GEOGRAPHERS
Biobibliographical Studies

VOLUME 10

Edited by

T. W. Freeman

on behalf of the
Working Group on the History of Geographical Thought
of the International Geographical Union and the
International Union of the History and Philosophy of Science

Mansell Publishing Limited

London and New York

First Published 1986 by Mansell Publishing Limited
(A subsidiary of The H.W. Wilson Company)
6 All Saints Street, London Nl 9RL, England
950 University Avenue, Bronx, New York 10452, U.S.A.

This volume forms part of the series *Study in the History of Geography* planned by the Working Group on the History
of Geographical Thought of the International Geographical Union and the Commission of the International Union of
the History and Philosophy of Science. *Chairman*, Professor David Hooson, Department of Geography, University of
California, Berkeley, California 94720, USA. *Secretary and Editor*, Professor T.W. Freeman, 2 Kingston Close,
Abingdon OX14 1ES, England. *Associate Editor*, Professor Geoffrey J. Martin, Southern Connecticut State
University, 501 Crescent Street, New Haven, Connecticut 06515, USA. *Full Members*, Professor Josef Babicz,
Institut d'Histoire des Sciences et de Techniques, Polska Akademia Nuak, Nowy Swiat 72, Warsaw, Poland;
Professor Anne Buttimer, Department of Social and Economic Geography, University of Lund, Sölvegatan 13, S 223 62,
Lund, Sweden; Professor Philippe Pinchemel, Institut de géographie, 191 rue St Jacques, Paris 75005, France;
Professor Hou Ren-Zhi, Geography Department, Peking University, Beijing, PRC; Dr Galina Vasil'yevna Sdasyuk,
Institute of Geography, Staromonetnyy 29, Moscow 109017, USSR; Professor Keiichi Takeuchi, Faculty of Social
Studies, Hitotsubashi University, Kunitachi, Tokyo, Japan. *Honorary Members*, Professor Manfred Buttner,
4620 Bochum, Kierfenweg 40, Federal Republic of Germany; Professor Clarence D. Glacken, Earth Sciences Building,
University of California, Berkeley, California 94730, USA; Professor Richard Hartshorne, Department of Geography,
Madison, Winsconsin 53706, USA; Professor George Kish, 3610 West Huron River Drive, Ann Arbor, Michigan 48103, USA;
Professor Emil Meynen, 53 Bonn-Bad Godesberg 1, Langenbergweg 82, Federal Republic of Germany; Professor O.H.K.
Spate, Research School of Pacific Studies, Australian National University, GPO Box 4, Canberra ACT 2601.   There
are approximately seventy corresponding members.

Enquiries concerning publications listed in the 'Bibliography and Sources' section of the biobibliographies should
be sent to the relevant publisher or journal, and *not* to Mansell.

British Library Cataloguing in Publication Data

Geographers : biobibliographical studies.--Vol. 10
  1. Geographers--Biography--Periodicals
  910'.92'2   G67

  ISBN 0-7201-1851-4

Printed in Great Britain

Contents

Introduction: The Real World of Geographers   *T.W. Freeman*   v

List of Abbreviations   x

Nikolay Nikolayevich Baranskiy 1881–1963   *Theodore Shabad*   1

Emrys George Bowen 1900–1983   *Colin Thomas*   17

Henry Charles Carey 1793–1879   *William B. Meyer*   25

Henry Chandler Cowles 1869–1939   *Garry F. Rogers and John M. Robertson*   29

Juan Dantín Cereceda 1881–1943   *Manuel Mollá Ruiz-Gomez*   35

William Gerard De Brahm 1718–1799   *Louis De Vorsey, jr*   41

John Dee 1527–1608   *Ian M. Matley*   49

Nevin Melancthon Fenneman 1865–1945   *Bruce Ryan*   57

Ernst Emil Kurt Hassert 1868–1947   *Gerhard Pansa*   69

Johann Gottfried Herder 1744–1803   *J.A.C. Birkenhauer*   77

Fridtjov Eide Isachsen 1906–1979   *Hallstein Myklebost*   85

John Scott Keltie 1840–1927   *L.J. Jay*   93

Johann Gottfried Otto Krümmel 1854–1912   *J. Ulrich*   99

Marguerite Alice Lefèvre 1894–1967   *J. Denis*   105

William John McGee 1853–1912   *Andrew S. Goudie*   111

Niels Nielsen 1893–1981   *N. Kingo Jacobsen*   117

Joseph Franz Maria Partsch 1851–1925   *Gabriele Schwarz*   125

Henri-François Pittier 1857–1950   *Jordi Martí-Henneberg and Anne Radeff*   135

Eduard Richter 1847–1905   *Josef Goldberger*   143

William Rosier 1856–1924   *C. Fischer, C. Mercier and C. Raffestin*   149

Index   155

Already Published

Volume 1 (1977)

Edited by T.W. Freeman, Marguerita Oughton and Philippe Pinchemel

Jules Blache, Isaiah Bowman, Alfred Hulse Brooks, Grove Karl Gilbert, Clement Gillman, Louis-Auguste Himly, Robert Ho, Vladimir Leontyevitch Komarov, Matthew Fontaine Maury, Simion Mehedinti, Hugh Robert Mill, Ernst Georg Ravenstein, James Rennell, Eugenuisz Romer, Franz Schrader, George Adam Smith, Jacques Weulersse and Naomasa Yamasaki

Volume 1 (1978)

Edited by T.W. Freeman and Philippe Pinchemel

Dimitry Nikolaevich Anunchin, Nirmal Kumar Bose, Albert Perry Brigham, Eugène Cortambert, Charles Andrew Cotton, George Davidson, Eratosthenes, Arthur Geddes, Patrick Geddes, Geoffrey Edward Hutchings, Bartholomäus Keckermann, Émile Lavasseur, Geoffrey Milne, Wincenty Pol, Carl Ortwin Sauer, Mary Somerville, Paweł Edmund Strzelecki, Camille Vallaux, George Vâlsan and Alexander Ivanovitch Vernadsky

Volume 3 (1979)

Edited by T.W. Freeman and Philippe Pinchemel

Jacques Ancel, Philippe Arbos, Wallace Walter Atwood, Augustin Bernard, Antoine D'Abbadie, Alexandre Dimistrescu-Aldem, Archibald Geikie, James Geikie, Edmund William Gilbert, Johannes Gabriel Granö, Andrew John Herbertson, Philipp Melanchthon, Sebastian Münster, Robert Swanton Platt, John Wesley Powell, Élisée Reclus, Nathaniel Southgate Shaler, Thomas Griffith Taylor, Alexey Andreyevich Tillo, Zachris Topelius

Volume 4 (1980)

Edited by T.W. Freeman and Philippe Pinchemel

Al-Muqaddasî, Henri Baulig, Constantin Bratescu, Jovan Cvijić, Vasily Vasilyevich Dokuchaev, Ludwig Von Höhnel, Llewellyn Rodwell Jones, Immanuel Kant, Alfred Kirchoff, Andrey Nikolaevich Krasnov, Jan Stanisław Kubary, Hermann Lautensach, Joachim Lelewel, William Vaughan Lewis, Georg Joachim Rheticus, Richard Joel Russell, William Scoresby

Volume 5 (1981)

Edited by T.W. Freeman

Lev Semenovich Berg, Lorin Blodget, Robert Capot-Rey, George Babcock Cressey, William Morris Davis, Johann Albert Fabricius, Johann Michael Franz, Henricus Glareanus, Andrei Alexandrovich Grigoryev, Arnold Henry Guyot, George Jobberns, Stanisław Lencewicz, Thomas Livingstone Mitchell, Ferdinand Jakob Heinrich Von Mueller, Hilda Ormsby, Carl Ritter, Percy Maude Roxby, Heinrich Schmitthenner, Johannes Stöffler, Pavle Vujević

Volume 6 (1982)

Edited by T.W. Freeman

Peter Apianus, Anton Friedrich Busching, Charles Carlyle Colby, Nicholas Copernicus, Mihai David, Ludovic Drapeyron, Charles Bungay Fawcett, Robert Gradmann, Alfred Hettner, Mikhail Vasilyevich Lomonosov, Takuji Ogawa, Nicolae Orghidan, Aleksei Petrovich Pavlov, Archibald Grenfell Price, Erwin Josephus Raisz, Gregor Reisch, Rollin D. Salisbury, Otto Schluter, Jerzy Smolenski, Vasili Nikitich Tatishchev, Louis Vivien de Saint-Martin, Leo Henrich Waibel

Volume 7 (1983)

Edited by T.W. Freeman

Pierre Camena D'Almeida, Paul Arqué, Henri Cavaillès, Walter Christaller, Jean De Charpentier, Erich Von Drygalski, James David Forbes, Alexander Nikolayevich Formozov, George Woodroofe Goyder, James MacDonand Holmes, Pyotr Alexeivich Kropotkin, Thomas Aiskew Larcom, David Leslie Linton, Jacques M. May, Ivan Vasylievitch Mushketov, Eugen Oberhummer, Albrecht Penck, Eugène Revert, Ferdinand Freiherr Von Richtofen, Johann Sölch, Jorge Leonides Tamayo, Pierre Teilhard De Chardin, Vladimir Ivanovich Vernadsky, Robert DeCourcy Ward, Clark Wissler

Volume 8 (1984)

Edited by T.W. Freeman

Ewald Banse, Robert Neal Rudmose Brown, Robert Eric Dickinson, James Fairgrieve, Alexei Pavlovich Fedchenko, John Forrest and Alexander Forrest, Henry Gannett, John Paul Goode, Oscar MacCarthy, Vintilă Mihăilescu, Helge Nelson, Sergei Semyonovich Neustruev, Hans Hugold Von Schwerin, Ellen Churchill Semple, Shigetaka Shiga, Wilhelm Sievers, Carl Troll, Alfred Russel Wallace, John Harold Wellington, Sidney William Wooldridge

Volume 9 (1985)

Edited by T.W. Freeman

Charles Frederick Arden-Close, Ralph Hall Brown, Philippe Buache, Vaughan Cornish, Charles Darwin, William Hughes, Emilio Huguet del Villar, Ragnar Hult, Halford John Mackinder, Anton Melik, Philipp Paulitschke, Johan Evert Rosberg, Peter Ivanovich Rychkov, Ludomir Ślepowran Sawicki, Wilfred Smith, Edward Louis Ullman, Eugene Van Cleef, Wilhelm Volz, Wang Yung

Introduction: The Real World of Geographers

Though having no precise definition the 'real world' is a phrase presumably having some meaning to those who use it. Negative definition could perhaps begin with the 'ideal world' and to an editor this would be one populated by eager authors who have carefully studied the conventions or 'house style' of the journal for which they write, who have checked all their references and have gone brightly forward to submit papers by promised dates or, still better, some time beforehand. And the editor's ideal world would include some who like the poet 'do but sing because they must' (the next line of the poem is tactfully omitted) and send unexpected contributions to an editor who may be surprised and delighted. Oddly there are such authors, of whom some are young but not all.

What actuates people to write is a mystery but it is a fact of the real world that the number of people who say they want to write is far in excess of the number who do so. It would be a futile psychological exercise to try to find out why this should be, but one spur for an author is the wish to share with others something that has proved to be interesting. Clearly in the case of this journal the authors will be concerned with the individuality of particular people, not only as worthy of study but also in their own circumstances of time and place. Some will have been known personally as teachers, scholars and friends but by no means all for some authors deal with geographers of past periods. Like historians, geographers may identify themselves with former periods with pleasure and profit. Nevertheless most of the essays

published over ten years have dealt with geographers born after 1800, though this volume is typical in having four out of twenty essays dealing with people born before 1800.

The modern advance of geography from the middle of the nineteenth century is a fact of the real world. Within that time the exploration and commercial development of the globe, the advance of cartography, the outpouring of new publications, the work of geographical societies, the liberalization of education, have all given impetus to the study of geography, as of other branches of learning. Historians of geography working within that period have an abundance of material, of which biographical material is just one strand. Every historian wonders how life seemed to people in past ages studied and naturally geographers will wonder what social, economic and political forces, among others, affected earlier workers. Authors working for *Geographers* are dealing with their own kind, with those who chose to be, or became, partially or wholly geographers, with those who--had they been born fifty or a hundred years earlier--they might well have encountered as colleagues. And some geographers of riper years will be writing of people they knew quite well.

It is here that difficulties may arise. Can an author be sure that his judgement is objective? Someone who has been known as a respected, perhaps much-loved figure who was successful as a teacher, encouraging as a tutor or thesis supervisor, gifted with pastoral concern for students, a committee man at once courteous, enterprising and constructive, a patient administrator building up a geography department from small beginnings, may not prove to be an effective subject for a paper in *Geographers*. The cynical may think that such a person as the one just simulated is a saint rather than a geographer and that only some saints are interesting. This journal has its own particular mandate, which is the general development of geographical thought, and therefore its concern is with figures who in their day and age made some clear contributions as authors as well as teachers and administrators. If the concern is with the history of particular departments of geography in universities, with geography in governmental and other official agencies, with developments in education, with the work of geographical societies, then people whose work was primarily administrative would certainly figure prominently. Such people are essential to the success of any organization as is shown, for example, in such works as R. W. Steel, *The Institute of British Geographers: the first fifty years*, London 1984. There is every reason to hope that at some time there may be more studies of geographical societies and of geography in particular universities, showing that much is owed to people who have given devoted service as well as to a smaller number of charismatic figures who have emerged from time to time. Unhappily many geographical societies established in England, France and other countries no longer exist and it is hard to accept the idea that their work was done. Why should they disappear? There may have been a failure to realize when some change of policy was needed or when some new opportunity existed. But the Institute of British Geographers appears to have found the ability to surmount crises and to attract devoted workers.

In considering geographers of past time fascination lies in relating their life and work to prevailing views of their period. This emerged prominently in a session, early in 1986, at a conference of the Institute of British Geographers with papers on Haushofer and other German political geographers. To what extent were they influenced, perhaps dominated, by current trends of political thought within Germany, not only as it existed as it might be with some change of boundaries and with more colonial territories? That the upsurge in French geography, notably seen in the establishment of numerous provincial geographical societies, was partly actuated by a desire to expand overseas after the disastrous Franco-Prussian war of 1870-71 is generally accepted and indeed confirmed by study of the contents of the journals of the then new societies. It is not difficult to find in early issues of the publications of the American Geographical Society expressions of the Monroe doctrine alongside admonitary comment on British imperialism. Very often the imperialist outlook of H.J. Mackinder (1861-1947) has been regarded as common to all British geographers and especially of the Royal Geographical Society but this society also had a welcome for Pyotr Kropotkin and others of varied political views. And in the nineteenth century the Russian Geographical Society was of markedly liberal outlook under an autocratic Czarist government. In short, there is no simple and clear relationship between a society, or an individual, and any corporate ethos of outlook. The individual has the power of choice, at least in theory, even if he is subject to powerful constraints in many countries at particular periods of history.

Geographers are concerned with the real world, however vivid their vision of an ideal world. In the real world they find their inspiration but to many this lies in the pursuit of the remote and rare and this is encouraged by the general public who regard a geographer as one who penetrates remote and possibly dangerous places bringing back scientific information of permanent value, as indeed Charles Darwin and many others did in the nineteenth century. Geography deals with the whole world and for that very reason is of vast scope but it also deals with the immediate environment and several of the essays in this volume show that the fascination of the homeland was revealed. In general the twenty papers now published fall into three groups. The first of these deals with geographers born before 1800, confronting problems that they found intriguing at their own distinct times. A second group deals with geographers whose main work was on their home countries, though not necessarily exclusively so, and the third group covers seven geographers whose main work was either in education or in some particular systematic aspect of geography. There is no unequivocal distinction between the three groups and the contrasts between the geographers of each group perhaps justifies the emphasis on individuality which is one theme of this essay.

Of the geographers in the first group the earliest in time is John Dee (1527-1608) whose active career almost coincided with the reigns of the two Tudor queens of England, Mary (1553-58) and Elizabeth I (1558-1603). Not surprisingly he was a product of the Age of Discovery, concerned with exploration and cartography, mingled with what would now seem to be an element of eccentricity in his faith in alchemy and astrology. He must be one of the few geographers to be suspected of witchcraft and the possession of magical powers, or indeed of possible impropriety in relations with royalty, in his case Queen Mary whom he allegedly bewitched; this suggestion may cast new light on the character of Queen Mary. Two centuries later William Gerard De Brahm (1718-99), after an Army career, settled in Georgia in 1750 where he acquired a large area of land, served for a time as surveyor-general and became at least incipiently a regional geographer, though his oceanographical work is now of most academic interest. Johann Gottfried Herder (1744-1803), also of German extraction, lived for part of his life in Riga, the Baltic Hansa town that ultimately became capital of Latvia, but ultimately settled in Weimar, known for its rich cultural life, and came to a broad conception of geography as part of a wide range of thought based on philosophy and religion. This had a prophetic element (he even saw the possibilities of space travel) and a concern for conservation. A true polymath, he was a writer who influenced von Humboldt and Ritter. Finally, of geographers born before 1800, Henry Charles Carey (1793-1879) was born in America and argued that the theory of diminishing returns from agriculture of the economist Ricardo was fallacious, for pioneer farmers first settled the poorer soils as the richer soils would be largely forested, probably difficult to drain, and only brought into cultivation when the demand of town populations would stimulate such enterprise. He held theories on the possible regional development of America as a local integration of enterprise in towns and rural areas but regarded any form of plantation agriculture as likely to mine the land and therefore harmful.

A second group of geographers found much of their inspiration in their homelands. The first in time was William John McGee (1853-1912) who was primarily a geomorphologist though also interested in anthropology and human geography. His working life was spent in government employment, which gave research opportunities at a time of great advance in geomorphology in the United States. As a geomorphologist he was concerned as much with 'process' as 'stage' and at all times he welcomed new ideas and methods. In fact he attributed the first use of the term 'geomorphology' to J.W. Powell, and gave it wide currency. Worthy still of consideration is his definition of conservation as the search for 'the greatest good to the greater number for the longest time'. Regrettably America had, he said in 1910, squandered its resources in the shortest time: this view led to comment on soil erosion. Another American considered here, Nevin Melanchthon Fenneman (1865-1945) is most remembered for his physical regionalization of the United States, published in two large books dated 1931 and 1938. As the bibliography shows a large number of papers supported his major works while others revealed his wider interests, notably in natural resources. He was a man of decided views, not least on school and university education. His contemporary, Henry Chandler Cowles (1869-1939), has been described as 'a pioneer in developing the dynamic view of the biological landscape'. Essentially he was a field scientist, in a broad sense an ecologist who was conscious of the human significance of his researches and known to geographers partly as a founder member of the Association of American Geographers under the chairmanship of W.M. Davis in 1904.

Two geographers born in the same year were Nikolay Nikolayevich Baranskiy (1881-1963) the Russian and Juan Dantín Cereceda (1881-1943) the Spaniard. Both lived through difficult times in their very different countries. Baranskiy was for much of his earlier life almost a professional revolutionary and only came fully into geography when forty-four years of age, though he had taught some economic geography from 1918 and by 1921 it had become a consuming interest. His work on the geography entries in the *Great Soviet Encyclopaedia* in 1925 was followed a year later by his first book on economic geography and then by his famous texts primarily on economic but also on physical geography. Established at Moscow university in 1929 he had the enterprise--and indeed the courage--to arrange for translations of books by geographers of other countries when outside cultural contacts were not favoured by the government. As a geographer he drew on his vast knowledge of Russia, acquired from childhood onwards, and he saw the economic and social revolution he had desired from his early years. His

concern with regions was practical for to him they were dynamic organisms, a synthesis of physical, economic and cultural phenomena. They were also entities for planning and management and his outlook crystallizes the concern of Soviet geographers with applied geography having the region as a human entity serving a practical purpose. Dantín-Cereceda (1881-1943) was a Spaniard of quieter mould, striving towards a convincing regional definition of Spain, aided by study of comparable efforts elsewhere, notably in France, Germany, Britain and America. He was especially interested in the arid and semiarid areas of Spain. His life was spent as a lecturer in agricultural colleges and his knowledge of rural settlement led to his selection as a member of the Rural Habitat Commission of the International Geographical Union, a body which stimulated vast rivers rather than streams of papers. This massive outpouring of data and assessment showed that virtually all the assumptions taught to students as general laws of rural settlement were fallacious.

Denmark, Belgium, Wales and Norway are represented here by geographers who found fascinating material in their own countries which they were eager to communicate to a wider circle beyond the frontiers. Although the Dane, Niels Nielsen (1893-1981), gained inspiration from scientific expeditions to Iceland and Greenland, he was a devoted exponent of the geography of his own country and eagerly led parties of students and others through its fascinating landscapes of glacial deposition. He was notably involved with the geomorphology of Denmark's west coast and through his enthusiasm drew a large number of field scientists into his researches, which in general were a fine contribution to applied geography. Marguerite Lefèvre (1894-1967) was perhaps the most eager traveller of the four and gave great service to the International Geographical Union though apart from her work on Central Africa it was her contribution to the geography of Belgium, notably to its National Atlas, which is best known. Her inspiration was drawn from French geographers but her career was impeded by the male dominance (indeed one could say chauvinism) in Louvain university during her lifetime. Happily she did not let this restrict her scholarly output and in the end she was given long overdue honour in her own country. Emrys Bowen (1900-1983) was also influenced by the work of Vidal de la Blache and his followers and in his own country developed a distinctive type of historical and regional geography which, especially in his later years, he eagerly and engagingly expounded to English audiences. Fridtjof Isachsen (1906-1979) was taken by his father on several occasions to Spitzbergen, Greenland and Iceland and retained an interest in Arctic lands throughout his life but he also found satisfaction in the study both of urban and of marginal rural areas in Norway as well as in its physical geography: eventually he was concerned with problems of conservation in Norway.

Finally, seven geographers considered in this volume were specialists in some aspect of geography, educationalists, or even in the case of John Scott Keltie (1840-1927) a professional writer. Keltie is often described as a journalist but his work was of a specialist and generally geographical character and at the age of twenty-one he joined the staff of the Chambers firm in Edinburgh to work on the first edition of their famous *Encyclopaedia*. Later he joined the Macmillan firm as subeditor of *Nature* and in 1883 he became editor of the *Statesman's Yearbook*. In his twenties he began to train as a Presbyterian minister but he abandoned the course as a matter of conscience and was described as having 'lapsed into literature'. In the geographical context he acquired fame for his investigation of school education for the Royal Geographical Society, of which the report published in 1885 was helpful in persuading Oxford university to appoint H.J. Mackinder as its Reader in geography. This initiated a period of marked expansion for geography in British universities, watched with quiet satisfaction by Keltie who spent the rest of his working life at the Royal Geographical Society. Oddly he had failed to take his degree while a student at Edinburgh and St Andrews universities but he shared that odd distinction with Sir Patrick Geddes (1854-1932) and also A. J. Herbertson (1865-1915), the first professor of geography at Oxford. But it is not intended here to give students the impression that the path to glory in geography begins with failure to take a degree.

Eduard Richter (1847-1905) was one of several geographers initially trained as an historian. He spent fifteen years as a grammar school teacher but from 1886 to his comparatively early death was professor of geography at Graz university. His interest in historical research was retained to the end of his life in his work for an historical atlas of the Austrian provinces. But it was for his work on glaciation that Richter was best known and in 1897 he became president of the newly-established International Commission on Glaciers.

Like Richter, Partsch (1851-1925) was not initially a geographer, for as a student he concentrated mainly on the classics and found his way to geography through the study of ancient history though he was also attracted by lectures on hydrology and geology. Also like Richter, Partsch became fascinated by glaciation, past and present, though he also wrote on the geography of classical times, notably on Greece. In his thirties work on Germany began, particularly on Silesia from his base in Breslau university where he served until his removal to Leipzig university in 1905. He was well suited too by the range of his education and interests to become a regional geographer and his work on Central Europe, published first in English in 1903 and in a fuller German edition a year later, became a standard work for many years. Very different from the life and interests of Partsch was that of Johann Krümmel (1854-1912), a student of geography with geology and botany at Leipzig university, who moved to Göttingen and then to Kiel where as professor of geography he was attracted to oceanography, then as now a specialization attracting a limited number of devoted workers.

Two Swiss geographers, Rosier and Pittier, born in consecutive years, had careers of marked contrast for one stayed at home and one went away. William Rosier (1856-1924) was at first a school teacher and a strong supporter of the Geneva Geographical Society, publishing textbooks on Switzerland and on general geography with various papers on education. Never having taken a degree at a university he became the first professor of geography at Geneva university in 1902. With this status he became a major figure in the Geneva canton and an enthusiastic worker for the development of geographical education, with travel as an occasional interlude. Henri-François Pittier (1857-1950), a field scientist of varied interests that were integrated by a geographical approach, spent his first thirty years in Switzerland, the next thirteen in Costa Rica, eighteen in the United States and finally over thirty in Venezuela. The range of his activities in these four countries was vast and practical for he used his scientific knowledge as a basis for economic advance. One notable feature of his outlook was a belief in the need for conservation, particularly in the economic exploitation of forests.

Finally Ernst Hassert (1868-1947), a major German geographer of his time who worked in Leipzig, Cologne, Tübingen and Dresden, had clear ideas which were firmly expressed on the position of Germany in the contemporary world. Any paper with the title 'the geographer as patriot' could dwell on his thought and career with advantage. He was firmly influenced by Friedrich Ratzel (1844-1904), who he regarded as 'the father of anthropogeography', and he also drew inspiration from Ferdinand von Richthofen (1833-1905), from whom he acquired a deep respect for physical geography as the essential foundation of any area from the home environment to the entire world. Hassert thought that students should have a world view and that geographers should meet this demand by providing texts even of areas they could never hope to visit. The earth was the home of man, but not the whole of it as Ratzel had shown in his discussion of the *ecumene*, the habitable world. Hassert wrote his first major publication on the Arctic limits of the habitable globe and also contributed fruitful studies on communications as a means of integrating the habitable world. But he attracted major attention by his work on colonial expansion as a basis for economic growth and emigration combined with his bitter condemnation of the loss of Germany's colonies following World War I. This was one reason for his warm support of the Nazi regime. He survived the Second World War only by a short time, having during his life provided books and papers welcomed by their readers not only for their geographical content but also for their political ethos.

In 1964 geographers formed a commission of the International Geographical Union on applied geography and at a series of meetings in various countries speakers from a wide range of countries dealt with local problems having a primary emphasis in planning. Especially valuable to geographers from the western democracies was the revelation of the use of geography in the socialist countries where new mineral, power and industrial resources were in process of development. Not less interesting was the consideration of industrial areas whose original resources such as coal and iron had been worked out or whose major industries had declined: similarly a conference in Brittany dealt vividly with the need for agricultural regeneration combined with appropriate industrial development in the towns. These were only aspects of a wide and interlocking range of problems including in several countries the redefinition of administrative boundaries to provide areas appropriate for modern governmental services.

Over the years the geographer deals with the landscape as it is, for he is concerned with the here and now, the landscape as he sees it. It is then, and only then, that the geographer may hope to be influential in public policy and one is compelled to comment that many geographers have underplayed their hands. They may fail to achieve very much, indeed anything at all, for of all human activities politics is the least predictable in judgement and action. The golden vision of today may be the disaster of tomorrow and it would only bore the reader to give examples as everyone is likely to know them already. But one point which remains valid emerges in several essays within this volume. We must first study our own home area for geography is concerned with place and only by knowing what is most accessible can we hope to understand the world. However remote and rare the environments eventually studied initial techniques are acquired in the homeland.

The landscape is a social, economic and historical as well as a physical entity. Some geographers have found so much to study in their own countries that they have not felt the need, or may not have been able to create the opportunity, to work elsewhere, as in the case of Bowen. Within their own countries they may have been absorbed by one major problem, such as the geomorphology of western Denmark for Nielsen. Through such an approach they have contributed to the national life, by inspiration or even to practical economic benefit, for example in land reclamation. Others have found unexpected interest at home, as in the case of Isachsen's pioneer urban studies. Larger questions arise in the case of work, in some cases of high scholarly quality, with distinct political implications, whether on German colonial territories or on contentious national boundaries. The use made of a geographer's findings is a matter for his own judgement, though he may have to face pressure from governmental authority: for the use made of a geographer's findings by others he cannot be responsible. All he can do is to show the real world as he sees it, but this does not preclude his vision of an ideal world, or at least a better world. Efforts to realise such a visionary world may have admirable fruit but may end in tragedy as Haushofer and others were to find in time.

With deep regret we record the death of Professor Preston Everett James, an honorary member of this Commission from 1982, on 5 January 1986. His work on the history of geographical thought is internationally respected and his integrity, warm hearted nature, happy temperament, sense of humour, tact and tolerance gave him the affection of a wide circle of friends.

T. W. Freeman

Note: Intending authors are asked to write to Professor T.W. Freeman, 2 Kingston Close, Abingdon, OX14 1ES, England, who will send a copy of *Notes for Authors* (revised edition) and other information.

List of Abbreviations

Abbreviations have been adopted from *British Standard 4148: Part 2, Word-abbreviation list.*

Abh. Geogr. Gesell. Wien   Abhandlungen geographische Gesellschaft in Wien
abstr.   abstract
Acta Geogr. Lovaniensia   Acta geographica Lovaniensia
Adv. Sci.   Advancement of Science
Allg. Dtsch. Biogr.   Allgemeine Deutsche Biographie
Am. Anthropol.   American Anthropologist
Am. Geogr. Soc. Res. Ser.   American Geographical Society Research Series
Am. J. Sci.   American Journal of Science
Am. Nat.   American Naturalist
Ann. Assoc. Am. Geogr.   Annals of the Association of American Geographers
Ann. géogr.   Annales de géographie
Ann. Hydrog. Marit. Meteorol.   Annalen der Hydrographie und maritime Meteorologie
Ann. Soc. Sci.   Annales de la Société scientifiques de Bruxelles
Arch. Sci. Nat.   Archives des Sciences Naturelles, Genève

Ber. Dtsch. Landeskunde   Berichte zur deutschen Landeskunde
Ber. Verh. K.-Sächischen Gesell. Wiss. Philolog.-Hist. Kl.   Berichte der Verhandlungen der Königlichen-Sächsischen Gesellschaft der Wissenschaften, Philologische-historische Klasse
Bol. Agric.   Boletín de Agricultura

Bol. Inst. Físico-Geográfico   Boletín del Instituto Físico-Geográfico de Costa Rica
Bol. Min.   Boletin Minero
Bol. Real Soc. Geogr.   Boletín de Real Sociedad Geográfica
Bol. Soc. Geogr. Nac.   Boletín de Sociedad Geográfica Nacional
Bol. Soc. Venez. Cienc. Nat.   Boletín de la Sociedad Venezolana de Ciencias Naturales
Boll. Soc. Geogr. Ital.   Bollettino Società geografica Italiana
Bot. Gaz.   Botanical Gazette
Bull. Geogr. Soc. Chicago   Bulletin of the Geographic Society of Chicago
Bull. Geol. Soc. Am.   Bulletin of the Geological Society of America
Bull. Illinois State Geol. Surv.   Bulletin of the Illinois State Geological Survey
Bull. Phil. Soc. Washington   Bulletin of the Philosophical Society of Washington
Bull. Soc. belge Et. Géogr.   Bulletin de la Société belge d'études géographiques
Bull. Soc. belge Géol. Paléont. Hydrog.   Bulletin de la Société belge de géologie, de paléontologie et de hydrographie
Bull. Soc. R. belge Géogr.   Bulletin de la Société royale belge de géographie
Bull. Soc. R. géogr. Anvers   Bulletin de la Société royale de géographie d'Anvers
Bull. Soc. Vaud. Sci. Nat.   Bulletin de la Société Vaudoise des Sciences Naturelles
Bull. U.S. Geol. Surv.   Bulletin of the United States Geological Survey

Bull. Wis. Geol. Nat. Hist. Surv. Bulletin of the Wisconsin Geological and Natural History Survey

C. R. Congr. Geol. Bull. Comptes Rendus, Congrès Géologique Bulletin

C. R. Congr. Int. Geogr. Congrès Internationale de Géographie

C. R. Int. Geol. Congr. Comptes Rendus, International Geological Congress

Congr. Venez. Med. Congreso Venezolana de Medicina

Cons. Perm. Int. Explor. Mer Conseil permanent international pour l'exploration de la mer

DNB Dictionary of National Biography

Danske Vid. Selsk. Danske Videnskabernes Selskab

diss. dissertation

Dresdner Geogr. Stud. Dresdner geographische Studien

Dtsch. Geschichtsblätter Deutsche Geschichtsblätter

Dtsch. Runds. Deutsche Rundschau

Est. geogr. Estudios geograficos

Festschr. ... Semin. Univ. Breslau ... Dtsch. Geogr. Festschrift ... Seminars der Universität Breslau ... Deutschen Geographentages

Forsch. Dtsch. Land.-Volksk. Forschungen zur deutschen Landes-und Volkskunde

Geogr. Abh. Geographische Abhandlungen

Geogr. Anz. Geographischer Anzeiger

Geogr. biobibl. stud. Geographers: biobibliographical studies

Geogr. J. Geographical Journal

Geogr. Jahrb. Verl. Geogr.-Kartog. Anstalt Gotha Geographisches Jahrbuch. Verlag Geographisch-Kartographische Anstalt Gotha

Geogr. Jahres. Geographische Jahresberichte

Geogr. Mag. Geographical Magazine

Geogr. Rev. Geographical Review

Geogr. Runds. Geographische Rundschau

Geogr. Teach. Geographical Teacher

Geogr. Tidsskr. Geografisk Tidsskrift

Geogr. Z. Geographische Zeitschrift

Int. Geogr. Congr. International Geographical Congress

Izv. AN (Akad. Nauk) SSSR ser. geogr. Izvestiya Akademiia Nauk SSSR, seria geograficheskaya

Izv. vses geogr. Obshch. Izvestiya Vsesoyuznogo Geograficheskogo Obshchestva

J. Cincinnati Soc. Nat. Hist. Journal of the Cincinnati Society of Natural History

J. Ecol. Journal of Ecology

J. Geogr. Journal of Geography

J. Geol. Journal of Geology

J. R. Anthropol. Inst. Journal of the Royal Anthropological Institute

J. Sch. Geogr. Journal of School Geography

J. Washington Acad. Sci. Journal of the Washington Academy of Sciences

Kieler Geogr. Schr. Kieler geographische Schriften

La geogr. La géographie

Min. Mag. Mining Magazine

Mitt. Dtsch. Gesell. Meeresforsch. Mitteilungen der Deutschen Gesellschaft für Meeresforschung

Mitt. Dtsch. Österr. Alpenver. Mitteilungen des Deutschen und Österreichischen Alpenvereins

Mitt. Dtsch. Schutzgebieten Mitteilungen aus den Deutsche Schutzgebieten

Mitt. Geogr. Gesell. Wien Mitteilungen der Geographischen Gesellschaft in Wien

Mitt. Gesell. Salzburger Landkd. Mitteilungen der Gesellschaft für Salzburger Landeskunde

Mitt. Inst. Österr. Geschichtsforschung, Erg. Mitteilungen des Institutes für österreichische Geschichtsforschung, Ergänzungsband

Mitt. Ver. Erdkd. Dresden Jahres. Mitteilungen des Vereins für Erdkunde zu Dresden Jahresbericht

Mitt. Ver. Erdkd. Leipzig. Mitteilungen des Verein für Erdkunde zu Leipzig

Mon.-Ber. Dtsch. Akad. Wies. Berlin Monatsbericht der Deutschen Akademie der Wissenschaften zu Berlin

Natl. Geogr. Mag. National Geographic Magazine

Norsk Biogr. Leks. Norsk Biografisk Leksikon

Norsk Geol. Tidsskr. Norsk Geologisk Tidsskrift

Ohio Geol. Surv. Bull. Ohio Geological Survey Bulletin

Österr. Alpen-Ztg. Österreichische Alpen-Zeitung

Österr. Runds. Österreichische Rundschau

Pan Am. Geol. Pan American Geologist

Petermanns geogr. Mitt. Erg. Petermanns Geographische Mitteilungen Ergänzungsheft

Philolog. Abh. Philologische Abhandlungen

Pop. Sci. Mon. Popular Science Monthly

Proc. Am. Assoc. Adv. Sci. Proceedings of the American Association for the Advancement of Science

Proc. Geol. Soc. Am. Proceedings of the Geological Society of America

Proc. Int. Congr. Plant Sci. Proceedings of the International Congress of Plant Sciences

Proc. Int. Geol. Congr. Proceedings of the International Geological Congress

Proc. Natl. Acad. Sci. Proceedings of the National Academy of Sciences

Proc. Washington Acad. Sci. Proceedings of the Washington Academy of Sciences

RGS Supp. Pap. Royal Geographical Society, Supplementary Papers

Rep. Br. Assoc. Adv. Sci. Report of the British Association for the Advancement of Science

Rev. Quest. Sci. Revue des questions scientifiques

Riv. Geogr. Ital. Rivista Geografica Italiana

Sch. Soc. School and Society

Schlesische Z. Schlesische Zeitung

Sci. Science

Sci. Mon. Scientific Monthly

Scott. Geogr. Mag.   Scottish Geographical Magazine
Sitz.-Ber. K-Preussischen Akad. Wiss. Berlin
      Sitzungs-Berichte der Königlich-Preussischen
      Akademie der Wissenschaften Berlin
Sitzungsber. K. Akad. Wiss. Wien math.-naturwiss. Kl.
      Sitzungsberichte der Kaiserlichen Akademie der
      Wissenschaften in Wien, mathematisch-
      naturwissenschaftliche Klasse
Smithsonian Inst. Bur. Ethnol. ... Ann. Rep.
      Smithsonian Institution Bureau of Ethnology ...
      Annual Report
Soc. Esp. Hist. Nat.   Sociedad Española de historia
      natural
Sven. Geogr. Årsbok   Svensk Geografisk Årsbok

Trab. Mus. Cienc. Nat. Madrid   Trabajos del Museo
      nacional de ciencias naturales Madrid
Trans. Acad. Sci. St Louis   Transactions of the
      Academy of Science of St Louis
Trans. Inst. Br. Geogr.   Transactions of the
      Institute of British Geographers
Trans. Iowa State Hortic. Soc.   Transactions of the
      Iowa State Horticultural Society

U.S. Bur. Am. Ethnol. ... Annu. Rep.   United States
      Bureau of American Ethnology ... Annual Report
U.S. Dept. Agric. Bur. Soils Bull.   United States
      Department of Agriculture Bureau of Soils
      Bulletin
U.S. Geol. Surv.   United States Geological Survey
Univ. Cincinnati Inst. Sci. Res.   University of
      Cincinnati Institute of Scientific Research

Verh. ... Dtsch. Geogr.   Verhandlungen des Deutschen
      Geographentags
Verh. Gesell. Erdkd. Berlin   Verhandlungen der
      Gesellschaft für Erdkunde zu Berlin
Verh. ... Int. Geogr.-Kong.   Verhandlungen ...
      Internationalen Geographen-Kongress
Veröff. Inst. Meereskunde Univ. Berlin
      Veröffentlichungen des Instituts für Meereskunde
      ... Universität Berlin
Vest. Mokskovsk. Univ. ser. Geogr.   Vestnik Moskovskogo
      Universiteta, seriya geografia
Vop. Geogr.   Voprosў Geografi

Wiss. Z. Univ. Rostock   Wissenschaftliche Zeitschrift
      der Universität Rostock

Z. Gesell. Erdkd. Berlin   Zeitschrift der
      Gesellschaft für Erdkunde zu Berlin
Z. Schulgeogr.   Zeitschrift für Schulgeographie
Z. Wiss. Geogr.   Zeitschrift für wissenschaftliche
      Geographie

Nikolay Nikolayevich Baranskiy

1881–1963

THEODORE SHABAD

It is probably fair to say that few geographers left
such a rich legacy to posterity as Nikolay
Nikolayevich Baranskiy of the Soviet Union.
Trained in law in his native Tomsk and later in
economics and statistics in Moscow, and long active
as a professional revolutionary, Baranskiy was a
latecomer to academic geography when he began to
devote himself full-time to teaching, research and
publication in the discipline in his forties.   But
he brought with him a baggage of practical knowledge
about the workings of the Russian economy, especially
in his native Siberia, that did much to shape his
approach to geography.   He spent his first decade
of teaching, in the 1920s, at Communist Party schools
that trained administrators and managers for the new
political and economic system and provided refresher
courses to experienced officials.   In 1929 he made
Moscow University his base of operations, establishing
a Department of Economic Geography that was to become
the Soviet Union's most prestigious training centre in
the discipline.   But in his broad-ranging activities
he was scarcely confined to his Moscow University
position.   Throughout his career in geography he held
parallel part-time teaching jobs in other
institutions, helped to found geographical journals,
edited encyclopedias, translated and edited foreign
geographical writings for Russian readers, took an
active interest in the reform of geography education
at the primary and secondary school level, and was
not above popularizing geography among the broad
public.   His contributions to professional geography
were many.   He developed a distinctive regional

school in economic geography, postulating the concepts
of the geographical division of labour and of the
economic-geographic situation of an area or a place.
He elaborated the principles and techniques of economic
cartography, contributed to the study of the economic
geography of cities and developed a separate subfield
of regional geography concerned with the complex
characterization of countries and regions.   But perhaps
one of Baranskiy's most memorable achievements was his
secondary school textbook on the economic geography of
the USSR, which served as a standard text for more than
twenty years, going through sixteen editions with
translations into thirty languages and introducing an
entire generation of Soviet teenagers to the economic
geography of their country 'the Baranskiy way', *po
Baranskomu*, as the Russians say.   Since his death in
1963 at the age of eighty-two, Baranskiy has remained
one of the most widely quoted authors in Soviet
geographical writing, and many young geographers have
sought to elaborate further some of the ideas he
advanced long before their time.

1.   EDUCATION, LIFE AND WORK

Nikolay Nikolayevich Baranskiy was born on 27 July 1881
in the Siberian city of Tomsk, into the family of a
teacher at a boys' gymnasium, or secondary school.
His parents had moved to Tomsk, then already a leading
cultural centre, in the early 1870s after having lived
for a number of years at Mogilev in western Russia.
Baranskiy's father, also named Nikolay Nikolayevich,
was the son of a priest but left home at an early age

and studied at the University of St. Petersburg. On
his graduation as a teacher, he moved to Mogilev, where
he taught in a gymnasium and earned extra income as a
private tutor, benefiting from the aura of having been
educated in the tsarist capital itself. Among those
whom Baranskiy Senior tutored was Olga Sergeyevna
Sokolova, a graduate of a private girls' boarding
school, where she had studied French, music, singing
and dancing. They were married and soon moved to
Tomsk, where the elder Baranskiy joined the staff of
the local gymnasium, teaching Russian literature and
geography.

By the time the younger Nikolay, known by the
Russian diminutive Kolya, was born the Baranskiys
already had three daughters, Yekaterina, Lyubov' and
Nadezhda. A second son, Dmitriy, was born five
years later, in 1886. The mother, Olga Sergeyevna,
was helped to bring up the children by her own mother,
Anastasiya Ivanovna, and by her sister, Valentina
Sergeyevna, who were also members of the household.
To augment his income to support such a large family
the father again did tutorial work in addition to his
regular teaching. The family environment in which
young Kolya was raised was fairly apolitical.
Although the father displayed a keen interest in
current events, he belonged to no party and did not
share the views of the Populism that was then widespread among the Russian intelligentsia as an
expression of the interests of the peasantry. In
fact, the elder Baranskiy often derided the Populist
movement as 'peasant socialism'. However, according
to the testimony of his son, he was always devoted to
the cause of public education and opposed to
bureaucracy, particularly the constant striving for
higher rank that preoccupied civil servants under the
tsars. It was a quality that also pervaded the young
Baranskiy who made the Russian word *chinovnik*,
functionary or bureaucrat, one of the most pejorative
epithets in his vocabulary. The elder Baranskiy
himself rose to the fairly respectable grade of state
counsellor, the highest of the middle-level rankings in
the civil service, short of the top four grades which
were bestowed for special services personally by the
Tsar. Not surprisingly, the Baranskiy household was
an enlightened and cultivated one. The father read a
great deal in his leisure time--books, magazines,
newspapers--and passed on this interest to the
children, encouraging them early to read and to
discuss what they read. The younger Baranskiy later
recalled that, starting at the age of six or seven, he
would join his father almost daily after dinner and
listen to him talk about developments in national and
world affairs. The names of the great political
figures of the time--Bismarck, Gladstone, Disraeli--
soon became familiar to the boy. All the children
learned French early on: they also did their share of
helping with the housework.

Tomsk in those days had a population of 45 000
and was a major commercial, transportation and
cultural centre of Siberia, with a considerable number
of merchants, civil servants and intellectuals, many
of whom had once been exiled and then remained in
Siberia after their release. As an administrative
centre for West Siberia, Tomsk had many government
agencies: four daily newspapers were being published

and a university was opened in 1888, consisting
initially only of a medical school.

In 1891, at the age of ten, Kolya Baranskiy was
admitted to the Tomsk gymnasium. He found the work
easy, consistently received high marks, and advanced
from class to class without having to pass
examinations. In those days boys' gymnasiums in
Tsarist Russia required the study of four languages,
French and German, Latin and Greek. Latin gave young
Kolya some difficulty, but he applied himself and was
soon able to read Cicero and Plutarch in the original.
Having long developed a penchant for self-study, he did
a great deal of work on his own. At the age of
thirteen he began to help his father with tutorial work
and in the process further enriched his own knowledge.

While still a student in the Tomsk gymnasium the
young Baranskiy was increasingly influenced by two of
his sisters, Nadezhda and Lyubov', who had joined
revolutionary student groups in St. Petersburg and
brought back illegal literature on visits to Tomsk.
Nadezhda was in fact arrested in 1897, but her father
managed to exert influence by virtue of his civil
service grade and succeeded in having her exiled back
home to Tomsk. Kolya also had an older friend, Pavel
Ivanovich Malinin, who was a student at Tomsk
University and introduced the young Baranskiy to the
writings of Marx and Engels at the age of fifteen.
These influences were not without effect. In 1897,
while he was in the sixth grade, Kolya started
attending meetings of an illegal student group and in
the following year he joined the newly formed Russian
Social Democratic Workers Party, from which the
Communist Party dates its official beginning.

On graduating from the gymnasium in 1899 at the
age of eighteen, with a gold medal for excellence,
young Baranskiy could have easily gained admission to
the nation's best universities, in Moscow or in St.
Petersburg. However he chose the newly opened law
school of Tomsk University. By this time Tomsk had
grown further in importance. The 1897 census had
reported a population of 52 000: a technology
institute and a teachers college had been opened, and
in 1901 the city was reached by a 50-mile spur from the
newly built Trans-Siberian Railway, which had by-passed
the city despite the protest of the Tomsk municipal
authorities. In the law school, Baranskiy continued
to take an interest in revolutionary affairs. A
tightening of university rules in face of such
political activity in turn led to a student strike and
in March 1901 Baranskiy, as one of the ringleaders, was
expelled from the university. He became interested in
statistics and was commissioned by the Society of Altay
Area Studies to conduct a study of the village of
Chistyun'ka, near Barnaul. Chistyun'ka was a typical
Siberian village that was growing rapidly as a result
of the resettlement movement then being fostered by the
tsarist authorities. Baranskiy's 1901 survey in effect
yielded a full-fledged social-geographic study that was
incorporated into a book by N. M. Tregubov, published
in 1907.

After this brief excursion into academic research
Baranskiy committed himself more seriously to
revolutionary agitation. In the summer of 1902 he
organized his own underground cell in Tomsk, the
'Siberian Group of Revolutionary Social Democracy',

which the following year merged with a broader regional revolutionary grouping, the Siberian Social Democratic Union. At first the Siberian revolutionaries followed a separate political line but in 1903, when the Russian Social Democrats broke up into two wings, Lenin's Bolsheviks and the Mensheviks, the Siberian revolutionary movement sided mainly with Lenin's faction. Baranskiy began travelling up and down the newly constructed Trans-Siberian Railway, helping to set up underground printing presses, organizing revolutionary cells, haranguing meetings of workers in factories and in the newly established railway repair shops. His travels took him from Tomsk to Krasnoyarsk, Irkutsk, Chita and even as far as Harbin in Manchuria, which was then a virtually Russian town developed in connection with the construction of the Chinese Eastern Railway. Baranskiy later calculated that his revolutionary travels during the first decade of the twentieth century must have covered a total of 200 000 km (125 000 miles). Like all revolutionaries he adopted a nickname, supposedly as an aid to eluding the tsarist police. His sobriquet, reflecting his massive physical build, was Nikolay Bol'shoy (Big Nikolay). Another political agitator, Nikolay Nikolayevich Suslov, was known as Little Nikolay for his slight stature, and legend has it that the two Nikolays, appearing here one day and there another, thoroughly confused the police on their trail and so they evaded capture.

The high point of Baranskiy's revolutionary days occurred in December 1905, when the Siberian Bolsheviks made him their delegate to a Bolshevik congress held in Tammerfors (Tampere), Finland. There he met Lenin and his wife, Nadezhda Krupskaya, both of whom are said to have recalled knowing Baranskiy's sisters from earlier revolutionary days in the 1890s. Back in Siberia early in 1906 Baranskiy found that the watchfulness of the authorities had made it more difficult for him to operate in his old haunts, and the underground reassigned him to European Russia. In Ufa he was recognised on the basis of a police 'wanted' poster but was freed from jail after three weeks on a technicality. He moved on to Kiev where he was again arrested, to be released after a month on lack of evidence. In early 1907 Baranskiy was back in Siberia and during May, in Chita, he was seized once more, this time spending almost a year in prison. He was released again for lack of evidence, for the police were persuaded that the man they really wanted was Little Nikolay, who by then had fled abroad.

On his release from the Chita prison in 1908 Baranskiy was prohibited from residing in Siberia and settled in Ufa, in European Russia, giving up his work as a political agitator. Two years later, at the age of twenty-nine he resumed his formal education on his admission to the Moscow Commercial Institute, a school of economics now known as the Institute of the National Economy named for Plekhanov. He graduated in 1914 with a major in economics and mathematical statistics. Soon thereafter, World War I broke out and Baranskiy went to work as an economist for a war relief society known by the Russian acronym Zemgor, which had been established to assist in the care of the sick and the wounded at the front and to mobilize small-scale craft industries to supply the army. It was the eve of the Revolution, seen first in the overthrow of the Tsar, Nicholas II, in March 1917 and then in the Bolshevik *coup d'état* in November, but Baranskiy was sidelined during these cataclysmic events by a serious disease of the throat affecting the larynx, from which he recovered only in 1918. At some point during the year 1917, according to official Soviet sources, Baranskiy shifted his political allegiance from the Bolsheviks to the so-called Internationalist wing of the Menshevik Party, a faction which, like the Bolsheviks, opposed the war and was in virtual alliance with them. It was not unusual for Russian intellectuals to cross these party lines in the years immediately after the Revolution, and Baranskiy reverted to the Bolshevik cause in 1920.

After recovering from his throat ailment in 1918 Baranskiy, while still a Menshevik, became a civil servant in the new revolutionary government. He was first employed as an economist in the Chemical Division of the Supreme Council of the National Economy, an agency that was then responsible for reconstruction of the war-damaged economy. In March 1919 he became an inspector of the newly formed People's Commissariat of State Control, a branch of government that was intended to root out inefficiency and corruption in the civil service and became a permanent fixture of the Soviet administrative system as a watchdog and auditing agency. One of the functions of the new agency was to train a new and reliable civil service in areas that came under the control of the new Bolshevik regime as it gradually extended its authority in the Civil War against various counter-revolutionary groups. Baranskiy, with his background in the Siberian underground, was considered to be a reliable official in the new government and was given responsible positions in the state control agency, which was then headed by Stalin. This connection may have served Baranskiy in good stead during the great purges of the 1930s, when he came under attack from his ideological enemies in geography but was saved from any dire consequences. The state control agency posted Baranskiy on the Eastern Front where the Red Army was seeking to gain control of the Urals and western Siberia from the forces of Admiral Alexander V. Kolchak, who had proclaimed himself supreme ruler of Russia with his capital in Omsk. In July 1919, the Urals city of Chelyabinsk fell to the Bolshevik troops, and Baranskiy was named as representative of the state control agency to supervise the change of administration in Chelyabinsk. After the Red Army seized Omsk in November 1919 he was assigned to the newly formed Siberian Revolutionary Committee as head of the Economics Division: he also served as deputy chief of the statistical office in the Bolsheviks' new Siberian Government. In 1921 Baranskiy was summoned back to Moscow to rejoin the headquarters staff of the state control agency, which by then had been renamed the People's Commissariat of Workers' and Peasants' Inspection. Baranskiy, who was to remain a member of the agency's collegium, or governing board, until 1925, later calculated that his work on various inspection trips had taken him over a total distance of 100 000 km, or 60 000 miles, during the five-year period from 1918 to 1923, when he began to lessen his governmental commitments.

While serving in these various capacities in the Bolsheviks' fledgeling administration in the years immediately after the Revolution, Baranskiy also made his first attempts at teaching the economic geography of Russia. In late 1918, when the new government was still fighting for its life within a confined segment of European Russia cut off from much of the nation's resources, Baranskiy was called on to lecture to workers about the perilous situation. He talked about the location of the southern wheat-growing areas, the cotton of Central Asia, the oil of Baku, the coal of the Donets Basin, all under the control of the counter-revolutionary forces. It was an economic geography of the utmost timeliness and practicality as he tried to explain the shortages that were besetting the Bolsheviks. Drawing on his own vast experience and his familiarity with the distribution of economic activity in Russia, he began to formulate some of the basic ideas he later perfected in more rigorous form concerning the geographical division of labour, economic location, and the distinctive characteristics of Russia's regions. During his brief stint as a Siberian economic administrator in Omsk, in 1920-21, Baranskiy further developed his approach to economic geography by lecturing at local military and party schools. By the time he was summoned back to Moscow in 1921, economic geography had become a consuming interest. In addition to his work for Rabkrin, as the workers and peasants inspection agency was known by its Russian acronym, Baranskiy accepted his first appointment as a teacher of economic geography. The appointment was at the Communist University, a party school that was a training ground for officials and also offered refresher courses for experienced administrators. There Baranskiy established a department of economic geography--one of several he was to set up in the course of his career--and further refined his approach to the subject in the challenging environment of teaching the discipline to informed adults, many of whom brought years of life experience to the discussion in the classroom.

Baranskiy later recalled that some time in 1923 he was summoned by Lenin, who was then already virtually incapacitated from a stroke, and was offered the post of deputy people's commissar in the Rabkrin control and inspection agency. By that time Baranskiy was already thinking of leaving government work and devoting himself entirely to research and teaching. Moreover he was about to write his first textbook on the economic geography of the USSR. He therefore declined the offer of a high-level appoint- ment and in fact proceeded to loosen his ties to the state control agency, from which he resigned entirely in 1925.

And so it happened that at the age of forty-four, without any formal training in geography, Baranskiy plunged into his new career as an academic geographer. He was at the height of his intellectual powers and brought to the field a pent-up energy, a tremendous love for work, and his life experience as a professional agitator and widely travelled government official. His first few years of full-time activity as a professional geographer were of extraordinary diversity, involving teaching and lectures at several institutions of higher learning, editing and writing

encyclopedia articles, reviewing of Soviet and, more importantly, of foreign publications in geography; editing and writing introductions to Russian editions of selected Western writings in economics and geography, and his own work on textbooks seeking to introduce his own regional school into the teaching of economic geography.

In education, in addition to his principal teaching appointment at the Communist University, he served briefly in 1925-26 as deputy rector and then as rector of another party school, the Communist University of Workers of the East, which trained officials specifically for administrative and managerial positions in the eastern regions of the Soviet Union. He also headed an economic geography section in the Communist Academy, a separate party-orientated teaching and research institution in the social sciences that was ultimately merged with the Soviet Academy of Sciences in 1936. In 1928 Baranskiy taught at the Industrial Academy, another party school of the period, for the training of factory directors and other industrial managers. Finally he also accepted an appointment as professor at what was then called the Second Moscow University (now the Moscow State Pedagogic Institute, a teachers college), where he was on the faculty from 1927 to 1930.

Among his many parallel activities, Baranskiy also became the geography editor of the Soviet Encyclopedia publishing house in 1925, a position he was to keep for twenty years. The publication of the first edition of the *Great Soviet Encyclopedia* was then just getting under way (it has since been followed by two more editions), and Baranskiy was responsible for the geography entries, both on places and on concepts, many of which he wrote himself. He became a frequent reviewer of geography publications, both Soviet and foreign, and was instrumental in having selected Western writings translated into Russian and for contributing introductions to the Russian editions. Among some of the classic American textbooks he reviewed in those years were *An Introduction to Economic Geography* by Wellington D. Jones and Derwent S. Whittlesey, and *North America* by J. Russell Smith. But it was probably the translations of important Western writings that had the greatest impact. Baranskiy edited and wrote introductions to Russian editions of Alfred Weber's *Über den Standort der Industrien* in 1926 and Alfred Hettner's *Die Geographie, ihre Geschichte, ihr Wesen und ihre Methoden* in 1930. These and other efforts by Baranskiy to bring significant Western writings to the attention of Soviet readers very nearly made life difficult for him when they were seized upon by his ideological enemies during the great purges of the middle to late 1930s, when any Western connections, especially with Germany, were considered to be highly suspect.

In the midst of all this activity Baranskiy published the first version of his economic geography of the USSR in 1926. Based on his lectures at the Communist University, the book took an entirely new approach to the subject by analyzing the USSR in terms of the new economic planning regions then being conceptualized by the State Planning Commission, known by the Russian acronym Gosplan. Baranskiy, in keeping with the methodology adopted by the economic planners, viewed the regions as integrated entities combining environment, population and economy into distinctive segments of the earth's surface. This principle of regional economic integration as the basis for analysis ran counter to the then prevailing sectoral-statistical approach to economic geography, founded on separate investigation of particular economic activities and commodities. The conflict between Baranskiy's regional approach and the entrenched commodity-based school was to pervade the Soviet geographical scene for many years to come. His regional method also differed from the traditional approach to regions as homogeneous entities by stressing the interactions and interconnections among physical, human and economic factors within the regional context and identifying a regional specialization within the country's geographical division of labour. Finally Baranskiy found himself in conflict with two kinds of determinism, one to the effect that environmental forces alone shape human activity, the other that human activity can overcome the constraints of natural forces. Baranskiy used every occasion and every forum to publicize his ideas and in the end prevailed over his ideological opponents, both because of his political connections and because his regional school conformed to the Soviet economic planners' basic philosophy of regional development. At a 1926 conference of economics teachers Baranskiy presented a paper on how to set up a course in economic geography along the lines of his regional approach. In 1928 he published a methodological article, 'A plan for the economic-geographic characterization of a Gosplan oblast', in which he offered a model for portraying an economic planning region in terms of spatial relations involving environment, historical heritage, population and economic activity. And in May 1929, at a national conference of geography teachers, he delivered two fruitful papers on concepts that were intimately related to his regional approach: the geographical division of labour, and the economic-geographic situation of a place.

Later in 1929 Baranskiy transferred his base of operations from the Communist University, with its emphasis on training Soviet officials and administrators, and the Second Moscow University, with its teacher training programme, to the Moscow University for thirty-five years, until the end of his days. Geography at Moscow University had a venerable history going back to Dmitriy Nikolayevich Anuchin, an anthropogeographer who had established a Department of Geography and Anthropology in 1887. He was followed after the Bolshevik Revolution by Aleksandr Aleksandrovich Kruber, another man-oriented geographer who was interested mainly in the impact of the physical environment on human activities and cultures. However, Kruber left teaching in 1924 because of illness, and by the time Baranskiy joined Moscow University the discipline there was limited largely to physical geography. A physical geographer, Aleksandr Aleksandrovich Borzov with whom Baranskiy, incidentally, was to develop excellent relations, headed what was at that time the Geography division of the Physics and Mathematics Faculty. (In 1933, geography was to be combined with soil science in a separate

faculty and five years later, in 1938, an independent Geography Faculty was established at Moscow University.)

On joining Moscow University, Baranskiy promptly organized his own Department of Economic Geography, to which he began to attract kindred scholars sharing his views about the nature of the discipline. They included Pyotr Nikolayevich Stepanov, the industrial geographer, who was to take over the reins of the department in the 1940s before being himself succeeded in 1948 by Yulian Glebovich Saushkin. Baranskiy also brought with him Ivan Aleksandrovich Vitver, who was to set up his own Department of World Economic Geography in 1934. And Baranskiy took particular pride in attracting Nikolay Nikolayevich Kolosovskiy, an economic planner concerned mainly with Siberian development, who became known in particular for his work on the economic regionalization of the USSR. Baranskiy himself, at one time or another, taught courses on the economic geography of the United States, economic cartography, and an introduction to economic geography. In addition to setting up his teaching department in economic geography, Baranskiy also introduced the discipline into the Geography Research Institute that had been established at Moscow University in 1922 and existed for thirty years until the university moved in 1953 from the city centre to the new skyscraper building in the Lenin Hills. Baranskiy used the institute, which was then concerned not only with research but also with graduate work, to organize so-called complex, or interdisciplinary, geographical field expeditions for instructors and students that often provided the material for graduate dissertations.

In moving his base of operations to Moscow University Baranskiy in no way diminished his many parallel activities. He lectured and established economic geography departments at what was then known as the Institute of World Economy and World Politics and at a succession of party training schools, and from 1936 to 1940 again headed the department at the Moscow State Pedagogic Institute, a teachers college. In 1933 he attended the first congress of the Geographical Society of the USSR in Leningrad, and the following year he was one of the Soviet geographers who participated in a congress of the International Geographical Union, the 14th in Warsaw in August 1934. On that occasion he delivered papers on the university training of economic geographers and on the development of economic cartography in the Soviet Union. Twenty-two years were to pass before Soviet geographers renewed their participation in the work of the International Geographical Union, at the 18th congress in Rio de Janeiro in 1956.

Baranskiy's first few years at Moscow University were a period of acute ideological struggle in economic geography, as he sought to introduce his regional approach in teaching and research in face of the entrenched interests of economists and economic geographers who favoured the traditional commodity-oriented statistical approach and considered economic geography to be the prerogative of economics. Responding to Baranskiy's injunction that economic geographers must consider the realities of the physical environment, his enemies accused him of geographical determinism. He, in turn, called them 'unnatural' geographers, just as he accused physical geographers who ignored the role of man of being 'inhuman' geographers. More seriously, perhaps, Baranskiy's adversaries seized upon his efforts to make classic Western writings, notably those of Alfred Weber and Alfred Hettner, available to Soviet readers, together with fairly open-minded introductions of his own to the Russian editions. Attacks on Weber and Hettner and on Soviet geographers who favoured some of their views continued through the 1930s. Baranskiy, by being called a 'Weberian' and a 'Hettnerian', became particularly vulnerable during the great Stalinist purges starting in the mid-1930s, when any association with the West was considered virtual treason. However Baranskiy appears to have survived that period virtually unscathed, possibly because of his early association with Stalin in the state control apparatus after the Revolution. In fact he was given a series of academic honours during the worst time of repression. In 1935 he was awarded a doctorate in geography, a degree considerably more advanced in the Soviet context than the Ph.D. in the West. Moreover the degree was awarded on the basis of published writings without the usual requirement of a special dissertation. In 1937-38, at the height of the purges, he served briefly as a member of the Academic Council of the Institute of Geography of the Academy of Sciences, although he did not find much in common with the institute, then headed by Andrey Aleksandrovich Grigor'yev, a physical geographer. And in 1939 Baranskiy was made a corresponding member, the lower of two categories of membership, in the prestigious Academy.

In contrast to the opposition he encountered in academic geography Baranskiy was eminently successful in introducing his views into the teaching of geography in secondary schools. At that level Baranskiy fought both for more systematic study of geography in general and for his regional approach to economic geography in particular. It was largely due to his efforts and his contacts with the party authorities and with educators that the position of geography was strengthened in Soviet schools. A party decree of August 1932 stressed the need for using maps in classroom instruction and for teaching both physical and economic geography, including basic concepts, world geography and the geography of the USSR, as a whole and by regions. Baranskiy was placed in charge of a group of professors of Moscow University and the Moscow State Pedagogic Institute who worked out a detailed geography curriculum from the fifth to the ninth grades. Baranskiy's victory was complete when a Soviet decree of May 1934 approved the curriculum and instructed Baranskiy to write a physical geography of the USSR for the seventh grade and an economic geography of the USSR for the eighth grade. Both textbooks appeared in 1935. The slim 120-page physical geography text went through seven editions, the last in 1943. But it was the economic geography, which like the physical geography was translated into all the major languages of the USSR, that established Baranskiy's reputation, going through as many as sixteen editions, the last in 1955. Over a period of twenty years, an entire generation of Soviet citizens studied the economic geography of their country 'the Baranskiy way' (*po*

Baranskomu) and his name became virtually a household word. Moreover his high standing with the nation's political authorities was attested in 1952 when the twelfth (1950) edition of the Baranskiy economic geography textbook was awarded the Stalin Prize.

The 1934 decree on the teaching of geography in the schools also authorized the publication of a new journal for geography teachers, containing both substantive material for classroom instruction and material on teaching methods. Baranskiy became its editor, a function he retained until 1947. Lest the May 1934 decree convey the impression that the Baranskiy way of doing things had gained official endorsement only in the secondary schools, it was followed two months later by a similar edict relating to geography education at the college and university level. The second decree, issued in July 1934 by the Presidium of what was then the Government's Committee on Higher Technical Education, specified not only that most of the content of economic geography of the USSR be taught on a regional basis but also that at least 70 per cent of the time be devoted to economic regions.

Baranskiy's professional activities, like those of most people in the Soviet Union, were disrupted during World War II. When the German forces stood at the gates of Moscow in October 1941 part of Moscow University, including the Geography Faculty, was evacuated to Ashkhabad, the capital of the Turkmen SSR in Central Asia, where the educational process resumed on 1 December. The following August the university moved to Sverdlovsk in the Urals, finally returning to Moscow in 1943. For reasons not apparent from the available sources, Baranskiy was not evacuated together with the rest of the Geography Faculty. Instead he left Moscow first for Kazan' and then for Alma-Ata, capital of the Kazakh SSR, where he established a geography sector within the Kazakh branch of the Academy of Sciences (the sector was elevated to a full-fledged Institute of Geography in 1983), headed the Geography Department of the Kazakh Pedagogic Institute, and supervised the writing of a geography of Kazakhstan.

On returning to Moscow in 1943 Baranskiy plunged again into his multifarious activities in geography, resuming his chairmanship of the Economic Geography of the USSR Department, setting up a Geography Department in the so-called Lenin Courses, a party training programme, and lecturing at the Moscow State Pedagogic Institute. In 1946, as he became increasingly involved in new activities, he gave up the chairmanship of the Moscow University Department, although he stayed on as a professor. Freed of the administrative duties of chairman, he was active in setting up a separate Geography Publishing House in Moscow and assumed the additional function of geography editor in the Foreign Languages Publishing House. In the latter capacity Baranskiy resumed one of his favourite occupations, that of making foreign geographical writings available in Russian editions. They included Henri Baulig's *Amérique septentrionale*, which Baranskiy translated himself, and Preston E. James's *Latin America*. Even after giving up the geography editor's position in 1951 Baranskiy remained instrumental in fostering the dissemination of foreign geographical writings in the Soviet Union, translating himself several chapters of the Russian edition of *American Geography, Inventory and Prospect* (1954; Russian edition, 1957) and editing a Russian edition of Gunnar Alexandersson's *Industrial Structure of American Cities* (1956; Russian edition, 1959). Another publishing activity that has left a lasting impact was his founding of the geographical serial *Voprosy Geografii [Issues in Geography]*, which began to appear in 1946 under the aegis of the newly established Moscow branch of the Soviet Geographical Society, which has its main office in Leningrad. While serving as editor for the first eighteen years--more than sixty issues of the serial appeared by the time of his death in 1963 and the number has since doubled--Baranskiy imposed his approach to geography on a combination of issues of mixed content and, increasingly, thematic issues that would long survive him.

Toward the end of his life, as his activities necessarily slowed down, Baranskiy's honours accumulated. In 1951 he was awarded the Semenov-Tyan-Shanskiy Gold Medal of the Geographical Society of the USSR for his contributions to economic geography. He was made honorary member of four geographical societies in Eastern Europe--the Bulgarian and Yugoslav societies in 1948, the Polish Society in 1954, and the Serbian society in 1960, the latter awarding him also the Jovan Cvijić Gold Medal, instituted for the great Balkan geographer who worked in the tradition of Paul Vidal de la Blache. In the Soviet Union, in addition to the 1952 Stalin Prize for his economic geography textbook, Baranskiy was awarded the title of Hero of Socialist Labour in 1962, a belated recognition of his eightieth birthday the year before. With the title came the third Order of Lenin (the first two having been awarded in 1946 and in 1953). He died on 29 November 1963 at the age of eighty-two and was buried in the famous Moscow cemetery adjoining the ancient Novo-Devich'ye Monastery, where many of the nation's notables lie.

2. SCIENTIFIC IDEAS AND GEOGRAPHICAL THOUGHT

Baranskiy's role in geography was all the more remarkable because he entered the discipline without any formal training in academic geography, for he was essentially a self-made geographer. According to Yulian Glebovich Saushkin, a close disciple who headed the Department of Economic Geography of the USSR at Moscow University from 1948 to 1981 (a year before his death),

Geography (and economic geography in particular) was for Baranskiy a logical continuation of his revolutionary party work. He saw in it one of the powerful tools for remaking the world, for building communism, for indoctrinating the people. Baranskiy was clearly aware of the role that geography played in the apprehension of the universe, and of the strength and party orientation of its scientific ideas.

It was probably largely this linkage between Baranskiy's approach to geography and the perceived ideological and pragmatic needs of the Soviet system that made him such a successful and revered figure in the Soviet Union. Saushkin, who wrote a biography and appraisal of Baranskiy's work, summarized his scientific ideas under seven headings: Baranskiy's view of the man-nature relationship, his concept of the geographical division of labour, his approach to the space-time relationship in geography, his emphasis on the role of economic regionalization in geography, his founding of the urban geography school in Soviet geography, the development of a distinctive Soviet geographical subfield known in Russian as *stranovedeniye*, a cognate of the German term *Länderkunde*, and Baranskiy's contributions to economic cartography.

On the man-nature relationship, he is presented as having taken a balanced approach to the role played by the physical environment in economic development from the standpoint of Marxist-Leninist theory. He was critical both of what were known in the Soviet Union as bourgeois theories of geographical environmentalism, emphasizing the impact of the natural setting on human activity, and what Baranskiy viewed as nihilistic under-estimation of the role of natural factors. He stressed man-nature interaction at a time when, for ideological reasons, a sharp distinction was being made between the physical and the socioeconomic subfields of geography, and talk of a unified geographical discipline was considered heresy or worse. However, he distinguished two aspects of the impact of the physical environment on human activity. One was the way natural factors have affected the evolution of human society as a whole as well as the transition from one type of social formation to another. This was an issue he left to philosophy and history to sort out. But when it came to the other aspect of the man-nature relationship, namely the impact of differences in the physical setting on differences in human activity from place to place within a particular social formation, this he considered to be at the very heart of the concerns of economic geography. In Baranskiy's view,

the natural environment is not the cause in the strictly logical sense (as a set of circumstances that inevitably give rise to a particular event) either of a given human activity or, all the more, of a given social system, but simply a factor that either promotes or retards; we cannot therefore speak of a causal relationship, but rather of a correlation. It is up to the economic geographer to correlate spatial differences, i.e. differences from place to place, in human activities with spatial differences in the natural environment and thus to identify certain regularities.

Baranskiy also felt that any analysis of the impact of the environment on economic activity must always consider natural factors from a historical perspective, in combination with other types of factors such as the social-historical and the transportation and marketing aspects. The ultimate aim, in his view, is to reduce the man-nature relationship to the level of cost calculations for 'The proposition that a particular set of natural conditions is more favourable or less favourable will always remain a generality unless it can be converted into the precise language of the ruble.'

The geographical division of labour was viewed by Baranskiy as the central, organizing concept of economic geography, involving in effect an entire system of concepts interconnecting both the types of economic activities and the economic regions, or what he considered to be the 'inventory' of economic geography. He defined the geographical division of labour as the spatial form of the social division of labour. His focus on the geographical division of labour was closely related to his interest in the regional approach since he considered the geographical division of labour and regional differentiation to be analogous processes. The great driving force of the geographical division of labour, both within a country and between countries, is the economic advantage that can be derived from its realization. It is a concept that finds increasing practical application in the economic life of the Soviet Union as policy-makers struggle with the optimal division of labour between the populous, economically developed western regions of the USSR and the underpopulated but resource-rich eastern (Siberian) regions. The concept has also become relevant to a growing tendency for the Soviet Union to take advantage of an international division of labour, importing goods that are either not available or too expensive to produce domestically.

In the process of developing his ideas about the geographical division of labour, Baranskiy came upon the proposition that in the course of time places (regions, countries) with a favourable or unfavourable economic-geographical situation may arise in space. He thus developed his theory of the economic-geographical situation, involving spatial linkages and relations as they change through history in the process of the geographical division of labour. Baranskiy attached great methodological importance to the analysis of the economic-geographical situation of places and regions. In Saushkin's view, Baranskiy's notion of geographical space has become fairly well accepted in geography as a science, and his theory of the economic-geographical situation of places has assumed renewed significance in the form of the theory of interaction within a spatial (geographical) system.

Another central focus in Baranskiy's approach to economic geography was his concern with economic regionalization. This concern was in keeping with the practical orientation of the early economic planners of the Soviet Union, who sought to devise a set of optimal economic regions for planning and development purposes. Baranskiy integrated his views on the geographical division of labour and on the economic-geographical situation of places and regions with this overriding interest in economic regionalization. The resulting, all-inclusive, theory dealt with the formation and differentiation of regions, a system of regions and their interregional linkages, regional territorial complexes, a typology and a hierarchy (taxonomy) of regions, and the techniques used by economic geographers in studying and characterizing regions. Baranskiy's followers credit him with having developed some of the principles of regional science long before Walter Isard conceptualized this new field of knowledge in the mid-1950s, pointing to Baranskiy's concern with regions as dynamic organisms, as a synthesis of physical, economic and cultural aspects, and as the objects of planning and management.

Like economic regionalization, Baranskiy's approach to the geography of cities was also in keeping with the ideological and pragmatic concerns of Soviet society. In the revolutionary scheme of things, with industrial workers (and hence urban centres) constituting the basis of the new political system, cities were traditionally regarded as the foci of political power. Baranskiy developed this proposition in geographical terms, presenting cities as the most active and creative elements, as the centres of economic regions of varying scale, and as the nodal points of transport networks. In one of his most widely cited aphorisms, Baranskiy said that

> from the economic-geographical point of view, cities plus a road net constitute the framework, the structure on which everything else hangs, the structure that forms a territory and endows it with a certain configuration.

On the role of cities as the key elements in the political and economic organization of the Soviet State, Baranskiy wrote that

> Cities may be said to represent the command structure of the nation, organizing it in all respects ... Like any command structure, cities have a hierarchy of their own, including a lower, middle and upper level command, as well as a supreme command.

According to his followers, Baranskiy's approach to cities as central places was broader than Walter Christaller's, in the sense that Baranskiy was less concerned with spacing and arrangement, focusing instead on systems of cities, their hierarchies and their functions as part of the integrated economy of countries and regions.

Baranskiy worked out the methodological principles of the geographical subfield known in Russian as *stranovedeniye* (*cf.* the German term *Länderkunde*), which he conceptualized as the integrated geographical characterization of countries and regions. At a time when physical and economic geography were separated in the Soviet scheme of things by rigid ideological barriers, he saw his integrated regional approach as a way of bridging the gap between what he called 'unnatural' economic geographers and 'inhuman' physical geographers. His principles of integrated regional characterization, combining physical, population, economic and cultural aspects, have been applied in Soviet geographical writing both to the portrayal of countries around the world and to the depiction of Soviet regions, notably in the various multivolume regional geographies of the USSR that have appeared over the years.

Finally, Baranskiy placed great emphasis on maps as a geographical tool and, as an economic geographer, on economic cartography as an integral part of the investigation of economic phenomena. Baranskiy always viewed maps as the distinguishing--and saving--feature of geography as a teaching and research discipline observing that

> geography is a very broad notion and has always been in danger of losing its identity and becoming diffused. Maps always have been and will continue to be the main guarantee that geography will retain its distinctiveness. They constitute a graphic and palpable criterion of what is 'geographical' and what relates to geography. The same applies to the relationship between economic maps and economic geography. It may well be said that economic maps constitute the alpha and omega, the beginning and the end, of economic geography.

At a time when the relationship between geography and cartography is being widely debated, Baranskiy's followers have been pointing to his views in support of the notion that cartography and geography are united by common goals, common tasks and common methods.

3.   INFLUENCE AND SPREAD OF IDEAS

Baranskiy's approach to geography in general and to economic geography in particular made itself felt in the Soviet Union at different levels and through a variety of means. His ideas probably had the deepest impact in geographical education. The vehicles in this area were not only his standard textbook on the economic geography of the USSR, used over a period of two decades in the eighth grade of all Soviet schools, but his writings on teaching methods and his work as founder and editor of the teachers' journal *Geografiya v Shkole*. His *Ocherki* (*Essays*) on the teaching of economic geography in schools first appeared in 1946 and went through two more editions, in 1954 and in 1955, before appearing in a completely revised version in 1960 under the new title *Metodika prepodavaniya ekonomicheskoy geografii* (*The Method of Teaching Economic Geography*). His guiding principle was that geography as a teaching discipline in schools should not be divorced from geography as a research science, and he saw to it that teachers were made aware of the basic methodological issues in academic geography. He dealt with the issue of the significance of the physical environment in economic geography, his basic concepts of the geographical division of labour and the economic-geographical situation of places and regions, and linkages between phenomena and processes. From the very beginning in the early 1930s, Baranskiy sought to counterpose his regional approach and his view of regions as integrated entities to what he described as the enumerative preoccupation of the commodity-oriented statistical school that preceded him. He made geography teachers aware of such issues as generalization and the use of concrete examples, the need for bold and eye-catching illustrations, the importance of maps in geography lessons, and the use of statistical data. Many of these themes have become regular features in the journal *Geografiya v Shkole*, which with a circulation of 130 000 reaches geography teachers all over the Soviet Union.

Another facet of his work that left a deep imprint was in the realm of geographical publications. He always devoted a great deal of time to monitoring the literature, reviewing and editing. Starting in the mid-1920s, when he began to devote himself to geography as a career, he published more than 140 reviews of the works of Soviet and foreign authors, edited and wrote introductory articles to more than fifty books, and translated, edited or wrote commentaries for eleven Russian editions of the works of British, American, French and German authors. It may well be repeated in this connection that his interest in the work of foreign geographers came at a time when the Soviet Union was largely isolated from the mainstream of world science, and any preoccupation with 'bourgeois' science involved certain political risks. Since the middle 1950s, Soviet translation and critical evaluation of geographical work abroad have become a more common (though still limited) occurrence, and Baranskiy's work may well be regarded as having set a precedent in this regard. The monument to his publishing activity undoubtedly remains the academic serial of the Moscow branch of the Geographical Society, *Voprosy Geografii*. Starting with Vol 66, on 'Cities of the World', the serial has been noting on its masthead that it owes its founding to the 'initiative and leadership of N. N. Baranskiy'. Now appearing at the rate of four thematic volumes a year, *Voprosy Geografii* may well represent one of the richest troves of geographical writing on varied themes. The four issues of 1984, for example, were devoted to theoretical aspects of geography; the economic geography of the countries of Comecon, the Soviet bloc's economic alliance from the standpoint of economic integration; natural complexes and agriculture; and oceanography.

Finally, Baranskiy exerted a profound influence through his lectures and his work with graduate students. Although he was a hard taskmaster, he is recalled fondly by anyone who attended his courses at Moscow University or had occasion to work more closely with him. His manner of lecturing alone has become a legend. Though always well prepared, he seldom referred to notes and preferred to improvise even while following a logical trend of thought. A big man with a booming voice, he was always at ease with his audience. He spoke simply and clearly, evidently reflecting his long years as a party propagandist. In lecturing he often thought out loud, taking his audience through his way of reasoning to the ultimate conclusion. It was his way of demonstrating what he called the geographical way of thinking. In his book on teaching methods, he wrote:

> The geographical way of thinking is a particular way of thinking that first, relates everything to a territory and expresses its reasoning in ways that can be mapped, and second is an interconnected, complex way of thinking that does not confine itself to one particular element or one particular human activity, in other words, a way of thinking that plays harmonious chords and does not go about pecking at keys with one finger.

Baranskiy was an adept coiner of phrases, and many of his expressions, passed from mouth to mouth, have become geographical aphorisms. Cautioning students in economic geography against disregard of physical factors, he would say, 'Chase nature out the door, and it will climb back through the window'. And in impressing audiences with the importance of using maps, Baranskiy would intone: 'A map is like litmus paper, telling you what is geography and what is not'.

Baranskiy spent an unusual amount of time advising graduate students and commenting on their work. Occasionally his remarks took up nearly as much space as the paper being criticized. He would number his comments consecutively, placing a number in a circle in the margin next to a particular passage, and then writing out his criticism in longhand on separate sheets of paper. There were times, his disciples recall, when a student would hand in a short paper of a few pages, and get back a folder bulging with Baranskiy's observations. Some of his comments would be sharp and biting, but rarely insulting. For example, he hated vague, general statements. 'Stop philosophizing,' he would say, 'Let's have more concrete facts and examples.'

From the testimony of those who knew him, it would appear that Nikolay Nikolayevich Baranskiy has carved an enduring niche for himself in the history of geography, not only through the energy with which he devoted himself to his chosen discipline and the occasionally unorthodox ideas he professed but, perhaps most of all, by the sheer force of his personality.

Bibliography and Sources

1. REFERENCES ON NIKOLAY NIKOLAYEVICH BARANSKIY
Five Soviet sources, published in Russian, are particularly useful for information on Baranskiy's life and work:
A list of N. N. Baranskiy's published works, compiled by O. R. Nazarevskiy in N. N. Baranskiy, *Ekonomicheskaya geografiya v sredney shkole. Ekonomicheskaya geografiya v vysshey shkole [Economic Geography in the Secondary School. Economic Geography in Higher School]*. Moscow, Geografgiz (1957), 303-26. This bibliography of about 400 items extending into 1956 appeared in one of the principal collections of Baranskiy's writings on the teaching of economic geography and served as a basis for subsequent updated compilations.
N. N. Baranskiy. 'My life in economic geography (excerpt from memoirs)', *Geografiya v Shkole [Geography in School]*, (1964), no 1, 8-12. This short fragment from memoirs that were not completed appeared in a geography magazine for schoolteachers that was founded by Baranskiy in 1934 and served as a rich source of material, especially of the personal and anecdotal variety, for later biographical studies.
Yu. G. Saushkin, 'Nikolay Nikolayevich Baranskiy

(1881-1963)' in N. N. Baranskiy, N. P. Nikitin, V. V. Pokshishevskiy and Yu. G. Saushkin (ed) *Ekonomicheskaya geografiya v SSSR, istoriya i sovremennoye razvitiye [Economic Geography in the USSR; Its History and Contemporary Development]*, Moscow, Prosveshcheniye (1965), 511-39. This biobibliographical article on Baranskiy, written by one of his leading disciples, is one of about fifty articles on pre-revolutionary and Soviet economic geographers included in this volume, which was originally conceived by Baranskiy and appeared after his death. The bibliographical portion, compiled by Ye. A. Stepanova, is of particular interest. It opens with a list of Baranskiy's works taken from the Nazarevskiy bibliography (see above) and checked off by Baranskiy himself as the most significant writings in his judgement. The bibliography is then updated to early 1964, and ends with a chronological listing of references on Baranskiy.
Nikolay Nikolayevich Baranskiy (1881-1963), Moscow, Nauka (1971), 119p. This little white-covered paperback is one of a series of biobibliographies of leading Soviet scientists published by the Academy of Sciences of the USSR. The Baranskiy volume, the fourth of a subseries devoted to geographers, includes a biographical section by Saushkin (8-36) and a bibliography compiled by G. N. Finashina and R. I. Kuz'menko (37-116). The bibliography consists of references on Baranskiy, arranged in alphabetical order, and works by Baranskiy, in chronological order, with separate listings of co-authors, titles of works, and authors whose works were reviewed by Baranskiy.
A. I. Solov'yev and M.G. Solov'yeva, *N. N. Baranskiy i sovetskaya ekonomicheskaya geografiya [N. N. Baranskiy and Soviet Economic Geography]*, Moscow, Prosveshcheniye (1978), 109p. This book, written in a simple and at times even chatty style, was intended as a teaching aid, introducing Baranskiy to secondary school students. The authors, a husband-and-wife geographer team, had a special relationship to Baranskiy. Aleksandr Ivanovich Solov'yev (1907-83), a physical geographer, was one of three students dispatched in 1929 by the Moscow University administration to invite Baranskiy to join the university staff full-time. Margarita Grigor'yevna Solov'yeva (b. 1909), was a student of Baranskiy, specializing in the economic geography of the United States.
Other references on Baranskiy are too numerous to be listed fully within the scope of the present study. Soviet scientists of prominence are honoured customarily with anniversary articles starting with the sixtieth birthday and continuing on so-called round dates (the seventieth birthday, the eightieth, and so forth). In an unusual twist of chivalry in this age of feminism, no ages are mentioned in articles honouring women. In Baranskiy's case, the sixtieth birthday, which fell in July 1941, was not marked for some reason, but beginning with the sixty-fifth, in 1946, he was the subject of anniversary articles in a wide range of publications at five-year intervals. He also had the honour of figuring in Soviet encyclopedias both during his lifetime and after his death. When he died, on 29 November 1963, at the age of eighty-two, his high standing in public life was demonstrated not only by commemorative articles in the Soviet geographical

journals, but by obituaries in the principal Soviet newspapers, *Pravda* and *Izvestiya*. The centennial of his birth, in 1981, was the occasion of another round of articles and events honouring his memory. Following, in chronological order, is a selection of some of the more significant references on Baranskiy in addition to the five special studies mentioned earlier:

1946 *Izvestiya Vsesoyuznogo Geograficheskogo Obshchestva*, vol 78, no. 2, 139-56. This issue of the journal of the Soviet Geographical Society honoured Baranskiy on his 65th birthday with an anniversary article (139-42) and articles on his role in economic cartography by A. I. Preobrazhenskiy (143-8), on his work at Moscow University by I. A. Vitver and I. M. Mayergoyz (149-52) and on his contributions to the teaching of geography in secondary schools by V. A. Raush, S. I. Nikol'skaya et al (153-6)

1951 *Voprosy Geografii*, vol 27. This thematic volume of a serial founded by Baranskiy in 1946 is devoted to economic geography and dedicated to his seventieth birthday. It contains a special anniversary article (5-16) on his life and work; a personal tribute by V. N. Sementovskiy (1882-1969), a Kazan' University physical geographer (17-30); an evaluation of Baranskiy's contribution to the teaching of geography in schools by A. A. Dometti and V. A. Raush (31-43), and a brief appreciation of Baranskiy's standard school textbook on the economic geography of the USSR by F. V. Stepukhin, an old rural schoolteacher (43-5). The volume also contains a brief report of a Moscow University meeting honouring Baranskiy on his seventieth birthday, at which Stepukhin, who had been a schoolteacher since 1899 in the northern Russian rural town of Ves'yegonsk, told how local collective farmers were using Baranskiy's economic geography text as an encyclopedic reference to current events when listening to the radio or reading newspapers

1956 I. M. Mayergoyz and Yu. G. Saushkin, 'Nikolay Nikolayevich Baranskiy (on his seventy-fifth birthday)', *Izvestiya Akademii Nauk SSSR, seriya geograficheskaya*, 1956, no 5, 56-60

1957 'On N. N. Baranskiy's seventy-fifth birthday', *Voprosy Geografii*, vol 41, 5-8

1961 V. A. Anuchin and V. I. Bykov, 'The first Marxist-Leninist in economic geography', *Geografiya i Khozyaystvo*, vol 10, 3-6

---- D. L. Armand, V. M. Gokhman et al. 'Nikolay Nikolayevich Baranskiy (on his eightieth birthday)', *Izv. AN SSSR, ser. geogr.*, 1961, no 5, 148-50

---- V. P. Korovitsyn, I. V. Nikol'skiy and A. N. Rakitnikov, 'Nikolay Nikolayevich Baranskiy (on his eightieth birthday)', *Izv. VGO*, vol 93, no 4, 292-6

---- I. M. Mayergoyz and Ye. A. Trofimovskaya. 'On the eightieth birthday of Nikolay Nikolayevich Baranskiy', *Izv. VGO Obshch.*, vol 93/4, 289-91

---- Yu. G. Saushkin. 'A great Soviet geographer (on the eightieth birthday of Nikolay Nikolayevich

Baranskiy', *Vestnik Moskovskogo Universiteta, seriya geografiya*, 1961, no 3, 4-7

1963 'Baranskii, Nikolai Nikolaevich', in: John Turkevich. *Soviet Men of Science*, Princetown, New Jersey, Van Nostrand (1963), 37-8 (book was reprinted in 1975 by Greenwood Press in Westport, Connecticut)

---- (obituary) *Pravda*, Dec 1, 1963

---- (obituary) *Izvestiya*, Dec 1, 1963

---- (obituary) *The New York Times*, Dec 1, 1963

1964 Fuller obituaries in Soviet geographical journals appeared in: *Geografia v Shkole*, 1964, no 1, 6-7; *Vestnik Moskovsk. Univ., ser. geogr.*, 1964, no 1, 56-60 (by A. I. Solov'yev): *Izv. VGO*, 1964, vol 96, no 1, 81-2 (by I. A. Vitver, V. M. Gokhman, I. M. Mayergoyz, A. N. Rakitnikov); *Izv. AN SSSR, ser. geogr.*, 1964, no 3, 130-2 (by V. M. Gokhman and O. R. Nazarevskiy); *Voprosy Geografii*, vol 65, 9-13 (by Yu. G. Saushkin et al.). In foreign geographical journals, there were obituaries in *Geogr. Rev.*, vol 54 (1964), 282-3 (by Chauncy D. Harris), in *Ann. géogr.*, vol 74 (1965), 195-7 (by Pierre George) and in *Petermanns geogr. Mitt.*, vol 108 (1964), 102 (by Alfred Zimm)

1965 I. M. Mayergoyz, 'Nikolay Nikolayevich Baranskiy and Soviet urban geography', *Voprosy Geografii*, vol 66, 3-5

---- A. I. Preobrazhenskiy. 'The scientific contributions of Nikolay Nikolayevich Baranskiy to the development of Soviet cartography', *Geografiya v Shkole*, 1965, no 4, 5-10

1966 Yu. A. Kolosova. 'Nikolay Nikolayevich Baranskiy's work on the geography of the United States', *Vestnik Moskovsk. Univ., ser. geogr.*, 1966, no 3, 74-6

---- A. M. Zimina. 'Nikolay Nikolayevich Baranskiy and school geography', *Uchenyye Zapiski Moskovskogo Gosudarstvennogo Zaochnogo Pedagogicheskogo Instituta, Kafedra Pedagogiki i Psikhologii [Scientific Papers of the Moscow State Correspondence Pedagogic Institute, Dept. of Pedagogy and Psychology]*, 1966, no 1, 109-21

1971 V. I. Bykov. *Nikolay Bol'shoy [Big Nikolay]*, Irkutsk, East Siberian Publishers (1971), 105 p. Account of Baranskiy's early life, based on his personal records and archival material.

1973 Z. G. Freykin, 'N. N. Baranskiy and school geography', *Geografiya v Shkole*, 1973, no 3, 33-9

1980 V. A. Anuchin, 'Striving for the wholeness of geography as a science', introduction to N. N. Baranskiy, *Izbrannyye trudy. Nauchnyye printsipy geografii [Selected Works. Scientific Principles of Geography]*, Moscow, Mysl' (1980), 3-17

1981 In this centennial year of Baranskiy's birth, Soviet geographical journals devoted large portions of an issue to the occasion, as follows:

---- *Izv. AN SSSR, ser. geogr.*, 1981, no 5, published an introductory speech given by I. P. Gerasimov at a meeting May 26 1981, marking the centennial, (37-8), an article by Gerasimov, O. A. Kibal'chich, G. M. Lappo and Ya. G. Mashbits on Baranskiy and so-called constructive geography

(38-44) and an article by L. S. Abramov and
E. M. Murzayev on Baranskiy and regional
geography (44-9)

---- *Izv. VGO*, 1981, no 3, published seven articles
on the occasion of the anniversary: I. D.
Papanin, 'The father of economic geography at
the Geographical Society' (193-4); Yu. G.
Saushkin, 'The centennial of the birth of
Nikolay Nikolayevich Baranskiy' (195-9);
O. I. Shabliy, 'The evolution of N. N.
Baranskiy's ideas in the concept of inter-
sectoral territorial complexes' (200-06);
P. V. Voloboy and V. A. Popovkin, 'N. N.
Baranskiy's ideas and economic-geographic
research in the Ukraine' (207-14); L. N. Karpov,
S. B. Lavrov and G. V. Sdasyuk, 'N. N. Baranskiy
and the geography of foreign countries' (215-20;
translated in *Soviet Geography*, June 1983, 423-
29); O. A. Konstantinov, 'N. N. Baranskiy's
struggle for the regional approach in economic
geography' (221-29); B. S. Khorev, 'N. N.
Baranskiy as an educator', (230-4; translated
in *Soviet Geography*, September 1984, 508-14)

---- *Vop. Geogr.*, vol. 116, a thematic volume devoted
to regional geography, contained articles by:
I. D. Papanin, 'N. N. Baranskiy and the Moscow
branch of the Geographical Society of the USSR
[which has main office in Leningrad]' (13-14);
L. S. Abramov and E. M. Murzayev, 'N. N.
Baranskiy's contribution to the development of
regional geography of the USSR' (14-28); Yu. K.
Yefremov, 'The logic and the method of regional
description (in light of N. N. Baranskiy's
ideas)' (28-36); V. A. Anuchin, 'Geography,
regional geography and the systems approach'
(36-50)

---- *Vestnik Moskovsk. Univ., ser. geogr.*, 1981, no 3,
contained two articles: Yu. G. Saushkin, 'The
lessons of Nikolay Nikolayevich Baranskiy'
(18-23; translated in *Soviet Geography*, December
1982, 736-43), and A. I. Alekseyev, 'N. N.
Baranskiy and the study of man in geography'
(29-32)

---- *Geografiya v Shkole*, the geography teachers'
magazine that Baranskiy founded in 1934, carried
centennial material in several issues.   In no 2,
A. I. Preobrazhenskiy, 'N. N. Baranskiy's work
in cartography' (19-21; no 3, Ye. S. Antonyuk
(a Moscow schoolteacher) 'N. N. Baranskiy's
ideas about teaching methods' (37-42); no 4,
M. G. Solov'yeva, 'N. N. Baranskiy's teachings
about the significance of the geographical
division of labour in economic geography' (20-4),
and under the overall title 'N. N. Baranskiy,
teacher of teachers', segments by V. P. Zamkovoy
(32-3), Z. G. Freykin (33-6) and K. I. Yurkevich
(a schoolteacher) (36-7)

Separate entries on Baranskiy can be found in
several Soviet encyclopedias, starting in the
1920s with the first edition of the *Bol'shaya
Sovetskaya Entsiklopediya [Great Soviet
Encyclopedia]*, in which he was the geography
editor and to which he contributed a number of
geography articles, including entries on cities

and states of the United States. Biographical
entries on Baranskiy appeared in the *Bol'shaya
Sovetskaya Entsiklopediya*, 1st ed, vol 4 (1926),
692; 2 ed, vol 4 (1951), 226-7; 3 ed, vol 2
(1970), 714.   Also in the *Malaya Sovetskaya
Entsiklopediya*, a smaller multi-volume set, 2 ed,
vol 1 (1933), 714, and 3 ed, vol 1, (1958),
799-800.   Other entries may be found in
Sibirskaya Sovetskaya Entsiklopediya, vol 1
(1929), 236; in the *Entsiklopedicheskiy slovar'
[Encyclopedic Dictionary]*, 1 ed, vol 1 (1953),
143, and 2 ed, vol 1 (1963), 95; in the single-
volume *Sovetskiy entsiklopedicheskiy slovar'*,
1 ed, (1979), 109, and subsequent revised
editions, and in *Kratkaya geograficheskaya
entsiklopediya (Short Geographical Encyclopedia*,
vol 5, 1966, 417.   Professor Georgiy
Nikolayevich Gekhtman (1870-1956) of Tbilisi
University in Soviet Georgia also included
Baranskiy in his biographical dictionary
*Vydayushchiyesya geografy i puteshestvenniki
[Outstanding Geographers and Travellers]*, Tbilisi,
Academy of Sciences of the Georgian SSR (1962),
28-9.   Chauncy D. Harris wrote an entry,
'Baranskiy, Nikolai N.' for the *International
Encyclopedia of the Social Sciences*, New York,
Macmillan and Free Press, vol 2, (1968), 10-12

2. SELECTIVE BIBLIOGRAPHY OF WORKS BY NIKOLAY NIKOLAYEVICH BARANSKIY

Many of Baranskiy's journal-length articles were
published, sometimes more than once, in book-form
collections of articles.   The following selective
bibliography includes only the most significant
journal articles as well as the collections of articles
and other books, written or edited alone or in
collaboration, and articles or books translated into
English

1907   [Description of the Siberian village of
Chistyun'ka based on a household census of 1901]
in: M. Tregubov.   *Sibirskoye pereselencheskoye
seleniye sela Chistyun'ka Barnaul'skoy volosti
Barnaul'skogo uyezda [The Settling of the
Siberian Village of Chistyun'ka in Barnaul
Volost of Barnaul Uyezd]*, Barnaul, 9, 10, 11,
14-15, 20-21, 32-33, 35-40, and Appendix on the
1901 census, 3-33 (separate pagination).
Baranskiy's 1901 census of this village,
generally regarded as his first attempt at an
economic-geographic study, was incorporated into
the Tregubov book together with the results of
censuses taken in 1897 and 1905 without
indication of Baranskiy's authorship.   In a
Foreword, Tregubov says that the 1901 census
was conducted by Baranskiy

1923   *V ryadakh Sibirskogo Sotsial-demokraticheskogo
soyuza (vospominaniya o podpol'noy rabote
1897-1908) [In the Ranks of the Siberian Social-
Democratic Union (Memoirs of Underground Work,
1897-1908)*, Novonikolayevsk [the present
Novosibirsk], 1923, 89p.; 2 ed, Tomsk, 1961

1926   *Ekonomicheskaya geografiya SSSR.   Obzor po
oblastyam Gosplana [Economic Geography of the*

USSR. Survey by Gosplan Oblasts]. Moscow-Leningrad, Gosizdat, 1926, 294 p.; 2 ed, 1927, 336 p. [Note: The 1926 edition was translated into Yiddish, Tatar and Finnish.]

---- (editor of Russian edition and author of Foreword) Alfred Weber, *Teoriya razmeshcheniya promyshlennosti*, Leningrad-Moscow, Kniga, 223 p. [Russian edition of: *Über den Standort der Industrien*], Baranskiy's Foreword, 3-13

1928 'Plan for the economic-geographic description of a Gosplan oblast', in: *Sotsialisticheskoye khozyaystvo [Socialist Economy]*, 1928, vol 1, 183-98

1930 (editor of Russian edition and author of Editor's Introduction) Alfred Hettner, *Geografiya, yeye istoriya, sushchnost' i metody*, Translated from the German by Ye. A. Torneus, Moscow-Leningrad, Gosizdat, 1930, 416 p. [Russian edition of: *Die Geographie, ihre Geschichte, ihr Wesen und ihre Methoden*], Baranskiy's Introduction, 3-6

1933 *Geografiya SSSR [Geography of the USSR]*, textbook for secondary schools. Part 1 (for 6th and 7th grades). Moscow, Uchpedgiz, 1933, 198 p.; Part 2 (for 8th grade). Moscow, Uchpedgiz, 1933, 176 p. (Part 1 was translated into sixteen Soviet languages and five foreign languages, including English: Part 2 was translated into seven Soviet languages and German.)

1935 *Fizicheskaya geografiya SSSR [Physical Geography of the USSR]*, 7th grade textbook. Moscow, Uchpedgiz, 1935, 120 p. (This textbook went through seven Russian-language editions, 1935-40, 1943, and was translated into thirty-one Soviet languages and seven foreign languages.)

---- *Ekonomicheskaya geografiya SSSR [Economic Geography of the USSR]*, 8th-grade textbook. Moscow, Uchpedgiz, 1935, 408 p. (This textbook went through sixteen Russian-language editions, 1935-39, 1945-55, and was translated into twenty Soviet languages and nine foreign languages, including English.)

1939 *Ekonomicheskaya kartografiya [Economic Cartography]*, vol 1, 'The Method of Mapping Economic Phenomena', Moscow, Moscow Engineering Institute of Geodesy, Aerial Photography and Cartography, 1939, 162 p., mimeographed

1940 *Ekonomicheskaya kartografiya*, vol 3, *The Process of Compiling Economic-Geographic Maps and Atlases, Map Reading and Analysis*, Moscow, Geodezizdat, 1940, 96 p.

1941 'The method for teaching a regional course on the economic geography of the USSR', *Izv. VGO*, 1941, vol 73, no 1, 16-22

1946 *Ekonomicheskaya geografiya SShA [Economic Geography of the United States]*, Part 1: *General Survey*, vol 1, Moscow, Institute of International Relations, Ministry of Foreign Affairs of the USSR, 1946, 92 p.

---- 'Generalization in cartography and in geographic textual description', *Uchenyye zapiski MGU [Moscow University Scientific Papers]*, vol 119, 1946, 180-205

---- 'On the economic-geographic study of cities', *Voprosy Geografii*, vol 2, 1946, 19-62;

translated in: George J. Demko and Roland J. Fuchs, ed, *Geographical Perspectives in the Soviet Union, a Selection of Readings*, Columbus, Ohio State University Press, 1974, 589-610

---- *Ocherki po shkol'noy metodike ekonomicheskoy geografii [Essays on the Method of Teaching Economic Geography]*. Moscow, Uchpedgiz, 1946, 227 p.; 2d ed, 1954: 3 ed, 1955

1954 *Istoricheskiy obzor uchebnikov geografii (1876-1934) [Historical Survey of Geography Textbooks (1876-1934)]*, Moscow, Geografgiz, 504 p.

1956 *Ekonomicheskaya geografiya. Ekonomicheskaya kartografiya [Economic Geography. Economic Cartography]*, collection of articles, Moscow, Geografgiz, 1956, 366 p.; 2d ed, 1960, 452 p.

1957 *Ekonomicheskaya geografiya v sredney shkole. Ekonomicheskaya geografiya v vysshey shkole [Economic Geography in Secondary School. Economic Geography in Higher School]*. Collection of articles, Moscow, Geografgiz, 1957, 328 p.

---- 'Introduction to Russian edition', in Preston E. James and Clarence F. Jones, ed, *Amerikanskaya geografiya. Sovremennoye sostoyaniye i perspektivy*, Moscow, 1957, 5-18 (Russian edition of: *American Geography, Inventory and Prospect*, Syracuse University Press, 1954, 590 p.). In addition to writing the Introduction to the Russian edition, Baranskiy also translated four of the chapters: Harold M. Mayer on urban geography, Harold H. McCarty on agricultural geography, Charles M. Davis on field techniques, and Arthur H. Robinson on geographical cartography

---- (editor, with N. P. Nikitin and Yu. G. Saushkin), *Otechestvennyye ekonomiko-geografy XVIII-XX vv. [Russian Economic Geographers of the 18th to 20th Centuries]*, Moscow, Uchpedgiz, 328 p.

1959 (editor, with N.Ye. Dik, Yu. K. Yefremov, A. I. Solov'yev and N. A. Solntsev) *Otechestvennyye fiziko-geografy i puteshestvenniki [Russian Physical Geographers and Travellers]*, Moscow, Uchpedgiz, 783 p.

1960 *Metodika prepodavaniya ekonomicheskoy geografii [The Method of Teaching Economic Geography]*, Moscow, Uchpedgiz, 451 p. Revised edition of *Ocherki ...*, 1946

1962 (author, with A. I. Preobrazhenskiy) *Ekonomicheskaya kartografiya [Economic Cartography]*, Moscow, Geografgiz, 1962, 286 p.

1964 'My life in economic geography (excerpt from memoirs)', *Geografiya v Shkole*, no 1, 8-12

1965 (editor, with N. P. Nikitin, V. V. Pokshishevskiy and Yu. G. Saushkin) *Ekonomicheskaya geografiya v SSSR. Istoriya i sovremennoye razvitiye [Economic Geography in the USSR: Its History and Contemporary Development]*, Moscow, Prosveshcheniye, 663 p.

1974 'A brief outline of the development of economic geography', in: George J. Demko and Roland J. Fuchs, ed, *Geographical Perspectives in the Soviet Union, a Selection of Readings*. Columbus, Ohio State University Press, 1974, 8-30: also includes 'Consideration of the natural environment in economic geography: formulation of the problem', *ibid*, 31-47 and 'On the economic-geographic study of cities', *ibid*, 589-610

1980 *Izbrannyye trudy. Stanovleniye sovetskoy ekonomicheskoy geografii [Selected Works. The Rise of Soviet Economic Geography]*, collection of articles, Moscow, Mysl'

---- *Izbrannye trudy. Nauchnyye printsipy geografii [Selected Works. Scientific Principles of Geography]*, collection of articles, Moscow, Mysl', 239 p.

3. UNPUBLISHED SOURCES

So far as can be ascertained, no effort has been made thus far to collect archival material on Baranskiy. Zakhar Grigor'yevich Freykin, a Baranskiy disciple, in the 1973 article 'N. N. Baranskiy and school geography', urged that a collection of Baranskiy material be undertaken. According to Freykin, Baranskiy conducted an active correspondence with a wide range of professional geographers, teachers and others, often voicing his views about issues in geography. For many years, Baranskiy sat on the academic councils of the People's Commissariat (now Ministry) of Education of the Soviet Union's Russian Federated Republic, of the Central Scientific Research Institute of Geodesy, Aerial Photography and Cartography, Moscow University, the Institute of Geography of the Academy of Sciences of the USSR. Freykin urged that these and other institutions with which Baranskiy was associated might undertake a search for Baranskiy material, collect it in one central place, and publish the most valuable papers.

Theodore Shabad is Editor of Soviet Geography

Chronology

1881	Born in Tomsk, Siberia, 27 July (14 July according to Russia's old-style Julian Calendar)
1898	Joined the Russian Social Democratic Workers Party, precursor of the Communist Party
1899	Graduated from the Tomsk Gymnasium (secondary school) with a Gold Medal and admitted to the Law Faculty of Tomsk University
1901	Expelled from Tomsk University for political agitation and barred from admission to other Russian universities
1901	Commissioned by the Society of Altay Area Studies to conduct a census among settlers in the village of Chistyun'ka, near Barnaul: his work in the village results in a social-geographical study, published in 1907
1902	Founded a revolutionary underground cell, the Siberian Group of Revolutionary Social Democracy, which in the following year became part of the Siberian Social Democratic Union
1903-8	Active as a professional revolutionary in Siberia, hunted by the tsarist police, and jailed three times
1905	Delegate of Siberian Bolsheviks at a conference of the Russian Social Democratic Workers Party in Tammerfors (Tampere), Finland
1908	After serving the third prison term, in Chita, was expelled from Siberia, abandoned revolutionary underground activity and settled in Ufa
1910-14	Studied economics and mathematical statistics at the Moscow Commercial Institute (now the Institute of the National Economy named for Plekhanov)
1915-17	Employed as an economist in Zemgor, a war relief society organized by the liberal bourgeoisie and Mensheviks to provide medical aid to the tsarist army in World War I
1917-20	Shifted his political allegiance temporarily to the so-called Mensheviks-Internationalists, a left-wing faction collaborating with the Bolsheviks
1918-19	Employed as an economist in the Chemical Division of the Supreme Council of the National Economy, an agency responsible for reconstruction of the war-damaged economy
1919	Became inspector for the People's Commissariat of State Control, an auditing agency, and began lecturing on economic geography in newly formed workers' study groups
1919-20	Headed the Chelyabinsk (Urals) branch of the State Control Commissariat
1920-1	Headed the Economics Division of the newly formed Siberian Revolutionary Committee in Omsk, and was appointed deputy chief of the committee's statistical office
1921-5	Summoned back to Moscow to work for the state control agency, which by then had been renamed the People's Commissariat

of Workers and Peasants Inspection

1921-9	Simultaneously accepted the first appointment in economic geography as chairman of department at what was then called the Communist University, a party school
1925-26	Served briefly as deputy rector and then rector of the Communist University of Workers of the East, a training school for officials for the eastern regions of the USSR
1925-45	Simultaneously geography editor of the Soviet Encyclopedia publishing house
1926	Attended a conference of party-school economics teachers and read a paper on how to set up a course in economic geography
1927	Attended an international congress of geographers and ethnographers of the Slavic countries in Poland
1927-30	Accepted a second teaching appointment as professor at what was then called Second Moscow University (now the Moscow State Pedagogic Institute, a teachers college)
1929	Attended the first national conference of geography teachers and read two papers, on the geographical division of labour and on the concept of the economic-geographic situation of a place or area
1929	Shifted his main teaching position to Moscow University, where he held the chairmanship of the Department of Economic Geography of the USSR until 1946, except for an interruption (1941-43) during World War II
1933	Attended the First Soviet Geographical Congress in Leningrad
1933-38	In addition to Moscow University appointment, headed the Department of Economic Geography of what was then the Institute of World Economy and World Politics
1934	Attended International Geographical Congress in Warsaw
1934	Founded the geography teachers journal *Geografiya v Shkole* and served as its editor until 1947, except for an interruption (1941-46) in World War II
1935	Awarded the degree of doctor of geographical sciences on the basis of

publications (without a dissertation)

1936-40	In addition to Moscow University position, headed the Department of Economic Geography at the Moscow State Pedagogic Institute
1937-8	Member of the Academic Council of the Institute of Geography, Academy of Sciences of USSR
1939	Became a corresponding member of the Academy of Sciences
1941	Served briefly as head of the Geography Department at the Higher Party School before leaving war-threatened Moscow
1941-3	During wartime evacuation in Alma-Ata, headed the Geography Sector of the Kazakh Branch of the Academy of Sciences, and the Geography Department of the Kazakh Pedagogic Institute
1943-6	In addition to resuming his Moscow University position on the return to Moscow, was in charge of the Geography Department of so-called Lenin Courses, a party study programme
1945	Awarded the Order of the Red Banner of Labour on the 220th anniversary of the Academy of Sciences
1946	Awarded the Order of Lenin on his sixty-fifth birthday
1946-51	Was editor for geography of the Foreign Literature Publishing House
1946-63	Founded the geographical serial publication *Voprosy Geografii* and served as chairman of editorial board until his death
1947	Attended the Second Soviet Geographical Congress in Leningrad
1948	Elected as honorary member of the Bulgarian Geographical Society and the Yugoslav Geographical Society
1951	Awarded the Semenov-Tyan-Shanskiy Gold Medal by the Soviet Geographical Society
1952	Awarded the Stalin Prize for the twelfth edition (1950) of his standard textbook on the economic geography of the USSR
1953	Awarded a second Order of Lenin for his work
1954	Elected as honorary member of the Polish Geographical Society

1960 Elected as honorary member of Serbian Geographical Society and awarded the Jovan Cvijic Gold Medal

1962 Awarded the title of Hero of Socialist Labour, the Soviet Union's highest civilian honour, and a third Order of Lenin on his eightieth birthday

1963 Died in Moscow 29 November

Emrys George Bowen

1900–1983

COLIN THOMAS

1. EDUCATION, LIFE AND WORK

Emrys Bowen was born on 28 December 1900 at Carmarthen where he was educated at Queen Elizabeth's Grammar School. His father had worked in the tinplate industry before becoming an insurance agent and, in characteristic South Walian fashion, he recognized that the further education he had not experienced would be the medium of his son's greater opportunity in life. The strict values of Nonconformist ethics, acquired in the local Baptist church, permeated family life and remained with Emrys and his sister throughout their days. Indeed that upbringing may be distinguished as one source not only of his character and behaviour but also of some of his academic work in later years.

Having taught as an assistant in a Carmarthen school in 1919-20 Bowen proceeded with the aid of a state bursary to the University College of Wales, Aberystwyth, an institution founded partly by mass public subscription during the great liberal democratic revival in Wales in the 1860s and 1870s. By the turn of the century the College had attracted students from beyond the confines of the Principality, one of whom was H. J. Fleure, a brilliant final-year zoologist at the time of Bowen's birth. It was Fleure who had been appointed to the newly endowed Gregynog Chair of Geography and Anthropology in 1917 and to which Bowen was to succeed in 1946. After graduating in 1923 with first class honours--the first geography student to do so--Bowen was awarded a teacher's diploma in 1924 and remained at Aberystwyth as a research student until taking up the Cecil Prosser Research Fellowship at the Welsh National School of

Medicine in Cardiff (1925-28). Already the intellectual influence of Fleure, which may be traced throughout much of Bowen's subsequent writing, was becoming apparent. His master's thesis, which was judged worthy of a distinction when the degree was awarded in 1926, was a study of the physical anthropology of the population of south-west Wales, while his pioneer essays on medical geography focused on the incidence of fibroid phthisis in the lead-mining area of north Cardiganshire. Relating the temporal and spatial aspects of industrially-induced tuberculosis to the economic geography of the communities in which it concentrated, and also to some of the anthropological characteristics of those who contracted the disease, Bowen published only five papers on this theme between 1928 and 1933. However they indicated that their author possessed a perceptive mind that integrated incisive originality of interpretation with systematic analysis of data and was not merely an image of those around him. Appropriately one of those essays was contained in *Studies in Regional Consciousness and Environment*, a collection of papers by former students which was presented to Fleure in 1930 as a token of their regard for his inspiration over two decades.

Bowen's teaching career began in 1929 when he was appointed to an assistant lectureship in geography and anthropology at Aberystwyth, where he became a full lecturer in 1933 and a senior lecturer in 1942. Partly because of his own poor health he took no active part in the Second World War, though in later years he sometimes referred impishly to his prowess as a physical geographer during the years 1939-44 when he served as a special lecturer in meteorology at the

Royal Air Force training establishment at Aberporth. For more than thirty years his singular contribution to the education of undergraduates and postgraduates lay in the realm of historical geography and in particular the interpretation of settlement patterns and culture regions in Wales.

The University College at Aberystwyth was a small and intimate community for most of the first hundred years of its existence, a feature emphasized by its physical remoteness from large population centres even within Wales. It was natural therefore that, irrespective of Fleure's interdisciplinary creed, geographers should interact frequently with scholars in both the sciences and humanities and Bowen never failed to remind his professional colleagues on both sides of Offa's Dyke that the Department founded in 1918 was the first in the British Isles to offer honours degrees in both the arts and science faculties. The origins of the College and the establishment in the town of the National Library of Wales presented it with an opportunity to develop a distinctively Welsh ethos and Bowen himself co-operated closely with a series of distinguished scholars and scientists, notably Sir J. E. Lloyd whose history of early medieval Wales had been published in 1911. Lloyd had been a lecturer at Aberystwyth before occupying the Chair of History at the University College of North Wales, Bangor, from 1899 to 1930 and one of the ventures he undertook during his retirement was to edit a two-volume history of Camarthenshire. That enterprise drew on talents capable of adding novel insights into the past by including a brief physiographic introduction by Bowen, who also joined Sir Cyril Fox, the dynamic director of the National Museum of Wales, to provide three further chapters which re-created a picture of the area in the Neolithic period, the Bronze Age and the early Iron Age.

While Bowen may have derived part of his interest in efforts to reconstruct the geography of the past from Fleure's breadth of vision, it is clear that he found a more direct stimulus from advances being made in archaeology during the inter-war period. O.G.S. Crawford had been appointed archaeological officer in the Ordnance Survey in 1920; R.E.M. Wheeler had initiated major excavations of Roman sites during his years at the National Museum (1920-26); Fox had published his expansive *Personality of Britain* in 1932 and V. E. Nash-Williams, also at Cardiff, had continued the discovery of relics of Roman and sub-Roman settlement in Wales.

Ten years before the collaboration with Fox on the Carmarthenshire volume, Bowen had declared an interest in the problems of deciphering successive horizons of historic and prehistoric landscapes in his first publication, 'A study of rural settlement in South-West Wales', which appeared in the *Geographical Teacher*, thereby also following Fleure in establishing what was to be a long connection with the Geographical Association. By 1935, too, he was beginning to mark out a research field that became essentially his own, namely a study of the evidence of culture regions in Wales during the Dark Ages. An unpublished paper read to the infant Institute of British Geographers reiterated the views outlined in an essay on 'Early

Christianity in the British Isles' which had appeared in *Geography* three years earlier and now heralded a spate of articles in the 1930s and 1940s, two of them in Crawford's journal *Antiquity*, which culminated in his book *The Settlements of the Celtic Saints in Wales* (1954). In similar vein there followed *Saints, Seaways and Settlements in the Celtic Lands* (1969) and, at the invitation of Professor Glyn Daniel, *Britain and the Western Seaways* (1972) in the series 'Ancient Peoples and Places'.

In 1946 Bowen was appointed to the Chair in Geography and Anthropology at Aberystwyth, and during his tenure of the next twenty-two years he sought to continue the traditions of a broad humanist education which had been laid down by Fleure and, more briefly, by his successor, the anthropologist Daryll Forde. Five years earlier he had published *Wales, a Study in Geography and History* as an introductory textbook for secondary school pupils and students in colleges or at first year University level. By the later 1950s other specialist appointments to his department made possible the production of a more lengthy, detailed and systematic text for university students, *Wales; a Physical, Historical and Regional Geography* (1958). As both the College and the Department expanded its number of students and staff, diversifying its range of undergraduate courses, strengthening and deepening its research effort across all branches of the subject, and sending out its graduates into an increasingly complex world, Bowen was always eager to keep in contact with former *alumni*. This he was able to achieve by active participation in the Old Students' Association, of which he was President in 1970, and by virtue of his position as Warden of the Guild of Graduates of the University of Wales from 1971 to 1974. In a characteristic style his address, *The Guild Today* (1973) recalled and illustrated the distinctly regional student catchments of the four constituent Colleges and the limited number of the Guild's subject sections, heavily biassed towards the Arts, in earlier generations. What was essential for the future, he argued, was a greater intensity of commitment to all facets of adult education and improved publicity between the administrative centre and the past and prospective students. Never a man to urge others to do what he would not do himself, Bowen had contributed towards that goal for many years by accepting invitations to speak either on local topics or wider issues to all manner of groups and societies throughout Britain. Within the opportunities provided by the history of his own department he was able to publicise with pride its past achievements and present specialisms on the occasion of its jubilee celebrations in 1968 in a volume of essays, *Geography at Aberystwyth*. Activities such as these, in addition to his personal academic status, led to the award of an honorary doctorate by the Open University in 1979, following a similar honour from the University of Wales in 1975.

Bowen's career and activities, even in his own discipline, were not confined to Wales and Welsh institutions. While he served on the governing bodies of the University and its constituent colleges, was a member of the University's Press Board (1955), and acted as Vice-President of the Cardiganshire (1956) and Carmarthenshire (1970) Antiquarian Societies, his qualities were also recognized further afield. In

1949 he was elected a Fellow of the Society of Antiquaries, London, and in 1954 he became a Fellow of the Royal Geographical Society which awarded him its Murchison Grant in 1958. Having been a member of the general committee of the British Association for the Advancement of Science in 1951, he presided over its Geography Section at the Association's meeting in Cardiff in 1960. Presidencies of the Institute of British Geographers (1958), the Geographical Association (1962) and the Cambrian Archaeological Association (1967) completed the honours bestowed on him by learned societies and professional bodies.

In a life where no set of related events or places seemed wholly coincidental, he spent his last years in the village of Llanbadarn Fawr, near the site of an ancient Celtic church and opposite the imposing outline of an Iron Age fort. Busy to the end he died at Aberystwyth, fifteen years after his retirement, on 8 November 1983.

2. *SCIENTIFIC IDEAS AND GEOGRAPHICAL THOUGHT*

In 1975 when Bowen's seventy-fifth birthday and the fiftieth anniversary of his first publication were marked by the presentation to him of a selection of his own writings, the editors of the volume, Harold Carter and Wayne Davies, suggested that 'the genealogy of his academic inheritance' seemed 'easily established' as being derived from Darwin via Ratzel. In reality nothing could be further from the truth. Apart from the fact that he understood very little German and probably read the French classical geographers largely in translation, Bowen regarded Ratzel as a key figure in the development of an unacceptable thesis of environmental determinism. According he persistently allied himself with the French school of Vidal de la Blache, admiring its members both for their stylish expressions of possibilism and also their coherent essays on regionalism in rural contexts. For many years students at Aberystwyth experienced something of that legacy through the teaching of Walter Fogg, a social anthropologist who had studied at the Sorbonne and transmitted that aura in his lectures on the regional geography of France. If any trace of German thinking was to be found in Bowen the least fanciful would have been that of von Richthofen's chorology, yet even that would imply that Bowen was a man of the nineteenth century, which he most certainly was not.

The essence of his philosophy, though he would never have used such an exalted term to describe his beliefs, was that geographers could do worse than to adhere to first principles. In turn that meant a sound appreciation of a three-dimensional world, subject to the effects of temporal variation as well as spatial contrast and interaction. To him it was virtually impossible to conceive a geography which excluded Man and his thoughts, therefore, were of *géographie humaine* rather than a vague *anthropogéographie* which all too often degenerated into unsubstantiated social or political theorizing. It may seem strange that one so immersed in his own culture, in no small part derived from his religious upbringing, should place such emphasis on material

things in his scholarship, but to Bowen the visible evidence of human creativity in the landscape was always the starting point for any journey of geographical discovery. In that he would readily be associated with the writings of Carl Sauer which he often quoted in his teaching of historical geography.

Bowen's research into settlement in the Dark Ages was greatly stimulated by Fox's *Personality of Britain* (1932) with its refinement and extension of Mackinder's general notion of the division of Britain into Highland and Lowland Zones. The interplay between Fox and Bowen was so close that it is impossible to determine in which direction ideas flowed, but the two were of one mind in their advocacy of the basic technique of plotting archaeological finds and other manifestations of past human activities as distribution maps. Whereas Fox concerned himself with regional patterns in prehistory, Bowen saw his task as the exploration of the proto-historic period. By its very nature that objective, while approached by the same methods, called for the critical use of different types of source material. In contrast to the more classical strategy employed by historians such as Lloyd, heavily reliant on uncertain literary evidence, Bowen turned to topography and toponymy as the basis for his studies of culture regions in post-Roman Wales. The spread of Celtic Christianity was seen to represent a central topic in that story and there appeared between 1932 and 1975 a series of papers which began by identifying from genealogies specific individuals or family groups reputed to have been active in spreading Celtic monasticism during the period from the fourth to the ninth centuries. Maps were compiled of distributions of churches and other holy places dedicated to those so-called saints, while their sites were examined and classified by location, altitude, slope and aspect. Interpretation of the emerging patterns did not rest, as some casual readers of Bowen's later books inferred, on intuitive and simplistic impressions drawn from the distribution alone. On the contrary, just as he had been scrupulous in sifting reliable data from archaic documentary sources, Bowen was equally judicious in relating the 'spheres of influence' of the religious cults to detailed archaeological evidence which was only beginning to come to light in the 1920s.

Having established the nuclear areas of these early cultural groupings in Wales, the search then began for their origins and diffusion. Historical tradition revealed two distinct, though parallel, links. The earliest connected some cults with the Romanised civil districts of the south-east and were enshrined in an atmosphere of residual sedentary scholasticism at well-known monasteries. The second strand was far more intricately woven and its diverse connections ranged over the whole length of the Atlantic seaboard from Spain, and more particularly Brittany, through Cornwall, Wales and Ireland to Scotland. The means whereby those separate personal and regional affiliations were articulated apparently comprised the skeletal network of roads in the Roman Military zone and especially the seaways, the significance of which in prehistoric times had been noted by Crawford and Fox. Such a framework was sketched in his introductory contribution to H. C. Darby's collaborative work, *An Historical Geography of England before A.D. 1800* (1936). The continuity of process implicit in the analogy, seized upon by Bowen as a major explanatory variable, proved to be complemented in numerous places by remarkable locational juxtaposition of finds from successive phases of occupation, all of which suggested continuity of settlement and culture.   Examples included the presence of early Christian churches within or adjacent to megalithic monuments, and of bilingual inscribed stones of the Dark Ages alongside Roman routes.

For more than forty years Bowen elaborated this thesis, developing his argument from the particular case of one saint to more representative regional clusters, eventually extending his carefully reasoned story to embrace the whole of Atlantic Europe.   As such his contribution to our understanding of the bridge between prehistory and history was widely acknowledged as being original and stimulating by archaeologists, who were regularly unearthing evidence which tended to support his views, while historians who were faced with conflicting evidence from textual analysis were sometimes less easily convinced of the precise validity of his tracery of connections.

An underestimated segment of Bowen's general scenario was his belief that developments occurring in the 'Age of the Saints' were responsible for outline settlement distributions in the medieval period.   By plotting graphically the altitudinal ranges of Celtic church sites and Iron Age forts and camps he demonstrated with his customary clarity that over a millenium there had been a redistribution of population foci, referred to as long ago as 1916 by Fleure and Whitehouse as the 'valley-ward movement', yet further detailed investigation of this dynamic three-dimensional view of settlement in Wales was left to his research students a generation later.

Another subtle variation on an old theme was his suggestion that closer analysis of cultural landscapes and their formative influences enabled us to distinguish not only the coarse dichotomy between Lowland and Highland Zones, but also within the latter the identification of a peninsular or Atlantic Zone of Britain, in which the 'mediterranean' character of the Irish Sea had consistently performed an integrative function.   At a lower order of scale Bowen did not embark upon regional studies on the French model, although he clearly believed that predominantly rural Wales was suitable for such treatment and in general terms clung to the view that regional geography was the most appropriate medium for expressing the unitary nature of the discipline.   In later years he modified his stance, as he had done in his progression of studies in historical geography, to stress the need for thematic selectivity of indicators of regionalism and in his native land it was logical that he took the changing distribution of the Welsh language to illustrate the complex relationships between the natural environment, economic, cultural and political forces.   He was averse to arid discussion of methodology for its own sake and publicly stood aloof from the debates that threatened to rupture geography in the 1960s, yet the core of his outlook was contained in his first book, published in 1941, in which he presented not only a summary of the physical and human geography of the country but also a concise

statement of opinion as to the purpose of historical geography and the part which the whole geographical training could play in education for citizenship. Appropriately that book concluded with a section on the fluctuating status and regional variations in the use of the spoken language.

When he returned to the theme in his Presidential Address to the Institute of British Geographers in 1959, and again in two papers written jointly with H. Carter in 1974-75, explanations of the linguistic patterns were formulated on lines very similar to those set out in 1941, despite the fact that new factors had come into play in the intervening period and no substantial research had been undertaken to analyse the processes of acculturation.   There survived, too, a perpetuation of the idea that the 'real Wales' could be defined by its retention of language, a cultural index that had been widely demonstrated in Europe to be highly volatile and oversimplified.   In Bowen's case its choice also appeared to be difficult to reconcile with his earlier commitment to landscape criteria for regional differentiation.   The closest he came to trying to resolve this dilemma was to write in 1968 that 'when no correlation exists between cultural and physical patterns it is futile to seek it', or even to describe both patterns side by side.   The historical geographer, like the student of present-day conditions, should not assume that the physical underlay is the essential pre-requisite for the understanding of every distribution'. Those sentences clearly use the term 'physical' as short-hand for 'the natural environment', but if its connotation is broadened to embrace the total visible landscape, including the handiwork of Man, then it does reveal how he was later drawn into the more intractable problem of trying to relate conflicting aspects of the material and non-material culture.

It should not be inferred from what has been said that Bowen's landscapes were immutable or that they were dominated by nature.   On the contrary, on every possible occasion he picked out the imprints of human activity: his were essentially 'landscapes with figures'.   They were, moreover, invariably human figures and he made few concessions to the modern trend towards quantification of a sophisticated sort, or indeed to nontraditional teaching methods.   Two of his favourite aphorisms were 'an ounce of insight is worth a pound of technique', and 'you cannot treat people merely as numbers', by which he meant that aggregate analysis, however refined and complex, tended to mask more than it revealed of the true nature of inter-personal and man-land relationships.   Against the power of that instinct, his presidential address to the Geographical Association, 'The Geography of Nations' (1963), also presented a paradox when he found himself arguing that politics at a national level held the key to a balanced geographer's view of the world, half way between the global strategic outlook that prevailed before 1914 and the inter-war regional methodology that was beginning to attract criticism in some academic quarters.

3.   INFLUENCE AND SPREAD OF IDEAS
The full extent of Bowen's impact on geography at large can only be measured by his personal qualities

and not by the quantity or quality of his writings, formidable though they may be. His books were not widely read or even reviewed outside Celtic Europe as far as is known he left the shores of the British Isles only in 1929, when he travelled to the meeting of the British Association for the Advancement of Science in Capetown, returning through Kenya. In any conventional way, therefore, on an international scale he was scarcely known in his own lifetime. Conversely, so active and practical was his single-minded dedication to upholding geography's role in education that he was a familar personality in many walks of life in Wales and was highly regarded as a guest lecturer in a wider sphere. Like the missionary saints, the cattle drovers, the oceanic explorers, or the pioneer industrial entrepreneurs of whom he spoke with such vivid enthusiasm, Bowen was a master of communication in every sense. His technique was simplicity itself: the use of chalk on blackboard to outline the carefully structured approach to a topic, the choice of example so sharply etched in its detail that it served as a memorable symbol of the whole theme, and presentation as punctuated with humour as it was perfectly timed.

While the impressions left in the countryside by this quiet, unassuming man were spasmodic, at Aberystwyth his presence was all-pervading and unbroken. Generations of students, not only geographers, were captivated by the stories he had to tell and the way of their telling. His frugal life style may have suggested that of an austere recluse, but it was universally known that he was always willing to interrupt his long working hours to share a conversation or attend to the worries of caretakers, undergraduates or colleagues alike. Ever increasing numbers of students were attracted to his department either directly by his magnetism or more especially on the recommendation of hundreds of former students who had entered the teaching profession throughout the British Isles.

Quality was never sacrificed merely to accumulate large student numbers, but many able graduates proceeded to programmes of research under his subtle supervision. Hardly any of those postgraduates embarked on investigations of the topics that he himself came to dominate, namely culture and settlement in the Dark Ages and prehistoric periods, but they frequently developed similar themes in other historic phases and diverse parts of Britain, shedding further light on our understanding of that continuity from a proto-historic substratum that Bowen had always sought to demonstrate. The most significant contribution in that direction came from G.R.J. Jones, blending together the results of archaeological discoveries with meticulous examination of medieval documentary sources in an essentially geographical treatment of rural settlement patterns and social organisation, work that also owed much to the historian T. Jones Pierce.

An overwhelming preoccupation with rural communities also gave scope for the stimulating explorations in social geography and sociology by Emrys Jones and Alwyn Rees respectively, in reconstruction of linguistic frontiers and religious affiliations in the eighteenth and nineteenth

centuries by G. J. Lewis and W.T.R. Pryce, and in David Thomas' re-creation of the agricultural geography of Wales in the Napoleonic era. Among the very few of his students who went on to develop his earliest interests in anthropology and medical geography were Eric Sunderland, R.M. Prothero and G.M. Howe. It was left very largely to a second generation to devise new analytical techniques, to carry Bowen's statements on distributions and morphology into another phase in which the functional relationships of both pattern and process could be unravelled, and devote greater attention to the problems of urban and industrial communities in the modern world. As one who fervently believed with Pope that 'the proper study of mankind is man', he never thought it necessary to proclaim in his writings any formal creed and was impatient with those who did. For him to bind together people and place was enough, and in that elemental faith lay his frailty and his strength. There can be little doubt that both geography and education in Britain were enriched by that long-term commitment.

Bibliography and Sources

1. REFERENCES ON E.G. BOWEN
Obituaries include those in *The Times*, 12 November 1983; in *Studia Celtica*, vol 38-9 (1983-4), 327-9 by H. Carter and in *Transactions of the Institute of British Geographers*, n.s., 9/3, (1984), 374-80 with a bibliography. A selection of his writings, edited and introduced by H. Carter and W.K.D. Davies, was published under the title *Geography, Culture and Habitat: Selected Essays (1925-1975) of E.G. Bowen*, Llandysul (1976). An appreciation by another former pupil, Professor David Thomas, appeared in *Cambria*, vol 3 (1976), 169-71.

2. SELECTIVE THEMATIC BIBLIOGRAPHY

a. Historical Geography
1932 'Early Christianity in the British Isles. A study in Historical Geography', *Geography*, vol 17, 267-76
1934 'The Travels of St. Samson of Dol', *Aberystwyth Studies*, vol 13, 61-7
1936 'Introductory Background: Prehistoric South Britain' in Darby, H.C. (ed), *An Historical Geography of England before A.D. 1800*, Cambridge, 1-29
1945 'The Settlements of the Celtic saints in South Wales', *Antiquity*, vol 19, 175-86
1954 *The Settlements of the Celtic Saints in Wales*, Cardiff, 175 p.
1966 'The Welsh Colony in Patagonia 1865-1885: a study in historical geography', *Geogr. J.*, vol 132, 16-27
1969 'The Irish Sea in the Age of the Saints', *Stud. Celt.*, vol 4, 56-71
1969 *Saints, Seaways and Settlements in the Celtic*

lands, Cardiff, 245 p. (revised edition 1977)
1972 *Britain and the Western Seaways*, London, 196pp.
1972 'The Geography of early monasticism in Ireland',
      Stud. Celt., vol 7, 30-44

b. Medical and population geography
1928 'The incidence of phthisis in relation to
      racial types and social environment in South
      and West Wales', *J. R. Anthropol. Inst.*, vol 58,
      363-98
1930 'A clinical study of miners' phthisis in relation
      to the geographical and racial features of the
      Cardiganshire lead-mining area', in Peate,
      I.C. (ed), *Studies in Regional Consciousness and
      Environment: Essays presented to H.J. Fleure*,
      Oxford, pp 189-202
1960 'Welsh emigration overseas', *Adv. Sci.*, no 67,
      1-12
1975 'The distribution of the Welsh language in 1971:
      an analysis', (with H. Carter) *Geography*, vol 60,
      1-15

c. Regional geography
1925-26 'A study of rural settlements in South-West
         Wales', *Geogr. Teach.*, vol 13, 317-26
1934 'The rural settlements of central Wales',
      C.R. Congr. Int. Geogr. Paris 1931, vol 3,
      205-13
1941 *Wales: a Study in Geography and History*,
      Cardiff, 183 p. (revised edition 1943)
1957 (Editor) *Wales: a physical, historical and
      regional geography*, London, 528 p.
1959 'Le Pays de Galles', *Trans. Inst. Br. Geogr.*,
      no 26, 1-23
1962 'Rural Wales' in Mitchell, J. (ed), *Great
      Britain: Geographical Essays*, Cambridge,
      247-64
1964 'The Settlement Pattern of Wales', in Steers,
      J.A. (ed), *Field Studies in the British Isles*,
      London, 279-93
1971 'The dispersed habitat of Wales', in Buchanan,
      R.H., Jones, E., and McCourt, D. (ed), *Man and
      His Habitat: Essays presented to E.E. Evans*,
      London, 186-201

Bowen also contributed to *Encyclopedia Britannica*
(14th edition, of which he was assistant editor in
1929) and the *New Universal Encyclopedia* (1961), the
Cambrian Archaeological Association's regional
handbooks (1970-74), and county atlases of Anglesey,
Caernarfon and Merioneth (1972-75)

d. Biographical essays and notes
1970 'Herbert John Fleure and Western European
      geography in the twentieth century', *Geogr. Z.*,
      vol 58, 28-35
1974 *David Samwell (Dafydd Ddu Feddyg) 1751-1798*,
      Cardiff, 117 p.
1978 *John Hughes (Yuzovka) 1814-1889*, Cardiff, 82 p.
1983 *Dewi Sant: Saint David*, Cardiff, 112 p.

*Colin Thomas is Senior Lecturer in Geography at the
University of Ulster, Coleraine, Northern Ireland*

Chronology

1900	Born at Carmarthen, 28 December
1912-19	At Queen Elizabeth's Grammar School, Carmarthen
1919-20	Assistant teacher at the Model and Practising School, Carmarthen
1923	Graduated with first class honours in Geography and Anthropology, University College of Wales, Aberystwyth
1924	Teacher's Diploma, University of Wales
1925-28	Cecil Prosser Research Fellow, Welsh National School of Medicine, Cardiff
1926	M.A. (Wales) with distinction
1928-29	Assistant Editor, *Encyclopedia Britannica* (14th Edition)
1928	Wrote the first of a series of papers on factors influencing the incidence of tuberculosis among industrial workers in Wales
1929	Made a brief field visit to South Africa before becoming Assistant Lecturer in Geography and Anthropology, at Aberystwyth
1932	Published the first of many papers on the geography of Celtic Britain in the early Christian period
1933	Appointed Lecturer in Geography and Anthropology at Aberystwyth
1935-39	Collaborated with Sir Cyril Fox and Sir J.E. Lloyd on the two-volume *History of Carmarthenshire*
1939-44	Special Lecturer in Meteorology, No. 6 Initial Training Wing, Royal Air Force
1941	Published *Wales: a study in geography and history* as an integrated text for senior school pupils and university students
1942	Appointed Senior Lecturer at Aberystwyth
1946	Elected Gregynog Professor of Geography and Anthropology at Aberyswyth, a

	position he held until his retirement in 1968
1949	Elected a Fellow of the Society of Antiquaries, London
1953	Member of Court, University of Wales
1954	Published *Settlements of the Celtic Saints in Wales* Elected a Fellow of the Royal Geographical Society
1955	Became a member of the University of Wales Press Board; member of Council, St. David's University College, Lampeter
1956	Vice-Chairman of the Cardiganshire Antiquarian Society
1957	Member of Court, University College of Wales, Aberystwyth
1958	With other colleagues in his department he wrote *Wales: a Physical, Historical and Regional Geography* Was awarded the Murchison Grant by the Royal Geographical Society for studies in the geography of Wales President, Institute of British Geographers
1960	President, Section E (Geography) at the Cardiff meeting of the British Association for the Advancement of Science
1962	President of the Geographical Association
1965	Member of Council, University College of Wales, Aberystwyth
1967	President of the Cambrian Archaeological Association
1968	Edited *Geography at Aberystwyth*, a collection of essays published to celebrate the jubilee of the department's foundation
1969	Published *Saints, Seaways and Settlements in the Celtic Lands*, synthesising the results of his research on those topics over the previous forty years Had the status of Emeritus Professor conferred on him by the University of Eales
1970	Awarded D.Litt. (Wales) Vice-President Carmarthenshire Antiquarian Society; President of the Aberystwyth Old Students' Association

1971-74	Warden of the Guild of Graduates, University of Wales
1972	Published *Britain and the Western Seaways*
1974-83	Published bilingual monographs on David Samwell, the doctor on Captain Cook's voyages; John Hughes, the nineteenth century founder of the Ukranian steel industry; and St. David
1975	Honorary degree of LL.D. conferred by the University of Wales
1979	Honorary degree of D.Univ. conferred on by him by the Open University
1983	Died at Aberystwyth, 8 November

Henry Charles Carey

1793–1879

WILLIAM B. MEYER

1. EDUCATION, LIFE, AND WORK

Henry Charles Carey was born in Philadelphia, Pennsylvania, on 15 December 1793. His father was Mathew Carey, one of the nation's most prominent publishers and a man active in the social, political and cultural life of the city. The family publishing business produced a wide range of fictional and non-fictional works, both American and foreign. It was to work in this firm that Henry left school at the age of twelve. He more than made up for his paucity of formal education, however, with a lifelong course of voracious if unsystematic reading, much of it done in connection with his duties at the publishing house.

Carey became a full partner in his father's business in 1814 and married in 1819. He retired from publishing in his early forties and thereafter pursued a double career as an entrepreneur and a writer on political and economic topics. Following some reverses, Carey prospered as a businessman. Successful speculations in the coal and iron lands of northwestern Pennsylvania permitted him to devote full time in later life to his work in economic theory. A resident, briefly, of Burlington, New Jersey, Carey returned to Philadelphia and established himself in his native city as a sage of enormous if somewhat remote prestige. Several times urged to run for office, he refused, but as an adviser to officeholders, including Presidents, he exercised considerable influence. Originally a Whig, Carey joined the new Republican Party in the 1850s, only to desert it several years before his death for the Greenback

movement, whose programme of inflation through the issue of paper money he found attractive.

2. SCIENTIFIC IDEAS AND GEOGRAPHICAL THOUGHT

In any study of the geographical thought of the mid-nineteenth century in the United States, Henry C. Carey deserves a place alongside such familiar eminences as Marsh, Maury, and Guyot. He was an acute observer and interpreter of the changing economic and social geography of the nation. Though he is usually identified as 'essentially' an economist or a sociologist, Carey showed a consistent concern with questions of location, land use, and environment that justifies his consideration by modern geographers.

The basis of Carey's economic thought was an inversion of the sequence of cultivation set up by the English classical school, particularly David Ricardo. The most fertile soils, Ricardo had argued, would always be the ones farmed first, and returns from the extension of cultivation would necessarily diminish. The fusion of this idea with Malthusian population theory created a pessimistic doctrine that many American observers found both unpleasant and unrealistic.

Though still a fairly faithful follower of Ricardo in his early works published in the 1830s Carey turned his back on the classical school in the next decade and began to elaborate a system of his own, derived in large part from the ideas of his father's circle and from the elements of American experience that fitted poorly into Ricardian doctrine. Cultivation, Carey argued in direct opposition to the

English orthodoxy, always began on the poorest soils and not the richest because, being the poorest, they were the easiest for a frontier population to clear and farm.   The earliest years of any settlement were always its least, not its most, prosperous.   Only with the increase of population, capital and technology at any site could the richer but more densely vegetated or swampy lands be brought into production.

Carey considered his sequence to be universal and he offered historical illustrations drawn from a wide range of times and places.   For the clearest depiction, however, he was fond of describing the progress of a hypothetical valley.   Its few original settlers would be limited to farming the relatively sterile and rockly land on the hilltops because of their inability to clear and drain the much more fertile land in the river bottom below.   As their numbers grew and as labour and technology became available and mere subsistence required less effort, the settlers would gradually move down the slope and bring more of the fertile soils under cultivation. Their wealth would outstrip their numbers. Eventually a town would grow up in the valley, manufacturing would spring up, increasing the nearby market for the farmer, and the comforts of civilization would make their way into the lives of the people.

This was not merely an idyllic sketch of economic history, for in Carey's hands it had much more profound consequences.   If cultivation proceeded as he outlined it the Ricardian picture of steadily diminishing returns to agriculture was incorrect.   The supply of food per capita, far from decreasing with the growth and concentration of population, would increase.   Should population decrease in an area the richest soils would be the first abandoned and the standard of life would sink. For Carey there were no inevitable laws of progress that political decisions could not repeal.   Policies encouraging the growth of a nation's communities increased its prosperity, and those that tended towards the spatial dispersion of the population reduced the level of subsistence and of civilization.

The 'safety-valve' theory of American institutions was one widely held among Carey's contemporaries.   It made democracy and social harmony in the United States dependent on the all but unlimited supply of 'free' land on the western frontier.   This frontier was thought to act as a safety-valve to release pressures building up in the overcrowded cities of the East; the availability of land allowed the discontented from the cities to begin a new life in the farming regions.   The corollary was that once the frontier disappeared, America would suffer the same social ills as plagued the older countries of Europe.   Consequently, westward expansion and conquest were often justified as extending the amount of free land available and hence the lifespan of the nation's institutions.

Carey, to be sure, shared the goal of limiting unrest, in particular by restraining the growth of large cities.   For the safety-valve model which encouraged the dispersion of population he substituted a gravitational analogy that has since become his major claim to the notice of historians of spatial theory.   He realised, as many of his contemporaries did not, that the frontier drew from the countryside and the small towns, not the cities. People, Carey suggested, other things being equal, exerted a mutual attraction analogous to gravitation in the physical world.   The population, or 'mass', of large cities could only be controlled, it followed, by the maintenance around them of lesser centres of attraction, whose pull on individuals would balance that of the larger but more distant cities.   Let the 'mass' of these smaller towns be reduced, as western migration tended to do, and the drawing power of the cities would increase in proportion, thereby further depopulating the countryside and swelling the urban areas.   For Carey westward expansion and unchecked urban growth were the two sides of a single coin.

This aspect of Carey's thought is usually taken in the geographical literature as a direct precursor of the later 'social physics' of John Stewart and William Warntz, who in the 1940s and 1950s elaborated such concepts of spatial interaction as 'population potential' and the 'gravity model'.   There are indeed some points in common, but the characterization does not do full justice to the variety of Carey's approach.   Fond of declaring the essential unity of all laws, physical and social, the Philadelphian was particularly fond of gravitational analogies.   In 1848 he spoke mainly of the wealth of cities as the element of 'mass' attracting population.   In the later *Principles of Social Science*, he outlined the demographic formulation of the analogy, in which population was treated as the element of mass. It was to this version that Carey appealed in his attack on the safety-valve principle.   In the *Principles*, however, he was most concerned with the attractive force exerted on migration by the spatial concentration of political authority.   The retention of much power in the state capitals, Carey observed, had kept Washington from growing into an enormous centre in the way Paris and London had done, while the pull of the state capital was likewise countered by the attraction exerted by county and town seats. Nothing but a political decision was needed to upset the balance of population and create a new and very different distribution.   If the lower units were destroyed, Carey argued, then the concentration of power would make the capital cities grow rapidly to enormous size.   He gave numerous examples to show that the centralization of authority through the destruction of local powers had always produced rampant urbanization, political tyranny, environmental degradation, and class divisions.

Carey sought spatial decentralization in the economic realm as well as in the political.   His goal for the American future was of a set of more or less self-sufficient regional economies.   Each would be internally diversified, producing agricultural and manufactured goods alike, and in consequence would be minimally dependent on other regions and other countries.   This did not mean isolation, for a flourishing, nonmaterial intercourse in ideas and inventions and cooperation in matters of common defence appealed to Carey as the best way of sustaining such independence.   For all material products he drew a distinction between 'commerce', of which he approved, and 'trade', which he abhorred. Commerce, or 'association', involved relatively short-distance transactions, and trade long-distance ones. The commercial town served its own surrounding area, the trading city a distant market or markets. Commerce bred home values and honesty, while trade fostered impersonal exploitation. A commercial system made the regional community master of its own affairs while trade rendered it vulnerable to the changing whims and fortunes of distant customers. Commerce, above all, meant regional self-sufficiency, while trade required regional specialization coupled with centralized direction. From the 1840s onward a high tariff discouraging the export of American products and the import of European ones was a cornerstone of Carey's thought. In this policy he followed the associates of his father who had elaborated the 'American System' of tariffs and internal improvements for their political hero Henry Clay, but the younger Carey provided a comprehensive theoretical justification that had before been missing.

Carey fortified his economic regionalism with frequent references to the cost and waste of transportation over long distances. Ricardian economics largely disregarded the role of space. From his earliest work Carey, like most American economists of the time, pictured the 'friction' of distance as a heavy burden on both agriculture and industry. One of his disciples, the economist and politician E. Peshine Smith, introduced the distance-based economics of von Thünen to the American literature in 1853. In the *Principles of Social Science*, Carey cited von Thünen in a discussion of the location of crop production. His own arguments for the dispersion of manufacturing establishments rested in large part on the wastefulness of transporting raw materials long distances to factories. It would not be too much to say that Carey brought to its apex of elaboration the traditional American concern with distance, mediated and defined by social factors, as an essential dimension in economics.

To Carey it was as much political as common advantage that maintained a worldwide trading system, dominated by Great Britain, in place of a healthier commercial one. The 'costs of transportation' he identified were not exclusively economic. Morality in dealings declined as the distance involved increased, while trade likewise inflicted grave damage on the natural environment. In the ideal decentralized economy, Carey argued, the soil would be recycled and replenished with the wastes and refuse of the consumers in the nearby towns, and no net 'export' of soil material would result. Long distance agricultural trade prevented this recycling and eventually destroyed the productiveness of the land. If the places of production and consumption were close to one another, such destructive 'mining' of the soil would not occur. As Carey saw the pre-Civil War United States, with a national low tariff policy and the emphasis in the Southern plantation system on commercial crops to the exclusion of balanced cultivation, were steadily exhausting the land. The notion of plantation agriculture as environmentally destructive was popular in the northern states at the time. To such

ravages, in turn, Carey attributed the depopulation of the tidewater South through westward migration which reduced the supply of labour, forced the steady abandonment of the rich soils, and prevented the draining of swamps and the eradication of malaria. This led to further depopulation and to calls for wars of territorial expansion for the acquisition of new lands. The entire dismal chain of events, Carey argued, could be broken only by a protective tariff, one that would encourage industry and discourage an agriculture oriented towards foreign trade.

The modification of the environment by human action was of greater concern to Carey than was environmental influence on human behaviour. Indeed regional self-sufficiency, his ideal, presupposed the fundamental similarity of all regions, although Carey did not allow for some interregional trade in items specific to certain natural zones. Many in the mid-nineteenth century United States thought that the South was destined by nature to the exclusive pursuit of plantation agriculture; Carey believed it equally suited to manufacturing and to other forms of cultivation, and urged it to diversify accordingly. The English historian Henry Buckle achieved prominence with environmental, largely climatic, explanations of the differences between nations; Carey dismissed his approach, arguing that it failed to account for most of the disparities in the wealth, character, and government of physically similar areas.

It is not difficult to trace at least some of Carey's prescriptions for the American future to his personal interests, and it is all the more reasonable to do so in light of his own attacks, often personal ones, on such exponents of the classical school of economics as Ricardo and John Stuart Mill. Carey saw English economic thought, particularly the doctrine of free trade, as a tool for maintaining the dominance of Britain in the world economy. If his own thought was indeed a rationalization of personal or national interests, it was nonetheless an impressively rich, coherent, and detailed formulation, and a worthy alternative to the system it sought to supplant.

3. INFLUENCE AND SPREAD OF IDEAS

Carey's influence on the thought and policy of his time is difficult to specify, but all accounts agree that it was far from negligible, particularly among lawmakers, editors, and other moulders and servants of public opinion. Yet even here what was usually accepted and acted upon was not Carey's closely woven system, but parts--such as tariff protectionism--extracted and justified according to the predilections of the user. Certainly the Philadelphian spent much of his life in opposition to policies dominant in the national government, such as low tariffs before the Civil War and hard money after it.

Carey's disciples formed a self-conscious 'American School' of political economy that branched off in a number of directions from the master's thought. Those most concerned with the geographical dimension included R.E. Thompson *Social Science and National Economy*, (1875), and Van Buren Denslow, a Chicago lawyer whose *Principles of Economic Philosophy* (1888) includes a remarkable systematization of the 'American' view of space as a factor in economics. Neither, however, had much academic influence. Carey's doctrines found little favour with the pillars of the emerging discipline of economics in the United States. Trained in the more rigorous formulations of Ricardo and Mill, academic economists tended to be hostile to the intrusion of such political and social questions--and, for that matter, geographical ones--as lay at the heart of Carey's thought, into their more abstract 'science of wealth'.

Recent historians of economics have at least accorded Carey some attention as an interesting and provocative rebel against the intellectual orthodoxies of his time. Some have found his work still pertinent to modern problems. If one finds the Philadelphian cited in geographical research, however, it is likely to be for his early, and relatively unimportant, use of gravitational analogy to explain human interaction. In fact Carey, in his constant attention to questions of space, location, and the modification of the natural world, is a worthy precursor to modern geography, and his concern with what would now be termed the relations of 'core' and 'periphery' in a world economy likewise prefigures much modern thought.

Bibliography and Sources

1. REFERENCES ON HENRY C. CAREY
Conkin, P. K., *Prophets of prosperity: America's first political economists*, Bloomington: Indiana University Press (1980)
Green, A. W., *Henry Charles Carey: nineteenth-century sociologist*, Philadelphia: University of Pennsylvania Press (1951)
Kaplan, A. D. H., 'Henry Charles Carey: a study in American economic thought', *Johns Hopkins Studies in Historical and Political Science*, vol 49/4 (1931)
McKinney, W. M., 'Carey, Spencer, and modern geography'. *The Professional Geographer*, vol 20 (1968), 103-06
Malin, J. C., *The contriving brain and the skillful hand in the United States*, Lawrence, Kansas (1955)
Morrison, R. J., 'Protection and development: a nineteenth-century view', unpublished Ph.D. dissertation, University of Wisconsin, 1965
Smith, G. W., *Henry C. Carey and American sectional conflict*, Albuquerque: University of New Mexico Press (1951)
Turner, J. R., *The Ricardian rent theory in early American economics*, New York: New York University Press (1921)

2. SELECTIVE BIBLIOGRAPHY OF WORKS BY HENRY C. CAREY
1835 *Essay on the rate of wages*, Philadelphia: Carey, Lea and Blanchard
1836 *The harmony of nature*, Philadelphia: Carey, Lea and Blanchard
1837-40 *Principles of political economy* (3 vols), Philadelphia: Carey, Lea and Blanchard
1848 *The past, the present, and the future*, Philadelphia: Carey and Hart
1852 *The harmony of interests*, New York: Myron Finch
1858-60 *Principles of social science* (3 vols), Philadelphia: J. B. Lippincott
1872 *The unity of law*, Philadelphia: Henry Carey Baird

William B. Meyer is a doctoral candidate in geography at Clark University, Madison, Wis., U.S.A.

Chronology

1793	Born in Philadelphia, 15 December
1806	Left school to enter the family publishing business
1819	Married to Martha Leslie
1835	Retirement from the publishing business
1840s	Conversion to tariff protectionism
1856	Joined the Republican Party
1859	Final trip to Europe
1863	Honorary Doctor of Laws, New York University
1872	Elected a delegate to Pennsylvania Constitutional Convention
1879	Died in Philadelphia, 13 October

Henry Chandler Cowles

1869–1939

GARRY F. ROGERS AND JOHN M. ROBERTSON

Recognized as a pioneer in developing the dynamic view of the biological landscape, Henry Cowles was also a key figure in the development of the field of ecology, and was widely liked and respected by biologists, geologists, and geographers. Cowles' major scientific contributions were two papers published in the *Botanical Gazette* in 1899 and 1901. In them, he eloquently described the spatiotemporal linkage between the processes of ecological succession, geomorphology, and climate change. His most elaborate achievement was in deciphering the interaction of geomorphological process and biological succession. Above all, however, Cowles was a great teacher who inspired a large proportion of the first modern generation of American plant ecologists and geographers.

1. EDUCATION, LIFE AND WORK

Henry Chandler Cowles, the elder of the two sons of Henry Martyn and Eliza (Whittlesey) Cowles, was born on 27 February 1869 in Kensington, Connecticut, USA. After graduating from the high school at New Britain, Connecticut, Cowles entered Oberlin College, and received the A.B. degree in 1893. In 1894 he taught Natural Science at Gates College, Nebraska, but in the following year he was given a fellowship to study in the Geology Department at the newly formed University of Chicago. There he studied with T.C. Chamberlin but, following a failed attempt to find and study what later proved to be a nonexistent geological formation in the western United States,

Cowles shifted his emphasis to botany, and eventually completed his doctorate (1898) in the Botany Department under J.M. Coulter.

Cowles remained at Chicago, teaching courses in plant ecology, until his retirement in 1934. He advanced steadily in rank to Professor in 1915 and succeeded Coulter as Chairman of the Botany Department in 1925, a position he held until his retirement. Cowles lectured on a variety of topics, but taught several courses consistently throughout his career. These included Physiographic Ecology, Ecological Anatomy, Geographical Botany, Experimental Ecology, Applied Ecology, and Field Ecology. Cowles was known locally as a supporter of efforts in conservation and preservation, and was one of the principal supporters of the fight to save the Indiana dunes. During his career Cowles participated in many organizations and received numerous honours and awards. He was revered nationally and internationally for his contributions to ecology, and following his death on 12 September 1939 his significance in the physiographic study of the American environment, a vital contribution to geographical research, was splendidly shown in a memorial article by C.C. Adams and G.D. Fuller (*Ann. Assoc. Am. Geogr.* vol 30 (1940), 39–43). Like others, these authors regret that he did not publish his researches more abundantly.

2. SCIENTIFIC IDEAS AND GEOGRAPHICAL THOUGHT

Cowles' dissertation research, published under the title 'The ecological relations of the vegetation of the sand

dunes of Lake Michigan' (*Bot. Gaz.*, 1899), and his
study of the interrelationship between geomorphic
processes and ecological succession in 'The
physiographic ecology of Chicago and vicinity'
(*Bot. Gaz.*, 1901), became instant classics among early
contributions to ecology. Cowles' work was
inspired by European developments in ecology,
particularly through the textbook and separately
published sand dune succession studies of E. Warming.
Cowles was well qualified to advance the work of
Warming. Because of his original intention to become
a geologist he spent the summer of 1895 in the field
working for the U.S. Geological Survey, and
subsequently extended his undergraduate training in
geology through graduate studies at the University of
Chicago. With his additional training in botany,
Cowles undertook his dissertation research with the
unique ability to perceive the linkages between
physical and biological landscape processes.

Cowles' 1899 and 1901 works provide exceptionally
clear expositions of the geomorphological and
biological interactions involved in the various phases
of succession. Throughout his analysis of succession
on the dunes, Cowles described the positive interaction
of plants and dunes in which each may alternately
control the character of the other. His chief
contribution in these works was his elucidation of the
various ways in which physiographic processes control
biological ones. Cowles saw topographical change as
a predictable, directional process of denudation,
deposition and, in the absence of crustal deformation,
planation to base level. Despite Cowles'
contributions to a synthesis of geomorphological and
biological processes, relatively few attempts have been
made to expand his work during the intervening eighty-
five years.

Cowles' work led him to devise a general
classification of succession based on clearly distinct
causes: chorographic (due to climatic change),
physiographic (due to geomorphological processes), and
biotic (due to the interactions of organisms). He
considered the first two to be principally of interest
to physical geographers because of the greater variety
of changes involved and the greater potential for
experimental control in their study. He considered
the potential for control to be due largely to the
rapidity with which the biological processes operated,
making it possible to study them directly during the
span of a human lifetime. 'If in their operation,
chorographic agencies are matters of eons, and
physiographic agencies matters of centuries, biotic
agencies may be expressed in terms of decades' (*Ann.
Assoc. Am. Geogr.*, vol. 1/1 (1911), 1-20). Although
Cowles' major contributions were in the area of
physiography, he acknowledged that the character of
current vegetation was largely due to biotic processes
and that biotic processes in turn influence
physiography. His classification of succession was
criticized, however, because its emphasis on
directional causality tends to lead to a de-emphasis
of biological interactions and the feedback effects of
biological or geomorphological processes.

The significance of Cowles' work is often
evaluated by comparing it to the work of F.E. Clements
(1874-1945), a graduate of the University of Nebraska

(Ph.D. 1898), and a major contributor to the formation
and development of ecology. Tobey (*Saving the
Prairies*, 1981) discusses the differing viewpoints of
Cowles and Clements at length and concludes that they
were part of a more general competition between the
Nebraska and Chicago schools for leadership of the new
field of ecology. Tobey traces the differences from
Humboldt through Grisebach and Drude to Nebraska, and
from de Candolle through Darwin and Warming to Chicago,
and emphasizes the contrasting philosophical traditions
behind the two schools. According to Tobey (*Saving
the Prairies*, 1981, p. 99), Cowles' work represented a
legacy of the mechanistic tradition in plant ecology
stemming from de Candolle, whose emphasis on the
individual plant as a mechanism responding to external
stimuli was a departure from Humboldt's concept of
broad vegetation groups responding cooperatively as
integrated units. Clements integrated parts of both
traditions, retaining the holistic formation concept
originated by Humboldt and passed on through Grisebach
and Drude.

It is true that Cowles disagreed with Clements'
holistic concept of succession. Clements viewed the
biological landscape as composed of complex organisms
formed by groups of interacting and interdependent
species that, like the cycle of landscape evolution
suggested by W.M. Davis, developed irreversibly from
youth to maturity. In his review of Clements' master-
piece *Plant Succession; An Analysis of the Development
of Vegetation* (1916), Cowles comments that 'Clements
states positively that "succession is inherently and
inevitably progressive". This reviewer is as
positive in his opinion as ever that succession may be
retrogressive as well as progressive, ...' (*Bot. Gaz.*,
vol 68 (1920), 478). According to Tobey this viewpoint
of Cowles was derived from Warming for

> Warming had made it clear in both the 1896 and 1909
> editions [of his textbook] that vegetational
> development was not directional and was not
> irreversible, both fundamental tenets of Clements'
> successional theory, and part of the reason that
> other English-speaking scientists, such as Cowles,
> spoke of vegetation as developing when they did
> not wish to endorse Clements' successionalism
> (*Saving the Prairies*, 1981, p. 104).

Despite some disagreements relations between the two
men never deteriorated to attacks in the literature.

For several reasons the competition involving
Cowles and Clements probably had little influence on
the contributions they made. First, it seems likely
that Clements faced stronger opposition elsewhere than
he did from Cowles. After 1901 Cowles published very
little more than short commentaries and reviews, and
when these involved Clements' work they were largely
positive. The few disagreements were based on
reasoned arguments. Clements, on the other hand,
published prolifically, including several major works
after 1901. The principal critics of his works, when
such existed, were as apt to be European as American,
and they were not at Chicago. P.B. Sears (*Fifty
Years of Botany*, 1958) suggests that both men were
stimulated by the cool reception they received from
other biologists for their efforts to introduce a new
discipline.

The tracing of separate schools of thought to Chicago and Nebraska provides an appealing synthesis, but its influence on the work of Cowles and Clements might be overemphasized. Although the conclusion in Tobey's *Saving the Prairies* (1981, pp.99-109) that Darwin's focus on the individual organism in his explanation of evolution led to Cowles' acceptance of the biological landscape as occupied by individual species rather than multi-species organisms seems reasonable, it is clear that Cowles was influenced more by Lamarck's definition of evolution than Darwin's. At several places in his dissertation Cowles states his intention to expand his studies of the plant ecology of the dunes with experimental analyses of the possible existence of acquired traits within the dune species. Cowles never carried out his experiments, and though he remained concerned with evolution he appears to have never fully understood Darwin's ideas.

Cowles' principal error in understanding Darwinian evolution appears to have been his misinterpretation that selection worked principally on species rather than on the traits of individuals (*Am. Nat.* vol 43 (1909), 361-2). This mistake, coupled with his low regard for jargon and teleological argument, led him to suggest abandoning the idea of adaptation and natural selection altogether. This view was strongly evident in his ecology text (*Ecology*, 1911), and was sharply criticized by Ganong (*Bot. Gaz.* vol 54 (1913)).

In many ways, however, Cowles' ideas closely corresponded to contemporary views on many subjects. He was repelled by the overly zealous efforts made by some to fit ecological justifications to morphological characteristics to the extent that it appeared that plants 'adapted' themselves consciously to features of the environment. At times it appeared that altruistic co-evolution of plants and animals was universal, and Cowles believed that as this was impossible to test, it was simply a teleological proposition. He doubted that all morphological features could be responses to the environment, and he called on ecologists to keep in mind that perfect adaptation was an illogical proposition simply because of the evidence that many species had become extinct in the past (*Am. Nat.* vol 43 (1909)).

3. INFLUENCE AND SPREAD OF IDEAS

By extending earlier work in Europe while simultaneously pioneering in a new American field of science, Cowles was assured of wide interest in his ideas. His personal relationships with leaders in the fields of geology and biology further facilitated the spread of ideas into other disciplines.

Cowles was not a prolific writer, but his long association, begun while a graduate student, as reviewer and editor of the *Botanical Gazette* provided opportunities for expressing his ideas. Cowles' reviews of ecological and botanical literature, frequently containing his own views on a wide variety of subjects, numbered literally in the hundreds. His close ties with the *Gazette* continued for thirty-six years until his retirement as editor in 1935.

Cowles' influence was also felt through his involvement in professional organizations. He was a member of the Committee on Organization which met in 1904 to found the Association of American Geographers under the chairmanship of W.M. Davis, and from 1904-08 was one of the three councillors who then formed its Council with a president, two vice-presidents, a secretary and a treasurer. In 1910 he served as President. He was similarly active in the Ecological Society of America, of which he became President in 1917. Cowles was vice-president of Section G of the American Association for the Advancement of Science in 1913, President of the Botanical Society of America and President of the section on Phytogeography and Ecology of the International Botanical Congress in 1930. Cowles also served as president of the Chicago Academy of Science and the Geographic Society of Chicago.

Though best known for his early crucial publication on succession, Cowles also influenced the development of American plant ecology through his teaching. He usually taught seven or eight courses per year, including extensive summer field classes composed of students, in-service teachers, and colleagues. Cowles' knowledge of plants and geological formations, coupled with his enthusiasm and warm personality, served to inspire those who accompanied him on his many field trips with greater appreciation and concern for natural processes. As an effective teacher, pioneering in a new field, Cowles' ideas and approach to ecology reached a large proportion of the first generation of American ecologists.

Bibliography and Sources

1. OBITUARIES AND REFERENCES ON HENRY CHANDLER COWLES

Adams, C.C. and G.D. Fuller, 'Henry Chandler Cowles, physiographic plant ecologist', *Ann. Assoc. Am. Geogr.*, vol 30 (1940), 39-43

Fuller, G.D., 'Henry Chandler Cowles', *Science* vol 90 (1939), 363-4

Clements, F.E., *Plant Succession, an Analysis of the Development of Vegetation*. Carnegie Institution of Washington, Publication No. 242, Washington, 1916, 512 p.

Cooper, W.S., 'Henry Chandler Cowles', *Ecology* vol 16 (1935), 281-3

Franklin, K. and N. Schaeffer. *Duel for the Dunes, Land Use Conflict on the Shores of Lake Michigan*. University of Illinois Press, Chicago, 1983, 278 p.

Ganong, W.F., 'The Chicago textbook', *Bot. Gaz.* vol 54 (1912), 73-5

Sears, P.B., 'Botanists and the conservation of natural resources', Steere, W.C. (ed) *Fifty years of Botany*, McGraw Hill, New York, 1958, 359-66

2. BIBLIOGRAPHY OF WORKS BY HENRY CHANDLER COWLES

1899 'The ecological relations of the vegetation of the sand dunes of Lake Michigan', *Bot. Gaz,* vol 27, 95-117, 167-202, 281-308, 361-91

1901 'The physiographic ecology of Chicago and vicinity; a study of the origin, development and classification of plant societies', *Bot. Gaz.*, vol 31, 73-108, 145-82

---- 'The plant societies of Chicago and its vicinity', *Bull. Geogr. Soc. Chicago*, no 2, 1-76

---- 'The influence of underlying rocks on the character of the vegetation', *Bull. Am. Bur. Geog.*, vol 2, 1-26

1903 'The contribution of Linnaeus and his students to phytogeography', *Science*, n.s., vol 17, 463-4

---- 'Contrasts and resemblances between the sand dune floras of Cape Cod and Lake Michigan', *Science*, n.s., vol 17, 262

1904 'The work of the year 1903 in ecology', *Science*, n.s., vol 19, 879-85

1905 'A remarkable colony of northern plants along the Apalachicola river, Florida, and its significance', *Eighth Int. Geogr. Congr.*, Abstr., 599

---- 'Importance of the physiographic standpoint in plant geography', *Eighth Int. Geogr. Congr.*, Abstr., 600

1908 'The response of plants to soil and climate', in Salisbury, R.D., *Physiography for High Schools*, New York, 462-73

---- 'An ecological aspect of the conception of species', *Am. Nat.* vol 42, 265-71

1909 'Present problems in plant ecology', *Am. Nat.*, vol 43, 356-6

---- 'Fundamental causes of succession among plant associations', *Rep. Brit. Assoc.*, Sect. K., 668-70

1910 'Charles Reid Barnes', *Science*, n.s. vol 31, 532-3

1911 'Ecology', vol 3, Coulter, J.M., Barnes, C.R., and Cowles, H.C., *A Textbook of Botany for Colleges and Universities*, New York. (Revised and enlarged in 1931 by G.D. Fuller), 485-964

---- 'The causes of vegetative cycles', *Bot. Gaz.*, vol 51, 161-83

---- 'The causes of vegetational cycles', *Ann. Assoc. Am. Geogr.*, vol 1, 1-20

1912 'A fifteen-year study of advancing sand dunes', *Rep. Brit. Assoc. Adv. Sci.*, 1911, 565

1915 'The economic trend of botany', *Science*, n.s. vol 41, 223-9

---- (with J.G. Coulter) *A Spring Flora for High Schools*, American Book Company, New York, 144 p.

1918 (with Carl O. Sauer and Gilbert H. Cady) 'Starved Rock State Park and its environs', *Bull. Geogr. Soc. Chicago*, vol 6, 1-148

---- 'Retrogressive and progressive successions in the Arkansas sunklands', *J. Ecol.*, vol 6, 95-6

1927 'Interactions between plants and their environment', in Newman, H.H. et al. (eds.) *The Nature of the World and of Man*, Univ. of Chicago Press, Chicago, 240-59

1928 'Ecology and human affairs', in Kunkel, L.O. et al. (ed.) *Mayo Foundation Lectures*, *Lectures on Plant Pathology and Physiology in Relation to Man*, W.B. Saunders Co., Philadelphia, 207 p.

---- 'Persistence of prairies', *Ecology*, vol 9, 380-2

1929 'John Merle Coulter'. *Bot. Gaz.*, vol 87, 211-17

---- 'The succession point of view in floristics', *Proc. Int. Congr. Plant Sci.*, 1926, vol 1, 687-91

---- (with O.W. Caldwell and others) 'John Merle Coulter (1851-1928)', *Science*, vol 70, 299-301

1932 'The ever-changing landscape', *Sci. Mon.*, vol 34, 457-9

Garry Francis Rogers holds a post at Columbia University but spent the academic year 1985-6 as Assistant Professor of Geography at the University of California, Los Angeles and John Michael Robertson is a doctorate student in the Department of Geography, UCLA

Chronology

1869   Born at Kensington, Connecticut, 27 February

1893   A.B. from Oberlin College

1894   Taught natural science at Gates College

1895   Given fellowship at University of Chicago

1897   Appointed Teaching Assistant in the Botany Department at University of Chicago; began extended summer field studies with students and colleagues that included most North American habitats, and continued yearly throughout most of his career

1898   Ph.D., University of Chicago

1899   'The ecological relations of the vegetation of the sand dunes of Lake Michigan'

1900   Married Elizabeth Waller

1901   'The Physiographic Ecology of Chicago and Vicinity'

1902   Appointed Instructor of Botany at University of Chicago; *Journal of Geography* formed with Cowles as phytogeography head

1904   Used the U.S. Subtropical Laboratory at Miami as a base of operations while studying the Florida Everglades

1905   Appointed first secretary/treasurer of newly formed 'Botanists of the Central States', American Association for the Advancement of Science delegate, with

Coulter and Shear, to the International
Botanical Congress in Vienna

1906   Spent the autumn and winter in Florida,
studying the Everglades under a grant
from the Carnegie Foundation

1907   Appointed Assistant Professor

1910   President of the Association of
American Geographers

1911   *Chicago Text Book of Botany*; promoted
to Associate Professor

1913   Vice President of Section G of the
American Association for the
Advancement of Science; led the
International Botanical Congress tour
of United States

1914   Participated in the formation of the
Ecological Society of America

1915   Professor of Botany, University of
Chicago

1917   President of the Ecological Society of
America

1922   President of the Botanical Society of
America

1923   Oberlin College conferred the honorary
Sc.D. degree

1925   Succeeded Coulter as chairman of the
Botany Department

1926   Editor of the *Botanical Gazette*

1930   President of the section on phytogeography and ecology of the
International Botanical Congress

1934   Honorary lifetime membership in
British Ecological Society; retired as
Professor Emeritus

1939   Died in Chicago, 12 September

Juan Dantín Cereceda

1881–1943

MANUEL MOLLÁ RUIZ-GOMEZ

Drawn from a photograph   by Olive E. Freeman

To a great extent Juan Dantín was a geographer whose emphasis on the recognition and definition of the natural region was characteristic of his time in many countries.   And there are many others who came to geography, as he did, with an initial training in the field sciences, bringing with them a firm conviction that geography must retain a firm link with geology, which was reinforced by observation of the French and German schools of geography and in his case also by admiration for W.M. Davis.   Perhaps the feature of his work most characteristic of his time was the constant emphasis on the need to recognise and define natural regions on the firm methodological basis of a thorough knowledge of both physical and human geography.   From this he never deviated in a career of active research and publication covering thirty years.   In his own country he found this satisfying and on this he was still working at the time of his death in his early sixties.

1.   EDUCATION, LIFE AND WORK

Juan Dantín Cereceda was born in Madrid on 25 December 1881.   During his high school years in Madrid he distinguished himself as an able student.   After his graduation from the high school he entered the School of Sciences at Madrid Central University, where in 1912 he obtained the Ph.D. degree in natural sciences. The title of his dissertation, which was awarded the highest honours, was *El relieve de la Península Ibérica.   Ensayo de un estudio geográfico-geológico sobre su constitución e interpretación (The relief of the Iberian peninsula.   An attempt at a geographical*

and geological study of its constitution and interpretation).   The director of Dantín's thesis (published in 1913) was the distinguished geologist Eduardo Hernandez Pacheco, with whom he worked on several later occasions.   Meanwhile in 1912 he went to the Sorbonne in Paris to study under Vidal de la Blache and Emmanuel de Martonne and in the following year he was sent to the newly acquired Spanish zone of Morocco as a member of the commission dispatched by the Real Sociedad Española de Historia Natural.

In 1913 Dantín became a teacher of agriculture at the Instituto de Bachillerato de Guadalajara, a provincial capital near Madrid.   There he was responsible for teaching pupils aged fourteen to seventeen and in 1922 he moved to a similar post at the Instituto San Isidro de Madrid.   He was deeply committed to the study of agriculture and one result of this work was his books, *Dry-Farming ibérico.   Cultivo de las tierras de secano en las comarcas áridas de España*, 1916 and *Agricultura elemental española*, 1935. As will be shown later, throughout his life Dantín had a particular interest in the problems of the arid lands of Spain.   He never held a university post but was always a dedicated researcher.

His main research interest was in geography, within which his contributions covered a wide range of subjects, ranging from methodological theory to physical and human geography.   Most of his researches were carried out at the Real Sociedad Española de Historia Natural, which was associated with the Museo de Ciencias Naturales de Madrid: there he had stimulating colleagues, who were field scientists including geologists but not geographers as such.

Geographers Biobibliographical Studies, volume 10 (1986)

This devotion to research work at the Sociedad de Historia Natural did not preclude his membership of other societies and a number of academic institutions, such as the Comité Nacional de Geodesia y Geofísica, the Real Sociedad Geográfica de Madrid, the American Geographical Society and the Sociedad Española de Antropología, Etnografía y Prehistoria. He was also a member of the Spanish national committee for geography, for which he acted as chairman of the section on biological geography and secretary of the section on human geography. These sections were in accord with those of the International Geographical Union's congresses, at which he was a prominent Spanish representative. At the Congress in Cairo in 1925 he attended sessions on the Rural Habitat with the support of the Real Sociedad Geográfica de Madrid. During the 1928 congress in Cambridge he was elected as one of the four ordinary members of the Rural Habitat commission: the other three were Renato Biasutti of Italy, H.J. Fleure of Britain and P.L. Michotte of Belgium, with Albert Demangeon as chairman. Dantín remained in touch with the Rural Habitat commission during the congress at Paris in 1931. At this congress he was an official representative of the Spanish government.

Out of Dantín's association with the Rural Habitat commission came his few researches in human geography. He agreed to carry out a study of the settlements of the rural population in Spain, with appropriate maps of a distributional character but this task was never finished. Nevertheless he wrote a number of papers on the subject of rural settlement, including 'Distribución geográfica de la población en Galicia' of 1925 and 'Geografía humana. Estado presente de la cuestión del hábitat rural. La población de la Mancha española en el centro de su máximo endorreismo' of 1932. Finally, in 1940 at a meeting in Oporto of the Congresso do Mondo Portuges he reopened the question of the *cañadas*, the routes followed by transhumant sheep on the Spanish *meseta*, on which he wrote two papers, published as no 79 (1940) and no 114 (1942) in *Publs. Real Soc. Geogr.* ser B on the *cañadas* of Leon.

In fact the most important work of Dantín was concerned with physical geography, though his contributions on methodological concepts of geography and especially on the natural region should also be mentioned. His methodological papers include 'Resumen fisiográfico de la Península Ibérica', 1912, in which he contended that one of the most significant objectives of modern geography is to discern the natural region. One year later he published his theoretical paper, 'Concepto de la región en Geografía' (1913), in which he mentioned the work of Vidal de la Blache, his famous paper in *Annales de géographie*, (vol 22, 289-99), 'Des caractères distinctifs de la géographie'. In 1915 Dantín provided another article, 'Evolución y concepto actual de la Geografía moderna', in which he emphasized his personal concept of geography and commented on the work done by other geographical schools, including the French school of Vidal de la Blache, the German school of Hettner and Penck, and above all the American school. Of this last Dantín emphasized the investigations of William Morris Davis, who combined the work of geologists and topographers to provide 'the newest geography', of which indeed he regarded Davis as the founder.

The fruits of Dantín's investigations on the concept of the natural region were most abundantly seen in 1922 with the publication of *Ensayo acerca de las regiones naturales de España*, volume 1. This work was divided into two parts. The first and general section dealt with the elements constituting the natural region, relief, climate, flora, fauna and man as well as the relation of one criteria to another. In the second part he dealt with two of the seventeen regions into which he had divided the Iberian peninsula, Galicia and Asturleonesa. The work was reprinted without alteration in 1942 but the second volume, which was to deal with the other fifteen regions, was not completed by the time of his death. The only completed regional description of Spain given by Dantín is the condensed account in the Spanish edition of Granger's *Nouvelle géographie universelle* (1929).

As noted earlier, Dantín's most intensive work was on physical geography and much of his energy was devoted to study of the relief and the climate of the Iberian peninsula. It may be observed that his work shows a clear sequence in time, for in the 1910s and the 1920s there is a predominance of investigations on relief, both on the peninsular and the wider regional basis while the work of later years is dominated by the analysis of aridity, clearly a very important problem in Spain. Significantly, his few works on human geography were devoted almost exclusively to the distribution of the population in arid and semiarid zones. In research Dantín went from largely geological to morphoclimatological investigations.

When he was teaching agriculture at the Instituto de Guadalajara he conducted an interesting investigation of the area, published in 1915 with the title 'Las terrazas del valle del Henares y sus formas topográficas'. Dantín was one of the first Spanish authors to use lithological criteria to determine the origin of materials. He was also one of the first authors to take account of the possible physiochemical alterations of alluvial land. He carried out some granulometric assessments, though he never conducted measurements of material. The edaphic criteria were important and a clear correlation between morphology, processes and vegetation emerged. In general the descriptions he gives include a lot of detail: his observations were conducted in a generally convincing manner and in some cases closely approach more recent interpretations.

During his later years Dantín made a notable contribution to the study of Spain's semiarid (endorreic) zones, from Andalucia to Aragon. One result was a study, with Antonio Revenga Carbonell, of the aridity index defined by de Martonne. In a paper of 1941, 'Las líneas y las zonas isóxeras de España según los índices termopluviométricos. Avance el estudio de la aridez en España', Dantín and Revenga criticized de Martonne's aridity index because the higher values are given to places with higher pluviosity and the lower values with places of lower temperature. They advocated an inversion of de Martonne's formula, done by dividing the temperature by the precipitation and then multiplying the result by 100 to prevent the quotient being less than one. They used this method to determine the relatively aridity of Spain's drier areas, which were found to have an index figure of between 3 and 6: any recording over 6 means a subdesertic zone.

Dantín travelled extensively. In part this was to congresses though on occasion he was asked to travel on behalf of societies or official commissions carrying out scientific expeditions, for example to the Alpine regions of France and Switzerland, but despite all his varied activities he maintained his educational work. He developed and elaborated a teaching plan for geography covering all the years of a child's education and also contributed frequently to daily newspapers in an attempt to spread geographical knowledge in an understandable form among the general public.

In the last years of his life, though in poor health, Dantín continued his research work relentlessly at the new Instituto de Geografía Juan Sebastián Elcano. This institute was founded on 24 November 1939 as a research centre of the Consejo Superior de Investigaciones Cientifica, a body independent of the universities. The Instituto de Geografía Juan Sebastian Elcano organized conferences and summer schools at which geographers working in universities and schools combined with the research staff. The geographer Amando Melón has noted that he last saw Dantín at the third summer Meeting on geographical studies held in Galicia. Though of weary appearance and in bad health, this did not prevent him from taking part in some arduous study expeditions to remote and hardly accessible places. He was working on material that was to be used in his last public appearance but he died shortly afterwards, on 23 October 1943 in Madrid.

2. *SCIENTIFIC IDEAS AND GEOGRAPHICAL THOUGHT*

Dantín believed that the limited development of geography in Spain resulted from its restriction to historical geography, political geography and statistics. Through these circumstances geography had become an auxillary rather than a major subject and therefore he advocated 'a new way for geography', which was to have a strong background in geology and to gain support and recognition from the Museo de Ciencias Naturales de Madrid. As a support and background for physical geography Dantín revered the work of some famous Spanish geologists, notably José Macpherson and Salvador Calderón and their successors Eduardo Hernández Pacheco and Ignacio Bolívar. The view of Dantín was that for geography the only acceptable division was between the physical and human aspects: with this he also argued that human geography is at once the basis for political geography and above and beyond it in scope. Alone, without the acceptance of human geography as dominant, political geography would not be anything more than a reflection of historically and constantly changing political circumstances.

In accordance with a number of other European authors Dantín argued that geographical studies must focus on two aspects: the description and interpretation of contrasts. Such studies require data from other sciences, particularly the physical and biological sciences though the high degree of specialization demanded of geographers by such studies might damage geography because its very essence, the capacity for synthesis, might be distorted. To avoid such a danger Dantín maintained that the geographer should collect all the materials that are susceptible of being integrated and then harmonize and synthesize them into a geographical whole. This he explained with an example: one worker may be a meteorologist who measures weather data with statistics of temperature, precipitation and other observations of places, while another is a climatologist whose concern lies in the interactions between the different but clearly related meteorological phenomena.

It should be emphasized that Dantín worked when positivist geography was favoured in Spain. This means that geography cannot be content with mere description but must go on beyond that and seek laws that take account of local similarities and differences. Later in this paper it will be seen that Dantín showed some penchant for determinism in his work, but only to a limited extent.

According to Dantín the study of places must be made with a full recognition of the past as well as the present. Consequently he disagreed with F. von Richthofen's idea of separating geography and geology as such a separation ignored the fact that surface relief is an expression of the geological structure cut and moulded by subaerial agents. If this separation from geology is valid for physical geography it is also valid for the separation of history from human geography, of which history will always be a valuable ally. This is not to say that Dantín, having made a distinction between determinism and possibilism, then placed exaggerated emphasis on relief in the study of the natural region, perhaps because of his background training in the natural sciences, though his book published in 1922 on the natural regions of Spain, *Ensayo acerca de las regiones naturales de España*, was criticized by Ortega y Gasset in a Madrid newspaper, *El Sol*, for its over-emphasis on the physical environment as an explanation of human behaviour. Apparently Dantín was considerably influenced by the book *Man and his Work: an Introduction to Human Geography* by A. J. and Frances D. Herbertson, originally published in 1899 and translated into Spanish by Palau Vera as *Geografía Humana*, Barcelona 1914.

Dantín was fully aware of the division between physical and human geography, but for him this did not mean that there were two different sciences. He constantly stressed the unity of geography when confronted by the idea of its duality. He quoted the comment from Ewald Banse's *Geographie* of 1912 that 'Geography is a science where the monistic idea should dominate, since the unity of physical geography implies the unity of human geography'. Certainly the geographer had to deal with a concatenation of phenomena, ranging from the interpretation of relief to climate, vegetation, fauna and also the influence of relief on man. Studying human geography independently of physical geography would make no sense for the study of effects must always be incomplete without the study of causes.

The complexity of human phenomena was well appreciated by Dantín and he clearly understood that in seeking reality the geographer must study human societies and phenomena as they exist, in inevitable connection.   Systems that are no more than statistical must be discarded for as disciplines geography and statistics have different objectives and principles.   Statistical data, rightly studied systematically, will only have geographical value if they are moulded into a synthetic whole sought by geographers as the one and only way of attaining a comprehensive knowledge of the factual situation.

Regional geography was to Dantín the clearest expression of the essential geographical concept for it involved the placing and description of individual phenomena.   However regional work was not the whole of the discipline for particular problems could well be studied on a local basis in detail and through such studies it would be possible to discern the 'distinction and separation of forms and types' and so to seek and hopefully to find some general laws.

Not surprisingly Dantín was willing to justify his recognition of the natural region as the fundamental support of his work as a geographer.   He faced squarely the question of exactly how it could be recognized and said that there were five components, relief, climate, vegetation, fauna and man.   The correlation between these components, established for any area of the world, would define a specific natural region, having its own distinction and individuality.   Before seeking a definition of natural regions it was necessary to establish criteria for any of the components, to discern their order of preference and to accept their concatenation as it would occur in any area studied.

On this general basis Dantín regarded relief and climate as the main elements, for all others depended on these two as basic.   In time this was modified by the view that relief was of greater significance than climate, 'the first and most fundamental of all constituent elements'.   At no time did Dantín regard the relation between the five elements as simple for he appreciated how strong and complex such relationships were.   He commented that

> The correspondence and solidarity that occurs between the constitutive elements of a region is of such a nature that the alteration of one of them will result not only in the alteration of the whole system, but also in that of any of the remaining elements taken individually. If the climate changes then relief, flora, fauna and man will change too.

After relief and climate the third element to be analyzed is vegetation.   The vegetation of the different natural regions is characterized by its fundamental adaptation to the environment, since the factor common to the different floristic species of any region is their physiological rather than their systematic affinities.   Here one may note that Dantín fully recognized that plant associations were part of the physiognomy of a regional entity.   The fourth element, fauna, is not considered in any detailed way in Dantín's work on the natural region,

though it too is treated in relation to other elements of the environment.

Man is the last element making up the natural region.   Dantín favoured a pragmatic way of studying man in the natural region by focusing attention on settlement and population distribution in any territory, and commented that 'the setting of human societies is the environment.   Human society and environment are united by such close ties that both must be studied simultaneously so as not to upset the natural balance: this approach will reveal the natural unity of the earth'.   In his study of the distribution of the population in the various natural regions he discerned in Spain Dantín followed this same principle and the results were of particular interest for the semiarid (endorreic) areas with which he was especially concerned.

Dantín maintained the traditional concept of the earth's surface as the theatre of human events, of the interactions between man and environment.   In his 1915 paper he mentioned the auxillary character of history though this was a statement of principles rather than one of practical reality for to him cause and effect relationships were so strong that they hardly allowed an historical interpretation of events.   When it came to anthropogeographical (he retained this term) considerations of human activities Dantín showed a main interest in two subjects: the first was the influence of physical conditions and the second the character and form of the population distribution.

Following this brief commentary on the constituent elements of a natural region the next stage is to define the natural region itself.   To Dantín

> the natural region is the product of the mutual reciprocity of the elements that make it up, and it should not be forgotten that these elements react on one another and change: it is a conflict between physical principles and biological ones that govern the world, with all their additions and interferences.

He saw the natural region as a closed landscape, defined by its morphological character.   The geological study is basic but the field of observation covers visible events since it is the results on the surface, itself the product of the interaction of several elements, that really matters.

In summary the following characteristics of Dantín's concept of the natural region may be noted. First, it is a geological definition, much modified by geomorphological processes.   The erosive component is as significant as the structural component, based on climate, and the conjunction of both defines the natural region.   Secondly, the living organisms that integrate themselves into the region are the result of relief and climate: their adaptation to the environment is a function of their physiological rather than their systematic affinities.   This gives a physiognomic view of living beings and these living beings and inorganic elements overlap within a visible phenomena which is the landscape.   Thirdly, the limits of the natural region are so precise that beyond them the interactions between the elements change and make up a landscape based on different attributes. Fourth, the positivist view that identifies geography with science remains valid. This promotes a deterministic approach to human behaviour. The natural region is governed by overriding laws that determine it and the task of the geographer is to discover such laws.

Finally one may summarize the view of Dantín by a short statement in which he emphasizes that the concept of the natural region shares the principles that govern other sciences:

> the natural region appears to be governed by the correspondence between the elements that make it up, and that of coordination between variants of each element. In effect every science is nothing but the result of relationships (of similarity or difference) and of coordination.

3. INFLUENCE AND SPREAD OF IDEAS

Dantín's extensive work, consisting of more than 400 works according to Amando Melón, was widely read and studied among Spanish geographers. Since his publication in 1913 of his paper, 'Concepto de la región natural en Geografía', studies of Spain's natural regions have been numerous. These papers show varied divisions of the Spanish territory as the several combinations used to define each region have yielded different results, though almost all the divisions have had a physiographical basis. It can be said that Dantín was the geographer who bridged the gap between Spanish geography before and after the Civil War and that for this reason he is, among the Spanish geographers born in the nineteenth century, one of the best known in Spain today. His thermopluviometric index, for example, is still to be found in books on Spain which deal with the arid zone. The widespread familiarity with Dantín shown by Spanish authors contrasts with the scant knowledge of him shown by other Europeans despite his membership of the International Geographical Union and active participation in international congresses during the first third of the twentieth century.

Bibliography and Sources

1. REFERENCES AND OBITUARIES ON JUAN DANTÍN CERECEDA

Aitkin, R., 'Juan Dantín Cereceda', *Geogr. Rev.*, vol 35/2 (1945), 330-1

Kant, E., 'Juan Dantín Cereceda 1881-1943', *Svensk Geogr. Arsbok*, no 23 (1947), 107-13 (French summary)

Melon, A. and de Gordejuela, R, R., *Est. Geogr*, vol 5/no 14 (1944), 5-20

Molla Ruiz-Gomez, Manual, *Juan Dantín Cereceda y la Geografía espagnol en el primer tercio del siglo XX (Juan Dantín Cereceda and Spanish geography during the first third of the twentieth century)*, thesis forthcoming

Praesant, H., 'Juan Dantín Cereceda', *Petermanns Geogr. Mitt.*, vol 90 (1944), 248-9

anon., *Bol. Soc. Geogr. Ital.*, ser. 7, vol 11/4 (1946), 238

A(itkin), R., 'The natural regions of Spain (an abstract specially prepared from *Regions naturales de Espana* by J. Dantín Cereceda, 2. vol, Madrid 1922), *Geogr. Teach.*, vol 11 (1921-2), 333-45, vol 12 (1922-3), 19-27, 82-90: Aitkin gives a map showing the areas having adequate rainfall and no problems of intermittent aridity with a description of the regions but no map of their approximate limits

2. SELECTED BIBLIOGRAPHY OF WORKS BY JUAN DANTÍN CERECEDA

1912 'Resumen fisiográfico de la Península Ibérica', *Trab. Mus. Nac. Cienc. Nat. Madrid*, no 9, 272-93

1913 'Concepto de la región natural en Geografía', *Bol. Real Soc. Geogr.*, vol 13, 507-14

---- *El relieve de la Península Ibérica*, Madrid, 99 p.

1915 'Evolución y concepto actual de la Geografía moderna', *Junta Amp. Est. Inv. Cient.*, vol 15, 289-317

---- 'Las terrazas del valle del Henares y sus formas topográficas', *Bol. Soc. Esp. Cienc. Nat.*, vol 15, 301-15

1916 *Dry-farming ibérico. Cultivo de las tierras de secano (Iberian dry-farming: the cultivation of the dry lands)*, Guadalajara, 148 p.

---- 'La zone espagnole du Maroc', *Ann. Géogr.*, vol 25, 366-73

1917 'Evolución morfológica de la bahía de Santander', *Trab. Mus. Cienc. Madrid*, ser. Geol., no 20, 43 p.

1922 *Ensayo acerca de las regiones naturales de España*, Madrid, vol 1, 397 p.; republished as *Régions naturales de España*, vol 1, 1942, 397 p.

1925 'Distribution geográfica de la población en Galicia', *Junta Amp. Est. Inv. Cient.*, vol 25, 40 p.

1929 'Localización de las zonas endorreicas de España', *Soc. Esp. Hist. Nat.*, vol 15, 829-36

1932 'Geografía humana. Estado presente de la cuestión del habitat rural. La población de La Mancha española en el centro de su máximo endorreismo', *Bol. Real Soc. Geogr.*, vol 72, 24-45 (also published earlier in *UGI Rapport de la Commission de l'Habitat Rural*, Cambridge, 1928, 3-23)

---- 'El dinamismo interno de la Geografía', *Revista de Occidente*, vol 9, 106-13

1934 'Contribución al conocimiento del 'habitat rural' en España', *C. R. Congr. Int. Géogr. Paris 1931*, vol 3, 61-72

1935 *Agricultura demantal espanola*, 247 p.

1936 (with A. Revenga Carbonell) (Geomorfometria de la Mancha: nota preliminar' in *Estudios geomorfométricos sobre España*, no 1, 131-40

1940 'Las cañadas ganaderas del reino de León', *Bol. Soc. Geogr. Nac.*, vol 76/8-12 August-December 1936 but published in 1940, 464-99

---- 'La aridez y el endorreismo en España', *Est. Geogr.*, vol 1, 75-118

1941   'Aspectos de la agricultura española', *Bol. Soc. Geogr. Nac.*, vol 77/nos 7,8,9, 491-525

----   (with A. Revenga Carbonell), 'Las lineas y las zonas isoxeras de España segun los indices termopluviometricos.   Avance al estudios de la aridez en España', *Est. Geogr.*, vol 2/no 2, 33-92

----   'El ciclon del 15-16 de Febrero de 1941 sobre la peninsula iberica', *ibid.*, 131-41

----   'La laguna salada de Gallocanta (Zaragoza), *Est. Geogr.*, vol 2/no 3, 269-301

----   'Distribución geografica de la escanda Asturiana', *Est. Geogr.*, vol 2/no 5, 739-97

1942   'El medio fisico aragones y el reparto de su poblacion', *Est. Geogr.*, vol 3/no 6, 51-162

----   'La reunion de geografos Europeos en Würzburg (Almenia)', *Est. Geogr.*, vol 3/no 7, 423-41

----   'Distribucion y extension del endorreismo aragones', *Est. Geogr.*, vol 3/no 8, 505-95

1943   'Aspectos geograficos de las vegas de Granada', *Est. Geogr.*, vol 4/no 11, 267-371

1944   Three posthumous papers were published in *Est. Geogr.*, vol 5/no 14 following the obituary by A. Melon and R. de Gordejuela (*v.s.*)
'Laguna salada endorreica de la Ratosa en la Alameda (Malaga)', 21-5
'Clima de la region Astureonesa', 27-44
'Tectonica del macizo galaico', 45-52

Manuel Mollá Ruiz-Gomez is a professor of geography at the Universidad Autónoma de Madrid

Chronology

1881   Born in Madrid, 25 December

1912   Became a Ph.D. of Madrid university for a thesis on the geography of Spain in relation to its geology: went to the Sorbonne in Paris to study with Vidal de la Blache and Emmanuel de Martonne

1913   Visited Morocco as a member of a commission from the Real Sociedad Espanola de Historia Natural: became a teacher of agriculture at the Instituto de Bachillerato de Guadalajara

1916   Published *Dry-farming iberico* ..., a book on aridity in Spain which was to be a lifelong interest

1922   Moved to the Instituto San Isidro at Madrid and in this year published his famous work on the natural regions of Spain

1925   Attended the Congress of the International Geographical Union at

Cairo and became a member of the Rural Habitat Commission by 1928, of which the President was A. Demangeon, the secretary H. J. Fleure and the ordinary members R. Biasutti and P. Michotte

1928   Was present at the IGU Congress in Cambridge, England, and remained a member of the Rural Habitat Commission

1931   Attended the IGU Congress at Paris and once more was actively associated with the Rural Habitat Commission

1936   Although he remained active as a geographer during the Civil War, much of his work could not be published until 1940 or later, when the new journal, *Estudios Geograficos*, was enriched by his contributions

1939   From this time he was closely associated with the new Instituto de Geografia Juan Sebastion Elcano, founded in November of this year

1940   Was present at a meeting in Oporto of the Congresso do Mondo Portugues, where he dealt with aspects of transhumance

1941   Published his paper criticizing the aridity index of E. de Martonne, with the collaboration of A. Revenga Carbonell

1942   Attended the Conference of European Geographers at Würzburg as an honorary member on the invitation of the German government in March, at which there were three other representatives from Spain, ten from Italy, and one each from Bulgaria and Finland.   He gave a paper on the aridity and agriculture of Spain, 'El clima árido de España en congruencia con las formas de su agricultura' and became a corresponding member of the Geographical Society of Würzburg

1943   Died in Madrid, 23 October

William Gerard De Brahm

1718–1799

LOUIS DE VORSEY, JR.

1. EDUCATION, LIFE AND WORK

Although a number of authorities have identified
William Gerard De Brahm as a citizen of Holland, the
records show that he was born in Koblenz, Germany on
20 August 1718. His father, Johann Phillip von
Brahm, served as court musician to the Elector of
Triers. As a member of the lesser nobility of the
day, young De Brahm secured an excellent early
education. His adult literary and scientific
performances exhibit a rich background doubtlessly
dervied from the serious study of classical and modern
languages, mathematics, history, geography,
literature, philosophy, and most importantly, the
burgeoning experimental sciences which were
influencing Europe so profoundly in the wake of the
Renaissance.

After establishing a successful career in the
Imperial Army of Charles VII, during which he rose to
the rank of captain engineer, De Brahm renounced the
Roman Catholic faith and resigned his army commission.
Thus, at the age of thirty he found himself and his
young wife, Wilhelmina, nearly destitute in the
troubled and fragmented Germany of the mid-eighteenth
century. Fortunately he was befriended by the
influential Samuel Urlsperger (1685-1772), Senior of
the Evangelical Ministry of Augsberg and one of the
two non-British trustees for establishing the Colony
of Georgia in America. Through Urlsperger's good
offices De Brahm was placed in charge of a contingent
of 156 German-speaking emigrants bound for new homes
in far-off Ebenezer, Georgia.

Once arrived in Georgia, De Brahm's talents and

abilities were recognized by those charged with the
administration of the depressed colony during the
twilight of the trusteeship period. James Habersham
(1715-75), a key figure in the trusteeship period and
sometime acting governor in the soon-to-be-formed
Royal Colony of Georgia, took time to express his
great satisfaction at the good health and condition of
De Brahm's party of newly arrived immigrants. He
wrote that

> The trustees I believe are not mistaken in
> Mr. von Brahm's abilities. He has been at a
> great deal of pains to view the country to fix
> on a settlement and has taken plans of all the
> places he has visited, and I look upon him to
> be one of the most intelligent men I ever met
> with, and will I doubt not make a very useful
> colonist.

There can be no doubt that Habersham was an excellent
judge of the human resources necessary in a raw
frontier colony. De Brahm's talents as a military
engineer and surveyor made him a much sought-after
individual in both Georgia and neighbouring South
Carolina. In 1752 he was summoned to Charleston and
commissioned to design and oversee the construction of
a comprehensive system of fortifications for the city.
It was at this time that the German surveyor-engineer
exchanged 'von' for 'De' as his surname prefix. Later
he served an interim term as that colony's surveyor-
general of lands and journeyed deep into the Indian
territory of what is today eastern Tennessee to design

and construct, in the heart of the Overhill Cherokee country, the ill-fated Fort Loudoun of which the garrison were slaughtered by the Cherokee Indians after an apparently successful escape.

In Georgia De Brahm busied himself in acquiring a large personal estate in landholdings in the vicinity of the German refugee settlement, Ebenezer, and near Savannah, Georgia's capital and chief city. Official recognition was not long in coming and he was named as one of Georgia's two first royal surveyor-generals. De Brahm and Henry Yonge (1713-78) were appointed jointly in 1754 to supervise this important function of the newly established royal government.

As tension with the French heightened in the mid-1750s, De Brahm was deeply involved in devising schemes of defence and building fortifications in Georgia. For Savannah he proposed a 'well pallisadoed (sic) Intrenchment to envelop the city, so as to make it a Receptacle and Shelter for all the Planters, their Families, Sales, etc.' Fortunately, no French invasion was ever mounted and Britain emerged from the Seven Years War as an unqualified victor. Among the territorial spoils won as a result of this victory was East Florida, the Spanish colony to the south which had represented an almost constant threat to Georgia's existence.

Realizing their virtual total ignorance of the geography of the new American dominions, now spreading from the Arctic to the Caribbean, George III's chief advisers recommended the establishment of two important new imperial districts which were to be divided by a line from the Potomac River westward. The northern and southern surveyor-generals were to undertake comprehensive detailed scientific surveys of their respective districts and employ the data collected in the construction of accurate charts and maps.

De Brahm enjoyed his finest hour when he unexpectedly received the king's commission to the post of surveyor-general of the southern district and surveyor of lands in the government of the newly formed colony of East Florida. In the early months of 1765 De Brahm left Georgia with his family to take up residence in St. Augustine and to begin his new duties in East Florida.

In a petition for a grant of land, which he submitted in the autumn of 1758, De Brahm had described his family as consisting of his wife Wilhelmina, and two children. Sadly, De Brahm's young son died in Georgia before the move to St. Augustine, the capital of East Florida. Death visited the small family again in September 1765, when his wife died in that former Spanish town. With a daughter to bring up and a household establishment which included at least four negro slaves, De Brahm lost no time in seeking a suitable spouse. In his search for a German-speaking second wife he wrote to the widow of a German Moravian living in far-off North Carolina to propose marriage. Not successful, he directed his attention locally and married the sister of John Row, a planter and surveyor in East Florida. Later, in July 1769, De Brahm's daughter married Lieutenant Frederick Mulcaster of the Royal Regiment of Artillery in St. Augustine.

After six years of arduous and productive surveys along the extensive Florida coasts he was ordered to London to answer charges which had arisen out of his rancourous feud with the colony's royal governor. He was eventually exonerated and, in 1775, returned to continue his southern surveys with the armed ship *H.M. Cherokee*.

Unfortunately he arrived in Charleston to see the outbreak of the American Revolution in 1775. His ship was commandeered for military purposes and he, stoutly loyal to his king, found himself a prisoner at large in Charleston. After a considerable time he was allowed to leave for France and eventually returned to England. Ill health kept him an invalid through the period of the active conflict in America. He returned to America and took up residence in the Philadelphia area in 1791 at the age of seventy-three. His last eight years of life were spent there. Much of that period was devoted to publishing a number of books on mystical themes with numerous biblical references.

De Brahm died during the early summer of 1799 at his estate known as Bellair near Philadelphia. An intriguing glimpse of De Brahm in his twilight years is provided by Plowden C. J. Weston (ob. 1864), who in 1856 wrote that 'I know nothing of De Brahm's life; but he lived within the memory of persons now alive, much addicted to alchemy, and wearing a long beard.'

2. SCIENTIFIC IDEAS AND GEOGRAPHICAL THOUGHT

Living in the American colonies of the eighteenth century, De Brahm's most numerous contributions to the field of geography are best described as applied and came in the form of geographical exploration, survey and cartography. This is not to imply, however, that he did not synthesize his observations, surveys and findings into a comprehensive written form worthy of being considered a regional geography in the modern sense of the term.

The most comprehensive geographical writing of De Brahm was unfortunately not published until recently so he has not enjoyed the notice which otherwise might have been afforded him by historians of science. This work was his manuscript 'Report of the General Survey in the Southern District of North America', which De Brahm had the privilege of personally delivering to King George III (ed de Vrosey, 1971). In this elegant document De Brahm exhibits the essential qualities of the geographer through his concern with the larger regional complex of which his individual sketches, plates, and observations were elements. De Brahm's active mind sought broad patterns and interrelationships in nature. He observed and commented on regional variations in patterns such as those created by the landforms, soils, flora, fauna, climate, and human activities he encountered in the American southeast.

Soon after his arrival he presented the administrators of Georgia with 'a map of that Part of Georgia, which I have had an opportunity of Surveying since my arrival here: which I flatter myself will speak in my behalf and be more satisfactory and agreeable than I could say in a long and tedious letter.' As a good geographer, De Brahm knew the eloquence which an accurate and well-designed map possessed. Needless to say he included a large number of such maps in his *Report*.

In some measure the content of De Brahm's *Report*

was pre-determined by his immediate supervisory agency, the British Board of Trade and Plantations. Although his orders were, by their own admission, only 'a rude sketch or outline', the Board made clear that they expected detailed descriptions of the southern provinces which would aid in forming 'a true judgement of the state of this important part of His Majesty's dominions.'

As he began to plan for this important task, De Brahm studied the writings of well-recognized world travellers and was guided by such figures as Pierre François Xavier de Charlevoix, Antonio de Ulloa, and Louis Antoine de Bougainville. As he stated in a communication to the Board of Trade, he had 'intentionally set no bounds to my Historical Report ... which to improve I studied by copying after the best I know who went before me upon similar Expeditions.'

Perhaps the best summary of De Brahm's *Report* is the one he provided in a paragraph on the title page:

> Delivered to the Board of Trade and Plantations in three separate Returns, and Sections entering with the History of South Carolina and Georgia; then proceeding to the History of East Florida, and Surveys, containing in general of said Provinces, the Climates, Beginnings, Boundaries, Figures, Contents, Cultures, Soils, Natural Products, Improvements, Navigable Streams, Rivers, Cities, Towns, Villages, Vapours, their Effect, and Remedies, burning of Forrests, Winds, how to preserve Health, Pathology, Materia Medica, Diet and Regimen, Ports, Bars, Numbers of Inhabitants, and Negroes, Exportation, Riches, Number of Trading Vessels, Cattle, Governments, Forces, Fortifications, of Fort Loudoun in particular; Indians and Apalachian Mountains, their Soil, Natural Produce, Air and Communications compiled from the Surveys, Voyages, Astronomical, Philosophical, and Chymical Observations and Experiments, Sea and Land Surveys of William Gerard de Brahm, His Majesty's Surveyor General for the Southern District of North America.

In an age when lines dividing the sciences were more plastic than formed, it is not surprising to find that De Brahm's scientific curiosity was soon directed to the ocean environment off Florida's extensive coast. His early reports, made during February and March 1765, included references to the dynamics of the Gulf Stream which flowed close along the shoreline of southern Florida. His sophisticated approach to the study of this powerful current is revealed in a report to the Board of Trade where he wrote, 'besides the geometrical Operations on shore I performed also Loxodromical Surveys with two Vessels and Boats upon the Coast, in order to obtain the Soundings, and, of the Florida Stream, its Settings to the North, and also its distance from the Shore.' In the extended descriptive text included on his 'Chart of Cape Florida according to surveys

made May 13 and 29, 1765', De Brahm first revealed his original theoretical explanation of the great current, 'The Florida, vulgarly called Gulf Stream, [is] whirled by the Trade winds round and out of the Gulf of Mexico along the southern and eastern coast of Florida.'

As De Brahm charted the Florida Keys he developed a highly original hypothesis to explain their evolution and fragmented morphology. His hypothesis is all the more interesting because he chose the cartographical medium in articulating it. Briefly, he was proposing that the keys and their paralleling line of reefs were the fragmented remnants of two paleo-peninsulas which had formerly extended southwest from the main bulk of southern Florida. In addition to supporting his hypothesis, his published maps of the 'before' peninsulas and 'after' islands and reefs served to make clear the sailing passage they enclosed. Named Hawk Channel it remains, in the words of the *U.S. Coast Pilot*, 'the navigable passage inside Florida Reefs and outside the keys from Cape Florida to Key West, a distance of about 127 miles.'

In 1771 De Brahm was recalled to London. The voyage enabled him to extend his observations of the Gulf Stream. He lost no time in acquainting the scientific community and maritime interests of his discoveries through the medium of a letter to Mr. Urban, editor of the widely read *Gentleman's Magazine*. Among other matters he wrote that he had

> traced the Florida, commonly called Gulf Stream, with all its windings from the Dry Tortugas, the westernmost of the Martiers [,] along the Atlantic coast to the Newfoundland bank: ... vessels bound from any part of America through the New Bahama Channel to Europe may take the benefit of that stream, which will not only guide them clear of all shoals projecting from the Capes on the coast of North America, but also accelerate their voyage in a near incredible measure from twice to sixtimes (sic) ... As I am convinced of the utility my discovery affords to the public, I would not lose a day to communicate it to your publication. I am, Sir, your most humble Servant, Wm. Gerard De Brahm, His Majesty's Surveyor for the Southern district of N. America.

De Brahm lost no time in following up on the letter to Mr. Urban with his book, *The Atlantic Pilot*, which was printed in London in 1772. It is a small but valuable work containing the first printed chart and empirically supported theoretical explanations of one of the earth's great oceanic phenomena--the Gulf Stream. Even a casual study of *The Atlantic Pilot* reveals the increasingly sophisticated refinement of De Brahm's theory concerning the current's cause. Whereas in 1765 he saw the Gulf Stream as originating in the Gulf of Mexico, in 1772 he wrote, 'but this stream is in reality carried into the gulf of Mexico by these trade-winds, and therein circulates at large.' Thus, fully sixteen years before Benjamin Franklin published his well-known *Maritime Observations* and map of the Gulf Stream, De Brahm had presented the current in terms of a complete North Atlantic surface gyre (rotary system).

In the summer of 1775 De Brahm, now equipped with

the armed ship *Cherokee*, sailed for America to resume the southern coastal surveys interrupted three years before.   During the lengthy voyage De Brahm conducted an ambitious oceanographic survey.   In his own words he had

> two objects in view, both equally interesting
> for Navigation and the Learned world.   The
> one was to discover the breadth and directions
> of the constant currents upon the coast of
> Europe and Africa, and whether [they cross]
> the Atlantic from Africa to America as I
> always supposed.   The other object was the
> preciseness of fixing the Line of No
> Variation.

The results of the 1775 *Cherokee* voyage were compiled in a manuscript titled 'The Continuation of the Atlantic Pilot' which remains as yet unpublished. While the largest part of this 160 folio page manuscript is devoted to De Brahm's determination of the direction and force of the North Atlantic surface current system, it also includes observations on 'the change of the weight of the Atmosphere' or barometric pressure differences and their effect on the changing weather.

Through his close relationship with Lord Dartmouth, the secretary of state for the Southern Department, De Brahm was exposed to the scientific avant-garde of London of the early 1770s.   One of the new areas of research then being reported dealt with the use of mercury barometers for determining elevation.   It was hoped that topographic surveyors would be able to obtain heights more accurately and rapidly than through the laborious trigonometrical methods which were usually employed.   Probably drawing on the work of J. A. DeLuc which had been published in Geneva in 1772, De Brahm prepared and had printed a small book with the title, *The Leveling Balance and Counter-Balance; or The Method of Observing, By The Weight and Height of Mercury, On any Place of Terra-firma on the Terrestrial Globe, The Exact Weight and Altitude of The Atmosphere Below and above the Place of Observation; Thereby to ascertain how much the Horizon of the Sea is lower than the Place whereon the Observation is made.*

Thus thoroughly grounded in the theory of atmospheric weight and pressure and equipped with 'Nairnes Marine Barometer and a Thermometer', De Brahm conducted careful observations during his voyage aboard the *Cherokee* in 1775.   He was able to relate the movement of low-pressure centres to storms at sea and note the generally lower atmospheric pressure of the tropics.   At one point in the *Continuation of the Atlantic Pilot* he wrote that he (on) 'the 31 of August foretold a Hurricane was approaching'.   In fact the Master's log indicates that there was a severe storm.

3.   *INFLUENCE AND SPREAD OF IDEAS*

It was indeed unfortunate that De Brahm's arrival in Charleston, South Carolina, on 7 September 1775 coincided with the outbreak of the American Revolution there.   The *Cherokee* was almost immediately withdrawn from his use and pressed into military duties by the British.   De Brahm, remaining intensely loyal to his King, was finally allowed to take ship for England in 1777.   Due to the distractions of a long and expensive war, the bureaucracy in London had little time or regard for De Brahm and his incessant claims for compensation.   He appears to have suffered some sort of breakdown and from 1778 until 1784 was, in his words, 'for the most part an invalid'.

During this period of convalescence, spent in Topsham, Devon, the tide of international politics once more swept Florida into Spanish hands after twenty years of British control, so any hope of a resumption of the survey there was dashed.   The evidence indicates that De Brahm embraced the ethos of the Religious Society of Friends during this period and turned his mind from science to metaphysical concerns.   In a letter he wrote to the King, De Brahm stated that 'my loyalty and faithfulness is now of a Spiritual Nature and not Subject to change or even to be disolved by Eternity'.

After a long period of petitioning the government for a pension and in less than robust health De Brahm left England to take up residence in Philadelphia in 1791.   From that time until his death in 1799, De Brahm occupied himself in publishing a number of books devoted to mystical themes with abundant biblical references.   An anonymous manuscript note found in a copy of one of these goes far in describing the whole corpus of De Brahm's mystical outpourings as 'very curious and madly mystical'.

There can be no doubt that the American Revolution combined with De Brahm's conversion to a life of mysticism in his later years seriously limited the influence and spread of his scientific ideas and contributions.   Among his contemporaries he was probably most well-recognized for his exploration and accurate charting of the Atlantic coastline from South Carolina to the Florida Keys.

De Brahm's large four-sheet *Map of South Carolina and a Part of Georgia* which was engraved and printed by Thomas Jeffreys in 1757, was described, by the authority William P. Cumming, as the first map of any large area in the Southern colonies to possess topographical accuracy based on scientific surveys. George Louis Le Rouge, the Paris-based cartographer, produced a French edition of the map in the late 1770s and a 1780 English edition with much new information was published in London by William Faden, Jeffreys' successor.   Both the French and English versions of De Brahm's Carolina-Georgia map were included in atlases and distributed widely in late eighteenth-century Europe and America.

Many other cartographers included De Brahm's work in their own compilations without acknowledging it. Occasionally, however, attribution to him was included such as the gloss on Emmanuel Bowen's *An Accurate Map of North America Describing and Distinguishing the British and Spanish Dominions on the Great Continent: According to the Definitive Treaty Concluded at Paris 10th Feb. 1763 ...*, which notes that 'Cape Sable is laid down according to Chabert ...   And South Carolina according to the Survey of William De Brahm'.   A similar mention of De Brahm's work is included on the 1770 sheet by William Fuller which contains charts of Amelia Island, the Entrance to St. Mary's River, and the Mouth of the Nassau River in northern Florida and adjacent Georgia. On the Amelia Island chart, it is noted that the latitude and longitude of the northern point of the island was 'Taken from De Brahm's Map of South Carolina and Geogria'.

During his period of Florida surveying and mapping De Brahm had been aided by a number of assistants, at least two of whom rose to fame as map makers in their own right. These were Bernard Romans and Joseph Purcell. Romans is best known as the author of the book *A Concise Natural History of East and West Florida*, published in 1775 in New York. In addition to this book Romans authored many manuscript and engraved maps and views as well as historical tracts.

Joseph Purcell, like Romans, espoused the Patriot cause and exercised his talents as surveyor and cartographer for the American side during the Revolutionary War. One of his notable post-war contributions was the map of the southern United States which was used by the first great American geographer, Jedidiah Morse in his highly successful book, *The New American Geography*. Through these two, as well as a cadre of other assistants he trained, De Brahm's work was made available to decision makers both during and after the American Revolution. It is probably not an exaggeration to assert that his surveys and the maps which flowed from them played some role in the war's outcome and the period of development which followed.

His book of sailing directions, *The Atlantic Pilot*, which had appeared in English in 1772, was published in a French edition in 1788, and more recently was reproduced as a part of Florida's American Revolution Bicentennial publishing project. His work was joined with that of Bernard Romans, George Gauld and others in several late eighteenth-century sailing directions which doubtlessly guided squadrons of navigators during the period before official hydrographic offices were established.

During the early nineteenth century De Brahm's contributions were acknowledged by figures such as the international boundary surveyor Andrew Ellicott and the Florida author James Grant Forbes. There are indications that copies of De Brahm's maps and *Report* provided valuable information for United States government figures concerned with the development of the area after it was purchased from Spain in 1821.

Most recently, workers interested in reconstructing the original natural environmental conditions of the area around Biscayne Bay and the modern urban centre Miami, found De Brahm's maps and surveys to be both accurate and valuable. His maps and *Report* have also been cited as evidence in two recent U.S. original jurisdiction cases heard by the Supreme Court. One of these dealt with the determination of Florida's seaward boundaries and the other the Savannah River boundary between Georgia and South Carolina.

Bibliography and Sources

1. *REFERENCES ON WILLIAM GERARD DE BRAHM*

Brown, Lloyd A., 'The River in the Ocean', *Essays Honoring Laurence C. Wroth*, Portland, Maine, (1951), 69-84

Brown, Ralph H., 'The De Brahm Charts of the Atlantic Ocean, 1772-1776', *Geogr. Rev.*, vol 28 (1938), 124-32

Chardon, Roland, 'A Best-Fit Evaluation of De Brahm's 1770 Chart of Northern Biscayne Bay, Florida', *The American Cartographer*, vol 9 (1982), 47-67

Corse, Carita D., 'De Brahm's Report on East Florida, 1773', *The Florida Historical Quarterly*, vol 17 (Jan. 1939), 219-26

Cumming, William P., *The Southeast in Early Maps*, 2 ed., Chapel Hill, North Carolina, (1962), 54-5

De Vorsey, Louis, Jr., 'William Gerard De Brahm, Eccentric Genius of Southeastern Geography', *Southeastern Geographer*, vol 10 (1970), 21-9

---- 'La Florida Revealed: The De Brahm Surveys of British East Florida, 1765-1771', in Ralph Ehrenberg (ed) *Pattern and Process*, Washington, D.C., (1974), 87-102

---- 'British East Florida on the Eve of the Revolution', in Samuel Proctor (ed) *Eighteenth-Century Florida and Its Borderlands*, Gainesville, Florida, (1975), 78-101

---- 'Pioneer Charting of the Gulf Stream: The Contributions of Benjamin Franklin and William Gerard De Brahm', *Imago Mundi*, vol 28 (1976), 105-20

---- 'A Colorful Resident of British Saint Augustine: William Gerard De Brahm', *El Escribano*, vol 12, (1975), 1-24

---- 'Hydrography: A Note on the Equipage of Eighteenth-Century Survey Vessels', *The Mariner's Mirror*, vol 58 (1972), 173-7

---- 'William De Brahm's Continuation of the Atlantic Pilot, An Empirically Supported Eighteenth-century Model of North Atlantic Surface Circulation', in M. Sears and D. Merriman (ed) *Oceanography: The Past*, New York, Heidelberg, Berlin (1980), 718-33

---- 'Wayward Ocean River', *Geogr. Mag.*, vol 52 (1980), 501-10

Goodheart, L. McCormick, 'Two Anglo-American Gold Medals Relating to St. Augustine, Florida and Sheldon, South Carolina', *The Numismatic Circular*, vol 64 (March, 1956), 111-12

Morrison, A.J., 'John G. De Brahm', *South Atlantic Quarterly*, vol 1 (July, 1922), 252-8

Mowat, Charles L., 'That Odd Being, De Brahm', *The Florida Historical Quarterly*, vol 20 (April, 1942), 323-45

Tate, Thad W., 'The Discovery and Development of the Southern Colonial Landscape: Six Comentators', *The American Society Proceedings*, vol 93 (1983), 289-311

2. SELECTED BIBLIOGRAPHY OF WORKS BY WILLIAM GERARD
DE BRAHM

a. *Scientific Books*
1772 *The Atlantic Pilot*, London (Facsimile Reprint
with Introduction and Index, Gainesville,
Florida, 1974)
1774 *The Levelling Balance and Counter-Balance;
Or, the Method of Observing by the Weight and
Height of Mercury, On any Place of Terra-Firma
on the Terrestrial Globe, The Exact Weight and
Altitude of the Atmosphere Below and Above the
Place of Observation; Thereby to Ascertain How
Much the Horizon of the Sea Is Lower Than the
Place Whereon the Observation Is Made*, London
1788 *Recherches faites par ordre de sa majesté
britannique, depuis 1765 jusqu'en 1771, pour
rectifier les cartes & perfectionner la
navigation du canal de Bahama ...
Traduite de l'anglois de W. Gerard De Brahm
ecuyer, hydrographe général au department du
sud de l'Amérique Septentrionale*, (Paris)
1849 *History of the Province of Georgia, With Maps
of Original Surveys*, Wormsloe, Georgia
(Privately printed and limited to 49 copies)
1856 *'De Brahm's Philisophico-Historico Hydrogeography
South Carolina, Georgia, and East Florida'*, *in*
Plowden C. J. Weston (ed) *Documents Connected
With the History of South Carolina*, London,
p. 155-227 (Printed for Private Distribution)
1971 *De Brahm's Report of The General Survey In
The Southern District of North America*,
Columbia, South Carolina (Edited and with
an Introduction by Louis De Vorsey, Jr.)

b. *Religious Books and Tracts*
1791 *Time an Aparition of Eternity*, Philadelphia
1792 *Voice of the Everlasting Gospel*, Philadelphia
1794 *Zeite Rechenschaft*, Ephrata, Pennsylvania
1795 *Apocalyptic Gnomon Points Out Eternity's
Divisibility Rated With Time Pointed At by
Gnomon Siderealis*, Philadelphia

*Louis de Vorsey is professor of geography at the
University of Georgia, Athens, Ga., USA*

Chronology

1718	Born in Koblenz, 20 August
1750	Count Seckendorf signs a testimonial to the abilities of De Brahm, 18 December
1751	De Brahm in charge of a party of 156 German Protestants bound for Georgia, 14 July
1751	Trustees for Establishing the Colony of Georgia recommend that De Brahm be granted 500 acres of land, 24 August
1751	The happy arrival of an embarkation of Germans under the prudent conduct of Mr. von Brahm reported from Georgia, 27 November
1752	De Brahm presents a map of part of Georgia to colony's authority with a request for some office of employment, 24 March
1752	South Carolina invites De Brahm to design and build a system of fortifications for Charleston, 17 July
1752	De Brahm announces intention to undertake a map of South Carolina with part of Georgia, 23 October
1754	De Brahm and Henry Yonge appointed joint Surveyors of Land in Georgia, 26 July
1755	De Brahm appointed as interim Surveyor General of South Carolina, 14 August
1756	De Brahm in charge of construction of Fort Loudon in Overhill Cherokee territory in what is now eastern Tennessee
1757-62	Involved in several fortification projects in Georgia
1764	Appointed Surveyor General of Southern District in North America and new colony of East Florida, 13 June
1765-71	Exploring and charting Florida's eastern coast
1770	Suspended from provincial office by Governor of East Florida, 4 October
1771	Sailed to England to answer governor's charges
1772	*The Atlantic Pilot* published in London
1773	First volume of *Report* presented to King George III, 2 April (now in the British Library)
1774	Board of Treasury exonerates De Brahm and he is reinstated in his East Florida office, 28 July; *De Brahm's Zonical Tables ...*, and *The Levelling Balance and Counter-Balance ...* published in London from press of T. Spilsbury
1775	Returns to America aboard *Cherokee* to undertake research for 'The Continuation of the Atlantic Pilot'

1775-77	Held as a prisoner at large by Revolutionary government in Charleston, S.C.: married Mrs. Mary Fenwick of Charleston, 18 February 1776
1777	Returned to England via France aboard the *Lincoln*
1778	Parliamentary allowance for conduct of Survey in the Southern District stopped
1778-84	Suffered breakdown and convalesced at Topsham near Exeter
1784	Attempted to claim his right to the Signory of Holzheim in the Bavarian Palatinate
1785	Returns to Charleston to recover wife's annuity
1788	Voyages to London to petition for compensation and pension, arrives 22 June; Writes to Queen regarding a cure for King George III's illness, 12 November
1798	Delivered personal copy of *Report* and *The Continuation of the Atlantic Pilot*, to Phineas Bond, British Consul General in Philadelphia, 1 November (now in Houghton Library, Harvard University)
1791-99	Last decade of life devoted to metaphysical concerns and publishing in Philadelphia area; died, 6 June 1799

John Dee

1527–1608

IAN M. MATLEY

A controversial figure even in his own day, John Dee combined the roles of geographer, mathematician, philosopher, antiquarian, astronomer, astrologer, alchemist and spiritualist at a time when science in England was still linked with magic. However, his contributions to geography and navigation were considerable, mainly through his part in the planning of many of the voyages of northern exploration of the Elizabethan age and his development of navigational techniques.

1. EDUCATION, LIFE AND WORK

John Dee was born at Mortlake near London in 1527 into a family of Welsh origin. His father was a minor official at the court of Henry VIII. Dee received his early education at the chantry school in Chelmsford and in 1542, at the age of fifteen, went to St John's College, Cambridge. In 1546 he graduated with a B.A. degree and was appointed as a Fellow of the newly founded Trinity College.

The following year Dee went to the University of Louvain to study under Gemma Frisius, who was cosmographer to the Emperor Charles V at his court in Brussels. Dee met Gerard Mercator, a student and assistant of Frisius, and became his close friend. Dee returned to Cambridge in 1548 and brought back a terrestrial and a celestial globe designed by Frisius. The terrestrial globe had been engraved by Mercator and showed among other features a passage to the north of the American continent, the *Fretum Trium Fratrum*. Dee's interest in the discovery of this passage probably dated from this initial contact with Frisius.

In 1548 Dee received his M.A. degree from Cambridge. His later title of 'Doctor' seems to have been an honorary one of doubtful origin. He returned to Louvain the same year, where his contact with Mercator was renewed. It was at this time that Dee's interest in alchemy and astrology may have begun. In 1550 Dee went to the University of Paris where his public lecture on Euclidian geometry was a great success. He was offered a position as a reader in mathematics in Paris, but declined and left for England in 1551. This success in France, along with the publication of two works, one on the use of the celestial globe (1550) and another on astronomy (1551) dedicated to Edward VI, resulted in the award of a royal pension. In 1553 he wrote a treatise, on the subject of tides, for the Duchess of Northumberland. In the following year he refused the offer of a lectureship in mathematics at Oxford University. At this time he was also called upon to cast horoscopes for the royal family: because of the accusations of an informant that Dee was attempting to enchant Queen Mary he was arrested and imprisoned, but he was able to clear himself of the charges of witchcraft and treason before the Star Chamber in 1555 and later was absolved of possible heresy. However, the stigma of magician remained with him for most of his life, especially in the eyes of the uneducated public. His astrological interests continued and he was requested by Elizabeth to choose the most propitious date for her coronation in 1559.

Geographers Biobibliographical Studies, volume 10 (1986)

Dee's contacts with the geographers of his time were expanding.   Apart from Mercator, he was also friendly with another pupil of Frisius, Pedro Nunes (Petrus Nonius), the Portuguese geographer who was professor of mathematics at Coimbra and Lisbon universities and was made cosmographer royal in 1529. Although Dee probably never met him, they corresponded and Nunes had a great influence on the development of Dee's theories on navigation.   When Dee was ill in 1558 he named Nunes as his literary executor.   At a later period he also knew Ortelius, whom he visited in 1571 in Antwerp and who visited him in England in 1577.

Dee's involvement in the planning and study of sea voyages began in 1553 with the proposed attempt by Richard Chancellor to discover the north-east passage to the Orient.   There is no direct evidence of Dee's involvement in the planning of this voyage, but he had a major part in the training of Chancellor, himself a trained mathematician, in the use of astronomical tables and other navigational aids. After the death of Chancellor the hopes of the Muscovy Company were transferred to Stephen Borough and Dee was asked to instruct him in navigational techniques for his unsuccessful voyage in 1556 to the north-east.   In fact Dee had taken Sebastian Cabot's place as adviser on navigational matters to the Muscovy Company and the trainer of their pilots.

In 1565 Dee married the widow of a London grocer and sometime before 1570 they went to live in his mother's house in Mortlake.   This house became his headquarters and there he started a library that was soon to become one of the richest collections of books and manuscripts, both contemporary and ancient, in England.   In 1571 the Queen sent Dee on a mission to the Duchy of Lorraine which has never been explained, though it was probably connected with Dee's interests in alchemy, which the royal court also seem to have encouraged in the hope that the secret of transmutation of base metals into gold would be an English discovery.   This trip has also been queried as a possible espionage mission and raises the question of whether Dee was a secret agent of Secretary Walsingham, with whom he was on friendly terms; but there is no clear evidence on this point. There is also little proof of a voyage by Dee to St Helena, although the island seems to have been of interest to him.

About this time Dee became interested in chorography which he saw as 'an underling and a twig of geography'.   As practised by Dee, it was a combination of geography and local history.   In 1573 he became interested in the reasons for the coastal erosion which had caused the disappearance of the port of Dunwich in East Anglia.   In 1574 he published a small book on a field trip he had made in Cheshire and North Wales.   He also produced a diagram showing ten places in England where treasure was buried, but nothing seems to have come of this knowledge.   In 1577 Ortelius visited Dee to enlist help for his plan in organizing a chorographical study of early Britain. Although Dee himself does not seem to have been interested in contributing to this work, he introduced Ortelius to his friend William Camden and Camden's *Britannia* was the result.   It was probably compiled largely with the aid of Dee's library of ancient manuscripts.

In 1575 Dee married for a second time, but his wife died during the following year.   He married for the third time in 1578, his new wife being in the retinue of Lady Howard, whose husband later became Lord Admiral.   This connection enabled Dee to strengthen his already considerable influence at Court and to keep in touch with persons who might help him with his various plans.

Dee's interest in the search for a route to Cathay (China) continued in the 1570s.   Dee believed that China was the repository of occult knowledge that would lead to the discovery of the philosophers' stone, the elixir of life and the formula for harmonious life among the peoples of the earth.   These goals were shared with other alchemists of his time, but Dee was unique in using his geographical knowledge to further the search.   He also wished to encourage an English expedition to discover the land of Beach or Locach, shown on the maps of Mercator and Ortelius as existing south of New Guinea.   At the same time Dee believed in the expansion of English sovereignty over the whole area of the North Atlantic.   His ideas on these subjects were contained in his work *General and Rare Memorials Pertayning to the Perfect Arte of Navigation*, published in 1577.   This essay was part of a larger work which was to include a section on *Famous and Rich Discoveries* which traced the voyage of Solomon to Ophir (close to the legendary Beach with its gold treasures), a set of navigational tables and a section which was later burnt by Dee for some obscure reason.   The aim of the whole work was to cover the history and theory of navigation and to promote the establishment of British rule in the North.   Dee was probably the inventor of the concept and term 'British Empire'.   He proposed to achieve these aims by the construction of a royal navy and by the establishment of English claims to various regions in the north Atlantic.

Due to his Welsh ancestry he was a great admirer of King Arthur and named his eldest son after him. The Tudor monarchs, themselves of Welsh origin, had encouraged this association between themselves and the early hero-king, and a cult of Arthur developed, based largely on the imaginative writings of Geoffrey of Monmouth, whose *Historia Regum Britanniae* (c. 1135) claimed that King Arthur and King Malgo had conquered not only Gaul and Ireland, but also Scandinavia and Iceland.   Dee extended Arthur's conquests to Greenland and even further north, as well as to all the northern isles as far as Russia.   These isles included the imaginary Frisland and Estotiland, where Dee thought that Arthur had established colonies. Thus England had a right to a large northern empire. England could also claim much of the east coast of America, called 'Atlantis' by Dee, based on the voyage of Madoc, the Welsh prince who was supposed to have established a colony there in 1170.   In 1577 Dee informed Queen Elizabeth of her titles to these lands and in 1580 he presented her with copies of these titles and a map on her request, but her adviser Burghley was doubtful of these claims and no practical action was taken to enforce them.

After the opening of trade with Russia by Chancellor, the lack of success of Borough's voyage and the cancellation of a planned voyage in 1568 to the north-east, interest in the north-east passage waned.

The idea of a north-west route was instead raised, particularly by Sir Humphrey Gilbert in 1576. He pointed to the 1564 world map of Ortelius which suggested a relatively easy passage to the north of the American continent. This short and quick route to the Orient would enable the English to sell their products there more cheaply than the Spaniards. Mercator's world map of 1569 and Ortelius' map of 1570 reinforced the belief in a 'Strait of Anian' between Asia and America which would provide a route to the South Sea (Pacific Ocean).

Although Ortelius and Mercator thought that the role of the English in the exploration of the time should be the further investigation of the north-east route, Gilbert and Dee thought otherwise. Preparations for a voyage to the north-west by Martin Frobisher and Christopher Hall were under way in 1576 and Dee stayed for a time in Muscovy House and visited Frobisher's ships to brief them on the voyage and instruct them in the use of navigational instruments. Frobisher returned from his first voyage in 1577 with the story of the discovery of the entrance to the 'passage to the South Sea', a sample of ore which it was hoped would contain gold, and a captive Eskimo, thought to be a Tatar. There is no real evidence that Dee was on board Frobisher's ship as has been claimed. The sample of ore excited enough interest for two further voyages. Dee had invested £25 in the first two voyages and was thus granted a free share in the third. However the ore finally was proved to be worthless and Dee's only consolation was to have some crags on Greenland named 'Dee's Pinnacles', although the name never lasted. Frobisher's reward was to have his name given to the wide straits shown on the 1578 world map in George Best's account of the voyages. The idea of a north-west passage was thus still very much alive.

While the first voyage of Frobisher was in progress secret preparations for a voyage by Francis Drake were taking place. It has not been proved that Dee took part directly in these plans, but there are some clues to his probable involvment. Drake left England in December 1577 and returned from his circumnavigation of the globe in September 1580. In the summer of 1580 Dee was busy drawing a chart and preparing navigational instructions for a proposed voyage by Arthur Pet and Charles Jackman to China via the north-east passage. It is probable that Dee was planning a meeting between Drake and Pet and Jackman somewhere in the north Pacific near the 'Straits of Anian' and they would all return to England via the route along the north coast of Asia. The fact that Drake sailed so far north along the west coast of North America, perhaps as far north as 48 degrees, suggest such a plan and Drake may have been briefed to look for the entrance to the 'Straits of Anian'. Mercator wrote to Ortelius after Drake's return and told him that he thought that the Drake and Pet-Jackman voyages were connected and that a plan for such a meeting had been made. In 1581 William Hawkins, who had been with Drake, called on Dee, presumably to report on the voyage. These pieces of evidence at least suggest a knowledge on Dee's part of Drake's plans. The failure of the Pet-Jackman voyage with the loss of Jackman and his crew, however,

dimmed Dee's enthusiasm for the north-east route, and he discussed a possible overland route to China with John Haller, a German, in 1581, directing him to note variations in compass readings along the way. Haller never undertook the journey. It is possible that Dee himself had hoped to make such a journey himself a few years earlier. By 1580 Dee's influence as a geographer was beginning to diminish. In that year the younger Hakluyt became so doubtful of some of Dee's ideas in connection with the Pet-Jackman voyage that he wrote in secret to Mercator to obtain his views. The letter arrived too late as the voyage had already started. Although he was still to act as adviser to another voyage, Dee's interests in the occult began to dominate his work.

In 1581 he entered a bizarre and unfortunate stage of his life. In that year he claimed to have had some sort of psychic experience and the next two years were spent looking for a medium to help him contact spirits who would aid him in his philosophical and other quests. In 1582 he met Edward Kelley, an unscrupulous charlatan who dominated Dee's life for the next six years. Dee never seems to have claimed to have contacted or seen spirits himself and all the messages and advice from the other world came through the medium of Kelley. As Dee refused to have any dealings with evil spirits, Kelley conveniently arranged for a series of interviews with angels who were sent to enlighten Dee on a variety of matters ranging from alchemy to geography.

By 1578 Humphrey Gilbert had begun to lose his enthusiasm for the north-west passage and had approached Queen Elizabeth about founding a model colony for English Catholics in North America. He does not seem to have given Dee any details of his plans and Dee assumed that he was sailing for the North. In 1580 Dee asked for and obtained from Gilbert the titles to all territories discovered by the expedition north of 50 degrees latitude. It was only in 1582 that Gilbert raised the necessary funds for his project, with the help of Sir George Peckham, a Catholic sympathizer, who came to Dee to enquire about the title to Norumbega (a so-called 'kingdom', probably the area around Penobscot Bay in present-day Maine). He was concerned that this potential site for the new colony might be subject to Spanish or Portuguese claims. He promised Dee 5,000 acres of land in the colony. Dee had thus a considerable stake in this venture.

Preparations for the voyage included the drawing up of a map in 1582-83 for Gilbert, almost certainly by Dee. It corresponds closely with Dee's map of the North Atlantic made in 1580. Dee refers to the gift of a map to an English sailor planning a voyage to the American coast. It was probably used as promotional advertising for the expedition. Gilbert's deed to the colony refers to the Dee River, probably Narragansett Bay in present-day Rhode Island. Sir Humphrey's voyage in 1583 was a disaster and the new colony never materialized.

During this period Dee was also active in planning what was to be his last navigational venture. In 1579 he had met Adrian Gilbert, the brother of Sir Humphrey, and John Davis, the navigator. Gilbert was interested in the occult and had influenced Davis in this direction, hence the attraction of association with

Dee.   Davis was the finest navigator of his time and Dee saw in him the man who could best further the exploration of the north-west passage.   He encouraged the pair and on occasions recourse was made to seances for information and advice from the spirits. In 1583 serious plans were made with the advice of Secretary Walsingham.   The idea was to explore further the approaches to the north-west passage and, if possible, to found a colony.   Dee was obviously interested in pursuing the title to the lands granted to him in 1580.   Later in 1583 year Dee held a seance with Kelley which Adrian Gilbert was allowed to attend. However, when an angel was contacted as to the prospects for the voyage and the geographical details of the territories to be visited he replied: 'These things belong not to my charge.   Thou knowest them which are sufficient, whom short time shall serve for the whole instruction'.   Other contacts with the spirits were equally unfruitful.   Although plans for the voyage were ready in 1583, Adrian Gilbert did not receive the royal patent for the voyage until 1585, by which time Dee had left the country.

In 1583 Dee's life entered a new phase in which geography and navigation played little part.   He had been visited by another dabbler in the occult, a Count Laski from Poland who invited Dee to visit him.   Dee left for Poland with Kelley who had become an indispensable companion.   Dee's interest in geography briefly reappeared in Krakow where he learned from an angel during a seance that the lands on the maps of Mercator and Pomponius Mela were accurately shown and that gold might be found in Greenland, along with some other bizarre information.   In 1584 Dee moved to Prague and seems to have thought that he was about to learn the great secret of the philosophers' stone. He even offered the King of Poland his services in making the stone with angelic assistance but was turned down.   In 1586, while in Bohemia, Dee received an invitation from Tsar Feodor I to come to Moscow to give advice on matters pertaining to the northern territories of Russia and also some 'other weighty occasions', probably alchemy.   It is probable that there were plans for a map of Russia to be drawn with Dee's help, but Dee declined the offer.

It has been suggested that Dee's stay on the continent involved espionage for Walsingham and that some of the carefully recorded conversations with the spirits at this time were in fact messages in code, but there is no proof of this theory.   Dee at that time seems to have been manipulated by Kelley who not only claimed to have made gold but also that the spirits wished Dee and himself to share all their property in common, even their wives.   However Kelley left Dee in 1588, was later imprisoned and was killed or died by accident in 1595.   In 1589 Dee returned to England.   Shortly after his departure in 1583 his house in Mortlake had been sacked by a mob who opposed his occult exercises and much of his valuable library was destroyed.   Short of money, he obtained the post of Warden of Christ's College in Manchester in 1596.   He arranged for Christopher Saxton to come and make a survey of the town and also finished a treatise on the Queen's sovereignty over the seas around the British Isles.   He still seems to have continued his interests in alchemy and in 1603

appealed to King James I to have his name cleared of charges of being a 'conjurer' and magician.   In 1605 his wife died and in 1608 he died, a poor man, at his old home in Mortlake.

2.   SCIENTIFIC IDEAS AND GEOGRAPHICAL THOUGHT

Dee's contributions to geography lie mainly in his encouragement of exploration in higher latitudes and his development and teaching of mathematical techniques of navigation.   Unfortunately Dee's advice to seekers for the north-east and north-west passages was less than useful and was in fact totally misleading, for his image of the northern world was no better than that of his contemporaries.   Too much credence was given to the works of early geographers, accounts of mythical voyages and the lies of seamen about their experiences in distant lands.   For example, the (probably fictitious) voyage of the Zeno brothers in 1380 resulted in a book and map, published in 1558, which showed the islands of Frisland and Estotiland, features which were to appear on maps of Mercator, Ortelius and Dee. Frobisher carried the Zeno book and map on his 1576 voyage and reported sighting Frisland (probably southern Greenland).   Belief in the Zeno map misled geographers and seamen for almost 100 years.

Dee was more optimistic than some of his contemporaries on the subject of the north-east passage. Mercator believed that a large promontory, known as Tabin (Cape Chelyushkin), extended far to the north and made it difficult to sail along the coast of northern Russia.   Apart from problems with ice, great variations of the compass in this region would create navigational problems.   He advised the use of one of the great Siberian rivers as a route to Asia.   Dee believed that Tabin did not lie as far north as Mercator claimed.   Once past Tabin the coast would turn more southerly and easterly toward China as the Arab geographer Abulfeda (Abu-al Fida) had shown.   In 1577 Dee corresponded with Mercator and Ortelius on this matter.   Dee's map of 1582, prepared for Humphrey Gilbert, shows a relatively easy route along the arctic coast of Russia once the North Cape had been rounded.

His views on the north-west passage were equally hopeful.   The idea of a through route to the 'Straits of Anian' and the Pacific Ocean was strengthened by Frobisher's 1576 voyage and the appearance of 'Frobisher's Straits' on the 1578 world map of George Best, the official reporter of the Frobisher expedition. Dee's map of 1582 shows somewhat the same outlines of the northern coast of America.   It is difficult to see any contribution which Dee made to understanding the configuration of the major land and ocean masses which places him ahead of the other geographers of his time and indeed it can be argued that he was inferior in this knowledge to others such as Mercator and Ortelius.

In the understanding of navigation Dee emerges as a pioneer on the English scene, though our knowledge of some of Dee's contributions is somewhat clouded by his secrecy.   He rarely made his ideas or inventions known except to those who sought his help.   Dee introduced the use of mathematics to English navigation which until that time   had   been conducted on an unscientific basis.   His introduction to the first

English translation of Euclid in 1570 laid the mathematical foundations for the future development of navigation. In particular Dee contributed to the understanding of navigation in the North. The problem of rhumb lines and their relationship to meridians and great circles had interested Dee's friend Nunes and Dee applied Nunes' ideas to the practical problem of navigation in high latitudes where the meridians converge towards the Pole. In the 1550s he invented 'a paradoxal compass' for this purpose and taught its use to the Boroughs brothers. Dee used the term 'paradoxal sailing' to describe a method of avoiding the navigational errors caused by use of the plain chart. This mysterious 'compass' seems to have been a circumpolar chart based on a zenithal equidistant projection, similar to his 1582 map for Gilbert. As such a projection was already used by the Portuguese and Spaniards, it is difficult to say if Dee devised it on his own or had heard of it, in spite of the secrecy maintained by England's rivals in these matters. That Dee thought this projection was best for navigation in northern latitudes is shown by his criticism of Mercator's famous world map of 1569. Dee, who seemingly did not understand the significance of Mercator's projection, thought that the distortions of scale in the north made it difficult for seamen to use. A study of Dee's *Canon Gubernauticus, An Arithmeticall Resolution of the Paradoxall Compas* has also led to the theory that the 'compass' was a curved line (a rhumb spiral and not a great circle) resulting from the plotting of rhumbs on a globe or polar projection. This idea may have been borrowed from Pedro Nunes. Dee has left us few clues about this and others of his inventions.

Dee also claimed the invention of 'compasses of variation'. It is not clear exactly what he meant as the variation compass was already well known since the 1530s as a means of comparing true and magnetic north by the altitude of the sun. Nunes had perfected such a compass and Dee may have learned about it from him. He certainly had such a compass and taught its use. Dee also designed a 'new sea compass' divided into degrees as well as points and also a table of departures, used to calculate changes in longitude caused by sailing one degree of latitude north or south on different courses. In 1576 he claimed to have developed a method for ascertaining the longitude and latitude of a position without recourse to astronomical observations. This mysterious technique may have involved noting changes in magnetic variation or possibly astrology. In 1575 Dee prepared a manuscript of a complete manual of navigation called *The British Complement of Perfect Navigation* which was never published and is now lost. References to it by Dee reveal that it was very mathematical and impractical for use at sea.

On the development of the theory and practice of mathematical navigation Dee was certainly ahead of his contemporaries and provided a base for future navigators on which to build. Although his knowledge of the earth may have been inadequate, his contributions to the science and art of navigation paved the way for future voyages and a greater knowledge of the geography of the globe.

3. *INFLUENCE AND SPREAD OF IDEAS*

Dee exercised a great influence on persons interested in furthering English exploration and the expansion of English trade and sovereignty in the North. Until 1583 he was in touch with scholars, seamen, travellers and officials at the court. Queen Elizabeth visited him several times and sought his advice on such varied subjects as astrology, comets and toothache. This royal favour helped Dee considerably in his schemes. His contacts with scholars such as Frisius, Mercator, Ortelius, Nunes and others led to an exchange of ideas on an international scale. For example, Dee was probably instrumental in supplying Mercator with a map of Great Britain, perhaps by Laurence Nowell, Dean of Lichfield, which Mercator engraved in 1564. His contacts with Nunes led to the further development and propagation of the latter's ideas in England by Dee. Dee also introduced new navigational instruments from the continent, such as the cross-staff and astronomical ring-dial (for telling time and finding latitude) designed by Frisius, as well as globes by Frisius and Mercator. Dee donated these instruments in 1584 to the Fellows and students of Trinity College, Cambridge, for study.

Dee's influence was probably greatest with other scholars and geographers of the time and least with the seaman, except as far as his enthusiasm for their schemes encouraged them. His mathematical techniques for navigation were beyond the intellectual capabilities of most of the master mariners of the period and, besides, were often unsuitable for use under shipboard conditions at that time. Frobisher was unable to use Dee's navigational techniques, while William Borough stated that he thought that scholars who had no experience at sea had little to offer. This fact does not detract from the important foundation laid by Dee in the study of problems of the use of charts and problems of compass variation, latitude and longitude. His work was to bear fruits in the next century when mathematical navigation finally came into its own.

Status of Research

Research on Dee's contributions to geography and navigation has been carried out mainly by Eva G.R. Taylor, who has devoted much space to Dee in her writings on Tudor geography, and by D.W. Waters, who has examined his navigational and cartographic techniques. There are references to Dee in many books on northern voyages and exploration, especially by Samuel Eliot Morison. Frank Rasky gives a more popular account of Dee's activities. His role in the voyage of Humphrey Gilbert is covered by David B. Quinn and other material on Dee occurs in Hakluyt and Purchas. A good general biography of Dee has been written by Charlotte F. Smith and various facets of Dee's thought and work are discussed by Peter J. French. Richard Deacon's book is somewhat sensational in its approach and says little about Dee's scientific work.

Bibliography and Sources

1. REFERENCES ON JOHN DEE

anon, 'The Gilbert Map of c.1582-3, *Geogr. J.*, vol 72 (1928), 235-7

Deacon, Richard, *John Dee: Scientist, Geographer, Astrologer and Secret Agent to Elizabeth I*, London (1968)

'Dee, John', *DNB*, vol 14, London (1888), 271-9

French, Peter J., *John Dee. The World of an Elizabethan Magus*, London (1972)

Halliwell, James O., *The Private Diary of Dr. John Dee*, London (1842)

Kendrick, T.D., *British Antiquity*, London (1950), 37, 43

Morison, Samuel Eliot, *The European Discovery of America*, The Northern Voyages A.D. 500-1600, New York (1971), passim

Quinn, David B., *The Voyages and Colonising Enterprises of Sir Humphrey Gilbert*, London (1940), vol 1, 31, 37, 52-3, 66-71, 96-8; vol 2, 280, 483-8

Rasky, Frank, *The Polar Voyages*, Toronto (1976), 121-51

Smith, Charlotte F., *John Dee (1527-1608)*, London (1909)

Taylor, E.G.R., *The Haven-Finding Art*, London (1956), 195-7, 202, 207

Taylor, E.G.R., 'John Dee and the Maps of North-East Asia', *Imago Mundi*, vol 12 (1955), 103-06

Taylor, E.G.R., 'John Dee and the Nautical Triangle, 1575', *Journal of the Institute of Navigation*, vol 8, no 4 (1955), 318-25

Taylor, E.G.R., *Late Tudor and Early Stuart Geography 1583-1650*, London (1934), passim

Taylor, E.G.R., 'A Letter Dated 1577 from Mercator to John Dee', *Imago Mundi*, vol 13 (1956), 56-68

Taylor, E.G.R., 'Master John Dee, Drake and the Straits of Anian', *Mariner's Mirror*, vol 15 (1929), 125-30

Taylor, E.G.R., *The Mathematical Practioners of Tudor and Stuart England*, London (1954), passim, esp. 170-1

Taylor, E.G.R., 'More Light on Drake 1577-80', *Mariner's Mirror*, vol 16 (1930), 134-51

Taylor, E.G.R., *Tudor Geography 1485-1583*, London (1930), 75-139

Taylor, E.G.R., 'Canon Gubernauticus, by John Dee (1558)', in Taylor, E.G.R. (ed), *'A Regiment for the Sea' by William Bourne*, Cambridge (1963), 415-33 (Hakluyt Society)

Ward, B.M., 'Martin Frobisher and Dr. John Dee', *Mariner's Mirror*, vol 12 (1926), 453-5

Waters, D.W., *The Art of Navigation in England in Elizabethan and Early Stuart Times*, New Haven (1958), passim, esp. 521-6

2. SELECTIVE BIBLIOGRAPHY OF WORKS BY JOHN DEE

1550 *De usu Globi Coelistis ad Regem Edwardum Sextum* (The use of the celestial globe, dedicated to Edward VI)

1551 *De nubium, solis, lunae, ac reliquorum planitarum, immo ipsius stelliferi coeli, ab infimo terrae centro, distantiis, mutuisque intervallis, et eorundem omnium magnitudine* (Treatise on astronomy, dedicated to Edward VI)

1553 *The True Account (not vulgar) of Floods and Ebbs* (Written for the Duchess of Northumberland)

---- *The astronomical and Logisticall rules and canons to calculate the Ephemerides by, and other necessary accounts of heavenly motions written at the request, and for the use of, that Excellent Mechanician Master Richard Chancellor, at his last voyage into Moschovia*

1556 *Inventum mechanicum paradoxum de nova ratione circumferentiam circularem; unde valde rara alia dependent inventa* (Reference to the 'paradoxal compass')

---- *Canon Gubernauticus: An Arithmeticall Resolution of the Paradoxall Compass* (Departure tables)

---- *De Annuli Astronomici Multiplicie: usu* (Use of the astronomical ring-dial)

---- *De Nova Navigatione Ratione* (A new system of navigation, probably with the 'paradoxal compass')

1560 *De Itinere subterraneo* (Methods of surveying underground mining tunnels)

1565 *Reipublicae Britannicae Synopsis* (Tract advocating English overseas expansion, the search for the north-west passage and the colonization of North America, written for Sir Edward Dyer, a favourite of the Queen)

1570 *Mathematical Preface* to Henry Billingsley, *Euclide*

1576-77 *The British Complement of the Perfect Art of Navigation. Vol. 1. General and Rare Memorials Pertayning to the Perfect Arte of Navigation: annexed to the Paradoxall Cumpass in Playne. Vol. 2. Queen Elizabeth her Tables Gubernatick* (Treatise on navigation by the 'paradoxal compass', great circles, compass variation and a method of finding latitude and longitude). *Vol. 3. Destroyed, thus contents unknown. Vol. 4. The Great Volume of Famous and Rich Discoveries* (King Solomon's voyage to Ophir and descriptions of the North)

1578 *Her Majesty's Title Royal to many foreign countreys, kingdoms and provinces* (Written at the Queen's request)

---- *Brytanici Imperii limites* (Treatise on English land titles, written at the Queen's request)

1580 *Atlantidis, vulgariter Indiae Occidentalis nominatae, emendiator descriptio, quam adhuc est divulgata* (Map of America with land titles, presented to the Queen)

---- *Instructions for the two Masters, Charles Jackman and Arthur Pet*

---- *Navigationis ad Cathayam per septentrionalia Scythiae et Tartariae Littera delineatio Hydrographica* (Chart to accompany the above instructions)

1583 *Hemisphaeri Borealis Geographica atque Hydrographica Descriptio* (Map on polar projection made for Humphrey Gilbert's voyage)

1597 *Thalattokratia Bretannike* 'Treatise on British sovereignty over the seas, written for Sir Edward Dyer)

3. ARCHIVAL SOURCES

Manuscripts, letters, notes and maps by Dee are located in the British Library, mainly in the Cotton, Harleian, Lansdowne and Sloane MS collections. Others are in the Bodleian Library, Oxford, in the Ashmole, Corpus Christi, Rawlinson and Smith MS collections. Dee's 1583 map is in the Free Library of Philadelphia. Many manuscripts are not complete. For example, of the four volumes of *The British Complement of the Perfect Art of Navigation*, volume 1 is in the British Library, volume 2 is lost, volume 3 was destroyed and volume 4 is in the British Library, badly damaged by fire. Several manuscripts and notes refer to inventions and navigational techniques which are not described in any detail. Some of these manuscripts are known only from Dee's own catalogues of his works.

Ian M. Matley is professor of geography at Michigan State University, Michigan, United States of America

1580	Planned the voyage of Pet and Jackman to the north-east passage
1582	Met Edward Kelley the medium
1582-83	Involved in planning a colonization venture with Adrian Gilbert and John Davis
1583	Left for Poland with Kelley. Library at Mortlake sacked by mob
1584	Arrived in Prague
1586	Refused offer to come to the Russian Court
1588	Kelley left Dee
1589	Dee returned to England
1896	Warden of Christ's College in Manchester
1605	Wife died
1608	Died at Mortlake

Chronology

1527	Born at Mortlake, near London
1542-46	Studied at St John's College, Cambridge
1547-50	Studied at the University of Louvain
1550-51	Visited the University of Paris
1553	Instructed Richard Chancellor in navigation
1555	Acquitted on charges of treason and heresy
1556	Trained William Borough as pilot for the Muscovy Company
1565	First marriage
1575	Second marriage
1576	Instructed Frobisher in navigation for his voyage to the north-west passage
1577	Propagated the idea of a northern British Empire
1578	Third marriage

Nevin Melancthon Fenneman

1865–1945

Courtesy of University Archives, University of Cincinnati

BRUCE RYAN

There is only one Fenneman in the history of American
geography, a famous and revered figure who died forty
years ago, but whose work has been cited continuously
ever since and whose two definitive volumes on the
physiography of the United States are still listed,
perdurable as Holy Writ, in the *Geographical
Bibliography for American Libraries* fifty years after
their original publication.   Appointed to the first
chair of geology in the University of Colorado at
Boulder in 1902, Fenneman became the foundation
professor of geology and geography at the university
of Cincinnati in 1907, where he remained as department
head until his retirement in 1937.   He was elected
President of the Association of American Geographers
in 1918, having been one of the thirteen intimates of
the inner circle who supported William Morris Davis in
founding the Association in 1904.   His was the double
distinction of also being elected President of the
Geological Society of America in 1935, in his
seventieth year.

1. EDUCATION, LIFE AND WORK

Fenneman's grandfather was Johann Heinrich Vennemann, a
Westphalian migrant who landed in Baltimore on
1 October 1840, journeyed to Pittsburgh by canal and
railroad, continued on to Cincinnati by river, and
eventually settled in Indianapolis in 1852.   One of the
sons who accompanied him across the Atlantic as a boy of
eight went on to study Calvinistic theology at
Heidelberg College in Tiffin, Ohio, was ordained as a
minister of the Reformed Church, and anglicized his name

to William Henry Fenneman.

Between 1859 and 1903, the Rev. Dr. W. H. Fenneman
ministered to a succession of rural, small-town
congregations in northern Indiana and Ohio.   They were
stern and law-abiding communities of the Protestant
Bible Belt where the Amish felt at home and today's
voters still begrudge the pleasures of the bottle.   In
Lima, Ohio, one of these towns, the Fenneman's third
child and only son was born on 26 December 1865.   He
was christened 'Nevin' after the contemporary American
theologian who had written *The Mystical Presence* (a
vindication of Calvinism), 'Melanchthon' after Philipp
Melanchthon (1497-1560) who had first formulated the
theology of Luther's Reformation (in the *Loci communes*
of 1521), and 'Fenneman' after the fen folk of the Low
Countries among whom his ancestors were numbered.   His
mother, Rebecca Oldfather (previously 'Aultvater'), was
descended from German and Irish pioneers of the
Shenandoah Valley in Virginia.   Despite her husband's
ambition that his son would also take holy orders she
had her more worldly way and guided him and his sisters
into teaching careers.   Fenneman was a downy-faced
youngster of seventeen when he graduated as A.B. from
Heidelberg College, Tiffin, in 1883.

For nine years Fenneman taught the higher grades
at the village schools of Wilkinsburg and Greensburg,
both near Pittsburgh.   Sarah Morrow Welty was the
second teacher at Wilkinsburg in 1884.   She and
Fenneman began a platonic friendship which lasted
through both their marriages, each addressing the other
as 'David' (Mrs. Welty) and 'Jonathan' (N.M.F.).
Mrs. Welty remembered Fenneman as being 'modest to a

fault', rather diffident about his 'humble beginnings', and the 'only alley-cat' (his own whimsical epithet) among a better-pedigreed set of young men who were the life of the village. At Greensburg, where Fenneman was headmaster from 1886, he taught mathematics and chemistry.

In 1892, at the age of twenty-seven, he moved to Greeley as Professor of Physical Sciences at the Colorado State Normal School, a cow college (that is, a small and not well-known college in a rural area) founded in 1889 with a two-year curriculum for the training of elementary schoolteachers. The term 'cow college' aptly describes Greeley which then had only one building. His responsibilities embraced geography and his conquests included a colleague, Sarah Alice Glisan from Fredonia, New York, whom he married on his birthday in 1893. Their first home was an eight-room wood-frame house with cottonwood trees in front and box elder behind. At that time, Horace Greeley's cooperative farming venture had not yet survived its first twenty-five years, and Turner's westward frontier of settlement had passed beyond Denver only in 1858. Fascinated though he remained by the Front Range of the Rocky Mountains for the remainder of his life, Fenneman's intellectual ambitions turned his attention eastwards, initially towards New York and Washington, D.C. During those eight years in Greeley he trod the stranger paths of banishment, taking care to map them indelibly in his memory. As Edgar L. Hewett inquired later, 'What would you have done without Greeley to practice on?'

What followed was quite literally a re-orientation, a turning once more towards the east. The first distraction was the summer course at Harvard in 1895 when Fenneman fell under the spell of William Morris Davis and adopted him--not all at once but over a forty year friendship--as mentor, confidant, and collaborator. The course at Harvard was an outgrowth of the National Education Association's foray into secondary school curricula and college entrance requirements. Much that Davis preached in 1895 was fastened upon by Fenneman, from the urge to lay the logical foundations for a more scientific geomorphology to the importance of geography everywhere in the curriculum, whether in the elementary school, the business college, or the War Department. Although Fenneman has been regarded by some as a disciple of Davis (verging at times on sycophancy), that was not his self-perception. The two men never agreed about the propriety of importing geological terms into geography, and Fenneman told Eliot Blackwelder that he 'always found both Davis and [Douglas W.] Johnson very much more interested in discussing types [of landforms] than in dividing up areas according to type'.

Nevin and Sarah Fenneman remained in Greeley eight years, but moved to Chicago in 1900. Fenneman had already passed his M.A. examination at the feldgling University of Chicago in March 1898, and had submitted his M.A. thesis eighteen months later. It dealt with 'The Laramie Cretacious Series' but was augmented by lacustrine field work for the Wisconsin Geological Survey, supervised by T.C. Chamberlin and C. R. Van Hise, whose lectures on the physiography of the United States Fenneman prepared for publication by Longman in 1903. His doctoral thesis, in emulation of Rollin D. Salisbury, was entitled 'The development of the profile of equilibrium of the subaqueous shore terrace'. He was awarded the Ph.D. degree in 1901 and returned to Colorado during the following year, but this time to Boulder, not Greeley, to become the first professor of geology at the University of Colorado. For twenty years, however, the University of Chicago remained the pivotal brokerage house in any negotiations he undertook, professional or academic. What it approved, he carefully considered; what it advised, he mostly did.

Its advice and approval prompted Fenneman to leave the University of Colorado after only one year, during which he nevertheless published over 200 pages of his graduate research. He was appointed professor of geology in the University of Wisconsin, Madison, at an annual salary of $2,000, joining an already distinguished department which promptly saw him elected to Fellowships in the American Association for the Advancement of Science (1903), the Geological Society of America (1904), and the Association of American Geographers (also in 1904, as one of the forty-eight original members). These were four years of professional ordination and increasing family responsibility. His father died in 1904 and Fenneman had to underwrite the mortgage of his parents' home. There was a summer abroad, at the Second Petroleum Congress in Belgium, and resource surveys in the Yampa coal field of Colorado, the oil fields of Texas, and in North Dakota. Fenneman earned another $100 per month ('plus expenses') in summer employment by the Wisconsin Geological Survey, and settled himself on the awkward, three-legged stool of corporate oil exploration, government field commissions, and academic research--preferring United States Geological Survey (USGS) assignments which were resource surveys rather than 'private work', because of the 'greater steadiness' of the former despite the 'larger remuneration' of the latter. 'I trust you will not entirely discard your official personality', advised C. W. Hayes of the USGS, 'but will be on the lookout for profitable fields for future investigation'. Another 400 pages of those investigations appeared in print during the four Wisconsin years.

Why Fenneman left the University of Wisconsin in 1907 is not transparently obvious. With such famous friends there as Frederick Jackson Turner (1861-1932) and Stephen Moulton Babcock (1843-1931), and with such easy access to the Wisconsin lakeland and the Chicago Sanhedrin, Madison should have seemed the ideal location. Yet as early as 1905 John W. Hall of the University of Cincinnati had tempted Fenneman with the lure of a $3,000 annual salary, and by 1906 Fenneman had sought unsuccessfully the chair of geology at the University of Michigan. They were turbulent, topsy-turvy times in the bearish academic market-place, with the closest of friends jockeying for the same positions. As chairman of the AAG nominating committee in 1906 Fenneman found himself advantageously placed to recommend others with characteristic fairness and responsibility, but not himself.

What tugged him to Cincinnati in 1907, into an ecological niche for which he was ideally adapted? Though attached to his work and home in Wisconsin,

Fenneman divulged in a letter to a Cincinnati ally the attractions of being more closely in touch with students in smaller classes, of the respect clearly accorded good teaching, and the now familiar 'pleasure of creating something'. 'I think too', he wrote, 'I should like the old time flavor of the Cincinnati Scientific Society'. Two 'ulterior motives' were also mentioned: the proximity of Cincinnati to his mother's home in Hamilton, Ohio, and his 'wish to get Mrs. Fenneman away from our rigorous Wisconsin winters'. The inducements were sufficient.

An offer from President Charles W. Dabney of the University of Cincinnati was accepted on 7 May 1907, and Fenneman resigned his Wisconsin position on 10 June. He was forty-one years old, at the threshold of a thirty-eight year association with Cincinnati that was to redound to the honour of both. Dabney had specified his wish for a 'scientific geologist, who is at the same time a good teacher and a good organizer', knowing that laboratories and a museum had to be established. He was equally cognizant that his Teachers College wanted a geographer and his Engineering College an economic geologist. Fenneman satisfied them all. 'Congratulations', wrote W. M. Davis. 'It is a beautiful district, a fine chance for your physiographic work'.

That chance, however, then seemed rather remote. In 1906 the U.S. Geological Survey had declined to support Fenneman's proposed synopsis of American physiography on the grounds that it lacked 'correlations'. One year later the *Journal of Geology* rejected a similar submission because it appeared to be, in T. C. Chamberlin's editorial view, 'an extension of an interesting principle far beyond the probable limit of its applicability'. Thus encouraged, Fenneman turned for consolation to building his new Department of Geology and Geography at the University of Cincinnati—the 'Taft institution', as one Colorado lawyer perceived it, the municipally-owned, civic university of 'one of the most highly cultured and most solidly established American cities'.

The University now enrolls 37 000 students, but in 1907-08 there were only 1,364. All departments in the College of Liberal Arts huddled under the common roof of McMicken Hall. On its fourth floor Fenneman assembled what were then the latest teaching materials and laboratory equipment including a Fuess petrographic microscope, lantern slides from the Smithsonian Institution, a stereopticon, donated journals, maps and map catalogues from the American Geographical Society, and the beginning of a geological museum which was eventually to house the Ordovician fossil collection for which Cincinnati is known around the world. Walter Bucher, whose own distinguished career in academic geology began when Fenneman appointed him museum curator in 1912, himself attached 8,160 labels to the S.A. Miller fossil collection. Rock specimens, many obtained through exchanges with Bryn Mawr College and other institutions, also enlarged the museum, as did mammoth tusks and teeth from Cincinnati sites. Within two years of Fenneman's appointment the University of Cincinnati had spent $16,000 in equipping its new Department of Geology and Geography, equivalent today to an initial outlay of $250,000.

In 1911 the Department moved out of McMicken Hall and became the sole tenant of the 'Old Technical' building, a red brick encrustation of the Holocene in which the university's geologists still silicify. A student guide book located the building at 'a stone's throw from dilipidation'. Cooperation with the Cincinnati Museum of Natural History, alternating with attempts to combine its collections with those of the university, occupied Fenneman until 1924 when he resigned from the museum committee, having fended off the Botany Department's six-year crusade to capture it, but weary, as he put it, of a city which consigned museums to attics and basements. Although the paleontological collection gradually eliminated the invidious reliance on better collections of Cincinnati fossils in Boston, Chicago, Washington, and Baltimore, and although Fenneman steered the museum through its 'natural history' phase into a 'laboratory' phase, some of its other exhibits attracted the jibes of condescending colleagues. One joked that the museum entrance was presided over by 'a decrepid (*sic*) elephant', for which the lack of budgeted maintenance occasioned Fenneman to rebuke his dean. 'Several years ago', he complained, 'we had to throw out and burn the giraffe for the same reason'. Almost from the day of his arrival in Cincinnati Fenneman had contended that 'the question whether we shall ever have a great geological department here depends in large measure on the closeness of our relations to that [Cincinnati's Natural History] museum'.

Among those appointed by Fenneman to the joint department were such notable geologists as Walter H. Bucher (who came from Heidelberg, Germany, and went to Columbia, N.Y., after thirty years in Cincinnati), the mineralogist Otto von Schlichten, Charles H. Behre, Jr. (who moved to Northwestern in 1928), John L. Rich (that rarity, as Fenneman described him, 'an intellectual among oil men'), George B. Barbour (a Scottish FRGS who became Dean of Arts and Sciences), and Kenneth E. Caster (President and Medallist of the American Paleontological Society). Caster had arrived in Cincinnati with the record Ohio River flood of 1936 and founded the 'Dry Dredgers', Cincinnati's organization of amateur fossil hunters. His wife, Anneliese S. Caster, drew or redrew most of the illustrations in Fenneman's *Physiography of Eastern United States*.

Of the geographers appointed by Fenneman to Cincinnati's joint department Earl C. Case taught from 1920 to 1957, Homer Martin from 1925 to 1930, Daniel Bergsmark from 1927 to 1945, and George F. Deasy from 1936 to 1940. A.E. Sandberg taught both geography and geology, as if to legitimize the academic union, and personally conducted Fenneman on a tour of Panama in 1940. Case and Bergsmark collaborated in writing *College Geography* (1932), for long the standard regional treatment of world economic geography, and their contributions to 'commercial education' (today's 'business administration') were highly valued by the pragmatic Fenneman. During the 1920s, in fact, over a third of the Department's students came from the very impure, very applied College of Engineering and Commerce. 'Are we training scientists or engineers?' Fenneman once inquired.

The Dean of that College was Fenneman's great friend and habitual confidant, Herman Schneider, the famed originator of the 'cooperative system of technological education' in 1906--a scheme for alternatively scheduling undergraduates in classes for one term and in a career-related job for the following term. From the University of Cincinnati the 'co-op system' has spread around the world. What his friendship with Schneider epitomozed was the trustworthiness, steadiness of purpose, authority, and influence that Fenneman attained in Cincinnati and its university. Geography and geology became protected birds in a feathered nest largely because Fenneman became an unflagging and almost avuncular source of sound advice. He served on two committees to find new presidents for the University, coming within an ace of securing the geographer Harlan H. Barrows from the University of Chicago in 1928. He chaired committees to find department heads for Economics and Botany and even after his retirement was consulted confidentially by the Board of Trustees about the fairest methods of staff retrenchment.

Herman Schneider's belief in meshing practical job experiences with university studies certainly also sustained Fenneman, who spent his summers in the field. His own letters, and those of his scientific correspondents, typically concluded with 'best wishes for a pleasant and successful [or profitable] field season'. That was the expectation. Fenneman worked almost every summer between 1900 and 1924 for the U.S. Geological Survey, or for various state geological surveys. After 1924, he returned to Montana or Colorado each summer for the ensuing twenty years, either as a participant in the Yellowstone-Bighorn Research Association's field camp at Red Lodge or to the University of Colorado's summer camp at Nederland. His field reports were published in a dozen USGS *Bulletins* and as the first editions of geological map quadrangles. His St. Louis and Cincinnati surveys were published with an eye to school use although he held his tongue when a Missouri educator taunted him with Oersted's aphorism, 'The laws of nature are the thoughts of God'. Fenneman may well have agreed. His efforts to reconcile science and religion were low-key but constant.

Fenneman travelled throughout the United States, exploring at first hand all the physiographic provinces which he was to describe so fully in the two volumes of his *Physiography*. His secretary, Lillian Smith, routinely made his railroad and sailing reservations, but Fenneman never yielded to the temptation to view his physiographic provinces from the sky and never boarded an airplane, even though Orville Wright and Admiral Richard E. Byrd were among his correspondents. Fenneman attended the 1908 International Geographical Congress in Geneva, offically representing the Association of American Geographers (of which he was then treasurer), the U.S. Geological Survey, and the Cincinnati Natural History Society. When he decided to join the Italian-French Excursion, which W. M. Davis called 'the geographical pilgrimage to Rome', an admirer of Fenneman's geomorphological divination wrote: 'I trust when you get to Naples, you will insist upon an eruption'. Fenneman also attended the International Geological Congresses in Stockholm (1910) and Toronto (1913), and the Pan-Pacific Congresses of 1923 (Australia) and

1926 (Japan).   On W.M. Davis's Transcontinental Excursion of 1912 Fenneman served as 'Journalist', fielding the questions from reporters in every town they visited throughout what Davis had predicted would be 'by far the biggest geographical event ever evented in the U.S.' in a letter to Fenneman dated 1 November 1911.

Although these gatherings brought Fenneman into contact with many European geographers, including Albrecht Penck, and although he lectured at Oxford for A.J. Herbertson in 1910 his research remained resolutely in focus on North American landforms. He corresponded with foreign scientists as friends and tour guides, not research associates. For Fenneman's generation perhaps the untimely First World War shattered any prospect of collaboration, especially with scientists from countries of the Pan-Germanic League. After the Armistice Fenneman sent gifts of books and money to Joseph Partsch at Leipzig, to the Rector of the University of Graz, and to colleagues in Vienna, but there were no proposals for joint research-- only inquiries about American studies from Penck and about charter forms of government from G.G. Chisholm of Edinburgh. Fenneman knew where he belonged. 'To offset the travels of a philosopher in Europe', he wrote to Van Meter Ames, the aesthete, 'I spent the week following Christmas in Oklahoma. The difference between Europe and Oklahoma is the difference between a philosopher and a geologist'.

Fenneman's magnanimity was widely appreciated. His beneficiaries ranged from depression victims who begged him for fifteen dollars to pay the rent to a niece whose incestuous husband abandoned her, in defiance of the White Slave Act. Fenneman served on the board of Cincinnati's Community Chest for the last five years of his life. The Children's Hospital received $6,000 as a memorial to his wife. Sarah Fenneman, the radiant hostess and blithe spirit who had pretended to be the frivolous fiancée of 'every eligible man in the country', had died in 1920, childless, leaving her husband another twenty-five years of solitary but not wholly inconsequential pursuits. Fenneman never remarried, but willed $73,000 of his trust assets to Heidelberg College, Ohio, and $50,000 to the University of Cincinnati.

It took a resolution of the Board of Trustees to force retirement upon him in 1937, at the age of seventy-one. He took it very hard--relinquishing the headship of his department during its thirtieth anniversary, surrendering his warm office on the ground floor of Old Tech, from which he could bark at the insolent Corryville boys when they lit fires in the trashcans or threw stones at the greenhouse. Though fearful that retirement would be a 'sentence to solitary confinement' as a 'pensioner and derelict', he was not forgotten. At the University's commencement exercises in 1940, when he gave the address to the graduating class, he was awarded the Honorary Degree of Doctor of Laws. He continued to teach his graduate seminar in physiography for all eight of his emeritus years and his Lincolnian portrait painted by Frank H. Myers, through subscription, still hangs in Old Tech.   In June 1945 a minor ailment put him in hospital but pneumonia set in and he died of a coronary occlusion on the Fourth of July.   He was buried in Fredonia, N.Y., beside his wife, reunited with the earth which had been his terrestrial abode and intellectual obsession, somewhere, appropriately, between the Rocky Mountains and the New England coast, between the Laurentian Shield and the Ohio Valley.

2.   *SCIENTIFIC IDEAS AND GEOGRAPHICAL THOUGHT*

Nevin Fenneman's life bridged the nineteenth and twentieth centuries--formed in one, fulfilled in the other.   Notionally this *fin de siècle* transition should have produced a dilettantish and despairing world-weariness, but instead, in Eliot Blackwelder's modest words, it produced 'a useful and well-rounded life'.   For all his starchy conservation Fenneman was a life-long initiator, especially at Greeley, Boulder, and Cincinnati.   He helped to liberate American geography from its early servitude in the House of Geology, while remaining an eminent and approachable figure in both fields.   Although he lived through two World Wars and the Great Depression Fenneman was too old to enlist and too comfortably off to be diverted from his physiographic pursuits.   He worked for peace, while investing in railroads and real estate.   Fenneman tuned out the partisan politics of his time or reduced them to satire or cynicism, routinely reading only the *Springfield Republican* (not the most cosmopolitan of newspapers) and acquiring a radio (as a gift) only to listen in on the Second World War.

But if the historical context was not crucial to his work, its regional context was.   Time and again Fenneman found his physiographic examples and inspiration in either his home locality or areas to which he returned summer after summer, often the Rocky Mountains of Colorado and Montana, and the glaciated landscape of Ontario and Hudson Bay, to which he made six camping and canoeing trips with his friend James Albert Green, Cincinnati's Public Librarian, together with Indian guides.   Fenneman filled forty-five notebooks with minutely pencilled descriptions of the landforms he saw out of train windows or from saddled horses, even using up the blank columns of his savings bank account book.   The site of Cincinnati also thrust under his nose all the ramifications, human and physical, of Ohio River flooding.   Although foreign lands did not distract him from American research he kept up an interest in Central America, the Manchurian and Japanese periphery of China, and Australia, where E.C. Andrews (Sydney), Charles Fenner (Adelaide), and E. de C. Clarke (Perth) corresponded about his work.

Fenneman's enduring monument is his immaculate division of the United States into twenty-five physiographic provinces (*Annals of the Association of American Geographers*, 1916, with revisions in 1921 and 1928), and his 1,200-page description of them, section by section, in *Physiography of Western United States* (1931) and *Physiography of Eastern United States* (1938). The task absorbed his mind, frontally or occipitally, for over forty years from the germinal inspiration of Major John Wesley Powell's *Physiographic Regions of the United States* (published in 1894 by the National Geographic Society), through his post-doctoral editing

of Van Hise's physiographic lectures in 1902, to the gold medal conferred on him in 1938 by the Chicago Geographic Society 'for eminent achievements in the Physiography of the United States'.   He treasured a photograph of Powell's statue in Arlington National Cemetery, Washington, and Powell's portrait came first among those displayed in the Department's museum.

The complementarity of regional analysis and synthesis has rarely been so snug or so compelling but it was not accomplished through divine inspiration alone.   Fenneman refined and multiplied the physiographic provinces employed in *Forest Physiography*, published in 1911 by Isaiah Bowman, whom he had met five years before, in Eckert's Cave, Illinois.   He assimilated W.L.G. Joerg's subdivision of the United States into 'natural regions' (1913) and had corresponded with A.J. Herbertson of Oxford about his 'major natural regions' (1905).   Fenneman chaired standing committees of the Association of American Geographers, which sponsored the project, and spent 1915-16 in Washington, D.C., on academic leave, plundering the U.S. Geological Survey for the unpublished field reports he needed.

Critics of his landform divisions included Cleveland Abbe (1838-1916), founder of the U.S. Weather Bureau (another Cincinnatian), who averred that North American climatological boundaries were 'better established', and Alfred H. Brooks, who wrote: 'You certainly started a fine row with your physiographic provinces'.   To geologists who accused him of rarely handling an actual rock or ever focusing on the earth processes responsible for his provinces, he retorted: 'I can't lay eggs either, but I know more about an omelet than any hen in the country'.

The encouragement he really desired came from Isaiah Bowman who wrote that 'You have set in motion a principle of research which will continue for many, many decades and will firmly establish the geographical idea in statistical work in the United States'. However environmentally deterministic, that was clearly Fenneman's hope: to define physiographic units that formed a 'proper basis for the study of statistics', and to establish the 'political base' (that is, counties or townships as enumeration districts) corresponding to the 'physical base'.   He proposed this next step to the U.S. Census Bureau and the U.S. Department of Agriculture, but in vain.   Louis C. Peltier rejected the degree of generalization implicit in Fenneman's provinces (and those of Bowman), finding them misaligned with the finer-grained 'phenomena of human occupance'--an unconformity of a different kind.   A more durable conclusion was reached by the *Geographical Review*: 'Dr. Fenneman's labors thus provide a solid physiographic framework for any serious study in the geography of our country'.   So they did and so they continue.

What is more the nomenclature he devised for the eighty-six physiographic 'provinces' and 'sections' has become coin of the realm, the accepted terminology used in virtually all subsequent regional studies. Detractors of Fenneman's 'mere description' casually overlook how precisely he defined his toponomy, how judiciously the sub-regions fit together into a logically and empirically valid whole.   He invented the alphabet, but not without contention.   G.K.

Gilbert and M.R. Campbell withheld his Boulder report from publication until Fenneman's so-called 'Culver sandstone' was better correlated with precisely designated members, while D.W. Johnson and F.E. Matthes long contested the propriety of 'older' and 'newer' Appalachians, and such coinages as 'vallemont'. Fenneman also took issue with W.M. Davis over banning the use of geological names such as 'Carboniferous' from regional geography. 'Why must we secrete the geologic map', he asked, 'as medieval priests secreted the Bible?'

To Fenneman topography was an end-product for the geologist but a point of departure for the geographer. For geography the 'areal relation' was crucial and studying specific areas comprehensively, however far they strayed from the singularly idealized Davisian landform 'types', was no less worthy than hunting down the geomorphological processes responsible for their topography. The two approaches, Fenneman thought, belonged to different sorts of minds: for him, 'the obligation to explain the actual spot seemed quite equal to that of elucidating and illustrating the principles'.

Fenneman made other contributions to science and geography. Erosion surfaces of various kinds attracted his attention sporadically over four decades. His doctoral dissertation on the profile of equilibrium of the subaqueous shore terrace matured into his presidential address to the Geological Society of America in 1935, 'Cyclic and non-cyclic aspects of erosion', which contended that normal erosive agencies, without reference to base level or cyclic processes, could produce landforms which his cocksure contemporaries were attributing to peneplanation. Fenneman styled himself 'a peacemaker on pediments', but the fiefdoms kept on squabbling.

During other phases of his career Fenneman also wore the hat of the economic geologist, surveying oil fields in Texas, Louisiana, and Colorado, mapping coal deposits, drilling (even divining) for artesian water, and assessing the commercial value of glacial gravels. In the vicinity of Cincinnati he reconstructed the preglacial river systems and advised the Ohio Flood Board, the U.S. Army Corps of Engineers, and the Central Inland Waterways Association, latterly (much too latterly) concerning 'commerce through a barge canal'. The 1920s, in fact, were swamped with commissioned resource surveys, many of them intrinsically geographical, which involved Fenneman's entire department. One was Herman Schneider's industrial resource survey for the Commercial Club of Cincinnati (now its Chamber of Commerce), an eighty-six page undertaking which Fenneman directed while protecting his flanks from free-loading hyenas in the Department of Chemical Engineering who believed they should be in charge. Bulging files were developed on the most proximate, suitable, or untapped raw materials for Cincinnati's manufacturing industries. Earlier, in 1914, Fenneman had prepared the case for locating a new Federal Reserve Bank in Cincinnati, but the building was erected in Cleveland. Few of today's dedicated location theorists are given such an opportunity to rearrange a nation's financial hierarchy.

In 1920 Fenneman again put 'science on the witness stand' (as he titled an address on the subject) by providing expert testimony for the United States Department of Justice, which hoped to show that Golden Lake and Young's Lake in Arkansas had not existed as water bodies when the area was originally surveyed about 1840. It was another case of riparian rights and 'quiet title' along the meandering and ever-shifting Mississippi. Litigation dragged on until 1923, when Fenneman and the ecologist, the forester, and the drainage engineer who assisted him failed to prove their case but learned the lesson that physiography is not quite an exact science.

All these resource surveys confirmed Fenneman's belief that geology and geography had a united role in practical affairs. As newly-elected president of the AAG in 1918, he was arranging a joint meeting with the Geological Society of America to consider the 'relation of our sciences to the War' when the Armistice intervened. By then he was already involved in President Woodrow Wilson's preparations for the peace commission negotiations. Many American geographers had been commandeered by Colonel Edward M. House to assist with his secretive but official 'Inquiry' into the international implications and options of the peace settlement. Fenneman's assigned responsibility was to gather and organize the African scientific data, since the political disposition of the German colonies had to be anticipated. His group was to be 'concerned primarily with the potentiality of the African tribes', including those considered to be 'from the start rather hopeless material'. He reported to Isaiah Bowman, whose field expediency was becoming a legend, but his own African 'team' included many notables. J. Paul Goode provided a hyposometric map of Africa 'with special reference to white settlement'. H.K.W. Kumm proposed a monograph on the 'Mohammedan advance in Africa', and prepared a tribal map. T.C. Chamberlin, Fenneman's old editorial nemesis, dispensed with geology and offered a paper on the 'question of colonial adjustments' including boundaries, international straits, free ports, 'outlets for inland peoples', and international rights of way. E.A. Hooton of Harvard prepared an ethnographic map of Africa showing the 'capacity for civilization' of each tribe and George Otis Smith, Director of the USGS, provided the mineral statistics. A monograph co-authored by H.L. Shantz of the U.S. Department of Agriculture and C.F. Marbut, *The Vegetation and Soils of Africa*, was published by the American Geographical Society in 1923. It was one of the larger, more scholarly outcomes of the Inquiry.

Fenneman turned for basic research assistance to 'women who have been trained in physiography and have good critical faculties ... who can read the various descriptive books and articles on Africa and see through them'. He recruited Aida A. Heine, Lila Thompson, and Mrs. Robert W. Jones. His own contribution, over and beyond organizing the project, was the *Report on Trade Routes in Africa*, which discussed 'hinterlands' (a term used in English only from 1890) and 'radial routes' which reflected 'the stage of exploitation'--a conceptualization usually thought to derive only from the labyrinthian search models of the 1960s.

There is a great deal more to Fenneman's scientific ideas and geographical thought than the physiographic provinces, nomenclature, resource surveys, and surficial geology, to use a modern term including soils, their parent materials and bedrock at or near the surface. Monumental contributions though they are, these were merely the alluvial cones of his mind. Fortunately he left a formidable record of the bedrock uplands from which those cones discharged. Fenneman's work was set in a well-formulated but rigid educational philosophy, and a clear sense of scientific mission. His most enduring and most-studied paper is surely 'The Circumference of Geography', the presidential address he gave to the Association of American Geographers in 1918. Depicting his notions in the now-famous Venn diagram, Fenneman showed how 'commercial geography' occupies the overlap between geography and economics, climatology the overlap between geography and meteorology, and so on. The central residual circle, he said, around which the overlapping sciences formed an encroaching ring, 'may represent regional geography'. This he defined as 'the study of areas in their compositeness or complexity'. He viewed it as the 'central core' of 'pure geography', containing the seeds by which the subject propogates itself and produces 'that *distinctive* geographic flavor which comes only when the various elements are studied in their inter-relations'. Much as he approved and practised G.K. Gilbert's 'scientific trespass' and 'cross-fertilization', he held that 'a science cannot be defined by its circumference' and that living 'too much on our borders and not enough in the center' allows other fields to 'claim us as a vassal'.

Fenneman was much intrigued by the nature, intellectual stature, and value of geography. In 1919 he circulated an outline for a hypothetical 'geography of Missouri' to sixteen American colleagues, requesting their reactions to his conspectus. Twenty years later, at the age of seventy-four, he presented a paper on the 'Development of Geography in America' to the Ohio Academy of Sciences, contending that the popular mind still saw geography as a comprehensive, non-technical, anthropocentric 'world picture', discouragingly dominated by commercial patterns and therefore an 'orderly assemblage of knowledge, but not a branch of science'. Three attempts to infuse it with 'scientific character'--Georg Gerland's 'geophysics', W.M. Davis's 'surface forms', and the disciples of 'environment and response'--had all failed. In the margin of his manuscript, Fenneman wrote: 'Not offered for publication'. As early as 1910 he was commending to Davis a paper on the 'geographic aspect of culture' by Captain Joshua Slocum, and in 1924 he reminded his Dean of Education why a staff geographer was needed: 'Speaking of *culture*, is there any subject in the range of Education in which one or two years of college study minister more to the *higher pleasures* than Geography. If it is not a "humanity", what is?'

Fenneman was well aware of his own scientific style. 'I am distinctly partial to hypotheses which involve the fewest unproved or extraordinary assumptions', he told Howard A. Meyerdorff in 1936, with reference to Appalachian drainage. In the margin of a letter from T.C. Chamberlin he pencilled 'No more

gathering of data but guided in doing this by multiple hypotheses. Any other "inquiry" is blind and wasteful and inefficient'.

Others remarked on the versatility of this professor of geology who could explain the land movements of the Japanese army in Manchuria and the geographical 'adjustments' of cities, but Fenneman's role models were not the polymathic geomorphologists with whom he corresponded--not Davis, Bowman, Salisbury, or Atwood--but those historic figures whose biographies he chose to write for his private edification: John Maynard Keynes, Malthus, Thomas Paine, and Jeremiah. When he wrote familiar essays about bronze statues, they were those of Woodrow Wilson and the Methodist pioneer, Bishop Francis Asbury, not the busts of Chamberlin or Salisbury, to which he nonetheless subscribed.

Despite the sprawling extent of his research and his determination to encompass regional entireties, he strove to simplify, 'casting the present account [as he put it] into large molds', deferring rather than denying the complications, and constantly questioning the degree of generalization 'allowable and advisable'. One postulate informed both volumes of his *Physiography* that 'the best scheme of division into provinces is the one which makes possible the largest number of general statements about each'. Where irreconcilable or merely divergent interpretations of landforms existed Fenneman rendered each impartially. He did so, Bucher said, because he 'hated polemics as much as dogmatism'--at least in others. In 1919, anticipating a lecture on 'Geology as a mode of thought', he jotted down the bare bones of an argument that seem, with hindsight, almost to define the circumference of Nevin M. Fenneman's 'thinking in terms of long time, of slow change, of ultimate destiny'.

3. INFLUENCE AND SPREAD OF IDEAS

What Fenneman tried to convey was aimed at distinctively and deliberately different audiences. He did not preach to the masses from the top of Mt. Monadnock. His students and university colleagues heard from Fenneman the Educator, his professional associates listened to Fenneman the Scholar, and his social circle knew Fenneman the Moralist. It is doubtful whether Fenneman's students loved him. As a teacher, he was gruff, old-fashioned, tyrannical, scornful, relentless, deliberate, withering, reverberative, tight-lipped, sarcastic, rigorous, resonant, mind-opening, and frightening. He deplored the educational 'drift toward easy and shiftless work' and assured Samuel Van Valkenberg, a European, that he looked upon American students 'with rather more dignity than they deserve'. His sixteen years of teaching in the grade schools and at the Colorado State Normal College (for teachers) left him with a life-long interest in the goals and practicalities of education. It unsettled his research-obsessed colleagues, who saw fit to separate the educational publications from his other papers when the time came to write his obituary.

Fenneman's educational antipathies were legion. In 1907, he chastised the Regents of the University of Wisconsin for wanting to hire professional coaches for intercollegiate athletics and in 1934 he opposed plans to enlarge the stadium at the University of Cincinnati for 'great football spectacles'. It alarmed him that grades in the Botany laboratories were so much higher on average than in the other laboratory sciences, that 'bootleggers' were visiting the fraternity houses, that 'girls' in the back row should chew gum, and that funds were being denied to schools that taught evolution as a result of pressure from the Princeton Theological Seminary. It did not disturb him that W.M. Davis's Transcontinental Excursion of 1912 was for 'men only', but it appalled him when 'girls' requested a 'loafing place' in McMicken Hall where they could smoke and mingle with 'boys'. 'We must face the question', he instructed President Schneider, 'what kind of girls are going to set the pace in this institution? ... Most people of culture want their daughters under better influence'.

Fenneman was no slavish reactionary who automatically dismissed any call for academic change, although he was by no means the 'extreme radical' for whom Richard E. Dodge mistook him in 1910. 'If I seem to express myself too positively', he told the Natural History Society, 'it is only the professor's habit, a kind of "mark of the beast"'. In his 'Letters to Advanced Students', the positive admonitions abounded: 'What do grades signify?', 'What grades do not signify', 'Interest', 'Excuses', and so on, including 'Love and Geology' for those who sat together in class--'a choice tidbit', one alumnus admitted, ranking with the 'lighter humor' of Tennyson, Browning, or Burns. Fenneman staunchly defended the place of geography in the school curriculum and insisted to successive Deans of Education, without much avail, that those who taught geography should be properly trained in it.

He also defended the university faculty against editorial abuse in the *Citizen's Bulletin* (Cincinnati) and chaired an investigation of Allegheny College in 1918 for the American Association of University Professors. He was then a member of its Committee on Academic Freedom and Academic Tenure but resigned from the AAUP in 1931, fed up with its 'labor union' manners. His openness and even-handedness made Fenneman an effective advocate for the faculty. He candidly told President Dabney in 1916 that the new policy of placing a Dean between himself and the faculty was deplorable, reminding him that 'the highest compliment a man can pay to his superior is to entrust his full case to him without reserve'. He also opposed the designation of 'Fellows' within the Graduate School, feeling that an unwarranted 'order of nobility' was being created, and resisting the 'assumption that our faculties are made up of "firsts" and "seconds"'.

If personal quirks coloured Fenneman's influence as an educator, he suppressed them when holding office in the national organizations that served American geographers and geologists. There he was the committee man *par excellence*, at once patient, receptive, organized and supportive. Whereas W.M. Davis could fulminate against the *National Geographic Magazine* ('the pictures are superb, as usual; but the geography is peculiar'), Fenneman could seal his lips. He even filed samples of 'courteous language', including an

attorney's excuse for being unable to address convocation: 'I am attending to an edifying case of co-respondency for a lady of consideration'.

For nine of the years between 1908 and 1923, Fenneman held office in the Association of American Geographers, as councillor, treasurer, and president. As late in his life as 1942 he served on its Atwood Award Committee with Carl O. Sauer and Vernor C. Finch, all three of them chagrined when their recommendation not to make the award was overruled by cronies of one applicant. Fenneman became a Fellow of the Geological Society of America in 1904, made five trips to New York as its president in 1935, and chaired its prioritizing Committee on Research Program from 1936 to 1939. He was an active member of the GSA's Committee of Past Presidents for five years. The American Association for the Advancement of Science elected him a Fellow in 1903, and Fenneman served as its AAG representative in 1926-27 and on its Committee on Grants in 1925-27. He organized its 1923 meeting in Cincinnati, but failed as prime promoter of W.M. Davis's candidacy for the AAAS presidency in 1926. In 1922-23, while Prohibition raged, Fenneman served a term as Chairman of the Division of Geology and Geography of the National Research Council. For so prestigious and influential as appointment he was granted sabbatical leave, and lived for the year in Washington, D.C., where he found 'a very decent group of people ... still in that larval stage which marine biologists distinguish as "free swimming", not yet attached to the bottom'.

These four professional affiliations made very different demands on Fenneman, from benignly approving research grants for the AAAS and determining with Solomonic finality that 'geophysics and ocean bottoms should receive immediate and special support' from the Geological Society of America, to breathing life into the almost still-born Association of American Geographers. As only the second treasurer of the nascent AAG, Fenneman of the broad shoulders received the apologetic resignations of those who could not pay their dues (including the unemployed Willard D. Johnson, who remained 'not unmindful of the honor of being an American geographer') and those like the intransigent Cleveland Abbe who simply found geography too peripheral to their main interests. From his Cincinnati heartland, Fenneman could also help resolve the 'east-west' factionalism brewing within the Association. There was despair among its founders when the National Geographic Society proposed to publish its own 'journal of technical geography' in 1912, and 'mutiny among the disadvantaged' members who questioned the 'exclusiveness' of the AAG and the credentials of its office-bearers. Fenneman proposed some ameliorative changes in the by-laws, but was not beyond scrutiny himself. Herbert E. Gregory wished to replace him as treasurer 'by a man who is nearer to the geographic type' and not a geographer 'only by a stretch of that term', but Albert P. Brigham knew better than to snub a friend in court. Brigham represented the devastated Association at the funeral of Ralph Tarr in 1912, and uncovered his feelings in a letter to Fenneman: 'I thought of you all--the inner circle of his friends-- as I looked at his coffin ... We will all draw up a little closer in our fellowship'.

Fenneman's circle of intimates, with whom he shared his worst fears and fondest hopes, comprised the sixteen correspondents who reacted to his outline 'geography of Missouri' in 1919, the thirty-one 'powerful individuals' expected to support W.M. Davis for president of the AAAS in 1926, and the colleagues whose criticisms of his manuscripts Fenneman sought and respected. There were the conduits of his influence. Prominent among them were Isaiah Bowman, W.M. Davis, Wallace W. Atwood, Herman Schneider, G.G. Chisholm, Harlan H. Barrows, E.C. Andrews, Ellen C. Semple, Eliot Blackwelder, and Albrecht Penck. In 1945, President Raymond Walters of the University of Cincinnati eulogized Fenneman in the AAAS journal, *Science*. He mailed 352 offprints to individuals whose lives had been touched by Fenneman's. Only three of them were relatives.

The larger context of Fenneman's enthusiasms can be gauged from his membership in the American Society of Naturalists, Sigma Xi, Phi Beta Kappa, the Community Chest, the Mount Auburn Presbyterian Church, the English-Speaking Union, the Foreign Policy Association, the City Club, the University Club, the Cosmos Club of Washington, D.C., and, above all else, the Literary Club. 'I like clubs with atmosphere', he would say. There he could promulgate calender reform or sing parodies of Gilbert and Sullivan, advocate Esperanto and its derivative Ido or tell tales about his outlandish friend, Edgar Hewett, whose second wife was the first lady sanctioned to sleep in the camp of the French Foreign Legion in Syria. Through those clubs he could focus the sympathy of Orville Wright and Charles Kettering on the plight of Dr. August Foerste, a Dayton schoolteacher whose retirement they made possible so that he might find the time, long overdue, to write up his local geological research. While puffing his Pittsburg 'stogies' in such company, the ageing Fenneman could behave, as George Barbour observed, 'like a rejuvenated stream in a land of senescence'. This was most true on the Monday nights between 1910 and 1945, when Fenneman gloried in his membership of the Literary Club. This was the oldest organization of its kind in the United States, a forum founded in 1849 for the purposes of civil conversation and the reading of specially written papers. Over the years, its members have included two Presidents of the United States, several Governors of Ohio, and many of Cincinnati's leaders in public affairs and learning. For Fenneman, it was a wonderful outlet for his only hobby, the writing of essays. He read altogether sixty-six of them to the club, and two more were read posthumously. They had, Van Meter Ames remembered, 'a wisdom and a thought-provoking humor about the great questions of morals and education, and a transforming interest in little things'. Fenneman was President of the Literary Club in 1924-25, during its 75th Anniversary, and his name was the answer to one of the questions on C.B. Firestone's Literary Club Quiz: 'What member used to terrorize the club by invoking "unwritten laws" known only to himself?'

Fenneman attained such dignity that he became the object of fun, celebration, and concern. Students and colleagues arranged dinner parties for his sixtieth and seventieth birthdays, and twenty-seven students expressed their 'deep sympathy' when 'Mrs. Fenneman slipped into the great beyond on Easter eve' in 1920, signing themselves 'Your loyal students'. Now there is an annual Nevin M. Fenneman Memorial Day Banquet, at which 'The Fenneman' (a trophy patched together from a sub-standard house-brick) is awarded for the 'Worst Geographical Thought'. Fenneman even served as the archetype for 'Mr. Mason', the prunes-for-desert character in *Scholars and Gentlemen*, an unpublished novel by Van Meter Ames. Admirers asked for his signed photograph, as E.C. Andrews did: 'Our writings are under our own control in great measure, they form an *exparte* statement but one's physiognomy is a kind of summary statement of all one's hopes and fears'. The Aetna Life Insurance Company recorded his height as 5 ft 11 inches and his weight as 180 pounds. The oval face pictured on his citizenship certificate in 1915 was itemized as having a regular forehead and chin, brown eyes and hair, dark complexion, straight nose, and a large mouth.

It has been said that 'you are not a genius until you have left a personal mark on your subject'. For all his modesty, Fenneman would have agreed. 'A university man has two duties', he once told his dean, 'one to his students and the other to his subject. His duty to the latter is not to leave it where he found it'. But was his subject geology or geography? Fenneman supplied his own answer: 'I am a geologist whose chief interest is in human beings'.

Bibliography and Sources

1. OBITUARIES AND REFERENCES ON NEVIN M. FENNEMAN

anon., 'Dr. Nevin Fenneman dies', *Cincinnati Enquirer*, 5 July 1945, 9

anon., 'A famous geologist passes', *Cincinnati Times-Star* (editorial), 6 July 1945, 6

anon., 'Obituary--Nevin M. Fenneman', *Geogr. Rev.*, vol 35 (1945), 682

Barbour, George B., 'In honor of times past', *The Isoline* (Department of Geography, University of Cincinnati), 3 (1967), 5-7

Bucher, Walter, 'Memorial to Nevin M. Fenneman', *Proc. Geol. Soc. Am. for 1945* (June 1946), 215-28

Decamp, John P., 'Nevin M. Fenneman', *New York Times*, 6 July 1945, 11

Green, James Albert, 'In Memoriam', Archives of The Literary Club, Cincinnati Historical Society, 26 September 1945

Lowrie, S. Gale, 'Resolution on the death of Nevin M. Fenneman', Minutes of the College of Liberal Arts, University of Cincinnati, 9 November 1945, 4 p.

McGrane, Reginald C., *The University of Cincinnati--A success story in urban higher education*, Harper & Row, New York, 1963

Rich, John L., 'Memorial to Nevin M. Fenneman', *Ann. Assoc. Am. Geogr.*, vol 45 (1945), 180-9

Ryan, Bruce, *Seventy-five years of geography at the University of Cincinnati*, University of Cincinnati, 1983

Walters, Raymond, 'Nevin M. Fenneman', *Science*, vol 102 (1945), 142-3

2. *SELECTIVE THEMATIC BIBLIOGRAPHY OF WORKS BY NEVIN M. FENNEMAN*

a. *Geography*
1899 'Climate of the Great Plains', *J. Sch. Geogr.*, vol 3, 1-13
1905 'Geography of Manchuria', *J. Geogr.*, vol 4, 6-11
1909 'Problems in the teaching of physical geography in secondary schools', *J. Geogr.*, vol 7, 145-157
---- 'Some anthropo-geographic effects of glacial erosion in the Alps', *J. Geogr.*, vol 7, 169-172
1913 'The Yellowstone National Park', *J. Geogr.*, vol 11, 314-320
1915 'The site of Cincinnati', *J. Geogr.*, vol 14, 10-12
1916 'Geographic influences affecting early Cincinnati', chapter 1 of *Citizens Book*, Cincinnati Chamber of Commerce
1919 'The circumference of geography', *Ann. Assoc. Am. Geogr.*, vol 9, 3-11. Also published in *Geogr. Rev.*, vol 7, 168-175
1921 'Ohio (Geography and Geology)', *The New International Encyclopedia*
1929 'Physical geography of the United States', *Encyclopedia Britannica*, vol 22, 714-723

b. *Physiographic Divisions and Processes*
1902 'On the lakes of southeastern Wisconsin', *Wis. Geol. Nat. Hist. Surv.*, *Bull*, 8, 178pp. (rev. ed., 1910, 188 pp.)
---- 'Development of the profile of equilibrium of the subaqueous shore terrace', *J. Geol.*, vol 10, 1-32
---- 'The Araphoe glacier in 1902', *J. Geol.*, vol 10, 839-51
1905 'Effect of cliff erosion on form of contact surfaces', *Bull. Geol. Soc. Am.*, vol 16, 205-14
1906 'Floodplains produced without floods', *Bull. Am. Geogr. Soc.*, vol 38, 89-91
1908 'Some features of erosion by unconcentrated wash', *J. Geol.*, vol 16, 746-54
1909 'Physiography of the St. Louis area', *Illinois State Geol. Surv.*, *Bull.* 12, 83pp.
1914 'Physiographic boundaries within the United States', *Ann. Assoc. Am. Geogr.*, vol 4, 84-134
1916 'Physiographic divisions of the United States', *Ann. Assoc. Am. Geogr.*, vol 6, 19-98
1917 'Physiographic subdivisions of the United States', *Proc. Natl. Acad. Sci.*, vol 3, 17-22
1922 'Physiographic provinces and sections in western Oklahoma and adjacent parts of Texas', *U.S. Geol. Surv. Bull.* 730, 115-34
1923 'Recent work in paleobotany', *Science*, vol 57, 44-45
1928 'Physiographic divisions of the United States', *Ann. Assoc. Am. Geogr.*, vol 18, 261-353 (rev. 3rd ed., with map)
---- 'Physical divisions of the United States', *U.S. Geol. Surv.*, map & table
1931 *Physiography of Western United States*, New York, 534 pp.

1932 'Physiographic history of Great Basin', *Pan Am. Geol.*, vol 57, 131-42
1936 'Cyclic and non-cyclic aspects of erosion', *Bull. Am. Geol. Soc.*, vol 47, 173-86 (also pub. in *Science*, vol 83, 87-94)
1938 *Physiography of Western United States*, New York, 714 pp.
1939 'The rise of physiography', *Bull. Geol. Soc. Am.*, vol 50, 349-60

c. *Resource Surveys*
1903 'The Boulder, Colorado, oil field', *U.S. Geol. Surv.*, *Bull.* 213, 322-32
1904 'Structure of the Boulder oil field, Colorado, with records for the year 1903', *U.S. Geol. Surv.*, *Bull.* 225, 383-91
1905 'The Florence, Colorado, oil field', *U.S. Geol. Surv.*, *Bull.* 260, 436-40
---- 'Oil fields of the Texas-Louisiana Gulf coast', *U.S. Geol. Surv.*, *Bull.* 260, 459-67
---- 'Geology of the Boulder district, Colorado', *U.S. Geol. Surv.*, *Bull.* 265, 101 pp.
---- 'Oil fields of the Texas-Louisiana Coastal Plain', *Min. Mag.*, vol 11, 313-22
1906 'Oil fields of the Texas-Louisiana Gulf Coastal Plain', *U.S. Geol. Surv.*, *Bull.* 282, 146 pp.
---- 'The Yampa coal field, Routt County, Colorado', *U.S. Geol. Surv.*, *Bull.* 285, 226-39, and *Bull.* 297, 7-81 (with H. S. Gale)
1907 'Clay resources of the St. Louis district, Missouri', *U.S. Geol. Surv.*, *Bull.* 315, 315-21
---- 'Stratigraphic work in the vicinity of East St. Louis', *Illinois State Geol. Surv.*, *Bull.* 4, 213-7
1911 'Geology and mineral resources of the St. Louis quadrangle, Missouri-Illinois', *U.S. Geol. Surv.*, *Bull.* 438, 73 pp.
1916 'Geology of Cincinnati and vicinity', *Ohio Geol. Surv.*, *Bull.* 19, 207 pp.
1925 'A classification of natural resources', *Science*, vol 61, 191-7
1927 *Resource survey of the Commercial Club of Cincinnati*, Univ. Cincinnati, Inst. Sci. Res., Ser. 2/1, 86 pp.

d. *Educational and Other Professional Publications*
1917 'The museum situation in Cincinnati', *J. Cincinnati Soc. Nat. Hist.*, vol 22, 28-41
1922 'Functions of the Division of Geology and Geography of the National Research Council', *Science*, vol 56, 620-4
1925 'What is a university for?', *Sch. Soc.*, vol 21, 393-7
---- 'Why we study', *Sch. Soc.*, vol 22, 196-201
---- 'The case for Latin', *Sch. Soc.*, vol 22, 639-44
1936 'The pupil and the student', *Sch. Soc.*, vol 23, 665-9
1936 'Presentation of the Penrose Medal to Reginal A. Daly', *Proc. Geol. Soc. Am. for 1935*, 48-53
1938 'A possible program of research in geology, A forum on the present needs of the Society', *Proc. Geol. Soc. Am. for 1937*, 143-56
1939 'Personally conducted', *Sci. Mon.*, vol 27, 451-4
1941 'Memorial address (Fellows of the Geological Society of America who died in 1940)', *Proc.*

Geol. Soc. Am. for 1940, 69-71
1943 'Memorial to Max Harrison Demorest', *Proc. Geol. Soc. Am. for 1942*, 173-7

3.   ARCHIVAL SOURCES

Fenneman died a childless widower.   All his personal and professional papers were willed to his colleague and successor, John L. Rich, whose widow, Nellie B. Rich, and whose son, Ralph Rich, donated the papers to the Archives and Rare Books Department of the Carl Blegen Library, University of Cincinnati.   Other departmental papers, manuscripts, and photographs were culled from inactive files by Richard Spohn, Geology-Geography Librarian, and were also placed in the Blegen Library.   The Fenneman papers now fill eleven small Hollinger boxes (7,000 items) and six large Paige containers (195 folders, 85 field maps, 45 notebooks, glass negatives, photograph albums, and memorabilia—including Fenneman's doctoral hood).   The relevant accession numbers are 15/Q9-D/35, UA-73-32, UA-81-35, and UA-84-15.   Fenneman's portrait hangs in the Department of Geology, which also inherited his library.   His rolltop desk and carved wooden owl repose in the Department of Geography.   The archives of The Literary Club have been deposited with the Cincinnati Historical Society, in Eden Park.   They contain, in the bound volumes of proceedings, all 68 papers that Fenneman wrote for the Club.

Bruce Ryan is Professor of Geography and Department Head, University of Cincinnati, Ohio 45221-131, U.S.A.

Chronology

1865	Born in Lima, Ohio, 26 December
1883	A.B., Heidelberg College, Tiffin, Ohio
1884-92	Taught mathematics and chemistry at grade schools in Wilkinsburg and Greensburg, Pennsylvania
1890	Fieldwork in southwestern Wisconsin
1892-1900	Professor of Physical Sciences, Colorado State Normal School, Greeley
1893	Married Sarah Alice Glisan of Fredonia, New York, 26 December
1895	Studied with William Morris Davis, Harvard University summer school
1899	M.A. in geology, University of Chicago; thesis—'The Laramie Cretacious Series'; first journal publication
1901	Ph.D. in geology, University of Chicago; thesis—'The development of the profile of equilibrium of the subaqueus shore Terrace'
1902-03	First Professor of Geology, University of Colorado, Boulder
1903-07	Professor of Geology, University of Wisconsin, Madison
1904	Charter member of Association of American Geographers; elected Fellow of Geological Society of America
1904-05	Travel in Europe; fieldwork in Colorado, North Dakota, Texas, and Louisiana
1906	Fieldwork in Missouri
1907-37	Foundation Professor of Geology and Geography, and Department Head, University of Cincinnati, Ohio
1908	Attended International Geographical Congress, Geneva
1908-12	Treasurer, Association of American Geographers
1910	Attended International Geological Congress, Stockholm; elected to membership in The Literary Club
1912	Journalist for American Geographical Society's Transcontinental Excursion
1913	Attended International Geological Congress, Toronto
1914	Presented claims of Cincinnati for a 'regional' [Federal Reserve] bank; published first version of 'physiographic boundaries within the United States'
1915-16	Sabbatical leave in Washington, D.C.
1918	President, Association of American Geographers; delivered the Presidential Address, 'The Circumference of Geography', at Baltimore meeting
1918-21	Headed African section of Colonel Edward M. House's 'Inquiry' re the peace settlement
1919	Field camp at Lake Superior
1920	Mrs. Sarah Glisan Fenneman died, 2 April
1920-23	Expert witness for U.S. Department of Justice re Golden Lake and Young's Lake, Arkansas

1922-23	Chairman, Division of Geology and Geography, National Research Council, Washington, D.C.
1922-27	Industrial Resource Survey for Commercial Club of Cincinnati
1923	Attended Pan-Pacific Scientific Congress, Australia
1924	Commonwealth Fund Conference on Social Values; hired Miss Lillian Smith as department secretary
1924-25	President of The Literary Club during its 75th anniversary
1925-27	Committee on Grants, American Association for the Advancement of Science
1926	Attended Pan-Pacific Scientific Congress, Japan, also visiting China and the Philippines; nominated W.M. Davis for President, American Association for the Advancement of Science
1927	Councillor, American Association for the Advancement of Science, representing the Association of American Geographers
1928	Third edition of 'Physiographic Divisions of the United States'
1930-41	Nine summer field seasons at Red Lodge, Montana, with Yellowstone-Bighorn Research Association
1931	Published *Physiography of Western United States*
1932	Vice president, Geological Society of America
1935	President, Geological Society of America; elected corresponding member, American Geographical Society
1936	President, Yellowstone-Bighorn Research Association; Chairman, Past Presidents of Geological Society of America
1936-39	Chairman, Committee on Research Program of the Geological Society of America
1937	Professor Emeritus, University of Cincinnati; succeeded as Department Head by Walter H. Bucher; 30th anniversary of Department of Geology and Geography
1938	Published *Physiography of Eastern United States*; awarded Gold Medal by Geographic Society of Chicago
1940	Awarded Honorary LL.D., University of Cincinnati; Commencement speaker
1941	Atwood Award Committee, Association of American Geographers; Hayden Medal Committee, Philadelphia Academy of Natural Science
1942	Portrait painted by Frank H. Myers
1945	Phi Beta Kappa address; died in Cincinnati, 4 July; buried in Fredonia, New York, 7 July

Ernst Emil Kurt Hassert

1868–1947

GERHARD PANSA

Kurt Hassert was a major German geographer of his time, concerned to provide a world view of his subject, to travel as widely as possible, to observe environments markedly different from his own homeland and, if this was not possible, to construct from available sources an explanatory account of distant lands he never saw, whether tropical or polar. He was clearly influenced by the work of Ratzel on human, including political, geography and accepted the ethos of colleagues who believed that world geography must be studied with local geography in education generally, and in the personal study and research of those who taught in universities and schools. His work on communications was particularly notable and much appreciated.

1. EDUCATION, LIFE AND WORK

The son of a lower middle class merchant, Kurt Hassert was well known for his works written during the first half of the twentieth century. Born on 15 March 1868 at Naumburg, he was educated at its Gymnasium, a church foundation, and became interested in the economic and political aims of the German ruling classes. This led him to study the European penetration and colonization of the African continent during his four years as a student in the universities of Leipzig and Berlin. At the former he became a devoted disciple of Friedrich Ratzel, who recognized his marked ability and regarded him as his *famulus*, in effect a student assistant, from the second semester of 1888. Hassert also profited from the lectures of Ferdinand von Richthofen in Berlin, particularly from the sound training in physical geography which became an

essential theoretical and methodoligical basis for his later researches. But it was Ratzel's view of anthropogeography that won his devotion and influenced all his later researches, though he was fully aware of the contribution of other scholars to a broadly based study of man and the land. He realized that geomorphological researches such as those of Richthofen needed geological understanding, and was also appreciative of the lectures of Ferdinand Zirkel. An understanding of botanical geography needed some fundamental study which he acquired from the lectures on botany of Wilhelm Pfeffer. On the more specifically human side an historical approach was an asset and therefore he attended lectures on history given by Wilhelm Maurenbrecher.

Hassert's first major publication was *Die Nordpolargrenze der bewohnten und bewohnbaren Erde*, a study of the Arctic limit of the habitable and inhabited earth, of 1891. This choice of subject was derived from the theories of Ratzel on the earth as the home of man (*oecumenie*) and Hassert was concerned to look at an area offering particular difficulties. The dissertation publication marked the end of his student career at Leipzig and Richthofen suggested that he should work on the western part of the Balkan peninsula, which from 1888 had become of great interest to Imperial Germany, then concerned with political expansion in the Balkans and the Middle East. The preparation for this enterprise included attendance at lectures by Albrecht Penck at Vienna university and later, in 1892-93, a six months course on military cartography and geographical survey.

Before taking the military course Hassert went on

Geographers Biobibliographical Studies, volume 10 (1986)

the first of several expeditions to the western Balkans in 1891 and spent five months walking in the mountains of Montenegro. In 1895 he received his doctorate and his *habilitation* as a unibersity teacher and began to lecture at Leipzig university. The first publications on the Balkans were on account of his tour of Montenegro of 1893, *Reise durch Montenegro nebst Bemerkungen ÜberLand und Leute* and a group pf papers on the physical geography of Montenegro, with special reference to karst formations, which appeared in *Petermanns Geographische Mitteilungen.*, (*Erg.* 115, 1895). His travels continued with a tour of Italy. He had already met Olavi Baldacci, the Italian geographer, on his tour of Montenegro in 1891-92 and was to visit Italy later on. In 1896 he went to the Central Plateau of France and in 1897 once again to Montenegro, Herzegovina and Albania, as well as to Italy.

Once settled as a *docent* in the Institute of Geography at Leipzig with Ratzel, Hassert began to show the breadth of interest characteristic of geographers of his period. His inaugural lecture was on the migration of Eskimos and their economy in relation to natural conditions. He also developed at this time an interest in communications and their economic aspects, partly through his teaching at the Leipzig High School of Commerce founded in 1898. A pioneering enterprise was in urban geography, shown in his study of the situation and development of Leipzig, on which he published a paper in 1899. This study of the town on the basis of natural features and historical growth indicated his wish that more attention should be given to local history and geography.

An abiding interest was German colonial policy, especially in Africa, and on this he lectured from 1895 onwards, both in the university and to a wider audience of the general public. His first book on this theme, *Deutschlands Kolonien*, appeared in 1899 and followed the established sequence of the physical geography, the use of the land by the native inhabitants and the economic opportunities available for new settlers. The recent acquisition of African colonies by Germany made such work welcome, but it was written before he was able to visit any of the colonies. His initial visit to Africa was made in 1905, when he attended the First Italian Colonial Congress at Asmara, Eritrea, through the encouragement of Baldacci, after which he visited the German colonies. In 1907-08 he was a member of an expedition to the Cameroons, financed by the German Colonial Office with orders to study economic possibilities. In 1903 he wrote on the new German colonies in the South Seas, which apparently he never saw.

In 1899 Hassert became a professor at Tübingen university, in succession to Alfred Hettner, having refused a similar offer from Jena university made on the recommendation of Alfred Kirchhoff and Ernst Haeckel. At Tübingen he began a study of rivers, seen in a general perspective but with a full understanding of their economic and political significance for Germany. Also at this time he wrote a monograph on the Polar regions, inspired partly by the Antarctic expedition led by Erich von Drygalski from 1901-03, which was supported by the German government's grant of 1.2 million marks. This was a time of active Antarctic exploration and his monograph,

Die Polarforschung, an historical study first published in 1902, achieved later editions and a Russian translation appeared in 1923. As on some earlier occasions, Hassert was writing on an area he had never seen in order to meet a popular demand.

On his home territory he wrote on Württemberg in 1903, with the definite purpose of giving readers, at that time drawn largely from the governing classes, knowledge that could both strengthen regional consciousness and improve local administration. Clearly Hassert was fully aware of the application of geography in political, economic and social life and in this he was following a practice much favoured by Albrecht Penck and also by Alfred Kirchhoff, who was a supporter of the Central Commission for the study of the geography of Germany and arranged that Hassert should be paid 300 marks annually as an editor and general worker for the Commission. The aim was to produce local regional studies and Hassert was the author of the first of three, on Württemberg. While working in Tübingen Hassert also produced a book on mapping in geography, arguing that teachers should in all fieldwork encourage pupils to record their findings on maps with the greatest possible accuracy.

A move to the High School of Commerce at Cologne in 1902 gave Hassert further encouragement to work on communications. The ethos of geography teaching at Cologne was that the political geography of the whole world should be studied on the basis of relationship between men, economic resources and commercial development, and communications. At the time Hassert was still greatly influenced by his friendship with Ernst Friedrich, the political geographer. remembered particularly for the phrase 'robber economy' and also with Max Eckert, who had worked with Ratzel from 1891 to 1899, when he became a professor at Aachen. Known as a leading commercial and political geographer, Eckert had founded the Colonial Association at Leipzig of which Hassert was a supporter.

The earlier work on rivers was extended to other waterways in 1902-04, when Hassert issued an edited and revised version of a thesis written by Martin Voss. In considering the Suez Canal Hassert saw that physical geography was of secondary significance. Essentially it was an artificial waterway, a triumph of modern engineering: far more significant were the economic and political aspects, which Hassert discussed in relation to Germany with a comparative treatment for other nations and especially for Great Britain in her trading relations with India. Both these major aspects were of great interest to German readers and Hassert was prophetic in discerning that the Suez Canal was to be of political significance in the future.

While at Cologne Hassert was able to make several journeys abroad, to Transylvania in 1902-03 and Italy in 1903, to the United States, Canada and Mexico in 1904 and again in 1910, and to Great Britain and Ireland in 1912-13. Mention has already been made of his attendance at the Italian Colonial Congress in 1905 and of his participation in the official German government expedition to the Cameroons in 1907-08. Perhaps oddly his published works from 1903-14 were on none of these areas and they included a general and economic geography of Australia (*Landeskunde und*

Wirtschaftsgeographie des Festlandes Australien), which he never saw. The publications show that his earlier interests continued into early middle age, except for the discarding of polar work. Writings of this time include in 1905 a paper dealing with communications on the frontiers of Germany, in 1907 a geographical study of towns and in 1911 a work on the Cameroon mountains. Like other fortunate authors he had to prepare new editions of earlier works, notably the second of his *Deutschlands Kolonien* in 1910, the second of his *Landeskunde des Konigsreiches Württemberg* in 1913 and the third of his *Die Polarforschung* in 1914.

Widely regarded as one of his major achievements, his work on communications, *Allgemeine Verkenhrsgeographie* was first published in 1913. To him communications were an essential aspect of human geography, to be considered in their own right as something that occurred in relation to the earth and man rather than as a minor consideration in a general treatment of economic geography. Hassert clearly understood that the communications system was in constant change, for the Industrial Revolution had made development essential and new means of communication had opened up the world as never before. He viewed communications in its causal relations to land and economic development and his work, for which there was no prototype in German, was widely regarded as a classic.

The death of Ratzel in 1904 was a severe blow to Hassert and on him he wrote two essays of which the first, in 1904, dealt with Ratzel's work in commercial geography, 'Die geographische Bildung des Kaufmanns', and the second was an obituary in the *Geographische Zeitschrift* of 1905. He regarded Ratzel as the founder of anthropogeography, conceived as human geography closely connected with physical geography in which climate and soil, having a natural influence on vegetation and possible agricultural activity, were of immense human significance. From Ratzel he also acquired his conception of the state as an earthbound organism, a human group with a definite organization and distribution of life on the surface of the earth. Throughout the work of Ratzel there is contact with the earth and Hassert approved strongly of such an approach, to be developed into the idea of *Lebensraum*, the living area of a state.

In July 1914 Hassert, though in his middle forties, became a captain in the Military Control Office of the 9th Army Corps but in 1915 he was invited to accept a chair at the Technical High School of Dresden. Owing to his military duties he was not able to take up this appointment until 1917 but characteristically his first lecture in Dresden was on 'J.J. Becher--the champion of German colonial policy in the seventeenth century'. He became an eager advocate of the colonial expansion of Germany as a basis for economic growth and emigration and on this theme wrote numerous essays in the *Deutsche Kolonialzeitung*, the *Informationen aus den deutschen Schutzgebieten* and the *Koloniale Rundschau*. His views were vigorously advocated in lectures at Dresden where after one year, in 1918, he became director of the Institute of Geography. As soon as he settled in Dresden he was invited to become vice-president of its Geographical Society and president a year later, and he held one of

these two offices with little intermission to 1928, when he became an honorary member, though he was to be president again from 1932-38, when his retirement was marked by the presentation of the Society's silver plaque. He persuaded the Society's committee to arrange that lectures on colonial geography should be given at least twice a year. Not surprisingly he was outraged by the provisions of the Treaty of Versailles in 1919 on the German colonies and when the Nazi movement developed he looked forward to the day when Germany would once again become a colonial power. He was proud of the continuing work of German explorers and scientists, especially geographers, in Africa and in 1941 published.his book on its exploration, *Die Erforschung Afrikas*.

Other works of his final period of university service, from 1917 to his retirement at the age of sixty-seven in 1936 included a paper on Montenegro and Upper Albania as theatres of war (*Geogr. Z.*, vol 22, 1916, 199-224), a book on Turkey in 1918 and in 1919 a paper on the nature and value of economic geography and also contributions on Africa, Australia, the Belgian Congo and Italy to Banse's *Lexikon der Geographie*. But it was in four books published at this time that Hassert showed the breadth of his regional interest: of these his work, *Die Wirtschaftsleben Deutschlands und seine geographischen Grundlagen* on the economic life of Germany appeared in 1923. The three books dealing with the New World were *Die Vereinigten Staaten von Amerika* (1922), *Australien und Neuseeland* (1923) and *Nordamerika* (1927). It was now that some of the observations made on his various travels were to be helpful. Hassert, like so many geographers of his time, was eager to cover a large part of the world in his publications as in his teaching: it was widely accepted that a university-trained geographer should have a comprehensive knowledge of the main features, physical and human, of the whole world.

Much of Hassert's energy was absorbed in building up the geographical work in Dresden but although he attracted large audiences to his lectures few presented geography at the university's examinations despite the Saxon government's recognition of the subject for degree work from 1920. In some years he was said to have as many as 200 students attending lectures but only five or less taking examinations. He was eager to confer with other geographers and in 1927 formed an informal group to discuss new writing in geography. Four years later he was one of the founders of the *Dresden Geographische Studien*, for which he wrote several papers. In volume 3 (1932) he published an essay which was a general introduction to geographical literature, and in volume 5 (1934), he wrote *Das Wirtschaftsleben Deutschlands*. Once more, as in the years before the First World War he was concerned with new editions of his earlier books, notably of his 1913 work on communications, *Allgemeine Verkehrgeographie*, of which a much enlarged and revised edition appeared in two volumes in 1931. This showed a marked advance in his thinking on geographical theory and methodology and is discussed later in this paper.

Following his retirement Hassert published little until his work on Africa appeared in 1941, but in 1939 his successor, Nikolaus Creutzburg, was called up for military service and Hassert returned to his teaching work at Dresden.   There he continued his work even though much of the Institute was destroyed in 1945. In April 1947 he moved to Leipzig as professor at the Institute of Geography of the University but during the summer he became ill and he died on 5 November.   He had worked almost alone in his last years under difficult conditions.   His work was impressive in quantity and quality and he made a rich contribution to German geography despite his eager acceptance of Nazi views, particularly on colonial policy.

2.   SCIENTIFIC IDEAS AND GEOGRAPHICAL THOUGHT

By 1899 Hassert had made it clear that like Ritter and Ratzel he thought that all geography had a scientific basis and therefore physical geography was of prime significance.   Phenomena of physical geography obey natural laws, of invariable application and independent of human life, and such phenomena determine the characteristics of the earth as the home of man in his progress through life.   Therefore man is the mirror of the environment and the variety of peoples is caused by the variety of natural conditions (1899).   Ignoring social classes in his work, Hassert argued that the fundamental influence on economic life and political decisions was the natural conditions.   But in the evolution of economic life based on capitalism the ruling classes had the dominant power, though the actual making of decisions on economic development could not be idiosyncratic.   Here lay the incipient challenge of social geography, of which one feature must be the sociological ideas of a bourgeois society.   In 1919 Hassert expressed his views on the 'character and educational value of economic geography' with more precision than in his earlier writings but he remained faithful to the need for study to be firmly rooted in physical geography.   This was in accord with his comment of 1898 in the paper on Leipzig that the future development of the city, or indeed of any place, depends on its geographical location.

This remained a fundamental concept of his work in human geography.   Though the human will actuated any development, such development was only made possible if natural conditions were favourable.   Settlements can prosper only if nature and human activity are in harmonious relationship.   Such an outlook is seen especially in Hassert's work on communications, notably in the book of 1913 with its fuller re-presentation of 1931.   In communications one could see 'the totality of ... relations between men, as a broad concept, and also in a narrower and pragmatic concept' as serving 'the spread of the products of the earth', subject to the constraints of distance in space and the expenditure of time.   In constructing his geographical theory of traffic flow Hassert recognized two fundamental considerations, of which the first is the actual means of transport and the routes followed and the second a broader view of the territorial and regional spread of communications as influenced by cause and effect in physical and human geography.   But he was not concerned with the visible effects of transport by road, rail, river or canal on the landscape, though he saw the need to study communications in their relationship to other branches of geography.

Especially communications must be seen as part of economic geography, which in its turn is one aspect of human geography.   Economic geography involves an understanding of the transport of goods as fundamental to commerce and industry and so worthy of close study in its own right and not merely as a minor or subsidiary aspect of economic geography.   In his long study of transport Hassert developed several related approaches.   Of these the first was the actual routes used, their character inevitably influenced by natural conditions.   This simple and basic consideration led to a second and more abstract treatment of the routes followed and then to a third aspect, the influence on routes of changing economic and social circumstances. A fourth consideration was the actual means of transport, greatly changed by technological developments, themselves part of a scientific advance of immense economic and social significance.   This brought forward the fifth point, the relative use of routes with the means available for conveying goods and people.   And all these five approaches were involved in the sixth and last, well illustrated by the revival of German industrial production in the 1930s after the World Economic Crisis which began in 1929, that the new economic strategy could only be successful with the careful planning of transport.   By 1931, when the revised version of the 1913 book appeared, Hassert had looked at all aspects of transport for their own interest but with a watchful eye for their relation to a broad human geography with its essential physical basis.

Hassert in all his work on communications concentrated on the German-speaking area, fortified by a strong understanding of geography in general.   In developing his theory and methodology he was well aware of the changing social and economic changes of his time.   His outlook was pragmatic, for he was always seeking a theory and methodology that in his view was confirmed by observation of what happened.   Never prone to imitate the theoretical conceptions of his colleagues, he was a dedicated and independent researcher and so did work of international significance, particularly for the German-speaking area of Europe.

3.   INFLUENCE AND SPREAD OF IDEAS

Politics were of abiding interest to Hassert and he lived through the crucial time when Europe saw the rise of Germany to great power status, the slaughter of World War I with the Treaty of Versailles to follow, the rise of the Soviet Union, the success of the Nazi movement in Germany and World War II with, for him, just two more years in a devastated homeland. He was not slow to express his political views, particularly in a social context.   He was a member of the *Alldeutscher Verband*, the Pan-German Association, a chauvinistically-minded body based on the idea of the *Herrenmensch* and the right of Germany to predominance in Europe.   His clear loyalty to such views remained through the imperial, republican and Nazi periods of German political history.

From the beginning of his professional career as a geographer Hassert supported the colonial policy of

Germany, and in a paper of 1899 he observed that their lack was an unfortunate consequence of the late achievement, in 1870, of the administrative unity of Germany. 'The young great power', he wrote, 'must become a world power and...a colonial power'. He was convinced that the 'lower, economically powerless classes' wanted colonial development as least as much as the political leaders and the captains of industry and in 1917 bitterly attacked the occupation of German colonial territories by the Allied Powers. In 1934 he argued that colonies were necessary for a great industrial state as sources of raw materials and markets for manufactures and in 1941 he was still commenting on 'the robbery of our protected territories'.

Associated aspects of his political outlook were clearly expressed and hardly surprising. Having made his 'great power' position clear, in 1904 he declared that 'the national interests of commerce...stand in the forefront of our policy' and in 1905 he added that Germany, surrounded by other great powers, needed strong military forces in the struggle for existence and later, in 1923 and 1934 he supported the idea that true patriotism for the Fatherland involved the mobilization of all workers for the support of the capitalistic economy. On his various tours abroad he regarded his task of observing and advocating German interests as compelling. In 1934 he gave full support to the Nazi regime established a year earlier and five years later in 1938 at the 75th anniversary of the Dresden Geographical Society he said that

> We enter the fourth quarter of our century... hoping that our Society will prosper as much as our Fatherland, which has been raised from the depths to new power and magnificence by one of its greatest sons...Adolf Hitler is the incarnation of faith in Germany.

After World War II, however, Hassert was willing to co-operate with the new rulers of Germany who held democratic and anti-Fascist views and he was in consequence able to share in reviving the geographical work at the Technical High School in Dresden and the university of Leipzig. To him the menace for Germany lay in France, Great Britain and the United States, as he showed in publications dated 1922, 1923 and 1934. His comments on the Soviet Union were equivocal but in 1922 he favoured the Treaty of Rapallo which had provisions for trading with Soviet Russia as he thought that such commerce could be valuable in the economic reconstruction of Germany. Constantly he said that geographers should 'help in solving the political and economic tasks of the Fatherland' (1919, 1923, 1928, 1934) and that the dissemination of geographical knowledge among the population, particularly the younger people, was a major responsibility (1903, 1922). Further, the teaching of economic geography should enlighten young people on the world position of Germany (1919).

Undoubtedly Hassert's views were those that many people were happy to read when they were written. Through career circumstances he had little opportunity of training future university teachers because he had no junior colleagues either at Dresden or Leipzig. As

his political outlook was in its day conventional it was the work on communications that attracted most attention among German colleagues. Of these Otto Schlüter, in a review of 1930, emphasized his support of the anthropological conceptions of Ratzel and disagreed with Hassert's view that the geography of communications was based on economic aspects: rather it was independent of economic geography and should be studied in relation to the 'cultural landscape' (*Kulturlandschaft*). This idea was supported by Alfred Hettner in his book on communications of 1952, though he recognized the strength of Hassert's views on the significance of production and consumption in studying communications. However he also thought that the voluminous treatment of particular means of transport was an unnecessary feature in a geographical work. Erich Otremba writing in 1957 agreed and he also thought that a pure geography of communications was the primary connection of traffic with natural features, as indeed Hassert had himself said in his work of 1899 on rivers. Criticism was also levelled at Hassert's separation of geography and commerce, which in Otremba's view should be taken together as they were clearly connected. At the same time he stressed that Hassert's work was a fine contribution to the methodology of communications and could indeed be regarded as methodologically faultless except that it was overloaded with some unnecessary material.

Writing ten years after the death of Hassert, Otremba ranked him with Hettner and also with Professor Otto Blum, an engineer by profession who produced some interesting works on geography, as writers on communications. Politically this was the least contraversial aspect of his work and his most sincerely academic contribution to geography. That he had great scientific diligence is beyond question, even though his views on colonial geography and his zeal in writing on a wide range of countries now seems to belong to a past phase in the modern development of geography. His political outlook, so frequently expressed, was not in accord with that of the post-1945 world, but in that he was a product of his day and age.

Bibliography and Sources

1. REFERENCES ON KURT HASSERT

Banse, E., 'Hassert, Kurt', *Lexikon der Geographie*, vol 1 (1923)

Siegel, J., 'Kurt Hassert: zu seinem sechzigsten Geburstag', ('... on his seventieth birthday'), *Geogr. Anz.*, vol 29 (1928), 76-9

Grunicke, E., 'Kurt Hassert's Verdienste um die deutschen Kolonien', ('The fine work of Hassert on German colonies'), *Mitt. Ver. Erdkd. Dresden Jahres.* (1936-8), 1-8

Reuter, M., 'Kurt Hassert, Leben und Werk', *Petermanns Geogr. Mitt.*, vol 94 (1950), 89-92

2.   *SELECTIVE BIBLIOGRAPHY OF WORKS BY KURT HASSERT*
Note: As his writings cover such a wide range of topics
they are here listed in sections, of which each is
identified by the date of its first appearance in his
writing.

a.   *Polar Geography*
1891   *Die Nordpolargrenze der bewohnten und bewohnbaren
Erde (The Arctic limit of the inhabited and
habitable earth)*, diss., Leipzig, Naumberg, 103p.
1902-04   *Die Polarforschung*, 156p., 2 ed (1907), 152p.,
3 ed (1914), 134p.
1956 (posthumous) *Die Polarforschung: Geschichte der
Entdeckungsreisen zum Nord- und Südpol (The polar
lands: a study of the exploration of the north
and south pole)*, Munich, 291 p.

b.   *The Balkan countries*
1893   'Eine Fusswanderung durch Montenegro' ('Walking
through Montenegro'), *Dtsch. Runds.*, vol 15,
97-106, 116-73
----   *Reise durch Montenegro nebst Bemerkunjen über
Land und deute (Tour of Montenegro with comment
on the land and people)*, Wien, Prest, Leipzig,
236p.
1895   'Die natürlichen und politischen Grenzen von
Montenegro' ('Natural and political frontiers in
Montenegro'), *Z. Gesell. Erdkd. Berlin*, vol 30,
375-405
----   'Beiträge zur physischen Geographie von
Montenegro mit besonderer Berücksichtigung des
Karstes' ('On the physical geography of
Montenegro with special reference to the karst'),
Petermanns Geogr. Mitt., *Ergänz.*, no 115, 174p.
1898   'Wanderungen in Nord-Albanien', *Mitt. Geogr.
Gesell. Wien*, vol 41, 351-79
1901   'Reise durch Montenegro im Sommer 1900', *Mitt.
Geogr. Gesell. Wien*, vol 44, 140-65
1916   'Montenegro und Oberalbanien als
Kriegsschauplatz' ('Montenegro and Upper
Albania as theatres of war'), *Geogr. Z.*, vol 22,
199-224

c.   *Urban Geography*
1898   'Die geographische Lage und Entwicklung
Leipzig' ('The geographical situation and
development of Leipzig'), *Mitt. Ver. Erdkd.
Leipzig*, 17-53 (publ 1899)
1907   *Die Städe geographisch betrachtet (Geographical
aspects of towns)*, Leipzig, 137 p.

d.   *Colonial Geography, especially of Africa*
1899   *Deutschlands Kolonien*, 332 p.; 2 ed 1910, 657 p.
1903   *Die neuen deutschen Erwerbungen in der Sudsee
(The new German colonies in the South Seas)*,
Leipzig, 111 p.
1910   'Forschungs-Expedition ins Kamerun-Gebirge und
ins Hinterland von Nordwest-Kamerun', *Z. Gesell.
Erdkd. Berlin*, 1-35
1911   'Das Kamerunegebirge', *Mitt. Dtsch. Schutzgebiete*,
vol 24, 55-112, 127-81
1917   'Beitrage zur Landeskunde der Grashochländer
NW-Kameruns' ('A contribution to the geography of
the grass highlands of northwest Cameroon'),
Mitt. Dtsch. Schutzgebietun, supp. 13, 144 p.

1926   'Das Kammerungebirge', *Geogr. Z.*, vol 32,
449-59
1941   *Die Erforschung Afrikas (The exploration of
Africa)*, Leipzig, 250 p.; 2 ed (1943), 259 p.

e.   *Communications*
1899   'Die anthropogeographische und politisch-
geographische Bedeutung der Flusse', ('The
anthropogeographical and political-geographical
significance of rivers'), *Z. Gewasserkunde*,
vol 2/4, 189-219
1903-04   revised and edited edition of Voss, M., *Der
Suezkanal und seine Stellung im Weltverkehr
(The Suez Canal and its position in world
communications)*, *Abh. Geogr. Gesell. Wien*,
vol 5/3, 1-76
1913   *Allgemeine Verkehrsgeographie (General geography
of communications)*, Berlin and Leipzig, viii +
494 p.; 2 ed in two vol, 1931, viii + 408 p.,
376 p.
1919   'Zur 400-jährigen Wiederkehr der ersten
Weltunseglung' ('The 400th anniversary of the
first voyage round the world'), *Die Umschau*,
vol 23/37, 577-83
1926   'Neuere Beiträge zur Geographie und Kartographie
der Eisenbahnen' ('A new contribution on the
geography and cartography of railways'), *Mitt.
Ver. Erdkd. Dresden, Jahres.*, (publ 1927),
94-111
1928   'Der neue Weltverkehr' ('The new world traffic'),
Meereskunde, vol 16/6, 1-27

f.   *Germany*
1903   *Landeskunde des Königreiches Württemburg
(Regional studies of the Kingdom of Wurttemburg)*,
Sammlung Goschen, Leipzig, 160 p.; 2 ed (1913),
139 p.
1923   *Das Wirtschaftsleben Deutschlands und seine
geographischen Grundlagan (The economic life of
Germany and its geographical foundations)*,
Leipzig, 127 p.; 2 ed (1934), 115 p.
----   'Deutschlands Lage und Grenzen in ihren
Beziehungen zu Verkehr und Politik', and 'Die
Schwabische Alb', in Grube, A.W., *Charakter-
bilder deutschenlandes und Lebens*, 17 ed, 1-19
and 341-52

g.   *Political Geography*
1904   review of Lukas, G.A., *Studien uber die
geographische Lage des österreichisch-ungarischen
Okkupationsgebietes und seiner wichtigeren
Siedlungen (Studies of the geographical situation
of the Austro-Hungarian territory and its major
settlements)*, *Geogr. Z.*, vol 10, 291-2
1905   'Deutschlands Lage und Grenzen in Ihren
Beziehungen zu Verkehr und Politik' ('Aspects of
communications and politics in relation to the
frontiers and position of Germany'), in
*Festschrift zur Feier des 70 Geburtstages von
Johann Justus Rein, zuglich 1.   Veröffentlichung
der geographischen Vereinigung zu Bonn*, Bonn,
120 p.
1922   *Die Vereinigten Staten von Amerika als politische
und wirtschaftliche Weltmach geographisch
betrachtet (The political and economic geography*

of the United States of America as a world power),
Tubingen, viii + 316 p.

h.   Educational and Biographical Geography
1904   'Die geographische Bilding des Kaufmanns' ('The geographical education of merchants'), in *In Memory of Ratzel*, Leipzig, 151-68
1905   'Friedrich Ratzel: sein Leben und Werken', *Geogr. Z.*, vol 11, 305-25, 361-80
1927   'Das Geographische Institut an der Tecynischen Hochschule Dresden', *Geogr. Anz.*, vol 28, 184-8
1932   'Einführung in die geographische Literatur' ('Introduction to geographical literature'), *Dresdner Geogr. Stud.*, vol 3, 1-89
1941   'Grosse deutsche Geographen: Friedrich Ratzel', *Atlantis*, no. 19, vol 8, 373-6

i.   General Regional Works
1918   *Das Türkische Reich*, Tübingen, 242 p.
1924   *Australien und Neuseeland: geographisch und wirtschaftlich*, Gotha, Stuttgart, viii + 178 p.
----   'Zur Wirtschaftsgeographie des australischen Staatenbundes' ('Economic geography of the Australian states'), *Geogr. Z.*, vol 30, 187-204

j.   Contributions to Encyclopaedias and Gazetteers
1922   'Afrika', 'Australien', 'Belgisch-Kongo', 'Italian', in Banse K., *Lexikon der Geographie*, vols 1, 2 Braunschweig
1924   'Afrika', 'Australien', *Meyer's Encyclopaedia*, vol 1
1927   'Nordamerika', in Andree, Heiderich, Sieger, *Geographie des Welthandels, eine Wirtschaftsgeographisches Beschreibung (Geography of the world's commerce...)*, Vienna, 4 ed, vol 2
----   'Australien und Ozeanien', 'Nordamerika' in Seydlitz, *Centenary Edition*, Breslau, vol 3
----   'Italien', *Meyer's Encyclopaedia*, Leipzig, vol 6

Dr Gerhard Pansa is senior lecturer at the Pedagogische Hochschule 'K.F. Wander' Sektion Marxismus/Leninismus, Wissenschaftsbereich Philosophische Probleme der Mathematik und Naturwissenschaften, DDR 8060 Dresden, Wigardstrasse 17

Chronology

1868	Born at Naumberg, 15 March
1874-87	Was a pupil at the church Gymnasium in Naumberg
1887-91	Studied at the universities of Leipzig and Vienna, where he was a *famulus* of Friedrich Ratzel from the second semester of 1888
1891-92	Went to Montenegro and spent six months

at Vienna, where he attended lectures by Albrecht Penck: met Olavi Baldacci, the Italian geographer, at this time

1892-93	Military service, during which he attended a six month course on military cartography and geographical survey
1895	Graduated as a Ph.D.: visited Italy
1895-99	Gave lectures at Leipzig university, particularly on colonial geography
1896	Visited the Central Plateau of France
1897	Made another visit to Montenegro, with Herzegovina and Albania: also went to Italy
1897-98	Gave a series of lectures for the general public at Leipzig
1898-99	Became the first lecturer in geography at the newly established High School of Commerce in Leipzig
1899	Succeeded Alfred Hettner in the chair of geography at Tübingen university and published a monograph on German colonial territories
1902	Moved to the High School of Commerce at Cologne and published a monograph on historical aspects of polar exploration
1902-03	Spent some months in Transylvania
1903	Visited Italy once again
1904	Toured the United States, Canada and Mexico
1905	With the encouragement of his friend Olavi Baldacci, took part in the first Italian Colonial Congress at Asmara, Eritrea
1907-08	Was a member of the Cameroons expedition of the Imperial Colonial Office, which was particularly concerned with economic development
1910	Made another visit to the United States, Mexico and Canada
1912-13	Visited Great Britain and Ireland
1913	Elected a corresponding member of the Dresden Geographical Society: issued his *General geography of communications* (revised edition 1931)
1914	In July became a captain in the Military Control office of the 13th Army Corps

1915	Invited to accept a chair at the Technical High School in Dresden but owing to military commitments was not able to take up this post until 1917
1917	At Dresden he began to foster geography enthusiastically
1918	Became a vice president of the Dresden Geographical Society, with which he was closely associated from this time
1926–27	Toured Yugoslavia
1934	Was given the Cvijić memorial award by Belgrade university
1936	Retired from his chair at the Technical High School of Dresden and was succeeded by N. Creutzburg: became an honorary member of the Leipzig Geographical Society
1938	Awarded the silver plaque of the Dresden Geographical Society
1939	When Creutzburg joined the Army Hassert returned to the Technical High School of Dresden
1945	Following the destruction of the Technical High School of Dresden, he worked assiduously to restore its work
1947	Invited to take up a chair at Leipzig university on 1 April, he became ill during the summer and died on 5 November at Leipzig

Johann Gottfried Herder

1744–1803

J.A.C. BIRKENHAUER

As will be seen, Herder deeply influenced both Alexander von Humboldt and Carl Ritter in their understanding of geography and, through them, geography as a whole subject. This is the reason to introduce Herder in this series.

1. EDUCATION, LIFE AND WORK

Johann Gottfried Herder was born on 25 August 1744 in the small town of Mohrungen in what was then East Prussia. Herder later called his birthplace 'the smallest of all towns in this barren countryside', a notion, however, prejudiced by some experiences in his youth. The inhabitants were mostly weavers, spinners, farmers. Cloth and potash were traded. Herder's grandfather was a small clothier, assisted in his business by his son, and his mother was a shoemaker's daughter. In order to have some income additional to his bare subsistence, Herder's father opened up a primary school and also served as sacristan and bell-ringer.

Both his parents were deeply religious, adherents of pietism, with a strong Lutheran touch. A favourite book of Herder's father (which through him Herder learnt by heart) was Johann Arndt's *Vier Bücher vom wahren Christentum (Four books of true Christianity)*. The kind of pietism preached in this book was a world-open one, in praise of God's fine and good creation, of the earth as a 'wonderful building' and God's 'great granary and treasury' for man. It is this good earth that offers to man all the means of his existence, but man has to make use of it in a careful way. The book had a strong influence on Herder's outlook on life, man and earth. Throughout his life Herder remained open to this kind of pietism, which is shown by his very warm response to Lavater's *Aussichten in die Ewigkeit (Prospects of eternity)* of 1768-69, by which Lavater soon became famous in Europe. In his childhood and youth Herder began to love the poetry of the Bible, and his love of nature and landscape was kindled. Beside Lavater's book, Herder welcomed a novel deeply merged in pietism, which appeared nearly at the same time. This was Sophie von La Roche's *Story of the young Lady Sternheim*, soon to be translated into English. In this book Herder found phrases dear to him and used later in his *Ideas* as key-words and core-phrases, such as 'the earth--our home', 'the motherly earth'.

From his mother Herder inherited a sanguine temperament (though occasionally he was made morose by his lifelong suffering from an eye-fistula which in spite of painful operations never could be removed), quick apprehension, versatility and a gift for language and rhetoric. According to his own words, it was his mother who taught him to pray, to feel and to think; his father taught him at home to write and to read and to sing the hymns in the Prayer Book. Herder then passed on to the small local grammar school where the headmaster, acting as a stern monarch, drilled his pupils in Latin, some Greek and a little Hebrew. Though Herder suffered from this man, he acquired sound foundations of learning, and by the age of fourteen he was able to assist his father in the elementary school.

Geographers Biobibliographical Studies, volume 10 (1986)

Happily for Herder in 1760, when he was sixteen, a new dean was appointed to Mohrungen; Dean Trescho possessed a very fine library, including the works of Rousseau and books of travels and geography. Herder's lifelong interest in books of geography and travels, especially those of the great explorers of his time, became established. Assisting the dean in his many literary endeavours, he was able to explore the dean's library freely. The dean even taught him to read and speak in French. In spite of Herder's obvious great gifts the dean advised Herder's parents against letting him become a *savant*. This advice, though well-meant in view of the very scanty means of the Herder family, was never to be forgiven by the infuriated young man.

Nevertheless, the day of great opportunity came when the German doctor of a Russian regiment (garrisoned at Mohrungen for several years during the Seven Years War) became a boarder with the dean and thereby began to know young Herder and his great talents. The kind doctor offered to take Herder with him to Königsberg (today, as Kalinigrad, part of the USSR) and to have him trained as a surgeon at its medical school. So in 1762, at the age of eighteen, Herder left Mohrungen and his parents, never to see them and his birthplace again. On the dean's recommendation, Herder was given a free boarding place in the *Collegium Fridericianum* in the city, because, as the dean wrote, such 'an ingenious youth, formed by nature for everything' should not remain without subsistence.

During his first sight of an anatomical section at Königsberg Herder fainted and therefore gave up the idea of becoming a surgeon. Instead, after a preliminary examination which he had to pass, he was inscribed in the faculty of divinity of the university. In the college where he boarded Herder's teaching talent was soon discovered; he was made a tutor for the students and a teacher in the college's elementary school, teaching nearly every subject, including French. These posts enabled him to earn some private money. Naturally, all these duties did not leave him much time for his own studies. These did not pertain to divinity alone, but included physics, astronomy and physical geography. It was Immanuel Kant, then at the beginning of his career as an eminent philosopher, who in his lectures on these subjects opened up new paths for Herder and created in him a lifelong understanding and love of geography and indeed of all subjects which today would be termed geo-sciences.

Because of the multiplicity of his duties and studies Herder forced himself to adhere to a rigid timetable set up by himself from seven in the morning to eleven at night. Self-discipline and independent work became two valuable and lifelong assets. Going along his own way was strengthened by the deep friendship with Johann Georg Hamann, fourteen years older than Herder. Hamann taught him English and Italian, the love of English and Italian poets, the love of the Old Testament, an understanding of language and poetry, and a critical attitude to any absolutisms in life, in politics, in science and in philosophy.

On Hamann's recommendation Herder was able to gain the position of teacher at the German cathedral school in the Baltic Hansa town of Riga (later capital of Latvia). Herder was appointed there at the age of twenty, after only two years of study. The years in Riga again became very formative for Herder's astute and versatile mind, for several reasons. Herder, again through Hamann's recommendation, soon had access to the leading intellectual and artistic social circles of the thriving city. Though under Russian supremacy the city was a republic enjoying a constitution and democratic rule of its own. Herder became an admirer and defender of democracy, which he considered to be the most humane form of government. He also developed a sense for the values of other nationalities and peoples whose customs, dances and songs he came to compare and to admire, including Germans, Poles, Russians, and Livonians, all of them citizens of Riga. The roots were here laid for his later work on the *Voices of peoples in their songs*, which made him a leader in the romantic movement and paved the way for the national revival of the Slavonic peoples. It was through him that the collection of native songs and fairy tales became widespread in Europe.

In Riga he was able to pass two more necessary examinations so that finally, in 1765, he could be ordained. Soon his Sunday services became much sought after because of his lively sermons. He found time to read the works of Buffon, Hume, Lessing, Leibniz, Montesquieu, Shaftesbury, all of which had a profound influence on his thoughts and later works. Yet in spite of his standing in the city Herder felt fettered. He wanted to see the world; so in 1769 he resigned from his duties, with great honour. Choosing a route through Copenhagen and Nantes he travelled to Paris where, among others, he met Voltaire, Montesquieu, d'Alembert, Rousseau and Diderot. While still on the ship he wrote the 'Journal of my voyage' (never published during his lifetime). In this 'Journal', at the age of twenty-five, he wrote down what he wanted to achieve in his life, notable reforms of schools and education and also the writing of a universal history of the world, including both the earth and its peoples. These were two ambitious aims that much later he was really able to achieve, in conformity with his high standards, in the *Ideen...Monschheit* of 1784-91.

From France he went on to the Netherlands and also to various places in northern Germany where he became tutor to the young Prince Holstein. Soon he accompanied the prince on his travels to southern Germany. During these travels, while at Darmstadt, he fell in love with his future wife, Caroline Flachsland, whom he married in 1773. While at Strasbourg, he had a considerable influence on Goethe, five years his junior. By various critical writings Herder's name had become fully known and his fame spread with his *Essay on the origin of language*. In this essay he made it clear that language is not of divine origin, opposing here, as always later on, any easy compromise between mind and relevation.

After these travels he first settled at the small court of Bückeburg in northern Germany, where he was appointed as chaplain to the count and countess and the chief minister of the town. He held this post from 1771 to 1776. In 1776 he was invited to become the religious superintendent-general of the Weimar dukedom and the chief minister of the town church of Weimar. As such he held two ranks: that of bishop and that of a minister of education, as we would say today. Except for a journey to Italy (1788-89) and various short visits to other places in Germany, from now on he stayed in Weimar up to his death. In Weimar, he began to execute what he had set himself to do in his 'Journal'. His main works began to appear and in them he proved himself to be one of the best stylists of the German language and, indeed, its best essayist. He won several awards from the Prussian and Bavarian Academies of Sciences and Fine Arts. His fame spread because of his progressive ideas on art, the human soul, on anthropology, ethnology, theology, and poetry. His most ambitious work, the *Ideas on a Philosophy of the history of mankind* appeared between 1784 and 1791, comprising twenty books (altogether nearly 600 pages in small modern print). He was befriended by the leading German poets: Goethe, Schiller, Wieland, all of whom were living at Weimar or in its neighbourhood.

By all these endeavours, Herder became the modern founder, or at least the chief innovator, of such various academic activities as psychology, ethnology, comparative linguistics and literature, history, philosophy of history, pedagogics, biblical criticism, archaeology, anthropology, ecology, and geography. Through his writings he prepared the way for the theory of evolution. Wherever he could he stressed the role of reason and experience against metaphysical principles. His influence was vast. Nisbet (1979) calls Herder 'a great innovator' and praises 'the novelty of his ideas'. Gillies (1944) referring to the *Ideas* finds them still 'fresh and modern', and that in a very surprising and remarkable way.

When Herder died, on 18 December 1803, not yet sixty years old, he had accomplished what thirty-five years earlier he had himself proposed to do; and yet he died in the knowledge that he still had not by far achieved what he himself considered as his life-work. As an admirer of Shakespeare and Milton since his early days in Königsberg, he could well have cited the lines in Milton's famous sonnet:

'When I consider how my life is spent
ere half my days in this dark world and wide'

2. SCIENTIFIC IDEAS AND GEOGRAPHICAL THOUGHT

In this context it is neither possible nor necessary to describe all of Herder's scientific ideas. Here it suffices to point out first some of his general ideas and then to proceed to his geographical views in somewhat larger detail.

His general ideas may be outlined as follows:

... distrust of metaphysics and myths, even of those in the Bible

... trust in what human experience and reason can achieve--in opposition to the platonic doctrine of the innate ideas ('Ghosts,' he said, 'are merely everything we do not know': *Ideas*, 204-05)

... man as the product of the generic energy of life on earth, making earth his home

... deep humanity

... energy (force, power) as the base both of mind and matter; on this he said: 'We do not know anything about a spirit that works without and outside of all matter, and, within matter, we perceive so many spiritual forces that any idea of a disunity and contradiction between the spiritual and the material appears to me to be entirely without proof' (*Ideas*, 134)

... unity of world history and mankind; the basic unity of space, time, energy (force); the monistic interpretation of the universe

... a broad concept of 'history', meaning and including the history of the earth and all the living things on earth as well as the history of mankind

... the concept of evolution, since earth, all the living things and mankind are held to be the products of a common process (According to Nisbet (1970, 72) Herder was one of the first men to propose the 'historical' description of an evolution in order to explain a particular subject matter. The *Ideas* helped to generate a favourable atmosphere for the later acceptance of the evolutionary theory (Nisbet, 1970, 221)

... the attempt for a universal theory of the cosmos, predicting that such a theory would soon become possible (as indeed it was: Laplace 1796), the look-out for common laws in nature

... man's liberty of instincts

... distrust of all anthropomorphism and anthropocentrism

... equality of all races, peoples, nations

The chief methods used by Herder and by him established as academic methods were those of (a) comparative studies; (b) genetic-processual studies; and (c) inductive reasoning. In turning to geography (and ecology) we are told by Nisbet (1970, 182) that 'Herder had an accurate and diversified knowledge of nearly all aspects of geography, and he used it as an explanatory commentary on universal history ...' (that is, according to Herder's sense of the universe and of history).

Herder's lively and lifelong interest in geography had originally been stimulated by listening to Kant's lectures on 'Physical geography' in 1764. Kant's (and Herder's) 'physical geography' would today be called 'systematic geography'. Kent chose the term 'physical' because he wanted to deal with the geographical aspects as ordered by time and space, these two being, according to Kant, the two physical qualities of the world. What made Herder interested in geography was Kant's attempt to get away from the mere enumeration of places and things and to show instead why conditions and phenomena on earth are varied and how these variations can be explained by rules and even by physical laws of land, water, atmosphere, positions of earth and sun and other natural laws. In order to show that the rules are valid Kant enumerated many examples, but in so doing in the end he himself became very encyclopedic and tedious. What Herder however gained from Kant was the zeal to search for underlying rules, to look for its scientific basis, but his own presentation is not based on Kant. Kant treats the earth as a mere theatre to be looked at or a quarry to be exploited. Herder, however, sees the earth as the constitutional home of mankind, to be considered and to be understood as such.

Herder's view of geography is expressed for instance in one of his numerous school addresses, with which he guided his school reforms. The title of one such address, probably from the year 1784, is 'On the amenity, usefulness and necessity of geography'. In one of the earlier passages in this address he asks the question: 'Who would not want to come to know this wonderful home in which we live ...?' He then goes on to explain its laws and its organizations, its natural history; he stresses the need to understand one's fellow men everywhere, to shake off all one's prejudices. He calls knowledge, especially historical knowledge that exists without the knowledge of geography, futile--a true cloud cuckoo land. A man who detests or abhors geography (and history) should not live on the earth's surface, but rather as the mole beneath it. All life is bound to this earth, our home, securing our existence within the universe; it is the very core of our existence.

In the parts of the *Ideas* on which he had worked in the previous year, he refers again and again to geographical matters: climate, atmosphere, mountains, their 'revolutions' (orogenies), how the life of plants and animals, and thereby that of man, is made possible by the interplay of the matters discussed; again and again he marvels at the astounding variety of living conditions on earth and the beauty of the landscapes and tries to explain their beauty and variety, looking out for the fundamental unity and searching for fundamental explanations. He deplores that at the time of writing he does not have solutions for the 'revolutions of the earth', its 'geogony', as he terms it, nor for the physical deduction of climatic variations. In both instances, however, is he sure that the human mind eventually will find explanatory and unifying rules. 'So, finally, we shall attain a geographical aerology and shall know that this great nature's glass-house, with all its thousand diversities works through just one fundamental law'. (*Ideas*, 55) He even envisages space travel and welcomes it, though he was sorry to live at a time where it was not yet possible.

Underlying this search, this hope of ultimate explanation is Herder's conviction of the unity of man, land, nature, culture, and history. Man is called the 'brother of all other organisations on earth' (*Ideas*, 56) which, however, man entered at a stage when it was already inhabited by these other 'organisations'. They guided him on his way for rule over plants and animals. 'How he achieved this (rule) is the history of civilisation ... the most interesting part of the history of mankind, ... and this history of his (human) civilisation is, to a very large extent, zoological and geographical' (p 73). It is a history in which place, situation, needs, circumstances, and opportunities interplay with each other; and it is this interplay of effect and counter-effect which makes this history so interesting and so stimulating. The sheer wealth in the manifold gifts of races and peoples is a result of this interplay, of effect and counter-effect of man and nature. The history of peoples and nations on earth, whether in Asia, in Africa, in the Americas or in Europe, is carefully examined by him in many long passages in the latter parts of the *Ideas* (e.g. 350-9) in order to illustrate this conviction and to point out some underlying rules for the interplay; for geography is an activity that does not merely describe the manifold potentials of the earth's varied regions and their equipment, but tries also to understand the reasons for these potentials and to research into them (p 186-7). It is Herder who first set this task for geography, and indeed it is still one of its essential tasks. Herder is truly concerned with the two modern trends (or 'paradigms') of geography, that is explaining the earth's regions and their varieties (the 'horizontal' paradigm) and stressing the interplay of man and nature (the 'vertical' paradigm).

Combined, both paradigms lead Herder on to a deep insight into ecological circumstances. Not only does he speak of the 'balance of regions', but he expressly mentions and refers to the chain of matter on earth (soil, rain, growth, decay, nutrition), closing this special passage with the remark, addressed to nature: 'Beneficial mother--how ecological (German: *haushälterisch*; of a household nature) is your circle and how thrifty' (p 66). And then he addresses man to draw the necessary conclusion of confining himself to a proper usage and not to a raping of nature, calling man the earth's responsible householder, whose task it is to keep the balance of this living household. 'Whenever a species, be it plant or animal, was exterminated its loss led in all except a few cases, to obvious drawbacks for the habitability of the whole region' (*Ideas*, 71)

Understandably, Herder does not use concepts and ideas with the same meaning as they are used today but his themes are those which are still the concern of geographers, as Table 1 shows.

3. INFLUENCE AND SPREAD OF IDEAS

According to Goethe (in *Talks with Eckermann*, 1828) the *Ideas* had become a work 'which had an unbelievable influence on the mind of the nation', and, indeed, Herder's ideas as such had become very popular in Germany and abroad, so much so that they became part of the common intellectual heritage. This explains that some of his thoughts today seem to be nothing but truisms. Yet this is so only because Herder's ideas have had a kind of anonymous effect, having in time become so widely accepted that they are deemed to be universally true. Nevertheless, as they have gratefully admitted, some of the leading psychologists, neurophysiologists, and biologists of the latter nineteenth century were led on to their discoveries by Herder (cf. Nisbet, 1970, 327).

Even greater was his profound influence on leading geographers of the nineteenth century, notably Humboldt, Ritter and Ratzel. Through them, all deeply and directly indebted to Herder, the two main paradigms

TABLE I

HERDER'S TERMS AND THEMES	GEOGRAPHY TODAY
+ The terms so marked were taken over by Herder from Kant	
alterations--world wide	categories of change (Lautensach: *Formenwandel*)
+ physical geography	systematic geography
mountains as catchers of rain	windward and leeward climates; high precipitation on the exposed parts of mountains, mediterranean mountains and the huertas at their feet, river oases in deserts
+ role of earth's ecliptic position	the same
mountainous crusts above the ocean (*Erdgebirg*)	hypsographical curve
organizations of the earth	natural regions, regions of similar landform types
Europe and the other continents, linked together by trade and commerce	regional division of labour, functional relations
'geogony'	orogeny, tectonics, plate tect.
geographical aerology	geography of climate, climatology, meteorology
frontiers of life on earth and their significance	inhabitable--uninhabitable world ('ecumene'--'unecumene')
Nature's household, 'circle'	ecology, self-containing circle, balance of in- and output
+ seasons and their significance	climate and weather, weathering, soils, morphodynamics
+ climate and its significance	the same
'worlds' of civilization	cultural realms
organization of regions	spatial organization
situation (in space)	geographical situation
Mittelstriche and their significance	the middle latitudes and their advantageous, various climates
organizations of peoples	geography of states, territories, ('regional geography')
spatial order	the same
'revolutions' of the earth	orogenies
+ ecliptic skewness	the same
+ insolation and its significance	the same
rivers as bestowers of waters	oases and huertas for which the water is provided by a river, e.g. the Nile
great rivers, mountain ranges, oceans and seas, frontiers as guiding lines	the same
relations and connections on the globe	horizontal dependence

HERDER'S TERMS AND THEMES	*GEOGRAPHY TODAY*
configuration of land-masses (continents) and oceans in its significance for climatic variation	the same, e.g. the eastern and western sides of the continents
global situation	functional relations--climatic coherence on the globe

of modern geography came to be established: the 'vertical' (man-land-relation) and the 'horizontal' (chorological distribution of phenomena on earth; 'spatial organization').

Alexander von Humboldt was directly influenced by Herder and not at all by Kant. The very title of his book when he planned his plant geography reveals this: *Ideas towards a future history and geography of plants* (1794). Before that time, in 1788, Humboldt had studied botany at the university of Frankfurt (on the Oder river) under the guidance of the gifted botanist Willldenow who himself was indebted to Herder, as his writings on plant geography show. Humboldt surely knew the *Ideas* himself, by him so tested in 1788 and 1789. Many passages in Humboldt's ambitious work on the *Cosmos* (1845-58) are nothing but echoes of similar phrases in Herder's *Ideas*. Even Humboldt's aims for his great travels in South America as well as the study of the very region itself was strongly influenced by Herder. (The conclusions in this preceding passage are drawn from the following sources: Meyer-Abich, 1967; Beck, 1959-61; two letters by Beck to the author, dated 19 May 1984 and 24 May 1985).

Turning to Carl Ritter it can be clearly established that he was profoundly in debt to Herder through the reading of the *Ideas*, which he so did thoroughly and painstakingly in 1811 (Beck 1979, Hoheisel 1980). It was from the *Ideas* that Ritter gathered his conception of geography as a science of spatial relations, both in the 'horizontal' and in the 'vertical' way. These adjectives were first and expressly used by Ritter (cf. Beck, 1979, 87, 125). 'Horizontal' here means the basic knowledge of spatial contiguity and ordering, 'vertical' the interplay of man and nature (Hoheisel, 1980, 70-1; Beck, 1979, 105). The earth is man's home (this became a central notion of Ritter's) and geography's task is to deal with its occupied space. Moreover, Ritter expressly adopts Herder's methods of comparison and of the generic-processual procedure. By comparing phenomena their variety can be reduced to the underlying and unifying pattern. Ritter's notion of cultural realms is based on Herder's conviction of the equality of all human civilizations (Beck, 1979, 91). Ritter's (and Mackinder's) opinion of what Europe owes to the configuration of land and seas and to its rivers for trade, wealth and cultural variety (see Freeman, 1979,

98) goes directly back to Herder's *Ideas*, where, indeed, he uses this notion at the very end of the last book. Freeman (1979) goes on to show how European geographers were influenced by Ritter, and that inevitably means indirectly by Herder.

Summing up one is really enabled to say that Herder was the author of a cognitive development which ultimately, through Humboldt and Ritter, gave birth to geography as a science in the modern sense. Sadly enough, this knowledge seems to have been lost in the twentieth century; yet in the second half of the nineteenth century it was still living knowledge. Two geographers of that time testify to it. The one is Guthe, the other Ratzel. Guthe, in 1868, published a *Textbook of Geography* which, on the title-page, was decorated with the portraits of three men, who were regarded by him as the founding fathers of modern geography: Humboldt, Ritter, and Herder. Ratzel too, according to Nisbet, was conscious of Herder's importance for geography and dedicated an essay to him; in his *Anthropogeography* of 1882 he mentions Herder several times. Yet in this book he turned the 'vertical' paradigm upside-down, reversing it from 'man--land' to 'land--man'. He wanted to express thereby the geodeterministic effects of land and nature on man (in order to make geography a true science). But Herder himself was not a geodeterminist at all.

Bibliography and Sources

1.   REFERENCES ON JOHANN GOTTFRIED HERDER
Clark, R.T., *Herder*, Berkeley, Calif., 1955
Gillies, A., *Herder*, London, 1944
Haym, R., *Herder nach seinem Leben und seinen Werken dargestellt (Herder represented through his life and works)*, Berlin, 1880-95, 2 vol
Kantzenbach, F.W., *Herder*, Hamburg, 1982
Nisbet, H.B., *Herder and the philosophy and history of science*, Cambridge, 1970

2. *SELECTIVE BIBLIOGRAPHY OF WORKS BY JOHANN*
    GOTTFRIED HERDER

1784? *Von der Annehmlichkeit, Nützlichkeit und*
    Notwendigkeit der Geographie (On the amenity,
    usefulness and necessity of geography): ed by
    B. Suphan in *Gesammelte Werke (Collected Works)*,
    30, 1930: School addresses. Reprinted 1967-68,
    Hildesheim

1784-91 *Ideen zu einer Philosophie der Geschichte der*
    Menschheit (Ideas on a philosophy of the history
    of mankind), 20 books in four parts, newly ed by
    G. Schmidt, Darmstadt, 1966 (pages and citations
    refer to this new ed)

3. *ARCHIVAL SOURCES*

Irmscher, H.D. and Adler, E., *Der handschriftliche*
    Nachlass Johann Gottfried Herder (The handwritten
    papers of Johann Gottfried Herder). This book
    gives a systematic list of Herder's papers kept
    in the Staatsbibliothek Preussischer
    Kulturbesitz in West Berlin (State Library of
    Prussian Cultural Possessions)

4. *OTHER BIBLIOGRAPHICAL SOURCES*

Arndt, J., *Vier Bücher vom wahren Christentum (Four*
    books on true Christianity), Bremen, 1706
Beck, H., *Alexander von Humboldt*, Wiesbaden, 1959-61,
    2 vol
Beck, H., *C. Ritter*, Berlin, 1979
Birkenhauer, J., 'Über den möglichen Ursprung der zwei
    Hauptparadigmen der Geographie bei Herder', ('On
    the possible origin of the two main geographical
    paradigms with Herder'). *Mitt. Geogr. Gesell.*
    München, 1986
Freeman, T.W., 'Carl Ritter und die britische
    Geographie' in Richter, H., *Carl Ritter: Werk*
    und Wirkungen, Gotha, 1983
Guthe, H., *Lehrbuch der Geographie (Textbook of*
    geography), 1869
Hoheisel, V., 'Kant--Herder--Ritter', in Büttner, M.,
    ed, *Carl Ritter*, Paderborn, 1980
Kant, I., *Schriften zur physischen Geographie*
    (Writings on physical geography), Leipzig, 1838
La Roche, S. von, *Geschichte des Fräuleins von*
    Sternheim (Story of the young Lady Sternheim),
    1771. Reprinted in Stuttgart, 1983
Lavater, J.H., *Aussichten in die Ewgkeit (Prospects of*
    eternity), 1768-69, 3 vol
Meyer-Abich, A., *Alexander von Humboldt*, Hamburg, 1967

Josef Birkenhauer is professor of education in
geography at the Geographical Institute of Munich
University

Chronology

1744	Born at Mohrungen, East Prussia (now part of Poland), 25 August
1762	Entered the university of Königsberg (now Kalinigrad, USSR): studied theology and geography and became a friend of Kant and Hamann
1764	Ordained as a priest in the Lutheran Cathedral at Riga
1769	Published critical essays on language, literature, theology and philosophy and in this year travelled to Paris where he met various supporters of the 'Enlightenment'; he also wrote a 'Journal' describing his travels
1770	Travelled to the Netherlands, and also in both northern and southern Germany. He met his future wife, Caroline Flachsland at Darmstad and during these travels went to the university of Strasbourg, where Goethe was then a law student: he established a strong friendship with Goethe at this time
1771	Became the chaplain to the Court at Buckeburg, a town on the Weser river. Published 'On the origin of language' and was given an award by the Prussian Academy of Sciences and Fine Arts in Berlin
1773	Married to Caroline Flachsland: wrote on Ossian and Shakespeare
1774	Published an essay 'On the oldest records of mankind': his eldest son was born, to be followed by six other sons and one daughter (to 1790)
1776	Appointed as 'Superintendent-General' for the duchy of Weimar with the rank of a bishop and also of a minister for education
1778	Presented some of his work on folk songs, of which a major collection was published posthumously in 1805-07
1779	Given an award by the Bavarian Academy, Munich, for his essay on the 'effects of poetry'
1780	Received further awards, from the Berlin Academy for his essay 'On the influence

of governments...' and from the Bavarian
Academy for his essay 'On the influences
of sciences'

1782   Wrote 'On the spirit of Hebrew poetry'

1784-91   Presented his main work, *Ideas on the
            philosophy of the history of mankind*,
            and travelled widely, including a visit
            to Italy in 1788-89

1794-98   Much of his writing was of a
            specifically Christian nature, and
            included contributions on the
            improvement of humanity

1799   Advocated 'Metacriticism', in effect
            opposition to the later philosophy of
            Kant

1803   Died at Weimar, 18 December

Fridtjov Eide Isachsen

1906–1979

HALLSTEIN MYKLEBOST

From early youth Fridtjov Isachsen was living in an atmosphere where discovery was venerated, and his own father was away for long periods of his childhood. In time Major Gunnar Isachsen, like many other fathers, was glad to introduce his son to the type of activity he himself enjoyed and Fridtjov eagerly seized opportunities of Arctic travel and residence, especially during holiday periods. Love of the Arctic remained with him throughout his life but as a geographer he found fascination in the places where he lived and worked. It is characteristic of his wide field of interests that he was concerned with forms of settlement, work and leisure both in the marginal high valleys and in the capital region of Oslo. Trained in the natural sciences as well as in the arts, he regarded an understanding of geology and geomorphology as basically necessary for good work as a geographer. In time widely known internationally, he was most devoted to Norway, enjoying a life of rich cultural interest, especially in music, literature and the graphic arts.

1. EDUCATION, LIFE AND WORK

Fridtjov Isachsen was born on 22 July 1906 at Haugesund during an era of active Polar exploration. In 1906 his father, Gunnar Isachsen (1868-1939), became leader of the first Spitzbergen expedition despatched by Prince Albert of Monaco. Gunnar was initially a cavalry officer but he acquired an interest in polar exploration and was a cartographer on the *Fram* expedition of 1898-1902 under the

leadership of Otto Sverdrup: he mapped large areas of the North American arctic archipelago and this work was recognized by the Royal Geographical Society's presentation of the Murchison Grant in 1903. From 1906 Gunnar Isachsen became famous as an explorer of Spitzbergen and in 1930-31, when over sixty years of age, he was leader of the *Norvegia* expedition around the Antarctic continent. From 1923-39 he was Director of the Norwegian Maritime Museum. Although his son was born at Haugesund, his mother's home town, his first five years were spent in Oslo and the family then moved to Asker, 20 km west of Oslo so Fridtjov grew up in rural surroundings, but with the help of his father he was able to spend long periods in Arctic lands.

Before Fridtjov Isachsen began his studies at the University of Oslo in 1924, he had sepnt 'five summers in Northern lands, three on Spitzbergen, one in Greenland and one in Iceland, and not as a tourist, I can assure you', according to *Geografien i Norge frem til 1940*, published in 1979. In Iceland he worked as a miner. In other words, he grew up in an environment characterized by familiarity with research, economic activity and the way of life in Northern areas. Alone or together with Gunnar Isachsen he prepared reports on Norwegian Arctic hunting and other economic activities as part of the government's evidence in the dispute with Denmark over the sovereignty of Greenland before the International Court at the Hague. In 1933 the Court ruled in favour of Denmark. Of particular interest is his attempt to determine how far the Norse Greenlanders

might have travelled in their hunting expeditions (1932)
(1932). He retained his interest in the Polar regions
throughout his entire life, and the last duty he
renounced was his deputy membership of the Polar
Council.

Fridtjov Isachsen's M.A. thesis--*The Geography of
Greater Oslo* (1929)--went in an entirely different
direction, and may be said to have heralded a new era
in Norwegian geography. It was the very first M.A.
degree in geography to be conferred in the country.

A series of publications on the form and function
of cities followed, inspired particularly by authors in
the German- and French-speaking countries, as well as
by the trend-setting works by Sten de Geer on Stockholm
and work in the United States on New York, Chicago and
other cities. His delimitation of the passenger and
goods traffic fields of East Norwegian towns and of the
traffic intensity of the railways are still of
interest. In 'Contributions to the geography of Oslo'
(1931) there are also interesting theoretical and
methodical features, presented in the simple and at the
same time elegant form which was to become a hall-mark
of Isachsen's work:

> This has to do with the fact that the
> geographical urban concept itself has been
> transformed during this development...until
> 1900 it was the compact and close settlement
> which characterised the town in contrast to
> the countryside. Today, on the contrary,
> the town dwellers are spread over a large
> area, where rural and urban elements are
> intimately mixed. The urban area has
> become more and more differentiated, and
> ever more people thus have their home and
> their work in entirely different places.
> The most important uniting feature of the
> metropolis is just the flow of people to
> the places of work in the inner city.
> (1931, 181)

Here we meet the regional city structure placed
in a correct historical set-up, and the emphasis on
the journey to work as 'the most important uniting
feature of the metropolis'. A generation later many
geographers were busy discovering that the difference
between town and country was no longer as distinct as
formerly. As far as the second point is concerned,
an extra generation may be needed before the
connection is generally understood. 'Contributions
to the geography of Oslo' ends with a chapter on
'City Geography and Home Region Studies in the High
Schools' which gives an even more fresh and less
conventional impression:

> It has been a general belief that the
> countryside is a better environment for
> teaching about the home community than
> are the towns. In the case of the high
> schools this is an entire misunderstanding...
> There are in a large city a number of
> subjects which are suited for that kind of
> study: the houses, the street system, the
> manufacturing areas, the traffic, the port
> conditions--all this may be used for object

lessons in the original sense. The geographer
does not concern himself with isolated features
in the home environment, but first and foremost
wants to draw attention towards a quantitative
perception of phenomena. This has to do with
the distribution and intensity of features of
importance to the visual landscape. Besides
providing a necessary point of support for the
geographical subject-matter generally, home
region teaching in the towns has the additional
advantage of being able to draw the interest of
the future citizen towards the economic and
social problems of his own town, problems which
he himself will once have to cope with, and the
home region subject thus even achieves importance
for the education for citizenhood.

Today few people will agree to a limitation of the
geographer's attention to phenomena which affect the
visual landscape. Those who believe that geography
until the 1970s consisted in learning the names of
Belgian towns should be able to improve their
knowledge from Isachsen's rejection of 'isolated
features' and his emphasis on the importance of
'quantitative perception' and on generalizing. The
final statement is a firmly-based plea for a
direction within human geography which at a much later
time was named *samfunnsgeografi* (geography concerned
with society) in Norwegian. The term is not easily
translated, but it implies that the concerns and
problems of man in society shall guide the selection
of themes relevant for study.

The question of delimiting major cities occupied
him also in the chapter on the environs of Oslo which
he contributed to a master plan of Greater Oslo,
published in 1934. He adheres to Hassinger and other
authors when he decides that 'smaller towns ... are
most easily delimited directly according to the extent
of settlement on the map, major cities, however,
preferably on the basis of traffic conditions'.

These mature statements were put on paper by an
author twenty-five years old, for that was Fridtjov
Isachsen's age when in 1931 he was appointed Associate
Professor (*dosent*) of Geography at the University of
Oslo. The position had been held by Aksel Arstal
(1855-1940) from 1915 until 1925, and had been vacant
since Arstal's retirement. Until 1947, when the
Department of Geography was given its first assistant
and Isachsen was appointed a full Professor, Fridtjov
Isachsen and Werner Werenskiold (1883-1961) were the
entire staff of the department. Werenskiold had a
cavalier attitude to administrative paper-work, so it
is not unreasonable to assume that the younger
colleague had to take care of administrative and other
time-consuming and less gratifying work.

Isachsen's interest in life in marginal rural areas
goes back as far as his interest in urban geography.
His article on the pedlars and cattle-traders from the
Uvdal community in Upper Numedal, with the subtitle 'A
contribution to our older cultural geography' was
printed in 1930. It is a fascinating account of the
organization and extent of the pedlar industry in a
mountain valley of East Norway, of its geographical
background and history, and not least of the human
aspects of the strenuous expeditions across the

mountain plateaux separating East and West Norway and further along the coast or across the fjords and peninsulas of West and even North Norway. It is a piece of research based on many interviews with people who knew the pedlars' life and activity from their own experience as well as on personal observation of the pedlars' environment.

Isachsen always retained a firm belief in the value of skilled observation obtained with your boots on, and he must have made his reflections on the discovery in a later age of the value of 'research on the individual level'. His article on the winter *seter* economy in Vågå (1938) is a work of the same type, where he studies a particular way of adjustment in another marginal area, this time the community of Vågå in a tributary valley Northwest of Gudbrandsdalen. Both studies remain works of lasting interest far outside the geographer's own circle. Studies dealing with themes from settlement geography, like 'A peculiar settlement form in Upper Numedal' (1932) and '*Seter* villages in Nordfjord' (1940) may be regarded as by-products, but they are valuable by-products.

In 'Deglaciation and the quaternary bases of settlement in the uppermost communities in Numedal and Hallingdal' (1933) Isachsen demonstrated that during his long walks he had always had an open eye for landforms and surface deposits. This open-minded interest in both main parts of geography resulted in a number of articles on quaternary and geomorphological themes during the 1940s. Even when he was tracing the Skagerrak moraine in Jaeren or fossil-carrying sediments on Karmoy, his awareness of signs of human activity was as keen as ever. Few of his publications make more absorbing reading than 'Long-ship boat-houses at Ferkingstad and the land-rise' (1940), where with the acuteness and gift of combination of an expert detective he arrives at a consistent explanation of the imposing sites found 5.50-6 metres above present sea level, a puzzle which has intrigued several people. Isachsen's interpretation implies that the sea must have been 1.90-2 metres higher when the boat-houses were built, basing his deductions mainly on the location of landing-places and signs of the route where the ships were hauled on land. Part of this early profile is an article on place-names on a map from Numedal, containing interesting observations on different terms used to denote *seter*, as well as a strong recommendation to the map-makers to give more thought to the correct rendering of place-names.

In Fridtjov Isachsen's contribution to a popular two-volume work *Norway Our Country* (1937) we may observe his versatility at its fullest extent. Landforms and vegetation, dialects and dress, architecture and popular poetry, all contribute to present the identity of the districts dealt with: Setesdal, Telemark, Numedal and Hallingdal. It is the writing of geography heightened to the level of an art, reminiscent of the French school of regional geography at its best. It is also a valuable analysis of Norwegian valley and mountain communities as they were before the thorough-going changes of the period after World War II.

We meet Isachsen as a writer of regional geography even in *World Geography* (1939), where he acted both as editor and as the author of the chapters on Black Africa, Australia and South East Asia. Here we are confronted with the geography centred around present-day problems of which Fridtjov Isachsen was such an articulate advocate.

In later years, from student cohorts whose readiness at drawing far-reaching conclusions does not always have an adequate backing in knowledge and understanding, a common reproach has been that geographers restrict themselves to describing the world as it is, or as they perceive it to be, without daring to leave their ivory towers to involve themselves in judgements on right or wrong or to express opinions. To quote Isachsen on what the world ought to be like

> It has been the pride of South Africa that the standard of living of Europeans is nowhere higher than here. But the economy of this society is based on exploitation of the native population and its mineral riches. It is a conqueror's economy, which sooner or later will have to be changed (1939, 216-17).

Isachsen's analysis of the case of South Africa was written ten years before *apartheid* was launched and many more years before the system was carried through to the extremes that it reached in the days of Messrs. Verwoerd and Vorster.

Other tasks of a similar character were the editing of a Norwegian edition of the four volumes of *The World's Lands and Peoples* (1949-50) and of a Norwegian edition of H. W:son Ahlmann's excellent text *Norway, Nature and Economy* (1957, together with H. Myklebost).

New and interesting work in urban geography resulted from the contact with the leader of Oslo's town planning office, H. Hals, and a promising young student, Tore Sund. Sund and Isachsen together published two reports on the holiday sojourn pattern of Oslo's school children, as well as *Homes and Work-Places in Oslo* (1942). The last-named work was strongly indebted to the successful Stockholm project in which H. W:son Ahlmann and W. William-Olsson in particular had been involved. It has had considerable influence, due to the vast amount of information on Oslo's inner differentiation which it presented in a clear and succinct form as well as to the useful opportunities for comparison with Stockholm which it offers.

Fridtjov Isachsen also made a significant contribution as editing a number of atlases, wall maps and maps in encyclopedias, alone or in cooperation with others.

One of the greatest services rendered to Norwegian geography by Fridtjov Isachsen was his work as editor of *Norsk geografisk Tidsskrift* from 1934 till 1960. He was in fact responsible for this important task for as many years as all the journal's other editors together. He was secretary of *Det norske geografiske Selskab* from 1936. His strong and persistent interest in the Society, where he was a highly treasured lecturer, led to his election as an honorary member in 1976, an acknowledgement already awarded him by a number of foreign geographical societies.

2. *SCIENTIFIC IDEAS AND GEOGRAPHICAL THOUGHT*

The war brought Isachsen as a refugee to Lund, where
from January 1944 to June 1945 he was the leader of
the Norwegian student office, a branch of the school
office of the legation in Stockholm. A number of
geography students of the war years were given an
opportunity to sit their examinations in Lund or
Uppsala. Fridtjov Isachsen's introductory lecture to
the new goegraphy students at the University of Oslo in
the autumn term of 1943, a short time before the
University was closed by the occupying power, was
printed in the Swedish periodical *Geographica* (1944).
It is an interesting document of the time, but first
and foremost valuable as the *credo* of a university
teacher:

> One should study in such a way that one
> learns to distinguish between what is
> important and what is insignificant, what
> is earnest from what is of no avail, what
> is the main thing from what is secondary.
> Do not accept uncritically what is presented
> to you in printed or spoken form! Consider
> thoroughly what is to you the aim of your
> studies, reach an opinion on what is really
> worth learning, do not regard the
> curriculum proposals in the study plan as
> fences which it is impossible to climb,
> but study first and foremost what you really
> feel absorbed by...

> No normal person is capable of being
> intellectually attracted by anything less
> than problems, i.e. questions that cannot
> be answered by everybody. A collection of
> unrelated facts has never been considered
> nourishing mental food...

> When you study problems that can be
> discussed and researched or have been the
> object of research, then you *have* contact
> with research, even if you will not make
> any independent contribution in the form
> of new understanding...

> A genuine University education will have
> provided its man with training for
> independent intellectual work, a store
> of knowledge with sense behind it and a
> critical attitude (1944, 491-2).

One might wish that today's university studies in
Norway had been arranged with greater weight on ideas
like these. Isachsen's attitude to the identity of
geography was at least decisive when a new plan of
study was brought forward in 1958 as part of the new
academic structure which was introduced at the Faculty
of Mathematics and Science. Now a new basic one
year's course was established, common to science and
art students, whereas the geography courses in the two
faculties formerly had differed from beginning to end.
Werenskiold had been strongly interested in
conservation and for a time was president in the
National Conservation League. The same was the case
with Isachsen, who took part in the committee work
which led to the Nature Conservancy Act of 1954. In
1955-60 he was a member of the State Nature
Conservancy Council. In 1963 he presented his view
that

> society by taking human reason systematically
> in its service must be able to lead the
> development towards more desirable conditions
> than we would have with the so called free
> 'interplay of forces'.

> If everybody gnaws from his corner, it will
> soon be a threadbare country for our children
> to inherit. This is a situation virtually
> crying for regional planning. Anything
> cannot take place anywhere--that is simply
> the basic law of ordered administration in
> the cultural landscape. (1963, 81)

This unequivocal public involvement, this
commitment to what he termed applied geography, went
back at least as far as to the cooperation with the
town planner Harald Hals in the mid-thirties. It
made it natural for him to contribute actively to the
process of public planning as a member of the Council
for Regional Planning and of the Railway Commission
of 1949.

3. *INFLUENCE AND SPREAD OF IDEAS*

The entire generation of Norwegian geographers who
went through a University education in the 1930s, 1940s
and 1950s, was strongly influenced by the ideas of the
contents, problems and methods of geography which they
received from Fridtjov Isachsen. Many of them
maintained a lasting contact with him. The present
author's obituary in the Norwegian Academy of Science
and Humanities brought a pile of letters from people
who wanted to tell how much the contact with Isachsen
had meant to them, intellectually as well as in human
terms.

Fridtjov Isachsen's last years as professor
brought him disappointments. He had always had an
open mind towards new directions in his subject, but
he was foreign to the uncritical worship of what at
any time was conceived as 'modern', which undeniably
was a widespread attitude for some time, particularly
among students. When the quantitative wave was at
its highest, he and his concept of geography was
rejected as out-dated by people who neither had much
knowledge of his concept nor knew anything about the
extent to which Isachsen had been a pioneer in the
use of quantitative methods. He did much to make
outstanding representatives of this trend known in
Norway. William-Olsson, Edgar Kant, David
Hannerberg and Torsten Hägerstrand were invited as
guest lecturers long before their names, theories
and methods became known outside the Norden countries.
The biographer's familiarity with the geography
subject at the University of Oslo goes back to the
autumn term of 1946 and comprises a vivid recollection
of how burning issues like Norwegian post-war
reconstruction, new enterprises and regional planning,
problems of underdeveloped countries (which they were
at that time called), world population growth, food

and resource problems were important parts of the study.

Fridtjov Isachsen retired in 1976, and died at Oslo just three years later on 11 August 1979, having been a University teacher for forty-five years. From 1969 he held the position of senior lecturer, at his own request, as a concession to his own wish for a less exacting period after a lifetime of professional duties. In later years he was less active as a geographer, but he still maintained his numerous contacts at home and abroad. He had particularly close ties with Swedish and French geography, and these went back to periods of study in Lund in 1927 where he met Helge Nelson, and to the Congress of the International Geographical Union at Paris in 1931 where he met Emmanuel de Martonne: their all-round geographical training was of lasting importance. He also went to IGU Congresses at Warsaw in 1934 and Washington in 1952. In the post-war years many Fulbright fellows came to Norway from the United States, and Fridtjov Isachsen himself made several journeys to America. But he did feel at home even in Britain, Germany and Italy. Norwegian and foreign colleagues who sent him a new publication did not receive the usual phrases, telling how it had been received with thanks and how the recipient looked forward to studying it more closely. Instead they would receive long letters which showed that the publication had been carefully studied. Through the years many people with highly different backgrounds were to receive advice and encouragement in this way. His readiness and resourcefulness in aiding present and former students went far.

Fridtjov Isachsen was not an uncomplicated man, but he was a person whom it was intellectually stimulating to know. His wide cultural background extended from the poets and philosophers of antiquity to modern painters and graphic artists. In the fields which caught his interest, he was no amateur, but a professional. He was a performing member of the Oslo Quartette Society and of the Student Orchestra and a member of the commission for music examinations at the University of Oslo. His wife shared his musical interests and became a member of the Oslo Philharmonic Society in 1921, just two years after its foundation. She accompanied her husband on various field-tours. Blessed by a happy home, Isachsen was a European in the old sense of the word. He was fluent in English, French and German and read Latin: in his later years he also learned Italian. His reading ranged from Virgil and Petrach--whose work he translated into Norwegian--to Eckermann's conversations with Goethe, and also from the Koran to the writings of St Teresa of Avila.

In a series of radio lectures Fridtjov Isachsen in 1953 presented his view on the problems and methods of geography:

> Geography shall teach us to know and understand the globe--and our own piece of it--so well that we are able to evaluate it--as a place (an environment) where we find our livelihood, shape our life, arrange our coexistence with other people. Our aim as geographers is to be able to judge with discernment our opportunities--and our limitations--in the geographical environment where destiny has placed us. Geography embraces nature as well as human life when it directs our attention towards the environments on earth. The discord between nature research and culture research which leaves such an unreasonable split in our intellectual life, looks more and more unreal when one concentrates on geography's fundamental questions. There is only an apparent contradiction when we assert that all geography is physical geography--and that all geography is human geography. Both are necessary; they are two aspects of the same thing.

Fridtjov Isachsen would hardly have had any objection to the reading of this as his testament as a geographer. The content of the subject will constantly change as new facts and new problems present themselves. Some of these changes will have lasting influence, others will turn out to be short-lived whims. To be able to profit from new impulses, to avoid becoming uncritical epigones, we need a professional identity that can enable us to 'distinguish between what is important and what is insignificant'.

Bibliography and Sources

1. *OBITUARIES*
Hertsberg, L.H., *Sven. Geogr. Årsbok*, vol 55 (1979), 100-02
Myklebost, H., *Det Norske Videnskaps-Akademi i Oslo, Årbok* (1980), 215-18
---- *Norsk Geogr. Tidsskr.*, vol 34 (1980), 1-8

2. *SELECTIVE BIBLIOGRAPHY OF WORKS BY FRIDTJOV ISACHSEN*

a. *General Works*
(i) *Content and Purpose of Geography*
1944 'Innledningsforelesning for ney geografistuderende ved Universitetet i Oslo 6. oktober 1943' ('Introductory lecture to new geography students at the university of Oslo ...'), *Geographica*, no 15, *Geografiska studier tillägnade John Frödin den 16. april 1944*, 488-500
1952 *Kulturgeografien i Norge (Human geography in Norway)*, Oslo
1953 *Geografiens arbeidsmåte og problemstilling (Methods and problems of geography)*, Naturforskeren i arbeid, Oslo
1967 'Hva er geografi?' ('What is geography?'), *Den høgre skolen*, no 16
1979 *Geografien i Norge frem til 1940. Innledningsord vod Norsk Samfunnsgeografisk*

*Forenings seminar på Gol 18. november 1978
(Geography in Norway to 1940. Introduction to
the Norsk Samfunnsgeografisk Forenings seminar
on Gol)* private mimeograph

(ii)   *Education, including Universities*

1930   'Geografiens stilling ved universiteter og
hoiskoler i Sverige og Danmark og hos oss'
('The place of geography in the universities of
Sweden and Denmark and with us'), *Norsk. Geogr.
Tidsskr.*, vol 3, 271-8

1933   (with A. Redse) 'Geografilaererne i Norges
høiere 1932-33, dares utdannelse' ('Geography
teachers in Norway's high schools; their
education'), *Norsk Geogr. Tidsskr.*, vol 4,
465-72

1934   'Sprogene i skolen' ('Languages in the school')
Den Holiere Skole, no 11, 1-12

----   'Quelques remarques sur un livre de M.
Zimmermann', *Norsk Geogr. Tidsskr.*, vol 5,
262-7

1946   'Geografistudentenes fagkombinasjoner' ('The
subject combinations of geography students'),
Norsk Geogr. Tidsskr., vol 11, 138-43

1957   'Gli studi geografici in Norvegia', *Boll. Soc.
Geogr. Ital.*, per 8 vol 10 (also given as vol 94),
165-71 (English abstract)

1967   'Geografisk institutt 50 år' ('Fifty years of
the Department of Geography'), *Univ. Oslo,
Arsberetning*, 47-54

(iii)   *Textbooks*

1939   *Verdens geografi (World geography)*, De Tusen
hjems bibliotek, Oslo

----   *Verdens geografi*, vol 2, 'Neger Afrika,
Australia, Sydost-Asia', 185-248, 348-62

1949-50   *Verdens land og folk (The world's land and
peoples)*, Oslo

1960   'Norden', in Somme, A., ed., *A geography of
Norden*, 11-17

1965-66   (with C.M. Mannerfelt, A. Schou, and H.
Hammerskiold) *Europa. Vår egen världsdel
utom Norden (Europe: our own continent
excluding Norden)*, Stockholm

(iv)   *Atlases and Cartography*

1935   Text and picture section of *Cappelens
verdensatals (Cappelen's world atlas)*

1936   'Quelques nouveautés de la cartographie
officielle norvegienne', *Mélanges de géographie*,
72-4

1951   *Cappelens store verdensatlas (Cappelen's Great
World Atlas)*, Oslo

1963   (with K. Gleditsch and A. Rohr) *Norge, Fjerde
Bind, Atlas*, Oslo

1965   *Fabritius Atlas*, Oslo

1970   *Cappelens internasjonale atlas (Cappelen's
International Atlas)*

(v)   *Obituaries*

1940   'Adkel Arstal', *Norsk. Geogr. Tidsskr.*, vol 8,
121-3

1947   'Minnetale over prof. dr. O. Solberg', *Det
Norske Videnskaps-Akademi i Oslo, Årbok*, 41-9

1953   'Werner Werenskiold', *Norsk. Geogr. Tidsskr.*,
vol 14, 1-5

1955   'Et livi fiell Gygda lars Olsen Grøtjorden 1855-
1941' ('A life in a mountain parish...'),
Norsk Geogr. Tidsskr., vol 15, 85-9

1976   '"Gunnar Holmsen", Minnetale', *Det. Norsk
Videnskaps-Akademi Årbok*, 163-9

1977   'Werner Werenskiold, 1883-1961', *Norsk Biogr.
Leks.*, vol 18, 463-70

b.   *Publications on Norway*

(i)   *Oslo and the Oslo region*

1928   'De geografiske hoveddrag ved Oslos
innenlandske distribusjonshandel' (French
abstract; 'Les traits géographiques principaux
du commerce de distribution d'Oslo dans
l'interieur'), *Sven. Geogr. Årsbok*, 91-116

1929   *The geography of Greater Oslo*, unpublished
M.A. thesis

1931   'Bigrad til Oslo geografi' ('A contribution to
the geography of Oslo. 1, A new study of
Oslo: 2, Oslo's building forms and their
development: 3, City geography and home regional
studies in the High Schools'), *Sven. Geogr.
Årsbok*, 166-87

1934   *Stor-Oslos omfatning. Omegnen (Greater Oslo.
The environs)* Stor-Oslo, Forslag til general-plan,
Oslo

1938   (with T. Sund) (Oslobarnas landopphold sommeren
1937' (Zusammenfassung: 'Der Sommeraufenthalt
der Oslo-Kinder 1937), *Norsk Geogr. Tidsskr.*,
vol 7, 101-08

----   'Oslobarnas landopphold sommeren 1938 og deres
besok hos slektninger' (Summary: 'Country
holidays of Oslo children, summer 1938, and their
stay with relations'), *Norsk Geogr. Tidsskr.*,
vol 7, 167-72

1942   (with T. Sund) *Bosteder og arbeidssteder i Oslo
(Homes and workplaces in Oslo)*, Oslo

(ii)   *Economy and Settlement of Mountain Areas*

1930   'Uvdølenes skreppehandel og dirftetrafikk' ('The
pedlar and cattle trading industry of Uvdal'),
Norsk Geogr. Tidsskr., vol 3, 165-84

1932   'En eiendommelig bebyggelsesform i Ovre Numedal'
('An extraordinary farm type in Upper Numedal'),
Norsk Geogr. Tidsskr., vol 4, 191-212

1933   'Fra Hallingskeid til Nordsfjord i
driftekarenes fotefar' ('From Hallingskeid to
Nordfjord in the path of the pedlars'), *Norsk
Geogr. Tidsskr.*, vol 4, 488-513

1936   (with W. Werenskiold) 'Naturog levevilkår i Øst-
Norges fjellbygder' ('Nature and living
conditions in the mountain districts of East
Norway'), *Universitetetets radioforedrag*, no 23,
Oslo

1937   'Setesdal-Telemark-Numedal-Hallingdal' in *Norge
Vart Land*, 97-168, Oslo

1938   'Vintersaeringen i Vågå' (Summary: 'La migration
hivernale du bétail a Vågå, Norvège'), *Norsk
Geogr. Tidsskr.*, vol 7, 203-39

1940   'Seterlandsbyer i Nordfjord' (Resumé: 'Montagnes
agglomérées dans le Nordfjord'), *Norsk Geogr.
Tidsskr.*, vol 8, 73-84

1949 'Raundalen mellem vest og-øst' ('Raundalen between west and east'), *Bergen Turlags Årbok*, 1-8

1953 'Numedal', in *Norske bygder*, 9-24, Bergen

1960 'Rural settlement in Norway' (IGU Norden Congress) 'Excursion E.N. 5 August 14-21, Oslo-Gudbrandsdal-Nordfjord Bergen', *Norsk Geogr. Tidsskr.*, vol 17, 187-96

1969 'Setringsintensiteten' ('Seter intensity') in Ouren, T. (ed), *Festskrift i anledning professor Axel Sommes 70-årsdag 19. april 1969*, *Ad Novas-Norwegian Geographical Studies*, no 8

(iii) Norway and the Arctic

1928 'Navene på Tunhovdbladet' ('The names on the Tunhovd map'), *Norsk Geogr. Tidsskr.*, vol 2, 371-89

---- 'Tidligere utforskning av området mellom Isfjorden og Wijdebay på Svalbard' ('The earlier exploration of the area between Isfjorden and Wijdebay on Spitzbergen'), *Norsk Geogr. Tidsskr.*, vol 2, 461-4

1932 (with G. Isachsen) 'Norske fangstmenns og fiskeres ferder til Grønland 1922-1931' ('Expeditions of Norwegian trappers, hunters and fishermen to Greenland...'), *Norsk Geogr. Tidsskr.*, vol 4, 21-74

---- 'Hvor lengt mot nord kom de norrøne grønlendinger på sine fangstferder i ubygdene?' ('How far north did the Norse Greenlanders travel in their hunting expeditions in the ecumene?'), *Norsk Geogr. Tidsskr.*, vol 4, 21-74

1933 'Verdien av den norske klappmyssfangst langs Sydøstgronland' ('The value of the Norwegian catch of hooded seals around southeast Greenland'), *Norges Svalbard- og Ishavsu ndersøkelser*, Medd. no 22

1935 'Svalbard, Norges besiddelse' ('Svalbard, Norway's possession'), *Cappelens verdensatlas*, 25-9, Oslo

1939 'Det nye norske land i Antarktis' ('The new Norwegian land in the Antarctic'), *Mellanfolkligt samarbate*, 133-7

---- 'The new Norwegian dependency in the Antarctic', *de Nord*, 67-78

(iv) Geomorphology

1934 'Terrassemålinger i Nord-Odal' ('Terrace measurements in Nord-Odal'), *Norsk Geogr. Tidsskr.*, vol 5, 33-52

1937 'Urdebøskredet i Rauland' ('The Urdebo landslide in Rauland'), *Norsk Geogr. Tidsskr.*, vol 6, 361-3

(v) Deglaciation

1933 'Isavsmeltningen og de kvartaegeologiske forutsetninger for bebyggelsen i Numedals og Hallingdals øverste bygder' ('The deglaciation and the Quaternary bases for the settlement of in the upper communities of Numedal and Hallingdal'), *Norsk Geogr. Tidsskr.*, vol 4, 428-41

1940 'Grefsenmorenens opbygning og fossilinnhold' ('Structure and fossil contents of the Grefsen moraine in Oslo'), *Norsk Geol. Tidsskr.*, vol 20, 253-62

1941 'Korrelasjon av strandmerker i løsmateriale og fast fjell' ('Correlation of shore marks in loose deposits and bedrock': Zusammenfassung: 'Zusammentstellung von Strandmarken in Locker-Ablagerungen und in festem Felsen'), *Norsk Geogr. Tidsskr.*, vol 8, 257-64

(vi) Geology

1939 'The breccia zone Bonnefjord-Halandspollen', *Norsk Geogr. Tidsskr.*, vol 7, 322-5

1940 'Kvartsporfyr i Åros, Royken' ('Quartz porphyry in ...'), *Norsk Geol. Tidsskr.*, vol 20, 263-5

---- 'Kvaedfjordkull fra Karmøy' ('Kvaedfjord coal from Karmoy'), *Naturen*, vol 64, 217-23

---- 'Fossilførende sedimenter ved Skudenes' ('Fossiliferous sediments at Skudenes'), *Norges Geol. Unders.*, no 155, 5-18

1942 'Permike gangspalter i grunnfjellet vest for Randsfjorden og deres betydning for utformingen av reliefet' ('Summary: 'Permian fissures in the Archaen west of lake Randsfjord, eastern Norway, and their geographic significance'), *Sven. Geogr. Årsbok*, 24-38

---- (with G. Horn) 'Et kullfund i Skagerrak-morenen på Jaeren' (Summary: 'A coal find in the Skagerack moraine of Jaeren, southwest Norway'), *Norsk Geol. Tidsskr.*, vol 22, 15-46

1949 'Forvitringsleire og blekejord på Karmøy' ('Weathered clay applicable for bleaching used lubricating oil, on Karmoy'), *Norsk Geol. Tidsskr.*, vol 27, 175-86

(vii) Historical

1940 'Langskibs-naustene ved Ferkingstad og landhevningen' (Zusammenfassung: 'Die Langschiffsschuppen am Ferkingstad'), *Norsk Geogr. Tidsskr.*, vol 8, 94-104

1942 'Grensen mellom Norge og Sverige ved Göta älv før 1658' ('The boundary between Norway and Sweden at Göta älv before 1658'), *Norsk Geogr. Tidsskr.*, vol 9, 126-8

(viii) Conservation and Planning

1956 'Selvstyrem og samarbeid' ('Self-government and co-operation'), *Samarbeid over Kommunegrensene. Regionplantomitéen for Oslo-området*, Brosjyre (pamphlet) no 2

1963 'Aktuelle naturvern-tiltak i Norge' ('Current conservation measures in Norway'), *By-og regionplanleggeng, Jursus ved Norges Tekniske Hogskole* (1962), Trondheim, 70-82

1974 'Naturvern', in Sollid, J.L., ed., *Vern og forvaltning av naturressurser. Noen utvalgte emner (Conservation of natural resources: some selected themes)*, Oslo, 1-4

Hallstein Myklebost is professor of geography at the University of Oslo

Chronology

1906   Born at Haugesund, 22 July

1924   Became a student at the university of Oslo

1927   Had the first of several study periods at Lund university and met Helge Nelson

1929   Awarded the Master of Arts degree of Oslo university

1930   Published a paper on rural geography

1931   Appointed as Associate Professor of Geography at Oslo university: attended the IGU Congress in Paris and from this time published several studies of the geography of Oslo and district
Married Sofie Kobro

1933   Wrote a paper on physical geography and settlement in mountain areas of Numedal and Hallingdal

1934   Became editor of *Norsk Geografisk Tidsskrift*: attended the IGU Congress at Warsaw

1936   Elected as secretary of the Norwegian Geographical Society

1937   Showed his wide range of interest and study in *Norway: our country*

1938   Edited a book of a general character, *World Geography*, to which he contributed the sections on Black Africa, Australia and South East Asis

1942   His continued interest in urban geography was shown in his work, *Homes and Workplaces in Oslo*, written in collaboration with Tore Sund

1944   Went as a refugee to Lund, where he worked for Norwegian students

1947   Appointed as a full professor in Oslo university

1952   Attended the IGU Congress in Washington, D.C.

1955   Appointed as a member of the State Nature Conservancy Council (to 1960)

1969   At his own request, and by arrangement with the authorities of Oslo university, he became a senior lecturer with less responsibility of an administrative nature

1976   Retired from university service

1979   Died at Oslo, 11 August

John Scott Keltie

1840–1927

L.J. JAY

The wind of change which carried new dimensions to the concept of geography in Britain during the closing decades of the nineteenth century blew vigorously across Edinburgh, where inspiration came from several quarters.   To name only a few, John Bartholomew the cartographer (1860-1920), Patrick Geddes (1854-1932) with his dual interest in biology and sociology, A.J. Herbertson (1865-1915) who later became professor of geography at Oxford, Hugh Robert Mill (1861-1950), the meteorologist and Sir John Murray (1841-1914) the oceanographer, each made a distinctive contribution. Also qualified to be included among these doughty Scottish pioneers with an Edinburgh connection was Sir John Scott Keltie, whose love of literature and journalism provided yet another facet to the many sides of the emerging discipline.

1.   *EDUCATION, LIFE AND WORK*
David Keltie was established as a builder and stonemason in Perth but it was during a brief visit which he and his wife made to Dundee that their son, John Scott, was born there on 29 March 1840.   Shortly afterwards the family returned to Perth where in due course young John went to school.   At the age of fourteen, when times were hard in his father's business, he became a pupil-teacher in the old school at the south end of Watergate.   After four years he did not proceed to a Training College but elected to enter St Andrews University, where he attended the normal first-year classes in Latin, Greek and mathematics.   In 1860 he transferred to Edinburgh University where he studied Arts subjects

intermittently for four out of the next seven years without, however, completing the requirements for a degree.   An earnest member of the United Presbyterian Church, he followed the course of training for the ministry at the Theological Hall in Edinburgh, but finding the aspect of theology which was then being demanded of him uncongenial he did not seek ordination at the conclusion of the course.   In later years he delighted to recount how the Calendar of his Theological College reported him as having 'lapsed into literature'.   This indeed proved to be his abiding interest, for as early as 1861 he had joined the editorial staff of the publishing firm of W. and R. Chambers in Edinburgh where he worked on the earliest edition of *Chambers' Encyclopaedia*; he also wrote articles for the *Edinburgh Evening Courant*. Throughout his student years he supported himself financially by his journalistic work.
    In 1865, in Edinburgh, he married Margaret, a daughter of Captain John Scott of Kirkwall.   Six years later he left the firm of Chambers and moved south to London to join the editorial staff of Macmillan's, where he became sub-editor of *Nature* under the guidance of Norman Lockyer and frequently contributed a quarter of the items to that weekly scientific periodical.   His literary output during these years included a new edition of the works of Daniel Defoe, an anthology entitled *The Works of the British Dramatists* and a revised version of Browne's *History of the Scottish Highlands, Highland Clans and Highland Regiments*, which became the standard work on that subject for many years.   His connection with *The*

Times began in 1875 through his possession of unpublished information on the island of Socotra which Great Britain had recently annexed; he contributed a series of highly topical articles to that journal on the subject of 'The Scramble for Africa'.   In 1883 he became editor of *The Statesman's Year Book*, and this appointment he retained for the rest of his life.

He was elected a Fellow of the Royal Geographical Society in 1883 and in July of the following year he made a successful application for the post of the Society's Inspector of Geographical Education which had been advertised.   Largely due to the insistence of Douglas Freshfield (1845-1934) the Society was seeking to improve the state of geographical teaching in Britain and wisely decided to investigate the extent of its imperfections before pressing for reform. Keltie spent an energetic year visiting schools and universities throughout Britain and in several countries on the Continent as well as eliciting evidence from informed correspondents in Canada and the United States.   His report on 'Geographical Education', presented to the Council of the Society in May 1885 and published in the first volume of *Supplementary Papers* of the Society a year later, has been rightly regarded as one of the landmarks in the development of the subject as an academic discipline. Two months before he submitted his findings to Council Keltie was appointed Librarian of the Society and he subsequently acknowledged the helpful advice he had received from the Assistant Secretary of the Society, H.W. Bates (1825-92), in the compilation of his report. Keltie was able to help Bates in editing the *Proceedings* of the Society and he often acted as his deputy in much of the committee work; it was an amicable partnership which sadly was dissolved by the death of the older man in 1892.   Keltie was the Council's automatic choice to inherit the mantle of Bates, and some of his most valuable work was achieved in his capacity as Secretary of the Society during the ensuing twenty-three years.

The Sixth International Geographical Congress was held in London in 1895 and the Royal Geographical Society assumed the chief responsibility for the organization of this gathering.   Keltie and Hugh Robert Mill (who had succeeded Keltie as Librarian) were appointed joint organizing secretaries and the advance planning of the Congress coupled with the careful supervision necessary to ensure that the entire programme passed off without a hitch left both men physically and mentally exhausted.   Two honours were accorded to Keltie in 1897; he was elected President of the Geography Section of the British Association for the Advancement of Science, held that year in Toronto, and he received an honorary degree of LL.D. from the University of St Andrews.

Keltie was deeply involved in the negotiations for the transfer of the Society's headquarters from Savile Row to Kensington Gore in 1913; in the same year he was President of the Geographical Association, choosing as the title for his Presidential Address 'Thirty years' progress in geographical education'. At the age of seventy-five, although still alert both physically and mentally, he deemed it expedient to resign his post as Secretary of the Society in favour of a younger man, whereupon he was awarded the

gold medals of the American, Paris and Royal Scottish Geographical Societies.   During his career he had been an Honorary Corresponding Member of almost every geographical society in Europe and of several in America.   He was decorated with the Orders of the North Star of Sweden, St Olaf of Norway, and the White Rose of Finland.   When he relinquished his post as Editor of the *Geographical Journal* in 1917, a publication which he had initiated twenty-four years earlier, the Royal Geographical Society gratefully awarded him its Victoria Medal and in the following year he received a knighthood for his services to education and geography.   His wife died in 1922 from shock caused by accidental burns and five years later, on 12 January 1927, he died in London following a sudden attack of bronchitis.

2.   *SCIENTIFIC IDEAS AND GEOGRAPHICAL THOUGHT*

In common with most of the British geographers who had studied at a university during the nineteenth century, Keltie's approach to the subject developed from an earlier interest which in his case was journalism. He had been in turn a pupil-teacher, an undergraduate and a theological student, yet at the age of twenty-seven he was neither schoolmaster, graduate nor minister of religion.   Literature absorbed him and the need to earn a living by his own exertions led him into journalism.   At the time when he moved to work in London events in Africa were very much in the public eye through the activities of explorers who were busily unveiling the physical features of that continent's interior, and the colonial ambitions of certain European nations.   In his quest for newsworthy copy that would attract the attention of educated readers Keltie focused his sights on affairs in Africa and it seems probable that by this avenue he became aware of the potentialities of geography in general.   His series of articles in *The Times* were both informative and opportune while his work as editor of *The Statesman's Year Book* gave him access to a mass of geographical information relating to all corners of the globe.

In formulating his own concept of geography he learned much from H.W. Bates for whom he had a deep admiration and affection.   He recalled this debt in the address he gave to the Royal Geographical Society on 5 February 1917 entitled 'Thirty Years' Work of the Royal Geographical Society'.

> When we talked about geography he (Bates) insisted that if we were to develop it and raise it from its low estate, man in his relations to his habitat must be made the centre, the culminating factor of the whole subject.

The extensive inquiries which Keltie conducted into the state of geographical teaching in schools and higher institutions of learning during 1884-85 revealed the deplorable condition of the subject in Britain in marked contrast to the situation in many countries on the continent where geography had attained a respected place in the curriculum. Keltie regretted the 'mischievous' subdivision of geography into physical and political sections in

British schools and was impressed by the unity of the subject he had observed in schools abroad. In particular, he admired the high standard of teaching and learning geography in Germany, especially the *Heimatskunde* whereby children were taught about their own neighbourhood in detail before their vision and understanding were extended to successively larger units of study in the province, the homeland and the continent. This was achieved, he observed, through placing emphasis on learning by actual observation (*Anschauungslehre*) and by using atlases and textbooks which were far superior to those available to teachers in Britain. Keltie admitted that the position of geography in the primary schools of Britain had improved considerably since the Education Act of 1870, but his inquiry was chiefly concerned with the public schools and universities. Here, he declared, the state of geographical education was very poor. Very little geography was taught in most public schools and when it was attempted it was given to the most incompetent master: in fact

> The only places where geography is systematically taught in England are the Training Colleges, male and female, and the National Board Schools; with now, and for the last few years, some few good High and Middle-Class Schools.

When he turned to examine the situation in the universities of Britain Keltie revealed that there was no professor of geography in any university or university college in this country, and he continued

> There can be little doubt...that if the Universities of Oxford and Cambridge recognised geography by giving it a real place in their examinations, it would have a powerful influence in exalting the position of the subject in schools.

In 1890 Keltie wrote a book of modest size entitled *Applied Geography: a preliminary Sketch*, in which he set out to illustrate some of the ways whereby geography has a bearing on human activity, with particular attention to history, colonization, industry and commerce. One reviewer considered the book to be 'great in scope, philosophical in treatment and going far to define the somewhat vague field of geography' (*Scott. Geogr. Mag.*, vol 7 (1891), 106). Keltie considered 'distribution' to be the keyword in geography but he asserted that at more advanced levels of the subject the task was to consider the interaction between humanity and environment. Incidentally he would appear to be one of the first geographers in Britain to employ the term 'applied geography'.

In his Presidential address to the Geography Section of the British Association meeting in Toronto in 1897 he referred to the 'excessively apologetic' language used by previous presidents when they attempted to define the field of geography, and Keltie expressed his opinion that this defensive attitude was no longer necessary (*Report of the British Association for the Advancement of Science* (1897) 700). Sixteen years later he collaborated with O.J.R. Howarth in writing a short *History of Geography* which gave a condensed outline of progress in geographical knowledge through the ages. After asserting that the whole structure of geography rested on the two massive pillars of exploration and measurement the authors admitted that the precise scope of the subject was not easy to define. Nevertheless, they continued, the study of distributions and environment had given a new dimension to the discipline in which regional subdivision had become a leading principle of research into both natural and human phenomena.

In an address which Keltie delivered to the Royal Scottish Geographical Society in Edinburgh in 1915, entitled 'A half-century of geographical progress', he declared that

> Personally it does not irk me whether geography is admitted to be a science or not. It is a department of investigation which deals with a field untouched by any other department--the earth as the home of humanity.

After describing geography as the mother of all the sciences he added a cautionary note. Geographers must not be too grasping, but should form a clear idea of the limits of their field of operations:

> The geographers in this country have been moving into a much more spacious mansion; we have hardly had time to put our house in order; we may find when we do so that we have not room for all the furniture that some of our friends would like to squeeze into it.

3. INFLUENCE AND SPREAD OF IDEAS

Keltie was one of the key figures in the revival of geography in Britain towards the end of the nineteenth century, for his report on geographical education touched off a train of momentous developments. In the years preceding his investigation the Council of the Royal Geographical Society had made a number of abortive attempts to stimulate the teaching of geography in schools. Annual cash prizes for outstanding work in geography by candidates for the examinations of the Royal Society of Arts were replaced by the award of gold and bronze medals to boys from selected public schools who were successful in specially designed examinations in geography, but both of these schemes failed to persuade more schools to introduce geography into the curriculum. In addition the Council had attempted, on more than one occasion, to induce the Universities of Oxford and Cambridge to appoint a lecturer in geographer, but without success.

To supplement the evidence marshalled by Keltie in his report he assembled an impressive exhibition of geographical appliances which were employed in continental schools such as wall-maps, atlases and textbooks, of a quality far superior to the material available in British schools. This exhibition attracted large crowds when it was first displayed in the Map Room of the Royal Geographical Society; it was subsequently taken on tour and seen by many people in London, Manchester and Edinburgh before it was eventually donated to the Teachers' Guild of Great Britain. Keltie himself travelled widely around the country, giving talks on the subject of his investigations.

Before the impact of Keltie's report had subsided Douglas Freshfield persuaded the Council of the Society to make yet another approach to the Universities of Oxford and Cambridge, and persistence was rewarded on this occasion; in 1887 Oxford agreed to establish a Readership in Geography for a trial period of five years at an initial salary of £300 per annum provided that the Royal Geographical Society would contribute one half of the cost.   The man to whom this post was offered was Halford John Mackinder (1861-1947) whose lectures on geographical topics to the widely dispersed centres of the Oxford University Extension System had recently been attracting large audiences and who in the early part of 1887 had delivered an impressive paper to the Society entitled 'On the scope and methods of geography'.   These events were recalled in the discussion which followed Keltie's Presidential Address to the Geographical Association in 1914 when Mackinder revealed how his own latent interest in geography as a boy had been stimulated in 1884, the year he graduated in modern history from Oxford.

> As a result of a notice in the paper I came up to London and went to a hall in Regent Street where I found a collection of maps, diagrams, books and Dr. Keltie.   I had never heard of Dr. Keltie at that time.   I knew very little of the Royal Geographical Society.   I got into conversation with Dr. Keltie and he was very kind to a young man who put some crude questions to him.   It was from that moment that my vague tendencies towards geography began to crystallize .... I am speaking in praise of my father in geography.   (*Geogr. Teach.*, vol 7 (1914), 226)

Thus not only was Keltie instrumental in persuading the University of Oxford to establish a lecturer in geography, but also he was the person who communicated to the future holder of that post the potentialities of the subject--no mean feat when one reflects on the significant part which Mackinder subsequently played in the revival of geography in Britain.

The tonic effects of Keltie's report were felt at all levels of the educational system.   Cambridge appointed a lecturer in geography in 1888, the University of Manchester followed suit four years later, and the trend gathered momentum in the early years of the twentieth century to produce a generation of graduates from a dozen different institutions who had studied some geography at university level; many of these subsequently became school teachers.   The Education Act of 1902 empowered local authorities to establish secondary schools within their respective areas and two years later the Board of Education issued regulations prescribing a minimum of two periods of geography teaching per week as an integral part of the basic four-year course in all secondary schools. The improved status of geography as an academic discipline was reinforced in 1893 by the creation of the Geographical Association, which henceforth acted as a medium for the dissemination of basic knowledge and practical techniques among teachers of the

subject.   Publishers were stimulated to produce improved specimens of atlases and wall-maps, while authors of the calibre of Herbertson and Mackinder wrote geography textbooks for schools which set a new standard in this genre; written in literary English they were liberally illustrated with photographs, diagrams and sketch-maps, providing a welcome replacement for the tedious works of reference which passed for school texts in the mid-nineteenth century.

Keltie brought to his journalism a flair for mastering complicated matters and expressing the essential facts of a topic in clear, concise terms. This skill was displayed in his contributions to *Nature* and *The Times*, and in his editorship of the *Statesman's Year Book*, which had been founded in 1864. G.G. Chisholm (1850-1930) testified to the immense and immediate improvement in the format and content of this annual work of reference which took place from the year Keltie became its editor.   Access to factual and reliable information about the countries of the world was the key to much geographical writing around the turn of the century.

Among the most valuable achievements of Keltie in the sphere of geography his report on geographical education must be set alongside the services he rendered as an official of the Royal Geographical Society.   When he became its Assistant Secretary in succession to Bates, Keltie was fifty-two although he looked much younger than his years, not least because he had recently shaved off his pointed auburn beard. Somewhat below the average in height he was always neat in appearance and precise in his speech and manner. One of his first moves was to cast a critical eye on the Society's serial publications.   Bates had revised the *Proceedings* in 1879, substituting a New Series which appeared monthly instead of at irregular intervals.   Within a year of entering upon his new duties Keltie changed the title from *Proceedings* to *Geographical Journal* and under his expert control it acquired international prestige among geographers. After 1892 Keltie modified his wide-ranging literary interests in order to concentrate on the task which henceforth became his main purpose in life, namely the promotion of the interests of the Royal Geographical Society, and in this direction his influence was positive, persuasive and permanent.   In June 1926, not long before his death, Keltie received a letter from Douglas Freshfield which ended with this observation:

> When I look at the Society's activities now and what they were in 1870 I think we may fairly say we between us broke a good many bad traditions and set up a lot of new activities--the *Journal* will be mainly your monument!

In 1893 Clements Markham (1830-1916) was elected President of the Royal Geographical Society and for the next dozen years he directed the activities of Council in a masterly fashion.   Steeped in the traditions of the Society (he had been an Honorary Secretary for a quarter of a century) Markham was a widely travelled man, knowledgeable and imperious, whose decisions reflected his vigorous likes and dislikes.   He supported with boyish enthusiasm the Society's interest in exploration and discovery but was out of sympathy with Freshfield's attempts to improve the status of geographical education, so Keltie found it prudent to move delicately in his dealings with Markham. This he achieved with such consummate skill that many of the measures which appeared to emanate from the President had been subtly suggested to him by the Secretary.

Keltie was a friend to all and a sympathetic listener to any who approached him with matters of routine or personal problems, whether they were titled members of Council, holders of minor posts on the staff, aspiring authors, returned explorers, or rank and file Fellows. He was a keen clubman whose kindness and hospitality became proverbial; at the Savile Row headquarters he continued to hold the daily lunch parties for callers which Bates had inaugurated, and after the founding of the Royal Societies Club in 1895 he made the big round table in the corner of its dining room the rendez vous for visiting explorers and foreign geographers. To supplement these daytime gatherings his daughter, Mrs. T.L. Gilmour acted as hostess for Sunday evening suppers at the home they shared, at which geographers and men of literature were equally welcome. Keltie had the gift of promoting conversation in company without dominating it and his views on controversial matters were rarely revealed; his success rested on the skilful exercise of tact, caution and intuition. This may have deceived some observers into thinking that he was facile but a close colleague of his once observed that Keltie had a store of passive resistance which was used to keep in check movements of which he disapproved; at the turn of the century, for example, King Leopold II of Belgium had tried in vain to win Keltie's support for the policy of exploitation in the Congo State. Keltie had a progressive outlook, encouraging developments from the great traditions of the past, resisting any reactionary tendencies which might appear but never advancing his own views or losing heart when they were temporarily thwarted.

During Keltie's youth geography was widely regarded as synonymous with exploration and had no standing in university circles. He belonged to the generation when social gatherings at luncheons or dinner-parties provided rare opportunities for the exchange of educated opinion on global topics outside the formal meetings of the Royal Geographical Society. He died a few years before the inception of the Institute of British Geographers which after a half-century of expansion now organises an impressive array of conferences, study groups, field meetings, international seminars and scholarly publications to meet the needs of its members. That these diverse forms of academic intercourse flourish today is due in no small measure to the achievements of the unassuming Scot from Perth.

Bibliography and Sources

1. REFERENCE ON JOHN SCOTT KELTIE

a. Obituaries
The Times, 13 January 1927
Mill, H.R. and Freshfield, D., *Geogr. J.*, vol 69 (1927), 281-7
Chisholm, G.G., *Scott. Geogr. Mag.*, vol 43 (1927), 102-05
anon., *Geogr. Rev.*, vol 17 (1927), 339-40
Mill, H.R., *Nature*, vol 119, 22 January 1927, 130-1

b. Other References
Brown, R.N. Rudmose, *DNB* (1922-30), 463-4
Mill, H.R., *The record of the Royal Geographical Society*, RGS (1930)
Mill, H.R., *An Autobiography*, London (1951)

2. SELECTIVE BIBLIOGRAPHY OF WORKS BY JOHN SCOTT KELTIE
1886   'Geographical Education', *R.G.S. Supp. Pap.*, vol 1 (1882-5), 439-594. This was also printed as a separate publication entitled *Report of the Proceedings of the Society in reference to the improvement of geographical education*, (1886)
1890   *Applied Geography: a preliminary sketch*, 169 p.
1893   *The Partition of Africa*, 498 p.
1913   (with O.J.R. Howarth) *History of Geography*, 154 p.
1914   'Thirty years' progress in geographical education', *Geogr. Teach.*, vol 7, 215-27
1915   'A half-century of geographical progress', *Scott. Geogr. Mag.*, vol 31, 617-36
1916   'Sir Clements Markham: biographical sketch', *Geogr. J.*, vol 47, 165-72
1917   'Thirty years' work of the Royal Geographical Society', *Geogr. J.*, vol 49, 350-76
1921   *The position of geography in British universities*, (Am. Geogr. Soc. Res. Ser. no 4), 33 p.

Keltie also edited numerous works, including
1883-1926   *The Statesman's Year Book*
1893-1917   *Geogr. J.*
1869   *The Works of Daniel Defoe*
1870   *The Works of the British dramatists*
1875   *A history of the Scottish Highlands, Highland clans and Highland regiments*
1890   *The story of Emin's rescue as told in Stanley's letters*
1903-06   *The story of exploration*

3. ARCHIVAL SOURCES
Apart from the minute books and out-letter books in the Administrative Records of the Royal Geographical Society, the archives of the Society also contain the

following material relating to Keltie in the
Correspondence Files:
27 letters written by Keltie to the Society, 1913-20
32 letters written by Keltie to Hugh Robert Mill,
1887-1926
A large collection of letters addressed to Keltie by
various people, including
75 letters from Col. and Mrs. P.H. Fawcett 1917-26;
22 from Douglas Freshfield 1885-1926; 29 from Sir
George Goldie 1893-1922, and 75 from Prince Peter
Kropotkin 1880-1917.

The author wishes to aknowledge the help received from
Mrs. Christine Kelly, Archivist of the Royal
Geographical Society, Dr. Robert Smart, Keeper of the
Muniments in the University of St Andrews and
Dr. J.T.D. Hall, Sub-Librarian of Special Collections
in the Library of the University of Edinburgh.

*L.J. Jay, formerly Senior Lecturer in Education in the
University of Sheffield, died on 24 January 1986*

Chronology

1840	Born in Dundee, 29 March
1854-58	Pupil-teacher in Perth
1859-60	Student at St Andrews University
1860-67	Attended Arts courses at Edinburgh University but did not complete the requirements for graduation
1861	Joined the editorial staff of W. and R. Chambers, Edinburgh
1865	Married Margaret Scott in Edinburgh
1871	Moved to London to work with Macmillan and Company
1873	Sub-editor of *Nature* (until 1885)
1875	Began contributing articles to *The Times*
1883	Became a Fellow of the Royal Geographical Society and Editor of the *Statesman's Year Book*
1884	Appointed Inspector of Geographical Education to the Royal Geographical Society
1885	Became Librarian of the Royal Geographical Society (March); presented report on 'Geographical Education' (May)
1892	Became Assistant Secretary to the Royal Geographical Society on the death of H.W. Bates
1893	Edited the first volume of the *Geographical Journal*
1895	Joint Secretary, with H.R. Mill, of the Sixth International Geographical Congress, held in London
1897	President of Geography Section of British Association (Toronto). Awarded honorary degree of LL.D. of St Andrews University
1898	Made a Commander of Swedish Order of North Star
1907	Awarded Order of St Olaf, Norway
1913	President of the Geographical Association; the Royal Geographical Society moved its headquarters from Savile Row to Kensington Gore
1915	Resigned as Secretary of the Royal Geographical Society: awarded gold medals of the Royal Scottish, American and Paris Geographical Societies
1917	Resigned the editorship of the *Geographical Journal*: awarded the Victoria Medal of the Royal Geographical Society
1918	Knighted for services to education and geography
1921	Awarded the Order of the White Rose, Finland
1922	Death of his wife
1927	Died in London, 12 January

Johann Gottfried Otto Krümmel

1854–1912

J. ULRICH

Oceanography, in Germany as in Britian and other countries, is now generally regarded as a subject in its own right, studied in a limited number of universities, but its modern development was part of the growth of the natural sciences in the second half of the nineteenth century in association with geography. Krümmel, as one of the pioneers of oceanography in Germany, worked in the University of Kiel until the last full year of his life as a geographer with his own specialization in oceanography. Since World War II oceanography, as an independent science in Germany, has had its main academic support at the universities of Kiel and Hamburg.

Johann Gottfried Otto Krümmel, known generally as Otto Krümmel, was one of the most prominent geographers and oceanographers at the turn of the century. His special achievement was to establish oceanography in Germany as a systematic part of geographical science. Through his valuable research work and teaching activities marine science was developed and acquired respect. During the first half of this century Otto Krümmel's main work, the *Handbuch der Ozeanographie* (1907 and 1911) was regarded as the best of the texts on oceanography at least in Germany, and Fridtjof Nansen called it 'the most important publication in geographical literature' of its period.

1. *EDUCATION, LIFE AND WORK*

Otto Krümmel was born at Exin, a small village in the former West Prussian district of Bromberg, on 8 July 1854. He was the eldest of twelve children, of whom

five died at an early age and only four outlived him. Otto's father, and also his grandfather, was a coppersmith who also had a small farm and his mother was also of farming stock. He grew up in a rural community and regarded this association with the land as a gift of providence.

Unfortunately this farm childhood was not to last for in 1863 Otto's father was obliged to sell his agricultural property and the family moved to Lissa, a small town with a long tradition of cloth making in the then Prussian province of Posen. There he attended the Comenius-Gymnasium, a famous secondary school where he was able to develop his talents to the full. His teacher, Dr Julius Töplitz, encouraged his enthusiasm not only for Latin, mathematics and physics but also for crystallography, chemistry and astronomical geographical in the upper forms of the school. Otto also found time to broaden his knowledge through private reading of botanical and geological works and the study of travel reports, including Alexander von Humboldt's *Kosmos*. He was familiar with the most important geographical journals before he went to the university.

Having successfully completed his school course at Lissa, Krümmel went to the university of Leipzig in the spring of 1873, where at first he studied medicine. He was attracted also to other fields of science, including ethnology and geography, and attended lectures by Oskar Peschel. It was Peschel who inspired him to study geography and from July 1874 he began serious study of the subject with cognate natural science subjects, especially geology and biology.

In April 1875 Krümmel moved from Leipzig to Göttingen, where he continued his study of geography under Eduard Wäppaus and Karl von Seebach and during the winter term of 1875-76 he followed lectures courses by Heinrich Kiepert and Adolf Bastian in Berlin.   Later he returned to Göttingen, where in July 1876 he took his Doctor of Philosophy degree with a thesis on *Die äquatorialen Meeresströmungen des Atlantischen Ozeans und das allgemeine System der Meereszirkulation (The equatorial currents of the Atlantic Ocean and the general system of the circulation of the Oceans)* under the supervision of Wappäus in geography and von Seebach in geology.

In April 1878, less than two years after he completed his Ph.D., he was appointed as lecturer in geography (*venia legendi*) at the university of Göttingen.   Also in 1878 he presented a *habilitation* thesis entitled *Versuch einer vergleichenden Morphologie der Meeresräume (An attempt at a comparative morphology of the oceans)* while holding the honorary position of a private lecturer.   He became known for his personal enthusiasm for oceanographical work and Georg von Neumayer, the first director of the German Naval Observatory, asked him to contribute, on an honorary basis, an oceanographical chapter to the *Segelhandbuch für den Atlantischen Ozean (Sailing handbook of the Atlantic Ocean)*.   Georg von Neumayer encouraged Krümmel's enthusiams for geography and they became close friends.

On 18 August 1883 Krümmel accepted an honorary appointment as an Associate Professor of Geography at the university of Kiel and in the following year, 1884, he became a professor of geography and also taught at the Imperial Naval Academy of Kiel.   In a letter to the hydrographer Friedrich Althoff (who was a 'Geheimer Rat', which meant a member of the secret council of the German Emperor) he expressed his enthusiasm for this new field of activity, saying that

> ... it would be difficult to take up another
> professorship with the same enthusiasm as
> that I have for the important tasks to be
> tacked in Kiel.   These duties will bring me
> so close to the sea, and my life's aim has
> been to explore the sea ...

In September 1884 Otto Krümmel married Helene Ludowieg, daughter of the mayor of Hameln.   Their only child, Heinrich, lived from 1885-1957 and is survived by his wife, Elsa Krümmel, who now resides in Münster, Westphalia.

The years at Kiel, when he was engaged in both teaching and research, were scientifically the most productive period of Krümmel's life.   He was not one of those academic lecturers able to fascinate students with powerful oratory but rather a presenter of systematically arranged material in a factual and clear manner.   He had a particular ability to discern essential scientific facts by the critical evaluation of available data and this precision of mind meant that both in his teaching and in his research he was able to present his conclusions in a clear and comprehensible way.

Although there were opportunities of sea travel for research purposes, Krümmel seldom undertook long journeys and spent most of his time in his own study. However, in 1889 he went on one of his few long cruises for research purposes and was a member of the famous German plankton expedition to the Atlantic on board the research vessel *National* under the leadership of Victor Hensen.   Later he joined several of the shorter and less ambitious expeditions on board the then *Poseidon* to the Baltic and the North Sea.

Initially oceanography at Kiel was included within Krümmel's teaching on *Allgemeine Geographie*, a general geographical course, but from the winter term of 1903-04 he gave a course of lectures on *Allgemeine Geophysik, Meteorologie und Ozeanographie*, so combining oceanography with geophysics and meteorology.   He was successful in attracting research students and of the thirty-five dissertations completed under his supervision sixteen dealt with oceanography.

Within the general field of oceanography the research contribution of Krümmel was of wide range. He made numerous contributions on the morphology of the sea floor and on the bathymetric conditions of the oceans.   Following the work presented in his research thesis on Atlantic ocean currents he welcomed the new data which appeared through various explorations and the new theories which such data made possible.   His career as a writer began with the publication of papers in various journals, including the famous *Petermanns Geographische Mitteilungen*.   He had become known to Professor Georg von Boguslawski who was then editor of the *Annalen der Hydrographie und maritimen Meteorologie* and had accepted papers by Krümmel.   Professor von Boguslawski, who was head of a special section of the Hydrographic Office of the German Imperial Academy in Berlin, had planned a work on oceanography in two volumes, of which he was only able to complete the first owing to a long period of illness leading to his death on 4 May 1884.   These volumes were included in a series of texts edited by Friedrich Ratzel (1844-1904), who asked Krümmel to continue the work on oceanography. Krümmel thought that he was not adequately qualified for so great a responsibility and suggested that it should be done by Professor Karl Zöppritz, but he died on 21 March 1885 and Ratzel turned to Krümmel once more and this time he accepted the offer, though still unsure of his adequacy for he was full of admiration for the vast knowledge of von Boguslawski.   Nevertheless he completed the work on the second volume originally planned by von Boguslawski by 1887 and both volumes were republished in 1907 and 1911.   By then a vast store of additional data on oceanography had been accumulated and the two volumes covered some 1,300 pages: the 1907 volume was a revision and development of von Boguslawski's volume 1 of the *Handbuch der Ozeanographie* and the second volume, in fact larger, of 1911 was a revision of Krümmel's own earlier work.   In these volumes the results of all the various deep-sea expeditions were summarized as never before.   Krümmel's smaller and more popular work, *Der Ozean*, first published in 1886, went into a second edition in 1902.

Soon after his appointment as a full professor at Kiel in 1884 Krümmel joined the *Kommission zur wissenschaftlichen Untersuchung der deutschen Meere*, the Commission for the scientific investigation of the German seas, which had been founded in 1870.

Increasingly it was recognized that oceanographical work should have an international organization and in 1899 meetings were held in Stockholm which led to the foundation of the International Council for the Exploration of the Seas (ICES), of which Krümmel was a member of the central committee.

As a writer Krümmel was highly productive and a complete list of his publications would include more than 100 titles: of these the more important items are listed in the bibliography. Initially he dealt with various branches and aspects of oceanography in scientific papers and these were the foundation of his main work, the *Handbuch der Ozeanographie*, which is the culmination of his life vocation for oceanography. Shortly after the publication of this book he received an invitation to be a professor at the university of Marburg and on 1 April 1911 he left Kiel to become the successor of Theobald Fischer at Marburg. Unhappily illness was already causing the loss of his creative powers and during the following year, on 12 October 1912, he died during a visit to Cologne at the age of fifty-eight.

2. SCIENTIFIC IDEAS AND GEOGRAPHICAL THOUGHT

Krümmel's significance for geographical science was well appreciated by Max Eckert, who in 1913 characterized him as 'one of our late geographers who still belong to the revival period of scientific geography which began in the second half of the last century.' Naturally it was as an oceanographer that Krümmel became best known, especially beyond Germany, and his view that oceanography was worthy of study because the seas cover two-thirds of the surface of the earth can hardly be disputed. He was fortunate in having a constant supply of new data readily available, including the observations made on the two research ships, the *National* and the *Poseidon*, on which he and even more his students made regular trips for study. Nevertheless like other geographers of his time, as well as those of a later time, as a university teacher he had to cover other aspects of geography, including the physical aspects of the subject, cartography, mathematical geography, ethnology and anthropogeography, on which he produced a map of the density of population in Europe, which was published in the *Brockhaus Encyclopaedia*, Leipzig, 1885.

As a student and admirer of Oskar Peschel, Krümmel had a sound knowledge of classic work in geography and was able to read the writings of Strabo and others in the original Greek. Of more recent authors he venerated Carl Ritter and Alexander von Humboldt and in 1904 he published his three-volume *Klassiker der Geographie* with the aim of introducing young scientists to the classical geographical literature. His first essay on regional economic geography was on the distribution of chernozem soils, 'Das Tschernosjom und seine Verbreitung' (*Dtsch. Geogr. Blatt*, Bremen, 1877). Three years later he published *Europäische Staatenkunde* (1880), a detailed study of the economic geography of Russia, Scandinavia and Great Britain but this work was adversely critized, unjustifiably it would now seem. Krümmel had hoped to develop some of the work of Peschel on various

aspects of geography but found his own main research interest in oceanography, though he published a few papers of economic interest, of which the last, a study of wind and water for power supply, was 'Die geographische Verbreitung der Wind--and Wassermotoren im Deutschen Reich' (*Petermanns Geogr. Mitt.*, 1903).

In his time Krümmel was known as a good general geographer and he received the recognition of honorary membership from various geographical societies. Clearly his main contribution of international significance was his work on oceanography. He became a member of the central committee of the International Council for the Exploration of the Sea and the GEBCO committee, which was responsible for the *General Bathymetric Chart of the Oceans*. This body was founded in 1903 by Prince Albert I of Monaco and among its members were some famous geographers, but oceanographers were few in number and therefore Krümmel's membership was highly valued. The work continues and the famous GEBCO international series of charts is now in its fifth edition. Krümmel's work for international bodies at times restricted the time available for research and writing though it also gave him the opportunity of two famous Norwegians, the oceanographer Helland-Hansen and the polar explorer Fridtjof Nansen, with whom Krümmel corresponded for a long time though they met only occasionally.

Krümmel wrote regular reports from 1895 to 1905 for the German journal, the *Geographisches Jahrbuch*, on the progress of oceanography. However he was more than a theoretical scientist for he occupied himself with technical problems of oceanographical research equipment, as shown in his 1890 paper 'Über die Bestimmung des spezifischen Gewichts des Seewassers an Bord' ('On the determination of the specific gravity of sea water on board') and in communications to the development of a workable areometer. His practical approach was also seen in the possible ways of discerning and measuring ocean currents, on which in 1908 he gave a lecture on 'Flaschenposten, treibende Wracks und andere Triftkörper in ihrer Bedeutung für die Enthüllung der Meeresströmungen' ('Bottles, wrecks and other drifting bodies and their importance for the investigation of ocean currents').

In all he was the author of some 150 reviews of works on oceanography and his critical and profound opinion on these publications was greatly respected. Undoubtedly his *Handbuch der Ozeanographie* was his triumph, for it gave him unlimited respect in Germany and within a short time other countries also. On 14 May 1907, after the publication of the first volume, Fridtjof Nansen wrote to Krümmel in these terms

... I send you the best and warmest congratulations on the completion of this volume of this work which will be of such great importance for the future of oceanography. It was a pleasure to read it and it is a work which has been wanted and needed and surely nobody could write it in a better way ...

When the second volume appeared in 1911 Hellend-Hansen wrote to Krümmel in similar terms. The earlier work of 1902, *Der Ozean*, intended as its subtitle showed to be 'an introduction to general marine science', was also warmly welcomed, not least because it showed the author's ability to simplify scientific language. In this work, as in other contributions made at various times, Krümmel was writing for the general educated public eager to know more of many aspects of science.

3.   *INFLUENCE AND SPREAD OF IDEAS*
The influence of Otto Krümmel's work remains to this day. There is a clear line leading from Krümmel's major works, mentioned above, to the well known textbook *Allgemeine Meereskinde* by the prominent late German oceanographer Gunter Dietrich, who taught oceanography in the university of Kiel from 1959 to 1972. This, the main work of Dietrich, of which the first edition was published in 1957 and the third in 1975, is the leading textbook in oceanography today. Indeed the work of Dietrich shows many qualities which could be described as 'Krümmel-like', including a critical approach to all scientific material, expressed with clarity of opinion, clearness of diction, fluidity of style, and brilliant handling of synthesis, all of which are admirable qualities not universally found among contemporary scientific writers.

Krümmel's morphological ideas on the relief of the sea bed and his first fundamental attempts to divide oceans formed a base for all the succeeding generations of marine scientists. His handling of the voluminous data into categories and divisions to give his work a constructive clarity has set a fine example for later workers. That he had aspirations for other lines of research is shown by the hope that in Marburg he would be able to write a much needed comprehensive geographical work about Germany. Death at a comparative early age removed this possibility but though much more could be said about the man and his work, the geographer Max Eckert described him with great sincerity as one with 'an ardent longing for knowledge which animated him: that is why he did not shrink from making great demands on himself and never spared himself. His whole nature was marked by great modesty'.

Bibliography and Sources

1.   *REFERENCES ON OTTO KRÜMMEL*
Brandt, K., 'Die beiden Meereslaboratorien in Kiel' ('The two laboratories for marine research at Kiel'), *Cons. Perm. Int. Explor. Mer, Rapport Jubilaire (1902-1977)*, Copenhagen, vol 47 (1928), 3-16

Eckert, M., 'Otto Krümmel', *Geogr. Z.*, vol 19 (1913), 545-54

Matthaus, W., 'Der Ozeanograph Johann Gottfried Otto Krümmel (1854-1912)', *Wiss. Z. Univ. Bostock*, vol 16/9-10 (1967), 1219-24

---- 'Die Berufung des Ozeanographen Otto Krümmel zum Ordinarius fur Geographie an der Universitat Kiel' ('The appointment of Otto Krümmel the oceanographer as a full professor at Kiel university'), *Mon.-Ber. Dtsch. Akad. Wiss. Berlin*, vol 9/6-7 (1967), 535-8

Meinardus, W., 'O. Krümmels "Handbuch der Ozeanographie"', *Geogr. Z.*, vol 18 (1912), 29-47, 98-111

---- 'Otto Krümmel+', *Petermanns Geogr. Mitt.*, vol 58 (1912), 281

Paffen, K.-H. and Kortum, G., 'Die Geographie der Meeres. Disziplingeschtliche Entwicklung seit 1650 und heutiger methodischer Stand' ('The geography of the sea. Historical development of the field since 1650 and contemporary methods of study'), *Kieler Geogr. Schr.*, vol 60 (1984), 293 p.

Ulrich, J., 'Otto Krümmel 1854-1912', *Mitt. Dtsch. Gesell. Meeresforsch.*, Hamburg, vol 3 (1984), 22-5

2.   *SELECTIVE AND THEMATIC BIBLIOGRAPHY OF WORKS BY OTTO KRÜMMEL*

a.   *General Oceanography and Regional Works*
1886   *Der Ozean. Eine Einführung in die allgemeine Meereskinde (The Ocean. A general introduction to oceanography)*, Das Wissen der Gegenwart 52, Leipzig and Prague, 242 p.

1887   'Die Fortschritte der Ozeanographie 1885 and 1886' ('Progress in oceanography ...'), *Geogr. Jahrb. Verl. Geogr.-Kartog. Amstalt Gotha.*, vol 11, 75-95

----   (with G. v. Boguslawski) *Handbuch der Ozeanographie*, vol 2, *Die Bewegungsformen des Meeres*, Stuttgart, 528 p.

1890   'Die Verteilung des Salzgehaltes an der Oberfläche des Nordatlantischen Ozeans' ('The distribution of salinity at the surface of the North Atlantic Ocean'), *Petermanns Geogr. Mitt.*, vol 36, 129-41

1891   'Die nordatlantische Sargassosee', *Petermanns Geogr. Mitt.*, vol 37, 129-41

1895   'Zur Physik der Ostsee' ('On the physics of the Baltic Sea'), *Petermanns Geogr. Mitt.*, vol 41, 81-6, 111-18

1902   *Der Ozean. Eine Einführung in die allgemeine Meereskinde (The Ocean: an introduction to general oceanography)*, 2 ed, Vienna and Leipzig, 285 p.

1904   'Die deutschen Meere im Pahmen der internationalen Meeresforschung' ('The German seas in the context of international marine research'), *Veröff. Inst. Meereskunde Univ. Berlin*, no 6, 36 p.

1907   *Handbuch der Ozeanographie*, vol 1, *Die räumlich, chemischen and physikalischen Verhältnisse des Meeres (Handbook of oceanography: the spatial, chemical and physical relations of the oceans)*, Stuttgart, 766 p.

b.   *Studies of Ocean Currents and Waves*
1876   *Die äquatorialen Meeresströmungen des Atlantischen Ozeans und das allgemeine System der Meereszirkulation (The equatorial ocean currents of the Atlantic Ocean and the general system of the sea)*, diss., Göttingen

1889 'Über Erosion durch Gezeitenströme' (On erosion by tidal currents'), *Petermanns Geogr. Mitt.*, vol 35, 129-38

1896 'Oberflächentemperaturen und Strömungsverhältnisse des äquatorialen Gürtels des Stillen Ozeans' ('Surface temperatures and ocean currents in the equatorial belt of the Pacific Ocean'), *Petermanns Geogr. Mitt.*, vol 42, 135-9

c. *Publications on the Relief of the Sea Floor and the Depths of the Ocean*

1878 'Die mittlere Tiefe der Ozeane und das Massenverhältnis von Land und Meer' ('The mean depth of the oceans and the general relations between land and sea'), *Verh. Gesell. Erdkd. Berlin*, vol 5, 258-63

1879 *Versuch einer vergleichenden Morpholie der Meeresräume (An attempt at a morphological comparison of oceanic space)*, Leipzig, 110 p.

1880 'Die mittlere Tiefe der Ozeane' ('The mean depths of the oceans'), *Z. Wiss. Geogr.*, vol 1, 40-6

1883 'Die Tiefseelotungen des Siemens'schen Dampiers "Faraday" im Nordatlantischen Ozean' ('The deep-sea soundings of the Siemens' steamer "Faraday" in the north Atlantic Ocean'), *Ann. Hydrog. Marit. Meteorol.*, vol 11, 5-8, 146-8

1899 'Die tiefste Depression des Meeresbodens' ('The deepest depressions of the sea floor'), *Geogr. Z.*, vol 5, 509-12

1901 'Die Einfuhrung einer einheitlichen Nomenklatur für das Bodenrelief der Ozeane' ('The introduction of a uniform nomenclature for the relief of the sea bottom'), *Verh. 7 Int. Geogr.-Kongr.*, *Berlin 1899*, vol 2, 379-86

d. *Methodological Contributions and Technical Reports*

1890 'Über die Bestimmung des spezifischen Gewichts des Seewassers an Bord' ('On the determination of the gravity of sea water on board'), *Ann. Hydrog. Marit. Meteorol*, vol 18, 381-95

1894 'Über einige neuere Beobachtungen an Aräometern' ('On some recent observations of areometers'), *Ann. Hydrog. Marit. Meteorol*, vol 22, 415-27

1896 'Professor Dr. Friedrich Dahls Aräometerbeobachtungen auf der Fahrt von Neapel nach Matupi' ('... observations with areometers of a voyage from Naples to Matupi'), *Ann. Hydrog. Marit. Meteorol.*, vol 24, 529-46

e. *Results of Expeditions*

1889 'Die Plankton-Expedition im Sommer 1889' ('The plankton expedition during the summer of 1889'), *Verh. Gesell. Erdkd. Berlin*, vol 16, 502-14

1892 *Reisebeschreibung der Plankton-Expedition (Cruise report of the Plankton expedition)*, vol 1A of *Ergebnisse der Plankton-Expedition der Humboldt-Stiftung (Results of the Plankton expedition of the Humboldt Foundation)*, Kiel and Leipzig, 370 p.

1893 *Geophysikalische Beobachtungen*, ibid., vol 1C, Keil and Leipzig, 118 p.

1902 'Über die ozeanographischen Ergebnisse der deutschen Südpolar-Expedition von Kiel bis Kapstadt' ('On the oceanographical results of the German Southpolar expedition from Kiel to Kapstadt'), *Ann. Hydrog. Marit. Meteorol.*, vol 30, 390-5

1904 'Die Fahrt der deutschen Südpolar-Expedition von Kerguelen in das südliche Eismeer und züruck nach Kapstad' ('The cruise of the German South Polar expedition from the Kerguelen to the southern ice-sea and back to Kapstad'), *Ann. Hydrog. Marit. Meteorol.*, vol 32, 11-20

f. *Other Publications in Geography and Oceanography*

1880 *Europäische Staatenkunde*, Leipzig

1888 'Die Temperaturverteilung in den Ozeanen' ('The distribution of temperature in the oceans'), *Z. Wiss. Geogr.*, vol 6, 31-41

1889 'Bemerkungen über die Durchsichtigkeit der Meerwassers' ('Remarks on the transparency of sea water'), *Ann. Hydrog. Marit. Meteorol.*, vol 17, 62-78

1903 'Die geographische Verbeitung der Wind--und Wassermotoren im Deutschen Reich', *Petermanns Geogr. Mitt.*, vol 49, 169-73

1904 *Klassiker der Geographie*, Kiel

3. ARCHIVAL SOURCES

Mrs Elsa Krümmel, of Münster, daughter-in-law of J.G.O. Krümmel, has kindly lent the author a collection of letters from Fridtjof Nansen.

Johannes F. Ulrich is Scientific Director and Custodian of the Institut für Meeresdunde at the University of Kiel

Chronology

1854	Born at Exin, 8 July
1863	His family moved to Lissa, where he attended the Comenius-Gymnasium
1873	Went to Leipzig university where he attended lectures by Oskar Peschel
1875	Moved to Göttingen university
1876	Successfully presented his doctorate thesis on oceanography
1878	Became a lecturer in geography at Göttingen university
1883	Appointed as an Associate Professor of Geography at Kiel university
1884	Became a professor of geography at Kiel university and began to teach at the Imperial Naval Academy: also in this year he became a member of the *Kommission*

zur wissenschaftlichen Untersuchung der Deutschen Meere: in September married Helene Ludoweig

1886   Published *Der Ozean* (2 ed 1902)

1889   Was a member of a marine expedition for the study of plankton

1899   Attended meetings in Stockholm which led to the foundation of the International Council for the Exploration of the Seas (ICES), of which he was a member of the central committee

1903-04   Began to lecture on the relation of oceanography to geophysics and meteorology

1907   Published *Handbuch der Ozeanographie*, a revision and development of the work of Professor G. von Boguslawski

1911   Added a second book on oceanography based on his own work: on 1 April left Kiel to be a professor at Marburg university

1912   Died at Cologne, 12 October

Marguerite Alice Lefèvre

1894–1967

J. DENIS

Belgium can claim a long tradition in studies which in various ways were the first fruits of geography. In cartography, for example, the work of Mercator and Ortelius gained world fame and the Ferraris map was the first to cover an entire country on a large scale. And in the middle of the nineteenth century the 1:20 000 map of Vandermaelen was the delight of eager amateurs. Both physical geography and geology have had scholars of international standing, among them many geologists who were stratigraphers who for particular periods have left names of Belgian origin, including Gedinnien, Couvinien, Frasnien, Famennien, Tourenaisien, Viséen, Dinantien and Namurien. Also many Belgians have explored Asia, Africa and the Americas and it was at Antwerp that the first International Geographical Congress was held in 1871.

Nevertheless a long and patient wait was needed before geography received full recognition in the Belgian universities, though some courses were provided for particular groups of students including engineering geologists, economists and historians. Only after the law of 21 May 1929 was passed could a full programme of courses be provided for the degrees of *licencié*, *agrégé* and *docteur* in geographical science. In 1929 Marguerite Lefèvre returned to Louvain from her long sojourn in Paris and became the assistant to Paul Michotte, mainly concerned with undergraduate teaching but with the clear ambition of bringing forward a new generation of scientifically minded geographers.

1. EDUCATION, LIFE AND WORK

When she returned from Paris in 1929 Marguerite Lefèvre was already a woman of considerable maturity as a geographer, for she was thirty-five years of age, having been born in Steenokkerzeel, Brabant, on 1 March 1894. To this stage her life had shown many of the paradoxes that mark human experience, notably that in her early years there was no apparent sign of the distinguished career in geography that was to follow. At the secondary school she attended geography induced a state of boredom, well known to writers such as G.S. Grosvenor who spoke of it as 'one of the dullest of subjects, something to inflict upon schoolboys and to avoid in later life': though only boys are mentioned here the growing demand for equality in the education of girls presumably meant that they too would be expected to accept the unacceptable in secondary schooling. Happily Marguerite had younger brothers, who were pupils at the Collège St Pierre in Louvain and told her of the teaching given by a master named Michotte, who showed that geography was something quite different from the normal encyclopaedic jumble of information. Her curiosity was aroused and, wanting to know more, she showed a combination of obstinacy and imagination that was to serve her well later on.

Louvain university, the oldest and most prestigious in Belgium, was traditionally minded, providing advanced education with dedicated purpose but only for male students. Marguerite Lefèvre had to wait until the end of the war in 1919 for its doors to be opened to women and even then there was no authorized course in geography. Anyone who wished to study the course was enrolled under history, economics or geology, so she entered as an *élève libre* (independent student) and with characteristic determination of mind followed not only all the courses given by Michotte but also others given in cognate subjects which might be useful in her general education. In addition she travelled regularly to Liège, to follow the courses in physical geography given by Paul Fourmarier. Resembling a hard working and diligent bee, she gathered honey from the finest flowers in the scientific garden.

Not surprisingly Paul Michotte watched with interest the progress of such an exceptional student. Well aware of gaps in his own education, for he was largely self-taught, he encouraged Marguerite Lefèvre to study under geographers of other countries, and so she went to Paris to broaden her knowledge of physical geography with Emmanuel de Martonne and of human geography with Albert Demangeon, under whose direction she wrote the thesis on the rural habitat in Belgium for which she was awarded the degree of Docteur de l'Université de Paris in 1926. Six years later, in 1932, she spent a semester at the Columbia University in New York, working with Douglas Wilson Johnson, a disciple of William Morris Davis.

From 1930, having acquired a grant as an associate from the Fonds National de la Recherche Scientifique Marguerite Lefèvre, in effect this time an assistant to Michotte, was able to devote much of her energy to her own research and publication. Her bibliography shows that forty of her contributions date from 1926 to 1938, almost as many as for the rest of her life. These years of intensive scientific production were only a small compensation for the fact that as a woman she had no hope of following a normal academic career as a university teacher, for in Belgium male chauvinism still survived in the academic world.

Or so it seemed at the time, but circumstances were to change. In 1938 Paul Michotte was incapacitated by illness and it was at once apparent that nobody was better suited to continue his work than Marguerite Lefèvre who for so many years had been his student and collaborator. The university authorities wisely decided that she should be given the responsibilities of Michotte on a temporary basis. In 1940, on the death of Michotte, this became a permanent appointment though she was not given full professional status but restricted to the position of 'chef de travaux', which in a British university would be a 'lecturer-in-charge'. However she became a 'full professor' in 1960, only a short time before her retirement and entry into the select company of emeritus professors in 1964. At last she was received by her peers and given a status entirely merited but long deferred, though three years later, on 27 December 1967, she died at Louvain after a short illness.

2. *SCIENTIFIC IDEAS AND GEOGRAPHICAL THOUGHT*

Study of the scientific work of Marguerite Lefèvre reveals the diversity of themes followed. It might seem at first to suggest a disparate approach to geography and even a disparate mind but this was not

so, for like other geographers of her time she was devoted to the idea of the unity of her subject and favoured the view expressed in the often quoted words of Vidal de la Blache on the indivisibility of nature, *à ne pas morceler ce que la nature rassemble*. She held the view that geographical unity was itself a challenge for a scholar to master in all its aspects and that such an aspiration could be achieved. Therefore she threw the resources of an ardent believer into both human and physical geography and cared as much for regional geography as for cartography and methodology. Should anyone fear that such a dispersion of interest might prejudice the quality of her work to her it was a firm principle that her researches must be complete and this is confirmed by study of her major works.

Four of these, published over a period of almost thirty years, may be cited in illustration. The first was *L'habitat rural en Belgique* (1926), a synthesis of her thesis written under the care of Demangeon: nearly ten years later *La Basse-Meuse, étude de morphologie fluviale* (1935), was published. Then followed, in 1946 shortly after the end of World War II, her work on *Principes et problèmes de géographie humaine* and finally in 1955, after a study period in Africa, a regional monograph entitled *La vie dans la brousse de Haut-Katanga*.

Marguerite Lefèvre's cartographical work is best seen in her substantial contributions to the *National Atlas of Belgium*, compiled from 1937 to 1964 and found to be a long and exacting task by the Belgian National Committee for Geography. Her work for the atlas was of wide range as she provided maps on hydrography, morphology, lithology, population movements and types of rural settlement. Each plate was accompanied by an explanatory text and in all she was responsible for 142 pages.

Nothing illustrates better her intellectual curiosity and relentless work than the bibliography published in the *Acta Geographica Levaniensia* (vol 3 (1964), 17-21), in her honour on retirement. In all there were 82 titles and six more were to be added in her remaining three years. Eager to follow the progress of geography she read voraciously, always with pen in hand to make notes. She also wrote forty-nine reviews, not all of them gentle in tone, for the *Bulletin de la Société belge d'études géographiques*.

Constantly Marguerite Lefèvre advocated walking as a means of acquiring geographical knowledge was in accord with what many of her contemporaries in other countries were saying. The students in Louvain found 'La géographie est une science qui entre par les pieds' to be almost her *leitmotif* so she led them forth into the living landscape and to relate theoretical concepts with the living scene. This she did not only in Belgium, of which she appeared to know every nook and cranny, but also in much of Europe. She was herself an eager traveller who made several visits to Africa and America: in 1956 she was visiting professor at the Université Lovanium, Léopoldville, Belgian Congo.

There can be few people who shared her record of attending every congress of the International Geographical Union from Cambridge in 1928 to London in 1964. She used the opportunity of seeing the countries in which the congresses were held and travelled on various tours arranged for members, as well as independently.   After some of the congresses she published papers on areas visited, for example on Madeira after the Lisbon congress of 1948, on the pediments of the Mojave desert in California after the congress in Washington D.C. of 1952, as well as notes on the Amazon, the Serra de Mar and the Itatiaia massifs after the Brazilian congress at Rio de Janeiro in 1956.

The International Geographical Union had no more devoted supporter than Marguerite Lefèvre and she was a member of several commissions, of which that on the rural habitat, founded in 1925, was clearly related to her post-graduate researches: for this commission she produced several reports of a synthetic character and acted as secretary from 1928.   She was equally active in the Commission on surfaces of erosion, formed in 1931, though her main work for this body appears to date from the 1938 International Geographical Congress at Amsterdam at which the geomorphological meetings, like those on colonial geography, were strongly supported.   (Incidentally she does not appear to have been attracted to political geography at that time.)   Later, in 1948, she turned to the Commission on Pliocene and Pleistocene terraces, which had existed from 1925 and in 1952 was united with the Commission on surfaces of erosion mentioned above. Of this new commission, on 'erosion surfaces around the Atlantic' she was one of the five full members from 1960.   She was aided in this international work by her linguistic ability, having French as her mother tongue and also fluency in Dutch, English and German.

Her devotion as a research worker in her own country and beyond it never lessened her concern for the Institute of Geography to which she had been drawn by Michotte.   For more than twenty-five years she carried a heavy load of teaching and made a strong impression on successive generations of students.   As head ('Directrice') of the institute she cared for every aspect of its welfare and this involved a constant stream of administrative and financial problems and duties.   Through all this she retained a passionate desire to educate young people and she communicated her enthusiasm with both scientific integrity and profound quality of character.   When in 1964 tribute was paid to her work one former student said that 'if it be true that for a teacher the greatest of all rewards is always to induce students to share the enthusiasm that has made possible the work of a lifetime, I am sure I can affirm, Mademoiselle, that your efforts have paid handsomely'.   (*Bull. Soc. belge Et. Geogr.*, vol 34 (1965), 219).

Marguerite Lefèvre's integrity of character and flair for organization brought responsibilities beyond the university of Louvain both in Belgium and internationally.   From 1931 to 1958 she was secretary of the Société belge d'Etudes géographiques, secretary of the society's editorial board from 1931 to 1962, member of the Belgian national committee for geography in 1945 and its vice-president in 1960, and first vice-president of the International Geographical Union from 1949 to 1952.   She accepted these responsibilities not through any unhealthy desire for status but as an opportunity of service to geography as the inspiration of her life.

The dominant idea of Mlle Lefèvre was that if the wish is to make geography a science it must advance firmly from the description of landscapes to explanation.   In this she was a convinced disciple of Michotte who wrote that

> The exploration of Belgium is now complete; its topography and hydrography, its vegetational aspect, its land use by man are all recorded on our military topographical maps with a clarity and distinction of expression that is the envy of many other countries: no, there is nothing more to discover but there is still the need to understand it all better.   (*Bull. Soc. belge Et Géogr.*, vol 1 (1931), 19).

Geography could only find its autonomy and direction as a discipline by seeking explanation and this view Marguerite Lefèvre explained clearly in the work on *Principes et problèmes de géographie humaine* of 1946 'written for students'

> Geography can only claim to be scientific if a completed description is followed by a rational explanation of the facts which have been observed and described.   Ritter and von Humboldt laid down as the scientific principle of geography that 'every fact studied should be related to a larger whole within which the parts are enlightened by the whole and within which also knowledge of all the parts is indispensable to knowledge of the whole'. This, the principle of synthesis, rests on the principle of analysis.
>
> Therefore, to understand any geographical fact, such as landforms, water circulation and associated vegetation with their essential scientific exactness it is not enough to indicate the altitude of the terrain, to describe the soil movement, to follow the course of a river, to make an inventory of the vegetation species found in the area studied. Beyond all this is the need to go further and seek the causes of the phenomena observed, to relate them to a general classification, to elucidate the conditions of their spatial localisation, to recognise their affinity of origin and evolution with that of comparable phenomena existing in other places (p 11).

Later in the same work Mlle Lefèvre wrote that

> Modern scientific geography presents itself as a rational study of the various countries of the earth.   It is an explanatory or reasoned description of the terrestrial forms in their regional diversity, modified by the complex forces which have combined to give variety to the earth's surface (*ibid.*, 16).

Mlle Lefèvre adhered firmly to this point of view while approaching with equal ease problems in human, physical or regional geography.   However in all her

publications she limited her view to what was actually
visible on the surface of the earth for which she
sought an explanation untinged by determinism, mindful
of the wholeness of a complex visible scene which had
the rich quality of marquetry in its unending colour
as an indication of the endless variety of the earth's
surface.   To her regional geography was the crown of
scientific geography and throughout her long career
she devoted her mind to this delicate and difficult
task with judgement and mastery.   The challenge lay in
determining the essential regional reality from the
apparent tangle of facts seen in a landscape, and
therefore to be considered and related--the integration
of individual phenomena with each other in an
environment of immense complexity as well as with other
phenomena, for all these relationships had to be
discerned and understood as part of the landscape whole,
itself an actual and visible reality.

Paul Michotte, commenting on Marguerite Lefèvre's
study of the regions of Belgium, said that it had

> the merit that it recognised at one and the
> same time the exigence reached by exact and
> detailed regional study and the combination
> of all these into a whole picture in which
> both the essential phenomena and the immense
> variations recognisable in a landscape, all
> dependent one upon another, could be
> explained.

This testimony shows the complete identity of outlook
shared by Michotte and Lefèvre, between master and
disciple.   It also reveals an intellectual integration
not unusual in the early part of this century,
especially in continental European universities.

3.   INFLUENCE AND SPREAD OF IDEAS

It is never easy to make a satisfying judgement on a
whole career and it is even more difficult to do this
with a few felicitous phrases for a personality of some
complexity.   Clearly Marguerite Lefèvre was a woman
of character, with decided opinions, indeed to such a
degree that she appeared to be intransigent.
Undoubtedly she maintained a deep loyalty, one might
almost say a filial devotion, to those she regarded as
masters of her subject, particularly Paul Michotte but
also Albert Demangeon, Emmanuel de Martonne, Henri
Baulig and Douglas Wilson Johnson.   Possibly this
devotion was excessive and even literal, for geography
was evolving and the ideas of her mentors were
developing in various ways: to some people it seemed
that her main conception of geography, derived from
Michotte, was marked by a certain rigidity of mind.
One colleague, Pierre Gourou, spoke of her outlook as
'somewhat Puritan' though he softened the comment on
her apparent restriction of outlook by adding that it
was 'certainly preferable to unlimited geographical
imperialism'.   Presumably in his mind there was a
place, perhaps deserving of special respect, for a
geographer who had found a satisfying way of working
and was willing to follow it rather than to seek to
adapt herself to every new academic fashion as it arose.
Whatever the judgement, Marguerite Lefèvre gained
the respect of her peers and an international

scientific reputation, based on her assiduous
research and her numerous publications.   George
Chabot, as President of the French National Committee
for geography, expressed this admirably.   In a
letter to Belgian colleagues on her death he wrote
that

> For a long time we have admired the geographical
> work of Mademoiselle Lefèvre.   She has made her
> mark with brilliance in all aspects of geography;
> morphology, human geography and cartography.
> And her approach has shown views based on fine
> critical judgement ...

On successive generations of students taught during
almost thirty years Marguerite Lefèvre left an indelible
impression.   One of them, J. Goffaux, expressed this
well at a gathering to mark her retirement.

> What impressed us as your students and what for
> us is a vivid memory, seen day by day, was your
> fine human qualities ... and with your rich
> personality, which meant so much to us during
> those years, may I also associate your
> scrupulous professional conscience, concerned
> only to seek the truth.   Devising and
> remodelling your courses constantly to seek
> the reality, firstly to share it with us, you
> were less concerned with giving us merely
> factual knowledge than with developing in us a
> scientific turn of mind, looking for the truth
> in each idea and so making us as people
> academically worthy of university education.

But that was not all for beyond scientific research
and the education of young geographers Marguerite
Lefèvre was eager to share her ideas and spread her
enthusiasm for geography to a wider public and she
expressed her missionary zeal in her *Principes et
Problèmes de géographie humaine* of 1946.   In this
book, she notes (p 2)

> I have also had in mind the cultivated people
> of our country.   Neither our secondary
> education, even if it lasts for as long as
> twelve years, nor our university education
> provides an understanding of the new
> orientation and scientific outlook of modern
> geography.   Unquestionably in several fields
> of study, history, economics, politics,
> geography cannot be neglected as a cognate
> science.   Historians, economists and
> politicians themselves prove this by their
> frequent reference to geographical
> influences on the major facts of history,
> economics and politics.   But their
> interpretation of geographical influence on
> many occasions is--or should be--denounced as
> inaccurate by geographers.
>
> In these days of upheaval of populations
> and of profound economic transformation of
> countries, explanation may be sought in
> geography in general and human geography in
> particular.   Geography can provide primary
> guidance on actuality and it is important

that there should be a thorough and reliable understanding of the method and principles of problems in both pure and applied geography and that people dominated by equivocal and misunderstanding attitudes towards their subject should definitely fade away.

Strong stuff, quite in character, written at a time when future academic developments seemed as uncertain as every other prospect for the future seemed questionable. But could it be said that Marguerite Lefèvre was better suited to consolidation and innovation?

Bibliography and Sources

1. OBITUARIES AND REFERENCES ON MARGUERITE ALICE LEFÈVRE

Polspoel, L.G., 'La carrière et l'activité scientifique de Mademoiselle M.A. Lefèvre, Professeur a l'Université de Louvain', *Acta Geogr. Lovaniensia*, vol 3 (1964), 5-21, with bibliography

---- 'Homage a Mademoiselle M.A. Lefèvre', *La Géogr.*, vol 16 (1964), 272-6

---- 'L'hommage au Professeur M.A. Lefèvre. Petit histoire d'un grand jubilé', *Bull. Soc. belge Et. Géogr.*, vol 34 (1965), 217-33

---- 'In Memoriam Mademoiselle Lefèvre', *Bull. Soc. belge Et. Géogr.*, vol 37 (1968), 27-31

Chabot, G., 'In Memoriam Marguerite Alice Lefèvre', *Ann. Géogr.*, vol 78 (1969), 472-3

Denis, J., 'Les maitres de géographie moderne'. Marguerite A. Lefèvre', *Florilège des Sciences en Belgique*, Bruxelles (1980), 528-32

---- 'Marguerite A. Lefèvre', *Biographie Nationale*, published by the Académie Royale de Belgique, Bruxelles, vol 42/2 (1982), 487-92

2. SELECTIVE BIBLIOGRAPHY OF WORKS BY MARGUERITE ALICE LEFÈVRE

a. Physical, Human and Regional Geography

1921 'Carte régionale du peuplement de la Belgique', *La Géogr.*, vol 36, 1-34

1923 'La densité des maisons rurales en Belgique', *Ann. Géogr.*, vol 32, 395-417

1926 *L'habitat rural en Belgique. Etude de géographie humaine*, Liège, Vaillant-Carmanne, 306 p.

1928 'Le cône alluvial de la Meuse', *Ann. Soc. Sci.*, vol 48/2, 121-38

---- *Commentaires de huit cartes échantillons-types des régions géographiques de Belgique*, Bruxelles, Office de Publicité, 40 p.

---- 'Les sites d'habitat', *Rev. Quest. Sci.*, 4e ser., vol 14, 39-63

---- 'Habitat rural et habitat urbain', *Bull. Soc. R. belge Géogr.*, vol 52, 113-21

1930 'La Poznanie rurale', *Bull. Soc. R. Géogr. Anvers*, vol 50, 1-29

1931 'Morphologie éolienne littorale entre Nieuport et la frontière française', *Bull. Soc. belge Et. Géogr.*, vol 1, 36-60

1932 'La Plaine Flamand', in *Mélanges offerts à R. Blanchard*, Grenoble, 337-51

1933 'La géographie des formes de l'habitat', *Bull. Soc. belge Et. Géogr.*, vol 3, 186-211

1934 'Compte rendu de l'excursions dans la vallee de la Basse-Meuse', *Bull. Soc. belge Géol....*, vol 44, 424-44

1935 *La Basse-Meuse. Etude de morphologie fluviale*, Louvain, Soc. belge Et. Géogr., mémoire no 1, 194 p.

1936 'Eustatisme et morphologie fluviale', *Bull. Soc. belge Et. Géogr.*, vol 6, 194-218

1937 'Le profil d'équilibre des rivières et son évolution', *Bull. Soc. belge Et. Géogr.*, vol 7, 55-65

1938 (in part) *Encyclopédie de la Belgique et du Congo*, ch. 3 'Le relief et les rivières de la Belgique', 41-54; ch. 5, 'Les régions géographiques belges', 99-111, Bruxelles

1940 'Carte des régions géographiques belges', *Bull. Soc. belge Et. Géogr.*, vol 10, 49-74

1941 *In Memoriam. Paul Lambert Michotte 1876-1940*, Louvain, 32 p.

---- 'L'hypothèse tectonique dans l'interprétation de l'origine du relief et du réseau fluviale de la Belgique', *Bull. Soc. belge Et. Géogr.*, vol 11, 3-32

1946 *Principes et Problèmes de Géographie humaine*, Bruxelles, 203 p.

1948 *Elements de Géographie générale*, Louvain, St Alphonse, 230 p.

1949 'Madère. Un milieu géographique jeune', *Bull. Soc. belge Et. Géogr.*, vol 18, 57-78

1952 'Carte morphologique de la Belgique', *Bull. Soc. belge Et. Géogr.*, vol 21, 41-68

1953 'Notes sur la morphologie du Katanga', *Bull. Soc. belge Et. Géogr.*, vol 22, 407-32

1955 *La vie dans la brousse du Haut-Katanga. Etude de géographie humaine*, Louvain, Soc. belge Et. Géogr., mémoire no 9, 182 p.

1957 *La Belgique et le Congo. Manuel de Géographie*, Namur, La Procure, 235 p.

1960 'Niveaux d'érosion. Les faits et leur interpretation', *Bull. Soc. belge Et. Géogr.*, vol 29, 21-46

1965 'Le concept de région géographique', *La Géogr.*, vol 17, 305-18

1967 'Historique de l'évolution de la géographie en Belgique', *Acta Géogr. Lovaniensia*, vol 5, 117-37

b. Contributions to the National Atlas of Belgium

1937 (with A. de Ghellinck and P.L. Michotte) *Carte oro-hydrographique de Belgique à 1:500 000*

---- *Notice sur la carte oro-hydrographique de Belgique*, Turnhout, Brepols, 30-63

1951-64 *Atlas de Belgique:* maps; 'Carte morphologique de la Belgique', plat 7 (1951); 'Carte lithologique de la Belgique', plate 9 (1952);

'Carte des mouvements de la population', plate
24 (1952); 'Coupes morphologiques de la Belgique',
plate 10 (1953); 'Carte des modes de peuplement
rural de la Belgique', plate 27 (1964);
Commentaires, plates 6, 7, 9, 10, 'oro-
hydrogéographie', 'morphologie', 'lithologie',
'coupes morphologiques', 56 p. (1956), plate 24,
'mouvements de la population', 46 p. (1959),
plate 27, 'modes de peuplement rural' (1964),
16 p.   All published by the Comité national de
Géographie, Bruxelles.

3.   UNPUBLISHED SOURCES

A complete set of offprints of works by Marguerite
Lefèvre, with manuscript notes of her courses and
other works, and also annotated syllabuses of courses,
are stored at the Institut de Géographie de
l'Université Catholique de Louvain, Bâtiment Mercator,
Place L. Pasteur, 1348 Louvain-la-Neuve.

*Professor Jacques G. Denis is head of the Department
at Namur university and secretary of the Belgian
National Committee for Geography
Translated by T.W. Freeman*

Chronology

1894   Born at Steenokkerzeel, Brabant, 1 March

1919   Became an *élève libre* at the Catholic
            university of Louvain and studied under
            Pierre Michotte

1922   Went to the university of Paris to work
            with Emmanuel de Martonne on physical
            geography and Albert Demangeon on human
            geography

1926   Awarded the *Docteur de l'Université*
            degree at Paris for her thesis on the
            rural habitat in Belgium

1930   Given a grant by the *Belgian Fonds
            national de la Recherche*

1931   Became secretary of the *Société belges
            d'Etudes géographiques* (to 1958) and also
            secretary of its editorial board (to
            1962)

1932   Spent one semester working with D.W.
            Johnson at Columbia University, New York

1938   During the illness of Pierre Michotte,
            gave his courses at Louvain university

1940   On Michotte's death became head of the
            Institute of Geography at Louvain

1945   Became a member of the Belgian National
            Committee for geography

1946   Appointed as a member of the Commission
            for the National Atlas of Belgium, to
            which she made a substantial contribution

1946-49   Was Secretary-General of the
            International Geographical Union

1960   Given the status of a 'full professor' at
            Louvain university

1962   Was president of the *Société belge
            d'Etudes géographiques*

1964   On retirement became an emeritus
            professor

1967   Died at Louvain, 27 December

W. J. McGee

1853–1912

ANDREW S. GOUDIE

Drawn from a photograph by Olive E. Freeman

William John McGee was a leading American scientist who, at the turn of the century, was significant both in terms of his influence on nascent institutions and of his substantive contributions to developing disciplines. His tastes were catholic, covering such major themes as geomorphology, geology, anthropology, hydrology, conservation and human geography.

1. EDUCATION, LIFE AND WORK

McGee, who was normally called 'Don' and always signed his name WJ, omitting the abbreviation marks after the initials, was born near Farley, Iowa, on 17 April 1853. His father was a strict Methodist of Scottish-Irish extraction from Armagh who came to the USA in 1831, while his mother, Martha Ann, was a Baptist. WJ was the fourth of their nine children, and the family lived in a lumber house made from the dismantled huts of the construction workers of the Dubuque and Sioux City Railroad Company (E.R. McGee, 1915). He was a very sickly, delicate child and the attending doctor said that he had 'too large a brain for the strength of his body'. The diagnosis was not totally correct: on his death an autopsy showed his brain to be smaller than that of his great contemporary, Major J. W. Powell. (Both McGee and Powell stated in their wills that their brains were to be measured and preserved in the Smithsonian Institution.) He attended irregularly the county district school until he was about fourteen years old, 'but as the school was confessedly of low grade, it is not probable that he advanced much beyond

the merest rudiments' (Knowlton, 1912, 19). Later, as a consequence of intense individual effort and the help and stimulus of various members of his family, he educated himself in a variety of disciplines, including Latin, German, mathematics, astronomy, surveying and law. Nonetheless he did not immediately progress into the remarkably rich and varied scientific career for which he was to become famous. Doubtless assisted by his great bulk, he became a blacksmith and engaged in the manufacture and sale of agricultural implements. His other occupation, which required the exercise of somewhat different skills, was to practise in courts of justice.

At some point in the 1870s he began to study geology, and embarked on an investigation of the superficial deposits of north eastern Iowa. His first scientific paper 'Relative positions of Forest Bed and Associated Drift Formations in north eastern Iowa' was published in 1878, and between 1877 and 1881 he prosecuted, as a private enterprise, a topographical and geological survey of some 12 000 square miles of that area.

In 1880 he began to work for the US Government, and was employed initially to prepare a report on the building stones of Iowa for the tenth census. In July 1883, after a study of the glacial phenomena of the upper Mississippi valley, Major John Wesley Powell attracted him to the US Geological Survey. Shortly thereafter he was placed in charge of the division of Atlantic Plain coastal geology and he stayed with the survey until 1893 (Darton, 1913). During this period he married his wife, the medical doctor Anita Newcomb,

on St Valentine's Day, 14 February 1888.   This was
a tremendously successful and academically productive
period of his life, and he helped Major Powell very
considerably in the development of the Survey and of
its maps.   He became a founding Committee member of
the Geological Society of America and the first Editor
of its *Bulletin*.   He was also intimately connected
with the National Geographic Society, both as a Vice
President and an Editor, and was a member of the select
caucus that established the Association of American
Geographers in 1904.

In 1893 Powell turned the Directorship of the
Survey over to Walcott and McGee left the Survey at the
same time, immediately taking up a position as
ethnologist with Powell at the Bureau of Ethnology, an
office he held until 1903.   Even thereafter he
continued his pioneering work in American ethnology as
Chief of the Department of Anthropology at the St Louis
Exposition of 1903.   Darton (1913, 106) has
highlighted the importance of this work:

> At this exposition he presented to the public
> a greater variety of natives of many parts of
> the world then had ever been assembled before,
> and illustrated their natural environments,
> customs and products in a most instructive
> manner.   It was one of the greatest lessons
> in geography that has ever been given and one
> that had the largest audience.

This anthropographic work has been identified as a
major stimulus to the development of Anthropo-
Geography in the USA (James and Martin, 1978).   In
1903 on the conclusion of the Exposition (known as the
'Fair') McGee was placed in charge of the St Louis
Public Museum.   He remained there until 1907, when he
entered another major period of his life: he joined
the Department of Agriculture as an expert in charge
of certain hydrological investigations.   During this
time he not only developed the new science of
hydrology (McGee, 1908), but also came to the forefront
of the conservation movement, having been appointed by
President Roosevelt as Vice-Chairman and Secretary of
the Inland Waterways Commission in 1907.

McGee died aged fifty-nine in Washington DC on
5 September 1912.   He had suffered from cancer since
1894, and recorded the progress of the disease in a
paper he caused to be published just a few days after
his death.   In spite of the disease he kept working
and producing.   He was a scientist to the very end,
bequeathing his body for the purposes of dissection.
One of his main memorials is his voluminous list of
publications.   Another is Mount McGee in Western
Greenland (*Natl. Geogr. Mag.*, 1898, vol 9, p 11), a
country in which he never worked: the naming was a
compliment to a revered scholar.

2.   SCIENTIFIC IDEAS AND GEOGRAPHICAL THOUGHT

McGee's bibliography gives an immediate indication of
the width of his interests in geomorphology.   He was
one of that great critical mass of scientists who came
together to revolutionise geomorphology in the USA
towards the end of the nineteenth century, and can
claim to be of the same calibre as his great
contemporaries: Dutton, Powell, Davis, Gilbert and
their ilk.

Among his notable contributions were the
establishment of the glacial history of Iowa, the
study of the evolution of the great coastal plain of
the eastern seaboard, the giving of publicity to the
threat of soil erosion, the analysis of isostasy and
other types of tectonic movement, discussion of the
causes of climatic change, examination of the scale of
glacial excavation, and recognition of the possible
role of sheetfloods in causing planation and pediment
formation in arid lands.   His approach was wide-
ranging and eclectic, and though living in the
Davisian era he was as much concerned with the
'process' aspects of geomorphology as with the 'stage'
aspects.   He was as much at home with formulae and
calculations as he was with graphic descriptions of
form.   He concerned himself with theory as much as
with empirical observation.

Given the range of his interests it is
impossible to discuss them all, though certain of his
papers have a particular interest.   For example, his
work on glacial valleys in the Sierra Nevada attempts
a theoretical explanation of U-shaped cross profiles
and the modification of pre-existing fluvial canons
(McGee, 1894b), and also seeks to explain upon the
observation that 'ice streams flowing upon plains are
deflected toward the sides upon which effective solar
accession is least' (1895, p 391).   In another paper
(McGee, 1893) he used calculations based on observed
rates of erosion and deposition to calculate the age
of the Earth, and came up with a most probable value
of 6,000 million years: this value is not wildly
different from present day estimates of 4,000 million
years.   He championed the Croll hypothesis that
climatic changes were brought about by orbital
fluctuations (McGee, 1881 and 1883), in opposition to
G.K. Gilbert, and he sought to show the importance of
isostatic adjustment in areas of rapid sediment
deposition such as the Mississippi Delta (McGee, 1892).

He is now perhaps best known, however, for his
views on pediment formation in the American south
west, for he had the good fortune to witness a great
sheetflood in the Sonoran Desert, which advanced at
'race-horse speed'.   He believed that such sheetfloods
could fashion the curious rock-cut surfaces abutting
the mountain fronts (McGee, 1897):

> Over dozens or scores of square miles in
> carefully examined localities, hard rocks
> like those of the mountains, and with no
> sign of decomposition, are planed off almost
> as smooth as the subsoil by the plowshare,
> with nothing either in configuration or
> covering to indicate that streams have flowed
> over them, and extended consideration has
> yielded no other suggestion as to the eroding
> agent than that found also in analogy with
> the observed sheetflood (p. 108-09).
> ... the general effect of sheet flooding in
> the Sonoran district is to carve baselevel
> plains, lightly veneered by the carving
> material ...; that these plains tend ever to
> retrogress into the mountains and thereby
> steepen their slopes and render them

exceptionally rugged (p 110).

Given McKee's research interests in superficial geology, it is not surprising that he had an interest in what we would now call geomorphology. The origins of the word itself are somewhat obscure, but it is possible that he was the first person to give the term wide currency. He himself, without giving a reference, attributes the invention of the word to Powell (McGee, 1888, p 27), though in 1893 (p 199) he goes into greater detail, and makes some claim to having himself originated the term. In any event he provided a definition, which if not *the* earliest is certainly one of the earliest:

The systematic examination of land forms and their interpretation as records of geological history introduces a new branch of geologic science, called 'physical geography' or 'physiography' by different writers, which has been designated 'geomorphic geology' by Powell and the 'new geology' or 'geomorphology' by the writer; but the 'geomorphy' first employed in a somewhat different connection by Sir William Dawson, though never extensively used with this meaning, is preferable.

Conservation, as we have seen, was one of the final interests to which McGee turned. He was one of the fathers of the conservation movement, and was the chief promoter of the great Conference of Governors which President T. Roosevelt convened at the White House in 1908. McGee was both the secretary of the conference and the editor of its proceedings. He defined conservation as being for 'the greatest good to the greater number for the longest time' (McGee, 1906), and in a later paper made a cogent and spirited attack on the way in which America had squandered its resources in the shortest time (McGee, 1910, quoted in McGee, 1915, p 93):

In all the world's history no other such saturnalia of squandering the sources of permanent prosperity was ever witnessed. In the material aspect, our individual liberty became collective licence, and our legislative and administrative prodigality grew into national profligacy; the balance between impulse and responsibility was lost, the future of the People and the Nation was forgotten, and the very name of posterity was made a by-word by men in high places; and worst of all the very profligacies came to be venerated as law and even crystallized foolishly in decisions or more questionably in enactments - and for long there was none to stand in the way of the growing avalanche of extravagance.

The roots of this interest in conservation of natural resources, however, can be traced back to his days in the Geological Survey (McGee, 1894) and his concern with potable water supplies:

The pioneer accepted the spring as a gift of bountiful providence or as the meed of patient search, to be enjoyed without profane scrutiny or ingrate question, and the origin of the water was a half-sacred mystery. Ignorant of the source of the crystal flood, he neglected the precaution necessary to preserve its purity, even to prolong its existence...half the springs known to the contemporaries of Daniel Boone and Davy Crockett have ceased to flow (p 11).

McGee expressed similar concern about declining groundwater levels in the central United States (McGee, 1912, p 490) and related the phenomenon to land use chance:

... The chief cause of lowering must be sought elsewhere than in actual consumption by man and animals, and passing over the consumption by plants (which on the average has probably diminished rather than increased in consequence of clearing and cultivation), it would appear that this dominant cause is the loss of storm and snow water through surface runoff unretarded by adequate cover, especially during the non-growing seasons--the same excessive run-off that leaches and impoverishes the soils of rolling country and initiates destructive soil erosion in hilly lands.

He returned to this theme in his posthumusly published monograph on *Wells and subsoil water* (1913), in which he analysed water levels in no less than 28,906 wells from forty-eight states (p 182):

The great fact is that under the delicate balance between soil and slope and cover toward which nature wrought during the ages before settlement, there was little surface runoff, practically all the storm and thaw waters entered the dust and soil to escape slowly by seepage into streams and springs, and the subsoil was normally kept surcharged with water often quite to the level reached by the roots of growing plants; while after clearing and under ordinary cultivation much of the storm water is permitted to run off the surface in freshets, not only leaching and eroding the soil on its way, but gathering in the streams in flood torrents and never entering either soil or subsoil to maintain that reserve stock required for full productivity.

McGee's interests in geomorphology, hydrology and conservation came together in his work on accelerated soil erosion (McGee, 1911), in which he presents a penetrating analysis of the processes of rainsplash, overland flow, subterranean piping, rill coalescence, gully incision, and siltation. The following extract is notable for its graphic language and for the appreciation of the way in which initial incision can set in train a sequence of erosion in sensitive parts of the landscape (p 21):

The careful observer, standing on an old field slope already denuded of soil and rent by gullies, and watching the work of a single heavy rain, sees the storm water gather in rivulets guttering and moving down the old channels and cataracts, sapping the banks at every bend, and so deepening and widening the trenches and undermining any protective sward on either side, yet always pushing most rapidly upslope. Each gully forms an open line of attack, occupied by a growing body of rushing, crushing, rending, grinding, scouring, sediment-bearing water; each water body is a monster of two-score arms, each ended in a hundred wriggling fingers, clawing into the humus, under the bordering sward, through the softening surface, slashing the soil into bits and separating these into the sand and the soluble and solid semi-organic grains of which the soil bed consists. Finally, the debris is sorted and scattered; the coarse sand is spread over the nearby bottom land and the fine sand is dropped in the stream channel, the silt is dumped in the neighbouring river, and the slime and soluble salts and organic matters are swept on towards the sea, muddying and befouling the waters on the way.

Along with Marsh and Shaler, McGee was one of the pioneer investigators of the threat that man-induced soil erosion posed to the nation's long term welfare.

3.   INFLUENCE AND SPREAD OF IDEAS

WJ McGee held an important place in the development of American geography, geomorphology, geology, ethnology, hydrology and conservation, and was a major influence in the development of certain major institutions and their publications, including the Geological Society of America, the National Geographic Society and the Association of American Geographers.   His life was mainly spent in government service so that he never produced a cadre of graduate students to perpetuate his name, but within the realms of government he had contacts at the highest levels, influencing decision makers at the State and National level.   He was one of the most productive, eclectic and wide-ranging of American geographers during the formative years of the discipline.

Bibliography and Sources

1.   REFERENCE ON WILLIAM JOHN MCGEE

Knowlton, F.G., 'Memoir of W. J. McGee', *Bull. Geol. Soc. Am.*, vol 24 (1912), 18-28

Darton, N. H., 'Memoir of W. J. McGee', *Ann. Assoc. Am. Geogr.*, vol 3 (1913), 103-10

McGee, E.R., *Life of W. J. McGee*, privately printed at Farley, Iowa (1915), 240 p.

James, P.E. and Martin, G.J., *The Association of American Geographers: the first seventy-five years 1904-1979*, Washington (1978), 5, 33, 37

2.   SELECTIVE BIBLIOGRAPHY OF WORKS BY W. J. MCGEE

a.   Anthropology and Archaeology

1878 'On the artificial mounds of northeastern Iowa, and the evidence of the employment of a unit of measurement in their erection', *Am. J. Sci.*, 3 ser. vol 16, 272-8

1888 'Palaeolithic man in America; his antiquity and environment', *Pop. Sci. Mon.*, vol 34 November, 20-36

1892 'Man and the glacial period', *Am. Anthropol.*, vol 5, 85-95

1895 'The beginning of agriculture', *Am. Anthropol.*, vol 8, 350-75

1896 (with W. D. Johnson), 'Seriland', *Natl. Geogr. Mag.*, vol 7, 261-5

---- 'The beginning of marriage', *Am. Anthropol.*, vol 9, 371-83

---- 'Expedition to Papagueria and Seriland', *Am. Anthropol.*, vol 9, 93-8

---- 'The Sioux Indians, a preliminary sketch', *U.S. Bur. Am. Ethnol.*, 15 Annu. Rep., 163-204

---- 'The science of humanity', *Am. Anthropol.*, vol 9, 241-72

1898 'Papagueria', *Natl. Geogr. Mag.*, vol 9, 345-71

---- 'The Sioux Indians', *Smithsonian Inst. Bur. Ethnol.*, 17 Annu. Rep.

1899 'National growth and national character', *Natl. Geogr. Mag.*, vol 10, 186-206

1901 'The old Yuma trail', *Natl. Geogr. Mag.*, vol 12, 103-07, 129-43

---- 'Primitive numbers', *U.S. Bur. Am. Ethnol.*, vol 19, 821-51

1902 'Anthropology at Pittsburg', *Am. Anthropol.*, n.s., vol 4, 464-81

1905 'Anthropology and its larger problems', *Sci.*, n.s., vol 21, 770-84

---- 'Anthropology at the Louisiana purchase exposition', *Sci.*, n.s., vol 22, 811-26

---- (with C. Thomas) *Prehistoric North America*, Barnie and Sons, Philadelphia, 487 p.

b.   Climatic Change

1881 'A contribution to Croll's theory of secular climatic changes', *Am. J. Sci.*, vol 22, 437-43

1883 'On the present status of the eccentricity theory of glacial climate', *Am. J. Sci.*, vol 26, 113-20

c.   Conservation, Soils, Soil Erosion and Applied Geography

1880 'The "laterite" of the Indian peninsula', *Geol. Mag.*, vol 7, 310-13

1881 'The geology of Iowa soils', *Trans. Iowa State Hortic. Soc.*, vol 15, 101-05

1882 'The relations of geology and agriculture', *Trans. Iowa State Hortic. Soc.*, vol 16, 202-23

1889 'The world's supply of fuel', *Forum*, vol 7, 553-66

1890 'Encroachments of the sea', *Forum*, vol 9, 437-49

1908 'The cult of conservation', *Conservation*, September, 469-72

1910 'The conservation of natural resources', *Mississippi Valley Historical Proceedings*, 1909-1910 (cited in E.R. McGee, 1915, p. 13)

1911 'Soil erosion', *U.S. Dept. Agric. Bur. Soils, Bull.*, 71, 1-60

d.   Geomorphology

1888 'The classification of geographic forms by genesis', *Natl. Geogr. Mag.*, vol 1, 27-36

1891 'Flood plains of rivers', *Forum*, vol 11, 221-34

1895 'Canyons of the Colorado', *Sci.*, vol 2, 593-7

1896 'Geographic history of the Piedmont plateau', *Natl. Geogr. Mag.*, vol 7, 261-5

1897 'Sheetflood erosion', *Bull. Geol. Soc. Am.*, vol 8, 87-112

e.   Glaciation

1878 'On the relative positions of the forest bed and associated drift formations in northeastern Iowa', *Am. J. Sci.*, 3 ser., vol 15, 339-41

1879 'On the complete series of superficial formations in northeastern Iowa', *Proc. Am. Assoc. Adv. Sci.*, vol 27, 188-231

---- 'Notes on the surface geology of a part of the Mississippi valley', *Geol. Mag.*, vol 6, 353-61, 412-20

1881 'On maximum synchronous glaciation', *Proc. Am. Assoc. Adv. Sci.*, vol 29 (Boston 1880), 447-509

1882 (with R.E. Call) 'On the loss and associated deposits of Des Moines', *Am. J. Sci.*, vol 24, 202-23

1885 'On the meridional deflection of ice-streams', *Am. J. Sci.*, vol 29, 386-92

1891 'The Pleistocene history of northeastern Iowa', *U.S. Geol. Surv. 11 Annu. Rep.*, part 1, 199-557

1894 'Glacial canons', *J. Geol.*, vol 2, 350-64

1895 'On the meridional deflection of ice-streams', *Am. J. Sci.*, vol 29, 261-5

f.   Hydrology

1894 'The potable waters of eastern United States', *U.S. Geol. Surv. 14 Annu. Rep. 1892-3*, part 2, 47 p.

1908 'Outlines of hydrology', *Bull. Geol. Soc. Am.*, vol 19, 193-200

1911 'The agricultural duty of water', *U.S. Dept. Agric. Bur. Soils, Bull. 71*, 1-60

---- 'Principles of water-power development', *Sci.*, vol 34, 813-25

1912 'Subsoil water of central United States', *U.S. Dept. Agric. Yearb. 1911*, 179-90

---- 'Field records relating to subsoil water', *U.S. Dept. Agric. Bur. Soils Bull.*, no 93, 40 p.

1913 'Wells and subsoil water', *U.S. Dept. Agric. Bur. Soils Bull.*, no 92, 185 p.

g.   Isostasy

1892 'The Gulf of Mexico as a measure of isostasy', *Am. J. Sci.*, vol 44, 177-92

1894 'The extension of uniformitarianism to deformation', *Bull. Geol. Soc. Am.*, vol 6, 55-70

h.   General Geology

1883 'On the origin and hade of normal faults', *Am. J. Sci.*, vol 26, 294-8

1885 'Methods of geological cartography in use in the United States Geological Survey', *C.R. Congr. Geol. Bull.*, 221-40

1887 'Tuscaloosa formation.   Summary of previous observations and opinions', *U.S. Geol. Surv. Bull.*, no 43, 247-55

---- 'The geology of the head of Chesapeake bay', *U.S. Geol. Surv. 7 Annu. Rep.*, 537-646

1887-9 'The field of geology and its promise for the future', *Minnesota Acad. Nat. Sci. Bull.*, no 3, 191-206

1888 'Some definitions in dynamical geology', *Geol. Mag.*, decade 3, vol 5, 489-95

---- 'Three formations of the middle Atlantic slope', *Am. J. Sci.*, vol 35, 120-43, 328-30, 448-66

---- 'Notes on the geology of Macon County, Mo.', *St. Louis Acad. Sci. Trans.*, vol 5, 305-36

1889 'The geologic antecedents of man in the Potomac valley', *Am. Anthropol.*, vol 2/3, 227-34

1890 'The southern extension of the Apponattox formation', *Am. J. Sci.*, vol 40, 15-41

1891 'Rock gas and related bitumens', *U.S. Geol. Surv. Annu. Rep. 10*, part 1, 589-616

1892 'The aerial work of the United States Geological Survey', *Am. Geol.*, vol 10, 337-79

---- 'The Lafayette Formation', *U.S. Geol. Surv. 12 Annu. Rep.*, part 1, 347-521

1893 (with G.H. Williams and N.H. Darton) 'Geology of Washington and vicinity', *C.R. Int. Geol. Congr.*, 5 session, 219-51

---- 'Note on "The age of the earth"', *Sci.*, vol 21, 309-10

---- Discussion in *Congr. Int. Geol.*, 5 ème session, *C.R.*, 198-207

i.   Miscellaneous

1891 'The evolution of serials published by scientific societies', *Bull. Phil. Soc. Washington*, vol 11, 221-45

1892 'Comparative chronology', *Am. Anthropol.*, vol 5, 327-44

1896 'The relations of institutions to environment', *Smithsonian Inst. Annu. Rep. for 1895*, 701-11

1898 'Geographical development of the District of Columbia', *Natl. Geogr. Mag.*, vol 9, 345-71

---- 'The geospheres', *Natl. Geogr. Mag.*, vol 9, 435, 447

1899 'The beginning of mathematics', *Am. Anthropol.*, n.s., vol 1, 646-74

1900 'The growth of the United States', *Natl. Geogr. Mag.*, vol 9, 377-83

---- 'Cardinal principles of science', *Proc. Washington Acad. Sci.*, vol 2, 1-12

1901 'Asia, the cradle of humanity', *Natl. Geogr. Mag.*, vol 12, 281-90

1911 'Prospective population of the United States', *Sci.*, vol 34, 428-35

A.S. Goudie is Professor of Geography in the University of Oxford

Chronology

1853	Born near Farley, Iowa, 17 April
1877	Started topographical and geological survey of Iowa
1880	Employed by U.S. Government to investigate building stones of Iowa
1883	Joined U.S. Geological Survey
1888	Married Anita Newcomb
1893	Became ethnologist in Bureau of Ethnology
1894	Contracted cancer
1903	Chief at Department of Anthropology at the St. Louis Exposition
1907	Hydrologist, U.S. Department of Agriculture
1912	Died in Washington D.C., 5 September

Niels Nielsen

1893–1981

N. KINGO JACOBSEN

In 1939 Niels Nielsen was appointed as a full professor of geography in Copenhagen university after serving for some years as a lecturer and this initiated a period of expansion for geography in Denmark. He was already well known by 1939 as a dynamic man with a fine combination of scientific, pedagogic and administrative gifts and his influence on colleagues was considerable from 1924, when he received his doctorate degree, to 1964 when he retired from university service.

Throughout his career he was a dedicated natural scientist and the design of his professional life emerged in 1917 when he took his master's degree and became a freelance researcher having free accommodation at the *Elers Kollegium* from 1917-22, while working on the production of iron in Denmark during the Iron Age. At that time he found in geography a compelling interest, for to him it was a field subject well suited to the study of the topography of Denmark. He possessed a clarity of mind that was tempered by a fine critical judgement and the wisdom to discern the core of a problem. He had the ability to arouse the interest of others in various research enterprises and to form working groups of which some were concerned with new researches: this aspect of his activity owed much to his fine teaching, engaging personality and unquenchable enthusiasm. He was clearly interested in people, whether colleagues, young research workers, or students and he was very successful as the senior president from 1943-45 of the *Studenterforeningen*, the students' union. This was a society for all types of students, guided by a praesidium composed partly of university teachers. A president was elected each

year, and one of his responsibilities was to act as chairman at weekly lectures and discussions dealing with current problems. Until 1968 the *Studenterforeningen* had a seven-floor house in Copenhagen and was in effect an academic centre at which students of the natural sciences, the humanities, engineering and other specialisms met: within it there were club facilities for various small groups, a library and a restaurant. Nielsen's interest in students was also shown in his work as warden of *Nordisk Kollegism*, a student hostel built in 1942 for 100 residents and providing board and lodging at a low charge: of this he was warden from 1945-68. He retained this position for four years after his retirement from the university chair in 1964.

Niels Nielsen was the fourth professor of geographer in Copenhagen after the subject entered its modern phase in 1883 and during his tenure of office geography flourished. The staff increased from two professors, one part time lecturer and one clerk in 1939 to seventeen full time teachers and seven technical and secretarial helpers by 1964. In January 1960 the Institute of Geography moved to new premises in the city at Kejsergade, which provided a fine background for the expanding work of a large staff. The position of geography at Copenhagen was discussed by Nielsen in two papers (*Geogr. Tidsskr.*, vol 51, 1951, 134-48 and vol 61, 1962, 1-78) and also in a work issued in 1979 dealing with the fifth centenary of the foundation of the university in 1479.

Geographers Biobibliographical Studies, volume 10 (1986)

1. EDUCATION, LIFE AND WORK

Niels Nielsen was born on 3 October 1893 in the farming household of Soren Peter Nielsen and his wife Karen *née* Rasmussen and throughout his life he preserved a strong attachment to his home area, the Sorder (south) Vissing parish, for there his interest in observing nature developed from childhood. In 1911 he was successful in the General Certificate examination at the Horsens Statsskole and matriculated at Copenhagen university, where he followed the obviously appropriate course in natural science with geography, graduating in 1917 with zoology as his major subject: this choice shows the influence and inspiration of Professor Hector F.E. Jungersen. Another professor of the time, H.P. Steensby, a man of fascinating personality, drew him towards human geography with a strong topographical and archaeological emphasis and this was seen in his research work on the evidence of iron age working in Jutland.

From 1915-17, before his graduation as a Master of Science, Nielsen was given a grant for residence at the old college, the *Regensen*. This was a Danish college founded in 1623 which gave free lodging and financial support to 103 students of particular merit after they had taken the Bachelor's degree. From 1917-23 Nielsen was a resident at the *Elers Kollegium* which had been founded in 1693 and provided comparable help for twenty students included three who were studying for the doctorate as Nielsen was at that time. While there he began his pedagogic career, teaching for a year at the Ingwarsen and Ellbrechts Skole from 1917-18 and then for a year to 1919 at the H. Alders Faellesskole. In 1919 he became a teacher at the Sortedam Gymnasium, where he stayed for twenty years to 1939 and was a much loved and respected member of the staff. Meanwhile, in 1924, he successfully presented his doctorate thesis on iron in Jutland. The material for this thesis was gathered largely during summer vacation periods and written up during the winter: Nielsen found the places where iron age objects had been recorded, mapped them and indicated the relation of the distribution of Iron Age artefacts to geographical circumstances. As a thesis his doctorate was a product of its time and predates by many years the considerable development of archaeological research on this topic in Denmark. And one may add that the thesis was a stage in his academic development rather than a revelation of his later career for at Copenhagen university he had received a broad scientific education which could have led him into a diversity of specialisms.

Though retaining his teaching post at the Sortedam Gymnasium to 1939, Nielsen--how having his doctorate degree--also lectured from 1933 at Copenhagen university before joining the staff on a full time basis in 1939. On his appointment Professor Martin Vahl and Professor Gudmund Hatt issued a joint statement in these words

> We think it is of very great value for the future of geography in this country that he who succeeds Professor Vahl shall be a representative of the natural science field within geography. The human geography field

is represented by Professor Hatt, who wishes to retain his chair in human geography but who fully recognises the importance of natural science for geography and therefore finds it of the greatest importance that the chair now vacant shall be given to a scientist whose main interest lies in the natural science disciplines.

A man of immense energy, Nielsen during these years supplemented his general scientific education by study travel in the other Scandinavian countries and also in France, Switzerland, Germany, Holland, Belgium, Britain and the Balkan countries: he also lectured in many universities and at various geographical societies. His administrative gifts were shown as secretary of the Royal Danish Geographical Society and as editor of the *Geografisk Tidsskrift* from 1931 to 1956 as well as of the two new series, the *Kulturgeografiske Skrifter* published from 1936 and the *Folia Geographica Danica*, which first appeared in 1940. Of these the *Geografisk Tidsskrift* is a journal published annually from 1877 but the two later series were monographs published occasionally. Finally Nielsen was responsible for the series of publications, *Meddelelser fra Skalling-Laboratoriet* from 1935, which were studies of work at the Skalling Laboratory in the southwest of Jutland.

Clearly Nielsen was one of those who became a geographer by gradual stages and was not finally committed to the subject when he received his doctorate in 1924, though this reveals the continued impact of his childhood interest in topography and archaeology: indeed the thesis could be regarded as a reaching forward to geography of a human aspect on an archaeological basis. However in 1924 he joined the first Danish Expedition to Iceland and in 1927 the second expedition. These expeditions gave him a fascinating view of the vast open landscapes of Iceland, and especially of the effects of vulcanicity and glaciation. The actual eruption of volcanoes with their impact of icefields and river drainage and the evidence of plate tectonics all provided abundant interest, not least as a revelation of geomorphological processes in action. However Nielsen wanted to work in Denmark, for it was also possible to see geomorphological processes in action locally and the investigations in the Skalling area were perhaps his major scientific contribution to the physical history of Denmark. Fortunately his broad understanding of natural science gave him the wisdom to draw various specialists such as botanists, zoologists and geologists into his enterprise.

Through circumstances he spent much of his working life as a grammar school teacher, notably for twenty years to 1939 at the Sortedam Gymnasium, and he was equally successful at the university. The pupils at Sortedam found 'Nilaus', as he was called, fascinating and enterprising and A. Noe Nygaard, later to become professor of geology at Copenhagen, wrote that

> During my first three years at the university when I was the only Danish student majoring in

geology, Niels Nielsen was indeed an inspiration, for he was the only teacher who gave a breath of fresh air from the world beyond the parish pump.

He followed the normal university routine of providing lectures and arranging seminars but his field excursions stand out as of supreme significance in his work. These he clearly enjoyed and so did the field party members, who included students, foreign study groups, members of congresses and others. He knew and loved the landscapes and made them understood by his audiences as he showed the continuing evolution of the physical features, by wind, water or other agencies. Like other successful teachers he was something of an actor eager to win over his audience. The volume presented by his students on his sixtieth birthday, conforming to tradition by having a somewhat humorous tone, contained drawings, articles from newspapers, even a list of remarks he had made on excursions, and gave a general impression of a large pedagogue who was a beloved and much discussed personality, a man of strong opinions with great working ability ready to sacrifice everything for a cause in which he believed. Though as a teacher he had to give clear instruction and a firm factual basis, based on evidence as he had carefully gathered it, at the same time he was aware of many of the uncertainties in geomorphological work and in discussion showed the way forward to an understanding of the core of a problem when the final solution appeared to be beyond the current stage of research enquiry.

That he was fascinated by his own country, and especially by Jutland, was apparent to all. He was a fine citizen of his country. In 1941, the year after the Nazi invasion, he gave a series of lectures on Denmark to large audiences, and this course was a forerunner of the work of the Folk University established in 1950. His work for the Student Union as president had obvious dangers during the period of Nazi occupation from 1941-45 and in 1943 he was sent to the internment camp at Horserod. After the war, from 1946-68, he was warden of the *Nordisk Kollegium*, which then had accommodation for over 100 students and also for four visiting Scandinavians with doctorate degrees: he shed these responsibilities in his middle seventies. He died on 15 September 1981 at Charlottenland, aged eighty-seven.

2. *SCIENTIFIC IDEAS AND GEOGRAPHICAL THOUGHT*

Nielsen's ideas emerge through a study of his works rather than from any methodological writings for he was one of those more eager to elucidate a problem than to offer advice and directives on how others should do their work. Therefore consideration is given here to his visits to Iceland and then to the work that began in Skallingen and in time was extended to a wider area ultimately to be recognized as a vital contribution to planning, and thirdly to the educational work of the Institute of Geography at Copenhagen.

a. *Iceland*

He first went to Iceland in 1924 and his next visit was in 1927 with Palmi Hannesson, Mag. scient. This second expedition provided an abundance of material, published in a report of more than 100 pages by the Royal Danish Academy of Sciences under the title 'Contributions to the physiography of Iceland with particular reference to the highland west of Vatnajokull' in 1933. The morphological complexity of Iceland, with its vulcanicity, earthquakes and glaciation made it a natural laboratory for physical geographers as field workers, and Nielsen fully appreciated its remarkable qualities. He dealt with five major themes:

(i) wind erosion as a moulding agent in landscapes, especially the blow-out conditions in relation to the vegetation cover: this line of enquiry showed the relation of force and direction of winds to the type and humidity of the soil;

(ii) the movement of the ice margin which had changed by some 800 metres in eighty years and therefore raised questions of glacial history;

(iii) the hydrological pattern, both in relation to moraines of various periods and also to volcanic activity: lava flows from uplands will use existing valleys and therefore dislocate the existing drainage patterns;

(iv) the lakes;

(v) tectonic conditions and volcanic activity: Nielsen suggested that Iceland will in time split in a north-south direction and so he was a supporter of A. Wegener's theories and of modern views on plate tectonics. Apparently Nielsen was a pioneer in applying such theories to Iceland, initially in papers given at the Natural Science meeting at Copenhagen in 1929 and at the international Arctic conference of 1930 at Greifswald.

For Nielsen one major consideration was whether the current state of the Icelandic ice masses, with the marked imprint of former glaciation on its landscapes, might help to elucidate the influence of former ice sheets on the Danish landscapes. He was working at a time when glacial history was closely studied in many countries and indeed a subject of vigorous international discussion by glaciologists. However it soon became clear that Iceland's morphological development was to a considerable extent a special case with unique, or at least rare, attributes for volcanic activity disturbed or at least modified ice movement. Nevertheless there were some possible correlations between Denmark and Iceland, as on the southern border of Hofsjokull where a recent ice margin was embedded in old moraine deposits of vast extent.

Understandably Nielsen became interested in vulcanicity and the third and fourth Danish Vatnajökull expeditions of 1934 and 1936 were given to the subglacial eruption of the Grimsvötn volcano. Nielsen tried to quantify the amount of heat energy released in a volcanic eruption by means of meltwater in a river course to the amount of ice known to have disappeared from a crater. This theory was of interest but at the time it was not known that Grimsvötn is a permanently active thermal area and that this causes the formation of a subglacial lake: therefore special conditions existed.   Later Sigdurur Thorarinsson (the Icelandic geographer) showed on the basis of studies of Skreidarajökull that the emptying of a subglacial lake could influence the internal pressure below the earth's crust and so have an effect on volcanic eruption. Finally, in 1947 Nielsen had the experience of visiting Iceland with A. Noe-Nygaard to see the eruption of Hekla.

b.   Skallingen Studies
These were at least to some extent inspired by studies of the desert landscapes of central Iceland and were designed to test prevalent theories on sand drift and dune formation.   The Skalling peninsula was chosen as the research area and in 1930 Nielsen and his wife went there with camping equipment, including food and water for a stay of four weeks.   Skallingen at that time was a barren and uninhabited peninsula with fine opportunities for the observation of geomorphological phenomena as well as of plant and animal life, in short an outdoor laboratory for the natural sciences. Characteristically Nielsen went straight forward with the planning of geomorphological, ecological and botanical research studies.   Also characteristically he acquired a base for the work, a laboratory first housed in the shepherd's cottage at Vagtbjerg on Skallingen, with the help of the Carlsberg Foundation. Wisely he wrote a paper on the enterprise in the *Geografisk Tidsskrift* (vol 36, 1933, 79-87) and also contributed another paper on a subject naturally much in his mind at the time, the joint ownership of land on Skallingen (*ibid*,. 88-119).   He was always eager to share his enthusiasm with others.

A fundamental task was to map and study the geomorphology of the area, with its varied ecological environments of dune, marsh and tidal waters (*wadden*), in all of which changes could be observed within a short time.   The basis of this geomorphological study was explained in a long paper published in 1935. 'Eine Methode zur exakten Sedimentationsmessung', later to be valued at other similar coastal stations. Several scientific institutions and individuals, both Danish and foreign, were drawn into the ecological work, which covered flora, fauna and landscapes and for many years Professor August Krogh and his colleagues were enthusiastic workers at the Laboratory.   Among those who spent some time there were various doctorate students in zoology, botany, physical and human geography and Nielsen continued to show his abilities as a leader and an organizer of the work.   Still at this time he was teaching in the Sortedam Gymnasium and giving lectures at the university.

Although various textbooks of the time gave general comments on such matters as sedimentation, Nielsen saw the need to work out the local conditions in relation to the tides, water movement, freshwater streams with tidal reaches, salinity, marshes and other features characteristic of the maritime environment.   He did a vast amount of work himself and discussed some aspect of the research in two papers published consecutively in *Geografisk Tidsskrift* (vol 40/2, 1938, 123-38, 139-46), of which the first dealt with the formation of marshland in the Danish Waddensea and the second with an instrument for measuring the transport of sand and mud in flowing water.   His aim was to carry out a study of sediment balance in Ho Bugt and later to cover the whole of the Danish Wadden Sea.   His work was greatly assisted when, in 1936, thanks to the friendly co-operation of the harbour authorities, a new laboratory was established in a former machinery house at Esbjerg. The Esbjerg laboratory became the centre for two large hydrological-sedimentalogical and ecological research investigations of the entire Danish Wadden Sea from 1938 and from 1941, when Nielsen became a member of the Wadden Sea committee under the Ministry of Public Works.   During the German occupation of 1940-45 coastal work was in abeyance and in 1945 the Germans removed the Skalling laboratory at Vagtbjerg.   It was re-established at Ho in 1950.

Following World War II the emphasis on regional planning in Denmark gave great significance to the work Nielsen had begun in the 1930s.   The building of the Romo dam between 1938 and 1949 had drawn on studies made of dunes, marshes and tidal areas before the German occupation and these continued at the new Ho laboratory from 1950, when test plots were laid out once more.   The year 1935 saw the foundation of *De Danske Vade-og Marskandersogelser*, the *wadden* (tidal flat) and marsh investigations under a co-ordinating committee of the Ministry of Agriculture: this was a working group for the Tonder Marsh which was given funds to finance a wide range of research tasks from 1953-72.

Nielsen at this time worked primarily as a geographer for planning gave opportunities to both physical and human specialists and the investigations covered the reclamation work along the southwest coast of Jutland with the physical and economic geography of the marsh area.   In effect it was an extension of the earlier pioneer work, with close study of the *wadden* (tidal flat) and marsh areas, the formation of coastal marshes both as a natural and a human feature, drainage and possible reclamation, and the influence of vegetation as an agent of land reclamation.   In 1958 Nielsen became chairman of the Committee on Land Reclamation at Hojer and on this he published two papers in *Geografisk Tidsskrift*, on the proposals for land reclamation along the Wadden Sea of the Jutland coast in 1958 (vol 55, 62-87) and on the organization of research work in southwest Jutland in 1960 (vol 59, 1-11).   Also in 1960 he published some detailed studies of wind erosion and sand drift in Denmark in *Jusk Landbrug*, vol 19.

One fine quality in Nielsen's work was his ability to attract younger workers, and by the time he retired in 1964 the publications in the *Meddelelser fra Skalling Laboritet*, initiated in 1935, numbered twenty volumes.   A new generation of colleagues attached to the *Vade-og Marokandersogelser* included three young scientists who presented their doctorate theses in 1963-64 and the *Rapporter fra Vade-og Marskandersogelser* series reached its ninth volume at this time.

c.   Work at Copenhagen University
When Nielsen became the full time professor of geography at Copenhagen university, he at once began the planning of the *Atlas of Denmark*, which in five

volumes was to describe its geography under the general headings of natural features, population, trade, industry and cultural background. This enterprise broadened the research work on all aspects of the geography of Denmark and new ideas were welcomed. It also gave an occupation to a number of young and unemployed research workers of whom several moved after the war into universities, colleges and schools as teachers. Financial support was given by the Danish government and by the Carlsberg Foundation and the work was virtually completed by 1944 when it was all destroyed by *Schalburgtage*, Nazi-inspired sabotage, against the publishers, the Allers Etablissement. From this time Nielsen and others managed to reconstruct some parts of the atlas, and in 1949 the first volume, Axel Schou's *Landskabsformerne* (landscapes), appeared. Efforts to restore the other sections of the work continued but many difficulties arose, including the lack of appropriate statistical material and the changed social conditions of the post-war world. Only one other volume, Aage Aagesen's *Befolkningen* (population) of 1961 was in fact published, partly because by that time further publication would be prohibitively expensive.

Plans to establish an institute of geography with adequate premises at Copenhagen were in mind before World War II and in 1938 Nielsen was invited to share in the discussions. The general view, warmly supported by Nielsen, was that the Geografisk Laboratorium, obviously too small for its activities, should join forces with the Royal Danish Geographical Society within a new institute having accommodation for both and the possibility of relating the work of the society to that of the university. Some space was given for the Society by the National Museum in 1932 and in 1956 the Gråbrodoe Annex was bought by the University. Finally the Institute was opened in its central city location at Kejsegade in 1960. The Society is still the owner of its library, map collections and other materials, which are under the care of a librarian from the Royal Danish National Library. All the reading rooms are open to staff and students of the university, the members of the Geographical Society and indeed to others. Through association with other societies the Geographical Society receives some 800 periodicals with many books, monographs, offprints, atlases and maps.

3. INFLUENCE AND SPREAD OF IDEAS

Essentially Niels Nielsen was a practical man and a fine organizer and his election as secretary of the Royal Danish Geographical Society was a crucial event in its history for in 1931 it appeared to be declining, partly because the period of great exploring expeditions, which for so long had helped to sustain geographical societies by attracting members, was near its end, even in Greenland. During his period of twenty-five years service as secretary he managed to stabilize its finances, to raise the level of papers in its journal, the *Geografisk Tidsskrift* which had been published from 1877, and to initiate two series of monographs, the *Kulturgeografiske Skrifte* in 1936 and the *Folio Geographica Danica* in 1940. The acquisition of new premises at a favourable rent in

the National Museum was helpful but the triumph of good sense was the final move to the Institute of Geography in 1960.

Denmark's political ties with Greenland perhaps gave impetus to the work of the Commission on Scientific Research which published monographs from 1879 onwards: in 1947 Nielsen, with N. Kingo Jacobsen as secretary, reorganized the work and re-established the exchange of publications with other institutions of comparable interests. As far as possible back numbers of the publications of its monographs, *Meddelser om Grønland*, were provided for European libraries which had lost their copies during the war. In 1959 Nielsen relinquished the editorial work, which was taken over by the newly established Arctic Institute.

Another practical enterprise of Nielsen was to provide material for TRAP, the gazetteer listed in libraries as J. P. Trap, *Danmark*, vol 1-10, 1958-67 in the case of the fourth edition. Nielsen became the chief editor for the fifth edition, of which publication was complete by 1972 and included, for the first time, material on the Faroe islands and on Greenland. His final work for TRAP was seen in the last volume of the fifth edition, which he completed in October 1970. By the Municipality Reform Act of 1970 the old local government system based on rural parishes and towns, which had been sued for the TRAP volumes, was replaced by a new administrative system with 277 primary municipalities and fourteen counties. Nielsen saw the need for a sixth edition of TRAP, which would obviously have to be completely recast but that he could not achieve in old age.

As a major figure in Danish geography Nielsen was involved in numerous enterprises as an adviser, such as the compilation of wall maps for the Geodetic Institute, designed and described by Axel Schou in *Geogr. Tidsskr.*, vol 52, 1952-3, 255-75. His advice was sought by writers of papers, and acknowledged gratefully by B. Fristrup who wrote two papers in *Geogr. Tidsskr.*, vol 52, 1952-3, 51-65 on 'Wind erosion within the Arctic deserts' and vol 59, 1960, 89-102, 'Studies of four glaciers in Greenland'. In the 1960 paper Fristrup notes that the investigations he carried out in Greenland from 1956-8 were a Danish contribution to the International Geophysical Year, 1957-8 and were sent out from the Geographical Institute of the University of Copenhagen under the joint direction of Nielsen and himself. Nielsen also served on the Danish National Committee for geography and was the leader of the Danish delegation at the congresses of the International Geographical Union at Amsterdam in 1938, Lisbon in 1949 and Washington D.C. in 1952. Naturally he was closely involved with the Norden Congress of 1960, both as leader of the Danish contingent and as part author of the *Guidebook to Denmark*. He presided over the Symposium on Agriculture at Horsens during the Congress.

Varied as his activities were it was perhaps as a field geomorphologist that Nielsen achieved his greatest success, especially with visiting groups of geographers from other countries. Waving a walking stick or even a cigar he enthralled them with his exposition of how Thyholm in Jutland came into existence or entertained them with stories of the reclamation of heath areas, whether by scientific improvement of infertile soils or even by adventures such as gipsy groups or others who came in furtively at night, the 'nightmen', seeking somewhere to 'live rough'.   That his work on Skallingen was eventually to be a major contribution to the post-1945 planning of Denmark shows that study of the physical landscape begun as an academic exercise may be a basis for modern applied geography of social and economic value.   Of farming stock, he knew that soils are made and remade by human effort, that the natural sciences must be based on the realities of weather and climate, of tides and currents, as well as on the history of the physical environment which on the surface in Denmark could be regarded as the effects of the Pleistocene glaciation.

His visits to Iceland showed that the physical environment is constantly changing and that understanding of such developments could be achieved only by enlisting the aid of a wide range of field scientists.   He had the ability to see the core of a problem and to ask pertinent questions in his own research, and also to appreciate the work of others in such enterprises as the planning of a national atlas. His social gifts contributed to his organizational success in such practical matters as the post-1945 rehousing of the geography department of Copenhagen university in association with the Royal Danish Geographical Society.   Niels Nielsen as a fine citizen of Denmark and one of its pioneer modern geographers is remarkable not only for what he did but also for what he enabled others to do.

Bibliography and Sources

1.   REFERENCES AND OBITUARIES ON NIELS NIELSEN
1924   *Festskrift udgivet af Københavns Universitet i Amledning af Universitets Aarsfest November 1924. Selvbiografi af Aarets Doktorer (Festschrift published by the university of Copenhagen on the occasion of the commemoration of the university: autobiographies by the doctors of the year)*, 122-3
1953   'Studier tilegnet Niels Nielsen' ('Studies dedicated to...'), *Geogr. Tidsskr.*, vol 52, xvi-xvii and 1-309
1981   Jacobsen, N. Kingo, 'Niels Nielsen 3-10 1893--15-9 1981', *Geogr. Tidsskr.*, vol 81, viii-ix
1982   Jacobsen, N. Kingo, 'Niels Nielsen...', *Sven. Geogr. Årsbok*, vol 58, 210-15
1983   Noe-Nygaard, Arne, 'Niels Nielsen...', *Nekrolog. Tale i Videnskabernes Selskabs møde 7 April 1983 (Obituary speech at the Royal Danish Academy of Sciences and Letters)*

2.   SELECTIVE BIBLIOGRAPHY OF WORKS BY NIELS NIELSEN

a.   Iceland
1925   'Foreløbig Beretning om den dansk-islandske Ekspedition til Islands indre Højland' ('Preliminary report on the Danish-Icelandic expedition to the central highlands of Iceland'), *Geogr. Tidsskr.*, vol 28, 32-9
1927   'Plan til en Ekspedition til den vestlige Del af Vatnajokull og tilgraensende Egne i Centralisland' ('Plan for an expedition to the western part of Vatnajokull and neighbouring regions of central Iceland'), *Geogr. Tidsskr.*, vol 30, 112-13
----   'Foreløbig Beretning om den 2. dansk. islandske Ekspedition til Islands indre Højland' ('Preliminary report on the 2. Danish-Icelandic expedition to the central highland of Iceland'), *Geogr. Tidsskr.*, vol 30, 250-7
1928   'Landskabet Syd-Øst for Hofsjokull i det indre Island' ('The landscape southeast of Hofsjokull in central Iceland'), *Geogr. Tidsskr.*, vol 31, 23-45
1933   'Contributions to The Physiography of Iceland with particular Reference to the Highlands West of Vatnajökull', *K. Danske Vid. Selsk. Skr. Naturv.-math. Afd.*, 9 R, IV, 5, 183-287, pl. I-XXXII, map 1-9
1934   'Foreløbig beretning om Vulkanudbruddet i Vatnajokull 1934' ('Preliminary report on the volcanic eruption of...'), *Geogr. Tidsskr.*, vol 37, 279
1936   (with Noe-Nygaard, A.), 'Om den islandske "Palagonitformation" Optindelelse' ('On the origin of the Icelandic "Palagonite Formation"'), *Geogr. Tidsskr.*, vol 39, 89-122
----   'Iagttagelser vedrørende den saakaldte "Palagonitformation" i Island' ('Observation on the socalled...in Iceland'), *Nord. naturforskarmotet i Helsingfors (Meeting of Nordic natural scientists in Helsinki)*, 414-20
1937   *Vatnajokull, Kampen mellum Ild og Is (... the fight between fire and ice)*, H. Hagerup, 1-124
----   'Traek af Islands fysiske Geografi og Geologi' ('Features of physical geography and geology'), *Grundrids ved folkelig Universitetsundervisning (Outline on popular university teaching)*, no 425, 1-16

b.   Studies of Skallingen, Wadden and Marsh
1933   'Den videnskabelige Undersøgelse af Halvøen Skallingen' ('The scientific investigation of the Skalling peninsula'), *Geogr. Tidsskr.*, vol 36, 79-87
----   'Jordf ae llesskabet paa Skallingen' ('The joint ownership of land on Skallingen'), *Geogr. Tidsskr.*, vol 36, 88-116
1935   'Eine Methode zur exakten Sedimentationsmessung', *K. Danske Vid. Selsk., Biol. Medd.* XII, 4, 1-98, Tafel 1-XVI
----   'Das Skalling-Laboratorium 1930-35', *Geogr. Tidsskr.*, vol 38/3-4, 202-09
1938   'Nogle bemaerkninger om Marskdannelsen i det danske Vadehav' ('Some remarks on the formation of marshland in the Danish Waddensea'),

Geogr. Tidsskr., vol 41/2, 139-46

---- 'Et instrument til Maaling af Sand- og
Slamtransport i strømmende Vand' ('An instrument
for measurement of transport of sand and mud in
flowing water'), Geogr. Tidsskr., vol 41/2,
139-46

1939 'Et stenalderfund under Ribemarsken' ('A find
from the Stone Age under Ribe marsh'), Geogr.
Tidsskr., vol 42, 156-8

---- 'Landvinding i Holland' ('Land reclamation in
Holland'), Geogr. Tidsskr., vol 42, 9-20

1954 'Vade og Marskproblemer i Danmark' ('Wadden and
marsh problems in Denmark'), Ingeniøren, vol 34,
682-8

1956 (with Jacobsen, Borge and Jensen, Kr. M.),
'Forslag til landvindingsarbejder langs den
sønderjyske Vadehavskyst' ('Proposal for land
reclamation along the Wadden sea coast'),
Geogr. Tidsskr., vol 59, 1-9

c.   Other Works

1924 *Studier over Jaernproduktionen i Jylland
(Studies of the production of iron in Jutland),*
doctorate thesis

1933 (with Fog-Heede, A.), 'Inholdsfortegnelse til
Geografisk Tidsskrift 1911-1930' ('Table of
contents...'), Geogr. Tidsskr., vol 36, 1-74

1936 'Et nyt vae vaerk om Stockholms økomomiske
Geografi' ('A new publication on the economic
geography of Stockholm'), Geogr. Tidsskr.,
vol 39, 35-40

---- 'Knud Rasmussen Fondet' ('The Knud Rasmussen
Foundation'), Geogr. Tidsskr., vol 39, 160-3

1942 'Tale ved Mindehøjtideligheden i Horsens for
Vitus Bering' ('Speech at the memorial
ceremony in Horsens for...'), Geogr. Tidsskr.,
vol 45, 1-9

---- 'Vitus Behrings Mindefond og Vitus Bering
Medaillen' ('The Vitus Bering Memorial
Foundation and the... medal'), Geogr. Tidsskr.,
vol 45, 10-14

---- 'En Oversigt over Danmarks Areal og dets
Udnyttelse' ('A survey of the area of Denmark and
its exploitation'), Sven. Geogr. Årsbok, vol 18,
195, 330-8

1944-5 'Den danske Kartografis Historie.
Publikationer fra Geodaetisk Institut' ('The
history of Danish cartography.   Publication from
the Geodetic Institute' no 1-4, 6, 7), Geogr.
Tidsskr., vol 47, 114-21

1945 'Nogle Bemaerkninger om Geografiens Stilling ved
Universitetet i de 18. Aarhundrede' ('Some
remarks on the position of geography at the
university in the eighteenth century'), Mat.
Tidsskr., B, 156-60

1946 'Martin Vahl' (obit.), Universitetets Festskrift,
144-51

1948-9 'Haslund-Christensen, Henning:
Ekspeditionsmanden og forskeren' ('..., explorer
and researcher'), Geogr. Tidsskr., vol 49, 1-4

1951 'Unifersitetets Geografiske Laboratorium',
Geogr. Tidsskr., vol 51, 134-48

1960 'Fristrup B.: "Studies of four glaciers in
Greenland"', Geogr. Tidsskr., vol 59, 89-102

---- (with Schou, A.), 'Den Internationale

Geografkongres i Norden 1960', Geogr. Tidsskr.,
vol 59, 226-40

---- contributions to *International Geographical
Congress XIX, 1960 Guidebook, Denmark*, symposia
and excursions, 1-372

---- 'Greater Copenhagen', Geogr. Tidsskr., vol 59,
184-203

---- 'Gudmund Hatt' (obit.), Geogr. Tidsskr., vol 59,
6-7

1962 'Københavns Universitets Geografiske Institut.
Et bidrag til dansk geografie hi stoire' ('The
Institute of Geography at the University of
Copenhagen: a contribution to the history of
Danish geography'), Geogr. Tidsskr., vol 61,
1-53; summary 54-7

1968 'Kronprins Frederik, Praesident for Det
Kongelige Danske Geografiske Selskab 1927-47',
Geogr. Tidsskr., vol 67/2, xi-xxi

1971 'Borge Jakobsen' (obit.) Festskr. Kobenhavns
Universitet, 256-61

1978 'Axel Schou' (obit.), Oversigt over det K.
Danske Videnskabernes Selskabs Virksomhed,
1977-8, 164-54

1979 (with Christiansen, S. and Kingo Jacobsen, N.),
'Geografi', in Københavns Universitet 1479-1979,
vol 13, 377-446, published by G.E.C. Gad,
Copenhagen

3.   ARCHIVAL SOURCES

Material in the Danish State Archives is inaccessible
for fifty years from the date of each item.

*Niels Kingo Jacobsen is professor of geography at the
university of Copenhagen*

Chronology

1893   Born at Silkeborg, 3 October

1911   Received the general certificate at the
            Horsens state school and matriculated at
            the university of Copenhagen

1915-17   Held the *Kommunitet* from an endowed charity
            dating back to the reign of Christian IV
            (*ob*. 1648) which gave him free residence
            at the *Regensen*

1917   Graduated as *Cand. mag.* (M. Sc.)

1917-22   Was in residence at the *Elers Kollegium*

1917-18   First experience of teaching, at the
            Ingwardsen og Ellbrechts Skole

1918-19   Taught at the H. Adlers FA Faellesskole

1919-39   Spent twenty years on the teaching staff

of the Sortedam Gymnasium

1924 Received the doctorate degree; was a member of the first Danish expedition to Central Iceland

1927 Took part in the second Danish expedition to Central Iceland

1927-28 Studied in Paris and also in Geneva with Professor Collet, author of a famous study on the structure of the Alpa

1930 Began the work on the Skalling peninsula with a stay of four weeks in a tent

1931 Continued the work on the Skalling peninsula and found a permanent base in a shepherd's cottage: became secretary of the Royal Danish Geographical Society

1932 Established the Skalling laboratory at Vagtbjerg Skallingen

1933-39 Was assistant professor of geography at Copenhagen university

1934 Took part in the third Danish-Icelandic expedition to Vatnajokull

1936 Went once again to Iceland with the fourth Danish expedition to Vatnajokull

1938 Was leader of the Danish delegation to the International Geographical Congress in Amsterdam: in Denmark carried out studies of the Gradyb tidal area with special reference to hydrological sedimentation

1939 Became a full professor at Copenhagen university

1941 Appointed a member of the Committee on the Danish Wadden Sea under the Ministry of Public Works, he continued his work based on hydrology, sedimentation and marine biology. In this year he lectured on the geography of Denmark to large audiences of the general public

1943 During this year he was detained by the German occupying power at the Horserød internment camp and the Skalling laboratory at Vagtbjerg was removed

1943-45 Served as president of the students' union, the *Studenterforeningen*

1945-46 Was dean of the faculty of mathematics and natural sciences

1945-68 Was warden of the *Nordisk Kollegium*

1946 Travelled for study in the United States as a representative of the University of Copenhagen

1947 With others studied the eruption of Hekla, Iceland

1947-59 Editor-in-chief of *Meddeleser om Grønland*, published by the Commission for Scientific Investigations in Greenland

1949 Acted as leader of the Danish delegation to the International Geographical Congress in Lisbon

1950 Achieved the re-establishment of the Skalling laboratory at Ho

1950-64 Was chairman of the Danish-Icelandic Foundation

1952 Was leader of the Danish delegation to the International Geographical Congress at Washington, D.C.

1953 Actively engaged with the Danish Wadden and Marsh Investigations under the Coordinating Committee on Tøndermarsken of the Ministry of Agriculture

1953-71 Was editor-in-chief of TRAP, fifth edition, Denmark

1956 Was vice-president of the Royal Danish Geographical Society

1957 Concerned with land reclamation works at Højer

1960 During the International Geographical Congress, Norden, his main task was the management of the symposium on agriculture at Horsens and he was also leader of the Danish delegation

1961-62 Was pro-rector of Copenhagen university

1964 Retired as professor emeritus

1981 Died at Charlottenlund 15 September

Joseph Franz Maria Partsch

1851–1925

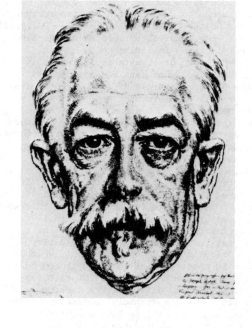

GABRIELE SCHWARZ

Joseph Partsch was a man whose life spanned a period of marked development in German geography and indeed throughout the world. Like many of his contemporaries he had only a limited training in the subject at school and university, but at both he was given a fine general education on which he built a distinguished career as a geographer. His life was firmly rooted in the cities and universities of Breslau and Leipzig and apart from his classical-philological studies, he gave most of his time to work on his immediate homeland, Central Europe, and European mountain areas moulded by glaciation. Gradually he advanced to a modern economic, political and applied geography and his great contribution led to the award of the Carl Ritter medal by the Berlin Geographical Society in 1923.

1. EDUCATION, LIFE AND WORK

Born on 4 July 1851 at Schreiberhau, Riesengebirge, Silesia, Joseph was the second son of Aloys and Emilie Partsch. His father was an official at the Josephine Glassworks (Josephinenhütte), established in 1842 and internationally known for its crystal glass. Aloys Partsch was a man of standing in his home area, a Roman Catholic anxious to see the maintenance of harmonious relations with the Protestant community. H. Waldbaur (*Joseph Partsch. Aus fünfzig Jahren*, 1927, p. 7), wrote that Joseph acquired valuable qualities from both parents, 'a particularly devoted sense of duty and an indefatigable eagerness to work from his father and a creative talent from his mother'.

Geographers Biobibliographical Studies, volume 10 (1986)

As a boy he showed an unusual capacity for observation, especially of the natural world, which was then beyond his powers of explanation, and this was to be valuable in his studies at the university.

Both parents were eager to give their sons the best possible education and Joseph went first to the Catholic Primary School at Schreiberhau, from which, like his elder brother Franz, he went to the nearest Catholic High School, St. Matthias, Breslau. In this connection, Partsch wrote (Chronicle, 184): 'In the first instance, it was a brave thought to send both of the eldest sons to High School, this only being possible by considerable personal deprivation. ... I think I am correct in saying that only by means of Whitman's (London) support--a business friend at the Josephine Glassworks--was it, at all, possible that we were able to be educated at such a school'. In 1860, schooling started at the above-mentioned High School and, particularly in the higher classes, Partsch praised the good instruction in Latin, Greek, German, history and geography. He also learned French fluently. Later he gained an adequate knowledge of English, was able to make himself fully understood in Modern Greek and Italian, and could read other European languages. He stayed at St Matthias for nine years until his matriculation.

The university stage of his education began with many personal questions. A course in law was considered but for this the financial resources were inadequate and he decided against theology, as at St Matthias High School he had been dubious about many points of Catholic theology and practice; notable is the discussion in the 1860s of the infallibility of the pope. In 1880, he became a member of the Evangelical Church, on the occasion of his marriage to Helene, *née* Doepke. At first in Breslau University, he concentrated on classical philosophy but he also attended other lectures including those of Carl Neumann (1823-80), who dealt with ancient history and geography at the time, when little geography was taught in the German universities following the death of Carl Ritter in 1859. Partsch was particularly impressed by Neumann's lectures on 'general hydrology' and attended them regularly, as well as the seminars on ancient history. He also followed several courses given by Ferdinand Roemer on geology.

Carl Neumann's proposed award-winning essay *Choreography, Topography and History of Administration ...of the Province Africa vetus* was tackled by him in his fourth semester and he received distinction for his work in 1872. He was able to use a part of this work for his doctorate thesis two years later, in 1874, *Africae veteris emandantur et explicantur*. At the end of this year he was successful in the public university examination, the 'Staatsexamen', which gave him authority to teach in high schools, though he did this only for a few months. Both at St Matthias and at the University he had given private lessons to relieve his father of the financial burden and before he passed his 'Staatsexamen' he became a private tutor with a respected Breslau family. This gave him time to prepare his habilitation thesis on the Representation of Europe in the works of Agrippa, *Die Darstellung Europas in den Werken des Agrippa*, (1875), showing the relations between ancient history and geography.

With his inaugural lecture on 'The Significance of Silesia for Prussia and Germany', his tendency towards Geography was clear. A year later, 1876, he became an associate professor (ausserordentlich) in ancient history and geography at Breslau University, and was then in a position to undertake study excursions in the Alps during semester holidays (until 1878). From 1880 onwards, he restricted his teaching activities to geography against the advice of Neumann and other friends, for the intention was that he should take over the work of Neumann in both geography and ancient history. However, his choice proved to be wise, for from 1880 he gave most of his time to geography and in 1884 he was given an *Ordinariat* Chair of geography, which meant that he became a full professor of the university.

His glacial-morphological works originated from his study excursions in the Alps. Already in 1878 he pointed out the Ice Age tracks in the Riesengebirge and subsequently, in 1882, the work *Die Gletscher der Vorzeit in den Karpathen und in den Mittelgebirgen Deutschlands* was published and was reviewed by Albrecht Penck in (*Verh. Geogr. Gesell. Berlin*, 1883, 320-5) in which he wrote (320/1):

It is certainly amazing that a man of learning, who has captivated his listeners for many years with lectures on Roman History, is to be thanked for the monograph mentioned on ancient glaciers in the Mittelgebirgen in Germany, and that this author must be even more praised for

revealing a thorough comprehension of relevant geological data. Professor Partsch's book is a work in natural science which early historians only revealed through the steady and conscientious treatment of controversies and which were handled in too much detail. However, this is precisely the point which distinguishes this book from various other works on related fields, which have often been too positively seen.

In 1884, Albrecht Penck asked Partsch to attend the Deutschen Geographentag meeting in Munich, at which problems of glaciation were to be considered (Letters 19/20). In 1887 Partsch and his friends founded the Breslau Section of the German and Austrian Alpine Club, the *Deutschen und Österreichischen Alpenverein*, and ten years later, the idea came from him and others to arrange an award-winning essay on the glaciation of the Austrian Alps, *Die Vergletscherung der österreichischen Alpenländer*, which subsequently led to the basic work by Albrecht Penck and Eduard Brückner, *Die Alpen in Eiszeitalter* (1909).

Nine years after the Munich meeting, in the summer of 1893, Penck, together with Eduard Richter and several students, went to study glacial phenomena in the Riesengebirge under Partsch's guidance. Moreover, he discovered that the improved road conditions enabled one to have a more accurate general view than ten years before. In the work on the glaciation of the Riesengebirge in the Ice-Age, 'Die Vergletscherung des Riesengebirge zur Eiszeit' (*Forsch. Landes- u. Volkskunde*, vol 8/2, 99-194), the results are represented by means of improved surveying of the Schneegrube, on the scale 1:10,000, a glaciation map on the scale 1:75,000, as well as numerous profiles and photographs, for which he was indebted to his younger brother Karl, later to become a well-known oral surgeon at the University of Breslau.

Joseph Partsch's interest in glaciation was to last to the end of his life. After his retirement, he published a work on the High Tatras during the Ice Age, *Die Hohe Tatra zur Eiszeit* (1923), for which he had gathered extensive material, including a glacier map on the scale 1:75,000, as well as numerous illustrations. The review of Albrecht Penck formed the introduction of his obituary (*Z. Gesell. Erdkd.* 1928, 81-7). It is a work which cannot be ignored by those involved in the study of glacial problems of the Donauländer.

Even though Partsch went his own way, he remained a friend and associate (which cannot be taken for granted) of Carl Neumann, who died in 1880, and this was made evident in several obituaries, of which the most detailed appeared in *Z. Gesell. Erdkd. Berlin*, 1882, 81-111. When the *Deutschen Geographentag* was held in Breslau in 1901, Partsch welcomed the members with a portrait of Carl Neumann and published his geographical inaugural lecture given at Breslau in 1863 (*Festschrift*, 1901, 16-27). Although Carl Neumann was well-known in Berlin for his public work, political outlook, and scientific aptitude his teaching work in Breslau had severely limited his time for writing, though much of his material was at least partially prepared for publication. It was, therefore, possible to publish some of this material afterwards. Of this Partsch chose the *Physical geography of Greece* ...(1885), a substantial volume of 475 pages. Apparently the editorial work of Partsch was considerable, for there had been great advances in the knowledge of climatology and geology since Neumann had written his studies. Even so, the work showed the affinity of Partsch with Neumann and the sharing of a compelling interest in the geography of classical Greece. Herrmann Wagner (*Göttinger Gelehrte Anseigen*, 1885, 953-64) 'did not know any other book which was written so much in the spirit of Carl Ritter' (954), and although written by two authors, Neumann and Partsch, 'the homogeneity of the work remained' (955).

There were two interesting results of the publication, of which the first was a grant of support for a visit to Greece by Partsch in 1885 by the Berlin Akademie der Wissenschaften and the second was an enquiry into the state of geography at Breslau university, by John Scott Keltie of the Royal Geographical Society in London. At that time Keltie was preparing his famous report on geographical education which included warm praise for German teaching of the subject in schools and universities, far more advanced in scope and scholarships than in Britain; this heralded the appointment of H.J. Mackinder at Oxford university in 1887, with the modern revival of British geography. As things happened, the paths of Mackinder and Partsch were to cross later.

Philological and historical studies continued to interest Partsch and led to the publication of a number of papers, including a study of Philipp Clüver (1580-1623), a 'pioneer of historical geography' (*Geogr. Abhand. Wien*, vol 5/2, 1891). Later, in 1909, he traced a medieval Latin reference back to its possible source, in an essay by Aristotle on 'the seasonal rhythm of the Nile'. *The Limits of the Ancient Classical World* followed in 1916, and *The River Bifurcation in the Argonaut Myth* in 1919. Finally, he published a paper on 'Palmyra, an historical-climatical study' (1922), in which he reaffirmed a view expressed in other papers, that there had been no change of climate in the Mediterranean region during historical times. All these studies appeared in the *Berichte der K. Sächsischen Gesellschaft der Wissenschaften, philologish-historische Klasse, Leipzig*. This element in his work shows the connection between geography and classical studies which was prevalent in Britain, as well as in Germany, in the last decades of the nineteenth century, indeed to such an extent that in German historical geography not only the conceptions of the classical world were apparent but from the early years of the twentieth century an historical approach was also applied to other and later periods.

Having made several visits to Greece from 1885 onwards, Partsch wrote monographs in the famous *Petermanns Geogr. Mitt. Erg.* series on *Korfu* in 1887, on *Leukas* in 1889 and *Kephollenia and Ithaca* in 1890. In 1891, he added an article on 'Zante' (*Petermanns Geogr. Mitt.*, vol 38, 1891, 161-74). Like other geographers, his first need was a suitable map, and for Korfu, Leukas and Kephallenia with Ithaca, he used a 1:100 000 map based on an English survey which he supplemented from his own measurements. Many other maps were used concerning the specific problems of the various islands, whether it was a geological map of Korfu, a map of the earthquake of 1867 on Kephallenia, or plans of reconstructions of older towns or the surveying of new towns. In this way, the studies vary in their internal structure. In the case of Korfu, he supplemented the 'natural history' with a section on the effect of malaria (53-5). As far as 'anthropogeography' was concerned, a division into 'situation', 'coasts' and 'interior' was made. With reference to Leukas, he suggests another method, where the 'lagoon' and the 'mountainous country' varied; nature and culture of the island were interwoven. Finally, in the case of Kephallenia, he supplemented the 'natural history' with a section dealing with malaria (36/7). The definition 'anthropogeography' had here been replaced by 'cultural geography', and the utilization of the soil took a special meaning from 'history of the Corinth' (100/6). E. Fels, in his article 'Korfu-Keffallenia-Ithaka' (*Mitt. Geogr. Gesell. München*, vol 20, 1927, 147-77), showed agreement with Partsch's view that the Kephallenians were the Prussians of Hellas, as compared with the people of Korfu. Later, H. Müller-Miny (*Verh. Dtsch. Geogr. Hamburg*, vol 30, 1955, 396-406) explained that the historical geography of Greece, in the modern sense, was only made possible by the work of Partsch. It was certainly admired in Greece and the Petermann monographs on Korfu and that of Kephallenia and Ithaca were translated into Greek in 1892. In 1912, Partsch received an honorary doctorate from the University of Athens.

Apart from methodological questions, because he had to comply with the geographical aspect treated, the monographs also received attention overseas 'so that an Englishman described them as exhaustive studies which could only be produced in Germany' (Chronicle, 285; original in the Partsch-Archive, Leipzig).

Work on Germany, and particularly on Silesia, was a concern of Partsch and from 1887 until shortly before his death he was a member of the *Centralkommission für wissenschaftliche Landeskunde*, the central commission on scientific geography. This body was founded by one of A. Kirchhoff's pupils and Kirchhoff was the first chairman from 1884-9 and also at various later periods. Every two years, at the meeting of the *Deutsche Geographentag*, a new executive committee was appointed, but re-election was possible. It was generally the case that the acting professor became the regional representative for the area in which his university was located and therefore Partsch's initial responsibility, from 1887 to 1905, was for Silesia and Posen but from 1905 to 1925 for Saxony. When Partsch moved to Leipzig in 1905 Siegfried Passarge became the regional representative for Silesia and in 1925, at the Breslau *Geographentag*, Partsch was replaced by Wilhelm Volz.

One useful idea was that a directive should be prepared for German army generals on the naming of places on topographical maps, of which Partsch had a profound knowledge. Also for military purposes, he prepared a study on the climate of Silesia which, as part of a geographical description, was praised by Field-Marshall General von Moltke in 1887 and his successor Count Waldersee in 1889.   From 1892 until 1900, in association with the Central Commission, he was involved with the bibliographical work in which comments were added to titles, the *Literatur der Landes und Volkskunde der Provinz Schlesien*; this work was highly praised by Kirchhoff (*Petermanns Geogr. Mitt.*, vol. 42, *Literaturbericht*, 23).   'Die Regenkarte Schlesiens und der Nachbargebiete mit einer Karte 1:1 000 000', in 1896, and in 1897, a paper on the Oder Basin, 'Der Oderstrom, sein Stromgebiet und seine Nebenflüsse' (*Petermanns Geogr. Mitt.*, vol 43, 37–41) formed a further basis of his later work.   Both these publications were basic material for the two-volume work, *Schlesien eine Landeskunde für das deutsche Volk*, of which the first volume appeared in 1896 and included a detailed general physical study of the environment, a shorter human section and a final chapter on Silesia as a war area 'Schlesien als Kreigsschauplatz', (398–420). The second volume was more concerned with anthropogeography and appeared in three parts-- Oberschlesien, (1903), Mittelschlesien (1907) and Niederschlesien (1911).   This work won praise from Norbert Krebs who said in a review (*Z. Gesell. Erdkd. Berlin*, 1912, 391–3):

> In contrast to the sparkling ideas of Ratzel, this work deals with realities and shows that anthropogeography can only be satisfying if it includes physical causes and also a profound understanding of settlement and economic history...Partsch, much attached to his native country, draws on many memories for his beautiful descriptions, his deep regard for many important places and both its tragic history and its present flourishing growth. He is an author who not only knows every corner of his own country but he also understands the joys and sorrows of its people and how to convey all this in his work.   Partsch's Silesia is a work for all the people of Germany and it has no rival.

In time it became known outside Germany, notably in France and Great Britain.

In 1899, Partsch became Chancellor of the Breslau university and gave as his address a study of 'Geographical research in the nineteenth century', included in the volume of 1927 by Waldbaur (op. cit., 35–45).   On this address, Kirchhoff ('Biography of Partsch', op. cit., 14) commented that 'no other study of whatever size to a large book gave so thoroughly the essence of the achievement of geography during the nineteenth century' (*Petermanns Geogr. Mitt.*, vol 42, 1896, *Literaturbericht* 23).   In this essay, he raised once again the idea of replacing the term 'anthropogeography' by 'cultural geography', which he had already used in his last works on the Ionian Islands.

> In my opinion, Ratzel's misinterpretation concerning the passive relation of other human beings to their dwelling place and mainly the behaviour of human beings towards their native country should be radically transformed.   It is so important, that I find the old term 'cultural geography' far more characteristic and by no means too narrow an expression (Waldbaur, ed., 1927, 42).

The use of 'cultural geography' was widely accepted, though in Anglo-Saxon countries some distinction was drawn between the use of 'cultural' and 'human' geography, the term drawn from the French *géographie humaine*.

By now a major figure in Silesia, Partsch spent part of his year as Chancellor organizing the *XIII Deutsche Geographentag* of 1901 in Breslau and this was the first congress where *Festschriften* were published.   In one of them he wrote a study of the geography of Breslau as an introduction to the meeting with other contributions on Breslau and Silesia.   He also organized a cartographical exhibition with the support of the Board of Mines, the Oder River Administration and the Ordnance Survey. The governor of Silesia, Prince Hatzfeld, himself countersigned all the letters asking for the valuable material presented, for example the Pless territory in 1635, a survey map of Oels by Field-Marshall von Moltke, which had a newly surveyed map next to it (Chronicle, 211/12).   In addition, there were successful excursions in Upper Silesia and the Riesengebirge, the latter led by Partsch himself.

Mackinder planned a series of twelve volumes as a world regional geography, of which his own, *Britain and the British Seas*, 1902, was the first published. In 1897, he asked Partsch to undertake the volume on Central Europe.   The definition of Central Europe was broad and, with reference to the physical conditions, described as 'The threefold belt of Alps, inferior chains, and northern lowlands (which) controls the landscape and scenery of Central Europe.   Whenever one of these elements dies out, Central Europe comes to an end' (*Central Europe*, 1903, p. 2).   Hilda Ormsby (*Scott. Geogr. Mag.*, vol 51/6, 1933, 338) observes that

> for Partsch, Central Europe extends from 42 to 55 North, from Ostend to Geneva, from Memel to Burgas, from Dunkirk to Sandomirz, and includes Teutons, Slavs, Romance and Low Altaic peoples, covering one-sixth of Europe and one-third of the European population. He includes the pre-war Empires of Germany and Austria-Hungary, together with a western fringe, comprising Switzerland, Belgium and the Netherlands, and a south-eastern fringe of Montenegro, Serbia, Roumania and Bulgaria. With Partsch, I think, Central Europe finds its maximum extension.

This was the definition given to Partsch by Mackinder.

The definition of Central Europe became a theme much discussed in political and regional geography later on and Partsch's work, with others, was widely read in the English edition.   However, the definition was bascially of a physical entity and this has remained acceptable to many German geographers. In Partsch's view the English edition of 1903 of his *Central Europe* had been emasculated, for Mackinder had insisted that technical terms and statistics should be avoided as far as possible, because the book should be regarded as for the general reader. In this, one may discern the influence, or at least the approval, of the commercial publishers. Partsch, therefore, decided to publish the more adequate German text which was issued by the Perthes firm in 1904; this included many revisions of the manuscript on which he had worked for the previous five years, a good deal of statistical data, more references to bibliographical sources and a number of sketches not given in the English edition.

The 1904 edition should be regarded as the crown of his work on Central Europe. Two major points he made were to be widely discussed later. Of these, the first was his view that in eastern and southeastern areas of Central Europe various peoples were increasingly conscious of their own nationality but were largely dependent on German as a medium of communication (1904, 173). The second point of consequence in political geography was the inclusion of a chapter on the geographical basis of national defence, 'Die geographischen Bedingungen der Landesverteidigung' (1904, 415-31). These two points are also included in the English edition.

During the last years before World War I, Partsch made two interesting journeys. In 1910 he attended the International Geological Congress at Stockholm which was followed by an excursion to Lappland on which he wrote a paper for the Leipzig geographical journal (*Mitt. Gesell. Erdkd. Leipzig*, 1911, 70-88) with several other papers that were read with great interest. In 1912, he was chosen together with Erich von Drygalski and Albrecht Penck, by the Berlin Geographical Society, to attend the diamond jubilee celebrations of the American Geographical Society and the famous trans-continental excursion which followed. In this he was most concerned with morphological problems, naturally prominent under the leadership of William Morris Davis. However, Waldbaur (op. cit. 1927, 18) shows that other aspects of the geography of the United States also engaged his attention and this was revealed in his account of the tour (*Z. Gesell. Erdkd. Berlin*, 1913, 249-73) and other articles.

Like many leading geographers of his time he was drawn into the preliminary work for the preparation of the 1:1 000 000 map, first suggested by Albrecht Penck 'who had flung' this object 'into the world' (Letters, 1891, 91/2) on the occasion of the International Geographical Congress in Berne in 1891. This led to international conferences concerning the World Map; 1895 in London and 1913 in Paris. Partsch, as representative of the Kingdom of Saxony, participated and was closely associated with Penck. Partsch wrote a paper on the history and progress of this enterprise (*Mitt. Gesell. Erdkd. Leipzig*, 1913, 80-98). At the Paris meeting he was well aware of the current tense international situation, not surprisingly for one who had studied the problems of national defence with particular insight.

Probably he was more perspicacious on the dangers

of war for Germany than most people but this did not deter him from giving lectures on the progress of the war to an audience in Leipzig. These included a lecture to 5,000 people in the Albert Hall in Leipzig on 'Germany's eastern border' during the first winter of the war in which he noted that the military situation was critical; this brought adverse comment from people who did not want such views to be held, still less mentioned. Significantly, the American *Journal of Geography* (vol 13, 1914-15, 102-10) published the section of his final chapter in the Central Europe volume on Poland in war, and this also appeared in *Geogr. Z.* (vol 20, 1914, 604-15, 670-88). The treatment of other war areas followed. He was active in war work and supported the enterprise in Leipzig of organizing a corps from East Prussia to defend the Austrian frontier in the Carpathians against Russian attack. In fact he was not surprised by the outcome of the war and was naturally troubled by the provisions of the Treaty of Versailles. He both spoke and wrote on the Silesian problem and held the view that a plebiscite would not conform to the provisions of international law.

During the war, Partsch wrote several obituaries of former students who had been killed and afterwards he did everything possible to help those who had returned to the university. In 1921, he organized the first post-war *Geographentag* at Leipzig, inevitably on a more modest scale than the gathering he had organized at Breslau in 1901. Given the status of emeritus professor on retirement in 1922, he completed his work on the glaciation of the High Tatra. He also worked on the material used at the Leipzig Commercial University on world trade, of which the first two parts were completed and edited by R. Reinhard, though to the regret of many geographers, the third section on world trade routes and sections never appeared.

At the beginning of 1925, he lost his eldest son Josef, who was a well-known lawyer of ancient and international law and, because of his vast knowledge, was the final person to be consulted regarding the difficult court case in Geneva. He had prepared a script concerning the French preliminary outline of the Treaty of Versailles. His old and bent father made it his duty to work on all the material which had been collected by his son until he, too, could not cope any more. During a course of treatment in Bad Brombach, in Vogtland, he died on 22 June 1925, after having had a stroke. His wife, who was also very ill, died several weeks later. Both were buried in Leipzig.

2. *SCIENTIFIC WORK AND GEOGRAPHICAL THOUGHT*
Joseph Partsch was a contemporary of Alfred Hettner (1859-1941) and Albrecht Penck (1858-1945), and his academic career covered fifty years, from the early seventies to the early 1920s. These three men have been grouped as the great pioneers of modern German geography. However, Penck was drawn to the subject from the natural sciences whereas Hettner, who originally came from the natural sciences, found his way to regional geography even though methodological questions were of more interest to him. The educational background of Partsch was in philology and

classical history.   He came to geography partly from
his interest in the character of the classical world
and had his reasons for being cautious with
methodological problems.

Looking backwards in time, Partsch had a link with
Carl Ritter who had taught both Carl Neumann of
Breslau and Heinrich Kiepert (1818-99), who was an
excellent cartographer and ran the Department of
Historical Geography in Berlin; Partsch wrote a study
of his life and work (*Geogr. Z.*, vol 7, 1901, 1-21,
77-94).   But as explained earlier, Partsch was eager
to broaden his education, for example by the study of
geology and of climate, and this was noted in
reviewing his work on the glaciation of the Central
German Mountains and the Carpathians (*Verh. Geogr.
Gesell. Berlin*, 1883, 320/1).   Problems of the Ice
Age remained an important field of study in his life.
With all this, he developed the interest in classical
geography which he had acquired from Neumann and he
showed not only a general geographical approach of a
regional character but also the occasional wish to
embark on some esoteric study such as a manuscript or
individual.   He edited and to some extent compiled
his first general work on Greece by 1885, using the
unfinished manuscript of Neumann before his own first
visit to Greece later in the same year.   From this
and later visits came the series of volumes that were
so warmly welcomed in Greece as well as in Germany.

It has been widely expected of geographers that
they should be well informed on their home area, and
Partsch's devotion to Silesia was unequivocal.   It
was clearly demonstrated from 1887, and as time went
on, he became increasingly aware of the danger
threatening Silesia in the event of war.   This was
one source of his strong interest in political
geography, further stimulated by his cartographical
and regional work.   This was most clearly demonstrated
by the publication of *Central Europe* in 1903, and the
more adequate German version in 1904.   As it happened
his fears for Silesia were prophetic but at the end
of his life he looked forward to a new economic
geography in his professional work.

3.   INFLUENCE AND SPREAD OF IDEAS

To a modern writer on geography, Partsch is remarkable
for the breadth of his scholarship; but he belonged to
a time when regional geography was venerated and many
of his exponents were competent in both its physical
and human aspects.   This was expected and, for
example, H.J. Mackinder in *Britain and the British
Seas* dealt not only with the British rivers but also
with theories of their origin.   In this one
influence was W.M. Davis with his explanatory
geomorphology as a basis for the presentation of
landforms in regional geography.   Partsch was in
accord with many leading geographers of his day in the
study of glaciation; he was also in tune with the
times in his work on the historical geography of
classical Greece, for such a work gave geography
academic status and also enriched the understanding of
history.

Although it is now clear that the English version
of his book on Central Europe is only part of a greater
whole, it is by this book that he is best known in the
English speaking world.   As a regional geography,
firmly rooted in the physical landscape, it is a work
with political and economic significance.   That
Partsch was much concerned with political geography
needs no emphasis but within the framework of the
lectures at the Commercial University in Leipzig, he
was seeking a new approach to economic geography, even
if only some were published afterwards.   This has been
well expressed by E. van Cleef (*Geogr. Rev.* vol 18,
1928, 171) in these words:

> Partsch was always a pioneer, but his modesty
> retarded a full appreciation of his work by
> his contemporaries.   Just as he was one of
> the first geographers to recognise the value
> of a regional treatment of his science, well
> developed in his *Central Europe*, still a
> standard work, so in his *Geographie des
> Welthandels* he blazes a trail by his
> departure from the method of treatment that
> characterises most of the commercial or
> economic geography texts.

More recently, in 1953, Overbeck emphasized the
'being ahead' quality of Partsch's regional geography,
however only concerning German geography.   His
assessment of his work is shared by R.E. Dickinson,
who in 1969 (93-7) wrote that the work of Partsch
speaks conclusively of the geographical tradition from
the founder Ritter, through Kiepert and Neumann.
Dickinson summarizes the life and influence of Partsch
in the statement that 'The development of the
discipline from its over-emphasis of classical
geography is evident in the career of this one man,
a great scholar of his time'.

A biography would be incomplete without dealing
with Joseph Partsch's personality.   He had contact
with all German university lectures concerned with
geography.   As can be seen from the friendly
relationship to classical philologists, historians and
historical geographers, he did not stand close to
Alfred Hettner but rather to glacial morphologists
like Eduard Brückner, Eduard Richter and above all
Albrecht Penck.   The latter characterized him very
clearly (Letters, 1918, 18):   it offers thanks for

> the thoughtful advice regarding the sureness
> of an opinion in questions concerning life
> and science.   You have the power to weigh up
> the possibilities and then to make the
> decision.   You have grown up on the soils of
> history and you have let your eyes roam over
> the entire world.   You find your roots in
> humanity and in your simple disguise, the most
> faithful and sincere heart beats.

Bibliography and Sources

1.   OBITUARIES AND REFERENCES ON JOSEPH PARTSCH
The main source is Waldbaur, H. (ed), *Joseph Partsch:*

Aus fünfzig Jahren, Verlorene Schriften, Breslau, 1927, which includes a biography, 7-45, a complete bibliography by Else Hauck (*née* Partsch), 170-84, with several papers of which some are listed below and marked with an asterisk.

Brückner, E., 'Joseph Partsch', *Petermanns Geogr. Mitt.*, vol 71 (1925), 179-81

Dickinson, R.E., *The Makers of Modern Geography*, London 1969, 89-93

Friederichsen, M., in *Schlesier des 19. und 20. Jahrhunderts*, Breslau (1926), vol 2, 380-9

Lehmann, F.W.P., 'Joseph Partsch', *Geogr. Z.*, vol 31 (1925), 321-9

---- 'Joseph Partsch', *Mitt. Geogr. Gesell. Erdkd. Leipzig*, (1923-5, publ. 1926), 5-7

Overbeck, H., 'Joseph Partsch', *Ber. Dtsch. Landeskunde*, vol 12 (1953), 34-56 (includes a list of doctorate theses made under Partsch in Breslau and Leipzig, 54-6)

Penck, A., 'Joseph Partsch', *Z. Gesell. Erdkd.*, vol 63 (1928), 81-98

Praesent, H., 'Joseph Partsch', *Mitt. Geogr. Gesell. München*, vol 19 (1926), 202-10

2. SELECTIVE AND THEMATIC BIBLIOGRAPHY OF WORKS BY JOSEPH PARTSCH

a. Classical-Philological Works with Historical Geography Research

(i) Classical Studies

1872 *Chorologie, Topographie und Geschichte der Verwaltung bis zum Einfall der Vandalen der Provinz Africa vetus (Chorology, topography and the history of administration in the Africa vetus province)*, Breslau, in manuscript form

1874 *Africa veteris itineraria emendantur et explicantur*, diss., Breslau, 70 p

1875 *Die Darstellung Europas in den Werken des Agrippa (The representation of Europe in the works of Agrippa)*, 80 p. (*habilitation* thesis, Breslau)

1879 *Corippi africani libri qui supersunt. Monumenta Germanica auctorum antiquissimorum*, Tomi 3, Pars 2, Berlin, 195 p.

1882-83 'Die Veränderungen des Küstensaumes der Regentschaft Tunis in historischer Zeit' ('The changing coastline of the regency of Tunis in historical times'), *Petermanns Geogr. Mitt.*, vol 29, 201-11

1885 (with C. Neumann) *Physikalische Geographie von Greichenland mit besonderer Berückichtigung auf das Altertum (Physical geography of Greece....)*, Breslau, 475 p.

*1888 'Geologie und Mythologie in Kleinasien' ('Geology and mythology in Asia Minor'), *Philolog. Abhand.Martin Hertz zum 70. Geburtstag*, 105-22

*1896 *Die Berbern in der Dichtung der Corippus. Satura Viadrina (The Berbers in the poetry of Corippus)*, special publication celebrating the 25th anniversary of the Philological Society of Breslau, 22 p.

1897 *Karte der Pisatis. Olympia*, ed Curtius and Adler, map volume; also guide to the map of Pisatis in text volume, ed Curtius and Adler, 1-15

1905 *Ägyptens Bedeutung für die Erdkunde (The importance of Egypt in the field of geography)*, inaugural lecture at the university of Leipzig, 39 p.

1909 'Des Aristoteles Buch "Uber das Steigen des Nils Nils". Ein Diagram' ('Aristotle's book "On the rising of the Nile". A diagram'), *Ber. Verh. K.-Sächsischen Gesell. Wiss.-Philolog. Hist. Kl.*, vol 27, no 16, 553-600

1916 'Die Grenzen der Menschheit', part 1, 'Die antike Oikumene' ('The limits of mankind', part 1, 'The ancient Oecumene'), *ibid.*, vol 68/2, 62 p.

1919 'Die Stromgabelungen der Argonautensage' ('River bifurcation in the Argonaut myth'), *ibid.*, vol 71/2, 17 p.

1922 'Palmyra, eine historisch-klimatische Studie', ('Palmyra, an historical climatic study'), *ibid.*, vol 74/1, 17 p.

(ii) Historical Geography and Historical Geographers

1882 'Zur Erinnerung an Carl Neumann' ('In memory of Carl Neumann'), *Z. Gesell. Erdkd. Berlin*, 81-111

1886 'Carl Neumann', *Allg. Dtsch. Biogr.*, 530-2

1891 'Philipp Clüver, der Begründer der historischen Länderkunde', ('Philip Clüver, founder of historical geography'), *Geogr. Abh.*, Vienna, vol 5, 47 p.

1892 'Die Entwicklung der historischen Länderkunde und ihre Stellung im Gesamtgebiet der Geographie' ('The development of historical geography and its place within geography as a whole'), *Das Ausland*, 401-03, 417-20

1901 'Heinrich Kiepart. Ein Bild seines Lebens und seiner Arbeit', ('Heinrich Kiepert. His life and work'), *Geogr. Z.*, vol 7, 1-21, 77-94

1922 *Carl Neumann. Schlesier des 19. Jahrhunderts (Carl Neumann. A Silesian of the nineteenth century)*, Breslau, 5 p.

b. Studies of Glacial Morphology

1878 'Gletscherspuren im Riesengebirge' ('Glacier tracks in the Riesengebirge'), *Mitt. Dtsch. Österr. Alpenvereins*, 241

1882 *Die Gletscher der Vorzeit in den Karpathen und in den Mittelgebirgen Deutschlands (Glaciers of the Carpathian mountains and the central chain of mountains in Germany in early times)*, Breslau, 198 p.

1894 'Die Vergletscherung des Riesengebirges zur Eiszeit' ('Glaciers of the Riesengebirge in the Ice Age'), *Forsch. Dtsch. Land.-Volsk.*, vol 8/2, 99-194

1904 'Die Eiszeit in den Gebirgen Europas zwischen dem nordischen und alpinen Eisgebeit' ('The Ice Age in the mountains of Europe between the northern and alpine ice areas'), *Geogr. Z.*, vol 39, 657-65

1923 *Die Hohe Tatra zur Eiszeit (The High Tatra in the Ice Age)*, Hirt, Leipzig

c.   *Regional Geography and Inter-Related Studies*
(i)   Silesia (excluding glacial morphology)
1889   *Kleine Landeskunde der Provinz Schlesien (Short
       geography of the Silesia province)*, F. Hirts
       Sammlung deutscher Landeskunder zur Ergänzung
       von der Schulgeographie von E. von Seydlitz,
       1 ed, 32 p.; 8 ed, 1918, 48 p.
*1891 'Die Schneedsecke als Bahn des Verkehrs' ('The
       snow cover as a way for transport'), *Das
       Wetter*, 150-63, 187-91
1892-1900 'Literatur der Landes-und Volkskunde der
       Provinz Schlesien in 7 Heften' ('Geographical
       and folklore literature in the Silesia province
       in 7 parts'), *Ergänzungshefte des Jahrbuchs der
       Gesellschaft für vaterländische Cultur*, no 69,
       70, 72, 73, 74, 75, 77
1895   'Die Regenkarte Schlesiens und der Nachbargebeite
       mit einer Karte 1:1 000 000' ('A precipitation
       map of Silesia and neighbouring areas'), *Forsch.
       Dtsch. Land.-Volsk.*, vol 9/3, 199-229
1896   *Schlesien, eine Landeskunde für das deutsche
       Volk*, part 1, *(Silesia, a geography for the
       German people)*, Breslau, 420 p. (see also
       1903-11 *infra*)
1897   'Der Oderstrom, sein Stromgebiet und seine
       Nebenflüsse' ('The river basin and tributaries
       of the river Oder'), *Petermanns Geogr. Mitt.*,
       vol 41, 37-41
1901   'Die Geographie an der Universität Breslau'
       ('Geography at the University of Breslau'),
       *Festschr. Semin. Univ. Breslau...13 Dtsch.
       Geogr.*, 1-37
----   'Lage und Bedeutung Breslaus' ('Situation and
       significance of Breslau'), in *Breslau: Lage,
       Natur und Entwicklung.   Eine Festgabe des
       13 Deutschen Geographentages*, 1-29
1903   'Schlesien an der Schwelle und am Ausgang des
       19. Jahrhunderts', *Festschrift der 100-Jahr-Feier
       der Gesellschaft für vaterlandische Cultur*,
       51-61
1903-11 *Schlesien* vol 2; 1, *Oberschlesien*, 1903,
       188 p.: 2, *Mittelschlesien*, 1907, 276 p.: 3,
       Niederschlesien, 1911, 220 p.

(ii)   Regional Geography of Greece
1886   'Bericht über die wissenschaftlichen Ergebnisse
       der Reise auf den Inseln des Ionischen Meeres'
       ('Report on the scientific results of the
       journey to the islands in the Ionian sea'),
       Sitz.-Ber. K.-Preussischen Akad. Wiss. Berlin,
       615-28
1887   'Die Insel Korfu.   Eine geographische
       Monographie' ('The island of Corfu: a
       geographical monograph'), *Petermanns Geogr.
       Mitt. Ergänz.*, no 88, 97 p. (also issued in a
       modern Greek translation by P. Begia in 1892,
       299 p.)
1889   'Die Insel Leukas.   Eine geographische
       Monographie' ('The island of Leukas: a
       geographical monograph'), *Petermanns Geogr.
       Mitt. Ergänz.*, no 95, 29 p.
1890   'Kephalenia und Ithaka.   Eine geographische
       Monographie' ('Cephalenia and Ithaka: a
       geographical monograph'), *Petermanns Geogr.
       Mitt. Ergänz.*, no 98, 108 p. (also issued in a

modern Greek translation by L.P. Papendreos,
       276 p., Athens, 1892)
*1902 'Auf der Insel der Pelops' ('On the islands of
       the Pelops'), *Schlesische Z.*, no 508, 514, 523,
       536, 541, 544

(iii) Regional Geography of Central Europe
1903   *Central Europe; with maps and diagrams*, Regions
       of the world series, London, no 3, 341 p.
1904   *Mitteleuropa.   Due Länder und Völker von den
       Westalpen und dem Balkan bis an den Kanal und
       das kurische Haff*, Gotha, 463 p.

(iv)   Studies in Economic and Political Geography
1914-15 'Der polnische Kriegsschauplatz' ('The Polish
       seat of war'), from chapter 20 of *Central
       Europe*, *Geogr. Z.*, vol 49, 605-15, 670-88:   also
       published in *J. Geogr.*, vol 13, 102-10 (in
       English)
1915   'Belgien', *Z. Geogr. Gesell. Erdkd. Berlin*,
       137-55
----   'Ostpreussen als Kriegsschauplatz' ('East
       Prussia as a seat of war'), *Geogr. Z.*, vol 50,
       22-32
----   'Der karpathische Kriegsschauplatz' ('The
       Carpathian seat of war'), *Geogr. Z.*, vol 50,
       177-94
*1921 'Oberschlsiens Schicksal' ('The destiny of Upper
       Silesia'), *Die Westmark, Rheinische
       Monatsschrift für Politik, Wirtschaft und
       Kultur*, 976-89
1927   (posthumous) *Die Geographie des Welthandels
       (The geography of world trade)*, ed R. Reinhard,
       348 p.

3.   SOURCES AND ARCHIVAL MATERIAL
The Partsch family chronicle, as well as other
important information was made available by Professor
Dr. Karl Hauck (Münster/Westphalis).   [Denoted as
'Chronik' in text.]
     The photograph of Joseph Partsch was made
available by Professor Dr. Franz Tichy (Erlangen/
Bavaria).
     A collection of local and foreign critical
reviews of works by Joseph Partsch was made available
by Professor Dr. Franz Tichy (Erlangen/Bavaria).
     The Partsch archive in the Deutsche Institut für
Länderkunde" in Leipzig (German Institute of
Geography), now called the Institut für Geographie und
Geoökologie, Akademie der Wissenschaften der DDR
(Institute of Geography and Geo-ecology, Academy of
Sciences of the German Democratic Republic).
[Denoted as 'Partsch Archiv' in text.]   The archive
is not at present available owing to reorganization.
     From this archive, letters written by Albrecht
Penck to Joseph Partsch were published in Engelmann,
G., 'Briefe Albrecht Pencks an Joseph Partsch',
*Wissenschaftliche Veröffentlichungen des Deutschen
Instituts für Länderkunde*, vol 17/18, Leipzig 1960 p
17-107.   Joseph Partsch's correspondence to Albrecht
Penck has been accepted as lost.
     Metz, F., Grosse Geographen des 19. und 20.
Jahrhunderts.   (Famous geographers of the 19th and
20th centuries), lecture 1958-59, was made available

by Professor Dr. Rudolf Metz.

The author offers grateful thanks to all those who have helped in the work.

Gabriele Schwarz is a retired Professor of Geography at the University of Freiburg i. Br.

Chronology

1851	Born 4 July at Schreiberhau, Silensia
1857	Started primary school in Schreiberhau
1860	Attended the Matthias High School in Breslau
1869	Matriculation at the University of Breslau
1869	Began studies in classical philology, incorporating the lecture by Carl Neumann 'Introduction to general hydrography'; from 1878 also including lectures in geology by Ferdinand Roemer
1874	Doctor's degree (Ph.D.) examination
1875	State examination in the subjects: Latin, Greek, history, geography, philosophical propaedeutics
1875	*Habilitation* in geography and ancient history
1876	Associate Professor of Geography and Ancient History
1880	Lecturing limited to the field of geography
1884	Appointed as Full Professor of Geography at Breslau university
1885	Refusal of a professorship in Königsberg
1899-1900	Chancellor of the University of Breslau
1901	Chairman of the local committee for the 13. Deutschen Geographentages (German Geographical Congress)
1902	Refusal of a professorship in Vienna
1904	Refusal of a professorship in Halle
1905	Acceptance of a professorship at the University of Leipzig
1908-09	Corresponding member of the Akademie München (Academy of Munich) Doctor honoris causa of the University of Geneva
1910	Participant of the XI. International Geological Congress, Stockholm
1911-12	Honorary member of the 'Verein für Geographie und Statistik' (Geographical and Statistical Society), Frankfurt a. M. Doctor honoris causa of the University of Athens
1912	Honorary member of the University of Greifswald
1912	Guest of honour at the American Geographical Society, New York
1921-22	Corresponding member of the 'Akademie der Wissenschaften Berlin, physikalisch-mathematische Klasse' (Academy of Sciences. Physical and Mathematical Section)
1922	Professor Emeritus
1923	Awarded the Carl Ritter Medal
1925	Died 22 June in Bad Brambach/Vogtland. Buried in Leipzig

Henri-François Pittier

1857–1950

JORDI MARTÍ-HENNEBERG AND ANNE RADEFF

Rather than a man who gave his whole life to geography, Henri Pittier would seem to be a geographer of encyclopaedic outlook who studied various specialisms at different periods of his life. In each of these several disciplines he followed the accepted geographical order and used his studies of form, relief, vegetation and other distributions as phenomena in space. Essentially he was a phytogeographer who defined regions in terms of the plants found within them, in contrast to a botanist unconcerned with geography who would classify plants without necessarily considering their location or their association in regional units. The logic of a whole and its parts, which sustains all geographical reasoning, dominated the thought of Henri Pittier.

1. EDUCATION, LIFE AND WORK

The life story of Henri Pittier would appear to be a series of daring choices, some of which were followed by failure. Disaster never quenched the spirit of Pittier, a man of quite extraordinary vitality who was born in the middle of the nineteenth century and died in the middle of the twentieth. He was a man of many specialisms, able to turn from one to another as occasion arose and perhaps dictated: essentially he was a product of the encyclopaedic phase of geography current in his time. Scientifically he was both in turn and at the same time a botanist, climatologist, cartographer or ethnologist while in his professional life he was at various times a teacher, researcher, landed proprietor, and much else besides. His life shows the versatility of knowledge and variety

Geographers Biobibliographical Studies, volume 10 (1986)

of activity that marked the careers of some savants during and for some time after the second half of the nineteenth century. It was long in duration and spent in four different countries, with his first thirty years spent in Switzerland, the next thirteen in Costa Rica, followed by eighteen in the United States and finally over thirty to his death in Venezuela.

Born at Bex in the French-speaking canton of Vaud, on 13 August 1857, Pittier was of humble origin. His father was a working carpenter at a saltworks some kilometres away and in 1858 the family became neighbours of the Thomas household at Dévens. This family were dealers in plants and devoted botanists and young Henri Pittier acquired his first knowledge of botany from the Thomas children in the Chablais area of the Vaud canton.

Fortunately Henri's artisan background did not preclude entry to the best educational facilities available and after five years at the local primary school in Fenalet he went to the *collège* for his secondary education in 1869 and then, in 1873, to the *Ecole normale* for intending school teachers. Two years later, in 1875, with the financial help of the Bex commune, he went to the *Académie* at Lausanne, which was the precursor of the university founded in 1890. He explained that he abandoned the course at the *Ecole normale* as he found that he lacked the vocation to become a teacher of general subjects. On 25 July 1877 he graduated as a Bachelor of physical and natural sciences.

Clearly he was fortunate in his opportunities for the Vaud canton had made higher education more readily

available for those of limited means since the election
of a radical government some forty years earlier.
During his childhood there were several events which
presaged an advance in geography teaching: in 1858 the
last volume of Alexander von Humboldt's *Kosmos* appeared;
in 1859 Charles Darwin's *Origin of Species* was
published and B. Studer's *Carte géologique de la Suisse*
was issued, to be followed in 1868 by E. Haeckel's
Natural History of the Creation. In Switzerland the
federal bureau of statistics was founded in 1860 and
the Alpine Club, of which in 1877 Pittier was to
become a member, was formed in 1863.

However there were delays in the full university
recognition of geography and on his graduation
Pittier wrote that though he had a marked partiality
for the subject from childhood he was unable to study
it at Lausanne for geography teaching had virtually
ceased in 1845. Nevertheless some of the courses at
the *Académie* were to be of value in his later work,
especially as a basis for physical geography. On
16 October 1877 he became a teacher of natural
sciences, history and geography at a *collège* run by the
Henchoz brothers at Château-d'Oex in the Vaud Préalps.
There he became a friend of other teachers, including
L. Dufour, J.B. Schnetzlar and L. Favrat. However he
was appointed on a provisional basis and his
appointment was not made permanent until 1883. He
did not regard this post as a life work and is known
to have sought other employment on at least two
occasions. His experiments in teaching included
fieldwork with his pupils; at this time he also
showed his enthusiasm for scientific organizations.
In 1877 he became a member of the *Société vaudoise des
sciences naturelles* and in 1879 he founded the
Institut météorologique de Château-d'Oex. He
retained his contacts with some of his former teachers
at the *Académie* in Lausanne, probably with the aim of
producing a thesis, which in fact he never wrote.
Pettier fielt that his varied abilities were not
adequately recognized by his colleagues at Château-
d'Oex: it may be that his political outlook was
regarded with suspicion, for he was the secretary of
the *Association pour le bien public*, which argued that
the banks were responsible for the virtual destitution
of some groups of people in Pays d'Enhaut. Other
adverse signs were his failure to be appointed to teach
history and geography at the *Ecole normale* and
discouragement at the *Académie*, where he was only
allowed to give one course of lectures though meanwhile
the authorities at the College Henchoz insisted that he
should be at Château d'Oex and not working in Lausanne.
Perhaps looking forward and outward, in 1881-2 he went
to the Stevens Institute of New York returning to
Château-d'Oex in the autumn.

Despite all his professional uncertainties Pittier
was married on 3 July 1883 to Jenny Hefti, by whom he
had three children, Mathilde (1884), Hans (1885) and
Rose (1886). With his wife and children he went to
settle in Costa Rica in 1887. This move, he explained,
was actuated by the need to find a warmer climate for
his family though that was not the only reason. One
attraction was his appointment as a teacher of natural
sciences in a secondary school at San José. At this
time Costa Rica was in the throes of developing a
system of lay education, fostered from 1885 by

President Bernardo Soto and also by Mauno Fernandez, a
government minister eager to attract teachers from
Europe. Trade recession from the 1870s onwards had
stimulated a trans-Atlantic migration which reached
its peak in the 1880s and to Pittier one attraction
was his escape from the class-ridden social structure
of Europe which prevailed despite conscious efforts
for equality of opportunity. On the economic side
Costa Rica at this time was gaining profit by the
export of coffee and of bananas by the United Fruit
Company.

Pittier went to Costa Rica as a teacher but soon
became deeply involved in various scientific
enterprises. These were contributory to a survey of
its geography and natural history. Pittier founded
the Meteorological Observatory and developed research
of wide range including geophysics, cartography,
history and ethnology. He was also concerned with
railway construction and the management of the rural
economy. Throughout his time in Costa Rica he
retained close connections with scientists in Europe
and published several books and papers and as
director of the *Instituto Físico Geografica* he
initiated the collection of a national herbarium. His
wife Jenny died in 1889 but two years later, on
18 April 1891, he married Guillermina de Fabrega, a
Columban, at San José. They had a family of three
children, Marguerite (1899), Emile (1900) and Maria
Teresa (1902).

At some time round the turn of the century, not
precisely known but probably early in 1900, Henri
Pittier, who could by now be regarded as an enterprising
scientist in early middle age, moved from Costa Rica to
the United States, following controversies with the
government in San José and personal financial problems
through loss of employment. At first he prudently
restricted his wide-ranging scientific interests and
became an applied botanist working particularly on the
possible timber and other resources of central America.
He was engaged by the United Fruit Company in 1902 and
by the U.S. Department of Agriculture in 1905 to
assess the economic value of tropical vegetation,
particularly in Columbia and Venezuela. In effect he
became a field worker and explorer having a directly
economic mandate and in 1902 he joined the Cosmos
Club, originally founded at Washington in 1878 by
J.W. Powell and others for those concerned with
practical exploration. In his tours of Central
America Pittier was especially concerned with forest
resources. To a marked degree his activities were
pragmatic and practical, for at that time Americans
were mainly interested in direct economic advantage,
notably in central American countries, rather than in
academic subjects like geography which did not in
themselves offer clear financial returns. In 1910
Pittier was engaged by the Smithsonian Institution to
make a botanical survey of the Panama Canal Zone, to
which he travelled frequently from 1911 to 1915. His
second marriage failed as Guillermina found life in
the United States uncongenial but following a divorce
he married Charlotte S. Falk in 1908 at Boston. There
were no children of this third and final marriage. He
retained his connection with geography throughout his
varied career. In 1896 he became an honorary
corresponding member of the Royal Geographical Society

in London and in 1917 he was given the Jane M. Smith Award of life membership of the National Geographic Society in Washington D.C. Miss Smith had bequeathed $5,000 to provide this award and in 1917 it was also given to Alfred H. Brooks, 'geologist', Hiram Brigham, 'historian', and George Kennan, 'authority on Russia'. Pittier was listed as an 'agriculturist' and it is curious that nobody was listed as a geographer though the award was to be a 'recognition of notable contributions to geography'.

From 1913 onwards Pittier was largely concerned with Venezuela, which he visited for long periods. Venezuela was then a dictatorship run from 1909 to 1935 by Juan Vicente Gómez, described by Preston James (*Latin America*, 1950, 44) as 'a man of tremendous ambitions, a ruthless fighter, one who ruled his country with an iron hand...(who)...completely eradicated the last vestiges of civil liberty, but with the aid of oil brought Venezuela through the difficult period after 1929 in a better condition than any other Latin-American country'. In 1919 Pittier acquired a property at Ingomar close to Caracas, partly attracted by a climate more suited to a man of sixty than that of Washington. This third major move, his last, was followed by an enlargement of his activities for though he retained his devotion to botany, both as director of the Commercial Museum from 1919 and in his private capacity, his studies henceforth included climatology, meteorology and ethnology, pursued with constant travel. He became director of the national observatory of Venezuela in 1933.

Venezuela during the 1920s, in its petroleum boom phase, was very different from Costa Rica in the 1880s but until 1933 when he fell out of favour Pittier had the support and personal friendship of Gómez, the country's dictator. After the death of Gómez in 1935, which was followed by a revolution, in 1936 Pittier was restored to official favour and became director of the botanical section of the Ministry of Agriculture. From 1935 he was troubled by attacks of pneumonia but he continued his work assiduously and from 1936 he was in charge of the national herbarium. In 1937 his eightieth birthday was celebrated by various institutions with which he had been associated, including the Ministry of Education in Costa Rica: in the same year he established the Rancho Grande national park which in 1953 was renamed the Henri Pittier park. In the winter of 1945-46 he moved to Caracas from his country home and in 1946 when the Venezuelan Ministry of Agriculture was reorganized he was appointed as research director for forestry. He died at the age of ninety-two at Caracas on 27 January 1950, having for many years astonished medical doctors by his resistance to illness. Almost thirty years later Francisco Tamayo, one of his closest associates and indeed the only geographer among them, wrote that Pittier was one of the pioneer thinkers who realized that in Venezuela, as elsewhere in the world, there would be tension on the degredation of the natural environment by the depredations of human activity. He fully understood the need for conservation.

2. SCIENTIFIC IDEAS AND GEOGRAPHICAL THOUGHT

Pittier's approach to science was empirical and deductive, with a full appreciation of the principle of evolution in the natural world. He presented both these aspects in a paper of 1880, titled 'Appel aux personne qui s'intéressent à l'étude de la météorologie dans la canton de Vaud', a study broader in scope than its title would suggest, and also commends a work by a Scottish philosopher, Dugald Stewart (1753-1828), *Elements of the philosophy of the human mind*, vol 1, 1792 and vol 2, 1814, as a guide to scientific methodology. Pittier in one of his early works, a treatment of the flora of the Pays d'Enhaut in Switzerland (1885, 20), shows the use of an evolutionary approach, then a relatively new paradigm. The range of his work, though seen in other workers of his time, was an expression of his boundless scientific curiosity, his compulsion to understand all the interpenetrating influences of the environment. This scientific approach was mentally deep-rooted and led him to view geography as a synthetic study, having a vision of one great environmental wholeness that could only be understood by grasping the significance of data revealed by other sciences.

An understanding of this view may be given by his own professional and scientific career, which was multidisciplinary and broad. His higher education at the *Académie* of Lausanne covered mathematics, physics and various branches of the natural sciences. Later, at Château-d'Oex, he taught not only geography and the natural sciences but also German, history and even gymnastics. His innate curiosity, his work as a teacher and the educational ethos of his time explain the diversity of his published work. Though it reveals a lack of specialist depth in any of the disciplines treated it also shows an approach to various problems that is fruitful and original.

Initially Pittier was drawn to botany by his association with the Thomas family, who had a fine knowledge of the local vegetation. They introduced him to the *Académie des Dévens*, a group of scientists whose associations was inspired by Jean de Charpentier (1786-1855), the geologist and pioneer glaciologist (*Geogr. biobibl. stud.*, vol 7, 1983, 17-22): several other Swiss naturalists supported this group. Its attraction may explain why Pittier, at an early age, chose to study the natural sciences rather than geography. His most ambitious work, the *Catalogue de la Flore vaudoise*, 1883-87, in collaboration with Th. Durand, shows his acceptance of botany as a study at a time when geography was little appreciated in Switzerland.

Even so, armed with an effective multidisciplinary education, Pittier felt the need to use other sciences to explain the distribution of vegetation.. This was especially true of climatology and in 1879 he founded the Meteorological Institute of Château-d'Oex. With the help of his pupils he made observations three times a day, and from 1880 eight times a day, of temperature, precipitation, air humidity (with the newly invented wet- and dry-bulb thermometer), sunshine, water content of the atmosphere, evaporation, cloud cover, winds and electrical storms. It was his aim to encourage the establishment of other meteorological stations in schools, all following the guidelines laid down by the International Meteorological Congress held at Vienna in 1873.

Pittier extended his study of the natural sciences in Switzerland to geology and zoology, and most of this work was published after his visit to the United States in 1881-82.

The work on the flora of the Vaud canton ranks as his major contribution at this time but his *Elements de la géographie du Pays d'Enhaut* of 1877 shows both his concern for the pupils he taught and his wish to integrate his diversified knowledge into a geographical text. Clearly a manual for students, it includes a summary treatment of the basic character of the region under the headings orography, hydrology, climatology, natural sciences, history (not generally included in Swiss manuals of the time), public institutions and customs. Pittier also wrote on his ideas of geography teaching in the bulletin for school teachers in French-speaking Switzerland (19). Here his attitude was in accord with his basic conception of geography as a synthetic subject and veneration for direct observation by pupils as an approach to reality.

Among those he admired for their work was d'H. Wettstein of Zurich, author of the German text, *Leitfaden für den geographischen Unterreich* and also Hölzen of Vienna. Both these writers wished to remove from geography the mere memorisation of much teaching at the time and stressed that pupils could only acquire a knowledge of the world that to them would have any meaning by direct observation of nature. This was in accord with the view of Pittier that pupils could only come to a knowledge of the larger world by possessing a basic knowledge of their own homeland. With this they could proceed from the known to the unknown by comparison: in fact the emphasis on local geography as a basis for world geography has become a commonplace of education method or at least theory. Pittier drew inspiration from the writings of other authors. The first view was expressed by various German writers who belonged to the group of 'philanthropes', notably J.B. Basedow and Ch. G. Salzmann, working in Germany at the end of the eighteenth century, who conducted school excursions as part of their teaching and had been influenced by Rousseau. H. Pestalozzi had favoured such methods while teaching at Yverdon from 1805 to 1825 and the general educational ideas of all these writers were also supported by W. Rosier and F. Geux, then principals of the teacher training colleges (*Ecoles normales*) at Geneva and Lausanne. Pittier's second conviction on the value of practical experience, for he remembered with gratitude the excursions arranged at the *Académie* of Lausanne as part of the course in botany by J.B. Schnetzler and in geology by E. Renevier.

Pittier's major contribution in Switzerland was the *Catalogue de la flore vaudoise* with Th. Durand, curator of the botanical museum in Bruxelles. As a mountain man Pittier dealt with the vertical distribution of species while Durand dealt with those of the plains. The zones of vegetation were considered and in addition to this already traditional approach attention was also given to the migrations, spread and evolution of species. This was a new approach in the 1880s, a new paradigm of the time which showed acceptance of the views of Charles Darwin.

Essentially, the work of Pittier and Druand dealt with the distribution of plant species of geographical regions in relations to climatic conditions and lithological circumstances.

Unfortunately in Switzerland Pittier lacked the institutional support that could have assisted his ambitious scientific projects and this actuated his move to Costa Rica, where within a year of his arrival, in 1888, he became director of the *Instituto Físico-Geográfico* and compiled the first meteorological map of the country. With this auspicious beginning the Institute's activity went rapidly forward. Scientifically Costa Rica was almost unexplored and an initial phase of cartographical mapping was followed by investigation of the flora, fauna, climate, volcanic and magnetic phenomena, folklore, in short studies contributory to a geographical synthesis. From the beginning of his work in central America Pittier showed the benefits that could accrue from scientific study, notably in relation to agriculture and this was fortunate as research institutions could justify government support on economic arguments.

During the next major phase of his life, in the United States, there was a surge forward in the study of tropical vegetation in the American continents with the support of various institutions. This was a crucial influence on Pittier, for he now concentrated primarily on botanical research. And this was also characteristic of the long and productive last phase of his life, in Venezuela, when his work was primarily on botany and its relation to agriculture. He made many significant contacts with other workers, including J. Cuatrecasas at the Smithsonian Institution and adopted his methods of geobotanical research, derived from Huguet del Villar in Spain and the northwest of Africa (*Geogr. biobibl. stud.*, vol 9, 1985, 55-60). Pittier maintained his scientific association with United States workers to the end of his life and frequently cited the work of Cuantrecasas, notably his ideas on the formation and *sinecia*, the environmental interdependence of various species, of vegetational associations: he also developed a growing interest in the relation of soils to vegetation. Conversely he ignored American academic geography, which in his view was dominated in the first half of the twentieth century by men brought up and educated in the Middle West who were excessively pragmatic in their approach.

From 1919, in Venezuela, Pittier continued his researches on botanical subjects, especially the distribution of species and their agricultural and medicinal value. To a great extent his work was on botanical geography, including plant associations, floristic regions and the influence of soil and climate on vegetation. A man whose scientific work was always related to the political and economic circumstances wherever he lived, he now had the fullest possible support from the government. His work included the direction of two research centres, the presentation of courses on botany, the founding of a museum and a national park. Through his research and activities he made a vital contribution to a developing country. It was his finest hour.

3. INFLUENCE AND SPREAD OF IDEAS

During a professional life divisible into four parts Pittier founded scientific institutions, established herbaria in several towns and attracted disciples and colleagues who were to become devoted scientific workers.

In Switzerland Pittier organized a network of meteorological stations covering the Pays d'Enhaut as part of a research project with L. Dufour, a professor at the Lausanne *Académie* and an English colonel, M.F. Ward. The meteorological station he founded at Château-d'Oex was of particular significance in the Vaud canton and its observations were published by the federal meteorological commission. While teaching at Château-d'Oex he established a splendid laboratory for physics and chemistry, according to a report of 1885, as well as natural history collections. He worked with Hans Schardt on the geology of the Pays d'Enhaut and also on the nomenclature and cartography of the Alps. For twenty years he collaborated with Colonel M.F. Ward on a catalogue of vertebrates and birds in the same area. He was similarly concerned with listing the distribution of botanical species in the Vaud canton with Th. Durand, L. Favrat, L. Leresche and J.B. Schnetzler.

The opportunities for research in Costa Rica were almost limitless and Pittier found several colleagues, mainly Europeans, who worked with him. Among the Europeans were P. Biolley, J. Rudin, A. Tonduz and Ch. Wercklé and the Costa Ricans included A. Calvo and P.N. Gutiérrez, all of whom worked at the *Instituto Físico-Geográfico*. With their co-operation Pittier built up a herbarium of 20 000 items, of which duplicate specimens are stored mainly in the Smithsonian Institution and in various European museums. Also in Costa Rica Pittier published maps from 1891-96 and studied varied aspects of the natural and human sciences as a basis for a geographical gazetteer, never in fact produced. An abiding legacy to Costa Rica was his meteorological organization, his development of cartographical work, the museum and the national herbarium.

Similarly in the United States Pittier left a vast herbarium of the tropical flora which, according to J. Cuatrecasas, is still in frequent use, as it was so admirably described and classified. The work on Panama was clearly of significance but Venezuela was to be the scene of his final maturity. Among the young workers and potential successors he trained were Dr. T. Lasser, Professor Francisco Tamayo and Professor Z. Lucas, and also Mademoiselle Artenza, who classified the botanical specimens. All of these worked at the Ministry of Agriculture during the 1940s but among them the only geographer was Tamayo, who compiled the geobotanical map published in the *Atlas de Venezuela*. Pittier's work expanded through the years but was always primarily in geobotany leading to the conservation of a richly abundant flora and to the rural economy. In 1946, when approaching ninety years of age, he was put in charge of the forestry research but he found the beaurocratic restraints unhelpful. He retained his association with European colleagues and those in the United States, in both of which museums and other research institutions have examples of his work. His scientific standing was such that he was widely known in the world's major scientific circles and in summary one may comment that he was a representative of European and American science in tropical America.

Pittier was a shining example of a general geographer at a time when such a vocation could only be contemplated with the basis of a broad education likely to produce a scholar of encyclopaedic mind. His career, though as a whole triumphant, was beset by many difficulties through changing circumstances, but as each new challenge arose he showed the wisdom to choose from his numerous specialisations the one which could be most academically fruitful in the changed circumstances. In giving his name to one of the lecture theatres in the Geography school of the Faculty of Letters at the Central University of Venezuela at Caracas, the authorities have recognized him as one of the fathers of geography in the country and in so doing have appreciated that his concept of the wholeness of geography was a source of unity in his scientific thought.

Bibliography and Sources

1. GENERAL REFERENCES

Glick, T.F., 'Antes de la revolución cuantitativa: Edward Ullman y la crisis de la geografía en Harward (1949-1950)', *Geo-Critica*, no 55 (1985), 1-45 (trans. Luis Urteaga)

Gomez P., L.D., 'Contribuciones a la pteridología costarricense. XI. Hermann Christ, su obra e inflenia en la botánica nacional', *Brenesia*, vol 12/13 (1977), 25-79

---- 'Contribuciones a la pteridología costarricense XII. Carlos Wercklé', *Brenesia*, vol 14-15 (1978), 361-93

Martí-Henneberg, J., 'E. Huguet del Villar', *Geogr. biobibl. stud.*, vol 9 (1985), 55-60

---- *L'excursionisme científic a Catalunya (1876-1900). La seva aportació a les ciències naturals i a la geografia*, doctorate thesis, Barcelona Univ. (1985)

Nicolas-O, G., *L'espace originel*, Bern (1984), 330 p., cf. 283 ff

Sala-Catalá, J., 'Conflictos y paradigmas en la biología de la segundamitad del siglo XIX', *Actas del II Congreso de la Sociedad Espanola de historia de las sciencas*, vol 3 (1984)

2. REFERENCES ON HENRI PITTIER

Jahn, A., 'El profesor Henri Pittier. Esbozo biografico', *Bol. Soc. Venez. Cienc. Nat.*, vol 4 (1937), 1-43

Cruz, B.L. et al., 'Henri Pittier', *Ceiba*, vol 1 (1950), 129-41

Chase, A., 'Henri François Pittier', *J. Washington Acad. Sci.*, vol 40, (1950), 240

Phelps, W., 'Henri François Pittier', *Geogr. Rev.*, vol 41 (1951), 341-2

Conejo-Guevara, A., *Materiales para una bio-bibliografía costarricense del Dr. Henri Pittier Dormand*, 2 vol., Licenciatura thesis (1972), Universidad de Costa Rica

Bernaldi, L., 'Vies parallèles de Mosé Bertoni et de Henri Pittier', *Les Musées de Genève*, no 201 (1980), 19-23

3. SELECTIVE BIBLIOGRAPHY OF WORKS OF HENRI PITTIER

a. Switzerland

1878 'Notes sur la nomenclature des Alpes du Pays d'Enhaut vaudoise', *Echo des Alpes*, vol 4, 249-58

1880 'Appel aux personnes qui s'intéressent à l'étude de la météorologie dans le canton de Vaud', *Bull. Soc. Vaud. Sci. Nat.*, vol 17

1881 'Contributions à l'histoire naturelle du Pays d'Enhaut vaudois.   I. Résumé des observations météorologiques faites à Chateau-d'Oex en 1879 et 1880 et à Cuves en 1880', *Bull. Soc. Vaud. Sci. Nat.*, vol 17, 397-421

1883-87 (with T. Durand) *Catalogue de la flora vaudoise*, vol 1 (1883), vol 2 (1885), vol 3 (1887)

1885 'L'enseignement de la geographie', *Educateur de la Suisse romande*, 21 year, no 1, 8010 and no 2, 22-5

---- *The flora of the Pays d'Enhaut (Switzerland), a botanical account*, Château-d'Oex, 14 p

---- (with H. Schard) 'La géologie de la vallée de la Grande Eeu et du Massif du Chamossaine', *Bull. Soc. Vaud. Sci. Nat.*, vol 19, Proc. Verb, I-III

1887 *Eléments de la geographie du Pays d'Enhaut*, Lausanne, 32 p

b. Costa Rica and Central America

1889 'Sur l'orographie de l'Amérique centrale et des volcans de Costa Rica', *Arch. Sci. Nat. Genève*, vol 22, 466-72

1890 *Apuntaciones sobre et clima y geografía de la República de Costa Rica*, San José de Costa Rica

1891 *Viaje de exploración al valle del Rio Grande de Térraba*, San José de Costa Rica, 140 p, 1 map

1892 'Note sur la geographie du Costa Rica. Lettre à Elisée Reclus', 'Nouvelles Géographies', supplement to *Tour du Monde*

---- (with C. Gagani) *Ensayo Lexicográfico sobre la lengua de Térraba*, San José de Costa Rica, 86 p

1895 *Nombres geográficos de Costa Rica*, San José de Costa Rica, 46 p

1899 'El caucho, las plantas que le producen y su cultivo', *Bol. Agric.*, 2 and 5

1901 'Primer ensayo de un mapa de la declinación magnética de Costa Rica', *Bol. Inst. Físico-Geografico*, vol 1, 10-31

1909 'New or noteworthy plants from Colombia and Central America', *Contributions to the U.S. Natural Herbarium*, vol 12, 171-81

1910 'Costa Rica-Vulcan's smithy', *Natl. Geogr. Mag.*, vol 21, 494-525

1912 'Little-known parts of Panama', *Natl. Geogr. Mag.*, vol 23, 627-62

1942 *Capitulos escagidos de la geografía fisica y prehistórica de Costa Rica*, San José de Costa Rica, 56 p

c. Venezuela

1920 'La evolución de las ciencias naturales y las exploraciones botánicas en Venezuela', *Cultura Venezolana*, no 14

---- *Esbozo de las formaciones vegetales de Venezuela, con una breve reseña de los productos naturales y agrícolas*, Caracas, with map

1925 'Flora venezolana: plantas medicinales', *IV Congreso Venezolana de Medicina*, vol 2, 167-72

1926 *Manual de plantas usuales de Venezuelas*, Caracas, 458 p

1928 'Contribuciones a la dendrología de Venezuela. I. Arboles y arbustos del orden de las leguminosas. II. Papilonáceas', *Bol. Min.*, no 4-7, 150-229

1933 'Contribuciones al estudio de la climatología de de Venezuela. I. Cuarenta años de observaciones pluviométricas en el observatorio de Cajigal', *Bol. Soc. Venez. Cienc. Nat.*, vol 2, 87-134

1936 *Apuntaciones sobre la geobotánica de Venezuela*, Caracas, 24 p

1937 'Classificación de los bosques', *Bol. Soc. Venez. Cienc. Nat.*, vol 4, 93-110

1942 *La mesa de Guanipa; ensayo de fitogeografía*, Caracas, 57 p; 2 ed, 1972

3. UNPUBLISHED SOURCES ON HENRI PITTIER

The authors are deeply indebted to Monsieur Henri Moreillon of Vevey, who has visited Venezuela and collected a vast amount of unedited material on Pittier, including numerous photocopies of his letters and those of his descendants.   The authors have worked on this material, which will be deposited in the Archives cantonales vaudoise.

The authors have consulted various materials in the Archives centrales vaudoise, filed under Ecole normale; Bdd 52 to 100; KXC; K XIII 24, 28, 30, 40, 107, 121.   On Costa Rica data has been obtained from material held by Dr David Henchoz, La Sarraz, Switzerland.   In Washington the Smithsonian Institution has more than 2,000 pages of Pittier's scientific correspondence.   Other archives include those at Caracas held by Tobias Lassar, director of the Botanic Garden: also at Caracas the authors have acquired useful data on Francisco Tamayo, one of the Venezuelan disciples of Pittier.

Jordi Martí-Henneberg is a Ph.D. in geography and Anne Radeff is lecturer in history at the university of Lausanne

Translated by T.W. Freeman

Chronology

1857   Born at Bex, Vaud, Switzerland, 13 August

1858	The Pittier family moved to Dévens and became friendly with the Thomas family, who were devoted botanists
1864	Henri went to the primary school in Bex and shared the botanical interests of the Thomas children
1869	Moved to grammar school in Bex
1873-75	Followed a two year course for intending teachers at the Ecole Normale, Lausanne
1875	Spent two years studying physical and natural sciences at the Académie de Lausanne and graduated on 25 July 1877
1877	On 16 October began teaching history and geography at Château d'Oex and during this year joined the Swiss Alpine Club
1878	Became a member of the Société vaudoise des sciences naturelles
1879	Founded the Institut meteorologique at Château d'Oex
1881	Went to the Stevens Institute in New York and returned in the autumn of 1882
1883	Married Jenny A. Hefti on 3 July: began the publication of his *Catalogue de la flora vaudoise*, 1883-85
1885	Applied without success for a post in history and geography at the Ecole normale: became secretary of the Société vaudoise des maîtres secondaires
1886	Gave a course of lectures on geography at the Académie
1887	Resigned from his post at Château d'Oex on 25 August and on 27 November left for Costa Rica to teach in a lycée: published his *Eléments de la géographie du Pays d'Enhaut*
1888	Founded a meteorological observatory and organized a network of rainfall recording stations: became director of the Instituto Físíco-Geográfico
1889	Death of his wife Jenny on 26 March
1891	Married to Guillermina J. de Fabrega on 18 April: began to publish a map of Costa Rica (to 1896) and at this time began to show his interest in nature conservation
1896	Became an honorary member of the Société vaudoise des sciences naturelles
1901?	At some time he moved permanently to the United States, probably in 1901, but the exact date is not known
1902	Worked for the United Fruit Company and carried out botanical surveys of Central America, Colombia and Venezuela for several years: became a member of the Washington Academy of Sciences, the Washington Botanical Society and the Cosmos Club of New York
1905	From 21 January worked at the United States Department of Agriculture
1909	Married Charlotte S. Falk, 25 June
1910	Appointed by the Smithsonian Institution to take part in the Botanical Survey of the Panama Canal zone and from 1911-15 made several visits to Panama
1913	Made several visits to Venezuela (to 1918)
1918	Bought an agricultural estate at Carabado, Venezuela
1919	Left the U.S. Department of Agriculture on 14 October and became director of the Museo comercial of Venezuela
1926	Published his *Manual de plantas usuales de Venezuela*
1927	Lausanne university made him an honorary Doctor
1933	Became director of the national observatory of Venezuela, but his friendship with Juan Gomez, dictator from 1909 to his death in 1935, came to an abrupt end
1935	Suffered his first attack of pneumonia, which recurred annually to the end of his life
1936	Following a change of government, which now included politicians who were his friends, he became director of the botanical section of the Ministry of Agriculture, established a national herbarium, planned measures for nature protection and the care of forests, gave a course of lectures for the Botanical service, and published his work on the geobotany of Venezuela, *Apuntaciones sobra la geobotanica de Venezuela*
1937	Founded the Rancho Grande national park, named after him in 1953: his eightieth birthday was marked by numerous bodies including the Ministry of Education in Costa Rica

1946   On the reorganization of the Ministry of
                Agriculture, Pittier became director of
                forestry research and visited the United
                States to recruit five civil engineers
                with an interest in forestry

1950   Died at Caracas, 27 January, aged
                ninety-two

Eduard Richter

1847–1905

JOSEF GOLDBERGER

On the Mönchsberg in the city of Salzburg Humboldt Terrace and Richter Höhe are the finest viewpoints. The fame of Alexander Humboldt as a naturalist abides and Eduard Richter is also an appropriate choice for commemoration as he lived in Salzburg from 1871 to 1886, taught at its *Gymnasium* and became the best known researcher on Alpine Austria before entering university service to continue this work and also to become a major figure in the academic society of Austria, with an international reputation as a glaciologist.

1. EDUCATION, LIFE AND WORK
Eduard Richter was born on 3 October 1847 at Mannersdorf, in Lower Austria at the foot of the Leitha mountains, the easternmost extension of the Alps. His father, an imperial civil servant, died in the following year shortly after his election to the all-German parliament at Frankfurt and his mother then returned with her two sons to her family home town of Wiener Neustadt. She was a well educated woman and had a considerable influence on the young Eduard, who was sent to a *Gymnasium* run by Cistercian monks. He was brought up in a helpful environment, with the resources of his grandfather's library, the artistic stimulus given by the family doctor and also the opportunity of discussing biological questions with an older family friend. He completed the school course in 1866 and was saved from taking the oral part of the final examinations as the school building had become a hospital during the Austro-Prussian war.

Geographers Biobibliographical Studies, volume 10 (1986)

Courtesy of Bildarchiv d. Öst. Nationalbibliothek

As a student in the university of Vienna from 1866 to 1871 Richter chose to read arts rather than science, as for science mathematics was one of the requirements and that subject, he said, was one for which he had neither the attraction nor the talent. Excursions in the Alps, including the nearby Hochschwab range and also the Ötztal range and the Ortler group of the South Tyrol, awakened in him a love of mountains. He was very active in the Institute of Austrian History, whose head was the famous historian Theodor von Sickel. Richter became a full member of the Institute and carried out an investigation into the estates of the Bavarian see of Freising within Austria, with the intention of seeking an academic career in history, but as his work was not approved by Professor von Sickel he turned his attention to the less ambitious but secure career of a grammar school teacher. With this aim he attended classes given by Professor Friedrich Simony, the first Austrian professor of geography, and established a close personal relationship with him. After a teaching test at a grammar school Simony assisted Richter to obtain a position at the *Gymnasium* in Salzburg.

During the summer of 1871, before the school opened, Richter toured the Hohe Tauern in Salzburg province and with the well-known Alpine climber Johann Stüdl carried out a first ascent of the Schliefer Spitzl in the Venediger range. At the *Gymnasium* in Salzburg Richter found ideal working conditions. The headmaster gave him the responsibility for all the geography teaching. At first, as he was still primarily an historian, he felt rather overtaxed but he soon became absorbed in the subject and his charming personality, his sense of humour and his quick wit earned him the esteem of his pupils.

Richter's first paper on glacial phenomena, 'Das Gletscherphänomen', appeared in 1873 in the annual report of the school.   He remained active as an historian: various articles on prehistoric, early and later historical periods of the Salzburg province led to his appointment in 1876 as editor of the periodical published by the Society for the study of Salzburg, the *Gesellschaft für Salzburger Landeskunde*. Unhappily at this time he suffered the loss of his wife in childbirth but he continued to write papers, including regional studies of the Salzburg province. And in time he remarried, and with his second wife, Louise, and their daughters enjoyed an exemplary family life.

Through his numerous excursions Richter became one of the leading authorities on the Eastern Alps. He acquired such high esteem that in 1883, when he was only thirty-six years of age, he was elected as president of the German and Austrian Alpine Society, the *Deutscher und Österreichischer Alpenverein*.   In this capacity he excelled as a promoter of the society's scientific activities, which included Alpine cartography, the study of glaciers and assistance in the establishment of a meteorological station on the summit of the Sennblick peak at an altitude of 3,105m. This was a time of active glacial research, stimulated by the publication in 1882 of Albrecht Penck's *Die Vergletscherung der Alpen*.   In 1876 Richter began to study the Unsterberg ice cave, Salzbrug and from 1880 he surveyed the Karling and Obersulzbach glacier of the Hohe Tauern, on which in 1883 he published a paper with a map in the *Zeitschrift des Deutschen und Österreichischen Alpenvereins* (vol 14).   This was followed by two comparable papers, on the Ötztal and Karling glaciers (vol 16, 1885 and vol 19, 1888) and by several general studies of glaciers in various journals.

In Richter's life at this time research was combined with administrative and editorial duties, as well as his school teaching and in 1885, fourteen years after completing his university studies he obtained his doctorate under Professor Simony.   One reason for seeking the doctorate at this time was that Simony was about to retire and this fact renewed Richter's hope of a university career.   In fact Professor Tomaschek of Graz moved to Vienna but Richter was appointed to the vacancy as a full professor of geography in Graz university on the recommendations of Simony and also of Ratzel.

When Richter arrived in Graz in 1886 he found that the facilities of the geography department were minimal and the work load enormous, for as the only teacher of geography in the university he had to cover the entire range of the subject.   His research during the first two years was exclusively directed to glaciology but subsequently he studied alpine lakes, carrying out measurements of their depths and temperatures, and so in time he became known as a limnologist.   The summers from 1884-94 were spent mostly on the shores of the Carinthian lakes. Meanwhile his university responsibilities were growing and in 1889 he was elected as Dean of the Faculty of Arts at Graz.

Initially his research shows the influence of Simony but during the last ten years of his life his geomorphological work shows his attraction to his younger friend, Albrecht Penck, whose *Die Morphologie der Erdoberfläche* was published in 1895. The new broadening scope of Richter's work involved several field study tours, to the Giant mountains (Riesengebirge) in 1893, to Norway in 1895, to the Black Forest and the Vosges in 1896 and finally a visit of considerable length to Russia in 1897. Meanwhile publication of his findings continued in Austrian and German journals: in 1896, in collaboration with Albrecht Penck, he published the *Atlas der österreichischen Alpenseen*.

When the International Commission on Glaciers was established in 1897 he was elected as its first president.   Two years later, at the International Geographical Congress in Berlin, he gave a paper on 'Die Gletscher-Konferenz im August 1899', published in the *Congress Proceedings* (vol 2 (1901), 269-78) and in 1900 he attended a meeting of the Congrès international de l'Alpinisme and gave a paper on 'des observations des glaciers et les associations alpines'. He was described in the *Compte rendu* of this congress as 'Président de la Commission internationale des glaciers, membre d'honneur du Club alpin francais'. Perhaps the greatest distinction was that he was asked to edit the *Historische Atlas der Alpenländer* by the Imperial Academy of Science, of which he became a full member in 1902.   However of all the honours bestowed on him the one that pleased Richter most was the honorary membership of the Berlin Geographical Society, the *Gesellschaft für Erdkunde zu Berlin*.   Having become dean of the Arts faculty at Graz in 1889, in 1898 he became *Rektor* of the university and in 1903 he was granted the title of *Kaiserlich-Königlicher Hofrat* (royal and imperial court counciller).

Also in 1903 he travelled in Sicily, Tunisia and Algeria and as his next academic task he planned a geography of Bosnia and Herzegovina, the two southern Slav territories that had been first occupied and then annexed by Austria.   He was unable to advance beyond the preliminary stage as he returned from Constantinople in 1904 seriously ill with a heart condition and a cure at Bad Nauheim brought no improvement.   At the end of January 1905 he had to abandon his academic work and shortly afterwards, on 5 February, he died at Graz.

2. SCIENTIFIC WORK AND GEOGRAPHICAL IDEAS

Two points are of special interest in Richter's career. Firstly, he came into geographical work with a thorough training in history, like several other geographers of his own time and later.   Secondly, having decided to become a geographer, Richter conformed to the general practice of continental Europeans that study and research must include physical aspects of the subject, partly as a basis for regional work.   In his case glaciation had a strong and clear appeal leading to a wider study of physical geography, including climate.   However, of him as of others it could be said that 'once an historian, always an historian', though his transition to geography could be discerned in his student research on the ecclesiastical see of Freising and also in his early publications for the German and Austrian Alpine Society on former wars and invasions in the Alps. A third consideration is that in a career of thirty-four years from 1871 Richter spent fifteen years as a *Gymnasium* teacher and was therefore fully conversant with the needs of school pupils and no less aware of the responsibilities of teachers under classroom conditions and, at least for geographers, on field courses. Finally, like many other geographers of Alpine lands, he was drawn to mountains for their academic as well as their scenic interest. Therefore his life work may be considered under four headings, as a glaciologist, historical geographer, teacher and methodologist, and alpinist.

a. *Glaciologist*

Richter became prominent initially as a glaciologist. In his first publication, 'Das Gletscherphänomen' of 1873, he followed closely in the steps of Agassiz, the Schlaginweit brothers, Sonklar and Tyndall. His observation of glaciers in the Hohe Tauern and Dachstein, with later visits to the Valais and Mont Blanc, gave him the incentive to carry out actual surveys so in 1879 he learned surveying under the instruction of an officer of the Imperial Army and in the following year he made a map of the Karling and Obersulzbach glaciers in the Salzburg province. Two years later, in 1882, he developed the use of profiles as illustrative material for his work and his 1883 paper, mentioned earlier as the first of three detailed studies with the general title of 'Beobachtungen an der Gletschern des Ostalpen', had the specific title of 'Der Obersulzbachgletscher 1880-82': it includes a carefully drawn map of the glacier lobe on the 1:50 000 scale. In this paper Richter was the first worker to prove the volume of a clagier geodetically. As causes of advance and retreat in glaciers he also analysed the meteorological elements. He also explained the reasons for the time lag between climatic oscillations and volume changes in glaciers and also the differences in changes between glaciers of the same area in terms of aspect and wind direction. In studies of the glaciers of the Ötztal Alps he paid particular attention to the greatly differing relationships between the supply and the wastage areas of individual glaciers.

Richter's main work, *Die Gletscher der Ostalpen*, appeared in 1888. It is one of the classics of glaciology. Based on the Austrian national survey on the 1:25 000 scale, carried out from 1870 to 1873, it contains not only the areal measurements of 1,012 glaciers but also a classification of glaciers into types with data on the snow lines of the various regions. Richter pointed out that the firn area and tongue are by no means identical with a glacier's feeding and wasting area. He fully recognized the importance of summer snow falls in glacier regimes. The most valuable part of the work is the final section in which he discusses the snow line in the Eastern Alps. On the time scale of glacier oscillations he recognized that the advances of 1820 and of 1850 were of approximately the same magnitude.

Together with S. Finsterwalder, Richter attempted a practical application of glaciology by investigating the causes of the annual flooding of the ice lake in the Martell valley. His historical training in the critical use of old sources stood him in good stead in research on the former advances and retreats of glaciers. On the glacial oscillations of the past three centuries he believed that he had found evidence conformable with the Brückner 35-year climatic cycle.

From the end of the 1880s Richter turned his attention to research on Alpine lakes and the atlas of Austrian alpine lakes, *Atlas der österreichischen Alpenseen*, initiated by Simony, was jointly edited by Richter and Albrecht Penck. Richter had himself surveyed the larger lakes of Carinthia and also the then Austrian part of Lake Garda, of which the greatest depth was 311m. For the measurement of depth he had developed a sounding apparatus accurate to a centimetre, and in the case of the Wörthersee, Carinthia, which has a tectonically complicated basin, he carried out 483 soundings. Fortunately the text on lake soundings, *Seestudien*, which accompanied the atlas, contributed greatly to the then young science of limnology. As early as 1891 Richter gave an account of his studies of convectional currents in lakes at the Congress of German Geographers in Vienna. He also discovered what he named the *Sprungschicht*, a layer of abrupt change of temperature, first observed in the Wörthersee in 1889.

It was through his glacial studies that Richter eventually became a geomorphologist. At that time there were sharp conflicts of view on the erosive power of ice, brought to a climax by the publication of Albrecht Penck's *Morphologie der Erdoberfläche*, especially in his work on cirque formation. Partly to seek further light in this controversy, Richter visited the so-called *Schneegruben* (snow hollows) in the Giant Mountains of Germany in the company of Joseph Partsch. But what finally convinced him of the erosive power of the Quaternary glaciers was his observations made during a visit to Norway in 1895, for there in the fjords he saw a morphological similarity to the trough shaped valleys of the Alps. These more general comparative studies were continued along with more local investigations which as a whole were a fine contribution to an understanding of the Quaternary local glaciation in the Eastern Alps.

b. *Historical Geography*

Richter's first research project, on the estates of the see of Freising in Austria, *Die Freisinger Besitzungen in Österreich*, during his student years, rank as a study in historical geography, and so do his early papers of 1875 and 1877 respectively in the publications of the German and Austrian Alpine Society, on the war in Tyrol of 1809 ('Der Krieg in Tirol im Jahre 1809') and on the Saracens in the Alps ('Die Sarazenen in den Alpen'). In 1877 also he produced a pamphlet on the teaching of historical geography, *Die historische Geographie als Unterrichtsgegenstand*, which stimulated considerable discussion.

Having worked on the see of Freising, Richter turned his attention to the archepiscopal see of Salzburg, on which he wrote a study of some 150 pages, 'Untersuchungen zur historischen Geographie des ehemaligen Hochatiftes Salzburg und seiner Nachbarg Nachbargebiete', published in the first issue of the *Mitteilungen des Institutes für Österreichische Geschichtsforschung* in 1885. The study included a series of maps of historical boundaries, supplementing the data on which the work was based, including legal documents and deeds. The results were crystallized in a masterly map on the 1:200 000 scale showing the development of the Salzburg territory, its former division into *Gaue (pays)* and the late medieval court districts. The interest in past boundaries was considerable at the time among historians and historical geographers and the work on Richter was much appreciated.

Later, after a break of ten years, Richter contributed a paper to a *festschrift* for Krones, an historian, in which he gave a plan for an historical atlas of the Austrian provinces. Sooner than he expected this idea was taken up by the Imperial Academy of Sciences, who entrusted him with the organization of the project. The work was to consist of thirty-eight maps on the 1:200 000 scale with the historical information superimposed on a relief background. Unfortunately Richter died before even the first part was in print.

c. *Teaching and Methodology*

Richter carried his great experience as a grammar school teacher into his university career and always cherished it. He published a textbook for the first, second and third forms of grammar schools, which became standard at more than two-thirds of these schools in the Austrian part of the Hapsburg monarchy. In addition he published a school atlas. In his opinion the scientific advance of geography had added to the value of the subject in school education. Provided that teachers were capable of meeting a challenge, the exposition of the world according to natural endowment and historical development was a very satisfying task. Conversely the most dangerous school tyrants were those whose knowledge barely extended beyond the contents of the textbook. School geography in Richter's view was a truly comprehensive subject in a liberal education.

Nevertheless Richter's textbook attracted criticism, for though the regional sections were appreciated it was also said that the sections on basic geographical concepts were nothing more than an accumulation of definitions. To this the author made the reply that the textbook was meant to be a précis of the lessons emphasizing their basic content, rather than a substitute for the teacher, who had the job of bringing the contents of the book to life. Though these criticisms diminished the author's satisfaction with the work it continued to be successful and reached a seventh edition after his death.

d. *Alpinism*

On Richter his friend Albrecht Penck commented that 'the geographer has grown out of the alpinist'. As the numerous reports of climbs which he published from 1872 in the journal of the *Alpenverein* show, in many cases the mountaineering interest preceded the scientific interest, though almost all his scientific work was linked with the Alps. *Die Erschliessung der Ostalpen (On the opening up of the eastern Alps)*,

published in 1894 by the Deutsche und Österreichische Alpenverein, was the work by which he was most widely known to the general public. He was its general editor and although there were many contributors it bears the mark of his work decisively. No mountain country in the world, not even Switzerland, has anything of comparable character. The introduction on the essential quality and the marked rise of alpinism was written by Richter himself and so too was the account of the Hohe Tauern. The account he gave of the climb of the Grossglockener, the highest peak in what is now Austria, is particularly noteworthy. As a postscript on the 25th anniversary of the foundation of the Alpenverein, in 1894 Richter published in the society's *Zeitschrift* a paper on the scientific discovery of the Eastern Alps, 'Die wissenschaftliche Erforschung der Ostalpen'.

But that was not all for as a paper asking rhetorically if the Alps are the most beautiful mountains in the world, 'Sind die Alpen das schönste Gebrige der Erde?' of 1885 shows, he also viewed mountains with the eye of the artist. To experience the beauty of the mountains was in his mind the greatest of rewards for all the effort involved, and one of the noblest activities of man, making him happy beyond words. Many readers will understand what he meant.

3. *INFLUENCE AND SPREAD OF IDEAS*

Little need be said of this but his influence was profound. One essential comment is that he opened up possibilities of research for others, especially on glaciation but also in geomorphology in general. His work on historical geography could well be studied again, for he saw many aspects of its significance in the development of administrative areas that were also of regional significance. As a teacher in school and university he laid foundations that were beneficial to the continuing growth of geography, especially in Austria, and so gave the subject abiding value in a liberal education. He stands out not only for what he taught and wrote, but for what he was and what he did.

Bibliography and Sources

1. *OBITUARIES ON EDUARD RICHTER*

Diener, K., 'Zur Erinnerung an Prof. Eduard Richter', *Österr. Alpen-Ztg.*, no 681, 2 March 1905

Jauker, O., 'Eduard Richter', *Geogr. Anz.*, vol 5 (1905), 49-52

---- 'Eduard Richter', *Z. Schulgeogr.*, vol 27/7 (1905)

Leukas, G.A., 'Eduard Richter', *Geogr. Z.*, vol 12 (1906), 121-35, 193-212, 252-77

Marek, R., 'Eduard Richters Leben und Wirken', *Mitt. k.-k. Geogr. Gesell. Wien*, vol 49 (1906), 161-255: this includes a complete bibliography of Richter's publications on p. 244-55

Marinelli, O., 'L'Opera scientifica di Edoaro Richter',

Riv. Geogr. Ital., vol 12 (1905), 274-83,
351-88

Mell, A., 'Eduard Richter', *Dtsch. Geschichtsblätter*,
vol 6, 186-9

Penck, A., 'Eduard Richter', *Leopoldina*, vol 41 (1905),
47

---- 'Eduard Richter', *Mitt. Dtsch. Österr. Alpenver.*,
no 3, 15 February 1905

Richthofen, F. von, *Z. Gesell. Erdkd. Berlin*, (1905),
152

Sieger, R., 'Eduard Richter', *Österr. Runds.*, vol 2/16
(1905), 148

Wutte, M., 'Eduard Richter', *Carinthia*, 1 and 2 (1905)

anon., *Ann. Geogr.*, vol 14 (1905), 178

anon., *Geogr. J.*, vol 25 (1905), 468

2. *SELECTIVE AND THEMATIC BIBLIOGRAPHY OF WORKS BY
EDUARD RICHTER*

a. *Physical Geography, especially Glaciology*

1873 'Das Gletscherphänomen', *Z. Dtsch. Österr.
Alpenver.*, vol 5, 1-56

1883 'Beobachtungen an den Gletschern der Ostalpen:
1, Der Obersulzbachgletscher 1880-2', *Z. Dtsch.
Österr. Alpenver.*, vol 14, 38-92

1884 'Über Beobachtungen an den gegenwärtigen
Gletschern der Alpen', *Verh. 4 Dtsch. Geogr.
München*, 9 p.

1885 'Beobachtungen an den Gletschern der Ostalpen:
2, Die Gletscher der Ötztaler Gruppe im Jahre
1883', *Z. Dtsch. Österr. Alpenver.*, vol 16,
54-65

1888 'Beobachtungen an den Gletschern der Ostalpen:
3, 'Der Karlinger Gletscher, 1880-6', *Z. Dtsch.
Österr. Alpenver.*, vol 19, 35-41

---- 'Die Gletscher der Ostalpen', *Handb. Dtsch.
Land.-Volks.*, vol 3 7 maps, 44 profiles, 306 p.

1889 'Der Gletscherausbruch im Martelltal und seine
Wiederkehr', *Mitt. Dtsch. Österr. Alpenver*,
vol 20, 231-3

1893 'Bericht über die Schwankingen der Gletscher
des Ostalpen 1888-92', *Z. Dtsch. Österr.
Alpenver.*, vol 24, 473-85

1896 'Die Gletscher Norwegens', *Geogr. Z.*, vol 2, 305-19

---- 'Beobachtungen über Gletscherschwankungen in
Norwegen 1895', *Petermanns Geogr. Mitt.*, vol 42,
107-10

---- 'Gemorphologische Beobachtungen aus Norwegen',
*Sitzungsber. K. Akad. Wiss. Wien, math.-
naturwiss. Kl.*, vol 40/1, 147-89

---- 'Aus Norwegen', *Z. Dtsch. Österr. Alpenver.*,
vol 22, 1-35

---- (with A. Penck) 'Atlas der österreichischen
Alpenseen', *Geogr. Abh.*, vol 6

1897 'Seestudien', *Geogr. Abh.*, vol 6/2, 72 p.

1899 'Neue Ergebnisse und Probleme der
Gletscherforschung', *Abh. k.-k. Geogr. Gesell.
Wien*, vol 1, 1-13

1900 'Die Gletscherkonferenz im August 1899',
Petermanns Geogr. Mitt., vol 46, 77-81

---- 'Geomorphologische Untersuchungen in den
Hochalpen', *Petermanns Geogr. Mitt., Erg.*,
no 132, 103 p.

1901 'Die Gletscher-Konferenz im August 1899', *Proc.
7 Int. Geogr. Congr.*, vol 2, 279-88

b. *Historical Geography*

1877 *Die historische Geographie als
Unterrichtsgegenstand, 27 Salzburger
Gymnasialprogram*, 25 p.

1885 'Untersuchungen zur historischen Geographie des
ehemaligen Hochstiftes Salzburg und seiner
Nachbargebiete', *Mitt. Inst. Österr.
Geschichtsforschung, Erg.*, vol 1, 1-152

1894 'Die wissenschaftliche Erforschung der Ostalpen
seit der Gründung des Österreichischen und des
Deutschen Alpenvereins', *Z. Dtsch. Österr.
Alpenver.*, vol 25, 1-94

1895 'Über einen historischen Atlas der
österreichischen Alpenländer', *Mitt. Wien. k.-k.
geogr. Gesell.*, vol 39, 529-40

1902 *Matthias Burgklehners Tirolische Landtafeln,
1608, 1611, 1620, Wien*

1903 'Die historische Atlas der österreichischen
Alpenländer', *Dtsch. Geschichtsblätter*, vol 4,
145-50

1904 *Historischer Atlas der österreichischen
Alpenländer, Erläuterungen zur Landgerichtskarte*,
Salzburg, Blatt 8, 9, 16 and 17

c. *Other Historical Publications*

1875 'Der Kreig in Tirol im Jahre 1809. Beiträge
zur Geschichte und Geographie der Alpen II',
Z. Dtsch. Österr. Alpenver., vol 6, 166-234

1877 'Die Sarazenen in den Alpen. Beitrag zur
Geschichte und Geographie der Alpen V', *Z. Dtsch.
Österr. Alpenver.*, vol 8, 221-9

1879 'Die Funde auf dem Dürenberg bei Hallein',
*Mitteilungen der k.-k. Zentralkommission für
Erforschung und Erhaltung der historischen
Denkmale*, 168

1881 'Verzeichnis der Fundstellen vorhistorischer und
römischer Gegenstände im Herzogtume Salzburg',
Mitt. Gesell. Salzburger Landkd., vol 21, 90-7

1889 'Salzburger Flachland und Pongau. Die Römerzeit.
Zur geschichte Salzburgs' in *Die Österreich-
ungarische Monarchie in Wort und Bild, Band
Oberösterreich und Salzburg*, 321-46, 382-424

d. *Regional Geography*

1881 *Das Herzogtum Salzburg, Die Länder Österreich-
Ungarns in Wort und Bild*, vol 5, 125 p.

1885 (with A. Penck) 'Das Land Berchtesgadan', *Z.
Dtsch. Österr. Alpenver.*, vol 16, 266-98

1887 'Neue wissenschaftliche Arbeiten über die Alpen',
Mitt. k.-k. Geogr. Gesell. Wien, 612-22

1898 'Die Karstländer und ihre Wirtschaft', *Z.
Schulgeogr.*, vol 20, 161-74

1906 'Bosnien. Aus dem literarischen Nachlasse E.
Richters, Lerausgegeben von G.A. Lukas',
Österreichische Revue, vol 6/69

e. *Educational Works*

1893 *Lehrbuch der Geographie für die I, II und III
Klasse der Mittelschulen*, Wien, 239 p.

1898 *Schulatlas für Gymnasien, Real- und
Handelsschulen, Lehrerbildungsanstalten sowie
sonstige höhere Lehranstalten*, Wien, Prague,
74 maps

1902 *Das Lehrbuch im Geographie-Unterricht*, Wien,
Prague, 18 p.

f.   Mountaineering

1871   'Wanderungen in der Venedigergruppe', *Z. Dtsch.
       Österr. Alpenver.*, vol 3, 275-316

1885   'Sind die Alpen des schönste Gebirge der Erde?',
       Mitt. Dtsch. Österr. Alpenver., vol 15, 1-2

1892   (ed) *Die Erschliessung der Ostalpen*, vol 1,
       Berlin, 1-19; also *ibid*, vol 2 and vol 3, 1894,
       130-223

1903   'Uber die Triebfedern der Bergsteigerei',
       Mitt. Dtsch. Österr. Alpenver., vol 23, 53-5

*Docent Dr. Josef Goldberger is a senior member of the
Geography Department staff at the university of
Salzburg
The German text has been translated by Professor
Karl A. Sinnhuber of the University of Vienna*

Chronology

1847   Born at Mannersdorf, Lower Austria,
           3 October

1848   On the death of his father, the family
           moved to Wiener Neustad

1866   Completed his school education at a
           Gymnasium run by Cistercian monks

1866-71   Was a student at the university of
           Vienna and concentrated on history in
           which, despite his major concern with
           geography, he maintained an interest
           throughout his life

1871   Became a master at the Salzburg Gymnasium
           with full responsibility for the
           teaching of geography

1873   Published his first paper on glacial
           phenomena

1883   Became president of the German and
           Austrian Alpine Society and continued
           to publish notable papers on glaciation

1885   Successfully presented his doctorate
           thesis

1886   Appointed as professor of geography in
           Graz university and continued to be an
           active worker on glaciation in the Alps
           with some visits abroad: with Albrecht
           Penck published the *Atlas der
           Österreichischen Alpenseen*

1889   Was Dean of the Arts Faculty at Graz
           university

1897   Elected as president of the newly-
           established International Commission on
           Glaciers

1898   Appointed as *Rektor* of Graz university

1899   Attended the International Geographical
           Congress in Berlin, at which he gave a
           paper on the 'Gletscher-Konferenz im
           August 1899'

1905   Following a tour of eastern Europe for
           further research, died at Graz,
           5 February

William Rosier

1856–1924

C. FISCHER, C. MERCIER AND C. RAFFESTIN

A geography teacher and politician, William Rosier was a man fully in tune with the spirit of the times in his political outlook, his general interests and his work for education as a practical teacher from the last decades of the nineteenth century. Although at his time the status of geography was uncertain, there was widespread interest in the areas newly opened by conquest and colonial annexation, and particularly in their commercial exploitation. It was the era of modern colonialism with favourable prospects for commercial and industrial expansion, and there was a widespread demand for more factual knowledge of the newly opened territories, for which the records of various travellers, administrators and settlers were obvious sources of data. The problem was to use this material intelligently to provide an understanding of the expanding known world and this challenge was clear to Rosier and his contemporaries working as geographers. It was not an easy task but German geographers had eagerly grasped the new opportunities, followed by the French, especially from 1870. In 1892 M. Dupuy expressed this well at the Berne IGU Congress in his paper on 'L'état de l'enseignement de la géographie en France' (*C.R. Congr. Int. Géogr.*, 289-95), notably in his comment that geography was a valuable contribution to the general intellectual life of young people, distinctive in its quality as part of their scientific studies and broad education.

In the University of Geneva there was only limited support for geography until a chair was established in 1902, of which Rosier was the first holder. Fortunately geography had already received some

recognition for during the 1890s the university had provided a few courses of lectures. Outside the university the *Société de géographie* was very active, as were similar societies in other countries. Indeed it could be said that at the end of the nineteenth century learned societies in many disciplines were powerful and influential, for though they were not official bodies they were virtually the only guardians of specialized knowledge, expounded at their meetings and in their publications. The geographical society of Geneva organized conferences and international congresses including those at Berne in 1895 followed by Geneva in 1908 (with the other Swiss geographical societies), published material of geographical interest (especially on its own area) and established friendly contacts with other societies of the world and also with individual geographers.

1. EDUCATION, LIFE AND WORK

Born on 26 September 1856 at Lancy in the Geneva canton, William Rosier was the son of a watchmaker and a milliner. Though of modest social origins he was able to continue his education beyond the normal school leaving age and in 1873 to enter a Gymnasium. This school offered two possibilities to its students, of which the first was the 'royal road', to a university through specialization in classical studies and the second was directly vocational and pedagogic for intending teachers. This second alternative was the choice of Rosier, and indeed the way more generally available for students of working class origin. In

Geographers Biobibliographical Studies, volume 10 (1986)

fact Rosier never went to a university for a degree
course.   Specialized students of geography of a later
period probably thought that his training as a
geographer was riddled with gaps yet it was rich and
personal in its impact for he absorbed the vivid
teachings of his master and mentor Paul Chaix, and he
acquired an aptitude for study that was to be
fruitful in his later researches.   As a geographer
he was largely self-taught, notably through reading in
the general and university libraries of Geneva and the
library of the city's geographical society.

Among writers of special value to him Rosier
mentions Vivien de Saint-Martin, Vigée, Malte-Brun and
Guyot, the geographer from Neuchâtel who was a
disciple of Carl Ritter and had emigrated to the
United States.   Elisée Reclus also seemed to have
influenced him to a considerable extent and they became
friends during Reclus' period of exile in Switzerland
from 1872.   From that time they maintained their
friendship by correspondence and perhaps Reclus
influenced Rosier's political outlook, of which one
expression was his growing concern for social and
economic welfare in his writings.

Paul Chaix encouraged Rosier to teach from the
time when he was only nineteen years old and in 1876,
when just twenty years of age, he gave a course of
lectures on the Atlantic Ocean, which in fact he had
never seen.   In the following year the material,
marked by minute description of physical features,
appeared as a monograph, *L'Océan Atlantique*.   He
found work readily as a teacher of history and
geography at a school for industrial workers, where
his courses in geography were given under the
headings of physical, economic and social.   He
developed a sense of social mission, expressed clearly
in a paper on the teaching of geography published in
the annual report of the *Association des sociétés
suisse de géographie*, (Berne, 1893).   In this he says
(p 25) that geography 'is a contribution to the
education of the true citizen, the patriot having at
once a wide intellectual vision and enlightened
judgement.   It shows him the importance of reasoned
effort and of active energy.

Other influences on Rosier were Adolphe Tschumi
(1856-1894) and Johann Heinrich Pestalozzi (1746-1827).
Tschumi, a contemporary of Rosier, finished his
classical studies before going to Germany as a tutor,
and later returned to Geneva convinced that scientific
development must be of benefit to the working classes.
In 1883 he wrote a pamphlet, *Routine et progrès*,
demanding radical changes in the educational system
of Geneva.   Instead of a humanistic outlook which
venerated dead languages and book learning he
favoured an integrated realism, a study of the material
world with which everyone inevitably had direct contact.
Through the study of the natural sciences, with the
emphasis on direct observation, the child could enter
the modern world.   Rosier adopted Tschumi's views and
believed that with such an education workers would be
able to respond to new circumstances at a time of
marked industrial and commercial development.

Rosier favoured Pestalozzi's view that a child
must progress from abstraction to known experience for
'teaching from the view, the appearance, the senses,
is the easiest way of giving young people an

understanding of the world when he acquired a direct
knowledge of the character of the territory by which
he is surrounded.

In 1893 (*op. cit. supra*, p. 25) Rosier gave
eloquent expression to his views in these words

> Both for its educational value and for its
> intellectual range, geography merits inclusion
> in the higher classes of gymnasia.   It
> describes the influence of the physical milieu
> on man but it also demonstrates, with a
> plenitude of examples, the powerful action of
> man on nature, the utility and indeed the
> necessity of ceaseless human effort.   We are
> shown man deriving from the earth his well
> being, knowledge and morality through
> indefatigable labour.   Therefore instead of
> constricting the minds of children with the
> sad and degraded history of human strife, the
> carnage and death of battlefields, we turn
> their attention to the consoling view of man
> striving against nature, his spirit aspiring
> to the triumph of mind over matter.

This apologia for geography was in accord with the
radical opinion of the time and with Rosier's own
political stance.   To him the replacement of
Geneva's fortifications in 1849 and the construction of
fine boulevards and parks on their former site was the
beginning of a new and progressive age.   The need was
to look outwards, to make a modern Geneva not only in
appearance but also in the minds of its citizens, and
he warmly approved the emphasis of radical thinkers on
democracy, progress and liberty.

Naturally in the years spent in school education
Rosier was mainly occupied with work in the classroom
but he was concerned with the whole world and gave
papers on a wide variety of subjects to the Geneva
Geographical Society.   In 1887 he spoke on Indian
railways on 23 February and two months later on the
lakes of Bavaria.   In April 1890 he gave a paper on
the then recent expedition of H.M. Stanley, noting its
origin, route and the vicissitudes of the journeys,
with some comments on colonialism.   In a paper of
1891 on the geographical exhibition during the IGU
Congress at Berne he spoke briefly on his views on
education, a theme more adequately developed in his
1893 paper to which reference has already been made.
Another paper of 1893 dealt with the *oecumene* concept
of Ratzel as the inhabited area of the earth and a
further paper of 1900 dealt more fully with the ideas
of Ratzel on geographical conditions and their
influence on the historical and social development of a
people.   During these years the published work of
Rosier was mostly of a directly educational nature,
including textbooks on Switzerland and on the world's
continents, on geography in general, with atlases, and
also papers dealing with various aspects of teaching
in schools.   The breadth of his knowledge was to be
more fully revealed in his tenure of the chair of
geography at Geneva university.

Meanwhile, from the 1890s, Rosier was emerging as
a political figure, partly through his professional
work and writing.   He acquired prestige partly through
his work as a member of numerous commissions and in

1898 was elected as a member of the *Grand Conseil*, the legislative body of the Geneva canton. In the following year he was chosen as a geographical expert by the Federal Council, the national body empowered to adjudicate between France and Brazil on the limits of the Guiana territory of South America. However in 1901 he ceased to be a member of the *Grand Conseil*, as a rule was made that none of its members could be in public employment.

In 1902 he became the first holder of the chair of geography at Geneva university and in the summer semester of that year began to teach the historical, political and economic aspects of the subject. Records of his courses show that he dealt with the political division of Africa during the nineteenth century and the economic geography of Germany, both subjects of contemporary interest. In the winter of 1903 a broader political study of Europe was offered with the emphasis on population and its distribution with races, peoples and nationalities, and at this time he also gave lectures on the economic geography of France with some general discussion of works on the political geography of Europe and on geographical methodology. The territorial area of his courses extended to cover emigration and colonial expansion on a world scale with the commercial policy and imperialism of England (he used the term 'Angleterre' rather than 'Grande-Bretagne'). From 1904 he dealt with Europe as an economic unit, with a comparative treatment of the economic capacity and its political implications possessed by the various European states, as well as with the growing power and significance of the 'new powers' in the Far East. In 1905 he dealt with the economic geography of the United States, along with the history of its discovery and exploration, and also at this time his teaching included what could be comprised as the *genres de vie* of hunters, fishers, nomadic pastoralists and sedentary farmers.

Through these years one may discern a steady deepening of learning on the geography of Europe, not only as an entity of political complexity but also as the economic and political heartland of a world in which colonial expansion and commercial development presaged a new global unity. Seeking the knowledge of the world that was favoured and desired at the time, he was well aware of the political implications of his teaching as well as of its historical basis. Finally in the winter of 1906-07 he included a course on the polar lands based on the history of their exploration and contemporary knowledge. This must have been greatly appreciated at a time when public interest had been aroused by several major expeditions.

From 1906 to 1918 Rosier was replaced by Paul Chaix's son, Emile, and returned once more to a political career, but he continued to publish works on geography teaching, atlases and textbooks in collaboration with Chaix (1908), Maurice Borel (1911), Paul Decker (1913) and Charles Biermann (1913 and later). However for twelve years he was Minister of State responsible for all public education in the Geneva Canton, and in this office he sought the highest possible standards for the whole community. In the University he constantly advocated the creation of a faculty for economic and social studies, within which geography could find a home. This was achieved in

1915, and he spent the last years of his life teaching and guiding the new faculty, and for two years from 1922 to 1924 was its Dean. Not until 1922 was he able to enjoy a long spell of fieldwork abroad, of several months in the Sahara. Unfortunately he was unable to complete the book based on his own observations. It is ironic that he never had the time to produce a work that could have been an expression of his view that geography was based on empirical data and on the interaction of both theory and experience, for he died on 16 September 1924 at Petit-Saconnex, Geneva from double pneumonia.

2. SCIENTIFIC IDEAS AND GEOGRAPHICAL THOUGHT

Rosier was a man of his time, a radical politician convinced that the modern world in which he lived was scientifically one whole and through its current development achieving a new unity in its economic, social and political life. To him, as to so many Genevois, the changes in Geneva meant growth and enrichment of its life, a New Age in which its old traditionally Protestant quality was modified by the immigration of people who were largely Catholic. Convinced that there must be change Rosier was eager to meet it and that in part explains why a large part of his working life was given to political activity. And to him the changes in his homeland area were part of a worldwide development of lasting insignificance.

At the same time geographical study of the whole world could only be satisfying when based on an analytical approach and Rosier was one of the many geographers who saw regional geography as the crown of the subject. Fully aware of the state as an entity in Europe, he regarded boundaries as one element in geographical study for geographers were concerned with any division of space. This could be expressed in many ways, including relief and river courses, local natural regions, towns, historical evolution, economic conditions, political and social qualities, all of which contributed to a study of natural regions. The aim was to discern a conformity, even an overlapping, of distributions which would give a credible result, so that a 'natural region' would not be merely physical or climatic but evident, perceptible, depending on a geomorphological foundation which imposed an attitude of life on the population.

This he illustrated by a consideration of the Basque province, in general mountainous, fertile and well cultivated. He found that history, language and political attitudes all served to identify the region, within which the names of rivers and mountains defined the parts. Firstly the mountains gave the region a coherence and identity which was visible and permanent, indeed recognizable with only the most rudimentary geomorphological study. Secondly, the fertility of the Basque region was a reality, existing long before anyone was there to appreciate it, but a reality also which in time made possible the varied economic, social and political characteristics of the area. Thirdly, the inherent value of the area for agriculture was an intrinsic quality which became part of the ethos of its people.

In the preface to his *Géographie générale illustrée. Europe* of 1891 Rosier gives a clear

exposition of his outlook on geography.   He explains
that there was a need for a work in Switzerland for the
higher forms of secondary schools and for the general
public, as the works published in France and Germany
were not suited to the ideas and general outlook of
Swiss readers.   He had been asked to write such a work
by the *Association des Sociétés suisse de géographie*,
and the volume on Europe was the first of three, of
which volume 2 on the remaining continents appeared in
1893 and volume 3 on physical geography in 1908, the
last in collaboration with Emile Chaix.   For Europe it
seemed reasonable to begin with Switzerland, the
continent's major watershed with rivers flowing and
opening in routes into neighbouring countries.   From
Switzerland he went on to Germany and Austria-Hungary,
linguistically Germanic, and then to France and Italy
with Romance languages, from which he turned to Spain
and Portugal.   This gave the idea of a circular ring
of countries around Switzerland, beyond which there
was a peripheral 'circle' made up of northwest Europe,
the Netherlands, Belgium and the British Isles, then
the Scandinavian states, Russia and the Balkan
peninsula.   In this arrangement one may see the
movement from the known, at least in some measure, to
the less known and the unknown, a long venerated
educational principle.

By 1891 a man of great experience as a teacher,
Rosier regarded the map as basic in geography
teaching, and for each state he first gave a general
view, including its area, boundaries, position in
Europe, shape, and then a study of relief and rivers
followed by a delimitation of its natural regions,
including the towns within them.   In his view study
of the provinces or other administrative divisions
was less 'scientific' than that of natural regions,
which he regarded as unchanging through time.   All
this study led to the map and was more in accord with
the approach of natural science than any other: no
doubt he knew the many dreary texts listing counties
and other local government divisions such as the
départements of France.   Rosier observed that he was
at variance with those who dealt with economic, social
and political geography before discussing towns.   His
emphasis on valley communications was widespread in
his time, notably among German geographers: perhaps
it is also an expression of his vision of Switzerland
but that thought raises the question of whether a
geographer looks at strange lands for their
similarities as well as their differences from the
home ground.   He was realistic in recognizing that
hopes of seeing all the countries adequately were
futile and that therefore he had to rely on map
evidence.

As a culmination of the geographical nature of a
country he moved forward to its economic, social and
political conditions, generally with a short
historical preamble.   Finally each state was seen as
a whole with the economic situation as an expression
of its natural physical characteristics, its social
conditions in relation to economic qualities, and
finally its political and administrative organization,
likely in some instance to be more subject to
modification than other aspects of regional geography.
Statistical information was provided for reference and
constant use of maps was necessary.   Rosier claimed
that

the different between this manual and other
short works does not rest on the inclusion
of a larger number of names but on the
stronger emphasis of a general view, on
comparisons, on the study of causes and
relationships between the physical nature
of countries and the material and spiritual
life of peoples, in general the social
wealth and civilization on the Earth.

Such an outlook may seem to be deterministic but
in the later editions of the 1891 book signs of any
earlier determinism diminished.   Rosier was familiar
with the writings of Ratzel but also with the
methodological controversies then current among French
geographers.   His outlook may be illustrated by a
brief treatment of his study of Switzerland.   A
section of some length deals with relief, for after
a short account of its geology several pages are given
to mountains, largely to summit heights.   The
geomorphological content is slight, as for example in
dealing with erosion as a sculpturing force of relief.
This is at variance with the prominence given to such
geomorphological data in classical French texts
dealing with natural regions.   Rosier uses rivers and
lakes as bases of natural regions and the various
upland basins with their hilly sloping ground emerge
as regional units, numbering forty-four in all.   The
emphasis on water courses remains, as for the Rhone
from its source to the Lake of Geneva, which is taken
as part of a major region with Lausanne as its major
centre though Geneva also appears, but without any
clear statement on its sphere of influence.   Then
follow other units such as Sarine, Orbe-Thièle,
Neuchâtel (the town here named the region), and the
valleys of the Jura with others such as the Aar valley
with Berne, the Grande Emme, the Reuss basin.

Rosier was concerned with the essential reality
of the scene and in general sought to describe areas
in simple ordinary terms, though he shows a reach
beyond this to a more scientific definition, as for
example in his statement that 'Berne stands on a
peninsula surrounded on three sides by the Aar river'.
This gives the essential features of the city's
fascinating site and situation but does not
specifically state that the river occupies an incised
meander and therefore the description, though
evocative of the reality, does not aspire to complete
explanatory scientific precision.   This is also true
of his comment on Bruxelles, where he comes close to
the idea of the city's centrality without explaining
it or succeeding in confirming it as a principle of
general application.   It is hard to say whether this
results from intellectual limitations or from current
fashion and it would seem that Rosier was more
interested in the factual content of any item of
information than in an explanatory statement of a
general structural nature.   His inclination was
towards an idiographic rather than a nomethetic use of
facts.

In dealing with population Rosier was more
concerned with social circumstances than with
demography.   Where the density of population was

perfectly clear he did not deal with the rate of growth but with the actual numerical increase year by year. It would be possible to give many examples of such a pragmatic approach, for example in his very modern approach of national budgets and external trade. This no doubt developed from his practical education which gave him a wish to establish a geography of direct utility and consequence, factual and practical, resting on his radical approach to political questions.

The use of sources by Rosier was comprehensive, particularly of those in French but also including German material and British for physical geography. Humboldt, Ritter, Guyot and Reclus figure prominently. Especially significant was the work of his friend Reclus, from whom he acquired abundant data, many views and documentary materials. The general attitude of Rosier was to acquire material whenever and wherever it might be found, seeking always to know and to make known. Naturally in his concern for political geography his views were influenced by the progress of World War I and historically moulded by the writings of French geographers of the time. He favoured the idea of national consciousness as a basis for a general study of political geography and was influenced in the distribution of languages, which he regarded as of more influence than race in national consciousness and in his view religious allegiance was also of significance in the emergence of nations. In fact there is no single attribute or combination of attributes that makes nationality but rather a collective will, a realisation of unity, what Richard Hartshorne was later to call a *raison d'être* that welds a people into a nation. This Rosier well understood, and in the last years of his life it was expressed in his book, *L'Europe nouvelle et le principe des nationalités*, a work which shows him to be modern in his outlook on political geography.

3. *INFLUENCE AND SPREAD OF IDEAS*

With others at the end of the nineteenth century Rosier shared fully in the 'geographical triumph' of the time, for then the whole world was seen more adequately than ever before. Even so, much of the discovery was recent, for in 1876 Elisée Reclus had written that 'in this century of the steamship, of the Press, of the incessant and feverish activity, the centre of Africa, part of the Australian continent, the beautiful and probably very rich island of New Guinea, and the vast plateaux of interior Asia still remain an unknown domain'. Reclus, born in 1830, twenty-six years before Rosier, was viewing the situation in his middle age and discovery continued avidly during the last quarter of the nineteenth century. Here was the newly revealed world that Rosier was eager to know and to describe. No longer was the main concern of the geographer with a world yet to be revealed but rather with a world to be carefully observed, explained in a systematic and orderly manner that could reveal as never before the face of the earth. In that lay the claim of geography as a study worthy of official recognition.

Consequently there was a clear task for the geographer of pragmatic mind, to deal with the newly revealed earth rather than to deal with philosophical and epistemological questions. Therefore it is useless to look for comparisons between the geographical conceptions of Rosier and of Kant, although each was concerned with the distribution of phenomena on the surface of the earth. Kant is never mentioned in Rosier's works: conceivably Rosier never read Kant's publications and encountered them only in the writings of other authors who had drawn inspiration from them. Rather Rosier saw the need for the geographer of his time to be pragmatic, empirical, to be primarily a clear exponent of geography more than the creator of a new geographical science. His strength lay in his vision of the world through his dedicated study and observation. Through that he gave others a way of sharing his understanding of the world. Many who came after him shared such aspirations but there were some who wanted more.

Bibliography and Sources

1. *REFERENCES ON WILLIAM ROSIER*
Geogg, E., 'William Rosier', *Le Globe*, vol 64 (1925), 23-6
Terrier, C., 'Hommage à la mémoire de William Rosier', *Bulletin de l'Institut national genevois*, vol 48 (1930), 1-18
anon., 'William Rosier', *La Patrie Suisse*, no 809 (1927), 243
Burky, Ch., 'William Rosier', in *L'histoire de l'Université de Genève*, vol 4, *Histoire des facultés et instituts*, (1956), 142

2. *SELECTIVE AND THEMATIC BIBLIOGRAPHY OF WORKS BY WILLIAM ROSIER*

a. *Works on Education and Methodology*
1891 'Des services que la géographie peut rendre dans les conflits économiques', *C.R. 5 Congr. Int. Géogr.*, Berne, 656-63
1893 *Le Congrès et l'Exposition de Geographie, Berne, Août 1891*, 30 p.
---- 'L'enseignement de la géographie dans les gymnases et la place de cette science dans le programme des examens de maturité', paper given to to the *Association des Sociétés suisses de géographie*, Berne, 13 p.
1894 'Le rôle intellectuel et moral de la géographie', *L'éducateur*, no 8, 118-20
1898 'L'enseignement de la sphère et de la lecture des cartes à l'école primaire', *L'éducateur*, no 2
1909 'Le domaine propre de la géographie considérée comme branche d'enseignement: utilité d'une résolution prise sur cette question par un congrès international', *C.R. 9 Cong. Int. Géogr.*, Geneva, vol 1, 275-84
1913 *L'éducation civique de la jeunesse*, Lausanne, 20 p.

b.   Hydrology and Oceanography
1877 *L'Océan Atlantique*, Geneva, 208 p.
1890 'Caractères généraux de l'hydrographie africane',
      La Globe, 34 p.

c.   Textbooks and Other Volumes
1888 *Premières lecons de géographie.   La Terre, sa
      forme, ses mouvements.   Lecture des cartes*,
      Geneva, 112 p., (also 1893, 1901, 1904)
1891 (with B. Dussand) *Géographie locale à l'usage
      des écoles primaires du canton de Genève*,
      Geneva, 27 p., (also 1902, 1914)
1892 *Géographie illustrée de la Suisse*, Lausanne,
      47 p.
1893 *Géographie générale ilustrée.   Asie, Afrique,
      Amérique, Océanie*, Lausanne, 339 p., (also
      1902, 1911)
1895 *Manuel-atlas, Cours Moyen.   Révision de canton
      de Genève, Suisse.   Premières notions sur les
      cinq parties du monde*, Lausanne, 111 p. (also
      1900, 1906)
1899 *Manuel-atlas.   Degré supérieur des écoles
      primaires.   Géographie des cinq parties du
      monde.   Révision de la Suisse*, Lausanne and
      Geneva, 180 p., (also 1900 and 1906)
1905 *Histoire illustrée de la Suisse à l'usage des
      écoles primaires*, Lausanne and Geneva, 200 p.,
      (also 1911 and 1926)
1908 (with E. Chaix) *Manuel de géographie physique
      destiné aux classes supérieures des
      établissements d'instruction secondaire*,
      Lausanne, 176 p.
1911 (with M. Borel) *Adaptation francaise de l'Atlas
      scolaire suisse 'Schweizerischer Schul-Atlas
      publié par la Confédération et les cantons*,
      Winterthur, XII + 136 p.
1913 (with P. Decker) *Manuel d'histoire suisse pour
      l'enseignement secondaire*, Lausanne, 224 p.
---- (with C. Biermann) *Manuel-atlas pour
      l'enseignement secondaires.   Géographie de la
      Suisse*, Lausanne, 163 p.
1914 (with C. Biermann) *Manuel-atlas, destiné aux
      classes inférieures des écoles secondaires.
      Géographie de la Suisse*, Lausanne, 100 p.
1920 (ed C. Biermann) *Cours de géographie générale à
      l'usage des classes supérieures de
      l'enseignement secondaires.   1, Europe*, 336 p.
      (see 1891 and 1923)
1921 *L'Europe nouvelle et le principe des
      nationalités*, Geneva and Lausanne, 18 p.
      (extrait de *l'Annoire de l'Instruction Publique*)
1923 (ed C. Biermann) *Cours de géographie générale à
      l'usage des classes supérieures de
      l'enseignement secondaire.   2, Amerique,
      Océanie, Asie, Afrique*, Lausanne, 352 p.
      (see 1893)

*Claire Fischer is Assistante in geography at the
university of Geneva, Claude Mercier is a teacher of
geography and history at a lycée in France and Claude
Raffestin is professor of geography at Geneva
university*
Translated by T.W. Freeman

Chronology

1856	Born at Lancy, canton of Geneva, 26 September
1873	Entered a Gymnasium
1876	Gave a course of lectures on the Atlantic Ocean
1877	Published a monograph on the Atlantic Ocean and taught geography and history
1887	From this time onwards he gave papers on a variety of subjects to the geographical society of Geneva
1893	His views on education were strongly expressed in a paper, later published, given to the *Association des Sociétés suisses de géographie*
1898	Became a member of the *Grand Conseil*, the legislative assembly of the Geneva canton
1899	Engaged, as a geographical expert, on the special commission adjudicating between France and Brazil on the limits of Guiana in South America
1901	Withdrew from the *Grand Conseil* as its members were precluded from public employment
1902	First holder of the chair of geography at Geneva university, where from this time he gave a wide range of courses
1906	Returned to a political career and became minister for all public education in the Geneva canton: during these years he continued to publish textbooks on geography
1915	His efforts to establish a Faculty in the university for economic and social studies, including geography, were at last successful
1918	Returned to university work and was particularly concerned with the development of the new faculty
1922	Spent several months on fieldwork in the Sahara: became Dean of the Economics and Social Studies faculty to 1924
1924	Died at Petit-Saconnex, Geneva, 16 September

Index

The index is divided into four parts:

1. *PERSONAL NAMES* as far as possible are given in full with the year of birth and death.
2. *ORGANIZATIONS AND RELATED REFERENCES* is subdivided into (a) *Colleges, Institutes, Institutions, Museums, Official and Research Organizations;* (b) *Scientific Congresses and Commissions;* (c) *Societies and Associations;* (d) *Universities.*
3. *SUBJECTS* cover concepts, geographical theories and specific research.
4. *CUMULATIVE LIST OF BIO-BIBLIOGRAPHIES* includes all the geographers listed in volumes 1-10 inclusive.

Page numbers in italic refer to the Bibliography and Sources, and to the Chronology sections of the biobibliographies.

1. *PERSONAL NAMES*

AAGESEN, Aage, 1956-   , 121
ABBE, Cleveland, 1838-1916, 61, 64
ABRAMOV, Lev Solomonovich, 1919-   , *12*
ABU-al FIDA, Ismael, 1273-1331, 52
ADAMS, Charles Christopher, 1873-1955, 29
AGASSIZ, Louis, 1807-1873, 145
AHLMANN, Hans Wilhelmson, 1899-1975, 87
ALEXANDERSSON, Gunnar, 1902-   , 7
ALTHOFF, Friedrich, 1839-1909, 100
AMES, Van Meter, 1898-1985, 60, 65

ANDREWS, Ernest Clayton, 1870-1948, 61, 65
ANUCHIN, Dmitry Nikolayevich, 1843-1923, *5, 11*
ANUCHIN, Vsevolod Alexandrovich, 1913-1984, *11, 12*
ARMAND, David L'vovich, 1905-1976, *11*
ARNDT, Johann, 1555-1621, 77, *83*
ARSTAL, Aksel, 1855-1940, 86
ATWOOD, Wallace Walter, 1872-1949, 65

BABCOCK, Stephen Moulton, 1843-1931, 58
BALDACCI, Olavi, 1901-1907, 70
BANSE, Ewald, 1883-1953, 37, 71
BARANSKIY, Nikolay Nikolayevich, 1881-1963, 1-16
BARBOUR, George Brown, 1890-1977, 59
BARROWS, Harlan Harland, 1877-1960, 60, 65
BARTHOLOMEW, John, 1860-1920, 93
BASEDOW, Johann Bernhard, 1723-1790, 148
BASTIAN, Adolf, 1826-1905, 100
BATES, Henry Walter, 1825-1892, 94, 96
BAULIG, Henri, 1877-1962, 6, 108
BECK, Hanno, 1923-   , 82, *83*
BEHRE, Charles Henry, 1896-   , 59
BERGSMARK, Daniel Rockman, 1898-1945, 59
BEST, George, ob. 1594?, 51
BIASUTTI, Renato, 1878-1965, 36, *40*
BIERMANN, Charles, 1875-1961, 151, *154*
BIOLLY, Paul, 1862-1908, 139
BIRKENHAUER, Josef A.C., 1929-   , 77-84
BLACKWELDER, Eliot, 1880-1969, 61
BLUM, Otto, 1876-1944, 73
BOGUSLAWSKI, Georg von, 1827-1884, 100

BOLIVAR Y URRATIA, Ignacio, 1850-1944, 37
BOREL, Maurice, 1860-1926, 151, *154*
BOROUGH, Stephen, 1525-1584, 50, *53*
BOROUGH, William, 1536-1599, 53, *55*
BORZOV, Alexander Alexandrovich, 1874-1939, 5
BOWEN, Emmanuel, ob. 1767, 44
BOWEN, Emrys George, 1900-1983, 17-23
BOWMAN, Isaiah, 1878-1950, 61, 62, 65
BRÜCKNER, Eduard, 1862-1927, 126, 130
BUCHER, Walter Hermann, 1888-1965, 59, 62, *68*
BUCKLE, Henry Thomas, 1821-1862, 27
BUFFON, Georges Louis Leclerc, 1707-1788, 78

CABOT, Sebastian, c. 1476-1557, 50, 108
CALDERÓN Y ARANA, Salvador, 1835-1911, 37
CAMDEN, William, 1551-1623, 50
CAMPBELL, Marius Robinson, 1858-1940, 62
CAREY, Henry Charles, 1793-1879, 25-8
CARTER, Harold, 1920-   , 20, *. 21*
CASE, Earl Clark, 1887-1978, 59
CASTER, Kenneth Edward, 1908-   , 59
CHAIX, Emile, 1855-1929, 151
CHAIX, Paul, 1808-1901, 150
CHAMBERLAIN, Thomas Chrowder, 1843-1928, 29, 58, 59, 62, 63
CHANCELLOR, Richard, ob. 1556, 50, *55*
CHISHOLM, George Goudie, 1850-1930, 60, 65, 96
CHRISTALLER, Walter, 1893-1969, 8
CLAY, Henry, 1777-1852, 27
CLEMENTS, Frederick Edward, 1874-1945, 30, 31
CLÜVER, Philipp, 1580-1623, 127

COULTER, John Merle, 1851-1928, 29, *32*

COWLES, Henry Chandler, 1869-1939, 29-33

CRAWFORD, Osbert Guy Stanhope, 1886-1957, 18, 19

CREUTZBURG, Nikolaus, 1893-1978, 71, 72

CUATRACAS, Josef, 1903-   , 138, 139

CUMMING, William P., 1900-   , 44

CVIJIĆ, Jovan, 1865-1927, 7

DABNEY, Charles William, 1855-1945, 59, 64

D'ALEMBERT, Jean Lerond, 1717-1783, 78

DANIEL, Glyn Edward, 1914-   , 18

DANTÍN-CERECEDA, Juan, 1881-1943, 35-40

DARBY, Henry Clifford, 1909-   , 19

DARWIN, Charles Robert, 1809-1882, 19, 30, 31, 136

DAVIES, Wayne Kenneth David, 1938-   , 19, *21*

DAVIS, John, c. 1550-1605, 51, 52

DAVIS, William Morris, 1850-1934, 30, 31, 35, 36, 57, 58, 59, 60, 62, 63, 106, 112, 129

DE BOUGAINVILLE, Louis Antonie, 1729-1811, 43

DE BRAHM, William Gerard, 1718-1799, 41-7

DE CANDOLLE, Alphonse Louis Pierre Pyrame, 1806-1893, 30

DE CHARLEVOIX, Pierre François de Xavier, 1682-1761, 43

DE CHARPENTIER, Jean, 1786-1855, 137

DE GEER, Sten, 1886-1933, 86

DE MARTONNE, Emmanuel, 1873-1955, 35, 36, 37, *40*, 106, 108

DE ULLOA, Antonio, 1716-1795, 43

DE VORSEY, Louis, Jr., 1929-   , 41-7

DECKER, Paul, 1882-1942, 151, *154*

DEE, John, 1527-1608, 49-55

DELUC, Jean André, 1727-1817, 44

DEMANGEON, Albert, 1872-1940, 36, *40*, 106, 108

DEMKO, George, 1933-   , *13*

DENIS, Jacques Georges, 1922-   , 105-10

DENSLOW, Van Buren, 1834-1902, 27

DICKINSON, Robert Eric, 1905-1981, 130

DIDEROT, Denis, 1713-1784, 78

DIETRICH, Gunter, 1911-1972, 101

DIK, Nikolay Yevgen'yevich, 1905-   , *13*

DODGE, Richard Elwood, 1868-1951, 64

DOMETTI, Aglaida Aleksandrovna, 1895-   , 11

DRAKE, Francis, 1543-1596, 51

DRUDE, Karl Georg Oskar, 1852-1933, 30

DRYGALSKI, Erich von, 1865-1949, 70, 129

DUFOUR, Louis, 1832-1892, 16

DURAND, Theophile, fl. 1888-1909, 137, 138

DUTTON, Clarence E., 1841-1912, 112

ECKERT, Max, 1868-1938, 70, 102

ELCANO, Juan Sebastián, 1474-1483?-1526?, 37

ELIZABETH I, Tudor Queen, born 1533, ruled 1558-1603, 49, 50, 51, 52, 53

ELLICOTT, Andrew, 1754-1820, 70, 101

FADEN, William, 1750?-1836, 44

FAVRAT, Louis, 1827-1893, 136, 139

FELS, Edwin, 1888-1983, 127

FENNEMAN, Nevin Melanchthon, 1865-1945, 57-68

FINCH, Vernor Clifford, 1883-1959, 64

FINSTERWALDER, Sebastian, 1862-1951, 145

FISCHER, Claire, 1957-   , 149-54

FISCHER, Theobold, 1846-1910, 101

FLEURE, Herbert John, 1877-1969, 17, 18, 21, *22*, 36, *40*

FOGG, Walter, 1899-1965, 19

FORBES, James Grant, 1769-1825, 45

FOURMARIER, Paul, 1877-   , 106

FOX, (Sir) Cyril Fred, 1882-1967, 18, 19, *22*

FRANKLIN, Benjamin, 1706-1790, 43

FREEMAN, Thomas Walter, 1908-   , 82, *83*, *110*, *140*, *154*

FRIEDRICH, Ernst, 1867-1939, 70

FREYKIN, Zakhar Grigor'yevich, 1904-   , *11*, *12*, *14*

FRISIUS, Gemma, 1508-1555, 49, 53

FROBISHER, Martin, c. 1539-1594, 51, 52, 53, *55*

FUCHS, Roland J., 1933-   , *13*

GANONG, William Francis, 1864-1941, 31

GAULD, George, 1732-1782, 45

GEDDES, (Sir) Patrick, 1854-1932, 83

GEKHTMAN, Georgiy Nikolayevich, 1870-1956, *12*

GEOFFREY OF MONMOUTH, ob. 1155, 50

GEORGE, Pierre, 1909-   , *11*

GERASIMOV, Innikentiy Petrovich, 1905-1985, *11*

GEUX, François, 1861-1918, 138

GILBERT, Grove Karl, 1843-1918, 61, 62, 63, 112

GILBERT, Humphrey, c. 1537-1583, 51, 52

GILLIES, Alexander, 1907-   , 79

GOETHE, Johann Wolfgang von, 1749-1832, 80, 89

GOLDBERGER, Josef, 1919-   , 143-8

GOODE, John Paul, 1863-1932, 62

GOUDIE, Andrew Shaw, 1945-   , 108, 111-16

GOUROU, Pierre, 1900-   , 108

GRIESBACH, August, 1814-1879, 30

GRIGOR'YEV, Andrey Alexandrovich, 1883-1968, 6

GUTHE, Hermann Adolf Wilhelm, 1825-1874, 82

GUYOT, Arnold Henry, 1807-1884, 25, 105, 153

HABERSHAM, James, 1715-1775, 41

HAECKEL, Ernst, 1834-1919, 70, 136

HÄGERSTRAND, Torsten, 1916-   , 88

HAKLUYT, Richard, 1552-1616, 51

HAMANN, Johann Georg, 1937-   , 82, *83*

HANNERBERG, David, 1900-1981, 88

HARRIS, Chauncy D., 1914-   , *11*, *12*

HARTSHORNE, Richard, 1899-   , 153

HASSERT, Ernst Emil Kurt, 1868-1947, 69-76

HASSINGER, Hugo, 1877-1945, 86

HATT, Gudmund, 1884-1960, 118

HAWKINS, William, ob. 1554, 51

HAYES, Charles Willard, 1859-1916, 58

HELLAND-HANSEN, Bjorn, 1877-1957, 101

HENSEN, Victor, 1835-1924, 100

HERBERTSON, Andrew John, 1865-1915, 37, 61, 93, 96

HERBERTSON, Frances Dorothea, 1864-1915, 37

HERNANDEZ, Pacheco Eduardo, 35, 37

HETTNER, Alfred, 1859-1941, 4, 6, *13*, 36, 129, 130

HOHEISEL, Karl Robert, 1937-   , 82, *83*

HOOTON, Earnest Albert, 1887-1954, 62

HOUSE, Edward Mandell, 1858-1938, 62

HOWARTH, Osbert John Radcliffe, 1877-1954, 95

HOWE, George Melvyn, 1920-   , 21

HUGUET DEL VILLAR, Emilia, 1871-1951, 138

HUMBOLDT, Alexander von, 1769-1859, 30, 77, 81, 82, 101, 107, 136, 143, 153

HUME, David, 1711-1763, 78

ISACHSEN, Fridtjov Eide, 1906-1979, 85-91

ISACHSEN, Gunnar, 1868-1939, 85

ISARD, Walter, 1909-   , 8

JACKMAN, Charles, ob. 1581, 51

JACOBSEN, Niels Kingo, 1921-   , 117-24

JAMES, Preston Everett, 1899-1986, 6, *13*, 112

JAY, Leslie Joseph, 1917-1986, 93-8

JOERG, Wolfgang Louis Gottfried, 1885-1952, 61

JOHNSON, Douglas Wilson, 1878-1944, 58, 106

JONES, Clarence Fielden, 1893-   , *13*

JONES, Emrys, 1920-   , 21

JONES, Glanville Rees Jeffreys, 1923-   , 21

JONES, Wellington D., 1886-1957, 4

JONES-PIERCE, Thomas, 1905-1964, 21
JUNGERSEN, Hector, 1854-1917, 118

KANT, Edgar, 1902-1978, 88
KANT, Immanuel, 1724-1804, 78, 79,
  80, 82, *83*
KARPOV, Lev Nikolayevich, 1928- ,
  12
KELTIE, (Sir) John Scott, 1840-1927,
  93-8, 127
KIBAL'CHICH, Oleg Aleksevich,
  1928- , *11*
KIEPERT, Heinrich, 1818-1899, 100,
  130
KIRCHOFF, Alfred, 1838-1907, 70,
  127, 128
KOLOSOVA, Yuliya Antonovna,
  1916-1984, *11*
KOLOSVSKIY, Nikolay Nikolayevich,
  1891-1954, 5
KONSTANTINOV, Oleg Arkad'yevich,
  1903- , 12
KOROVITSYN, Vasily Petrovich,
  1907-1983, *11*
KREBS, Norbet, 1867-1947, 128
KROGH, August, 1874-1909, 128
KRUBER, Alexander Alexandrovich,
  1871-1945, 5
KRÜMMEL, Johann Gottfried Otto,
  1854-1912, 99-104
KUMM, Hermann Karl Wilhelm,
  1874- , 62

LAMARCK, Jean Baptiste Pierre
  Antoine de Monnet, 1744-1809, 31
LAPPO, Georgiy Mikhaylovich,
  1923- , *11*
LASSER, Tobias, fl. 1915-1950, 139
LAVATER, Johann Kaspar, 1741-1801,
  77
LAVROV, Sergey Berisovich, 1928- ,
  12
LE ROUGE, George Louis, fl.
  1740-1770, 44
LEFÈVRE, Marguerite Alice,
  1894-1967, 105-10
LEIBNIZ, Gottfried Wilhelm,
  1646-1716, 78
LERESCHE, Louis, 1808-1885, 139
LESSING, Gotthold Ephriam, 1729-1781,
  78
LLOYD, (Sir) John Edward, 1861-1947,
  18, *22*
LUCAS, Zoraida, fl. 1915-1950, 139

MACKINDER, (Sir) Halford John,
  1861-1947, 82, 96, 127, 128, 130
MACPHERSON HEMAS, José, 1829-1902,
  37
MALTE-BRUN, Conrad, 1775-1826, 150
MARBUT, Curtis Fletcher, 1863-1935,
  62
MARKHAM, (Sir) Clements, 1830-1916,
  96, 97
MARSH, George Perkins, 1801-1882,
  25, 114

MARTI-HENNEBERG, Jordi, 135-42
MARTIN, Geoffrey John, 1934- , 112
MARY, Tudor Queen, born 1516,
  ruled 1553-1558, 49
MASHBITS, Yakov Grigor'yevich,
  1928- , *11*
MAURENBRECHER, Wilhelm, 1838-1892,
  69
MAURY, Matthew Fontaine, 1806-1873,
  25
MAYER, Harold M., 1916- , *15*
MAYER-ABICH, Adolf, ob. 1968, *83*
MAYERGOYZ, Yakov Grigor'yevich,
  1928- , *11*
MCCARTY, Harold H., 1901- , *13*
MCGEE, William John, 1853-1912,
  111-16
MELA, Pompanius, fl. 43 A.D., 52
MELANCHTON, Philipp, 1497-1560, 57
MELON Y RUIZ DE GORDEJUELA, Amando,
  1895-1977, 37, 39
MERCATOR, Gerard, 1512-1594, 49, 51,
  52, 53, 105
MERCIER, Claude, 1951- , 149-54
MEYER, William Bruce, 1961- ,
  25-8
MICHOTTE, Paul Lambert, 1876-1940,
  36, *40*, 105, 106, 107, 108
MILL, Hugh Robert, 1861-1950, 93, 94
MILL, John Stuart, 1806-1873, 27, 28
MOLLÁ RUIZ-GOMEZ, Manuel, 1954- ,
  35-40
MONACO, Prince Albert I of,
  1848-1922, 85, 101
MONTESQUIEU, Charles Louis,
  1689-1755, 78
MORSE, Jedediah, 1761-1826, 45
MULLER-MINY, Heinrich, 1900-1981,
  127
MURRAY, (Sir) John, 1841-1914, 93
MURZAYEV, Eduard Markovich,
  1908- , *12*
MYKLEBOST, Hallstein, 1923- ,
  85-92

NANSEN, Fridtjof, 1861-1930, 99, 101
NASH-WILLIAMS, Victor Erle,
  1897-1955, 18
NAZAREVSKIY, Oleg Rostistlarovich,
  1910- , *10, 11*
NELSON, Helge, 1882-1966, 89
NEUMANN, Carl, 1823-1880, 126, 127,
  130
NEUMAYER, Geog von, 1826-1909, 100
NIELSEN, Niels, 1893-1981, 117-24
NIKITKIN, Nikolay Pavlovich,
  1893-1975, *10, 13*
NIKOL'SKIY, Igor Vladimirovich,
  1907- , *11*
NOE-NYGARD, Arne, 1908- , 118,
  119
NOWELL, Laurence, Dean of Lichfield,
  fl. 1559-1576, 53
NUNES, Pedro (Petrus Novius),
  1492-1577, 50, 53

ORMSBY, Hilda, 1877-1973, 128
ORTELIUS, Abraham, 1527-1598, 50,
  51, 52, 53, 105
OTREMBA, Erich, 1910- , 73

PANSA, Gerhard, 1933- , 69-76
PAPANIN, Ivan Dmitriyevich,
  1894-1986, *12*
PARTSCH, Joseph Franz Maria,
  1851-1925, 99, 101, 125-33
PASSARGE, Siegfried, 1867-1958, 127
PELTIER, Louis C., 1916- , 61
PENCK, Albrecht, 1858-1945, 36, 65,
  69, 70, 126, 129, 130, 144, 145
PESCHEL, Oskar, 1826-1875, 99, 101
PESTALOZZI, Johann Heinrich,
  1746-1827, 138, 150
PET, Arthur, fl. 1560-1580, 51
PFEFFER, Wilhelm, 1845-1920, 69
PITTIER, Henri-François, 1857-1950,
  135-42
POWELL, John Wesley, 1834-1902, 61,
  111, 112, 113, 136
PREOBEAZHENSKIY, Arkadiy Ivanovich,
  1907- , *10, 11, 12, 13*
PROTHERO, Ralph Mansell, 1924- ,
  21
PRYCE, William Rees Thomas,
  1925- , 21
PURCELL, Joseph, fl. c. 1754-?, 45

QUINN, David Beers, 1909- , 53,
  54

RADEFF, Anne, 135-42,
RAFFESTIN, Claude, 1936- , 149-54
RAKITNIKOV, Andrey Nikolayevich,
  1903- , *11*
RATZEL, Friedrich, 1844-1904, 19, 69,
  71, 73, 80, 100, 128, 144, 150, 152
RECLUS, Elisée, 1830-1905, 153
REES, Alwyn David, 1911-1974, 21
REINHARD, Rudolf, 1876-1946, 129
RENEVIER, Eugène, 1831-1906, 138
REVENGA CARBONELL, Antonio, 36, *39,
  40*
RICARDO, David, 1772-1823, 25, 26, 28
RICH, John Lyon, 1884-1956, *65, 67*
RICHTER, Eduard, 1847-1905, 126, 130,
  143-8
RICHTHOFEN, Ferdinand von, 1833-1905,
  19, 37, 69
RITTER, Karl, 1779-1859, 77, 80, 82,
  101, 107, 130, 153
ROBERTSON, John Michael, 1958- ,
  29-33
ROBINSON, Arthur H., 1915- , *11,
  13*
ROEMER, Ferdinand, 1818-1891, 126,
  133
ROGERS, Garry Francis, 1946- ,
  29-33
ROMANS, Bernard, c. 1720-?1784, 45
ROSIER, William, 1856-1924, 138,
  149-54
ROUSSEAU, Jean Jacques, 1726-1778, 78

RYAN, Bruce Kenneth, 1936-   ,
    57-68

SALISBURY, Rollin D., 1838-1922, 58
SALZMANN, Christian Gotthilf,
    1744-1811, 138
SANDBERG, Adolph Engelbrekt,
    1898-   , 59
SAUER, Carl Ortwin, 1889-1975, 19,
    32, 65
SAUSHKIN, Yulian Glebovich,
    1911-1982, 5, 7, 8, *10*, *11*, *12*
SAXTON, Christoper, c. 1542-1611, 52
SCHLAGINWEIT, Adolf, 1829-1856, 145
SCHLAGINWEIT, Bruder Hermann,
    1826-1882, 145
SCHLAGINWEIT, Robert, 1833-1885, 145
SCHLÜTER, Otto, 1872-1959, 73
SCHNEIDER, Herman, 1872-1939, 59,
    60, 62
SCHNETZLER, Jean Balthasar,
    1823-1896, 136, 139
SCHOU, Axel, 1902-1971, 121
SCHWARZ, Gabrielle, 1914-   ,
    125-33
SDASYUK, Galina Vasil'yevna,
    1931-   , *12*
SEARS, Paul Bigelow, 1891-   , 30
SEEBACH, Karl Albert von,
    1839-1880, 100
SEMENTOVSKIY, Vladimir Nikolayevich,
    1882-1969, 11
SEMPLE, Ellen Churchill, 1863-1932,
    65
SHABAD, Theodore, 1922-   , 1-16
SHABILY, Oleg Ivanovich, 1935-   ,
    12
SHAFTSBURY, Anthony Ashley Cooper,
    3rd Earl of, 1671-1713, 78
SHALER, Nathaniel Southgate,
    1841-1906, 114
SHANTZ, Homer LeRoy, 1876-1958, 62
SHEAR, Cornelius Lott, 1865-1956, 33
SICKEL, Theodor von, 1826-1908, 143
SIMONY, Friedrich, 1813-1896, 143,
    144
SINNHUBER, Karl Aemilian, 1919-   ,
    148
SLOCUM, Joshua, 1844-?, 63
SMITH, Erasmus Peshine, 1814-1882,
    27
SMITH, George Otis, 1871-1944, 62
SMITH, Joseph Russell, 1874-1966, 4
SOLNTSEV, Nikolay Ivanovich,
    1935-   , *12*
SOLOV'YEV, Aleksandr Ivanovich,
    1907-1983, *10*, *11*
SOLOV'YEV, Margarita Grigor'yevna,
    1909-   , *10*, *12*
SONKLAR, Carl von, 1816-1885, 145
STANLEY, Henry Morton, 1840-1904,
    145
STEPANOV, Pyotr Nikolayevich,
    1921-   , *5*
STEWART, Dugold, 1753-1828, 137
STEWART, John, 1894-1972, 26

STUDER, Bernard-Rudolf, 1794-1887,
    136
STUDL, Johann, 1839-1925, 143
SUND, Tore, 1914-1965, 87
SVERDRUP, Otto Neumann, 1854-1930,
    85

TAMAYO, Francisco, 1902-1985, 137,
    139
TARR, Ralph Stockman, 1864-1912, 64
TAYLOR, Eva Germaine Rimington,
    1879-1966, 53, *54*
THOMAS, Colin, 1939-   , 17-23
THOMAS, David, 1931-   , *21*
THOMPSON, Robert Ellis, 1844-1924, 27
THORARINSSON, Sigurdur, 1912-1983,
    120
THÜNEN, Johann Heinrich von,
    1783-1850, 27
TOBEY, Ronald C., 1942-   , 30
TOMASCHEK, Wilhelm, 1841-1901, 144
TSCHUMI, Adolphe, 1856-1894, 150
TURKEVICH, John, 1907-   , *11*
TURNER, Frederick Jackson, 1861-1932,
    58
TYNDALL, John, 1820-1893, 145

ULRICH, Johannes, 1925-   , 99-104
URLSPERGER, Samuel, 1685-1772, 41

VAHL, Martin, 1869-1946, 118
VAN CLEEF, Eugene, 1887-1973, 130
VAN HISE, Charles Richard,
    1857-1918, 58, 61
VIDAL DE LA BLACHE, Paul,
    1845-1918, 7, 35, 36, *40*, 106
VIGÉE, Louis-Jean, 1758-1820, 150
VITVER, Ivan Aleksandrovich,
    1891-1966, 5, *11*
VIVIEN DE SAINT-MARTIN, Louis,
    1802-1896, 150
VOLTAIRE, François Marie Arouet de,
    1694-1788, 78
VOLZ, Wilhelm, 1870-1958, 127
VOSS, Martin, ob. 1903, 70

WAGNER, Hermann, 1840-1920, 127
WALDBUR, Harry, 1888-1961, 128, 129
WALSINGHAM, Francis, c. 1532-1590,
    50, 52
WAPPEUS, Eduard Johann, 1812-1879,
    100
WARMING, Eugenius, 1841-1924, 30
WARNTZ, William, 1922-   , 26
WATERS, David Watkin, 1911-   , 53,
    54
WEBER, Alfred, 1868-1958, 4, 6, *13*
WEGENER, Alfred Lothar, 1880-1931,
    119
WERCKLÉ, Charles, 1860-1924, 139
WERENSKIOLD, Werner, 1883-1961, 86,
    88
WETTSTEIN, Heinrich, 1831-1895, 138
WHEELER, (Sir) Robert Eric Mortimer,
    1890-1976, 18
WHITEHOUSE, Wallace Edward, 1882-?, 20

WHITTLESEY, Derwent Stanithorp,
    1890-1956, 4
WILLIAM-OLSSON, William, 1902-
    87
WRIGHT, Orville, 1871-1948, 60

YEFREMOV, Yuri Konstantinovich,
    1913-   , *13*
YONGE, Henry, 1713-1778, 42, *46*

ZAMKOVOY, Valentin Polikarpovich,
    1911-   , *12*
ZENO, Antonio, 53
ZENO, Niccolo, 1326?-1402, 52
ZIMM, Alfred, 1926-   , *11*
ZÖPPRITZ, Karl Jacob, 1838-1885, 100

2. *ORGANIZATIONS AND RELATED REFERENCES*

a. *Colleges, Institutes, Institutions, Museums and Research Organizations*

Allhegeny College, 64
Arctic Institute, Denmark, 121
Bavarian Academy of Sciences and
    Fine Arts, 79, *83*
Berlin Academy of Sciences, *83*
Berlin Akademie der Wissenschaft, 127
Board of Education, GB, 96
Bodleian Library, Oxford, *55*
British Board of Trade and
    Plantations, de Brahm and, 43
British Library, London, *55*
Bryn Mawr College, 59
Carlsberg Foundation, Denmark, 120
Centralkommission für wissenschaftliche
    Landeskunde, 127
Chicago Academy of Sciences, 31
Cincinnati Museum of Natural History,
    59
Colorado State Normal School, 58, 63
Comité Nacional de Geodesia y
    Geofísica, 36
Committee on Land Reclamation,
    Denmark, 120
Communist 'University', USSR, 4
Consejo Superior de Investigaciones
    Cientifica, 37
Danish Geodetic Institute, 121
Denmark,
    National Museum of, 121
    TRAP gazetteer, 121, *124*
Gates College, Nabraska, 29
Germany,
    Board of Mines, 128
    Central Commission for the study
        of, 70
    Imperial Academy, Hydrographic
        Office, 100
    Naval Observatory, 100
    Oder River Administration, 128
    Ordnance Survey, 128
Heidelberg College, Tiffin, 57, 60

Imperial Academy of Science, Vienna, 143
Imperial Naval Academy, Kiel, 100
Industrial Academy, USSR, 4
Institute of Austrian History, 143
Institute of World Economy and World Politics, USSR, 5, *14*
Institut météorologique de Château-d-Oex, 136, 139, *141*
Instituto de Bachillerato de Guadaljara, 35
Instituto de Geografía Juan Sebastian Elcano, 37, *40*
Instituto Físico-Geográfico, Costa Rica, 136, 138, *141*
Instituto San Isidro de Madrid, 35, *40*
International Council for the exploration of the seas, 101, *103*
International Court of Justice, The Hague, 85
Kazakh Pedagogic Institute, 6, *15*
Kommission zur wissenschaftlichen Untersuchung der deutschen Meere, 100
Moscow,
    Commercial Institute, 3, 14
    Communist Academy (Soviet Academy of Sciences), 4
    Communist 'University' (of workers in the East), 4
    Foreign Languages Publishing House, 6, *15*
    Industrial Academy, 4
    State Pedagogic Institute, 6, *15*
Museo de Ciencias Naturales de Madrid, 35, 37
Norway,
    Council for Regional Planning, 88
    Maritime Museum, 85
    National Conservation League, 88
    Nature Conservation Act, 88
    Railway Commission 1949, 88
    State Nature Conservancy Council, 88, 92
Norwegian Academy of Science and Humanities, 88
Ohio Academy of Sciences, 63
Ohio Flood Board, 62
Philadelphia Free Library, *55*
Prussian Academy of Sciences, *68*, 79
Royal Danish Academy of Sciences, 119
Royal Danish National Library, 121
St Louis Exposition, Department of Anthropology 1903, 112
St Louis Public Museum, 112
Smithsonian Institution, 111, 136, 138, *140*
Stevens Institute, New York, 136, *141*
United States of America
    Army Corps of Engineers, 62
    Bureau of Ethnology, 112

Conference of Governors 1908 (on conservation), 113
Department of Agriculture, 62, 112, 136, *141*
Department of Justice, 62
Geological Survey, 30, 58, 60, 111, 112, 113, 114
Inland Waterways Commission 1907, 112
National Research Council, 64
Weather Bureau, 61
Union of Socialist Soviet Republics
    Academy of Sciences, Kazakh branch, 4, 6, *10*, *15*
    Committee on Higher Technical Education, 6
    Institute of the National Economy, 3, *14*
    People's Commissariat of State Control, 3, *14*
    State Planning Commission (GOSPLAN), 4, 5
    Supreme Council of the National Economy, 3, *14*
Wales,
    National Library of, 18
    National Museum of, 18
Wisconsin Geological Survey, 58

b. *Scientific Congresses and Commissions*

Arctic Conference, Greifswald, 1930, 119
Conference of European Geographers, Würzburg, 1942, *40*
Congress of German Geographers, Vienna, 1891, 145
Congresso do Mondo Portuges, 1940, 36, *40*
International Botanical Congress, Vienna 1905, *33*
    USA 1913, *33*
    Vienna 1930, 31, *33*
International Commission on Glaciers, 1897, 144
International Congress of Slavic geographers and ethnologists, Poland 1927, *15*
International Geographical Union
    Commission on Pliocene and Pleistocene terraces, 107
    Commission on the rural habitat, 36, *40*, 107
    Commission on Surfaces of Erosion, 107
    Congresses (including pre-1922)
        1 Antwerp 1871, 105
        5 Bern 1891, 129, 149, 150
        6 London 1895, 94, 129
        7 Berlin 1899, 144
        9 Geneva 1908, *67*, 149
        10 Paris 1913, 129
        11 Cairo 1925, 36, *40*
        12 Cambridge 1928, **36**, *40*, 106
        13 Paris 1931, 36, *40*, 89

        14 Warsaw 1934, 5, *15*, 89
        15 Amsterdam 1938, 107, 121
        16 Lisbon 1949, 107, 121
        17 Washington, D.C. 1952, 89, 107, 121
        18 Rio de Janeiro 1956, 5, 107
        19 Norden (Stockholm), 1960, *67*, 89, 121
        20 London 1964, 106
International Geological Congresses, Stockholm 1910, 129
    Toronto 1913, *67*
International Meteorological Congress, Vienna 1873, 137
Italian Colonial Congress, Asmara, Eritrea 1905, 70, *75*
Pan-Pacific Congresses,
    Australia 1923, *68*
    Japan 1926, *68*
USSR (first) Geographical Congress 1933, Leningrad, 5, *15*

c. *Societies and Associations*

Academie des Dévens, 137
Alpine Club, 136, *141*
Altay Area Studies, Society of, 2, *12*, *14*
American Association for the Advancement of Science, 31, *32*, *33*, 58, 64
American Association of University Professors, 64
American Geographical Society, 62, 94, 129
    Transcontinental Excursion, 1912, 60, 64, 129
American Palaeontological Society, 59
Association of American Geographers, 31, *33*, 57, 58, 61, 63, 64, 112, 114
Association pour le bien publique, 136
Association des sociétés suisse de géographie, 150, *154*
Berlin Geographical Society, 7, *15*, 125, 144
Botanical Society of America, 31, *33*
British Association for the Advancement of Science, 19, 21, 94, 95
British Ecological Society, *33*
Bulgarian Geographical Society, 7, *15*
Cambrian Archaeological Society, 19, *22*, *23*
Cardiganshire Antiquarian Society, 18
Carmarthenshire Antiquarian Society, 18, *23*
Central Inland Waterways Association, USA, 62
Chicago Geographic Society, 31, 61, *68*
Cincinnati Scientific Society, 59
Commercial Club of Cincinnati, 62
Cosmos Club, Washington, D.C., 65, 136, *141*
Deutschen Geographentag, 126, 127, 128, 129

Dresden Geographical Society, 71
Ecological Society of America, 31, *33*
Geographical Association, 18, 19, 20, 96
Geological Society of America, 57, 58, 62, 64, *68*, 112, 114
German and Austrian Alpine Society, 126, 144, 146
Institute of British Geographers, 18, 19, 20, 97
Leipzig Colonial Association, 70
Leipzig Geographical Society, 70, *76*
Literary Club, Cincinnati, 65, *67*
National Conservation League, Norway, 88
National Education Association, USA, 58
National Geographic Society, 61, 64, 112, 114, 137
Norwegian Geographical Society, 87, 92
Pan-German Association (Alldeutscher Verband), 72
Paris Geographical Society, 94
Polish Geographical Society, 7, *15*
Real Sociedad Española de Historia Natural, 35, 36, *40*
Real Sociedad Geográfica de Madrid, 36
Royal Danish Geographical Society, 118, 121, 122
Royal Geographical Society, London, 19, *23*, 94, 96, *97*, *98*, 136
    educational work of, 94, 95
Royal Scottish Geographical Society, 94, 95
Royal Societies Club, London, 97
Royal Society of Arts, 95
Serbian Geographical Society, 7, *16*
Sociedad Española de Antropología, Etnografía y Prehistoria, 36
Société belge d'études géographiques, 107
Société de géographie de Genève, 149
Société vaudoise des sciences naturelles, 136, *141*
Society of Antiquaries, GB, 19, 23
Teachers' Guild of Great Britain, 95
USSR Geographical Society, 5, 7, 9, *15*
Washington Botanical Society, *141*
Würzburg Geographical Society, *40*
Yellowstone-Bighorn Research Association, 60, *68*
Yugoslav Geographical Society, 7, *15*

d. *Universities*

Athens, 127
Belgrade, *76*
Berlin, 69
Breslau, 125, 126, 127, 128
California, Los Angeles, *32*
Cambridge, 19, 49, 53, *55*, 95, 96

Caracas, Venezuela, 139
Chicago, 29, 30, *32*, 58, 60
Cincinnati, 57, 58, 59, 60
Coimbra, 50
Cologne, 70
Colorado, 58, 60
Columbia, New York, *32*, 106
Copenhagen, 117, 118, 119, 120, 121, *122*, *123*
Dresden, 71, 72, 73
Edinburgh, 60, 93
Folk University, Denmark, 119
Geneva, 149, 150, 151, *154*
Georgia, Athens, USA, *46*
Göttingen, 100, *103*
Graz, 60, 144
Hamburg, 99
Harvard, *47*, 58, 62
Jena, 70
Kiel, 99, 100, 101, *103*
Königsberg, 78, *83*, *133*
Lausanne (*Académie* to 1880), 135, 136, *140*, *141*
Leipzig, 60, 69, 70, 72, 99, *103*, 125, 129
Léopoldville, 106
Liège, 106
Lisbon, 50
Louvain, 49, *55*, 105, 106, 107, *110*
Lund, 88, *92*
Madrid, 35, *40*
Manchester, 96
Marburg, 101, *104*
Michigan, 58
Michigan State, *55*
Moscow, 1, 2, 5, 6, 7
Moscow pedagogic, 4, 5
Munich, *83*
Namur, *110*
Nebraska, 30, 31
New York, *28*
Open University, England, 18
Oslo, 85, 86, 88, 89
Oxford, 49, 60, 95, 96
Paris, 19, 35, *40*, 49, *55*, 105
St Andrews, 93, 94, *98*
St Petersburg, 2
Tomsk, 2, *14*
Tübingen, 70
Ulster, Coleraine, Northern Ireland, *22*
Uppsala, 88
Vienna, 60, 69, 143, 144
Wales
    Aberystwyth University College, 17, 18, *22*, *23*
    Guild of Graduates, 18, *23*
    National School of Medicine, Cardiff, 17
    Press Board, 18, *23*
Wisconsin, Madison, Wis., 58, 63

3. *SUBJECTS*

Africa

    Hassert on, 69, 70, 71
    Keltie on, 94, 97
    Lefèvre on, 106
    Rosier on, 151
agriculture, Carey on, 25, 26
alchemy, 42, 49, 50, 51, 52
American Revolution 1775, 42, 44, 45
anthropogeography
    Dantín-Cereceda on, 38
    Hassert on, 69
    McGee on, 112
    Partsch on, 127, 128
    of Ratzel, 69, 71
anthropology
    Herder on, 79
    physical, 17
archaeology
    Nielsen on Danish, 118
    of Wales, 18, 19, 20
astrology, 49, 53

Balkan peninsula, work of Hassert on, 69, 70, 71
Belgium
    Lefèvre on geography of, 106, 107, 108
    National Atlas of, 106
botanical research
    of Cowles, 29, 30, 31, *32*, *33*
    of Pittier, 135, 136, 137, 138, 139
British Empire, Dee and, 50

Cameroons, German expedition to, 1907-08, 70
cartography, Baranskiy and, 8, 9
Cathay (China), 50
Central Europe, Partsch on, 128, 129, 130
chorography, Dee on, 50
climate of Spain, Dantín-Cereceda on, 36, 37, 38
coasts, geomorphology of, Dee on, 50
colonial geography, 70, 71, 72, 73
Columbia, 136
Comecon, 9
communications, Hassert on, 70, 71, 72
conservation
    Cowles on, 29
    McGee on, 113, 114
    Pittier and, *141*
Costa Rica
    education in, 136
    Pittier on, 136, 138, 139
'Cultural' geography, Partsch on, 128

Denmark
    National Atlas of, 120, 121
    physical history of, 118, 119, 122,
    Skalling laboratory, 118, 119, 120, 122, *124*
    TRAP gazetteer of, 121
determinism
    Baranskiy and, 5, 6

Bowen on, 19
Dantín-Cereceda and, 37, 39

ecology, Cowles on, 29, 30, 31, *32*
economic cartography, Baranskiy on, 7
economic geography,
    Baranskiy on, 1, 4, 5, 6, 7, 8, 9, 10
    Carey on, 26, 27, 28
    Krümmel on, 101
    Rosier on, 151
economic theory, Carey on, 25, 26, 27, 28

forestry, Pittier's work on, 136, 137, 139, *141*

General Bathymetric Chart of the Oceans, 101
geographical education
    in Belgium, 105, 106, 107, 108
    Keltie on, 94, 95, 96
    Pittier on, 138
    Richter on, 143, 144, 146
    Rosier on, 150, 151, 152
    and textbooks of Baranskiy, 1, 4, 6, 9
geography, views on
    Baranskiy, 5, 6, 7
    Bowen, 19, 20, 21
    Dantín-Cereceda, 36, 37
    Fenneman, 61, 62, 63
    Hassert, 69, 70, 71, 72, 73
    Herder, 79, 80, 81, 82
    Isachsen, 86, 87, 88
    Lefèvre, 106, 107, 108, 109
    Partsch, 128, 129
    Rosier, 151, 152
geology, Fenneman on, 58, 59, 60, 62
geomorphology
    of Denmark, 118, 119, 120
    McGee on, 112, 113
    origin of term, 113
Georgia, Royal Colony of, USA, 41
German Colonial Office, 70
Germany
    Central Commission on scientific geography, 127
    Commission for scientific investigation of the German seas, 100
    colonies of, 69, 70, 71, 72, 73
    frontiers of, 71
    geographical education in, 95
glaciation
    of Alps, 126, 130
    of Denmark, 118, 119, 122
    McGee on, 112
    of Norway, 87, 145
    Richter on, 144, 145
Greece, work of Partsch on, 127, 130
Greenland, 50, 51, 52, 85, 86
Gulf Stream, 43, 44

historical geography
    Bowen on, 19
    Dantín-Cereceda on, 36, 37
    Richter on, 144, 145, 146
    work of Partsch on, 127, 130
human geography
    Bowen on, 19
    Dantín-Cereceda on, 36, 37
    Isachsen on, 86, 87
    Lefèvre on, 106, 107

Iceland
    Danish expeditions, 1924, 1927, 119
    Hekla eruption 1947, 120
    Isachsen in, 85
    Vatnajokull expeditions, 1934, 1936, 119
'Inquiry' for Paris Peace Conference 1919, 62

Länderkunde, concept of, 8
Lebensraum concept, 71

medical geography in Wales, 17, 21, *22*
meteorology and oceanography
    general, 100
    De Brahm on, 44
    Pittier on, 136, 137
Muscovy Company, 50, *54*

navigation, 49, 50, 52, 53
North America, proposals for the settlement of, 51
North Atlantic, map of, 51
Northeast passage, 50, 51, 52
Northwest passage, 51, 52
Norway, geography of, 86, 87

oceanography
    De Brahm on, 43, 44
    Krümmel on, 99, 100, 101
oecumene
    Hassert on, 69
    Rosier on, 150

Panama Canal zone, 136, *141*
physiography
    of Cowles, 30
    of Fenneman, 58, 60, 61
    of McGee, 112, 113
phytogeography, 135, 137, 138
plate tectonics, 119
Polar regions
    Hassert on, 70
    Isachsen on, 86, 87
    Rosier on, 151
political geography
    Carey on, 26, 27
    Dantín-Cereceda on, 37
    Hassert on, 71, 72, 73
    Rosier on, 151, 152, 153
positivism in geography, 37, 38, 39

quantification in geography, 38, 88

regional geography
    Baranskiy on, 5, 8, 9
    Bowen on, 20
    Dantín-Cereceda on, 35, 36, 37, 38, 39
    Hassert on, 71, 72
    Isachsen on, 87
    of Partsch, 127, 128, 129, 130
    Rosier on, 151, 152

sea level, changes of, in Norway, 87
Siberia, 1, 2, 3, 4, 8
Silesia, work of Partsch on, 127, 128, 130
soil erosion, 51, 52
Spain
    aridity of, 35, 36, 38, *40*
    natural regions in, 36, 37, 38, 39
    physical geography of, 36
    rural settlement in, 36, 38
    transhumance in, 36, *40*
Spitzbergen, Isachsen on, 85
'Straits of Anian', 51, 52
Suez Canal, 70
Switzerland, Rosier on, 150, 152

terrestrial unity, Herder's concept of, 79, 80
Tomsk, 1, 2, 3, *14*
Trans-Siberian railway, 2, 3
Treaty of Versailles 1919, 71, 72, 73, 129

United States of America
    in 18th century, 25, 26, 41, 42, 43
    coastal and marine survey of, 43, 44, 45, 46
    De Brahm survey of, 42, 43
    frontiers of settlement in, 25, 26
    physiographic regions of Fenneman, 61
    regional geography of, Carey on, 26
    state capitals of, 25
    tariff policy of, 27
    urban geography of, 26
urban geography
    Baranskiy on, 8
    Carey on, 26
    Isachsen on, 86

Venezuela, Pittier in, 136, 137, 138, 139, *141*

Wales
    geography of, Bowen on, 18, 19, 20
    languages of, 20

4. *CUMULATIVE LIST OF BIO-*
    BIBLIOGRAPHIES

AL-MUQADDASĪ, *c*.945-*c*.988, vol 4
1-6

ANCEL, Jacques, 1882-1943, vol 3,
1-6

ANUCHIN, Dmitry Nikolaevich, 1843-
1923, vol 2, 1-8

APIANUS, Peter, 1495 or 1501-1552,
vol 6, 1-6

ARBOS, Philippe, 1882-1956, vol 3,
7-12

ARDEN-CLOSE, Charles Frederick,
1865-1952, vol 9, 1-13

ARQUÉ, Paul, 1887-1970, vol 7, 5-9

ATWOOD, Wallace Walter, 1872-1949,
vol 3, 13-18

BANSE, Ewald, 1883-1953, vol 8, 1-5

BARANSKIY, Nikolay Nikolayevich,
1881-1963, vol 10, 1-16

BAULIG, Henri, 1877-1962, vol 4,
7-17

BERG, Lev Semenovich, 1876-1950,
vol 5, 1-7

BERNARD, Augustin, 1865-1947, vol. 3,
19-27

BLACHE, Jules, 1893-1970, vol 1,
1-8

BLODGET, Lorin, 1823-1901, vol 5,
9-12

BOSE, Nirmal Kumas, 1901-1972,
vol 2, 9-11

BOWEN, Emrys George, 1900-1983,
vol 10, 17-23

BOWMAN, Isaiah, 1878-1950, vol 1,
9-18

BRATESCU, Constantin, 1882-1945,
vol 4, 19-24

BRIGHAM, Albert Perry, 1855-1929,
vol 2, 13-19

BROOKS, Alfred Hulse, 1871-1924,
vol 1, 19-23

BROWN, Ralph Hall, 1898-1948,
vol 9, 15-20

BROWN, Robert Neal Rudmose, 1879-
1957, vol 8, 7-16

BUACHE, Philippe, 1700-1773, vol 9,
21-7

BÜSCHING, Anton Friedrich, 1724-
1793, vol 6, 7-15

CAMENA d'ALMEIDA, Pierre, 1865-1943,
vol 7, 1-4

CAPOT-REY, Robert, 1897-1977, vol 5,
13-19

CAREY, Henry Charles, 1793-1879,
vol 10, 25-8

CAVAILLÈS, Henri, 1870-1951, vol 7,
5-9

CHRISTALLER, Walter, 1893-1969,
vol 7, 11-16

COLBY, Charles Carlyle, 1884-1965,
vol 6, 17-22

COPERNICUS, Nicholas, 1473-1543,
vol 6, 23-9

CORNISH, Vaughan, 1862-1948, vol 9,
29-35

CORTAMBERT, Eugène, 1805-1881,
vol 2, 21-5

COTTON, Charles Andrew, 1885-1970,
vol 2, 27-32

COWLES, Henry Chandler, 1869-1939,
vol 10, 29-33

CRESSEY, George Babcock, 1896-1963,
vol 5, 21-5

CVIJIĆ, Jovan, 1865-1927, vol 4,
25-32

D'ABBADIE, Antoine, 1810-1897,
vol 3, 29-33

DANTÍN-CERECEDA, Juan, 1881-1943,
vol 10, 35-40

DARWIN, Charles, 1809-1882,
vol 9, 37-45

DAVID, Mihai, 1886-1954, vol 6,
31-3

DAVIDSON, George, 1825-1911, vol 2,
33-7

DAVIS, William Morris, 1850-1934,
vol 5, 27-33

DE BRAHM, William Gerard, 1718-1799,
vol 10, 41-7

DE CHARPENTIER, Jean, 1786-1855,
vol 7, 17-22

DEE, John, 1527-1608, vol 10, 49-55

DICKINSON, Robert Eric, 1905-1981,
vol 8, 17-25

DIMITRESCU-ALDEM, Alexandre, 1880-
1917, vol 3, 35-7

DOKUACHAEV, Vasily Vasilyevich,
1846-1903, vol 4, 33-42

DRAPEYRON, Ludovic, 1839-1901,
vol 6, 35-8

DRYGALSKI, Erich von, 1865-1949,
vol 7, 23-9

ERATOSTHENES, *c*.275-*c*.195 B.C.,
vol 2, 39-43

FABRICIUS, Johann Albert, 1668-1736,
vol 5, 35-9

FAIRGRIEVE, James, 1870-1953, vol 8,
27-33

FAWCETT, Charles Bungay, 1883-1952,
vol 6, 39-46

FEDCHENKO, Alexei Pavlovich, 1844-
1873, vol 8, 35-38

FENNEMAN, Nevin Melancthon, 1865-1945,
vol 10, 57-68

FORBES, James David, 1809-1868,
vol 7, 31-7

FORMOZOV, Aledander Nikolayevich,
1899-1973, vol 7, 39-46

FORREST, Alexander, 1849-1901, vol 8,
39-43

FORREST, John, 1847-1918, vol 8,
39-43

FRANZ, Johann Michael, 1700-1761,
vol 5, 41-8

GANNETT, Henry, 1846-1914, vol 8,
45-49

GEDDES, Arthur, 1895-1968, vol 2,
45-51

GEDDES, Patrick, 1854-1932, vol 2,
53-65

GEIKIE, Archibald, 1835-1924, vol 3,
39-52

GEIKIE, James, 1839-1915, vol 3,
53-62

GILBERT, Edmund William, 1900-1973,
vol 3, 63-71

GILBERT, Grove Karl, 1843-1918, vol 1,
25-33

GILLMAN, Clement, 1882-1946, vol 1,
35-41

GLAREANUS, Henricus, 1488-1563, vol 5,
49-54

GOODE, John Paul, 1862-1932, vol 8,
51-55

GOYDER, George Woodroofe, 1826-1898,
vol 7, 47-50

GRADMANN, Robert, 1865-1950, vol 6,
47-54

GRANÖ, Johannes Gabriel, 1882-1956,
vol 3, 73-84

GRIGORYEV, Andrei Alexandrovich,
1883-1968, vol 5, 55-61

GUYOT, Arnold Henry, 1807-1884,
vol 5, 63-71

HASSERT, Ernst Emil Kurt, 1868-1947,
vol 10, 69-76

HERBERTSON, Andrew John, 1865-1915,
vol 3, 85-92

HERDER, Johann Gottfried, 1744-1803,
vol 10, 77-84

HETTNER, Alfred, 1859-1941, vol 6,
55-63

HIMLY, Louis-Auguste, 1823-1906,
vol 1, 43-7

HO, Robert, 1921-1972, vol 1, 49-54

HÖHNEL, Ludwig von, 1857-1942, vol 7,
43-7

HOLMES, James Macdonald, 1896-1966,
vol 7, 51-5

HUGHES, William, 1818-1876, vol 9,
47-53

HUGUET DEL VILLAR, Emilio, 1871-1951,
vol 9, 55-60

HULT, Ragnar, 1857-1899, vol 9, 61-9

HUTCHINGS, Geoffrey Edward, 1900-1964,
vol 2, 67-71

ISACHSEN, Fridtjov Eide, 1906-1979,
vol 10, 85-92

JOBBERNS, George, 1895-1974, vol 5,
73-6

JONES, Llewellyn Rodwell, 1881-1947,
vol 4, 49-53

KANT, Immanuel, 1724-1804, vol 4,
55-67

KECKERMANN, Bartholomäus, 1572-1600,
vol 2, 73-9

KELTIE, John Scott, 1840-1927, vol 10,
93-8

KIRCHOFF, Alfred, 1838-1907, vol 4, 69-76

KOMAROV, Vladimir Leontyevitch, 1869-1945, vol 1, 55-8

KRASNOV, Andrey Nikolaevich, 1862-1914, vol 4, 77-86

KROPOTKIN, Pyotr (Peter) Alexeivich, vol 7, 57-62, 63-9

KRÜMMEL, Johann Gottfried Otto, 1854-1912, vol 10, 99-104

KUBARY, Jan Stanislaw, 1846-1896, vol 4, 87-9

LARCOM, Thomas Aiskew, 1801-1879, vol 7, 71-4

LAUTENSACH, Hermann, 1886-1971, vol 4, 91-101

LELEWEL, Joachim, 1786-1861, vol 4, 103-12

LENCEWICZ, Stanislaw, 1899-1944, vol 5, 77-81

LEVASSEUR, Emile, 1828-1911, vol 2, 81-7

LEFÈVRE, Marguerite Alice, 1894-1967, vol 10, 105-10

LEWIS, William Vaughan, 1907-1961, vol 4, 113-20

LINTON, David Leslie, 1906-1971, vol 7, 75-83

LOMONOSOV, Mikhail Vasilyevich, 1711-1765, vol 6, 65-70

MacCARTHY, Oscar, 1815-1894, vol 8, 57-60

McGEE, William John, 1853-1912, vol 10, 111-16

MACKINDER, Halford John, 1861-1947, vol 9, 71-86

MAURY, Matthew Fontaine, 1806-1973, vol 1, 59-63

MAY, Jacques M., 1896-1975, vol 7, 85-8

MEHEDINTI, Simion, 1868-1962, vol 1, 65-72

MELANCHTHON, Philipp, 1497-1560, vol 3, 93-7

MELIK, Anton, 1890-1966, vol 9, 87-94

MIHAILESCU, Vintila, 1890-1978, vol 8, 61-7

MILL, Hugh Robert, 1861-1950, vol 1, 73-8

MILNE, Geoffrey, 1898-1942, vol 2, 89-92

MITCHELL, Thomas Livingstone, 1792-1855, vol 5, 83-7

MUELLER, Ferdinand Jakob Heinrich von, 1825-1896, vol 5, 89-93

MUNSTER, Sebastian, 1488-1552, vol 3, 99-106

MUSHKETOV, Ivan Vasylievitch, 1850-1902, vol 7, 89-91

NELSON, Helge, 1882-1966, vol 8, 69-75

NEUSTRUEV, Sergei Semyonovich, 1874-1928, vol 8, 77-80

NIELSEN, Niels, 1893-1981, vol 10, 117-24

OBERHUMMER, Eugen, 1859-1944, vol 7, 93-100

OGAWA, Takuji, 1870-1941, vol 6, 71-6

ORGHIDAN, Nicolai, 1881-1967, vol 6, 77-9

ORMSBY, Hilda, 1877-1973, vol 5, 95-7

PARTSCH, Joseph Franz Maria, 1851-1925, vol 10, 125-33

PAULITSCHKE, Philipp, 1854-1899, vol 9, 95-100

PAVLOV, Alexsei Petrovich, 1854-1929, vol 6, 81-5

PENCK, Albrecht, 1858-1945, vol 7, 101-08

PITTIER, Henri-François, 1857-1950, vol 10, 135-42

PLATT, Robert Swanton, 1891-1964, vol 3, 107-16

POL, Wincenty, 1807-1872, vol 2, 93-7

POWELL, John Wesley, 1834-1902, vol 3, 117-24

PRICE, Archibald Grenfell, 1892-1977, vol 6, 87-92

RAISZ, Erwin Josephus, 1893-1968, vol 6, 93-7

RAVENSTEIN, Ernst Georg, 1834-1913, vol 1, 79-82

RECLUS, Elisée, 1830-1905, vol 3, 125-32

REISCH, Gregor, c.1470-1525, vol 6, 99-104

RENNELL, James, 1742-1830, vol 1, 83-8

REVERT, Eugène, 1895-1957, vol 7, 5-9

RHETICUS, Georg Joachim, 1514-1573, vol 4, 121-6

RICHTER, Eduard, 1847-1905, vol 10, 143-8

RICHTHOFEN, Ferdinand Freiherr von, 1833-1905, vol 7, 109-15

RITTER, Carl, 1779-1859, vol 5, 99-108

ROMER, Eugeniusz, 1871-1954, vol 1, 89-96

ROSBERG, Johan Evert, 1864-1932, vol 9, 101-08

ROSIER, William, 1856-1924, vol 10, 149-54

ROXBY, Percy Maude, 1880-1947, vol 5, 109-16

RUSSELL, Richard Joel, 1895-1971, vol 4, 127-38

RYCHKOV, Peter Ivanovich, 1712-1777, vol 9, 109-12

SALISBURY, Rollin D., 1858-1922, vol 6, 105-13

SAUER, Carl Ortwin, 1889-1975, vol 2, 99-108

SAWICKI, Ludomir Slepowran, 1884-1928, vol 9, 113-19

SCHLÜTER, Otto, 1872-1959, vol 6, 115-22

SCHMITTHENNER, Heinrich, 1887-1957, vol 5, 117-21

SCHRADER, Franz, 1844-1924, vol 1, 97-103

SCHWERIN, Hans Hugold von, 1853-1912, vol 8, 81-6

SCORESBY, William, 1789-1857, vol 4, 139-47

SEMPLE, Ellen Churchill, 1863-1932, vol 8, 87-94

SHALER, Nathaniel Southgate, 1841-1906, vol 3, 133-9

SHIGA, Shigetaka, 1863-1927, vol 8, 95-105

SIEVERS, Wilhelm, 1860-1921, vol 8, 107-10

SMITH, George Adam, 1856-1942, vol 1, 105-06

SMITH, Wilfred, 1903-1955, vol 9, 121-7

SMOLENSKI, Jerzy, 1881-1940, vol 6, 123-7

SÖLCH, Johann, 1883-1951, vol 7, 117-24

SOMERVILLE, Mary, 1780-1872, vol 2, 109-11

STÖFFLER, Johannes, 1452-1531, vol 5, 123-8

STRZELECKI, Pawel Edmund, 1797-1873, vol 2, 113-18

TAMAYO, Jorge Leonides, 1912-1978, vol 7, 125-8

TATISHCHEV, Vasili Nikitich, 1686-1750, vol 6, 129-32

TAYLOR, Thomas Griffith, 1880-1963, vol 3, 141-53

TEILHARD DE CHARDIN, Pierre, 1881-1955, vol 7, 129-33

TILLO, Alexey Andreyevich, 1839-1900, vol 3, 155-9

TOPELIUS, Zachris, 1818-1898, vol 3, 161-3

TROLL, Carl, 1899-1975, vol 8, 111-124

ULLMAN, Edward Louis, 1912-1976, vol 9, 129-35

VALLAUX, Camille, 1870-1945, vol 2, 119-26

VALSAN, Georg, 1885-1935, vol 2, 127-33

VAN CLEEF, Eugene, 1887-1973, vol 9, 137-43

VERNADSKY, Vladimir Ivanovich, 1863-1945, vol 7, 135-44

VIVIEN DE SAINT-MARTIN, Louis, 1802-1896, vol 6, 133-8

VOLZ, Wilhelm, 1870-1958, vol 9, 145-50

VOYEIKOV, Alexander Ivanovich, 1842-1916,
  1842-1916, vol 2, 135-41
VUJEVIĆ, Pavle, 1881-1966, vol 5,
  129-31

WAIBEL, Leo Heinrich, 1888-1951,
  vol 6, 139-47
WALLACE, Alfred Russel, 1823-1913,
  vol 8, 125-33
WANG YUNG, 1899-1956, vol 9, 151-4
WARD, Robert DeCourcy, 1867-1931,
  vol 7, 145-50
WELLINGTON, John Harold, 1892-1981,
  vol 8, 135-40
WEULERSSE, Jacques, 1905-1946,
  vol 1, 107-12
WISSLER, Clark, 1870-1947, vol 7,
  151-4
WOOLDRIDGE, Sidney William, 1900-
  1963, vol 8, 141-9

YAMASAKI, Naomasa, 1870-1928,
  vol 1, 113-17

THE FOREIGN OFFICE
AND FOREIGN POLICY, 1919–1926

The Foreign Office
and Foreign Policy, 1919–1926

Ephraim Maisel

Foreword by
Martin Gilbert CBE

Preface by
Zara Steiner

sussex
ACADEMIC
PRESS

First published 1994

Sussex Academic Press
18 Chichester Place
Brighton BN2 1FF, United Kingdom

British Library Cataloguing in Publication Data
A CIP catalogue record for this book is available from the British Library.
ISBN 1–898723–04–4

Copy-edited and typeset in 10 on 12 Palatino
by Grahame & Grahame Editorial, Brighton, East Sussex
Printed and bound in Great Britain
This book is printed on acid-free paper

Contents

_____

Acknowledgements   vi

Foreword by *Martin Gilbert*   vii

Preface by *Zara Steiner*   ix

1   Organizational Changes at the Onset of Peace   1

2   The Division of Functions between the Foreign Secretary
    and his Assistants   31

3   Co-ordination of Foreign Policy in the Cabinet   60

4   European Policy of the Prime Minister's Office   89

5   The First Labour Government: MacDonald and the Foreign
    Office   130

6   The Second Baldwin Government: J. Austen
    Chamberlain and the Foreign Office   155

Appendix 1: The Establishment of the Department of
Overseas Trade   189

Appendix 2: The Formation of the Middle East Department   204

Abbreviations   228

Notes   230

Bibliography   300

Index   312

Acknowledgements

The Publisher and the Family of Ephraim Maisel wish to acknowledge the following for assistance accorded the author whilst he researched the book: The First Beaverbrook Foundation; University Library, The University of Birmingham; The British Library; The Archivist, Churchill College, Cambridge; Cambridge University Library; the Record Office, House of Lords; the India Office Library and Records; Ms Aileen Kent of the *Journal of Contemporary History*; The Most Honourable the Marquess of Lothian; the Library, University of Newcastle upon Tyne; the Public Record Office; the Scottish Record Office.

The sad circumstances under which this book has been published mean that many individuals and institutes that played a part in the author's research may not have been formally acknowledged; the Publisher requests their understanding in this respect.

Material in Appendix 1 was first published in the *Journal of Contemporary History*, vol. xxx (1989).

The text was keyed from the author's handwritten manuscript and card indexes by Mrs Rivka Elron, and edited and indexed by in-house staff at Sussex Academic Press. The Publisher wishes to thank Eric Goldstein, University of Birmingham, for providing additional text material on the Political Information Department; staff at the Library and Records Dept., Foreign & Commonwealth Office; and all those individuals who have assisted in the editorial and referencing tasks.

The cartoon texts are reproduced by kind permission of Routledge, London, from Roy Douglas, *Between the Wars, 1919–1939: The Cartoonist's Vision*. Permission for the use of the two cartoons by David Low is granted by Solo Syndication, London. In any case where the Publisher may have failed to clear copyright of a cartoon, the necessary arrangement will be made at the first opportunity.

Foreword

Martin Gilbert CBE

Twenty years have past since Ephraim Maisel was my pupil at Tel Aviv University. But the impact of his personality is still vivid. 'Rammy', as we knew him, and as he signed himself, was the keenest member of a keen seminar, in which the controversies of twentieth-century British history impinged continually on the controversies of the Middle East. For British historians, the Balfour Declaration is an important diplomatic curiosity. For Rammy and those who debated with him, it was a living instrument encapsulating their aspirations, and from which their contemporary status had sprung.

Rammy was thoughtful and full of questions. The idea that archival research could shed new light on old questions was one that intrigued him. After each seminar, we would sit together in the Tel Aviv University café, discussing the policies and personalities of the inter-war years. He was drawn to the immediate post-war years, to a study centred on the British Foreign Office, and to Britain. In 1975 he was in Oxford, pursuing his research. 'I spent this morning in the Bodleian Library,' he wrote to me that October, 'and collected some material from the Fisher and Milner papers. Oxford is really a lovely place and has an atmosphere of its own, which I enjoyed very much, and also the little tour that I made in some of the colleges.'

Within a year, Rammy was deep into his work, steadily mastering the voluminous files, and through them the complex but important stories that he pieced together from them. In September 1976 he wrote to me in mid-toil: 'With me everything is fine including the work, although at times I feel like a midget trying to take over Atlas's job with all this huge material.'

Having graduated from the Aranne School of History at Tel Aviv University, Rammy devoted himself to his researches with an intensity and passion that sustained him through almost two decades. In 1989 he published 'The Formation of the Department of Overseas Trade, 1919–26' in the *Journal of Contemporary History*. In the journal's note about him, he was described as 'an independent scholar', and this he was, without a university affiliation, but with all the rigour and devotion of a scholar's calling. In that article, it was also noted that Rammy was 'currently working on a book, *The Foreign Office and Foreign policy, 1919–1926*'.

Rammy and I had several talks about his book when we spent some time together in Israel in 1989, when Rammy selflessly helped me with the rigours of an index. On his sad and untimely death in 1991, his researches, which had been remarkably wide ranging, were almost completed. Yet, as in all creative works, the last lap is a demanding one, and Rammy did not live to complete it. In the process of transforming his handwritten manuscript and copious card index system into a printed book, he has been remarkably well served by the labours of Mrs Rivka Elron and Dr Zara Steiner, each of whom has made it possible for that last lap to be completed. Just as I am proud that Rammy was my pupil, so he would be proud that these devoted hands have made it possible for his life's work to be published.

Martin Gilbert
Honorary Fellow
Merton College, Oxford
May 1994

Preface

Zara Steiner

The 1920s are very much in fashion among historians. There has been in the last decades an outpouring of books both on countries and problems, security, reparations, disarmament, to mention but a few. The opening of new archives, particularly in France, but also the explosion in the contours of traditional diplomatic history, resulted in new interpretations of the decade which had been viewed for so long in terms of the origins of the Second World War. Possible parallels between the two post-war decades stimulated interest not only in financial and economic diplomacy but also in American relations with the European continent. It was not without significance that the German foreign minister, Genscher, should have had a portrait of Stresemann hanging in his office. British historians were somewhat late in joining the stream of monographic literature but soon were producing their own accounts of British diplomacy during and after the Peace Conference and under Lord Curzon, Ramsay MacDonald and Austen Chamberlain. They began, too, to look at British relations with other European countries, the Soviet Union and the United States. Studies of British diplomacy in the Far East and Middle East took the Great War as a starting point, altering the traditional foreign-policy perspective. What was refreshing about much of this work was its multinational approach with regard both to archival research and centres of interest.

Parallel to the expansion of the fields of enquiry, a response both to the nature of international history in the 1920s and to the new tools of historical research, was a renewed interest in the decision-making process and in those who formulated and executed foreign policy. It was clear that this was a revolutionary decade in the evolution of diplomacy and the means used to conduct inter-state relations on a global scale. The number of actors dramatically increased as did the subjects of international negotiation. The creation of the League of Nations affected the conduct of diplomacy as did the increasing involvement of statesmen and non-professional diplomats in the many multinational conferences of the period. What was the impact of these changes on the traditional machinery of policy-making and how did the traditional foreign ministries react to the challenges of a changing environment, both at home and abroad?

This study of The Foreign Office and Foreign Policy, 1919–26 by the late Ephraim Maisel provides some of the answers to these questions. It is based on a wide reading in the Foreign Office archives but looks outside the Office walls. It is concerned with the administrative changes of the post-war period and with the senior permanent officials who advised the foreign secretary and carried out his policies. The later chapters look at the policies of these foreign secretaries to illustrate how foreign policy was actually made. The author's objective was not to write a diplomatic history of these years but to give examples of the foreign policy machine in action as seen from King Charles Street. Mr Maisel brings together a good deal of scattered material on the organization of the Foreign Office but has gone far beyond existing accounts in describing changes which took place both during and after the Great War. He has a good deal to say about the views of the main Foreign Office figures, the foreign secretary, the permanent under-secretary and those immediately under them. The task he has set himself is not an easy one, particularly for the Lloyd George years, because of the uneasy relationship between 10 Downing Street and the Foreign Office. The question of influence is always a difficult one for the historian and there can be no final answers. But Mr Maisel has good historical antennae and his picture of the relations between foreign secretary and senior officials commands attention and respect. The studies of the MacDonald and Chamberlain foreign secretaryships contain an impressive amount of new material not found elsewhere.

The publication of this book has involved special problems. Mr Maisel died in 1991 and left an incomplete manuscript. Ideally, it should have been expanded; it needs an introduction and conclusion that only the author could have written. Those of us who read the manuscript felt, however, that there was so much new material here which other historians of the Foreign Office and of the immediate post-war period would find useful, that it was worth publishing in the form in which it was left. It addresses important questions and breaks new ground. The work has been meticulously done and the quotations and citations are relevant and memorable. Hopefully its appearance will encourage others to take up the task of writing the inter-war history of the Foreign Office, a task still to be completed. One of the main difficulties in publishing this study arose from the number of books that have appeared since these chapters were written and which have a direct bearing on its contents. I have taken the liberty of incorporating some of this material where I felt omission detracted from the utility of the book and have tried in the notes to indicate where recent studies and articles have dealt in depth with the topics covered or offered different interpretations of the views and actions of the men described here which require consideration. I have not touched the essential argument in any of the chapters and have cited throughout the sources from which the new evidence comes. I know this would have been a somewhat different book had the author lived to complete it. I feel, nonetheless, that it can be published as a fitting tribute to the talents of a highly-gifted scholar.

Zara Steiner
New Hall, Cambridge
February 1994

1

Organizational Changes at the Onset of Peace

The Great War had a mixed impact on the Foreign Office. At a time of war, diplomacy was 'the handmaid of the necessities of the War Office and the Admiralty' and Sir Edward Grey fully recognized the changed order of priorities. The replacement of Asquith by Lloyd George, a man who thought 'diplomatists were invented simply to waste time', was guaranteed not just to magnify the power of 10 Downing Street in the making of foreign policy but to create divisions of responsibility that would continue after the fighting ceased. The Prime Minister, conducting a total war, pursued his diplomatic goals without consulting or even informing Arthur Balfour, Grey's successor, or the Foreign Office, while making known his contempt for traditional modes of diplomatic intercourse. Balfour, an experienced if detached senior political figure, was not a member of the War Cabinet and though present at over half its meetings was not necessarily included in the final stages of decision-making. He parried rather than faced Lloyd George and though he could be stubborn and determined, more often he gave way to his dynamic and opinionated Prime Minister.

If the purely political work of the Foreign Office shrank in volume and importance as the war progressed, it was hoped that preparations for peace-making would restore the standing and prestige of the department. Individual Foreign Office officials did make a major contribution to the peace process but it was Lloyd George who put his personal imprint on the Treaty of Versailles and who dominated the proceedings in Paris. Foreign Office representatives were often left in ignorance of what the Big Four had decided and even those officials actively engaged in the influential territorial committees complained bitterly about the lack of order in the Prime Minister's conduct of business. It was a fitting commentary on the often strained relations between Prime Minister and Foreign Office that there was a sharp clash between Lloyd George and the efficient and effective Eyre Crowe, the Under-Secretary who became head of the British Delegation and its representative on the Supreme Council when the politicians returned home. The Prime Minister demanded Crowe's recall and it was only strong resistance from Lord Curzon, the newly-appointed Foreign Secretary, and Lord Derby, Lloyd George's own appointee as ambassador to France, that thwarted the Prime Minister's unjustified action.

Nor did the Foreign Office enjoy the hoped-for resurgence of influence once the peace conference ended for Lloyd George engaged in a form of summit diplomacy that allowed him both to exclude the Foreign Office and to ignore its advice. There were hundreds of Allied and international conferences during the 1919–22 period; the Foreign Secretary attended most of them but his officials were present at less than half. One must not exaggerate the degree of Foreign Office impotence. Lloyd George's interventions were sporadic and focused on a few key questions. The Foreign Office continued to handle the bulk of a swelling load of global diplomatic business. This was as true of European affairs, including the French and German questions on which Lloyd George concentrated, as it was of the Near and Middle East where Curzon's expertise was rarely ignored.

Nonetheless, the strained relations between Lloyd George and Curzon, and the former's devious methods of conducting business, did little to correct the impression that there were two centres of authority in the making of foreign policy and that Downing Street was the more powerful. Curzon was experienced in diplomacy and accustomed to having his own way. The Prime Minister's slippery methods enraged the highly sensitive Curzon. There were instances, as in February 1922, when the Prime Minister's intrigues brought the Foreign Minister to boiling point. Lloyd George bullied Curzon and enjoyed the latter's discomfort, for, to the Prime Minister, Curzon was the symbolic representative of the upper-class aristocrat he so detested. There was little Curzon could do except threaten to resign. Instead, he stayed in office, his pent-up rage expressed in private letters and offers of resignation never sent.

Lloyd George's negative attitude towards the Foreign Office must be seen against the more general public distrust of professional diplomacy. The very outbreak of war represented a defeat for the diplomatists. President Wilson's attacks on secret diplomacy and the demand, not confined to the political left, that the diplomatic services should be subjected to 'democratic' oversight and control and should be recruited from a far broader and more representative section of the population, served to undermine the traditional respect paid to the diplomatic profession. At the same time, both during and after the war, the diplomatic canvas expanded and other ministers and departments took a more active interest in foreign affairs than previously. The War Office and Admiralty, the India Office, Board of Trade and Treasury insisted, on grounds of expertise, that they had the right to handle questions over which the Foreign Office claimed jurisdiction. In some instances, as over the Middle Eastern Department and the Department of Overseas Trade, the Foreign Office put up a fierce struggle to maintain its primacy against that of the India Office and Board of Trade respectively. Appendices 1 and 2[1] detail how these battles often resulted in a Foreign Office defeat or in unsatisfactory compromises. In other cases, as with Treasury control over reparations, the Foreign Office case went by default. Treasury officials took the initiative before the peace conference opened and, at Paris, the Prime Minister's appointees, and Lloyd George himself, determined British policy. After the war, Sir John Bradbury, the permanent British delegate to the Reparations Commission, reported to the Treasury and not to the Foreign Office, and the latter had to struggle to keep itself properly informed. As reparations became the central battleground between Britain, France and Germany, the exclusion of the Foreign Office, which lacked competence in financial matters but which understood the political implications of the question, became a major issue. The Foreign Office was no longer *primus inter pares* in a whole range of questions with a direct bearing on the country's international relations.

The Foreign Office's war-time experience was not entirely negative. New departments and sections were added and Foreign Office officials found themselves doing unaccustomed work and coming into contact with individuals far removed from the diplomatic inner circle. It was Eyre Crowe's driving energy and Alwyn Parker's extraordinary administrative talents that first turned the Contraband Department and then the Ministry of Blockade into the effective instrument of war which played such a crucial part in the British victory. Though its staff increased and included non-Foreign Office members, career officials were the key to its success. The diplomats found that they could handle commercial work of a highly technical nature. It is not an exaggeration to suggest that the Ministry of Blockade had become the real heart of the war-time Foreign Office though separate from it. Awareness of the important role of commerce and trade gave added weight to the arguments of men like Eyre Crowe that the Foreign Office needed to strengthen its commercial side. The defeat of the Office in its battle with the Board of Trade when much of the work of the pre-war Commercial Department, despite the strong disapproval of the Foreign Office, was handed over to the new Department of Overseas Trade, represented a check to progress in this direction.

During the 1920s, Foreign Office officials did not feel that there was a separation between foreign policy, economy and commerce. The head of the Western Department, Gerald Villiers, observed that the political departments now had a fair working knowledge of economic questions.[2] Waterlow, too, considered that the danger that political steps would be taken whilst economic considerations were disregarded, was greatly reduced.[3] It was only in the 1940s that Tyrrell admitted that the dissolution of the Commercial Department and the establishment of the Department of Overseas Trade effectively deprived the Foreign Office of one of its main functions, which was to handle economic, commercial and financial matters that had political bearings.

> It had perhaps subconsciously the effect of leaving the members of the Foreign Office more and more disinterested in any of these questions. I repeat, in the light of after events, that this separation, if not divorce, between the Foreign Office and the Departments of State that look after trade, commerce and finance grew. I remember being struck after the war by the fact – and it is a fact – that reparations which undoubtedly were financial in their nature, but which had a political aspect far transcending the financial merits of the question, were outside the domain of the Foreign Office, which occasionally was called in, in a consultative capacity.[4]

The war also alerted the Foreign Office to the importance of news, propaganda and intelligence and in this sphere as well war-time changes might well have had more far-reaching effects were it not for Treasury parsimony, an innate suspicion of outsiders that proved hard to shake, and an inherited dislike of intelligence operations as something quite alien to the traditional world of diplomacy. Whereas before the war it was primarily the Foreign Secretary's private secretary who handled relations with the press, it soon became apparent that something far more organized was necessary once war began. A separate News Department was created in 1915. It became a general co-ordinating body for all the committees, bureaus and ministries involved in reporting the war, influencing opinion and censoring undesirable news. It was a department far more comfortable with information than with propaganda and some form of reorganization became necessary as the battle to win over the neutral powers intensified. Nonetheless, the Foreign Office fought hard to keep control over the 'war of words' and it was not until 1918, with the creation of Lord Beaverbrook's Ministry of Information, that the Office was defeated.[5]

The Permanent Under-Secretary, Lord Hardinge, made the battle against Beaverbrook one of personal causes and when it became apparent that he might lose the bureaucratic struggle decided to 'poach' a whole department of experts – the Intelligence Bureau of the Department of Information which was responsible for weekly country reports that were circulated to the War Cabinet and other offices. The creation of this new Political Intelligence Department at the Foreign Office in the spring of 1918 was part of Hardinge's effort to counter Lloyd George's 'Garden Suburb'.[6] Beaverbrook was presented with a 'fait accompli' and ten former Intelligence Bureau officials soon to be joined by five more experts from other departments were recruited for the PID. The 'Ministry of All Talents', as it was called, was staffed by 'outsiders' marked by their catholicity of backgrounds, education and experience. William Tyrrell, a Foreign Office man, was named to head the department 'to protect them from the more irascible elements in the Office' who objected to their presence and their influence. The PID officials continued to provide reports on the political situation in a wide variety of countries; these reports were circulated to ministers but the PID's major function, seen by Hardinge as a way to move the Foreign Office into the centre of diplomatic activity, was to prepare memoranda, 71 in all, on every important topic and country that might be the subject of negotiation at the Peace Conference. Several members of the PID went to Paris and left their mark on the territorial clauses of the peace treaties. The PID made a real place for itself in the Office, creating a new repository of information, gathered and interpreted by temporary civil servants many of whom were recruited from the academic world.

Neither the PID nor the Historical Section, another war-time creation intended to strengthen the research side of the Foreign Office Library, survived the return to peace. The Treasury, anxious to cut expenditure, demanded the closure of all temporary departments and, with the appointment of Hardinge as Ambassador to France, no one took on the fight against the dismemberment of the two departments. Older members of the service were not unhappy about their dissolution. Some PID members, such as the Leeper brothers, became career diplomats. J. W. Headlam-Morley, picked by Lloyd George to deal with Polish questions in Paris, was appointed on a personal basis as Historical Adviser to the Foreign Office. His numerous and valuable memoranda were often a reminder of the work done previously in the disbanded PID until its revival at the outbreak of war in 1939.

The Press Department, another victim of Treasury parsimony, managed to survive albeit on a much reduced scale.[7] Old assumptions and prejudices against publicity lingered on in an institution that both prided itself on its professionalism and resented the interference of politicians and outsiders in its conduct of affairs. Eyre Crowe, appointed Permanent Under-Secretary in 1920, could not understand the need for such a department, asking Arthur Willert, who joined its small staff in 1921, 'why diplomacy had to hold the press in such consideration?'[8] Much of the work was still done, as before the war, by the private secretaries and through informal channels of communication.

The Great War had also seen a vastly expanded intelligence effort and a major breakthrough in cryptography.[9] The 'Decyphering Branch' was closed in 1844 depriving the pre-war Foreign Office of the most valuable form of diplomatic intelligence gathering. Secret service funds were under the control of the Foreign Secretary and audited by the Permanent Under-Secretary. The war radically changed the intelligence scene and in early 1919 a secret service committee was set up under the chairmanship of Lord Curzon, an old hand at the Great Game on the North-Western frontier, to reorganize the intelligence services for peace-time purposes. The threat posed by the Bolshevik revolution proved to be an important catalyst in this process. By 1922 the Foreign Office had assumed responsibility both for the newly-created Government Code and Cypher School and for the Secret Intelligence Service, the government's counter-espionage agency. It also ran the Passport Control Officer system, which became the Office's main cover for secret service work abroad. As visa and other fees paid for the covert as well as the public side of passport control, this proved a useful and inexpensive way to fund Foreign Office activities. The SIS was small and its organization informal. In 1921, following the military and naval examples, the Foreign Office appointed a liaison officer to head one of the three 'circulating sections' that assessed intelligence reports; this individual decided what to pass on to the respective departments and advised on priorities for intelligence collection.

The GC&CS was also in its infancy and ministers did not know how to interpret or to use decrypted intelligence, mainly focused on the Soviet Union. Lord Curzon studied the Soviet intercepts 'with almost obsessional interest' though it was the decrypt of a telegram from the French Ambassador in London reporting Curzon's criticism of cabinet policy that forced the Foreign Office to take over direct responsibility for the GC&CS from the Admiralty in April 1922. Convinced of major Soviet subversion at home and in India, Curzon was not always sensitive to the quality of his intelligence, which led to the debacle of 1921 when secret information proving that the Soviet government was breaching the terms of the 1921 trade agreement proved to be forgeries. Nor had ministers yet appreciated the need to protect the source of their intelligence. In April 1923, in the so-called 'Curzon ultimatum', the Foreign Secretary not only quoted from the Russian intercepts but taunted the Soviets with the British cryptographic success. It was through the decrypts that the much shaken Curzon learnt not only of Poincaré's disloyal dealings with the Turkish nationalists but of the French premier's intrigues to have Curzon replaced by a more Francophile Foreign Secretary. Curzon was outraged; it was 'the worst thing that I have come across in my public life . . . '[10] During his last months in office, it was the perfidious French rather than the Russians, whom the decrypts showed to be loyal to the terms of the Anglo-Soviet trade agreement, who became the special object of Curzon's fury.

Changes in Structure

Nevertheless, the process of making these reorganizations effective was slow. The problem was not so much the lack of enthusiasm at the top for the proposed changes as the difficulty in translating them organization-ally, and the fact that they were called upon to implement them even before the war had reached its conclusion. The confusion created by the complicated chain of command within the Office did not help matters. The Foreign Secretary, Balfour, was an ageing man, inclined to be lazy about paper work and uninterested in the workings of the Office. By making Cecil his deputy, the de facto existence of two Foreign Secretaries resulted in overlapping jurisdictions, and made rapid decision-making difficult.[11] Equally problematical were the differences of opinion between Cecil and the Permanent Under-Secretary, Hardinge, over the questions of appointments and internal reform; this situation was compounded by their uneasy working and personal relations. Cecil was critical of Hardinge's management of affairs,[12] while Hardinge thought Cecil a 'crank'.[13] And as both had overbearing natures, it gave rise to frequent clashes between them.[14]

The opening of the Paris Peace Conference made it even more difficult for the various parties to get on with the job of reorganization. The Foreign Office was split into two: one part was in Paris under Balfour, Hardinge and Crowe, and its other part in London, where Curzon sub-stituted for Balfour until officially made Foreign Secretary on 25 October 1919. The final complication concerned the Foreign Office's negotiations with the Treasury over the provision of the requisite expenditure for carrying out the proposed reforms. The proposed expenditures were at variance with the Treasury's demand for cuts and economy; the inevi-table outcome was further delay in the introduction of organizational change.[15]

On setting itself to deal with the organizational issue, the Foreign Office embarked, first of all, upon the process of dismantling its war-time structure and replacing it with a structure that would enable it to function once more as an independent department servicing its minister fully, representing him to other departments, formulating major policy for ministerial decision, implementing it after decision; and acting as the government's interpreter and adviser upon all matters pertaining to the significance of foreign developments and the steps called for on the part of Britain.[16]

The War Department was dissolved on 7 October 1920, and in its place the old Eastern and Western Departments were re-established, as well as two new ones – the Central and Northern Departments. The American (for a short while known as American and African) and the Far Eastern Departments continued as before.

The Ministry of Blockade was disbanded in May 1919, and likewise the Commercial Department of the Foreign Office.[17] Like his opposite numbers – Phillipe Berthelot at the Quai d'Orsay, and F. Edmond Schüler at the Auswärtiges Amt[18] – Hardinge came to the conclusion that the pre-war functional separation of political from commercial departments should be abandoned in favour of a geographical system. He proposed grouping political and economic questions according to regions – for the purpose of adjusting the Office organizationally to the growing linkage between foreign policy, and economy and commerce. The functions of the Commercial Department were to be transferred in part to the Department of Overseas Trade and in part to the political departments. In the words of Hardinge, as he noted to Curzon:

> The scheme involves a new departure in the fusion of the Commercial and Political work of each geographical department of the Foreign Office which it is hoped will result in a better coordination of the Political and Commercial interests involved in each country without administrative inconvenience. It is thought that by this means the staff of each Department will obtain a wider grasp of commercial questions and the importance of promoting trade in unison with British political interests abroad.[19]

His intention was well meant, but in hindsight it misfired.[20] The Political Intelligence Department, as mentioned earlier, was wound up in October 1920 and the Supreme Economic Council and the Rhineland Commission, yet more war-time creations, were placed under the Western Department.[21]

Changes in Hierarchy

At the same time, a new division of responsibilities was made among the Foreign Office hierarchy. After consulting the Chief Clerk of that time, John Tilley,[22] Tilley's assistant, C. Hubert Montgomery,[23] and Ronald Campbell,[24] his own Private Secretary, Hardinge came to the conclusion that if they wished to reduce significantly the number of senior grades (an assurance given to the Treasury) and not to complicate the chain of command, which according to him was the one thing that exasperated Curzon more than anything else, then unlike the pre-1914 period, he would have – as Permanent Under-Secretary – to be content with one Assistant Under-Secretary of State, namely Crowe. It was suggested that Crowe would superintend the African (shortly disbanded), American, Consular and Far Eastern Departments, and be responsible, also, for all the work connected with the League of Nations and the

Department of Overseas Trade; while Hardinge himself would handle the Northern, Western and Eastern Departments.[25] In fact, originally, Hardinge intended Crowe to superintend the Eastern as well,[26] but as Curzon insisted that he would have to see the papers of the department, being a Near Eastern specialist, he retracted.[27] Hardinge's division of responsibility reflected his wish to handle most of the political work of the Foreign Office.

Instead of the practice instituted in 1906 according to which all papers going to the Foreign Secretary passed first through the hands of the Permanent Under-Secretary, Hardinge decided that henceforth Crowe would send his papers direct to Curzon.[28] In a letter to Crowe, who, in his capacity as British Minister Plenipotentiary for the completion of the peace settlement, remained in Paris, he explained that even though Curzon insisted that all the papers brought to his attention be read in the first instant by the Permanent Under-Secretary, if Crowe (who would be responsible for certain departments) would sort out for him the papers he thought should be read by the Foreign Secretary, Hardinge would be able to pass them on, merely affixing his initial. According to Hardinge, this would save valuable time:

> My idea is, and it seems to meet with the approval of others whom I have consulted, that you should deal with certain Departments and only send up to me, for transmission to the Secretary of State, the most necessary and most important papers. This would relieve me practically entirely of control of certain Departments, while it would enable me to concentrate more on those of which I have greater experience and more technical knowledge.[29]

At the same time it was determined that, in addition to the Permanent Under-Secretary and Assistant Under-Secretary, there would be eight Heads of Departments with the grade of Assistant Secretary (subsequently renamed Counsellors) corresponding to the former Senior Clerks, to whom – given the intensification of the pressure of work in the Office – would be delegated greater authority than in the past,[30] and whose function would be to handle on their own, low and sectoral questions, and, as far as possible, to avoid swamping the Permanent Under-Secretary and Assistant Under-Secretary with routine matters,[31] thus freeing them to deal with the most pressing issues on the agenda.

These arrangements, which were designed to restore to the Foreign Office its habitual functionality, were approved by Curzon, but their introduction entailed continual difficulties. First, there was the shortage of Higher Division staff, aggravated by the reluctance of Hardinge to press the Treasury on this point, in the face of severe cutbacks of excess bureaucratic staff.[32] Secondly, there was the problem of the cramped premises of the Foreign Office building – a troublesome issue since

1913, but now particularly difficult given the substantial expansion of departments and personnel during and after the war. Thus, for instance, the Head of the News Department (formed in 1915) received the Press in a small cubicle inside the building; the Cables and Wireless Section was housed in a hut located in the yard of the building, while the clerical and typing staff were put in a tottering structure on the roof. The clerks of the Central Department were divided between several rooms; files and documents of the Registry were stored in the corridors, where any passer-by had access to them; and the heads of the various departments did not have rooms of their own, but worked in a huddle, together with their staffs, and received persons in a small, dark and stuffy room at the bottom of the first-floor stairs.[33] To alleviate the acute shortage of space, in 1924 the Office of the Works began to erect a new storey on top of the existing building. Construction was completed in October 1925,[34] but only thanks to the intervention of the Prime Minister and new Foreign Secretary, J. Ramsay MacDonald, who responded to the appeal of Crowe[35] and secured for the Foreign Office the requisite budget.[36]

The new division of work at the top of the organizational hierarchy, as determined in 1919, was far from efficient. In 1920, Crowe, Tyrrell and Montgomery pointed out that the practice according to which papers from all departments in the Office reached the Foreign Secretary via the Permanent Under-Secretary, placed in doubt the ability of this individual to give matters passing through his hands the attention they might deserve.[37] Hardinge complained that the reduction in the number of Assistant Under-Secretaries from three to one, and the upgrading of the heads of departments, only served to increase the burden that fell on his shoulders:

> Sometimes between 50 and 60 boxes reach me during the day, some containing only a few papers, but others tightly packed. This, together with interviews, creates a heavy strain upon the Permanent Under-Secretary. But bad as this may be, it is worse when an Assistant Secretary is absent on leave or on account of illness, for then the permanent Under-Secretary has not only his own work to do, but also that of the absent Assistant Secretary.[38]

New Appointments

Thereupon the trio – Crowe, Tyrrell and Montgomery – with the support of Hardinge, suggested that the number of Assistant Under-Secretaries of State be increased once more, by making Tyrrell a second Assistant Under-Secretary. The recommended division of functions between them would be: Hardinge – Eastern and Northern; Crowe – Central and Western; and Tyrrell – American, Far Eastern and News Departments.

The trio recommended, with the object of relieving the pressure on the Permanent Under-Secretary and streamlining the system, that the Assistant Under-Secretaries would be entitled to submit papers directly to the Foreign Secretary.[39] To this, however, and in contrast to 1919, Hardinge objected on the grounds that it would hinder the co-ordination of the activities of the various departments. He assured Curzon that with the appointment of a second Assistant Under-Secretary he would no longer have any difficulty in acting as the 'neck of the bottle' through whom all papers were passed to the Foreign Secretary.[40]

The recommendations of the quartet about the appointment of an additional Assistant Under-Secretary, and the division of labour that would be between him and his colleagues, were adopted by Curzon. He did not even press Hardinge to give up the 'neck of the bottle' practice, so long as it did not upset the flow of work.[41] This chain of command continued until 1922, when the status of the Chief Clerk, by that time Montgomery, was raised to Assistant Under-Secretary, and he was allotted the superintendence of the Chief Clerk's, Consular, King's Messengers and Communications (formed in 1919), Library and Treaty Departments – all of them departments that were, hitherto, under the direct control of the Permanent Under-Secretary, Crowe.[42] This development took place against the background of the controversy which had ensued between the Treasury and the Foreign Office as to who would be responsible for the Foreign Office accounts. The Treasury, aware of the Public Accounts Committee criticisms about the inadequacy of the system of financial control within the Office,[43] adopted its recommendation, which it itself brought up in February 1918,[44] to make the Permanent Under-Secretary the Accounting Officer, and to assign to him, on behalf of the Treasury, an Assistant Secretary who would handle financial details.[45] The Treasury saw a golden opportunity to plant in the Foreign Office a man of its own and to enforce, through him, the budgetary and organizational standards of the Home Civil Service upon the Foreign Office, which hitherto had acted as a closed system.

The Foreign Office, quite understandably, desired to continue its way of conducting financial matters, and in response suggested at first that the Chief Clerk, who in any case took care of the establishment and financial business, be made the Accounting Officer, and as such be upgraded to an Assistant Under-Secretary.[46] Later the post would be given to a particular person, who together with the Chief Clerk – to whom he would be subordinate and who would be raised to an Assistant Under-Secretary – would take care of the Foreign Office accounts and all financial problems.[47] As far back as 3 December 1920, Crowe had explained to the Permanent Under-Secretary to the Treasury, Sir Warren Fisher, that the pressure of work on the Permanent Under-Secretary of the Foreign Office was so high that, if this individual was also to become an Accounting Officer, the appointment would be a sham. It would not only be impossible to give financial questions detailed consideration,[48] but, even if the question of putting somebody in charge of accounts had not come on the agenda, the Foreign Office would have had to ask the Treasury to sanction the promotion of the Chief Clerk to an Assistant Under-Secretary, given the burden of work at the top.[49] Nevertheless, two years of negotiations and constant bickering, which typified the Foreign Office's relations with the Treasury, were to elapse before the later suggestion of the Foreign Office was accepted and Sir Frederick Butler, formerly of the Department of Overseas Trade, was appointed, on 1 May 1922, an Accounting Officer.[50] Three months passed before the upgrading of the Chief Clerk was sanctioned by the Treasury,[51] and yet another month before Montgomery was officially made Assistant Under-Secretary.[52]

The Need for an Expanded Outlook

Throughout the period, and in fact well before the transition from war to peace, the Foreign Office had to contend with the more fundamental problem of modernizing its home and overseas services. This issue delayed, more than once, the organization of the peace-time structure and brought to the surface acute questions regarding the future shape and character of the diplomatic establishment.

In spite of the evolutions through which the Foreign Office and Diplomatic Service had passed prior to the First World War, their failure to catch up with the twentieth century had remained relatively unaltered. The criticism made so forcefully by the members of the Liberal Foreign Affairs Committee and other radical backbenchers in the Commons, and before the Royal Commission on the Civil Service of 1914, when they observed that Foreign Office and Diplomatic Service personnel were recruited from too narrow a social and educational stratum,[53] was still as valid as ever.[54] Just as marked was what they described as the central wrong of the two services,[55] namely that they tended to move in narrow social and professional circles, and to think in political rather than economic terms.[56] Furthermore, the slowness with which the 1905–6 reforms – i.e. the division of labour between intellectual and mechanical work, and the devolution of work from the Permanent Under-Secretary to the junior clerks[57] – were introduced into the Diplomatic Service, and the rise of the Foreign Office against the decline of the Diplomatic Service, were the makings of a gap between the managerial and advisory capabilities of the two services.[58] It is true that after 1906, in contrast to the period which preceded it, there was a considerable interchange between the Foreign

Office and the Diplomatic Service, in spite of the fact that officially the two services remained separate. This interchange, however, was on a voluntary, non-systematic basis and had to be privately arranged.[59]

In 1914, the Royal Commission, like its predecessor, the Ridley Commission of 1890,[60] recommended that the Foreign Office and Diplomatic Service be amalgamated on the grounds that a single Foreign Service would allow a systematically free interchange between the two services, supply the Office at home and the missions abroad with men of more varied training, facilitate the choice of the best man for each vacant post, and standardize the professional requirements for both services.[61] Representative bodies such as the Union of Democratic Control, an influential group of radical MPs and academics under the leadership of E. D. Morel and Arthur Ponsonby, insisted during the period 1915–17, like their French and German counterparts,[62] that the Foreign Office and Diplomatic Service be made more representative of British society and more responsible to Parliament.[63] They were joined by Sir Arthur Evans, Lewis B. Namier, Eustace Percy, Sir George Young and other public figures, who, in a series of articles published in *The New Europe* – a weekly journal of international affairs, edited by R. W. Seton-Watson and George Glasgow – urged administration improvements capable of encouraging initiative and talent, greater systematization of promotion, and a more broadly-based approach to the problems of recruitment and staffing.[64] All were suggestions that bore striking similarity to reforms pressed on other foreign ministries world-wide.[65]

Within the Foreign Office itself, there was a group who clearly realized that the very dimensions of foreign policy were being expanded, and that the Foreign Office would have to develop new competencies if it were to re-establish its primacy in the post-war world.[66] Together these elements proved to be a formidable pressure group, and in 1918 Sir Eric Drummond, Private Secretary of Balfour, warned that unless the Foreign Office took the initiative and implemented reforms suitable to it, they would be forced, by others, to reform in a much more extensive manner.[67]

Updated Financial Structure

The outbreak of the First World War and the replacement of the peace-time structure with a war-time one prevented the pre-war recommendations being put into practice. Nevertheless, they were not altogether forgotten and came on the agenda occasionally during the war. Thus, for instance, on the basis of the Royal Commission recommendations that salaries and allowances be raised with a view to abolishing the existing property qualification for admission to the Diplomatic Service (i.e. the possession of a private income of at least £400 a year) and the two-year unpaid attachéship,[68] thus enabling members of the Service both at home and abroad to live on their pay,[69] the Foreign Office submitted to the Treasury, on 29 September 1916, a scheme for improving emoluments.[70] The Treasury responded that improving the salaries of the Foreign Service would not do by itself, and that the rest of the Royal Commission recommendations ought to be adopted as well. The Treasury suggested that representatives of the two departments should meet to discuss the question.[71] Balfour acquiesced in the idea,[72] but after a meeting between the two sides on 24 May 1917 it became clear that the Treasury was, for the moment, not prepared to provide the requisite funds for carrying out the reforms suggested by the Royal Commission.[73] For a while, Foreign Office officials hoped that Cecil, who took up the handling of the question at the suggestion of Hardinge,[74] would succeed in getting the requisite funds, thanks to his intimate relations with Bonar Law, then Chancellor of the Exchequer. But this hope was not realized either.[75]

The crux of the problem was the contrary organizational trends of the Foreign Office and the Treasury. On a number of points, such as abolishing the property qualification, assimilating the entrance examination for the Foreign Service to the Class I scheme of the Home Civil Service, and raising the emoluments of those serving abroad, an agreement was reached.[76] However, the Foreign Office strove to detach the newly amalgamated service from the rest of the Home Civil Service, while the Treasury (which acted on behalf of the government in whatever pertained to the integration of the Civil Service – something that was recommended by several commissions)[77] strove to sweep away the existing quasi-independent status of the Foreign Office and bring it into line with the rest of the Civil Service.[78]

Foreign Office vs. Home Civil Service and Treasury

These contrary trends did not stem solely from financial calculations, but resulted also, and perhaps principally, from the conflicting perceptions as to the character and functions of the Foreign Office and Civil Service. The Foreign Office maintained that foreign affairs, unlike the standardized questions of domestic affairs, was a special field unto itself[79] that required years of specialization before a person was able to cope successfully with its problems; and that, therefore, the existent inter-departmental interchangeability between the staffs of the domestic departments was not applicable to it, for if its staff should depart from the Foreign Office it would cease to be specialist, while new recruits would most certainly lack the requisite skills for coping with foreign questions.[80]

The type of men required for the Foreign (or 'Imperial') Service is the man who by previous education, personality and growing experience gained in his career should be fitted to understand and interpret foreign governments and foreign people . . . In other words a specialist. The Civil Servant is in administration . . . To transfer him when in London to a Civil Service Department would be to change the type of his administrative work while removing him *absolutely* from his specialist work for so long as he is employed in the Civil Service Department.[81]

Other points on which, in the opinion of the Foreign Office officials, the two were distinguished from each other were the conditions of service (namely the post-amalgamation requirement to serve abroad, which did not fall on home civil servants, and the liability to incur, on being transferred to and from the United Kingdom, greater expenditure over housing, the education of children, etc., than officials who resided permanently in Britain and could plan their lives on a long-term basis), and the particular kind of gentlemen who would have to be recruited to the new service. Differentiation between Foreign Office and other departmental emoluments so as to compensate its officers for extra expenses arising from transfers, and entrance qualifications other than the standardized ones for the Home Civil Service, were points strenuously advanced by the Foreign Office.[82] Finally, it was pointed out that Civil Service work did not meet the social expectations of the members of the Foreign Office, who regarded the Foreign Service as ranking socially higher.[83] In reality, this limited the social contacts between the Foreign Office staff and that of the Civil Service, turned the Foreign Office into a closed system, and deterred, still further, Foreign Office staff from considering favourably the tearing down of the frameworks between them and the Civil Service.[84]

The Treasury, by contrast, held that the best mode of ensuring that a person would perceive the full implications of the problems which confronted him, and adjust policy if needs be – in short, to be a good civil servant with a universal approach – was by obliging him to serve in as many departments as possible.[85] The implication was that interchangeability between Foreign Office staff and the staffs of the other departments should be constant; that staffs of the various departments should be generalists and not specialists, as the Foreign Office claimed; and that, in order to accomplish this end, the Foreign Office ought to be integrated within the Civil Service as much as possible, with uniform standards regarding entrance and pay. In short, what was desired was the full amalgamation of the Foreign Office with the Civil Service.[86]

Stanley Baldwin, then Financial Secretary to the Treasury, explained in February 1918 to Cecil that:

The idea was that both sides would gain by an admixture of the two strains – the Foreign Office acquiring officers with a knowledge of Civil

Service administration and traditions, and the other Departments officers with diplomatic experience.[87]

Warren Fisher, Permanent Under-Secretary to the Treasury (from 1 October 1919), asserted for his part that interchangeability between the Permanent Under-Secretaries of the various departments would pose no problem, since that post would go to men of such breadth of experience that they would soon find themselves picking out the essential points and applying wide general knowledge to the subjects they dealt with.[88] Edward Bridges, a Treasury temporary official at that time, and, years later, its Permanent Under-Secretary, enlarged this point, noting that:

> Nearly every problem bears a family resemblance to something which the experienced administrator has handled before. To him . . . it is the old wolf in a new-look sheepskin . . . His experience tells him where the point of entry will probably be found; or warns him of the difficulties ahead which others without his training, would not see.[89]

The irreconcilable viewpoints between the sides prevented them from reaching a quick settlement on the question of the amalgamation of the Diplomatic Service within the Foreign Office. No progress was made in negotiations during 1917. Wherefore Hardinge formed, in February 1918, with the approval of Cecil,[90] a six-member committee and charged it with finding a solution.[91] The committee members, and particularly Crowe,[92] who was not over-enthusiastic about the idea of the amalgamation on the grounds that not every diplomatist was fit to be a Foreign Office official and vice versa,[93] reiterated old stands.[94] The committee was supported on this point by other persons in the Foreign Office. It was mooted that the main object was not amalgamation, but the salary rises expected in its wake, and that, should that object not be accomplished, they had better abandon the amalgamation scheme. They would try instead for an all-round increase of Civil Service salaries, in conjunction with the other departments[95] – a view shared by one of the members of the aforementioned committee, Ronald Campbell.[96] The Assistant Clerk, George Warren, ventured an opinion that, given the Treasury refusal to accede to Foreign Office financial demands, they ought not to drop the property qualification for the Diplomatic Service. Furthermore, he opposed the very idea of amalgamation on the grounds that:

> The fact of the matter is that the type of man required for the two services is so entirely different that it is hardly an exaggeration to say that the very qualities which make a man suitable for the one render him unsuitable for the other.

He warned that should there be constant interchangeability between the two services – as would follow from the amalgamation – the staff working at the Office would become transient in character, officials would lack the requisite knowledge for the proper conduct of the business of the Foreign Office, and tend to pass the solid work to Staff Officers, who would be the sole permanent officials in London.[97] Wellesley, for his part, pointed out that if members of the Foreign Office were to be called upon to serve from time to time anywhere in the world, then their salaries ought to be on a scale commensurate with the extra expense which such liability entailed.[98] Even Hardinge, who was favourably disposed towards the amalgamation, began, under the pressure of the growing criticism within the Office, to doubt whether there was any point in continuing the negotiations with the Treasury, which he suspected of intending to gain control over the Foreign Office. He suggested to Cecil that the whole question be handed over to the arbitration of a fresh committee.[99]

However, Cecil, who had pledged himself in the House of Commons to carry out the amalgamation, and who (according to him) had devoted to the question a considerable amount of his time and energy ever since he was asked by Hardinge and others to bring about the implementation of the reform, was not prepared to give up. He instructed the committee to moderate the tone of its report.[100] On 16 April 1918, he met with representatives of the Treasury and expressed his willingness to agree to the principle of transfers – in exceptional cases only – between the Foreign Service and other departments of state, he agreed to abolish the property qualification for the Diplomatic Service, accepted the recommendation of the Royal Commission of 1914 to broaden the Board of Selection, and he offered to reduce the number of Assistant Under-Secretaries to one. An Inter-Departmental Committee would be created to investigate the ways and means of rationalizing the manner in which the Chief Clerk managed Foreign Office financial affairs.[101]

The Treasury, however, refused to make similar concessions. It rejected Cecil's petition that the Foreign Office be given a higher salary than the rest of the departments of state on the grounds that it could not be justified to Parliament, and that the pay-limits of the Civil Service would be breached. At the most, the Treasury was willing to increase the number of archivists, and to create special allowances for those serving abroad.[102]

Problems of Recruitment

This brought about a renewal of the controversy between the two sides. Cecil and his assistants cautioned the Treasury that, should the members of the Service, both at home and abroad, not be provided with a proper (high) living wage, it would not be practicable to amalgamate the two services unless the property qualification was reinstated. Otherwise Foreign Office officers would not be able to serve abroad for want of financial means.[103] Balfour, too, emphasized to Bonar Law that the Diplomatic Service was not the only one which was entitled to a differential treatment, but also the Foreign Office, both so as to attract able young men, and to compensate them for their inability to make long-term financial arrangements while temporarily employed in London.[104] But Bonar Law continued to maintain the sufficiency of the proposed allowances to meet the extra expenses of the Foreign Office officials in foreign posts, and observed that even if the pay rises suggested by the Foreign Office were to be realized, he did not believe that they would attract young men.[105] The controversy continued throughout the summer and in its course the Foreign Office officials began, given the Treasury refusal to temper its stand, to look for a redeeming formula which would save the situation, but in vain.

If that was not enough, the dissensions over the amalgamation issue were compounded by questions of recruitment. Namely, whether the best method of determining the moral character and ability of candidates was an interview by the Board of Selection,[106] or literary examination.[107] And secondly, whether the Foreign Office system of selection through limited competition – that is competition restricted to those who had previously secured a nomination at the Foreign Office – should be retained, subject to the modifications recommended by the Royal Commission,[108] or replaced by open competition.

The growth of these uncertainties owed something to the findings of the Royal Commission,[109] which lent support to the suspicions that the process of selection was biased in favour of those born in aristocratic and gentry stations, possessed of independent means, and educated in favoured public schools and universities (mostly Eton, Oxford and Cambridge).[110] It owed something, also, to the rising need of ensuring a more democratic representation in the ruling élite – from which the Service was drawn – as the electorate expanded and as social and geographical mobility increased.[111] But underlying these wider explanations there were disagreements between the Foreign Office and the Treasury about the applicability of open competition to the diplomatic establishment. The Foreign Office alleged that direct-entry recruitment through open competitive examinations without recourse to the Board of Selection, as suggested by Baldwin,[112] would end up by substituting character, which was so important for overseas representation, with either academic skills or strange and unsuitable personnel.[113] The Treasury, on the other hand, held, like Lord Macaulay in his day,[114] that academic examinations were also indicative of general character, and pointed out that open competition produced a high average level of ability among entrants.[115]

In an effort to resolve this deadlock, Cecil agreed to adapt in full the Royal Commission recommendations as regards recruitment.[116] The

Treasury, for its part, was prepared to concede that the peculiarly confidential and sensitive nature of the Foreign Office's work required it to take special precautions when recruiting clerks, and that the Foreign Secretary should be allowed to retain a veto on the admission of entrants.[117] The Committee on Conditions of Admission into the Foreign Office and Diplomatic Service was set up, on 28 June 1918, under W. W. Palmer, Second Earl of Selborne, former President of the Board of Agriculture and Fisheries, with Sir Stanley Leathes, the First Civil Service Commissioner, Ian Malcolm MP, John Tilley, the Chief Clerks and O. Theophilus V. Russell, the Diplomatic Secretary. Committee discussions revealed that the divisions between the sides, as well as within the Foreign Office itself, still cut deep. Russell, for instance, had no doubts that the old Board of Selection should be retained and that they would have to find ways, 'How to exclude Jews, coloured men and infidels who are British subjects'.[118] Tilley feared lest open competition would admit into the service elements with negative characteristics.[119] Even Crowe put up a number of arguments against the recruitment of foreign-born British subjects.[120] Drummond, by contrast, favoured a more liberal entrance procedure. In response to the arguments of Tilley, he noted that open competition was essential, for the future of the Foreign Service depended greatly on recruitment from all sections of the population.[121] Eventually, a compromise was reached along the lines agreed between Cecil and Baldwin. It was concluded that the Board of Selection would be broadened in its composition and henceforward would include the First Civil Service Commissioner and two MPs, in addition to Foreign Office representatives; that candidates would first take the Civil Service written examination (which was being remodelled), and a viva voce (a major innovation)[122] only after they had appeared before the Board of Selection. Hitherto, appearance before the Board had been the first step.[123]

Towards Agreement on Wages and Conditions

No progress was made on the question of giving Foreign Office staff the same privileges as regards living wages as diplomats, and detaching the Foreign Office from the Home Civil Service. Consequently, the argument of those who held all along that the amalgamation between the Foreign Office and the Diplomatic Service was undesirable strengthened. Henceforth, the Foreign Office concentrated on bringing its salary scale into line with the Treasury, whose officials earned considerably more than their Foreign Office counterparts. In a meeting that Cecil held with Bonar Law on 3 September 1918, he expressed his willingness to consider bringing the number of Foreign Office grades into line with the Civil Service, and in return Bonar Law offered him higher living wages for the members of the Foreign Office.[124] This proposal did not satisfy the Foreign Office, and particularly not Crowe, who rejected the Treasury claim that thereby the salaries of the two departments were equated, as pulling the wool over their eyes.[125] However, after prolonged discussions within the Foreign Office, and between it and the Treasury – in which the driving force was Tilley, who considered that with slight modifications it would be possible to adapt the Treasury proposals, including the reduction of the number of Assistant Under-Secretaries of State to one[126] – a scheme finally materialized which was favoured by Hardinge (on 11 October 1918),[127] and approved enthusiastically by Cecil (on 16 October).[128]

As it turned out, although deliberated at length between the Foreign Office and the Treasury, this scheme was not final. Subsequently, on 28 December, Hardinge reported that in his meeting the day before with Sir Thomas Heath, Joint Permanent Under-Secretary to the Treasury, they had reached a general agreement that the Assistant Under-Secretary of State grade was to be amalgamated with that of the Heads of Departments; that Tyrrell and Graham, who took many years to reach that level, were to continue to be salaried on the same scale as the Assistant Under-Secretaries, even though they ceased to be such.[129] Grades in the Foreign Office were to be brought into line with the Civil Service, namely, Permanent Under-Secretary, Assistant Under-Secretary, eight Heads of Departments, and First, Second and Third Secretaries and the salary of the Diplomatic Service was to start at £300 per annum, and be supplemented by living and rent allowances. Existing Foreign Office staff were to receive the maximum pay of each grade while newcomers were to receive the initial pay, in order to avoid a situation whereby senior officials who were on the verge of being promoted and receiving maximum salary would, all of a sudden, become juniors receiving initial salary, and have to wait a number of years until they were, once again, entitled to the maximum salary of their grade, and to a promotion.[130]

But here an unexpected hitch occurred. Heath and Sir Robert Chalmers, the other Joint Permanent Under-Secretary, to whom Hardinge had sent a copy of his letter, announced that the distinction between seniors and juniors was contrary to Civil Service regulations, and that all of them would have to start at the new initial scale.[131] Cecil, who was about to leave the Foreign Office for the Peace Conference in Paris, seethed with anger. For months and months, he wrote to Bonar Law, he had struggled to reform the Foreign Service, something that he had pledged himself to carry out in Parliament. He had thought that at their last meeting, in September, a general agreement was reached, and that all that was left was to fill in a few details. 'Now at the last minute that intolerable creature Heath, and that pompous ass Chalmers, have come down with new proposals which absolutely destroy the scheme.'[132] But his outburst was to no avail.

This hitch – together with the departure of Cecil, Hardinge and Crowe to Paris – brought about a pause in the negotiations. Only in April 1919, on the entrance of Curzon as the deputy of Balfour, and his intervention with the Chancellor of the Exchequer, did the Treasury announce that it had no intention of going back on the understanding reached with Cecil.[133] This led to a fresh round of proposals and counter-proposals, until, on 23 June, the sides reached a final agreement according to which the emoluments of the Foreign Office staff were to be brought into line with those of the Treasury; rent and other special allowances were to be introduced for diplomats; the property qualification was to be eliminated; and it was determined that the entrance examination was to be assimilated to that of the Civil Service – except that candidates would have to demonstrate proficiency in French and German, and appear before the Board of Selection.[134]

Towards Amalgamation

This final settlement paved the way for Foreign Office action. The composition of the Board of Selection was altered to include the First Civil Service Commissioner and MPs. A new career ladder and new pay scales were introduced, and a joint seniority list of the two services was created, something which implied that persons in one branch would be replaced within a fixed period by their counterparts in the other branch. But when they embarked upon putting the principle of interchangeability into practice they found out, according to Akers-Douglas, the new Diplomatic Secretary, that the shortage of manpower and the constant interchangeability meant that there were not enough experienced men in the various missions.[135]

A year after the introduction of the amalgamation – in the course of a meeting of the Promotion Board consisting of Hardinge, Crowe, Tyrrell, Montgomery, Tilley and Akers-Douglas – strong doubts were raised as to the extent of the success of the principle of interchangeability. To consider these doubts, Hardinge appointed a four-man sub-committee, which submitted its conclusions on 4 August 1920.[136] According to Crowe, Tyrrell and Montgomery, the experience acquired during the previous year did not bear out the assumption that members of the two services were equally well-suited for employment at home and abroad.

> The result of the present system of amalgamation must be either (a) to put into posts abroad, on promotion from the joint list, men who would be doing much more useful work in the Foreign Office, and to bring

home men who would be more usefully employed abroad, or, (b) to pass over men for promotion at home or abroad, because they are useful where they are.

They suggested that interchangeability should be carried out up to the First Secretary grade, when it would be decided by the Foreign Secretary – on the basis of past experience – whether the particular individual was fit to serve at the Foreign Office or in the Diplomatic Service.[137] Tilley, however, did not agree with this approach. He protested to Hardinge that the cases in which interchangeability did not prove itself were isolated, and that the non-unification of the two services, to which they had pledged themselves in Parliament, would wreck the Foreign Office's reputation and rouse the resentment of Foreign Office officials at being robbed of the chance they always dreamt of, namely serving abroad.[138]

Interestingly, Hardinge, who had favoured amalgamation from the start, now sided with the majority. He confirmed that first-rate Foreign Office officials proved themselves as comparative failures in Embassies and vice versa and that this stemmed from the fact that the life and duties of the two services were entirely different. Very few individual officers were equally useful at home and abroad. Should the system not be altered, then officials such as Gerald Villiers, who were much better suited for employment at the Foreign Office than at the Diplomatic Service, but whose promotion was dependent on their serving abroad, would have either to go abroad and fail, or to stay in London while losing promotion. The inevitable result would be that younger men would by-pass officials who served abroad even though such postings might not be entirely suitable. He therefore recommended to Curzon that the majority opinion be adopted, with one modification. Hardinge suggested that at the end of ten years' service, the choice of service should be made by the individual.[139]

Curzon warmly approved the majority opinion. According to him, after a year and nine months' experience of the Foreign Office under the new scheme, he himself perceived that the constant interchangeability was fatal to the acquisition of departmental knowledge and experience. He complained that:

> During my short time, there is scarcely a single official in the same department as he was when I arrived. The result is that the notes are sometimes uninformed and inadequate, and that much additional labour is thrown upon the S. of S.

He rejected, however, the proposal of Hardinge concerning leaving the choice of service to the individual, and insisted on the decision being made by the Foreign Office.[140] Thus the committee's recommendations were put into operation, and the joint seniority list of the amalgamated Foreign Service was reduced to a joint seniority list of Third and Second Secretaries.

The attempt to implement in full an extensive organizational reform, which turned in due course into a tussle between the Foreign Office (wishing to maintain itself as a closed system) and the Treasury (desiring to bring the Diplomatic establishment into line with the home Civil Service), constituted a classic example of a widespread phenomenon in public administration, namely a distortion of goals whereby the final product is different from the one envisaged by the reformers. The Foreign Office decision to turn the amalgamation into a partial one constituted a certain degree of distortion, though of the displacement kind, since the means of achieving the main end – i.e. the rise in pay scales – turned for some of the Foreign Office officials into the principal goal, while the amalgamation became a means to an end.[141]

In the late 1920s a considerable number of the members of the Service abroad – especially in the case of Heads of Missions – still required private means, that officially had been eliminated, in order to cover their expenses.[142] Candidates desiring to enter for the examination for the Combined Service had first to qualify at an interview before the Board of Selection[143] – a complete reversal of the order of selection agreed on 1 July 1918. The examination, it is true, was the same as for the administrative class of the Home Civil Service, with a special arrangement whereby candidates were obliged to include modern languages among their subjects.[144] Yet, despite reform, the examination was not nearly as effective for assessing the personal suitability of candidates as one might suppose.[145] Moreover, its vast scope was, by the same token, in itself self-selecting. 'The fact, however, that candidates were still expected to have reached a standard equal to a University Honours standard, and that at the same time they were obliged to pass a very severe qualifying test in French and German', wrote Harold Nicolson, who served at the time in the Foreign Office, 'implied, none the less, that on leaving school they would have to spend at least three years at a University and that on leaving the University they would have to maintain themselves on their private means, while studying foreign languages abroad. This long and expensive period of preparation in itself narrowed the range of candidates to those who could support themselves up to the age of twenty-two by private means.'[146]

All the same, the extent of the defeat for the reforms should not be exaggerated. Interchanges between the two services noticeably increased; it has been established, for example, that by 1930 only 37 out of 1933 diplomats and 25 out of 74 Foreign Office officers had not interchanged – a very different picture from the one that prevailed before the war.[147] Promotions for both the Foreign Office and the Diplomatic Service were transferred from the Private and Diplomatic Secretaries – it was thought they had too much personal say in the matter[148] – to a Promotion Board.[149] Professional families provided a greater number of entrants to the Service,[150] and the educational background of successful candidates became more varied. The overwhelming majority, it is true, still attended either Oxford or Cambridge. Between 1919 and 1929, however, the proportion of Etonians fell from 67 per cent (1908–1914)[151] to 24 per cent; the number of other public schools represented increased, and a handful even came from state grammar schools, including William Strang, who was to become the Permanent Under-Secretary to the Foreign Office.[152] In the event, complete amalgamation of the Foreign Office and Diplomatic Service was accomplished only in 1943, when the Consular Service was brought in from the cold after the introduction of the Scott–Eden reforms.

Keeping Records

Another departmental problem that preoccupied the Foreign Office was the need to reorganize the registry. Between 1914 and 1917 the number of papers handled by the Office increased from 24,586 to 68,119,[153] while there was only a slight increase in clerical staff, due to the Treasury's parsimony, and the absence of improvements in the registration system. As a result the old Central Registry became, not surprisingly, a giant bottle-neck: dockets were often inadequate, indices were almost always in arrear, and there were constant delays in producing previous papers.[154] Alwyn Parker, the Foreign Office Librarian, was asked in 1917 to examine the filing and registration systems of other governmental departments, and to submit suggestions for reorganizing the Foreign Office system, a task accomplished by him in summer 1919.[155]

The implementation of Parker's scheme was conditional on the willingness of the Treasury to provide the requisite funds. The Sub-committee of the Committee of Staffs reported to the Treasury, on 29 July 1918, that it doubted whether the Parker scheme was an adequate answer to the problems of the Registry and that, in any case, major alterations were, at present, not possible, given the general demand for a reduction in the Foreign Office organization.[156] The report gave rise to a wave of protests in the Foreign Office, and in particular aroused the wrath of Crowe, who was one of the chief initiators of this project. Crowe complained that:

> This growing system of enquiring into other people's conduct by unqualified outsiders instead of entrusting the proper administration of an office to its own responsible head is going to introduce more and more anarchy into the whole service . . . Royal Commissions used to be bad enough. These mushroom committees are worse.[157]

Despite the general resentment, the Foreign Office became apprehensive lest the Treasury be driven to veto the whole reform. Consequently, it was decided – on the suggestion of Tilley,[158] concurred by Crowe,[159] and approved by Hardinge[160] – to adopt the proposal of Bradbury, the Third Joint Permanent Under-Secretary to the Treasury, that an Inter-Departmental Committee be established to examine Parker's new registration system.[161] The committee was duly formed in September 1918, and submitted its report on 14 November, in which it endorsed (with certain modifications) the recommendations of Parker.[162] The proposed system was adopted by the British Delegation to the Paris Peace Conference, revised considerably as a result of its experiences,[163] and approved, in its final form, by the Treasury on 12 January 1920.[164] Similarly, the Treasury granted the Foreign Office's request for additional clerical staff,[165] and, after prolonged negotiations, the establishment of a Chancery Service, aimed at relieving the members of the Diplomatic Service of clerical duties.[166] Whereupon, the Foreign Office embarked upon applying it department by department.

The new system involved the dismantlement of the three autonomous Sub-Registries of the old Central Registry, which were found to occasion too much duplication and delay, and the establishment of a new, four-branched Registry. Each incoming paper was sent to the Opening Branch, where it was opened, placed inside docket sheets, and sent to the appropriate division of the Archives Branch. There it was encased in its own jacket and given a full registry number – comprising the departmental designation, the paper number, the file number and the index number. A synopsis was made of the contents and the whole passed, together with the related file, to the appropriate department. The action taken and the subsequent movements of the paper were recorded on its précis jacket. The Despatch Branch was responsible for preparing and despatching circulars to posts abroad, and making up the Foreign Office bags; while the fourth branch, the Main Index Branch, compiled and printed, from the original papers and minutes, an index of the official correspondence annually.[167]

Yet, for all its modernity, this system was not free of a number of difficulties. First, there was the chronic lack of clerical staff, caused by the steady increase in the volume of Foreign Office business[168] and poor conditions of service.[169] Secondly, there was Curzon's insistence on introducing into the Foreign Office the file system which had been obtained by him in the Foreign and Political Department of the Government of India. According to Curzon's 'system', all the more important telegrams and papers were submitted to the Secretary of State in special subject files made up in the departments by removing relevant previous papers from their files, encasing them in the same folder, and flagging them alphabetically by means of metal tabs with pins. These pins were strong and sharp, and they were apt to pierce the fingers of Foreign Office officials. Harold Nicolson, who served as Second Secretary (1919–20) and then First Secretary (1920–5), remarked humorously that 'Curzon's reform of the file system was not unaccompanied by blood and tears'.[170] Nevertheless, Curzon not only refused to recognize the merits of the new system but also assailed his staff for not acquiring the mastery of the 'Curzon files'. On one occasion, after studying one of the files, he wrote peevishly:

> This is a particularly stupid file. It is about the trial of ex-Ministers. All papers showing who they are have been carefully excluded . . . This is a reversion to the worst type of file – which I thought I had killed 3 years ago.[171]

However, in the long run, after Curzon retired, the new system triumphed, and indeed it rationalized the Registry considerably.

Releasing Information

A further administrative change that took place in the Foreign Office was the establishment of information and cultural services. As Lord Beaverbrook, Minister of Information, and Sir Campbell Stuart, Deputy of Lord Northcliffe, Director of Department of Enemy Propaganda – the heads of Britain's First World War propaganda machinery – observed, the cessation of hostilities did not eliminate altogether the founded interdependence between foreign affairs and publicity.[172] Nor did they consider it wise for the Foreign Office to take over responsibility for the work of the Ministry of Information, which was wound up on 31 December 1918.[173] Opinion within the Foreign Office concerning the desirability of continuing propaganda varied greatly. Though acknowledging its war-time effectiveness as an instrument of power over opinion, Cecil, who had been involved in propaganda in one way or another since 1915,[174] probably spoke for many when he said that now, with the arrival of peace, he 'disliked the idea of propaganda in foreign countries on general principles, apart from commercial propaganda which was the real line to follow'.[175]

On the other hand, a small but active and articulate group of former war-time propagandists, like S. A. Guest, for example, pointed out that, 'One of the chief lessons to be drawn from the experience of the war, and the events leading up to it, is that Diplomacy by itself is not enough for the maintenance of satisfactory international relations.'[176] George Beak, Consul-General in Zurich, added that one of their main objects in future would be the protection of the masses from half-truths.[177] Victor

Wellesley went so far as to suggest that the peace-time character of British overseas propaganda should fall into political, commercial and cultural categories, so as to encourage, maintain, develop and render respectively, (a) smaller states' political orientation towards Britain and the latter's prestige in all parts of the globe, (b) British commerce in all available directions, and (c) 'all that is best in British life and education'.[178]

To settle differences of opinion over the propaganda issue, a departmental conference was proposed by Cecil Harmsworth, who was now Parliamentary Under-Secretary.[179] The conference was presided over by Curzon, who for all his aversion to the media found it necessary, in the light of new strategic techniques, to lay down at the start two fundamental principles, namely: (1) that propaganda should continue, and (2) that it should continue strictly in relation with, or under the control of, the Foreign Office. Discussion revealed that organizationally there was room for a 'Propaganda Department' in the post-war Office; and that the minimum budget required was estimated to be £100,000. It was decided to accept the suggestion of the Chairman to appoint a permanent official.[180]

Four days after the departmental conference, Curzon offered Tyrrell, who had headed the Political Intelligence Department during the war, the position of head of the new department. Tyrrell accepted and it was agreed between them that structurally the department would be called the News Department, have its own establishment abroad[181] and be linked with the Political Intelligence Department (which was wound up in 1920).[182] Functionally, the News Department was to fight anti-British propaganda as well as strive to improve the international standing of Britain by collecting information from, and supplying it to, foreign opinion.[183] It would work in co-ordination with the Secret Intelligence Service, the Home Propaganda Organization and the press.[184] Departmentally, the News Department was to be divided into three broad sections: (1) Cables and Wireless Division; (2) a National or Administrative Division with seven geographical sub-divisions, responsible for collating information; (3) a Facilities Division, which would also be responsible for receiving the press.[185] Financially, its expenditure was not to rise above the minimum level agreed upon at the departmental conference.[186]

Accordingly, on 2 April, the estimates were submitted formally. But here they met with severe opposition. Mindful of the need for financial retrenchment, the Foreign Office worked hard to keep the estimates as low as possible – a reduction, in fact, of 75 per cent on the £2 million spent on all forms of propaganda during the final year of the war.[187] To the more critical Treasury, determined to cut costs wherever possible, £100,000 on overseas propaganda appeared quite unduly large, if justified in peace-time at all. There began a series of fruitless negotiations culminating in Gasele, Harmsworth and Tyrrell attending, on 14

May, an inter-departmental conference at the Treasury under Baldwin to resolve this deadlock. They conceded to the Treasury that the proposed propaganda programme 'must be regarded as tentative and experimental and that for the present the Foreign Office must to a great extent feel their way'. It was noted that 'calling attention at once to statements relating [to] foreign affairs which required contradiction or correction' was an essential prerequisite for diplomacy, and assurances were given by the Office that propaganda of the 'Corpse Factory' type would be avoided. Matters were left, pending a decision of the Chancellor of the Exchequer.[188]

On 31 May the Treasury gave its reply. While recognizing that it was not 'at the present moment practicable to terminate altogether the system of propaganda and the expenditure which it involves', the Treasury did expect either its future discontinuance or gradual conversion 'from a purely political aspect to largely commercial lines', and subsequent transfer, to a great extent, to the Department of Overseas Trade. Cultural propaganda was taken as being already out of place. The approved budget was reduced by £20,000.[189] Alterations were made in the planned structure and functions of the new creation: the proposed Facilities Division was done away with; its Press functions were taken over by the Private Secretary to the Secretary of State[190] until spring 1921, when they were placed under a freshly set-up Press section; and some duties were incorporated into the Government Hospitality Fund. The Administration and Cables and Wireless Divisions continued in operation, in a shrunken form, and as an amalgamated section. The first was run by Sir Arthur Willert, former Washington Correspondent of *The Times*, who was recruited to take over the press work by Tyrrell in 1921.[191] The second was handled by Percy A. Koppel, First Secretary, shortly (early in 1921) to succeed Tyrrell as head of the department, until being transferred, in the 1925 reorganization (i.e. the merger of the two sections),[192] to the Dominions Information Department. He was replaced, on 17 November 1925, by Willert, who was appointed a temporary Counsellor in the Foreign Office. The end result was that the Treasury failed to carry the case for elimination of propaganda machinery, but the Foreign Office was required to reduce the organization of the News Department.

Codes and Ciphers

Another post-war reorganization was that of the cipher department. Prior to 1914 the function of coding and decoding telegrams was undertaken in the Parliamentary Department.[193] After the war, however, the Foreign Office found it increasingly difficult to return to the practice whereby cipher work was carried out by regular staff of the diplomatic establishment. For one thing, the workload of its officials had increased greatly, and for another, the number of ciphered despatches had grown too voluminous for the Foreign Office staff.[194]

The obvious solution was to pass the Parliamentary work to another department and man the cipher department with skilled staff.[195] Knowing, however, that the Treasury was against any increase in the number of Foreign Office staff, Tilley suggested that the King's Messengers division of the Chief Clerk's Department – charged with the transmission of despatch bags to and from diplomatic and consular posts abroad – and the clerical staff made responsible for the cipher work during the war, be combined organizationally and functionally. As he pointed out, this would provide the prospective department with the necessary manpower and vary the work of its members, who, in between assignments, would perform cipher work, thereby making it a tolerable career.[196] The subject was brought up before Hardinge, whereupon Cecil Dormer, George Clerk, Montgomery and Russell were detailed to form a committee under Tilley, to enquire into the matter.[197] In its report of March 1919, this committee endorsed the recommendations of Tilley, and suggested that the new cipher department be known as King's Messengers and Communications Department, so as to camouflage its occupation. The other major recommendations of the committee were that the combined service be composed initially of fifteen men, aged between thirty and forty, who served in one of the branches of the army; and that the entrance examination should be after the pattern of the Foreign Messengership examination, but of somewhat more advanced character.[198]

The proposed scheme received Treasury approval on 19 August 1919,[199] a minute on the organization of the Communications Department was issued,[200] and after a period of trial detailed instructions for the new service procedure were issued in October.[201] At the same time the Parliamentary Department was dissolved and the work of editing papers for Parliamentary publication was passed to the Library.[202] Finally, in 1921 the Code and Cypher School, that was responsible for the construction of ciphers for the various departments as well as its other functions, was transferred from the Admiralty to the Foreign Office.[203]

Other Innovations

Other miscellaneous new departments were the Passport Office, which issued and renewed passports for British subjects;[204] the Passport Control Department, which granted visas to foreigners;[205] the Egyptian Department, formed in 1924, to which were transferred Egypt (from the Eastern

Department) and Abyssinia (from the American Department); and the Dominions Information Department, established in 1926.[206] Also, the war-time Historical Section was retained and made the responsibility of the Librarian.[207]

The post-war Foreign Office was administratively and physically different from its pre-war predecessor. Its organization was altered; its responsibilities increased. Also, its home and overseas staffs expanded – the Office establishment had risen from 185 in 1913–14 to 880 in 1925–26;[208] the Diplomatic Service from 200 to 522[209] and the Consular Service from 299 to 413.[210] The close-knit family feeling was diminished. The long-term effects of the reforms on the Foreign Office were far less clear, for the broadening process was a slow one and new recruits took on the ethos of existing staff. The majority were still recruited from the upper classes of the country, mainly educated at public schools and either at Oxford or Cambridge.

2

The Division of Functions between the Foreign Secretary and his Assistants

The Role of Lord Curzon as Foreign Secretary

The appointment of Lord Curzon, on 29 October 1919, as Foreign Secretary gave double validity to the trend revealed during the war, according to which the Foreign Office ceased to be a close-knit family and turned gradually into a bureaucratic department of state. The system according to which every incoming paper was handed over to the study of all the grades, from the Third Secretary upwards, for the purpose of minuting their comments, created group consultation, thus limiting the danger that the Heads of the Foreign Office would be unable to perceive the full implications of one development or another.[1] Moreover, the rational system of formulating and implementing foreign policy included a wide range of preliminary tasks – such as collecting and analysing information about events and trends abroad, defining the various goals, examining the alternative means by which they could be achieved, predicting the likely reactions at home and abroad to each alternative, anticipating the political steps that would follow, etc. Because they were so numerous the tasks could not be left to the sole care of the senior grades.[2] Ironically, despite the modernization of the Foreign Office in the days of Curzon, foreign policy remained vulnerable to errors and inconsistencies. This stemmed from a series of outer factors that will be examined in due course; but it had, also, inner factors, the most prominent of which were the personalities and political views of the Foreign Office officials in general and those of the Foreign Secretary in particular.

Character and Abilities

George Nathaniel Curzon, First Marquess Curzon of Kedleston (1857–

1925), remains an elusive figure. The haughty aristocratic figure, drawn in graphic colours in various memoirs, must be compared with the charming social and somewhat childish man who hid behind it. Although he had a good memory, was proficient in French, Italian and Latin,[3] and excelled in his ability to reduce complicated situations to their principal components and define clearly the essence of the problem,[4] he was more an intellectual than a man of action.[5] It was not that he lacked experience in public administration. A descendant of a family of country gentlemen whose founder came over from Normandy with William the Conqueror,[6] he was raised in the tradition of landowners giving public service to their immediate surroundings. He went through a difficult childhood, in which his governess emphasized the Calvinistic doctrine that success in life was the distinguishing mark of the chosen people. This early experience instilled in him a fervent desire to prove to the world that he was more than just the son of a Derbyshire squire.[7] He was educated at Eton and Oxford; entered Parliament at the age of twenty-seven; became, in 1885, Parliamentary Under-Secretary to the Foreign Office; and from 1898 until his resignation in 1905 he served as Viceroy of India, where he is remembered favourably to this very day for his public works. For the next ten years he was in the political wilderness, but in 1915 he was co-opted to the First Coalition Government of Herbert Asquith as Lord Privy Seal, and in 1916 he became a full member of the War Cabinet of the Second Coalition Government under Lloyd George.[8]

Foreign affairs were not a new field for him. His travels in 1881–8 and 1892–3 around the world; his books on Persia, Central Asia and the Far East; and his preoccupation with antique collecting, restoring old castles, and writing about their history and architecture, gained him wide knowledge in different spheres, on general and international developments and Asiatic matters in particular. Even his opponents saw him to be a man possessing rare knowledge. Thus, for instance, in 1916 Lloyd George observed about him that:

> He had great knowledge – information of a sort which is uncommon amongst British politicians. He knows foreign countries; he has travelled widely. He is dogmatic and often unreasonable but he brings something to the general stock which is very valuable.[9]

But despite this background, Curzon was more a policy selector than an initiator.[10] He had difficulty in envisioning the future,[11] and expected his assistants to submit him suggestions as to what steps should be taken.[12] As a result, his foreign policy was a great disappointment to his colleagues.

There were further inconsistencies in his character which blur an accurate portrait. He was described alternately as strong and weak,[13] able and superficial,[14] unfeeling and sensitive,[15] arrogant and humble,[16] reason-able[17] and controversialist.[18] He was capable of creating an impression of self-control, yet would 'collapse' when the strain became too great. He was almost child-like in his belief that it was possible to determine political moves with mathematical accuracy on the basis of facts and precedents, and that two and two in diplomacy were four. Nonetheless, he was sufficiently intelligent not to trust international institutions such as the League of Nations or the 'diplomacy by conference' system,[19] and more than once showed political shrewdness and the ability to negotiate. Sir Andrew Ryan, the Chief Dragoman of the staff of the British High Commissioner at Constantinople, said of him:

> He shone particularly in the kind of oratory required at an international conference. It was an intellectual joy to sit among the experts behind his chair and to see with what consummate artistry he handled the points they suggested to him. He did not snatch at them crudely . . . but rejected some and wove the rest into an ordered theme delivered in an exquisitely modulated voice.[20]

And, indeed, this was perhaps his main forte.

Working Methods

Curzon inherited something of the capacity for work of Robert Castle-reagh and George Canning.[21] Despite the large number of incoming papers to the Foreign Office, Curzon insisted that before taking any decision he should see the maximum number of relevant files. He would spend long hours minuting the jackets of the files, drafting memoranda and writing letters, although he suffered from spinal weakness and had difficulty writing.[22] He met his daily schedule despite his health problem, which often made him very weary, and even though he had a greater bur-den of work than his predecessors. In illustrating his daily programme, Curzon complained to Hardinge that Cabinet meetings took his morning hours, and a whole series of further duties the rest of the day:

> many of which I haven't time to record in the afternoon – the House of Lords now and then turning in – occasional speeches at public dinners. Of the ceaseless stream of boxes – each pregnant with a crisis till 3 a.m. – are too much for any man – all the more that I so rarely get a holiday.[23]

Some ministers solved the workload problem by passing down some of the files to the sole care of their assistants. But Curzon was not good at delegating functions despite his dependence on his assistants' advice. He had no faith in the ability of anybody to summarize properly the contents of a political memorandum or do justice verbally to his views.[24] Moreover, he had a compulsive impulse to work. The hours in-between sessions of the Lausanne Conference were utilized by him for writing his memoirs.[25] And it is a fact, and not legend, that at the height of the Ruhr crisis he sat up to the early hours of the morning carefully adding up his domestic accounts or writing indignant memoranda on the incidence of cost in the matter of the coal-shed at Hackwood.[26]

His relationship with his senior officials did not reflect his technical dependence on them. Against the strengthening position of the Foreign Secretary's counsellors on foreign affairs, in the light of the bureaucratization of the Foreign Office, Curzon took pains to ensure that his authority would remain unchanged by examining their work minutely and keeping them from acting on minor issues as they saw fit.

As a rule, in government, officialdom does its best to ensure that it knows its minister's mind; sectoral and low questions are decided and implemented below ministerial level in order that the Secretary of State be able to concentrate on truly important issues without being submerged in day-to-day detail.[27] But when the Secretary to the Cabinet, Sir Maurice Hankey, proposed to the Permanent Under-Secretary at that time, Crowe, to transfer the League of Nations section from the Cabinet Secretariat to the Foreign Office, the Permanent Under-Secretary had to admit that he was not authorized to discuss it and he could not say what were Curzon's intentions.[28]

Curzon not only refused to delegate responsibility to his subordinates, but also would not treat them as equals. As one who demanded perfection from himself and from others, he watched over their work minutely, and when mistakes occurred occasionally – such as printing despatches on flimsy paper,[29] non-inclusion of all the required papers in the files submitted to him, misprints in outgoing despatches,[30] non-expression of a stand by one official or another,[31] and forgetting to draw the blind in his room in the afternoon hours – he became irritable and offensive.[32] With strangers he endeavoured to create an impression of a calm, and even amiable figure. But in his contacts with his staff he was far from being cool-headed; he avoided giving praise on the grounds that 'if you praise your butler for the state of your plate, there is a scratch on your best salver next day'.[33] His nervousness and tantrums, which stemmed to no little degree from his physical pain, created an electrified atmosphere.[34] While it is true that the ongoing rationalization at the Foreign Office created its own tensions in the work of its staff, Curzon's leadership attitude affected morale,[35] and undermined confidence.[36]

Policies

In his political ends Curzon was the symbol of tradition. Although a good deal of his and the department's time was devoted to various global problems, top of his order of priorities – against the background of the collapse of pre-war arrangements – was the necessity to build a new political system that would secure the territories of the British Empire. According to him there were three British Empires: first, the African Empire, populated by black barbarians, whose position was pretty stable; second, the established part of the Empire, the White Dominions that made Britain the greatest colonial power in the world; and third, the most vulnerable part, to which he devoted most of his attention, the Eastern Empire that stretched from Suez to Hong Kong.[37] At the centre of the latter Empire stood India, and it was axiomatic for him that what ought to guide British policy in the Middle and Near East was the traditional necessity for an Asiatic Pax Britannica which would secure Central Asia, India and interests in the Far East, such as Hong Kong.[38]

But if his ends were old-fashioned then the means by which he endeavoured to achieve them were, in the Near East at least, pretty modern. As an alternative to the informal co-operation with the Ottoman Empire, which placed itself on the side of the Central Powers, and to the Anglo-Russian Convention of 1907, which divided Persia and was abrogated during the war, Curzon desired to utilize the doctrine of self-determination in order to establish, in the Caucasus, Trans-Caspia, Persia and Mesopotamia, independent states that would act as a buffer between India and any hostile element. According to him a British initiative in this direction would not only ensure that these states be pro-British but also put an end to the Sykes–Picot Agreement and to the Russian presence, thus securing the monopoly of British influence in the region.[39] For this purpose, and on account of the awareness that some time would elapse before these buffer states were able to protect themselves, Curzon demanded that the British presence remain in the Caucasus, southern Mesopotamia and Persia;[40] recommended that Britain take on the mandate in Palestine, on the grounds that it protected the Suez Canal;[41] and signed, on his own initiative, on 9 August 1919, the Anglo-Persian Agreement in which Britain undertook to respect the independence and integrity of Persia and restore her economy.[42]

In the Far East, Curzon sought to secure British local interests and prevent a threat to India from this flank by maintaining collaboration with the traditional ally in the region, Japan. This position stemmed not from his estimation that, being of a higher standard than other coloured races, the Japanese were natural partners to all that pertained to the execution of Britain's divine mission in Asia.[43] Curzon perceived that, should the Anglo-Japanese Alliance be abolished, the British would lose one of the main mediums that had enabled them, hitherto, to temper Japan's ambitions, and that they would no longer be able to prevent her from pursuing an aggressive policy towards China.[44] Another advantage which he saw in the Alliance was that if necessary it could constitute a counterbalance to a Russo-German alliance in the Far East.[45]

Finally, so as to remove any renewed Turkish threat, Curzon held that they ought to dismantle the Ottoman Empire, turn the Turks out of Europe, and deprive them of control of the Straits (which would enable Britain to create discord among the nations in the area and expel the Sultan from Constantinople – the historical symbol of the rule of its owner over the Moslem world – and thus to cancel one of the main foundations for the right of the Sultan to be the Leader of the Faithful).[46] In order that the Turks would resign themselves to this policy manipulation, Curzon proposed to give them sovereignty over Asia Minor, thereby satisfying their national aspirations.[47]

Inevitably, the question arises as to what place Europe took in Curzon's outlook on the world. His position was that the economic recovery of Europe in general, and Britain in particular, was dependent on the succession of the conditions of war by conditions of peace. For Curzon, the re-establishment of Germany as a stable state in Europe was one of the main essentials for peace; he noted that 'any idea of obliterating Germany from the community of nations or treating her as an outcast is not only ridiculous but is insane'. He even favoured British involvement in the mainland of Europe for the purpose of safeguarding Britain's firm position and thwarting any attempt to undermine the peace.[48]

Nevertheless, he did not consider the problems of the Continent as comparable in importance to the security of India and her communications. It appeared inconceivable to him that Lloyd George could risk a general explosion in the Middle East in order to compensate France for making concessions in Europe. In his opinion:

> Germany though assuredly destined to recover, cannot for many years be a military danger to Europe, or even to France alone.[49]

He mentioned this as early as 2 December 1918.[50] Instead of concentrating on secondary issues, such as the Rhineland and reparations,[51] Britain ought, in his opinion, to place at the top of her priorities the creation of a chain of buffer states in the Near East, and to compensate, elsewhere, those nations that were ready to make concessions to Britain in Asia. In particular he opposed giving France a foothold in the former territories of the Ottoman Empire. He told the Eastern Committee:[52]

> I am seriously afraid that the great Power from whom we may have most to fear in the future is France, and I almost shudder at the possibility of putting France in such a position.

Finally, Curzon renounced, at least in theory, the 'splendid isolation' policy in Europe, and defined co-operation with France as the pivot around which European peace revolved.[53] However, in practice, he was opposed to an alliance with France or with Belgium,[54] and preferred an informal collaboration that would not tie Britain to the Continent,[55] and that at the same time would enable her to restrain France.[56] This was an old-fashioned order of priorities that caused, more than once, frictions between him and the French, and also between him and his colleagues.

Europe was not the only region where his calculations were to prove erroneous. Thus, for instance, he did not appreciate correctly how limited was Britain's ability to enforce arrangements favourable to her in the Near East; he did not anticipate objective difficulties, such as the refusal of Persia to avail itself of Britain in order to consolidate independence; and he did not take into account the positions of his colleagues and those of other Powers which had interests in the region. The result was that, up to the end of 1922, his plans regarding the Eastern Empire collapsed like a house of cards.

Assistants and Advisers

The knowledge that he had no alternatives to his original plans, or ability to find solutions to international crises on his own, forced Curzon to turn to his assistants. Foreign policy had become so complicated and the volume of information regarding it so great, that it was beyond the scope of a single human mind to apprehend the full implications of one development or the other. He recognized that he now had to rely on group analysis. All incoming papers were passed down to be studied by all grades, from the Third Secretary upwards, in order that they would minute their comments, in this way limiting the danger of errors of judgement.[57] Officials who in the nineteenth century performed clerical functions now administered the Foreign Office, consulted the neighbouring departments, related between incoming pieces of information and previous developments,[58] and narrowed the range of considerations down to one or two broad alternatives.[59] The increasing bureaucratization of the Foreign Office, and Curzon's insistence on being consulted in routine matters as well, did not, it is true, bring about a growth in the scope of the formal powers of the officialdom. But this did not eliminate, altogether, the existence of a certain scope for discretion, or the fact that, informally, officialdom had a far-reaching influence upon the Foreign Secretary.

The individuals who served Curzon, MacDonald and Chamberlain differed from each other in their personalities and abilities. In Curzon's first year at the Foreign Office a change of guards had taken place:

Graham was appointed British Minister to the Hague; Hardinge, who twice acted as Permanent Under-Secretary, retired in 1920 to the Paris Embassy. Their places were taken by Crowe and Tyrrell, who longed to turn the Office into a professional body that monopolized advising the government on foreign affairs. As a result, First and Second Secretaries became specialists to whom the heads of the departments and the Foreign Secretary turned with questions such as the state of disarmament and reparations. They also drafted, on their own initiative, memoranda on various subjects, in an open attempt to influence their superiors, and although their memoranda were not adopted automatically, they were read attentively by the senior officialdom.

Hardinge, Crowe and Tyrrell, like other less prominent personalities, provided an excellent reservoir of talent for the Foreign Secretary. Their judgement regarding Europe was sound and their advice always practical. The vast experience of this trio, the greater part of whose pre-war career had been devoted to finding solutions to the German threat to the position of Britain,[60] convinced them that Germany remained a potential adversary. As early as 7 August 1916, Tyrrell, together with Sir Ralph Paget, an Assistant Under-Secretary, referred to the possibility that Germany might remain the country that threatened to upset the balance of power after the war.[61] Hardinge believed, as he wrote on 28 April 1921 to Curzon, that France was right to fear Germany, for 'Germany is like a leopard which cannot change its spots, certainly for some years to come'.[62] Crowe, for his part, went so far as to assert in December 1921 that Germany had not given up her plans for waging a war of revenge against the Allies,[63] an opinion shared by Tyrrell.[64] Even the Second Secretary in the Central Department, J. M. Troutbeck, became convinced, following his jaunt of May 1924 in Germany, that the day was not far off when Germany would re-endanger Britain no less than in 1914. According to him:

> The General Staff made a pertinent observation when in a recent memorandum they spoke of France as being a buffer between this country and Germany, and it needs no great stretch of imagination to foresee the day when Germany will buckle on her armour once again to make the world safe for monarchy.[65]

The German menace did not seem to them immediate,[66] but they were sufficiently sensitive to see, in the non-execution of various Articles of the Peace Treaty by the Germans, proof of true intentions. They noted the significance of the consistent German attempts to conceal arms and maintain military frameworks under the guise of civilian bodies, and interpreted the delays that took place in the reparation payments, and the constant fall of the mark, as an attempt by German industrialists to hoard wealth and money, avoid tax-payments and bring about the cancellation of the reparation payments.[67] In their opinion, cancelling the reparation payments and granting a financial loan to Germany created a danger that, thanks to her inflationary policy, Germany might become the only state in Europe to enjoy the advantages of remission of external and internal debts.[68]

To what extent did the trio's apprehensions determine what would be British policy towards Germany? Policy was ultimately decided by the Foreign Secretary and the Cabinet; moreover, the broadening spheres of foreign policy turned the neighbouring departments into active partners in all that pertained to the formulation of policy towards Germany. Thus the trio was kept, from the beginning, from concentrating on the complex of German issues and forcing positions on this point. If, however, it is taken for granted that the top operated within a defined area of responsibility, then by examining the personalities of the individuals that staffed it, their views and functions, some generalizations about the policy-making process in the Foreign Office can be reached.

The Role of the Permanent Under-Secretary

The most senior official in the Office was the Permanent Under-Secretary. His duties were to advise the minister in all that pertained to foreign policy[69] and to ensure that the Foreign Secretary went into Cabinet and Parliament properly briefed in terms of what the Office thought he ought to know.[70] He was responsible to the Foreign Secretary for the administration of the Diplomatic Service and the smooth functioning of the Foreign Office. As chairman of the Promotion Board it was his function to make recommendations to the Foreign Secretary on all senior appointments and promotions. Up until 1922 he was also the Accounting Officer, personally responsible to Parliament for the proper employment of funds provided for the Foreign Office. A good deal of the papers dealt with by the Permanent Under-Secretary went to the Foreign Secretary, and he therefore had to make sure that the papers were in a fit state to be placed before the minister for a decision, and, if not, to put them right. He had to make it his business to know where the minister was to be found, day and night, and to see to it, should he not be in the vicinity of the Office, that the various files were sent to him for his scrutiny. The Permanent Under-Secretary's orders of the day included dealing each morning with files that came in during the night; meeting British ambassadors and foreign dignitaries, and attending departmental and inter-departmental committees and Cabinet meetings. The primary aim of his efforts was to lighten the burden of work on the Foreign Secretary so that he could concentrate on truly important issues. Finally, his duties included dealing with files which amassed during the day and began to flow in the evening hours, thus obliging him to take the files to his home and work until after midnight.[71] Altogether, it was a most loaded programme.

Charles Hardinge

To perform this role competently required special qualifications and vast experience. Charles Hardinge, First Baron Hardinge of Penshurst, came to the post with all the advantages of a long career. He was familiar with Foreign Office administration, had sound judgement,[72] was specialized in high policy matters, and always formulated his advice with the politically possible in mind.[73] Although Curzon's contemporary, Hardinge was more experienced than him in diplomacy. In the course of his career he had served in Constantinople, Berlin, Washington, Sofia, Bucharest, Teheran and St Petersburg; between 1906 and 1910 he had held the post of Permanent Under-Secretary at the Foreign Office, a period in which Foreign Office influence reached its peak; and in the six years after that he had acted as Viceroy of India.[74]

Nevertheless, after 1916 Hardinge found it difficult to function as Permanent Under-Secretary. There were three principal reasons. First, he found the changes that had taken place in the Office too sharp for him. The growth in Foreign Office staff meant that it was no longer possible to know everyone and his work, as he used to in the past.[75]

The broadening foreign policy spheres forced him to familiarize himself with new fields, such as economy and finance, in which he had no interest.[76] The increased paperwork was a burden to which he was not accustomed, and made it difficult for him to implement the bottle-neck system that he had devised, according to which all papers intended for the Foreign Secretary went through him, and which contributed so much to the magnifying of his personal influence. On 9 April 1920, he confessed:

> As for myself, I am tired from overwork, but very well in health. By June I shall have completed forty years [of] service, and I think the time is approaching for me to hand over my job to a younger man. The F.O. is a very great tie.[77]

Secondly, it was unfortunate that his return to the Foreign Office coincided with the transition from the methods of the old diplomacy to the methods of the new diplomacy. The increasing intervention of outer elements, such as the Prime Minister and 'Garden Suburb' in matters such as ambassadors[78] and advising the Government on foreign affairs,[79] gave him the feeling that the diplomatic functions and traditions of the Foreign Office were being undermined by amateurs, a feeling that reached its climax during the Paris Peace Conference, where he failed to have the central role he anticipated. Hardinge predicted that the Peace Treaty, as framed by the 'Big Four', was bound to cause disastrous results.[80] This episode repeated itself, according to him, at the beginning of 1920 when, during the negotiations on the framework of the peace treaty with Turkey, they ignored Near Eastern experts like himself and instead gave weight to the views of cranks and enthusiasts.[81]

Finally, and perhaps most important of all, there was Hardinge's shaky relationship with Curzon, who had not forgiven Hardinge for reversing the partition of Bengal in 1911, a policy act of which Curzon was proud. Hardinge, for his part, was hurt following the critical report of 1917 on the manner with which he conducted the Mesopotamian campaign. Curzon had asked him, in the name of the War Cabinet, to resign in order to 'ease the situation of the Government and to avoid hostile criticism of the Foreign Office in future'.[82] Hardinge's refusal to give up his office annoyed Curzon.[83] A constant source of irritation to Hardinge was Curzon's haughtiness and criticisms about the contents of files passed down for scrutiny.[84] 'G.N.C. is very impossible, and having served under Lansdowne, Grey and Balfour I feel the difference. However, I hope it will not be for long', wrote Hardinge to Ronald Graham on the eve of the San Remo Conference.[85]

It seems that the grudge was borne more on the side of Hardinge, who found himself subordinate to a man whom he detested, than on the side of Curzon. In his grievances towards Hardinge, Curzon's attitude was no different from the attitude he revealed towards the rest of officialdom. And in his own way, Curzon even saw in Hardinge the only person who approached him socially, for both were ex-Viceroys of India, both were peers, both had a soft spot for India, and both had similar ideas regarding the policy that ought to be taken by Britain in the Near East. This special treatment was manifested particularly after Hardinge became British Ambassador to France. Curzon commenced writing him frequent letters, in which he poured out his heart's discontent and again and again sought his advice.

Nor should it be forgotten that after 1916 Hardinge was by no means the Hardinge of 1906–10. He had been badly shaken by the assassination attempt made on him in India in 1912, followed in 1914 by the death of his wife and the death of his eldest son in battle. He had always tended to be imperious and his Indian years exaggerated this tendency. The transition from being Viceroy to being subordinate to the will of others required a considerable psychological adjustment, difficult for a man of Hardinge's temperament.[86]

Hardinge himself recognized it, and a short while after his return he came to a decision not to prolong his stay at the Foreign Office over the war period.[87] At the end of 1916 he turned to Grey and received a promise that he be appointed Britain's next Ambassador to France, after the retirement of Francis Bertie.[88] But Bertie was in no hurry to leave the post, and when he was forced to retire, by Lloyd George, the Prime Minister insisted that his successor be one of his confidants, Edward G. V. Stanley, Seventeenth Earl of Derby. Consequently, Hardinge remained, for the time being, at the Foreign Office and waited patiently for his turn.[89]

Hardinge's main areas of concern were the Northern and Eastern Departments, in which Curzon had a special interest. Unlike Curzon, who was relatively indifferent to Europe, Hardinge did not refrain from formulating, already during the war, a number of ideas about the desirable arrangements in post-war Europe. In his opinion, any arrangement had to involve the continued independence of Holland and Belgium in order to secure the coast-line opposite the British Isles.[90] In Central Europe, he did not see the dismantlement of the Austrian Empire as something serving British interests. He did not believe in the ability of national elements in Central Europe to function as independent states,[91] and thought that the dismantlement of the Empire would prevent the attainment of one of Britain's war aims, the creation of a barrier to the German *Drang nach Osten*. Consequently, he did not advocate the full dismantlement of the Austrian Empire.[92] The Treaty of Versailles failed to incorporate these ideas, much to Hardinge's bitter disappointment. In Western Europe, he urged Curzon to found the balance of power on the *entente cordiale* with France.[93]

If Hardinge was consistent in his belief that Germany remained a potential menace, then his ideas regarding the means by which it ought to be neutralized were vague, and at times even fraught with inconsistencies. He agreed that in order to attain viable peace German militarism should be uprooted,[94] but he did not accept the cold logic of the French that this necessitated the non-recovery of Germany. Only years later did he admit that this was a mistake.[95]

He thought that in order to put an end to German ambitions in Western Europe, she should be weakened by the amputation of Alsace-Lorraine.[96] At the same time he rejected completely the detachment of the Saar and Silesia on the grounds that it would prevent Germany from functioning as a political entity, an indispensable condition for the creation of any European balance of power.[97] British policy, in his opinion, ought to have been 'to show a united front to the Boche, and to act as the closest ally of the French'.[98] But he opposed turning the *entente cordiale* into a formal alliance,[99] which would act as the policeman of Europe,[100] and even toyed with the idea of turning Germany into a buffer to Bolshevik Russia.[101]

Similar inconsistencies were revealed in his attitude towards the reparations problem. On the one hand, he sympathized with France's position that the fulfilment of the Articles of the Treaty of Versailles on this point constituted a test of the credibility of Germany and that she must compensate the Allies for the damages caused them during the war.[102] On the other hand, given the figures provided to him by the Treasury, Bradbury and the Central Department officials, he could not see how Germany would be able to pay reparations. Thus, for instance, in July 1920 he approved the conclusions of Sidney P. O. Waterlow, the First Secretary of the Central Department, who asserted that the proposed amount of reparations was above and beyond what Germany was capable of paying, and that if it was not moderated it would create a chain reaction beginning with an internal economic crisis in Germany and ending with a *coup d'état* by the Right. The result, according to Waterlow, would bring about a Communist counter-revolution and the creation of a German–Russian bloc that would genuinely endanger Western Europe.[103]

According to Hardinge there was but one way of uprooting German militarism, and that was to disarm Germany. Once this was achieved, and the danger of militarism was removed, he saw no reason why Britain should not do its utmost to help them improve their economic state, and incidentally restore Europe's economy.[104] As in the pre-war period, Hardinge still thought in terms of a free hand. At most, and for the purpose of overcoming political obstacles, he was willing to consider giving France a guarantee against German aggression. A limited commitment such as this, together with the *entente cordiale*, would constitute, in his opinion, a security insurance policy and win French support for Britain's policy in the Middle East without unnecessary British involvement in French wars. It would dispel, in the long run, France's apprehensions of Germany, thus making it possible to moderate the Reparation Articles and to start the economic restoration of Germany and Europe. From his viewpoint, there was another ominous aspect of the question to be considered. Hardinge was aware that the main impulse in French political and military policies was inspired by her fear of a German war of revenge.[105] But he could not help wondering whether French sanctions, such as the occupation of German cities, would not give birth to expansionary aspirations,[106] which would badly affect the European balance of power. Another issue that troubled him was the identity of the nation against which the French submarine and aeroplane fleets were being constructed, was this not Britain, given that the German navy and air force had ceased to exist with the ending of the war?[107]

Hardinge's views were interpreted by Curzon as an affirmation of his own position, according to which the desirable state was minimal British commitments – as opposed to involvement – in the Continent of Europe.[108] Moreover, Curzon took satisfaction in the fact that in contrast to the disagreement which existed between them concerning the importance that should be attached to European problems, Hardinge fully supported the Near Eastern policy proposed by Curzon. Hardinge affirmed Curzon's assertions that the liquidation of even one threatened region from British presence, before it had a chance to consolidate its independence, would lead to its falling to the Bolsheviks, thus initiating a chain reaction that would bring about the downfall of other neighbouring regions up to the fall of India.[109] He approved Curzon's position that, even though the Ottoman Empire ought to be dismantled, Asia Minor should not be divided among the Allies and Greece, as was done by the Powers in the course of the Paris Peace Conference and after it.[110] And he criticized the Cabinet's decision of 6 January 1920 not to expel the Sultan from Constantinople.[111] Hardinge stuck strongly to these beliefs and his decisiveness undoubtedly encouraged Curzon to continue to fight for their application, despite the opposition he met in the Cabinet.

The Assistant Under-Secretaries of State

Apart from Hardinge, two other men in the Foreign Office attained positions of power: they were Sir Eyre A. B. W. Crowe and Sir William G. Tyrrell. As Assistant Under-Secretaries of State it was their function to handle routine matters on their own and to simplify as much as possible the material reaching the Permanent Under-Secretary and Foreign Secretary. Each superintended the work of several Heads of Departments and was responsible in his own sphere for elaborating political recommendations and co-ordinating political moves with the other concerned parties. Their recommendations were usually forwarded to the Foreign Secretary through the Permanent Under-Secretary, with whom they consulted before forming their final positions. This chain of command, however, did not deny them a good deal of authority. The burden of work of the Foreign Secretary and Permanent Under-Secretary was so overwhelming that they were forced to share with the Assistant Under-Secretaries of State all that pertained to the co-ordination of policy with other departments, attendance at inter-departmental committees, interviews with various people, etc. Moreover, it happened, more than once, that the Permanent Under-Secretary was hard-pressed or otherwise occupied. And in that event the Assistant Under-Secretaries of State reported directly to the Foreign Secretary, thus creating an immediate link between the two grades and requiring them to know the minister's mind no less than the Permanent Under-Secretary.[112]

Crowe and Tyrrell were one of the most successful combinations that ever arose in the Foreign Office, not because they worked as a team but because the one was the opposite of his companion. Each complemented the characteristics of the other. Harold Nicolson, who worked with both of them, says:

> Sir Eyre Crowe believed in facts; Sir William Tyrrell believed in personal relations; the former relied upon lucidity; the latter upon atmosphere; the minutes of Sir Eyre Crowe were precise and forcible; the conversations of Sir William Tyrrell were intangible but suggestive; the former concentrated his energies upon penetrating the matter in hand without regard to collateral contingencies; the latter, who kept aloof from the machinery of office life, excelled in examining the outer radius of international problems.[113]

Another distinction which all those who knew them remarked upon was that Crowe was the embodiment of the perfect civil servant, while Tyrrell was the symbol of the non-conventional diplomat. Crowe, a bureaucrat in the broadest sense,[114] devoted himself wholly to the work of his office and held in suspicion and distrust the rest of the members of the political establishment, whom he considered as meddling in a sphere not theirs. He was both loved and admired by his subordinates.[115] Tyrrell, on the other hand, cultivated the laziness which Talleyrand enjoined on diplomats,[116] and was more gregarious. He went everywhere and knew everybody; he had a wide range of acquaintances in and out of the various departments of state, and what he did not know about what went on behind the scenes was not worth knowing.[117]

It is surprising that Crowe, an agnostic, and Tyrrell, a Roman Catholic, should have worked so well together and been such good friends, but such was the case. Tyrrell paid frequent visits to the house of Crowe,[118] and the latter accepted without demur the judgement of Tyrrell in those spheres in which he excelled.[119]

Sir Eyre Crowe

The biographies of the two are a fascinating story in themselves. As the third son of Sir Joseph Crowe, the Commercial Attaché for all Europe, and Asta, daughter of Gustav von Barley and Eveline von Ribbentrop, Crowe was born, on 30 July 1864, at Leipzig.[120] He was educated in Germany and France, and only at the age of seventeen did he first set foot on English soil in order to cram for the Foreign Office examination, which he passed successfully in 1885. From the start, the distinction between him and his contemporaries was evident. While his contemporaries passed the time dealing with routine matters, Crowe devoted the first ten years of service as Resident Clerk to exploring past records and reading all the despatches entering or leaving the Foreign Office. In his spare time he made an intensive study of modern European history; read avidly, books on economics, the military, classic and contemporary fiction, subjects in different languages, and thus acquired a rare knowledge of the essence of modern diplomacy. His annual bibliographies are astonishing in their range and variety of subjects. The diversions of his snobbish contemporaries had no appeal for him. Instead, he gave himself to playing the piano, composing music and participating in the training of the First Volunteer Battalion of the City of London. It was typical of this extraordinary figure that he should volunteer for active service during the Boer War, but he was turned down after having failed to pass the medical examination.

First in the Consular and then in the African Department, Crowe impressed his superiors with his industry and qualifications to such an extent that, though still only an Assistant Clerk, he was asked in 1905 to join the group preparing the reforms of the Foreign Office. The decisive role played by Crowe in the introduction of the reforms increased his reputation for brilliance and, in 1906, brought his promotion to Senior Clerk and appointment as Head of the Western Department.

Crowe became an active partner in formulating Office policy towards Europe. His analytical ability and knowledge gave his minutes and memoranda an authoritativeness unusual for a person of his grade. Another factor that made him an expert on Europe in general and Germany in particular was his correspondences with senior members of the German establishment, such as Admiral von Hozendorff, the Commander of the German High Sea Fleet (1909–13), to whom he was related through his mother and cousin Clema, widow of Eberhardt von Bonin, whom he married in 1903.[121]

Against this background it is not surprising that of all the Foreign Office members it was Crowe who drafted, in 1907, the famous 'Memorandum on the Present State of British Relations with France and Germany', in which he traced the history of those relations, defined Britain's political and strategic interests, and showed that the trend of German policy might ultimately bring about the military domination of Germany over Europe. However, this did not make Crowe *éminence grise*, for Crowe was one amongst many who were concerned with Germany and although his razor-sharp analyses were read with great attention by his superiors, they were not adopted by them automatically, and more than once were even rejected.[122]

The advance of Crowe on the hierarchic ladder, albeit swift, was not achieved the easy way. His German origins, interests and inability to benefit from connection to a titled family or old school ties so indispensable in the Foreign Service, made him an outsider. He was not admitted into that select circle from which was composed the Foreign

Office hierarchy before the war, whose members corresponded among themselves, exchanging ideas, thoughts and gossip, on the basis of which the foreign policy was formulated.[123] Hardinge in particular, while recognizing that it was only a matter of time, disliked the thought of his ever becoming Permanent Under-Secretary. 'Much as I admire Crowe's ability I shall be sorry if he becomes head of the Foreign Office. It will lower the prestige of the office, as he is so palpably German and his wife quite unrepresentable. Further, I mistrust the soundness of his judgement', wrote Hardinge, on 8 June 1913, to Sir Arthur Nicolson.[124]

The wish of a small minority to see Crowe removed from the centre of power was almost fulfilled with the outbreak of the First World War. The failure of the government to recognize that this was a total war that would require the mobilization of all British resources in order to win it, on the one hand, and the continued business-as-usual policy towards Austria and Turkey, and the avoidance of destroying the German fleet in the Far East for fear that it would affect Anglo-Japanese relations, on the other hand, were received by Crowe with clear impatience.

In the autumn of 1914 Grey and Crowe clashed not only on major policy issues, but also on the need to develop an economic warfare machinery, on which Crowe was an acknowledged expert. He was transferred at the beginning of 1915, to the relief of the élitists, from the War Department to the Contraband Department.[125] At the same time Crowe fell victim, in 1915–18, to an anti-German wave which washed over the press and led to repeated demands for the investigation of the pro-German tendencies of persons with a German background, such as Crowe and Prince Louis Battenberg. So serious was the campaign against him that both Cecil and Grey had to defend him in Parliament, and this meant that the question of his advance was taken off the agenda so long as the war continued.[126]

Nonetheless, Crowe was too gifted to disappear off the map. Instead of the Contraband Department becoming for him a wilderness, it grew, in part, through his industry and talents, into the Ministry of Blockade, which contributed decisively to the war effort. The appointment of Cecil as Minister of Blockade and Deputy of the Foreign Secretary won Crowe a new patron. Cecil considered him a key-man at the Foreign Office.[127] Crowe's position was finally rehabilitated with the opening of the Paris Peace Conference. Because of his many abilities, members of the British Delegation to the Conference considered him their leader, and the one responsible for co-ordinating their actions,[128] though officially, of course, Hardinge was Head of the Delegation. Derby[129] and Herman Norman, the First Secretary,[130] sent the Foreign Office enthusiastic reports on the manner with which Crowe represented Britain in the Supreme Council, convened after the conclusion of the Peace Conference for the purpose of formulating the final details, and how he managed by the force of his personality and knowledge to dominate that assembly and make it accept proposals suitable to Britain. Harold Nicolson, who was also included on the British Delegation to the Peace Conference, wrote reverently:

> Immediate to me and incessantly controlling, this man of extreme violence and extreme gentleness almost became an obsession. He was so human. He was so superhuman . . . 'Crowe', said Clemenceau (who had an eye for value), 'c'est un homme à part.' . . . It is difficult to speak of Crowe without lapsing into the soft ground of sentimentality. Yet here, if ever, was a man of truth and vigour.

For these reasons, no doubt, Nicolson decided to dedicate his book on the Peace Conference to Crowe.[131]

In September 1919, Ronald Graham, Hardinge's hoped for heir apparent, was appointed Britain's Minister to the Hague. It seemed, by this appointment, that the way was finally paved for Crowe to succeed Hardinge as the next Permanent Under-Secretary at the Foreign Office. But, as it quickly turned out, this was not quite the case. The petition of Curzon to the Prime Minister that he should give his sanction to the appointment of Hardinge as Britain's next Ambassador to Paris, and to the appointment of Crowe as Permanent Under-Secretary, met with sharp opposition on the part of Lloyd George.[132] Lloyd George, who distrusted the Civil Service in general, wanted to continue appointing to the Paris Embassy one of his own protégés, and as Permanent Under-Secretary in the Foreign Office he wanted Sir Maurice Hankey, the Secretary to the Cabinet. At the same time, he never ceased, for some reason, denouncing the manner with which Crowe represented Britain in the Supreme Council. Only after a long struggle, assisted by the King's veto on the transfer of Hankey from his present post,[133] did Curzon manage to secure the desired appointment,[134] and, on 27 November 1920, Crowe took the place of Hardinge, who went to Paris as Permanent Under-Secretary to the Foreign Office.

Immediately upon entering his post, Crowe set about rationalizing the division of functions at the top, so that the Office would function more smoothly. The increase in the number of international crises, the growth in the number of independent states, the broadening of the spheres of foreign policy, the increase in the range of incoming and outgoing papers to the Foreign Office, and the speed with which international developments took place, convinced him that the number of grades dealing with any problem, before bringing it to the Foreign Secretary, would have to be curtailed if the matter to hand was to be executed speedily. He therefore abolished Hardinge's bottle-neck system and laid down that the Assistant Under-Secretaries of State henceforth send their papers direct to the Foreign Secretary.[135] Crowe endeavoured to improve the legibility of handwriting and demanded that his subordinates formulate their meaning clearly. He observed that writing froth signified, in the case where a writer meant it, rot, and in the case where he did not mean it, the wasting of the Permanent Under-Secretary's time.[136]

Anyone who saw Crowe in action could not help being impressed by his prodigious industry and knowledge. His order of the day was to work in the morning in the Foreign Office, a short break for luncheon at his club, and to return home, usually at some ungodly hour, for dinner.[137] He read the copies of almost every incoming and outgoing despatch pertaining to his department,[138] and there is hardly a major memorandum which does not carry his minutes, which were always relevant and formulated lengthily and clearly. His minutes and memoranda, particularly those pertaining to Europe, are staggering not just because of their number, but also because of their very content, which combines historical and strategic analyses, and an ability to see the situation for what it was. Once, on the eve of the Conference of Geneva (20 June – 4 August 1927), one of the Under-Secretaries asked Sir Austen Chamberlain, the Foreign Secretary, what he should do if the Germans tried to steal a march on them; 'read Crowe's minutes to them' was the reply.[139]

His personality, too, captured the hearts of many of those who served under him. Some, it is true, accentuated in their memoirs his violent temper and Prussian mentality.[140] But all of them stressed his personal charm. In one of the finest descriptions about Crowe, the then First Secretary of the Northern Department said:

> Anyhow, we always got on very well, and I loved him dearly. What I remember best is not his sensibility and good nature though these were enough to make him a good friend; not his intelligence and flawless integrity though this would have made him a valued member of any society. There was besides these things such a heat in his spirit that knowledge of history and contemporary politics, acute judgement and power of exposition ran together with a kind of incandescence which lit up everything on which his mind and feeling and words were directed.[141]

Maurice Peterson, Second Secretary to the British Embassy at Washington, narrated for his part that his crippled father, in the last few months of his life, wrote to Crowe, whom he did not know personally, to congratulate him on his promotion to be Permanent Under-Secretary. That drew from this awkward, abrupt man a letter of such cordiality and sympathy as have seldom been seen. For this alone, said Peterson, he held his memory dear.[142]

Yet, despite the fact that his administrative attributes, his willingness to place himself at other people's disposal[143] and his regard for the feelings of others, excited feelings of fidelity and admiration, and made him a model for imitation among his subordinates,[144] Crowe, paradoxically, had to work in a new diplomatic environment. The conditions in which foreign policy was formulated and implemented were entirely different from those that had existed till 1914. The broadening of the spheres of foreign policy, and its having become, more than in the past, a subject that pertained to several departments, reduced considerably the monopoly of the Foreign Office and Permanent Under-Secretary over this sphere.

Unfortunately for Crowe, his first three years in office were passed in very awkward conditions. His good nature had to contend not only with his poor health, but also with the tantrums and nervousness of Curzon, who believed that the way to achieve competence was to be high-handed. Instead of praising Crowe he made him the object of his daily complaints about the functioning of the Foreign Office,[145] and deliberately used to ask for him on the telephone from his London residence or Kedleston at times when Crowe could not reasonably have been expected still to be at the Office, instruct that he be fetched from his home,[146] and be plaintive and peevish about it the following day.

> 'Can't the man realize', Crowe used to say, 'that long after he has gone home in his Rolls-Royce, I have to catch a No. 11 bus for Elm-Park Road and sup off sardines or cold sausages before dealing with the evening's telegrams?'[147]

The devotion of Crowe and his ability to meet all the assignments that were imposed upon him were used by Curzon to heap more and more work on his willing shoulders. In the opinion of observers, this accelerated his death.[148]

Since Crowe was not only Permanent Under-Secretary but also super-intendent of the Western and Central Departments, which included Germany, his opinions merit special attention. He was, undoubtedly, the most logical and consistent believer of all Foreign Office members in the need to turn the European balance of power primarily against Germany and to a lesser degree against Russia. As one who had been brought up since childhood to appreciate correctly the sources of strength and weakness of German society, he was convinced that so long as it had the slightest hope, Germany would strive systematically for a war of revenge.[149] The presumption that the various machineries established for the purpose of disarming Germany deprived her of the requisite means to threaten the peace of Europe did not commend itself to him. In order to disprove it, Crowe reviewed the history of European countries' evasions from disarming, underlining the capability of Germany to build an army within the boundaries of Russia. The assumption of the General Staff, Crowe argued, that the First World War had demolished German nationalism and militarism was erroneous; and the view that rehabilitating Germany and refraining from plaguing it would end them once and for all was naïve.

> I think that on the contrary, when Germany returns to more normal and settled conditions, there will be a steady growth of the feeling, traditional in German thought, that national spirit and military strength are the pride of a healthy and self-confident state. There is likely to be a harkening back to the idea of general national service under arms, and the more normal and prosperous the German State, the stronger will become that feeling.[150]

Crowe did not claim that the danger of German recovery was immediate – although he did consider it inevitable – but that Britain ought to take several precautionary steps, which, if carried out, would mean that a long time would pass before Germany was able to re-endanger France and Europe.[151]

These appreciations raised the question of whether it would not be better for Britain to resign herself to the inevitable and make Germany her main ally on the mainland of Europe. To that, Crowe responded with an emphatic 'No'. Resignation to Germany's becoming number one power in Europe meant German expansion at the expense of other countries and the non-fulfilment of the Treaty of Versailles,[152] which in Crowe's opinion were contrary to the British fundamental interest of preventing war in Europe, creating conditions that would allow an increase in inter-European trade for the purpose of restoring to Britain a healthy economy,[153] and opposing the domination of one or a group of anti-British powers on the Continent. The need to insist on the principle of the balance of power was dictated not only by common sense, given the danger that German expansion would eventually affect Britain, but also by the fact that its naval and imperial superiority were tolerated by the various powers, thanks to the endeavours of Britain to preserve the independence of other nations. If the independence of these states was challenged by a single power, it was almost a law of nature that Britain should oppose the aggressor.[154] For better or for worse, added Crowe, European peace was founded on the Treaty of Versailles.[155] If the treaty did not endure, Germany would be encouraged to think that none of its Articles was sacred.[156] The attempt to create an *entente* with Germany, in place of the one that existed with France,[157] would precipitate an Anglo-French crisis, which was not in Britain's interests.

The need to weaken Germany, the dismantlement of the Habsburg Empire and the turning of Russia, with the outbreak of the Bolshevik revolution, into a hostile, loathsome element, removed from the agenda the possibility of restoring the pre-war balance of power. By contrast, the possibility that a chain of Slavic states – Czechoslovakia, Hungary and Poland – be established in Central Europe, which would replace Russia as a counterbalance to Germany and check the Bolshevik revolutionism in the East, seemed a very attractive policy.[158] Therefore, in contrast to Hardinge, as a partial solution Crowe joined those who held, on the eve and in the course of the Paris Peace Conference, that they ought to found the European balance of power, henceforth not on existent states but on national nuclei.[159]

Another solution put forward by Crowe, in response to the suggestion that these states were far too weak to constitute a counterbalance to Germany or Russia, was to abandon the free-hand policy, adhered to by Curzon, in favour of an Anglo-French alliance. Such an alliance, contended Crowe, would consolidate the peace in Europe for a long time to come, since there were not at that moment, in Europe, states capable of opposing this combination; make it possible to embark upon the recovery of Europe; put an end to Anglo-French controversies; and, in the long run, enable some moderations to the Treaty of Versailles.[160]

Crowe was not anti-German in the narrow sense of the word. Politically, he did not favour putting off the economic recovery of Germany or undermining her internal stability, as the French desired.[161] Nor did he consider the Treaty of Versailles an ideal document. Personally, he appreciated the administrative capabilities and cultural achievements of the Germans, noting on one occasion that the world would be poorer but for German ideas, methods and character.[162] But from a review that he made of German past behaviour he deduced that she would attempt to restore the territories that were taken away from her in the Treaty of Versailles, and thought that any change in the *status quo* would eventually affect Britain. Crowe justified the various parts of the Treaty of Versailles not just on the grounds that it was based on the peace, but also on the fact that reparations were necessary for France.[163] If they had to choose, as occurred in Paris, between leaving territories that consisted of a million Czechs in Germany and leaving three million Germans in Czechoslovakia, naturally the choice had to fall in favour of their allies.[164] Therefore it was the duty of Britain to observe that the Versailles arrangements be honoured and the *entente* with France maintained. The Germans understood only strong language and they ought to be treated firmly if the Allies wished to prevent flagrant infractions of the Treaty.[165] The desirable situation for Britain, concluded Crowe, was the present division of powers; if change was, indeed, required, then it would come out of the dynamism of the Peace Treaty.[166]

Crowe was entirely consistent in his attitude towards Germany. He saw in the non-fulfilment of Articles, disarmament, reparations and putting war criminals on trial, a deliberate German effort to break down the Treaty of Versailles. His vast experience and knowledge of German mentality convinced him that there was no substance to the fears that disarming Germany would enable the Communists to make a revolution there without hindrance;[167] or that the reparations policy would bring Germany to the brink of starvation[168] and bankruptcy;[169] nor that should they continue to treat Germany firmly it would lead to the creation of a Russo-German alliance that would forcibly break down the Treaty of Versailles.[170]

It seemed to him that underlining the Communist danger was part of a German effort to persuade the Allies to refrain from disarming Germany;[171] while the grim descriptions about the position of the German economy and continuous fall of the mark were designed to bring about the cancellation of the reparations.[172] These appreciations did not correspond to the secondary importance attributed by Curzon to Europe, and though he did not dismiss Crowe's analyses, he did not always follow his advice, as, for instance, on the subject of turning the *entente cordiale* into a formal alliance with France.

While at the same time supporting *entente* with France and even pressing for the guarantee French leaders sought, Crowe was not unaware of possible French threats to the balance of power. He believed that only the support of Britain would provide that sense of security which would encourage the French to make the concessions to Germany necessary to establish a stable equilibrium and to compromise their many differences with Britain, above all in the Middle East, where the two countries were following different policies. Crowe believed that apart from the question of Smyrna, Britain should give every possible support to Greece, believing that it was in British interests to maintain and support a friendly Greece. However, if such a policy were to succeed, it could only do so if Britain gained the full co-operation of France, which was more and more inclined to support the nationalists. At the time it seemed to Crowe that the only way to secure French co-operation on the Graeco-Turkish War was to offer France a guarantee of security against a German attack on the Rhine. Crowe attempted, over a long period, to induce Curzon and the Cabinet to consider this approach, but when they did so it was on terms unacceptable to France.

There were, however, disagreements between Curzon and Crowe with regard to Middle Eastern policy. Crowe wished to put an end to the Pan-Islamism menace and secure British interests in the Black Sea region by means of enforcing a peace treaty upon Turkey no less harsh than the one imposed upon Germany – he recommended splitting Asia Minor amongst the powers – and by setting up a Great Greece in Europe and Asia Minor, which would protect the freedom of the Straits and replace Turkey as Britain's ally in the region.[173] To that, Curzon and Hardinge were strongly opposed. But this policy was imposed on them by Lloyd George, thus creating, ironically, an unholy alliance between him and Crowe. However, in hindsight, in contrast to European issues, Crowe may have shown less flexibility than his chief.

Crowe was closer in sympathy to Curzon's student anti-Bolshevism thanks to Lloyd George's far more pragmatic approach to the Russians. Nor did his antipathy diminish after the Anglo-Soviet trade agreement of March 1921 or when MacDonald took office intent on offering them some recognition. Crowe clearly felt that the Soviet Union was not fit to be included in the community of civilised nations.

William Tyrrell

Second-in-command at the Foreign Office was William Tyrrell, the *alter ego* of Crowe. Even before the war, Tyrrell became conspicuous for his ability to influence his fellow men. His superiors recognized his unusual diplomatic talents, and advanced him rapidly, until in 1907 he became Edward Grey's Private Secretary. And in the opinion of many, the one who was responsible more than anybody else for the decisions taken by the Foreign Secretary.[174] He went everywhere and knew everybody.[175] For foreign diplomats he represented one of the few Englishmen capable of understanding their point of view.[176] For politicians at home, Members of Parliament and journalists, he was an inexhaustible source of information regarding what went on behind the scenes; a person who always knew how to give advice at the right place and at the right time, and a charming companion.[177] At least up to the Conference of Lausanne even Curzon was not immune to his charms.[178] His ability to make dignitaries from different political streams and nationalities dependent on his advice, and capture their hearts by his technique of setting things forth so that people subconsciously adopted positions favourable to him as if they were their own, thus enabling him to smooth over differences of opinion, was renowned. Gaston Palewski, Chef de Cabinet of Paul Reynaud and General Charles de Gaulle, described this technique as 'verbal fencing'.[179] The breakdown which he suffered in 1915, due to overwork, the fall of his younger son in battle, and his drinking,[180] temporarily brought his career to an abrupt stop. He returned, however, in 1916, when he was asked to define the war aims of Britain; in 1918 he became Head of the Political Intelligence Department; in 1919 he was co-opted to the British Delegation to the Peace Conference; and, with the introduction of the organizational reforms in the Foreign Office, he became an Assistant Under-Secretary of State.[181]

As Assistant Under-Secretary he was assigned to superintend the work of the Heads of the American, Far Eastern and News Departments. He was head of an inter-departmental committee that accumulated all the data regarding Russian subversion against the British Empire,[182] and so must have had connections with GC & CS. Tyrrell joined many of the special committees designed to submit political recommendations to the Permanent Under-Secretary and Foreign Secretary (for example, the Foreign Office committee which recommended, on 21 January 1921, the abolition of the Anglo-Japanese Alliance),[183] put forward organizational suggestions, such as the composition of the British Delegation to the Washington Conference (November 1921).[184] He received ambassadors and ministers and would deputize in the absence of Curzon and Crowe.[185] All of these were typical Assistant Under-Secretaryship functions, designed to lighten the burden of work of the Permanent Under-Secretary and Foreign Secretary.

However, Tyrrell was far from being a typical Assistant Under-Secretary. He shunned the drudgery of departmental drafts, was selective in reading Foreign Office files, and had a fundamental objection to committing his views to paper. Instead, he preferred to rely on personal contacts. Once, one of the files was returned to him with the remark: 'This matter requires your decision', with the intention that he should minute on the jacket his position on the question in point. Tyrrell, however, circumvented the need to commit himself in writing one way or another by confirming in his spidery handwriting that the matter, indeed, required his decision.[186] These methods, according to his contemporaries, however unconventional, did not affect his position in the least. The positive results he achieved in the scope of his daily work earned him a reputation as a fairly good Assistant Under-Secretary, and his intimate relations with various public personages, including Stanley Baldwin, the Prime Minister (1923–4, 1924–9, 1935–7), turned him into one of the outstanding figures in the Civil Service. Finally, he did not suffer from the querulousness of Curzon, like Crowe did. Baldwin once remarked to Thomas Jones, Deputy-Secretary to the Cabinet:

> Curzon is difficult – very bad tempered in the forenoon, better as the day advances. He gets on Eyre Crowe's nerves but Sir William Tyrrell has humour and can handle him better.[187]

As a Roman Catholic born in India, on 17 August 1866, to a mother whose own mother was a daughter of a Hindu Vizier, Tyrrell, like Crowe, had been educated in Germany and had spent much of his early years in the home of Prince Hugo Radolin, who had married into his family. Again, like Crowe, Tyrrell spoke fluent German and knew intimately many German leaders.[188] His war-time experiences heightened his fears about German intentions after the war, and more than once he expressed his opinion that Germany was planning a war of revenge.[189] On 7 August 1916, together with Ralph Paget, he proposed to contain Germany in the West by means of an Anglo-Franco-Belgian alliance, which would be checked in the East by means of a series of national states – Poland, Yugoslavia and others – which would replace the Habsburg Empire and act as a buffer between Germany and Russia.[190]

In Tyrrell's opinion, German reasons why they were unable to pay reparations were not convincing. He argued that Germany was deliberately undermining her economy in order to evade the reparation payments.[191] However, the need to put an end to Germany as a military menace, after the war, was not interpreted by him as identical with her internal and economic disintegration, and he criticized the French moves in this direction as clumsy and foolish.[192] In order to check Germany on the one hand, and contain France's extreme leanings on the other, Tyrrell agreed with Crowe that the solution lay in giving a guarantee to France against German aggression,[193] or signing an alliance with France.[194]

Moreover, Tyrrell pointed out that the international conditions that developed after the First World War,[195] together with the advent of the aeroplane and the submarine, made Britain an integral part of Europe.[196] In order, therefore, that there would be a proper defence to the Channel, England had to turn the Rhine into a frontier the crossing of which constituted an act of war.[197]

Sir Ronald Lindsay

The other Assistant Under-Secretary was Sir Ronald Lindsay (1877–1945), brother of the twenty-seventh Earl of Crawford. Originally he was a member of the Diplomatic Service, and except for a three-year period from 1908, when he was transferred to London, he filled various posts abroad, including, from 15 September 1919, the post of Counsellor at the Washington Embassy and later at the Paris Embassy, to which he was assigned at the request of the new Ambassador, Hardinge.[198] However, a short while afterwards, Crowe requested Curzon to appoint Lindsay an Assistant Under-Secretary. Curzon acquiesced on the grounds that he needed a good man who would put together the chaos left behind by Tilley, who had been sent to Rio de Janeiro.[199]

Hardinge's protests that Lindsay was the right man in the right place at the Paris Embassy were to no purpose,[200] and on 1 January 1921 he was transferred to London and promoted to Assistant Under-Secretary of State. His functions were to superintend the Northern and Eastern Departments, and one can discern from the minutes that his superiors were pleased with the manner of his work. However, unlike Crowe and Tyrrell, Lindsay confined himself to the matters of his department and tended to sit on the fence as to the question of whether Britain should support Turkey or Greece.[201]

Other Advisers and Problem-Solvers

The Foreign Secretary and his senior assistants were not the only ones in the Foreign Office who directly influenced the process of foreign policy formulation. Collecting information, piecing it together with previous pieces of information, elaborating political principles, evaluating situations, simplifying the incoming material and submitting suggestions regarding the necessary political tactics, were performed, for the most part, by the Heads of Departments and their staff. This stemmed, first of all, from the fact that there was a limit to the capacity of the Foreign Secretary and his immediate assistants.[202] Even Curzon was aware of this limitation, and parallel to his violent outbursts as to why they did not seek his approval before taking one step or another, he used to ask the members of the Central Department, 'But what is the next step?',[203] or complain to the members of the Eastern Department that he had difficulty understanding what their position was from their minutes.[204]

The 'bounded rationality catch' gave birth to a group consultation among all the grades, which was manifested in the following way: an incoming paper would be handed first to the most junior member of the department – the Third or Second Secretary – who minuted on the jacket of the file the outstanding points in the paper. He would make the first suggestion concerning any action to be taken and pass it up for study by the grade immediately above him. If the paper was of small importance and this individual felt sure of his ground, he would submit it for approval to the Head of the Department or draft a reply to the incoming paper. If, however, the paper was of major importance, the Head of the Department, and after him the Assistant Under-Secretary, would elaborate the minutes of the junior members, specify what political tactics could be employed and which of them were favoured, and amend, as necessary, the draft reply submitted by their subordinates. These minutes would then be passed to the Permanent Under-Secretary, who would raise points not mentioned previously and submit his final suggestion to the Foreign Secretary; on the basis of all this, the Foreign Secretary would give his final decision on the matter.[205]

The Foreign Secretary and his immediate assistants could not, for want of time and burden of work, attend to all departmental matters themselves. Consequently, they devolved some of their responsibilities to the Heads of Departments and their staffs, and allowed them to deal with sectoral and low-level subjects on their own account.[206]

Thus, for example, a study of the Western Department files reveals that the Head of the Department until 1921, Charles Tufton, and his successor Gerald Villiers, and their staff, dealt with a good many of the problems themselves – despite the profusion of countries that comprised the Department. This picture was repeated in the American Department.

The seondary importance of the American problems enabled, more than once, the Head of the Department, Rowland Sperling, and his four-man staff, to split the handling of North, Central and South America, Ethiopia (until 1924), and a series of miscellaneous subjects which nobody else wanted (such as pollution at sea and whales)[207] among themselves. In the more important departments, such as the Central, Far East and Northern, files regarding taxes and railways, and similar low-key subjects, were dealt with by the junior grades. The Heads of the Departments, therefore, had to be capable of judging when the handling of one subject or another be left to their discretion, and at what point they should pass it down to the decision of their superiors, thus turning the middling officialdom from passive to active. The Heads of Departments discharged their duties so competently that juniors who were not familiar with the essence of the senior grades' work used to wonder what was the use of Assistant Under-Secretaries.[208]

Although the capability of the Heads of the Departments and their staffs to intervene in the affairs of the neighbouring departments was limited owing to the fact that they were organized as autonomic cells,[209] and could not become key figures, like Crowe and Tyrrell, in the departmental framework, the junior grades could have some influence on the decisions taken.

Much more complicated was the situation in all that pertained to long-term appreciations and moves towards Germany. Here the views of the young officialdom contradicted those of the 'old top', but not those of the Foreign Secretary, Curzon. Thus, for instance, the First Secretary of the Central Department, Waterlow, dismissed airily the assumption that the more normalization was resumed in Germany the greater the danger of militarism.[210] He contended that precisely the non-moderation of the disarmament and reparation clauses, and the imposition of sanctions upon Germany on account of her inability to fulfil those Articles, would create a Rightist revolution and perhaps even a Bolshevik counter-revolution.[211] At the same time, the Second Secretary of the Central Department, J. C. Sterndale Bennett, the First Secretary, Harold Nicolson, and the Head of the Department, Miles Lampson, rejected the General Staff's assertion of 28 March 1924 that an Anglo-French war was as impossible as an Anglo-German war was inevitable, and that the danger would become tangible from the moment that the Rhineland was evacuated in 1935.[212] Bennett observed coolly that historically the facts contradicted these assumptions;[213] Lampson noted that it was not possible to accept such assumptions as axiomatic;[214] while Nicolson said, concerning a further memorandum in the same vein, of the General Staff,[215] that he, too, doubted the correctness of these assumptions.[216] Moreover, as a long-term policy, they proposed to found the European balance of power on Germany and France, which would neutralize each other, thus enabling them to reduce to a minimum British involvement on the Continent.[217] This young group was a powerful lobby. Nevertheless, a confrontation between it and the 'old top' was averted, partly owing to the fact that most of the time deliberations over short-term crises, and items that cropped up daily, prevented any discussion on the question of the long-term policy of Britain towards Germany; and partly thanks to the willingness of the young guard, notwithstanding its views, to accept the position that peace had, for the time being, to be founded on the Treaty of Versailles and the *entente cordiale* with France, thus placing Germany on the other side of the fence.

The German question cropped up occasionally in other departments as well. In the Northern Department the need to prevent the establishment of a Russo-German bloc was discussed, and within this framework the department accentuated the importance of Poland as a buffer zone between the two countries.[218] In the Far East Department, Wellesley defined Germany as a negative element that had imperilled 1914 Europe and constituted, later on, a very bad example to Japan. Nevertheless, given the current post-war position of Germany, and the long time that would pass until the recovery of Russia, in his opinion there was no fear of the establishment of a German–Russian–Japanese alliance, which under normal circumstances might have inflicted a mortal blow upon the Western democracies.[219]

In the Western Department it was agreed that Britain ought to maintain the *entente* with France, not, however, because of the German peril, but in order to moderate French political moves and propaganda,[220] which the department viewed as anti-British, and concerning which they had bitter complaints.[221] These grievances were repeated, only with double force, by Eastern Department officials, against the background of Anglo-French controversies over nearly every subject in the region. Finally, the American Department contended that the refusal of France to assume limitations over the issue of submarine building turned it into Britain's immediate adversary in the world. This attitude was sharpened by Curzon, who noted that if, in addition to its total land and air domination of Europe, France should also build a fleet of submarines, then she would be able to impose upon Britain an air and naval blockade. The formidable French armies, in his opinion, were bound to lead to an arms race and perhaps, even, to an Anglo-French war. He cited Germany's arrogance, bred by a sense of overwhelming military superiority and by the desire to destroy every incipient challenge to military supremacy before the war, as pertaining to France's current position.[222]

3

Co-ordination of Foreign Policy in the Cabinet

Loss of Foreign Office Independence

In the period following the First World War, changes took place in the domestic and international settings in which the foreign policy of the developed European countries was elaborated and conducted. On the domestic level, against the background of the demands of various pressure groups (to restore the war-damaged economy, obtain a greater share of social wealth, ensure employment, etc.), governments ceased to be merely regulative agencies in a 'night-watchman' state. Instead, greater state centralization, with predominance given to domestic economic objectives, was fostered for the purpose of providing for the physical and human needs of the inhabitants of the state; the relative priority of external goals was decreased.[1]

On the international level there was increased interdependence amongst the various European states, and between them and other modern societies. Governments discovered that they could no longer satisfy some of their principal domestic needs nor achieve a number of their most valued foreign-policy objectives without reciprocal action from others.[2] For instance, the economic recovery of the states of Europe appeared to be dependent on the economic recovery of Germany;[3] and in order to prevent a naval arms race and to rehabilitate China, the Washington Conference – convened on 12 November 1921 and lasting until 6 February 1922 – attempted to secure the co-operation of Belgium, England, France, Japan, Portugal and the United States. The creation of the League of Nations added a new dimension to multilateral diplomacy.

These domestic and international transformations brought about a number of far-reaching developments in all that pertained to the form of national foreign policies. First, the traditional distinction between foreign and domestic policies became blurred. Previously it had been assumed that the divorce between foreign and domestic affairs was paramount, since the former aimed at national rather than particular goals and did not confine itself to the utilization of legitimate means as did domestic policy, and it was oriented towards a decentralized anarchic milieu over which the state maintained little control, as opposed to the centralized domestic order in which the state had a monopoly of the instruments of social order. However, the more the levels of interdependence among the various powers strengthened and the governments' economic and social functions expanded, the greater the necessity to revise these assumptions. The repercussions that the domestic measures of one country (in monetary policy, gross agricultural product, constitutional reforms, technological innovations, etc.) had on other countries, and the repercussions that resulted from the sacrifice of domestic policy goals for foreign political reasons (reduction of resources allocated to economic growth, welfare, etc.), gave foreign policy a domestic dimension and domestic policy a foreign dimension so that the two coincided in their timing and goals. This is not to say that a complete uniformity ensued between the two spheres. But what did happen was that they began to affect each other to an extent not experienced in the past.[4]

Secondly, the content of foreign policy widened. If, in the past, foreign policy had been confined to attaining high policy goals as a means to maintaining the integrity of the state (security, defence) or to enhancing some attribute of the state (territory, fulfilment of national ideas), now it included issues which in the past had been considered pure domestic matters. Moreover, even the definition of the essence of traditional goals as power and security altered. These were no longer identified with territory and population alone, but also with wealth and welfare, given the increasing demands of the citizens of the state that their government provide them with economic and social services. Thus the distinction between high and low policy became less important than in the past, and foreign policy was harnessed to achieving domestic goals just as the needs of domestic policy were sacrificed occasionally to attain external goals.[5]

Thirdly, it became increasingly difficult to centralize all aspects of foreign policy in one department. The blurring of the distinctions between foreign and domestic policies not only meant that decisions taken in one field had significant repercussions in others, but also made it difficult to keep a sharp division of functions between the Foreign Office and the neighbouring departments. More than anything else, it was economic factors that pressured the Cabinet to increase co-ordination among the various departments. In this respect, noted Edward Bridges, a Treasury official,

No government can today discharge its responsibilities unless it has a

coherent economic policy and such a policy must be framed after bringing together the view of the separate Departments while its execution
demands constant consultation between them.[6]

The inevitable result of increased co-ordination was an increase in
governmental bureaucratization, which of course put in jeopardy the
monopoly of the Foreign Office over foreign affairs, given the fact that the
increased spheres with which neighbouring departments were charged
all of a sudden acquired a foreign dimension.[7] The loss of Foreign
Office influence in the context of these bureaucratic developments was
compounded during the first three post-war years by a number of
other factors and incidents, which combined to weaken the Office still
further.

A major factor was the great suspicion in which the public held the
Foreign Office in 1919. Revelations regarding the character and methods
of pre-war statesmen convinced various sections of the public that the
secret diplomacy concerning the United Kingdom's various international
agreements, and indirectly the professional diplomats who carried this
diplomacy out, had caused the outbreak of the First World War. Liberals,
Labour politicians, as well as the Union of Democratic Control called
on Parliament to put an end to the balance-of-power policies and to
expand the Parliamentary controls over the conduct of foreign affairs.
There were those who demanded that the government should bring all
treaties and major commitments to Parliament for ratification. It would
be a mistake, warned one of the Radicals, to leave foreign policy in the
hands of the British Junkers, i.e. the professional diplomats; they had
proved themselves to be old-fashioned and incompetent and represented
aristocratic interests rather than democratic ones.[8] 'I want no diplomats . . . diplomats were invented simply to waste time',[9] summed up
Lloyd George, who had similar grievances towards the pre-war Foreign
Office.[10]

These criticisms were, in part, responsible for the transition from the
old diplomacy to the system known as diplomacy by conference.[11] The
high hopes pinned on this system were not realized. It failed to give
expression to the advantages latent in it, such as settling problems
pertaining to several powers, that could not rapidly be secured by
the ordinary methods of diplomatic communication, enabling heads of
states to conduct negotiations themselves, and to come to know and even
befriend foreign statesmen.[12]

The holding of no less than twenty-three international conferences
between 1920 and 1922, accompanied by resounding publicity and in
which the Foreign Office played a secondary role, did not add to its
prestige. The fact that in Curzon's view continental problems were not
comparable in importance to imperial questions,[13] gave Lloyd George

ample opportunity to act as his own Foreign Secretary.[14] The Foreign Office was therefore already in a weakened position when Lloyd George decided to leave control over foreign policy in the Prime Minister's Office, despite the disbandment of the War Cabinet in October 1919.[15] It was the Prime Minister's intention that his Private Secretariat should simplify for him the data and suggestions arriving from the Foreign Office, give the Foreign Office, in his name, appropriate directives, and be his adviser on foreign affairs. The principle on which the division of work between the Prime Minister and the Foreign Office was based, in his words:

> . . . that great questions should be discussed between principals, meeting alternately in London, Paris, Italy, and that details should be settled by communications between the Foreign Offices.[16]

The result was that the Foreign Office was deprived of its monopolistic position as adviser to the Prime Minister. Indeed, as Foreign Secretary, Curzon attended most Cabinet meetings, and also took his seat along with Lloyd George or even represented him in the vast majority of international conferences. Crowe, too, occasionally attended Cabinet meetings, and the same applied to Curzon's Private Secretary and other officials who accompanied him to international conferences. But in spite of the continued activities of the Foreign Secretary, the reduction of Lloyd George's dependence on Curzon inevitably weakened his influence and therefore that of his Office.[17]   Moreover, although in theory the Cabinet had the authority to veto the Prime Minister's actions, Lloyd George continually by-passed his colleagues by creating temporary committees of ministers to replace the whole Cabinet. He had, noted Churchill,

> a habit of picking his colleagues for any preliminary decision so as to have a working majority of those who were favourable to his view. One set for one phase of a question and another for its complementary part![18]

Confirmation of this situation can be found not only in the files of the Departments of State, but also in the private papers of persons who took an active part in the process of foreign-policy formulation. This is not to say, however, that by the end of the war the Foreign Office suffered from a total eclipse. With Curzon, Hardinge, Crowe and Tyrrell it continued to play a part in the elaboration of Britain's foreign-policy goals. A good deal of international business was conducted through British diplomats and was handled by the Foreign Office. Foreign Office representatives met with officials from other departments and although the growing involvement of various others in foreign policy made it more difficult for the Foreign Office to carry out its advisory and diplomatic functions, it did not accept this situation without fighting back. It even succeeded in reversing the trend once Lloyd George left the diplomatic scene.

Lloyd George's Policies

One of the chief sources of embarrassment to the Foreign Office during the three years after the war was the Prime Minister, David Lloyd George, who had little respect for the traditional establishment. He was prepared to ignore the Foreign Office when it suited him to do so.[19] To take just one flagrant example, from the time of the Paris Peace Conference, when he persuaded Clemenceau and Wilson that the Greeks should be invited to occupy Smyrna for the Allies,[20] Lloyd George set the parameters of British policy towards Greece, with Curzon following in his wake.

In London, under the influence of Eleutherios Venizelos, the Prime Minister of Greece, Lloyd George managed to pass a decision according to which the dominance over the western coast of Asia Minor and Thrace be accorded to Greece's hands.[21] In San Remo he overcame the second thoughts of the French, and particularly the Italians, in all that pertained to according territories to Greece in Asia Minor, and it was, in the main, due to his influence that the final text of the peace treaty presented to the Turks on 11 May 1920, and signed by them on 10 August 1920 at Sèvres, was very similar to the draft drawn during the Conference of London.[22] His motives for doing so were to put an end to the might of Turkey, which had stood during the war at the side of the Central Powers, and to create conditions that would enable Greece to take the place of Turkey as the guardian of the Straits and as Britain's ally in the eastern basin of the Mediterranean.[23] Given previous international treaties, the ethnic composition of the Smyrna population, the present balance of power between the Greeks and Allies and the Turks in western Anatolia, and the past record of Venizelos, in Lloyd George's opinion this policy was not only feasible but also had the force of reality.[24] According to Lloyd George, the Turks were 'obsolete' while the Greeks were the up and coming nation.[25]

Foreign Office appreciation of the problem was that the Turks would resign themselves to the loss of the Ottoman Empire but not to the loss of their historical homeland, Asia Minor. Instead of doing away, once and for all, with the 'Sick Man of Europe' problem, the Greek presence would inflame Muslim passions all over the Near East and bring Asia Minor to ruins. The main body of opinion in the London Foreign Office was almost unanimously against the policy of Lloyd George. Hardinge wrote in March 1920 that:

The Turks will never agree to the handing over of Adrianople and Smyrna to the Greeks whom they hate and despise. This is not my personal view only but I believe it to be the view of all who have lived any time in Constantinople and know both the Turks and the Greeks.[26]

Robert Vansittart, the Private Secretary of Curzon, claimed that everybody realized that Lloyd George's policy was a mistake.[27] Curzon, who warned against this policy all through the Paris Peace Conference, confirmed that the Turks would not bow their heads humbly to kismet.[28] Likewise, the 'easterners' of the Foreign Office were united in their opinion that the Greeks did not possess the military capability to secure Asia Minor for any length of time. From this point of view, any initial Greek presence was entirely superfluous.[29]

Despite these decisive views, Lloyd George flew in the face of the main body of Foreign Office opinion. In a meeting held between him and the Prime Minister of France and the President of the United States on 7 May 1919, a green light was, under his influence, given to the Greeks. And on 15 May, together with Italian forces, the Greeks occupied Smyrna. When Lloyd George's behind-the-scenes activities became known, stupefaction at such a policy blunder, among the 'easterners', was universal. Vansittart complained that the decision was made without any reference to those who knew the country, and who had been unanimous against such an action.[30] Hardinge wrote of the political chaos that ensued and foresaw that it would create crises in the region.[31]

I am afraid that much of the muddle is due to the ignorance of our leading statesmen as regards foreign countries and their unwillingness to consult those who really know.

Although assigned the task of negotiating with the delegations of the Allies on the precise text of the peace treaty, and even though Curzon attended all the deliberations of the London and San Remo Conferences and his suggestions were those which enabled them, not once, to bridge over the differences of opinions among the parties, there was little the Foreign Office could do about the situation. Curzon complained that in the six months he had been writing memoranda, minutes and despatches to the British Delegation to the Peace Conference, in which he warned of the results that would ensue from according territories in Asia Minor to foreign elements, he had not been favoured with a single reply.[32] His own recommendation, and that of Hardinge, to expel the Sultan from Constantinople, was rejected on 6 January 1920 in a majority vote by the Cabinet (though it had been supported by Lloyd George).[33] Finally, at the London and San Remo Conferences, Curzon found himself forwarding a Middle Eastern settlement and giving his support to solutions with which personally he was not in accord. Later on he claimed that the main features of the Turkish Peace Treaty were determined at a meeting of the Supreme Council, which was run by Lloyd George:

> in which he had forced a pro-Greek solution upon Miller and Nitti – my committee in the F.O. being merely left to fill in the details – while the final form of the Treaty had been debated and decided upon, again in his presence and carrying his influence at San Remo a little later in March 1920.[34]

Hardinge, who was no less stunned by Lloyd George's disregard of the Foreign Secretary's advice and the final text of the peace treaty with Turkey, wrote:

> Nothing could have been more mismanaged and we are only at the beginning of our trouble with Turkey over the conditions of peace. All those with experience and knowledge of Turkey and of Near Eastern policies have been ignored and the views of cranks and enthusiasts adopted. The merest tyro [tyrant] who has lived in Turkey would know that the Turks would never agree to give up Smyrna and Adrianople to the Greeks whom they both hate and despise.[35]

Another instance in which Lloyd George circumvented the Foreign Office was on 16 January 1920, during the Paris Conference. Unaccompanied by any Foreign Office representative,[36] Lloyd George met the French and Italian Premiers, ostensibly to discuss the theoretical possibility of reopening trading relations with Russia.[37] He initiated the taking of this decision, which was announced publicly on 17 January.[38] His motives in doing so stemmed in part from the belief that this was the best way of moderating Bolshevism;[39] in part from the desire to put an end to the armed intervention of the Allies in Russia; and in part because Russian food and raw materials resources were, in his opinion, essential to the economic reconstruction of Europe.[40] Lloyd George was also guided by the assumption that in the foreseeable future Russia would cease to constitute a military menace to her neighbours. In sum, therefore, it would not be taking an unnecessary risk to bring about a normalization in relations.[41]

Because of his unwillingness to make any step that might contribute to the recovery of Russia, and restore her traditional might,[42] Curzon was opposed to the idea of reopening trading relations, though not to the withdrawal of British forces. As he wrote on 14 November 1920:

> I firmly believe that the renewed lease of life which the agreement if concluded will give them, will be consecrated to no purpose more unswervingly than to the subversion and destruction of the British connection with the Indian Empire.[43]

For similar reasons he vetoed any aid to the White Russians.[44] The sole acceptable solution was, in his opinion, to leave Russia in a state of weakness. In order to accomplish that end, contact should be avoided, and a chain of buffer states should be encouraged that would seal and restrict the Bolshevik disquiet and hostility to the boundaries of Russia.[45]

Despite Curzon's views, Lloyd George, armed with a memorandum of E. F. Wise, a Food Ministry official and the British representative on the Supreme Economic Council, regarding the advantages latent in forming trading relations with the Russians,[46] pursued an opposite policy. When the Paris Conference reconvened, on 16 January 1920, Lloyd George would not allow Curzon or any other Foreign Office representative to attend the deliberations, nor did he give them in advance any hint of his intentions[47] on the grounds that it was not the affair of the Foreign Office, but that of the Food Ministry.[48]

At the Foreign Office, confusion prevailed when it became known that it was decided to begin trade with Russia. Hardinge telegraphed Curzon, who was participating in the Paris Conference, requesting that he bring the Foreign Office up to date on this point.[49] In reply Curzon had to admit to Hardinge that he, too, was in the dark, as the account of the proceedings was being withheld. Curzon noted that 'Answer [to your] question should therefore be sought from Hankey and Wise.'[50] Only a month later, when the Cabinet Secretariat circulated copies of the Paris Conference proceedings, did the Foreign Office learn precisely what had taken place.[51]

It is necessary to emphasize that the Prime Minister's disregard of the 'easterners'' position on the question of Asia Minor, and of Curzon's on the question of Russia, did not mean that his policies on these two issues were vetoed by all Foreign Office officials. Harold Nicolson, Eric Forbes Adam, Second Secretary to the Eastern Department, and, above all, Crowe, sided during the Paris Peace Conference with the pro-Greek policy of Lloyd George.[52] Likewise, J. D. Gregory, Head of the Northern Department, and Hardinge agreed with Lloyd George in the middle of February 1920 that peace with Bolshevik Russia was inevitable and that they ought to resign themselves to it.[53] The trouble was that there was no uniformity between these views and the Foreign Office's recommended position, as expressed by the Foreign Secretary. Another problem was the Prime Minister's penchant for first acting and only thereafter troubling, if at all, to bring the Foreign Office up to date. These circumstances created functional difficulties for the Foreign Office and meant that, more and more, foreign policy was conducted via the Prime Minister's Office.

It has previously been mentioned that no representative of the Foreign Office was present at the meeting of the Council of Four of 7 May 1919 and, later on, at the conference held between the Allies on 14 January

1920 at Paris. Lloyd George's choice of fellow-delegates on these occasions was, however, extremely significant. In addition to Henry Wilson[54] and Wise,[55] the occasional experts, he was accompanied by Maurice Hankey, Head of the Cabinet Secretariat, and Philip Kerr, one of his Private Secretaries. The presence of these two persons gives some indication of the activities, during the 1919–22 period, of the organizations which they represented, and of the position of these two bodies *vis-à-vis* the Foreign Office. For in the Private Secretariat, commonly known as the 'Garden Suburb', and the Cabinet Secretariat, the Foreign Office faced potential rivals in its advisory and diplomatic spheres of activity.[56] On the reorganization of the Private Secretariat after the war,[57] one of its members described its tasks as follows:

> . . . to keep an eye on the Government departments, to maintain a liaison with the heads of these departments and keep the Prime Minister informed of important items . . . It was not that he was inaccessible to his Ministers, but that a great deal of preliminary work in arriving at decisions was done beforehand by his Secretariat, who sifted the facts and prepared memoranda and often conveyed messages from him to his Ministers.[58]

Understandably, the 'Garden Suburb' aroused resentment because the Prime Minister was riding roughshod over precedent and procedure. Curzon, in particular, was far from being over pleased and when, due to Lloyd George, J. T. Davies, his Principal Private Secretary, a former schoolteacher who was not even a civil servant, was awarded a knighthood, Curzon's indignation knew no bounds. 'Surely nothing so absurd has happened since Caligula made his horse a pro-consul!', he cried.[59]

Influence of the Private Secretary for Foreign Affairs

The member of the Secretariat specifically responsible for foreign and imperial affairs was Philip Kerr, Eleventh Marquess of Lothian. His functions were divided into two. The first consisted of liaison functions between the Prime Minister and the Foreign Office on questions of policy and diplomatic contacts, giving instructions to the Foreign Secretary in the name of the Prime Minister and establishing a direct link between Britain's representatives overseas and the Prime Minister.[60] The second was to advise the Prime Minister on foreign affairs, a function he had begun performing already during the war. Lloyd George's affection for private advice, and the vast knowledge Kerr acquired of what went on behind the scenes, brought Lloyd George to ask more and more for his advice. By the time the Peace Conference convened, Kerr was established as the Prime Minister's private adviser and confidant. He also began to take an active part in the elaboration of foreign policy.[61] He attended Cabinet meetings, took his seat along with the Prime Minister in the course of international conferences, and maintained frequent contacts with foreigners.

Even though officially he was just one of the Prime Minister's assistants, in practice such broad powers were delegated to him by Lloyd George that from the respect of his hierarchical standing *vis-à-vis* the Foreign Secretary he became, more than once, the Prime Minister's deputy in all that pertained to foreign affairs. A famous example is Kerr's reply to Balfour's enquiry as to whether a certain memorandum had been read by the Prime Minister: 'I don't think so, but I have.' 'Not quite the same thing is it, Philip – yet?' said Balfour mildly.[62]

Curzon, too, suffered from Kerr's closeness to Lloyd George. Most of his contacts with the Prime Minister were, according to him, conducted through Kerr or his successor, Sir Edward Grigg,[63] He had virtually no personal contact with Lloyd George, who rarely answered his letters promptly, or replied only after a long lapse of time (and sometimes never).[64] Curzon was also annoyed by Kerr's penchant for assigning missions to Foreign Office members. On one occasion Kerr had telephoned the Foreign Office asking that a certain despatch should be telegraphed. Crowe reported to Curzon, who minuted acidly: 'This is the first I have heard of it. But the authority of the P.M.s Priv Sec appears to be sufficient.'[65]

Kerr, and Grigg also, performed diplomatic missions which might have been expected to go to the Foreign Office. For instance, it was Kerr and not the Foreign Office who, on 28–29 April 1921, held talks with the counsellors of the German and French Embassies, each one separately. Kerr warned the Germans that French occupation of the Ruhr, without affording Germany every opportunity of withdrawing from her intention to turn down the Allies' proposals on the reparation issue, would put an end to the *entente cordiale*.[66] By contrast, Kerr gave notice to France of the likely dangers to Germany if the present moderate French Government fell because of the reparations issue and in its stead there ascended an intransigent government.[67] A day earlier, Kerr had observed that should the French occupy the Ruhr it would weaken Germany significantly, the more so as it would be extremely difficult to get them out.[68]

The Foreign Office viewed these contacts and private communications of the 'Garden Suburb' with British ambassadors as undermining its main function, i.e. to be the sole communication channel between domestic and foreign spheres. As the Foreign Office was not always informed of these meetings or, indeed, what was said in them, the situation was regarded as particularly intolerable. Once, when it became known to Curzon that Grigg had held a meeting with the Greek representative and discussed such questions as Greece's position in Asia Minor, and the financial aid it would require, he turned to him asking that in future he be kept informed of such contacts, noting that it was very difficult for him to conduct a policy for which he was largely responsible when things were said and done of which he and the Foreign Office were unaware.[69] On another occasion, when the Foreign Office learnt that the German Embassy had delivered directly to Grigg a copy of the German Chancellor's letter to the Pope, in which he asked for his aid in the context of his struggle against the Treaty of Versailles, without delivering a similar copy to the Foreign Office, Ralph Wigram minuted that:

> If importance is still attached to the maintenance of the Foreign Office as the channel of communication with foreign governments through foreign missions here, I submit that the time has arrived when a definite ruling must be obtained and adhered to, that no British official whatever is to receive communications from foreign missions except with the sanction of the Secretary of State.[70]

Kerr's position and intimate relations with the Prime Minister made him, from the Foreign Office's point of view, a potential threat in another sphere that hitherto the Foreign Office had regarded as its exclusive domain, namely advising the Prime Minister on foreign affairs. It is known, for example, that it was Kerr who wrote, together with Lloyd George, the famous Fontainebleau Memorandum, which described the post-war aims of Lloyd George's government and highlighted the need for a moderate peace.[71] Another striking example was a paper of 2 September 1920, wherein Kerr laid out before Lloyd George his ideas regarding the desirable policy in Europe, in which he suggested, in contrast to the Foreign Office, that in lieu of British political intervention in Europe, 'Great Britain must deliberately draw in its horns in the matter of foreign policy.' Kerr explained the need to return to the splendid-isolation policy, and to concentrate on solving domestic and imperial problems, by arguing that England had neither the time nor the ability to solve Europe's problems, particularly as he did not believe that British public opinion would be willing to allocate resources, troops and time for the solving of problems such as Upper Silesia and reparations.[72]

Europe was not the only region in which Kerr's views clashed with the Foreign Office's recommended position. For instance, in contrast to Curzon, he favoured Greece's presence in Smyrna,[73] and criticized the Foreign Office for the lack of understanding of the need, for Russia, of a settlement. He noted that the Foreign Office 'had no conception of policy in its wider sense and did not in the least understand what L.G. was driving at'[74] – a criticism he continued to hold although later on he himself recommended that the Russian mission be expelled from Britain because of its subversive activities.[75]

The advice of Kerr and the 'Garden Suburb' on various subjects was numerous and variegated. Nevertheless, it is difficult to evaluate its precise influence on the Prime Minister's policies not only because of the similarity between views, but also because contact between the Prime Minister and his Secretariat was by word of mouth. All that can be said with certainty is that, in the eyes of many, the Private Secretary for Foreign Affairs – Grigg, and particularly Kerr – was considered a key-person,[76] Churchill even went so far as to describe him as the unofficial Foreign Secretary of Britain.[77]

Responsibilities of the Cabinet Secretariat

The most influential member of the Cabinet Secretariat was its creator and manager, Sir Maurice Hankey. During 1919–22 Hankey and his staff performed secretarial duties – recording the proceedings of the Cabinet, preparing the Agenda Papers, circulating the Cabinet's decisions to the relevant Departments, and so forth.[78] But, at the same time, they took over from the Foreign Office the responsibility for several other functions – organizing international conferences, providing the secretariat for the British delegations, circulating the proceedings and resolutions of the various conferences (copies that more than once were late in reaching the Foreign Office), and handling relations with the League of Nations.[79] Hardinge opposed the assignation of this last function to the Cabinet Secretariat, saying that all questions to be brought before the League of Nations would have foreign aspects. Crowe argued that, should the Foreign Secretary cease to be the sole channel of communication through which the British representative to the League of Nations received his instructions, he would lose his control over foreign policy. If it was desired that the Foreign Secretary and his assistants be the advisers of the Prime Minister and Cabinet on foreign affairs it was essential that this be executed via a sole communication channel.[80] But their protests were of no avail. Kerr, who conceived the idea, and Lloyd George[81] insisted on it that contact between the Prime Minister[82] and the Dominions' representatives and the League of Nations was of far greater importance, and that therefore responsibility ought to be transferred to the Cabinet Secretariat, where both had representation.[83]

This position was finally approved in Cabinet on 10 November 1919. When it transpired, at the end of 1920, that the Foreign Office was continuing to handle the affairs of the League of Nations in the framework of a special department, it was decided to transfer and amalgamate it with that of the Cabinet Secretariat and to place at its head a Foreign Office official, given the close linkage that existed between foreign affairs and League of Nations matters.[84] Only at the end of 1922, with the advent of Andrew Bonar Law as Prime Minister, were the functions pertaining to international conferences and the League of Nations returned to the Foreign Office.[85]

Parallel to their participation in the administrative side of foreign-policy work, Hankey and his organization also performed advisory functions. As Head of the Cabinet Secretariat, Secretary to the British Delegation to Imperial and international conferences, and as one who attended various Cabinet meetings and accompanied the Prime Minister everywhere, Hankey was qualified to offer Lloyd George advice on foreign affairs – though on the whole he was more a spectator than the Foreign Office's rival in this sphere.[86] In June 1920, for instance, Hankey submitted suggestions to Lloyd George regarding the line the Allies ought to take towards Germany in the forthcoming Spa Conference. Other examples were his constant pressures upon the Prime Minister to implement a non-intervention policy in Europe and his suggestion in June 1921 regarding the policy that ought to be pursued in order to solve the Greco-Turkish conflict.[87] He also took it upon himself, in the course of international conferences, to perform functions that might have been expected to be performed by the Foreign Office representatives to the conferences.

During the San Remo,[88] Hythe (the first)[89] and Genoa Conferences,[90] for instance, it was Hankey who kept London informed of their proceedings. More than once the Foreign Office was kept in the dark as to what was going on. Hankey also acted as a connecting link between Lloyd George and foreign representatives. Thus, on 9 March 1921, while the Third Conference of London was taking place – called by Lloyd George for the purpose of putting an end to the Greco-Turkish conflict – and whilst the Foreign Office was putting in strenuous efforts to find compromising formulas that would be acceptable to both parties, Hankey, under instructions from Lloyd George, told Greek Prime Minister Nikolaos Kalogerpoulos that it was of vital importance to the safety of the Greek army to strike a blow at Mustapha Kemal, thereby encouraging the Turks to reject the peace terms that he himself had suggested to them. This statement was received by the Greeks with a sigh of relief, and they, as a result, hardened their position.[91] The Foreign Office learned of it for the first time from the Intelligence Services, which happened on transcripts of messages sent by the Greek Embassy to Athens.[92] Later on, confirmation was received from Hankey,[93] but there was nothing the Foreign Office could do about it. The London Conference ended without a result, and on 21 March 1921 the Greeks embarked on an offensive to the hinterland of Anatolia which eventually brought about their military collapse and expulsion from Asia Minor.[94]

Against this background it is easy to understand Curzon's resentment against the Prime Minister and his assistants, and his complaints that:

> He, Hankey and Philip Kerr ... ruled the country and under the convenient aegis of Philip Sasoon's hospitality at Lympne menaced or sought to menace the Foreign Affairs of the Continent.[95]

As intimated earlier on, too much should not be made of the picture of Lloyd George railroading the Foreign Secretary and his Office along paths which they either did not care about or else positively detested. On Franco-German questions, negotiations with the United States, and the extension of the Anglo-Japanese Alliance, the Foreign Secretary consistently endorsed the main lines of the Prime Minister's policies.[96] Over the handling of relations with Russia, too, Curzon backed down.[97] And even over Turkey – despite his growing frustration, hoping against hope that once the parties fought to a standstill the Greeks would retire of their own accord from Asia Minor – Curzon still looked for a general settlement, on the basis of the Treaty of Sèvres, which would allow Britain to retain her wartime gains.[98]

Relations between Foreign Office and Neighbouring Departments

The argument between the Foreign Office and the Prime Minister's Office went beyond the wounded ego of Foreign Office officials, who focused more and more around questions of the essence of function and vocation of their respective offices. The Foreign Office maintained that, as a specialised department, its function was to centralize the administration of foreign affairs and act as the sole channel of communication between foreign and domestic spheres. While the 'Garden Suburb' and Lloyd George held that the blurring of the distinctions between foreign and domestic affairs, the increasing intervention of the various departments in foreign affairs, and the widening range of information that had to be gathered for decision-making, necessitated the centralization and processing of the whole complex of foreign issues in the Prime Minister's Office.[99] It would, they maintained, increase the rationalization and consistency of policy-making and reduce the bureaucratic process. The Foreign Office held that its function was to advise the Prime Minister and Cabinet on foreign affairs, whereas *their* function was to take decisions. If they took decisions without first taking its advice they might decide matters without knowing the full background of detail, of opportunity and danger, for the Foreign Office was the sole body that had the professional qualifications to present subjects clearly, succinctly and objectively.[100] Lloyd George, on the other hand, who disliked the conventional methods of the Foreign Office and mistrusted its advice, preferred to rely on the private advice of his assistants. He maintained that with a little organization the Prime Minister's office could supply relevant information much more rapidly and efficiently than the Foreign Office.[101]

The very same argument went on between the Foreign Office and Cabinet and the neighbouring departments. The Foreign Office saw itself as the principal adviser to the Prime Minister and Cabinet on foreign affairs, and as the exclusive communication channel between foreign and domestic affairs. Neighbouring departments insisted on their right to be full partners, if not more than that, whenever issues dealt with by them had foreign repercussions. The Foreign Office held that the tasks of departments of state should be grouped according to their subordination to a major goal.[102] This meant that anything that had a relation, even indirectly, to foreign policy had to be handed over to the department whose staff was committed to the pursuit of that goal, namely the Foreign Office.[103] The Cabinet, on the other hand, maintained that the issues dealt with by the various departments were, in fact, different aspects of the same problem. In order to ensure that steps taken by one governmental body did not contradict, but complement, those taken by other bodies, it insisted that inter-departmental consultation be expanded and that the proficiency of the various departments in what took place in other spheres be deepened. The Foreign Office claimed that the participation of others in the process of administering foreign policy created frictions and delays in the decision-making process and that the best way to simplify and strengthen the chain of command was to pass to it the sections of the various departments dealing with foreign aspects. The Cabinet, on the other hand, maintained that the changes that were taking place in the domestic and foreign settings in which foreign policy had hitherto been conducted necessitated the turning of departments of state into a co-operative body.[104]

If the rest of the departments had been willing to co-operate with the Cabinet in creating a political co-operative, it is possible that an end would have been put to the controversy. But the departments were unwilling to concede part of their autonomy for some common purpose. The Prime Minister's Office also preferred to circumvent the bureaucratic system and act independently of it. The result was that the various departments began to treat issues such as control, co-ordination and superintendence less as technical means by which it was possible to attain some goal and more as prestigious goals that ought to be fought over on the grounds that their achievement would preserve (and hopefully increase) the independence of the department. This created a process that Crozier defined as a vicious circle of displacement of goals.[105] It led to inter-departmental struggles over means instead of ends and created power struggles over the control of one sphere or another in the course of which neighbouring departments endeavoured to improve their position at the expense of the Foreign Office. Whenever the question of whether a certain subject constituted part of domestic or foreign policy was in dispute, it was inevitable that the particular department would act less as the Foreign Office's partner and more as a rival.[106]

Role of the War Office

Among the departments that threatened the position of the Foreign Office was the War Office. Already by the end of the Paris Peace Conference it had begun to voice reservations to the Foreign Office over the political stands of its officials. At the time this did not worry the Foreign Office. 'I see no necessity to discuss with the W.O. the terms of Sir E. Crowe's Des. to the S. of State', minuted Hardinge.[107] But soon, in the course of 1920, the War Office began to interfere in the conduct of foreign policy – not, perhaps, a surprising development in view of the fact that, owing to lack of manpower, the British Embassy in Berlin, for instance, was dependent on the military representatives in all that pertained to gathering political information, interviewing German politicians, and writing reports on the goings-on in the various districts.[108] The question of disarming Germany, which was the focus of the international deliberations, and in which the War Office had an interest no less than the Foreign Office, also gave legitimation to its increasing intervention in foreign affairs.

British diplomats abroad were frequently perturbed by the activities of War Office representatives. Francis Lindley, Chargé d'Affaires to Vienna, and George Clerk, Chargé d'Affaires to Prague, for instance, were alarmed by the inaccuracy of the War Office representatives' political reports and feared lest Whitehall be misled into believing that they gave a faithful picture about the goings-on in Austria and Czechoslovakia.[109] In London, Foreign Office officials were vexed by the turning of the War Office into the prime information supplier regarding Central Europe, something that in their opinion ought to have been the exclusive function of the Diplomatic Service.[110] They were annoyed by the refusal of the War Office representatives abroad to confine themselves to military issues, and by their persistence in attempting to dictate to the Foreign Office, through political analyses, appreciations and conclusions, what policy it should pursue.

Richard Haking, the Commanding Officer of British military forces in Upper Silesia, sent in a memorandum in which he noted that in order to save Europe from anarchy, and Germany from Bolshevism, the insane policy of attempting to crush Germany completely should be abandoned; the Treaty of Versailles should be moderated; and aid in foodstuffs and finance should be granted for the purpose of ensuring that the regime established would be democratic.[111] In response to Haking's memorandum, Wigram asserted that the General should be instructed to stop talking about the revision of the Treaty of Versailles and forcefully told that it was not British policy to crush Germany.[112] Eric Phipps, of the Central Department, recommended that the General should also be told that there was no justification for his criticisms or for his action in seeking to dictate to the Government what policy it should pursue,[113] Crowe summed up bitterly:

> Our generals, like the officers in the War Office, who lay down the law about foreign policy, are a terror. But as there is no real discipline in the upper ranks of the army, and as every general, whether serving or retired, freely expresses his views about the policy of H.M.G., we should gain nothing by complaining to the War Office, who rather like this state of things.[114]

On 16 August 1920, Winston Churchill, the War Secretary, circulated among his colleagues a memorandum (6 August) by the General Staff which described in bleak colours the political state of Germany. The memorandum noted that the Germans were being forced to choose between resigning themselves to their complete repression by France and forming an alliance with Bolshevik Russia, for the purpose of subverting the Treaty of Versailles. It analysed the military reaction of the Germans to each of these possibilities, and indicated that in its opinion Germany would prefer signing an alliance with Russia, despite their aversion to Bolshevism, rather than give way to France and the Treaty of Versailles. The General Staff recommended desisting from vindictiveness against Germany but treating it, rather, as a Great Power. It proposed that economic aid be given and that the German armed forces be increased so as to check the Bolshevik threat. It advocated signing a defensive agreement with France and Belgium in order to soften their attitude towards the Treaty of Versailles and warned that, without such measures, a German–Russian alliance would emerge that would destroy all international agreements, including the Treaty of Versailles, turn Europe towards Bolshevism, and trigger off a new European war.[115]

This memorandum elicited particularly indignant responses from Foreign Office officials. Waterlow observed that, although he agreed with the General Staff's stand that they ought to rehabilitate Germany and that what prevented the implementation of a conciliatory policy were the anxieties of the French and Belgians, in his opinion the War Office exaggerated the extent of the danger latent in a Russo-German combination. According to Waterlow, even if such an alliance did emerge, it was likely that a considerable number of Germans would oppose it and that Germany would be split politically into two. And if, indeed, that should take place, then the ability of the alliance to threaten Western Europe would be diminished greatly.[116] Crowe minuted indignantly that though the memorandum was drawn at the beginning of August and considered, no doubt, long ago by the War Office, it was brought to the knowledge of the Foreign Office only at the end of September. Worse than that, the military authorities' persistence in attempting to play the role of political advisers, he asserted, belonged to the type of usurpation of functions at a political level that had contributed, perhaps more than any other factor, to the impelling of Germany towards the First World War.[117] Hardinge, for his part, observed scornfully that:

> All soldiers regard themselves as Heaven born diplomatists and much prefer diplomacy to military strategy . . . As a matter of fact they are as a rule regularly short sighted and we have an excellent instance in their attitude towards the situation in the Caucasus and Northern Persia which might have been very different if they had had a wider outlook last January. The present paper is full of fallacies which are hardly worth criticizing as it had apparently cut no ice with the Cabinet.[118]

Curzon minuted that there was no need to get over-excited as it was one of those memoranda which neither attracted the attention of the Cabinet nor was discussed there. He had not, therefore, thought it worthwhile to send it in on the day of its appearance.[119]

Nevertheless, the Foreign Office managed eventually to restrain the War Office. Being convinced that, on the German question, the Cabinet tended to rely more on the opinion of the Army than on the opinions of the Foreign Office and British Embassy officials in Berlin,[120] it fought to change this relationship. On 14 November 1920, Curzon approved, with the support of the Central Department officials and Crowe, the request of Lord D'Abernon to increase the manpower of the Berlin Embassy in order to put an end to the dependence of the political establishment in London on the military representatives with regard to the gathering of political information.[121]

After discussions with the Treasury, the Foreign Office was given the requisite expenditure. The staff of the Embassy was increased and in consequence political information from Germany began to flow, once more, through conventional channels. Although the penchant of British officers for holding interviews with German officials without the knowledge of the Embassy, or submitting to their superiors political analyses, was not totally eliminated,[122] they did begin to confine themselves to reviewing the rate of Germany's disarmament, and this fact brought about a significant decrease in the extent of the War Office's intervention in foreign affairs.

Political issues apart, the Foreign Office was still fully dependent on War Office appreciations regarding the extent of Germany's disarmament. For instance, it was the General Staff that indicated, on 5 November 1920, to the Foreign Office that many years would pass before Germany could re-threaten the Allies militarily.[123] On 5 May 1921 it was again the General Staff who asserted that Germany had fulfilled the vast majority of the Inter-Allied Military Commission of Control's demands. And on 16 July the War Office announced, again, in the name of the British representative to the Commission of Control, that the rate of disarmament was progressing satisfactorily.[124] The War Office, it is true, did not go as far as D'Abernon, who asserted on several occasions that the work of the Commission of Control was terminated.[125] But, generally, it agreed, on 4 July 1922, that Germany was disarmed.[126]

Against this background it is not surprising that the assumption that Germany no longer constituted a military threat became a cornerstone in the mind of the young guard at the Foreign Office. Waterlow did not miss any opportunity to express his satisfaction at the level of Germany's disarmament. By August 1921 he had written that most of the military articles of the Peace Treaty, in all that pertained to the disarmament of Germany, were executed; and that the continued presence of the Commission of Control in Germany was superfluous and harmful,[127] and ought to be replaced with a much more contracted body.[128] This opinion was also shared by Wigram, Lampson and Curzon. And in 1922, a year later than Waterlow, they reached the conclusion that, generally, Germany was disarmed;[129] but, in contrast to Waterlow, they were in no hurry to withdraw the Commission of Control from Germany so long as it did not announce the completion of its work.[130] All the same, they regarded this latter point as a technical detail. Lampson reported, in the name of the War Office, that the replacement of the Commission of Control with a much more contracted body would not impair the capability to ensure that Germany remained disarmed.[131] This judgement was also affirmed by Wigram, who specialized on the issue of disarmament.[132]

The only people who were not prepared to accept the War Office reports at their face value were Tyrrell,[133] and particularly Crowe, who rejected the General Staff's assumption that the disappearance of the German threat was dependent on restoring normalcy to Germany.[134] Similarly, he treated with scepticism D'Abernon's optimistic reports about the prospective dissolution of private military organizations by the German government, minuting that he had every reason to believe that they were unable to do so.[135]

The information that Germany was evading the execution of the disarmament clauses was received by him with exceptional severity, and in April 1922 he still maintained that one could understand, under these circumstances, the reluctance of France to disarm as well.[136] In contrast to Wigram[137] and Lampson,[138] Crowe had the impression that there was substance to the claims of Brigadier-General J. H. Morgan, Deputy Adjutant-General on the Commission of Control, that Germany had managed to preserve the structure of the Imperial Army and the armament industry. She was gradually building a large new army and, when she was ready, she would unleash a war of revenge.[139] Crowe minuted that Morgan's arguments, and the facts reported by him, were exceedingly convincing, the more so as personally he had no confidence in the judgement of the Chief of the British Section to the Commission of Control, Major-General Sir (cr. 1918) Francis Bingham, who saw everything in rosy colours and through intensely pro-German spectacles.[140] However, despite this information, and his close familiarity with the German character, Crowe had difficulty in producing tangible statistics that would contradict the assumption that Germany was, indeed, disarmed. And eventually he had to agree that, when the Commission of Control pronounced the completion of its work, the Allies would no longer have the right to prolong the stay of the Commission in Germany.[141]

Therefore, when, in 1924, the War Office announced that an Anglo-French war was as impossible as an Anglo-German war was inevitable, and that the danger would become tangible from the moment that the Rhineland was evacuated in 1935,[142] Bennett treated it as sheer nonsense. 'What ground is there for assuming that for years to come renewed German aggression is the greatest possible danger that faces us?' he asked.[143] Similarly, Bennett did not accept the General Staff's assumption that the continued French presence in the Rhineland did not affect the security of Britain, minuting that this view contradicted the stands of the Air Ministry and the Admiralty.[144] Lampson, too, considered that the War Office's assumptions were unfounded,[145] and Nicolson held likewise.[146] Indeed, it is true not only that the Commission of Control was still operating in Germany until the end of 1926, but also that the Foreign Office had not made any demand that it should be withdrawn, in view of the running reports of the War Office and the Commission of Control that Germany was not yet totally disarmed.

However, the Foreign Office refused to link the disarmament issue with the evacuation of French forces from the Ruhr.[147] It held, at the end of 1925, that the process of disarming Germany had reached such an advanced stage that it was possible to evacuate the Cologne district, notwithstanding that the Germans were not meeting their obligations.[148] Only in September 1926 did the Foreign Office reach the conclusion that it was possible to withdraw the Commission of Control from Germany.[149]

Even Austen Chamberlain, then Foreign Secretary, who insisted on the execution of the disarmament clauses in full, dismissed the French opposition to this last step as standing on trifles.[150]

Interventions of the Prime Minister

But if the Foreign Office managed to hold its own against the intervention of the War Office in foreign affairs, there was little it could do to combat the diplomatic activities of the Prime Minister. On 8 December 1921 Lloyd George met Louis Loucheur, French Minister of Reconstruction, at Chequers,[151] and, between 19 and 20 December, a French delegation headed by Aristide Briand, the Premier, at London, in order to deal with a number of issues face to face. In these talks Lloyd George came up with a scheme that was supposed to solve, once and for all, the reparations problem and the European economic slump. Lloyd George asserted, as a fundamental premise, that the reparations, economic slump and detachment of the Russian market from the European system were different aspects of the same problem. He proposed to give Germany a moratorium on the reparations, open up the markets of Central and Eastern Europe to German goods in order to increase economic gain and hence their long-term capability to pay reparations, and establish an all-European organization that would take over the handling of the reconstruction of Europe.[152] He looked to reintegrate Russia into Europe by means of economic aid reconstruction; and recognize Russia *de jure* should she assent to grant the all-European organization executive powers within Russian boundaries and recognize the right of private ownership.[153] In return for French willingness to co-operate with Britain in all that pertained to the implementation of these proposals, Lloyd George expressed his readiness to guarantee the frontiers of France against external aggression.[154]

At the time, Lloyd George kept secret from the Foreign Office and Foreign Secretary the particulars of the negotiations with the French. The guarantee idea was first brought up by Comte Auguste F. de Saint-Aulaire, French Ambassador to London, in a conversation that he held with Curzon on 5 December 1921.[155] But they did not know the 'return' terms that Lloyd George demanded. In Curzon's opinion there ought to be, for instance, the liquidation of Anglo-French dissensions in Morocco and the Near East in a manner satisfactory to Britain.[156] Moreover, the Foreign Office had no intimation that Lloyd George intended to offer the French at the forthcoming Cannes Conference (6–13 January 1922) the Guarantee Pact.[157] The only persons amongst Government members who were familiar with the details of the contacts with France were Sir Robert Horne, Chancellor of the Exchequer, and Sir Worthington-Evans, Secretary of State for War, and his companions at the Cannes Conference.[158] In this Conference Lloyd George and Briand reached a number of conclusions, including the summoning of a European economic conference with the participation of Germany and Russia. They also concluded that practical steps be taken in order to integrate Russia into the European system, and that a British guarantee be given to France against any external aggression.[159] However, all these achievements dissolved with the resignation of Briand from office, on 12 January, in response to a revolt against his policies.[160]

On 16 December 1921 Curzon protested against Foreign Office exclusion on the grounds that at an earlier meeting of the Cabinet the Government members refrained deliberately from discussing the proposals Lloyd George intended to bring up before Briand. Curzon asked to have his view recorded that the Foreign Office should be consulted before a decision on this point was taken and that he reserved his opinion as to diplomatic recognition of Bolshevik Russia.[161] At the end of December 1921 he became more and more conscious of the fact that he was not versed in the details of Lloyd George's contact with the French, and on the eve of the Cannes Conference even considered taking his name off the list of the British delegation to the conference for fear lest his ignorance be exposed before Briand.[162]

In January–February 1922, as a follow-up to the above, Lloyd George negotiated with Briand's successor, Raymond Poincaré, on the ratification of the Cannes Resolutions. In view of the intransigence displayed by the latter, on 9 February Lloyd George expressed his wish to use Eduard Beneš, Premier of Czechoslovakia, as an intermediary between him and Poincaré. Curzon was none too happy over the idea but resigned himself to it, observing, in connection with this, to Hardinge that, 'As you know he thinks that nothing is ever done except in a private and secret conference between principals, i.e. between himself and a second party.'[163] However, what he did not know, until Poincaré revealed it to Hardinge, was that Lloyd George's talk of using Beneš was merely a blind, and that all the time he maintained a flowing contact with Poincaré via the French Ambassador. For instance, Poincaré's three demands, that Hardinge telegraphed to London after his meeting with him – that the forthcoming Genoa Conference would not undermine the Treaty of Versailles, not impair the Covenant of the League of Nations and not interfere with the reparations question – had already been submitted to Lloyd George by the French Ambassador and accepted by him unreservedly. Also, the two personages had agreed, through the Ambassador, to meet in Boulogne. Hardinge, who reported all this to Curzon, observed that:

> After he told me all this, I really felt thoroughly sickened by what I have heard and I said to myself, what is the good of diplomacy or working hard in the interests of one's country when one does not know what is going on and things are done behind the backs of those who are responsible.[164]

On the basis of Hardinge's report, Curzon summoned the French Ambassador for a clearing talk and was stunned to hear from him that he was merely a 'post-box' and that the individual who gave him messages for Poincaré in Lloyd George's name – and received from him, the Ambassador, the replies of the French Prime Minister – was Lord Derby. The outraged Curzon promised Hardinge that he would have it out with the Prime Minister, for 'without telling me of it L.G. had employed an ex Amb.ʳ to go behind our backs to the French Premier'.[165]

However, neither did he have a clearing talk with Lloyd George nor did he send him the draft of his protest letter.[166] Instead, he settled for having a meeting, on 1 March 1922, with Derby, who told him that Lloyd George had assured him that Curzon was aware of his mediatory mission. This statement, concluded Curzon, was a most flagrant lie.[167]

The Power to Make Appointments

Besides its advisory and diplomatic functions, there was a third sphere of activity in which the Foreign Office encountered interference during the 1919–22 period. This was in the appointment and dismissal of Ambassadors, a process that until 1916 was the concern of the Foreign Secretary and Permanent Under-Secretary, though both the Sovereign and the Prime Minister could be concerned. But from 1916 onwards, the Prime Minister forced on the Foreign Office appointments and dismissals they disliked and opposed.

For instance, in 1916–18 Lloyd George brought about the substitution of Derby for Lord Bertie as Britain's Ambassador to Paris. He also took an active, though not exclusive, part in the replacement of Britain's ambassadors at the Hague, St Petersburg and Washington with his own men.[168] On 30 January 1920, a Cabinet committee took a decision – endorsed on 11 February by the plenary – stating that the appointment or removal of senior officials in the various departments should be confirmed by the Prime Minister before taking effect.[169] A short while afterwards, Lloyd George pressed Curzon to appoint Baron Edgar Vincent D'Abernon, a businessman, as Britain's Ambassador to Germany. The deliberation between the two on this point was done without consulting the Permanent Under-Secretary, Hardinge, who, as it turned out after the fact, wanted the appointment to be given to

Sir Horace Rumbold.[170] However, Curzon did not object personally to D'Abernon for he was aware of the special qualifications that would be required from whoever was appointed British Ambassador to Berlin, and he accepted that D'Abernon answered them. Whereupon, on 25 June 1920, he officially offered D'Abernon the post, and on 2 July handed him his letter of accreditation to the President of Germany.[171]

The appointment of Auckland Geddes, President of the Board of Trade, as Britain's Ambassador to Washington was, however, purely the work of Lloyd George. At the beginning of 1920, Arthur C. Murray, Third Viscount of Elibank (1879–1962), relayed to Curzon the recommendation of Lord Grey, temporarily acting as Ambassador, to appoint to the post Herbert A. L. Fisher, MP. Curzon agreed that Fisher would make a good Ambassador but told Murray that, if he wished to secure for him the appointment, he would have to convince the Prime Minister. Murray replied that he did not understand, for he presumed that the appointment of Ambassadors was in the hands of the Foreign Secretary. 'I can only repeat', replied Curzon, 'that the appointment will be made by the Prime Minister and that your message ought to be delivered to him.'[172]

And, indeed, the vocal protests of the Foreign Office and Diplomatic Service members were of no avail,[173] for on 25 March 1920 Geddes became British Ambassador to Washington. The outcome was that, throughout his years in the American capital, the Foreign Office treated him with mistrust.[174] On one occasion, Geddes, who loved to flaunt his medical knowledge, reported that he diagnosed Charles Evans Hughes, the American Secretary of State, as exhibiting symptoms of mental instability. In response, Curzon wrote to the Prime Minister that he was by no means sure that it was not the British Ambassador who was afflicted with a mild form of mania.[175] Lloyd George's interference in the process of appointing various personages to senior positions in the Foreign Service did not end there. In mid-1920 he opposed Curzon's wish to appoint Hardinge as British Ambassador to France and make Crowe his successor; only with great effort did Curzon manage to secure for them the longed-for appointment.[176]

An Enduring Relationship

During the 1919–22 period the Foreign Office thus encountered several obstacles which made it difficult for it to function in the same way as it had in the past. Against some of these obstacles – the encroachments of the War Office, for example – it was able to take a stand. But against others – such as the activities of the Prime Minister and his two Secretariats – it was difficult to take a stand. This situation resulted in part from the changes that occurred in the domestic and international settings in which foreign policy was elaborated and conducted, but it arose, too, from the personalities and interrelations of the Prime Minister and the Foreign Secretary.

The relationship that prevailed between Lloyd George and Curzon was exceedingly complex. In character, origin, education, outlook and working methods it was difficult to find persons more diverse.[177]

Lloyd George was aware, however, of Curzon's political value and his vast knowledge.[178] He knew, too, of Curzon's admiration for the skill with which Lloyd George guided the Empire through the war[179] and for his political powers of persuasion.[180] These factors enabled the two men to bridge over the differences in their working relationship despite being an odd team of contrasts. Yet one cannot help suspecting that one of Curzon's main virtues, for the Prime Minister, was his extreme pliability as Foreign Secretary. Although Curzon objected more than once to the methods and political moves of the Prime Minister, he did not bring things to a head and always bowed ultimately to the will of the Prime Minister.[181]

Curzon's lack of firmness stemmed from several factors. One factor was his poor state of physical health. In May 1922, for instance, he was absent from the Foreign Office for two whole months because of an attack of phlebitis. Another factor was his difficulty in taking decisions, something attributed by Harold Nicolson to his being an historian rather than a man of action.[182] But above all it emanated from his disinclination to resign. During his first three years in office Curzon had many good reasons to leave, both because of the erosion in his standing and because of Lloyd George's humiliating treatment of him. This last phenomenon found expression in Lloyd George's penchant for imitating in public Curzon's pompous manner.[183] He complained that Curzon constituted an obstacle to the signing of the Trade Agreement with Russia and he publicly criticized his Foreign Secretary.[184] Nor were relations improved by Lloyd George's taunts about Curzon's aristocratic origin and outlook.[185] Yet, in spite of all that, Curzon did not tender his resignation. It is true that he threatened more than once to do so,[186] and constantly used to send protest notes to Lloyd George via Vansittart. But he loved his work too much[187] and thought it his duty to remain in Government.[188]

> 'I suppose' the Prime Minister would say, 'that you have the usual message from your chief', and when I boggled he would grin. 'Consider it said.'[189]

All that was left for Curzon was to bear his humiliation in silence and pour out occasionally his heart's discontent to persons such as Austen

Chamberlain, in the hope that they would, one way or another, bring about the longed-for change.[190]

Changes of Prime Minister Renew the Role of the Foreign Office

In 1922–3 governments changed. In October 1922 Lloyd George surrendered his office to Andrew Bonar Law and the Conservatives – following trenchant public criticism at the Prime Minister's handling of the Chanak débâcle and the decision of the Conservatives, who for months had been chafing under the dominance of Lloyd George, to withdraw their support from the Second Coalition.[191] Seven months later Bonar Law, who was dying from cancer, made way for Stanley Baldwin. The change was significant because it marked not only the end of the Lloyd George era, a fact which would have seemed incredible at the time, but also a turning point in the fortunes of Curzon, who stayed on as Foreign Secretary.

In matters of formulation and execution of policy, neither Bonar Law nor Baldwin believed in trying to be their own Foreign Secretary. Under their premiership the 'Garden Suburb', whose activities had been such a thorn in the side of the Foreign Secretary, was abolished.[192] Tasks were assigned back to the Foreign Office. The Cabinet Secretariat was reduced in size;[193] extra-diplomatic functions were farmed out to the two Secretariats as League of Nations work;[194] and communication with ambassadors,[195] drafting answers to Parliamentary Questions on foreign affairs for the Prime Minister,[196] and updating reporters on foreign issues, were all now safely back in Foreign Office hands.[197] It was the two Prime Ministers' intention that the machinery of government should revert to its former decentralized mode, thus leaving the Secretaries of State to do their own jobs.[198] In Bonar Law's own words, the work of the Prime Minister should be likened to 'that of a man at the head of a big business who allows the work to be done by others and gives it general supervision'.[199]

On the personal level, too, the Foreign Secretary – for all his mortification at having been passed over for the Premiership by the King in favour of Baldwin[200] – found things easier. Both Bonar Law and Baldwin had watched the way in which Lloyd George had treated Curzon with evident distaste and were resolved not to act in a similar way.[201] The masterly way in which Curzon led and dominated the First Lausanne Conference (20 November 1922 – 4 February 1923), convened to negotiate a new European Peace Treaty with Turkey,[202] and his handling of the 1923 confrontation with Russia over the latter's anti-British propaganda, which included outrages against British subjects and trawlers, and religious persecution,[203] did not go unnoticed. His triumph in both bringing the Lausanne negotiations, on 24 July 1923, to a successful end,[204] and compelling the Russians to give way,[205] earned him tributes that went a long way towards re-establishing his reputation and self-confidence.[206]

These developments did not mean that the conditions under which the Foreign Office and Diplomatic Service had operated before 1914 were restored. For one thing, Curzon's absence from the country, at Lausanne, made it necessary for Bonar Law to act temporarily as Foreign Secretary, head the British delegation to the London (9–11 December 1922) and Paris (2–4 January 1923) Reparation Conferences, and approve, together with the Cabinet, matters of policy.[207] Further, the closeness with which foreign policy was interlocked, since 1914, with other spheres forced the intervention of the two Prime Ministers and their Cabinets in the details of policy to a greater extent than in the past. As Chancellor of the Exchequer (until August 1923) as well as Prime Minister, Baldwin was obliged to superintend personally the unravelment of the tangle arising out of the continued Franco-Belgian occupation of the Ruhr, within complex political, economic and military intra-European issues.[208] Another result of the blurring of the distinction between foreign policy and other fields was that the day-to-day secretarial functions of the Cabinet Secretariat, and its role in co-ordinating government policy, linking together the Cabinet's proposals and the departments' execution of these proposals, had to be left intact.[209] Finally, unlike the pre-1914 period, in 1922–4 foreign policy was a subject on which almost every member of the Cabinet felt qualified to offer an opinion.

Nor, for that matter, were frictions between the Foreign Secretary and his colleagues altogether eliminated. Without informing the Foreign Office, Lord Derby, the Secretary of State for War, 1922–4, sent Major-General J. T. Burnett-Stuart, Director of Military Operations and Intelligence, over to Paris in November 1922 to discuss with the French a War Office paper on the question of peace with Turkey, despite many of the proposals being counter to those advocated by the Foreign Secretary and his staff.[210] Equally unhelpful practices included Derby himself venturing out '[to attempt] to correct the blunders of the British Ambassador and Foreign Secretary',[211] and opposing in Cabinet the tactics employed by Curzon for resolving the Ruhr crisis, which gave rise to an acrimonious exchange of letters between Derby, Curzon and the Prime Minister.[212] Likewise, the insistence of Robert Cecil, the Lord Privy Seal, 1923–4, on being responsible for the League of Nations affairs, was brushed aside by Curzon[213] while his failure to consult the Cabinet with regard to foreign affairs in general and the League in particular was highly annoying.[214] There was an irritable exchange of letters between 'My dear George', and 'My dear

Bob', in June and July 1923 over Cecil's conversations with the French in Geneva over the question of the Ruhr deadlock. The multiplication of occasions on which members of the Cabinet attempted to tell Curzon how to conduct his business proved yet another constant irritant to the Foreign Secretary. It was to combat this phenomenon that Curzon, in a burst of irritation that autumn (1923), wrote to his Prime Minister:

> But I must confess I am almost in despair as to the way in which Foreign policy is carried on in this Cabinet. Any member may make any suggestion he pleases and the discussion wanders off into helpless irrelevancies. No decision is arrived at and no policy prepared. Do please let us revert to the time-honoured procedure. I am at any time at your disposal for discussion. I have no fear we shall not achieve harmony. But we must act together and the P.M. must see his F.S. through.[215]

Further, there was the episode of the Baldwin–Poincaré meeting of 19 September 1923 in Paris, and the issuing of a press communiqué to the effect that the French and British Prime Ministers had been 'happy to establish a common agreement of views, and to discover that on no question is there any difference of purpose or divergence of principle which could impair the co-operation of the two countries, upon which so much depend the settlement and the peace of the world'.[216] 'Lord Curzon, on reading this communique', Harold Nicolson testified as Second Secretary at that time, 'was aghast'. He regarded it as others regarded it, as a repudiation, by his own Prime Minister, of the policy of strict neutrality between France and Germany for which he had been responsible since February. From that moment his relations with M. de Saint-Aulaire, one of the organizers of the meeting, 'became strained to the point of rupture. His resentment of Mr. Baldwin's apparent act of disloyalty was extreme.'[217] And he refused thereafter either to speak to Tyrrell, Baldwin's companion and author of the communiqué,[218] or appoint him Ambassador to Berlin as Crowe had wanted.[219]

To cap it all, during his final months in office Curzon became almost obsessed by the belief that he was being plotted against. According to intercepted telegrams exchanged between Poincaré and Saint-Aulaire, the French hoped, through the intermediary of H. A. Gwynne, Editor of the *Morning Post*, to persuade Baldwin to replace Curzon by a more Francophile Foreign Secretary.[220] Curzon could no longer bring himself even to meet the French Ambassador, whom he henceforth 'declined to see . . . on one excuse or another'.[221] 'This is the worst thing that I have come across in my public life . . . I had not realized that diplomacy was such a dirty game.'[222] Baldwin, too, though secretly anxious to replace Curzon,[223] confessed himself 'unaware that such dirty things were done in diplomacy'.[224]

Yet to say all this is not to deny that, overall, the Foreign Secretary and his Office enjoyed during the 1922–4 period a greater measure of independence. The relations between the Foreign Office and the Prime Minister's Office were more cordial; and despite the dramatization of the aforementioned cases by Curzon, who understandably was still smarting from the eccentricities of Lloyd George, in the day-to-day business of diplomacy the Foreign Secretary had, broadly speaking, a much freer hand. This change derived in part from the Foreign Secretary's recent diplomatic successes. More important, both Bonar Law and Baldwin attempted to interfere as little as possible in the administration of foreign affairs. The morale of the Foreign Office undoubtedly improved, once more traditional relations were established between King Charles Street and Downing Street.

4

European Policy of the Prime Minister's Office

Attitudes towards Germany and France

The end of the First World War altered inevitably the system of relations that prevailed between Britain, France and Germany before the War. The armistice and the Treaty of Versailles put an end to the splitting of Europe into two hostile camps; against the background of the Allied victory this created a new military and political situation, and necessitated embarking upon the task of the economic and social reconstruction of Europe. Gradually, the realization crystallized that, in order to ensure a durable peace, the co-operation of all European powers was required,[1] and this necessarily led to a decline in the measure of importance of the European pre-war system of alignments.

Although a change in the traditional principles that had guided the British Administration on the eve of the war was inevitable, a number of factors and developments brought about a rapid weakening of pre-war assumptions. First and foremost was the retrospective public revulsion at war as a means of settling international difficulties in the twentieth century. The legacy of the decimation of a generation of young men by the holocaust of 1914–18, the horrors of trench warfare on the Western Front and the passionate hope, indeed demand, that the experience of the First World War should not repeat itself, were strong public demands.[2] Secondly, although the Allied contribution to the defence of Britain and Europe from Germany's deliberate attempt at conquest had been decisive, closer collaboration did not mean that the prospects of the *entente cordiale* being converted to an alliance (as desired by the French) struck a chord in Britain after the war. On the contrary. Public persons expressed the opinion that the Continental giants that previously had had the capacity to upset the *status quo* in Europe were fast disappearing, given the engulfment of Russia by revolution and anarchy, the military defeat of Germany, her disarmament and Allied occupation of the Rhineland, and the heavy blows inflicted upon France during the war.[3] It was therefore proposed that Britain could resume her traditional role as arbiter of Europe.

The bitter condemnation of the Continental commitment as an unwise and unnecessary departure from an allegedly traditional maritime strategy, in which small military expeditionary forces played a valuable but subsidiary role, by critics such as Captain Basil Liddell Hart, tended to obscure among military men and statesmen how long Britain's security had been intimately related to the security of Western Europe, and how much the subduing of a powerful Continental Power through maritime supremacy and the financing of allies on the Continent had, under the conditions of the twentieth century, become impossible without substantial British military support.[4]

Equally vibrant opinions were centred on the social and economic pressures. For instance, John Maynard Keynes's stance that the enforcement of the Versailles Treaty would create poverty and starvation in Germany, increase her hatred of the Allies, strengthen extremist groups, and perhaps even bring them to power, found sympathy amongst the public. 'If we aim deliberately at the impoverishment of Europe vengeance, I dare predict, will not limp', he wrote.[5] Furthermore, Keynes asserted, the repression of Germany would hamper the economic recovery of Europe.[6] His views regarding treating Germany as a hostile element won widespread public support. The industrial unrest that prevailed in Britain; the protests of the British garrison on the Continent against the slowness of demobilization; the view that unemployment in Britain was consequent upon the collapse of international trade, which in turn stemmed from the economic depression in Central Europe, together created a tendency to link the German question with the solving of Britain's domestic problems.[7] The conception that the pre-war system of alignments was to blame for the outbreak of the First World War – no less than Germany – did not encourage the Government to insist on the maintenance of the European pre-war blocs either.[8] Nor, for that matter, did the Ten Year Rule – according to which it was to be assumed, for framing the estimates of the armed forces, that the British Empire would not be engaged in any great war during the next ten years, and that no Expeditionary Force would be required for this purpose.[9] If anything, the combination of the campaign against 'old diplomacy', British defence policy based on the assumption behind the Ten Year Rule, the multiplication of Imperial problems, and the growing number of domestic crises, provided together a powerful discouragement to any British military involvement in Europe.[10]

Moreover, from 1919 onwards, the pro-French leanings of the Foreign Office were seriously diminished following the appointment of Curzon as Foreign Secretary. It was not that he had a particular predilection for the German mentality, which he portrayed as 'at once the most formidable and the stupidest in Europe'.[11] Nor did he reject out of hand the approach that the Germans were harbouring feelings of revenge. He saw them, however, as a defeated element, impotent and conscious of their impotence.[12]

France, on the other hand, according to Curzon, emerged from the war victorious and as the most formidable Power on the Continent. As a nation she was, in his opinion, no better than Germany, given the fact that her citizens excelled in perfidy and her policy in hypocrisy.[13] Curzon admitted openly that he did not understand such people.[14] 'What treacherous dogs they are!', he complained.[15] He feared, however, that owing to her dread of Germany, France was liable to take extreme steps that might lead Europe to a renewed war. His strongly held opinions meant that Curzon did not advocate that Britain should trust France as in the past.[16] The difficulties and conflicts of interest that arose between Britain and France in the Balkans, Bavaria, Egypt, Hungary, Mesopotamia, Morocco, Silesia, Tangier and Turkey increased his scepticism as to the advisability of the *entente cordiale* in its old form.[17] In this connection Curzon noted that:

> It is difficult in the present transitional state of international relation-ships . . . to anticipate by what enemies we may be threatened in the future, or with whom we may find ourselves at war.[18]

Lloyd George's Strong Views on Europe

One of the chief obstacles to the Foreign Office during the first three post-war years was the Prime Minister, Lloyd George. He had his own ideas as to the means by which the attainment of Britain's three main external goals – the command of the seas, the balance of power in Europe, and the defence of the imperial frontiers and communications – could be secured.[19] Lloyd George did not think that the second aim would be attained if Britain kept pitching the *entente* with France as a counterbalance to 'this miserable thing called Germany'.[20] On 30 December 1920,[21] he turned down Churchill's proposals to moderate France's attitude towards Germany and the Peace Treaty. In his diary, the Cabinet Secretary concluded that the desirable policy, in Lloyd George's view, was:

> our traditional policy of aloofness from the affairs of Central Europe, only intervening at long intervals when our safety compelled it, as in the late war and in the Napoleonic wars.[22]

Lloyd George was of the opinion that the signing of an alliance would merely spur France not to reckon with, to ignore or challenge, later on, Britain's international interests.[23]

Lloyd George was not alone in his opposition to the alliance concept. Together with Balfour and Bonar Law, Curzon, for similar reasons, favoured the Prime Minister's stand. For instance, Crowe's rather indulgent attitude towards post-war Anglo-French discord was not accepted by Curzon. Similarly, following the occupation of five German towns by the French (in response to the entrance of German troops into the demilitarized zone for the purpose of quelling the Spartacists), despite the decisive opposition of Britain, he shelved Crowe's suggestion of 6 April 1920 to reinforce the *entente cordiale* and promote better relations between the two Powers. Crowe wanted to utilize the present opportunity – i.e. the fact that the French had put themselves hopelessly in the wrong and that Britain had the right and the power to make things very disagreeable for them in consequence – to have Anglo-French problems in Turkey, Syria, Palestine and Tangier out with France and thus clean the slate.[24] Curzon asserted that:

> After so flagrant a case as this however I am not very enthusiastic about 'kissing again with tears'.
>   It should be possible to point out that the idea of a British guarantee for the protection of France, or of Belgium with France, is rendered well nigh impossible by such action.[25]

Also, on 30 June and 30 December 1920, Curzon sided with Lloyd George against Churchill and Austen Chamberlain in turning down the idea of a Treaty of Alliance or Guarantee with France and Belgium, noting that no danger was imminent,[26] and that the anti-British policies of France in the Near East, Poland and Washington did not entitle France to any reward.[27] Finally, Curzon rejected the opinion of Crowe, Hardinge, Tyrrell and Vansittart, as put forward in a memorandum by Crowe on 12 February 1921, that if a definite British security was to be given to the frontiers of France, such as the kind that was rescinded in 1919, then it might be expected with some confidence that the French would be disposed to prove far more conciliatory not only in Middle Eastern matters, but also towards Germany. For France's intransigence, in questions of disarmament, reparations and control of the Rhineland, stemmed from her apprehension that within a measurable time there would arise on the other side of the French frontier a resurgent, revengeful Germany with whom France, exhausted by war and with a stationary population, would be powerless to contend. Curzon was unimpressed by Crowe's memorandum detailing Britain's isolation in Europe, her strained relations with the United States and the Bolshevik threat in Asia:

> If we could at this juncture reconstitute, or if possible fortify, the solidarity of the *entente* with France, the whole situation would be materially changed to our advantage. The stability effect of settling on a strong basis of common action and policy Anglo-French dealings

with Germany, cannot be overrated: it would probably settle the attitude of French Governments definitely for a long time to come; it would make for peace and harmony among all the new States made or enlarged by the Treaty of Peace; our relations with America could only be improved . . . and a solution of the Eastern question in a sense favourable to British interests and political aims would clear the atmosphere in all the regions extending from the Balkans to Central Asia and India.[28]

He observed that he did not believe that the French would be prepared to make to Britain concessions such as these.[29]

For all that, the Foreign Secretary and the Prime Minister differed on one basic point. Curzon insisted vigorously that, if Britain wished to control France and incidentally to prevent a renewed war in Europe, everything should be done to maintain the *entente* with France. According to Curzon:

We are at the present moment, believe me, the only moderating influence in respect of France. We go about arm in arm with her, but with one of our hands on her collar . . . and if we were now to relax it, I doubt very much whether the peace of Europe would be maintained for another five years.[30]

Lloyd George's tendency, on the other hand, was gradually to dissolve the *entente* with France,[31] and he even envisaged a day in which the European balance-of-power system for Britain would re-include Germany as a leading partner.[32] The motive was twofold: on the one hand, to create a counterweight to the French, whom he suspected of Napoleonic aspirations,[33] which were, in his opinion, the real danger to the peace of Europe;[34] and, on the other hand, to form the requisite conditions for the reconstruction of Germany and, *en passant*, Europe. Lloyd George feared that if Britain was to treat Germany harshly, and prevent her from taking her proper place in the European community, feelings of resentment would accumulate in the hearts of her citizens, and that this would drive Germany to conclude an alliance with the second European outcast – Russia. The fact that Germany was, at that time, at the mercy of France was not, in his opinion, significant. With Russia's extensive natural resources and the German organizational capability, between them they could raise and equip vast armies that would in all likelihood sweep Europe into a fresh round of war. Lloyd George maintained that the weakening of Germany did not contribute to the promotion of peace, the economic reconstruction of Europe, nor the turning of Germany into a buffer against Bolshevism. Nor did it serve French interests. On the contrary:

France, in pursuing this policy of driving the Germans into fierce but suppressed hatred, is guilty, I think, of the greatest act of folly which any race has ever perpetrated. It has produced the destruction of Germany, and unless France pursues a much wiser cause in future, it may very well end in a more complete destruction of France than anything that has ever been inflicted upon her.[35]

Because of the appreciation that the European peace was founded upon the *entente* with France,[36] and that German nationalism and militarism would not be checked but enhanced as conditions in Germany reverted further and further to their former selves,[37] top officials at the Foreign Office unanimously opposed the dissolution of the partnership with France, or its substitution by a partnership with Germany. Crowe wrote in November 1921:

> Some of the Treasury and other Downing Street tendencies are towards the substitution of an *entente* with Germany in the place of that with France. This is a chimera under present conditions and must remain so for a long time to come. What we, and what the whole world wants, is peace. Peace must for the present rest on the execution of the peace treaties. These would hardly survive a breach between England and France at this moment.[38]

Tyrrell, in one of his few minutes, seconded Crowe's words. Highlighting Germany's evasions from fulfilling her obligations under the Peace Treaty,[39] and the moderating role that the *entente* with France had,[40] Hardinge, for his part, stressed the importance of the *entente* as a means of securing world culture and freedom.[41]

Despite Foreign Office views, Lloyd George, armed with Treasury statistics and fortified by War Office appreciations, flew in the face of these Foreign Office opinions. Officials were themselves aware of the need to restore German prosperity. Thus, Crowe warned that restoring the Germans to settled and normal conditions would merely revive their nationalism and militarism. At the same time he admitted that for political and economic reasons the economic reconstruction of Germany was mandatory.[42] Although a number of officials acknowledged that Germany was deliberately avoiding paying reparations, they agreed in principle with Lloyd George that the sum of reparations demanded by the French was not realistic and had to be scaled down.[43] Most of all, however, it was the post-war Anglo-French controversies that impaired the bargaining power of Foreign Office officials, who shared some of the critiques of the French position made by the Cabinet and so increased sympathy towards Germany. This came to light during the Ruhr occupation, which Crowe condemned as illegal and unlawful.

Trouble with France

The first incident in the chain of Anglo-French incidents in the 1920s occurred against the background of the Kapp Putsch, which took place on the night of 12–13 March 1920. The attempt of right-wing groups to assert their authority in Germany failed within four days. But in the course of the putsch it caused a Communist counter-revolution in the Ruhr which lasted even after the German Government had returned to Berlin. On 16 March, the German Government turned to the Allies with a request that it be permitted to send troops into the neutral zone for the purpose of putting down the revolt.[44] The Foreign Office, with the backing of Lloyd George and Curzon,[45] tended to approve this request and reject the French counter-proposal that it should be done by Allied troops.[46] Waterlow expressed the opinion that under the present circumstances the German solution was the best,[47] for it was feared that the French entrance into the Ruhr would be used by them in order to detach the Rhineland from Germany.[48] Crowe explained that, given the need to restore order in Germany from within, and at the same time avoid taking steps which would imperil the present regime and unite all German factions against the foreign invaders, the quelling of the revolt had better be done by the Germans rather than by the Allies.[49]

Hardinge minuted that he approved Crowe's stand;[50] Lloyd George and Curzon, who had similar views, pointed out, on 18 March, to the French Ambassador that in the opinion of the Government of Britain it was desirable to treat the whole case as the internal affair of Germany, permit the German army to enter the Neutral Zone, accept an undertaking that this presence would be temporary, and intimate that, should the Germans not meet this latter condition, the Allies would be forced to act.[51] Given the British opposition, the French premier, Alexander Millerand, came up, on 21 and 23 March, with a new suggestion to the effect that the Germans might be allowed to enter the demilitarized zone, and that, in return, the Allies would occupy Frankfurt and Darmstadt;[52] the Foreign Office objected. Crowe asserted that such a step would only cause greater disturbances than at present. He observed that the Allied warning that, unless the Germans withdrew after the restoration of order, they would occupy the neutral zone, constituted on appropriate response to the French contention that the whole affair was a German ploy designated to re-perpetuate their presence in these territories.[53] Hardinge minuted that, 'It is quite clear that what the French desire, is the permanent occupation of Frankfurt and Darmstadt.'[54] Curzon, too, utterly disapproved of the French proposal.[55] Only Phipps expressed the opinion that, given the urgent need to allow the Germans to suppress the revolt, it would be better to accept the French suggestion.[56] Finally, after a secret negotiation between France and Germany, Millerand agreed to the entrance of German forces, provided that five further towns would be occupied by the French if there was no German withdrawal within two or three weeks.[57] But immediately afterwards, under pressure from right-wing circles, he reneged on his intention to agree to a time limit and demanded in return an immediate occupation.[58]

Under the pressure of events the Germans agreed to let France occupy any town whatever as long as they, the Germans, were allowed to enter the neutral zone.[59] Thus, when the German Chancellor announced the entrance of the Reichswehr, on 6 April, the French occupied Frankfurt, Darmstadt and Hamburg, and two other towns, without even notifying London.[60] This move gave rise to sharp protests from the British Administration. Bonar Law[61] and Curzon[62] maintained that France, which promised one thing and the day after did something else without any consultation with Britain, could not be an ally worthy of the name. Lloyd George expressed a fear that actions such as these would drag Britain, in the wake of France, into European wars for which she had no desire, or would cause her to repudiate the Alliance with France,[63] and charged France with having helped to bring about the Spartacists' revolt because of her intransigence.[64] Even Austen Chamberlain, the most pro-French of all the members of the Cabinet, regarded the French step as intolerable.[65] Nevertheless, when Lloyd George proposed that, in addition to the formal protest, Derby be withdrawn from the Ambassadors' Conference in Paris, Curzon objected on the grounds that it would create a hostile atmosphere on the eve of the San Remo Conference. As to Lloyd George's words, that he had read most of its works and that this body was worthless and ought to be wound up, Lloyd George cut him short, saying that he had read every line of them and that was why he remembered them better than Curzon or Crowe.[66] The Prime Minister's stand was approved by the Cabinet members at the same meeting, in April, and, parallel to the formal protest sent to the French, Derby was withdrawn from the Ambassadors' Conference in Paris.[67] Indeed, following Britain's response, the French withdrew from the Ruhr. But the damage to Anglo-French relations had already been done.

Successive Conferences

The next link in the chain of frictions between the two states was their deliberations within the framework of the various international conferences.[68] At San Remo (19–26 April 1920), Lloyd George called for the opening of negotiations with Germany in order to determine the total of the reparation payment; while France demanded that they should not establish hard and fast rules on this point but increase the amount annually in direct proportion to the recovery of the German economy.[69] At the First Hythe Conference (15–16 May), Lloyd George advocated the maximum reduction of the sum of reparations; while the French insisted on its enlargement and on their receiving from it the lion's share.[70] At the First Conference of Boulogne (21–22 June), experts on both sides estimated the payment capacity of the Germans at 69 milliard marks.[71] But with delaying tactics the French succeeded in bringing about the postponement of the Geneva Conference, which was supposed, according to the Spa Conference (5–16 July), to fix officially the total of German reparations.[72] France's motive was the wish that this issue should be adjudicated by the Reparation Commission, in which they had, together with the Belgians and Italians, a majority. The Commission was supposed to submit its report on that subject by 1 May 1921. Lloyd George, who wished to scale down as much and as soon as possible the sum of reparations, objected to the French position. He also negated the French suggestion that a conference should be held at Brussels instead of at Geneva, where the members of the Reparation Commission would attend as experts, and that the report of the conference be sent as a proposal to the Reparation Commission, which would rule finally on this point.[73] In an attempt to defeat the French manoeuvrings and enforce his stand, Lloyd George withdrew his demand for a Geneva Conference. But at the same time he met, on 22–23 August, at Lucerne, with the Italian Prime Minister, and on 11 October, in London, with the Belgian Prime Minister. He received from the first a promise to co-operate with him at Brussels against France,[74] and from the second an understanding that Belgium would not enter the Ruhr with France, even if the report of the Brussels Conference regarding Germany should prove to be negative.[75] Eventually, Lloyd George's stand won the day, and on 11 November the French agreed that the Brussels Conference should be composed of Allied and German technical experts.[76] However, this did not avert tension between the two sides, for on the major question – the total sum of reparations Germany would be required to pay – France and Britain would not budge from their respective positions. In quoting the various experts, Lloyd George observed that even the most confirmed optimist amongst them did not believe in Germany's capacity to pay 12 milliard marks per annum.[77]

> We must not give the impression of always insisting on our 'pound of flesh', to use a well-known English expression.[78]

The French, however, continued to propose that Germany was able, and also ought, to pay large sums of money in order to enable France to reconstruct the regions devastated during the war. Thus, when the Brussels Conference convened, on 16 December, and ruled, on 22

December, under the inspiration of the British Delegates, D'Abernon and Bradbury, that it would be necessary to reduce the scale of France's financial demands on Germany, France would not accept the Conference's report.[79]

At the First Hythe and Boulogne Conferences Lloyd George had begun – in part because of the need to win French co-operation in Turkey, and in part because of Germany obduracy – to come around to the French critical stand as regards the slow rate of German disarmament.[80] On 22 June he added his signature to a note in which the Allies protested over the slowness and non-compliance of the Germans.[81] At the Spa Conference the delegates concluded how the sum of reparations would be split among the Allies. An agreement was signed concerning the amounts of coal that would be delivered by the Germans, and it was decided that, should, thereafter, the disarmament progress at a slower pace than that determined by the Allies, further German territories would be occupied, including the Ruhr.[82] At the Second Conference of Paris, held from 24 to 29 January 1921, the Allies crystallized a schedule and an order of magnitude for the reparation payments in the following years, which were submitted as a proposal to Germany.[83] Later, at the Third Conference of London (21 February – 14 March), the Germans rejected the Paris Schedule of Payments and the compromises suggested to them at London, and threatened to suspend the payments altogether if it was decided not to leave Upper Silesia in their hands.[84]

On 7 March, Lloyd George accepted the French demand to break off the negotiations and impose sanctions on Germany, which included the occupation of Duisburg, Dusseldorf and Ruhrort, and a customs cordon between Germany and the Neutral Zone. Lloyd George did not, however, believe that these steps would prompt the Germans to pay, something that retrospectively proved to be true.[85] At the Fourth Conference of London (30 April – 5 May), the Reparation Commission's report, which fixed Germany's total reparation liability at 132 milliard gold marks, was approved, and the Conference reached an agreement on the basis of the Paris Proposals regarding the schedule of payments of this sum.[86] Lloyd George, convinced that he would no longer be able to stop the French from occupying the Ruhr should the agreement be rejected by the Germans, urged the Germans to adopt it, warning that 'the choice before Germany is between accepting the Paris terms and giving France a strangle hold on her throat'.[87] As a last step, intended to stave off the French occupation of the Ruhr – to which he and the Cabinet were opposed[88] – Lloyd George prevailed upon the French to send an ultimatum to the Germans.[89] It was presented on 5 May; on 11 May the Germans accepted the decision of the London Conference in full.[90]

Die ausgeplünderte Germania. *Simplicissimus*,
Munich, 21 April 1920

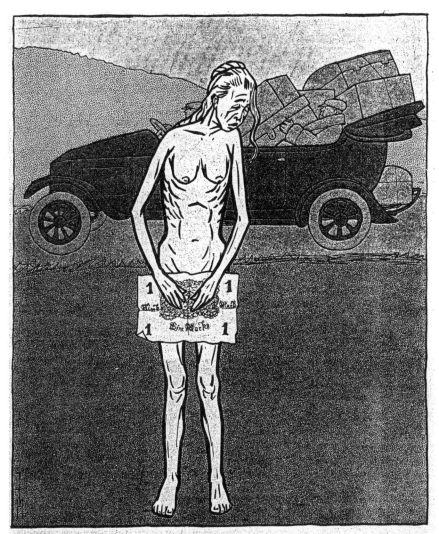

„Alles haben mir meine Söhne verschoben. Nur eine lumpige Papiermark ist mir geblieben, um meine Blöße zu bedecken."

The cartoonist is complaining both about reparations and about the depreciation in value of the mark. Before the war, it had been backed by gold; but wartime and post-war inflation have left merely the 'tattered paper mark'.

Stop!

(Karl Arnold)

"Bevor du einen neuen anfängst, bezahle erst deinen alten Krieg!"

Antanta. *Pravda*, Moscow, 6 September 1922

Антента – entente cordiale (по-русски сердечное согласие)

Two periodicals with very different political tendencies note the rapid deterioration of France's relations with her erstwhile Allies during the period of Poincaré's ministry, which commenced in January 1922.

In the *Simplicissimus* cartoon, Poincaré, in military uniform and with drawn sword, is mounted on a war-horse. He is confronted by Uncle Sam, with a long list of France's war debts, who tells him, 'Before you start another, pay for your last war!' The implication is that Poincaré's policy involves a serious risk of starting a new European war.

The *Pravda* cartoon, '*Entente*', explains to its readers the meaning of the term '*Entente Cordiale*' in Russian. Lloyd George and Poincaré are in furious battle, while the space around them is littered with memoranda, notes and questions relating to their many differences.

Lloyd George, and other British ministers who came into contact with Poincaré, were on bad personal terms with him. Anglo-French relations did not improve significantly in October 1922 when Lloyd George's Coalition Government fell, and was replaced by the Conservative administration of Bonar Law.

Dettes de guerre. *De Amsterdammer*, Amsterdam, copied in *L'Europe Nouvelle*, Paris, 9 December 1922

DETTES DE GUERRE

L'Allemagne supporte la France, qui supporte l'Angleterre, pendant que l'oncle Sam domine cet échafaudage branlant.

(Extrait du « De Amsterdammer.»).

The French insistence on reparations, despite the appalling plight of Germany, is not wholly attributable to greed or to 'revanchism'. The French were under pressure to pay war debts to Britain and the United States. Britain in turn was heavily indebted to the United States.

It is easy to understand how each creditor, conscious of pressure from others, was unwilling to relax its claim against debtors. The Americans, who were the general creditors, tended to argue that they had been brought into the European war, in which their own interests were scarcely involved, and received little gratitude from the Europeans, who now sought to avoid obligations which they had freely incurred. The maximum ill-feeling was generated, which made all forms of cooperation difficult, and easily tempted all countries into economic policies of narrow nationalism.

The vulture. *Star*, London, 11 January 1923

THE VULTURE: "Righto, Poinc., I'll take your message, but I'll come home to roost."

A remarkably prophetic cartoon, which appeared the day after the French government had informed Germany of its intention to occupy the Ruhr.

Poincaré despatches the vulture 'Revanche' – 'Revenge' – in lieu of a pigeon to Berlin. The vulture agrees to take the French message, but promises to return to roost. That promise was amply fulfilled.

The French decision generated enormous bitterness, not only in Germany but in Allied countries. In years to come, many people in Britain and other countries who might have been expected to sympathize with France's military and diplomatic problems in face of the rising threat of Hitler's Germany recollected the Ruhr invasion, and were quite willing that the balance between France and Germany should move in favour of the latter.

And now the next step! *Star*, London,
1 December 1925

In the mid-1920s, there was considerable optimism in Europe. The
Locarno agreements, concluded towards the end of 1925, were generally
welcomed as important moves towards general peace.

In this British cartoon, Europa has already advanced from the Dawes
Plan (1924) to the Locarno Treaties, and is now prepared for the more
difficult, but not impossible, next step, disarmament. The governments
represented at Locarno accepted the idea that they would now work out
plans for general disarmament, in preparation for a World Disarmament
Conference which would be held at an unspecified date.

Bessil'ny maestro. *Pravda*, Moscow,
3 December 1925

Бессильный маэстро.

Рис. МАЙКЕЛЯ.

This Soviet cartoon, 'Powerless maestro', takes a very different view of
Locarno. The despairing British chess-master, Foreign Secretary Sir Austen
Chamberlain, sets his pawns – Poland, Czechoslovakia, France and Germany – against the much more impressive piece 'USSR', whose proletarian
face regards his opponents with visible contempt.

The implication is that the Locarno agreements were designed primarily
to damage the Soviet Union, but that they were bound to fail. The first
part of that judgement, at least, seems to be erroneous. There is little, if
any, evidence that European governments of the mid-1920s were either
plotting to destroy the Soviet Union, or particularly fearful of its likely
effects upon their own countries in the foreseeable future. Chamberlain
was much more concerned to resolve disputes between France and Germany.

As for the roles of Czechoslovakia and Poland, the British Foreign
Office showed no enthusiasm whatever for the eastern alliances which so
interested the French. Those alliances, indeed, were designed essentially
to protect Germany's neighbours against a German military recrudescence,
not with reference to the Soviet Union.

The Scale of Reparation Payments

The cliffhanger negotiations and stances did not alter essentially the reparation and disarmament situation *per se*, nor did they constitute the beginning of the elimination of Anglo-French dissension, for within a short while the London agreements went to waste. Although, in the summer of 1921, Germany met the first cash instalment, her acquiescence stemmed, to no little degree, from the customs cordon of the Allies that was still in force. But after the abolition of the cordon Germany paid scanty amounts (in November 1921 and early in 1922), and from then on the payments virtually ceased altogether, given the loss of the mark's value.[91] Even though a moratorium was granted to Germany at the end of 1921, it helped neither to stabilize her economy nor to stop the galloping inflation. At this point the paths of England and France once again diverged.

In December 1921 Lloyd George was prepared to consider a scaling down of the reparation payments, but in France the Prime Minister flinched from the very idea for fear lest public opinion would depose him from office. In May 1922 Lloyd George referred at length to the need to reconstruct Germany economically before making any further claims upon her,[92] while Poincaré, who in January 1922 had replaced Briand, spoke passionately about Germany's duty to meet her financial liabilities.[93] Lloyd George was convinced that:

> any policy involving the handing over of Europe to the tender mercies of M. Poincaré and the French militarists would be directly contrary to the traditional policy of Great Britain and would be fatal to the reconstruction of Europe and highly dangerous to the British Empire.[94]

While Poincaré held, with no less measure of vigour, that it would be an even greater blunder to yield to the Germans, he was prepared, in March 1922, to grant a partial moratorium. Subsequently, on 31 May, the Reparation Commission granted the moratorium, minus the preconditions previously rejected by the Germans. But when, on 12 July, the German Government applied for a full relief from the reparation payments not only for that year but also for the coming years,[95] the French saw it as an affront and were not prepared to consider it.[96]

In all these proceedings, and despite the Foreign Secretary's attendance at most of the conferences, the Foreign Office was a sideline spectator. The dominant figure in the contacts with France and Germany was the Prime Minister, whilst the Treasury took responsibility for the handling of politico-economic issues. Against this background it was inevitable that the Foreign Office would become increasingly dependent on the interpretation of the British Treasury in all that pertained to the economic repercussions of political developments, and that a considerable decline would take place in Foreign Office capability to criticize, from a political point of view, economic or financial moves suggested to it by the Board of Trade or the Treasury. For instance, it was the Treasury that made it clear to the Foreign Office, in the course of 1920–22, that the implications of reparation collection were hindering the recovery of Germany, that the non-recovery of Germany would hinder a European recovery, and that a European non-recovery would hinder the recovery of the British economy.[97] Lord D'Abernon, who was equipped with greater knowledge of currency and economics than anyone in the Foreign Service,[98] made it clear, in December 1920, to the Foreign Office that:

> Germany is practically bankrupt, and that any idea of getting, for the present, large payments over and beyond what she is already doing is fantastic.[99]

Britain's Delegate to the Reparation Commission, Sir John Bradbury, added his voice:

> The present total of 132 milliards (or rather something like 70 milliards – the nearest estimate which can be made of the present value of the Schedule of Payments liability) has the advantage of being ridiculous as well as absurd . . . but if the Allies really set themselves to attempt to recover 40 milliards we should be passing from pure to applied lunacy.[100]

In 1923 the Board of Trade and the Treasury proposed to the Foreign Office that, for the sake of Europe in general and the British economy in particular, the Ruhr ought to be evacuated at the earliest date.[101] When, in 1924, the Office wished to make out the economic ramifications of the French presence in the Ruhr, it turned, on this point, first to the Department of Overseas Trade and then to the Board of Trade, which, in a detailed memorandum, noted that the French control over the mining industry in the Ruhr would bring about the cutting-off of British industrialists from these raw resources. It would give France a distinct commercial predominance in Central and Western Europe; and hinder the recovery of Germany, for the Ruhr and the Saar constituted a quarter of the whole of German industry.[102] Finally, with the publication of the Dawes Plan, Crowe minuted that it would be useless on the part of the Foreign Office to suggest lines of action so long as it did not know whether the plan was workable or acceptable, and that the Foreign Office ought to wait until the Treasury had crystallized its stand on this point, and also to consult with Bradbury.[103] Likewise, the Foreign Office turned to the Board of Trade with a request that it would analyse for it the commercial repercussions of the Dawes plan.[104]

On the basis of the statistics and other information supplied to them, Foreign Office officials came to the conclusion that the reparations resulted in despair, bankruptcy, inflation and the complete ruin of the middle class, which alone can 'give spiritual stability in an industrial age'.[105] J. W. Headlam-Morley, Historical Adviser to the Secretary of State for Foreign Affairs, criticized the Allied reparation policy as something impractical, as it was bringing Germany to the verge of starvation.[106] His opinion was joined by the Head of the Central Department, Lampson, who asserted that the reparation policy would be revised, as the Treasury advocated.[107] For his part, Waterlow added that unless the reparation policy was revised, as the Treasury was advocating:[108]

> The German Government may be expected under economic and other pressure, internal and external, to swing gradually to the right. In time a reactionary coup d'etat, or even the peaceful access of a reactionary Government, to power, might well provoke a communist revolution, which would prepare the ground for a bankrupt and desperate Germany to 'join the Bolshevists' in earnest.

In Waterlow's opinion, that would be too heavy a price for the Allies,[109] the more so as he did not believe that France truly needed reparations for the purpose of her reconstruction.[110] Bennett, too, remarked that one should not pursue a policy that would prevent Germany from regaining her power.[111]

These officials, as well as their overlords,[112] accepted the Treasury–Bradbury–D'Abernon stand that the problem was not Germany's willingness to pay reparations, but her capacity to do so. Lampson admitted that the Foreign Office lacked the requisite qualifications to assess Germany's measure of capacity to pay reparations,[113] and accepted without demur the economic departments' assertions that any attempt to extract large payments from Germany would lead to German bankruptcy,[114] which would affect Europe in general, and Britain in particular. Thus, when the Reparation Commission fixed, on 27 April, and the London Conference approved, on 5 May 1921, the total German liability at 132 milliard gold marks, Foreign Office officials were appalled, for they knew that the defeated Germany could not possibly meet these demands. The facts that from the summer of 1921 to the end of 1924 the Germans paid very little in cash reparations; that they frequently failed to meet the full quotas of deliveries in kind; that in the course of 1922 the mark's depreciation had become acute; and that there was a conjunction between reparations deadlines and dramatic inflationary falls of the mark,[115] were taken by the Foreign Office as a further proof that the reparation payments were destroying the German currency.[116]

Only a few suspected Germany's true intentions, the most prominent of whom were Tyrrell[117] and Crowe, who minuted that:

The Germans actively connive at, if not promote the fall in the exchange, which they regard as a weapon with which to force France to revise the Treaty of Versailles![118]

Crowe also realized that, having spent billions on reconstruction in the hope of huge payments from Germany, the French need for reparations was real, as it seemed politically impossible to use taxes for these purposes. Should reparations not be paid, the French would not be able to reconstruct their economy.[119] In spite of this, it is not surprising that Poincaré's refusal, in November 1922, on 9–11 December 1922 and on 2–4 January 1923, to accept compromise solutions on the basis of granting a full four-year moratorium on all reparation payments, was perceived by the majority of officials as superfluous stubbornness.[120] France's and Belgium's entrance into the Ruhr, on 11 January 1923, on the pretext that Germany was behind schedule in delivery of wood, coal and telegraph poles, appeared to Foreign Office officials as an unreasonable and illegal move that would cause a disaster. Headlam-Morley concluded:[121]

It was now generally recognised that the Reparation Clauses could not be defended. They can be indeed explained, condoned, excused, but an attempt to defend and justify them would almost certainly fail.

These officials were so impressed by the statistics supplied to them by the economic departments that they overlooked Germany's offer, on 24 April 1921, to pay a sum that exceeded considerably the one they were ordered to pay on 5 May. Also, they were unaware of what was an open secret in the Treasury, namely that the sum Germany was called to pay according to the London Schedule of Payments did not even approach 132 milliard gold marks. The Schedule of Payments determined that bonds of series A be issued at a grand total of 12 milliard gold marks and bonds of series B at a grand total of 38 milliard gold marks; and that these be redeemed by the Germans in annual payments of 2 milliard gold marks plus 26 per cent of the export's profits, which together amounted to 3 milliard gold marks.

The experts, however, anticipated that the Germans would not be able to meet their first payments; therefore it was decided that any default would not be interest-chargeable. Also, it was decided that the percentage of the export would be paid only when Germany was able to do so, and in case of a delay she would not be finable. It is evident that this arrangement constituted an incentive for the Germans to falsify export figures and to lag behind on interest payments, as indeed they did. The bulk of the German obligation, however, was contained in bonds of series C, at a grand total of 32 milliard gold marks, and it was determined that as long as Germany did not redeem the previous bonds – something that the experts estimated would not take place in the foreseeable future

– the C bonds were not to be issued and Germany would not be obliged to provide coupons for them.

The likelihood that Germany would provide coupons of her own free will was slight, and so, of course, was a swift redemption of the A and B bonds, that was a precondition to the issuing of C bonds. A situation was thus created in which the British Treasury designed chimerical bonds with the clear knowledge that the Germans would not be obligated to pay them. Furthermore, even the reduced London Schedule of Payments held out until early 1922; and from then till 1924 Germany paid very little in cash payments, arguing that the constant fall of the mark was not enabling her to honour her liabilities. Even the suspicions of Crowe and Tyrrell, and the estimation of a number of British and French experts, that Germany was deliberately ruining the mark, partly to avoid budgeting and currency reform, but primarily to escape reparations, did not convince those who accepted the German claims that the cause of inflation was the payment of reparations.

Nowadays there is no doubt that Crowe's and Tyrrell's suspicions were sound. The Reich Chancellery archives of the Weimar Republic indicate that in 1922 and 1923 German leaders deliberately chose to postpone tax reform and currency-stabilization measures in the hope of obtaining substantial reductions in reparations. Also, the assumption that the French occupation of the Ruhr would break Germany down economically, and cause heavy damage to British industry, proved inaccurate. The inflation enabled the German Government to pay off its domestic debts, including the war debt and those of state enterprises, in worthless marks. German industrialists who had close ties with the Government made handsome profits until hyper-inflation set in. The ailing British economy also benefited considerably from the disruption of German exports, but British officials would never acknowledge this fact, even to themselves, convinced as they were that this was something temporary, which did not certify that the French occupation of the Ruhr was something favourable. They did not cease from urging France to evacuate it.[122]

The inability of Foreign Office officials to criticize the financial and commercial appreciations of the economic departments meant that the latter did not realize the political significance of the German default in reparation payments. Instead of seeing in the default a German attempt to reverse the military verdict that was attained against them in 1918, and an unwillingness to recognize the Treaty of Versailles, they focused upon Germany's capacity to pay, and that too on the basis of dubious figures.

On 5 December 1922, Joseph Addison, the Counsellor to the British Embassy in Berlin, reported that the Germans had not given up their intention to avoid reparation payments. Their request that they be granted a full moratorium on reparation and foreign loans reminded him of the attitude of the Sultan in Hilaire Belloc's novel *Mercy of Allah*:

> Readers of this book will remember that the Sultan having issued with great benefit to himself pieces of paper on which was inscribed a promise to redeem in gold, was finally led by the success of this plan to issue pieces of paper on which appeared the words 'This is gold', whereupon a camel dealer, whom the Sultan had sent a piece of paper in payment for a camel, sent in return another piece of paper with the words 'This is a camel'.[123]

In a memorandum of the Central Department, in which was summarized the reparation position to December 1922, it was observed that if Germany did not pay in future a large sum of money there was a danger that eventually she would pay nothing.

> She would then, owing to her inflationist proceedings, be the one great industrialist power in Europe enjoying the advantages accruing from the extinction of both foreign and internal debt.[124]

Like the rest of their colleagues in the Foreign Office, Crowe and Tyrrell, in 1922, reached the conclusion that since Germany had managed to wreck her economy she should be granted a moratorium on reparation payments. It was agreed with the Treasury that tactically one ought to strive for the evacuation of the Ruhr, for the purpose of preventing a perpetuation of the French presence there, and to adopt the Dawes Plan of 9 April 1924 as a means of reconstructing Germany and renewing the reparation payments.[125]

Conflict over Upper Silesia

The next link in the chain of German issues which caused Anglo-French discord was Upper Silesia. Under the Versailles Treaty the fate of this province, to which both Germany and Poland laid claim, was to be determined by a plebiscite. However, when the Allies embarked upon executing this resolution, in August 1920, the British Commissioner on the Inter-Allied Administrative and Plebiscite Commission reported that disarming the Germans but not the Poles gave encouragement to armed Polish bands to assault and murder German citizens. The French, who constituted the main garrison force in the province, hoped to tip the plebiscite in favour of the Poles.[126] At the same time the British District Controllers reported their wish to resign their commission.[127] Phipps and Waterlow recommended dismissing the French representatives on the grounds that only this action would restore credibility to the Commission.[128] Waterlow stressed that allowing the French representatives, whom he described as arch-criminals, to remain would imply a continuing bias towards the Poles. Perforce, it would create a local crisis that was liable to sweep in its wake the whole of Europe.[129]

Crowe, on the other hand, favoured more moderate steps, and was also not prepared to take the facts as read. He pointed out that the British representatives had made no attempt to have it out with their French colleagues and observed that as long as it had not been proved that the French force was, indeed, inadequate to enforce order, and that it did not encounter the same problems facing Britain in Ireland, one should not have grievances against them.[130] Furthermore, he did not believe that the French and the Poles were exclusively responsible for the current situation. And when it was reported to the Foreign Office that large caches of arms had been found in the possession of German citizens in Upper Silesia he minuted: 'The seizure of the German arms shows that the Germans are not so innocent as Col. Percival has tried to make out.'[131] Hardinge changed his position a number of times but eventually sided with Crowe.[132] Curzon, however, had no doubt from the beginning on whom the blame should be laid, and even Crowe's observations would not convince him.[133] In the end it was agreed at the Foreign Office and, later on, at Cabinet level, to urge Paris to refrain, henceforth, from any bias. Subsequently an assurance to that effect was given to Derby by the French.[134]

All the same, Waterlow was still not satisfied. The steps taken were interpreted by him as a British failure to restrain France, and he called on Britain to lay publicly the full responsibility for what was taking place in Upper Silesia on the French and to wait until they mended their ways, or withdraw altogether from the plebiscite.[135] To that, Crowe vigorously objected. He minuted that a withdrawal from the plebiscite would bring about a breach with France, encourage the Germans to object to the outcome of the plebiscite, should it prove to be against them,[136] and be a blatant violation of the Peace Treaty, all the more so because it had been Britain at Paris who had introduced and supported the original idea of plebiscite. It was partly the fault of Britain that there were not sufficient forces in Upper Silesia capable of enforcing order, for she had reneged on her assurance to send troops there. Crowe suggested a solution: reinforce the French garrison with British troops; insist on the removal of the Polish violence-mongers;[137] and if all else failed, threaten British withdrawal – a step that would, in his opinion, deter the French.[138] His stand was approved by Curzon, with the exception of his proposal to send troops,[139] to which the War Office was also opposed.[140] On 30 December 1920, the Cabinet decided to tell the French that Britain was prepared to send four battalions to Upper Silesia.[141] Yet, when the French responded to the proposal in the negative, Curzon minuted, with a sigh of relief, that by that Britain had done her duty and she would no longer have to send troops.[142] Crowe did not, however, give up. And against the background of the continued unrest in Upper Silesia, and the need to accord weight to the stands of the British representatives on the spot, he once again suggested, on 12 May 1921, that British troops be stationed in the area.[143]

Curzon continued to oppose any British involvement on the Continent of Europe, and so did Churchill and the Chief of the Imperial General Staff.[144] But this time Crowe's recommendations were accepted by the Cabinet, which approved them on 24 May, exposed the fact of the Prime Minister's decision to send four battalions, but also instructed the War Office to send an additional two.[145]

These deliberations brought to the surface the fundamental question of to which of the sides Britain preferred that the district be ceded. Curzon and Lloyd George were in favour of assigning Upper Silesia to Germany on the grounds that there was a German majority there,[146] and that it was vital to Germany economically – a consideration that was of decisive importance as far as the Prime Minister was concerned.[147] This was also the opinion of D'Abernon, John Thelwall, the Commercial Secretary to the Berlin Embassy, and Waterlow. D'Abernon emphasized the loss of prestige that would be caused to Britain in Germany should any other decision be taken,[148] while Thelwall highlighted the grave economic consequences that would ensue should Upper Silesia be ceded to Poland.[149] Crowe, on the other hand, thought that Britain should remain absolutely neutral on this point.[150] He also dismissed appreciations regarding the devastating consequences that would result from the cession of the district to Poland, minuting that:

> . . . complete disaster immediately overcomes German territory ceded to Poland is not altogether borne out by such parts of the former German Empire, as have been actually incorporated in Poland since the Peace.[151]

A similar opinion was also expressed by Hardinge, though his reasons were different. Referring to the distinctions between British and French policy in Upper Silesia he observed that:

> they do not regard the question from a point of justice or right, but simply as one of policy. They cannot understand that we should consider ourselves in such a matter, bound by our word given and registered by our signature of the treaty, and their avowed policy is by hook or by crook, to obtain possession for the Poles of the whole of the industrial area of the province. They regard it as their only safeguard against future war and at present will listen to no compromise.[152]

As might have been expected, the Prime Minister's stand triumphed on this point.[153] In the event, this had a secondary importance, for Britain did not have the necessary dominance in order to decide the fate of the province as she saw fit. The results of the plebiscite, which was held on 20 March 1921, were far from being unequivocal, and did not justify the cession of the whole of Upper Silesia either to Poland or to Germany.[154] Consequently, Allied representatives on the Plebiscite Commission recommended the partition of the province between the two states,[155] an idea taken up by Waterlow.[156] However, the Commission was unable to arrive at an agreement as to where precisely the German–Polish frontier be delineated,[157] a problem that recurred during the prolonged negotiation between the various governments – with Poland's claims upheld by France and Germany's by Britain and Italy.[158] Finally, at a Paris meeting held by the Allied Supreme Council on 12 August, it was agreed to refer the question to the adjudication of the League of Nations.[159] And, on 12 October, the League submitted a sketch of the new frontier, which was accepted on 19 October by the Conference of Ambassadors. Upper Silesia in general, and the industrial area in particular, were partitioned between the two states – with Germany being assigned two-thirds of the territory, while Poland was assigned the greater share of the industrial area.[160] This decision left Germans feeling very bitter.

Further Disagreements with France and Some Moves towards Reconciliation

Although the German tactic of letting Lloyd George fight their battles for the revision of the Treaty of Versailles contributed to the driving of a wedge between Britain and France, there were additional developments that fouled the relations between the two even further. For instance, the hostile attitude of the French Press and public more than once drew protests on the part of Britain.[161] France's desertion of Britain and Greece, in the midst of their campaign against the Turkish Nationalists, provoked indignation in Britain.[162] Finally, there was the French refusal, at the Washington Conference on Naval Disarmament and Far Eastern Affairs (12 November 1921 – 6 February 1922), to limit their Continental and air forces, and refrain from an extensive development of a submarine fleet, which together with the constant frictions in Europe raised gloomy thoughts as to the continued worthiness of the *entente* with France, which, it seemed, was becoming a far greater potential threat to Britain than any German menace.[163]

In order to halt this deterioration in relations, Saint-Aulaire, the French Ambassador, met with Curzon in his room at the Foreign Office on 5 December 1921. The Ambassador proposed to him privately to put an end to France's fears of Germany, to her reluctance to limit her forces, and to the rivalry between her and Britain, by means of an Anglo-French defensive alliance, an idea that had been discussed secretly between him and the now Lord Privy Seal, Austen Chamberlain.[164] However, the Ambassador held that the alliance was to become operative not merely in case of an invasion into France proper, but also if France's Allies in Eastern Europe were to be attacked by Germany or Russia. In return, Britain would gain a French guarantee of the frontiers of India, the settlement of their disputes in the Far East, and a French readiness to make concessions to Germany, such as the obviation of her opposition to the admission of Germany to the League of Nations, to a French partaking in Lloyd George's plans to reconstruct Russia, and to a considerable reduction in the scope of France's land armaments.[165]

The idea did not fall on deaf ears in Paris, and although Briand was not aware in advance of the substance of this initiative, he welcomed it.[166] On the one hand, it corresponded with his intense desire, as expressed on 4 December, to find an overall solution to the Anglo-French disputes, so that it would be possible to return to the days of the *entente cordiale*.[167] On the other hand, it constituted for him an honourable solution to his current dilemma, namely the need to choose between the maintenance of the *entente* with Britain, the implication of which was a more flexible line towards Germany, and his perseverance with a hard line towards Germany, the implications of which were an almost certain break with Britain, something that he wished to prevent at all costs. In London, the British Administration, for its part, was also prepared to consider the idea seriously.

Lloyd George, who had previously opposed the idea, was now ready, against the background of his failure to promote the economic and political reconstruction of Europe, given the French opposition, to admit that under certain circumstances an Anglo-French pact was preferable to the present deadlock.[168] Churchill and Chamberlain saw in the proposal a confirmation of their opinion of old, that Britain's European policy had to be founded on co-operation with France, and that this was the best way of bringing a moderation in France's attitude towards Germany.[169]

Even Curzon was prepared, under certain terms, to modify his opposition to a formal tie-up with France. In the first place, he was attracted by the idea of putting an end, once and for all, to the French threat to Britain's interests in Africa and the Middle East. Secondly, he saw in it an opportunity to check the continued heading of the two states towards a rupture, which he regarded as a calamity,[170] noting in this connection that:

Lord Lansdowne thought that he had disposed of the main causes of Anglo-French misunderstanding when he signed the Agreement of 1902. Long before the war there was a fresh and fertile crop. The war was intended to effect another clearance; it has produced a new germination.[171]

After Briand submitted a formal proposal in meetings that he held with Curzon and then Lloyd George, on 21 December,[172] the British Administration was inclined to consider it favourably.

Nonetheless, there were differences of opinion as regards the form of the alliance and the return that France would be required to pay. Hardinge, Britain's Ambassador to Paris, thought, as he alluded in the course of his meeting with Briand, that Britain should restrict herself to guaranteeing the frontiers of France; for otherwise Britain would be liable to find herself entangled in wars in Eastern Europe, and France might be turned into the landlord of the Continent.[173] Crowe, on the other hand, answered the Foreign Secretary's question as to what the Foreign Office stand on this point was, observing in a long and inspiring memorandum (which the officialdom found on their return from the Christmas holidays!),[174] that generally it was possible to accede to the French appeal not to settle for a guarantee pact after the form of 1919, but to conclude a comprehensive defensive alliance. Crowe pointed out that Britain had, in any case, a moral obligation to guarantee France's frontiers, given the fact that this was the return that was promised France in 1919 for her consent to refraining from annexing the Rhineland. The more important reason, however, for this step stemmed, according to Crowe, from the fact that the achievement of Britain's European goals was contingent on the co-operation of France, and that that co-operation was dependent on an alliance.

> The one great preoccupation of the French mind is the danger of a German war of revenge. That Germany will prepare for it systematically, relentlessly so long as any hope of success remains, is indubitable. This danger clouds the French outlook like a nightmare. If there is anything open to France that can exorcise it, so precious a loan should be worth paying for even at a high price.

Practically speaking, observed Crowe, the proposed alliance would assist in promoting Britain's goals on the following subjects: first and foremost it would ensure that a new and prolonged war would not break out, given the fact that there did not exist a power, or a bloc of powers, capable of opposing this alliance in the foreseeable future. At the same time it would bring about the removal of French opposition to the economic reconstruction of Germany and the reduction of reparation payments, without which it was not possible, in the opinion of experts, to restore industrial and commercial prosperity to Germany; and, as the familiar argument ran, in consequence, to Britain (and Europe):

> There is ground for believing that France's exacting attitude towards Germany in the reparations question is – at least partly – inspired by the same considerations which actuated her in originally demanding

the middle-Rhine frontier; namely the desire to hold Germany down for fear of her becoming strong enough to attach France successfully. If the menace of a German invasion were conquered, as it would be, by an effective Anglo-French alliance, it might be presumed that France would become less bent on preventing or retarding Germany's economic recovery.

Finally, it was thought that with the pact in her pocket France would be willing to accommodate Britain's wish to start disarming and resolving the differences of opinion between them in various places all over the world, to the satisfaction of Britain. Crowe was aware of the danger that the alliance was liable to entangle Britain in wars that were not hers. But in his opinion this could be overcome by restricting the British pledge to go automatically to war to the case of a direct attack on France.[175]

Curzon found Crowe's arguments most convincing. However, like Hardinge, he did not agree with him that the form should be of a definite alliance. Instead, he advocated, on 28 December, in a memorandum which he circulated among his cabinet colleagues and that was based largely on Crowe's notes, that Britain should be content at this stage with giving a security to France's frontiers against an unprovoked German attack, in the form of the Guarantee Pact offered to them in 1919.[176]

As was later to become evident to Curzon and the Foreign Office that this was a vain debate. The Foreign Office did not know that in the course of Lloyd George's meeting, on 21 December, with Briand, the Prime Minister had already decided that what would be offered to the French was merely a British guarantee,[177] and that the return that France would be required to pay would be, first and foremost, the adoption of the Prime Minister's economic schemes for the reconstruction of Europe, a *de jure* recognition of Russia, and co-operation at Genoa.[178] Nor was the Foreign Secretary informed that the Prime Minister intended discussing the pact question in his forthcoming meetings with Briand at Paris and Cannes, as indeed he did.[179] All that the Foreign Office knew was that a fundamental decision was taken by the Government to adopt the Foreign Secretary's recommendation to restrict the British commitment to a guarantee pact only.[180]

These circumstances placed, from the outset, the success of the Paris talks (26 December 1921) and Cannes Conference (6–13 January 1922) in jeopardy. Briand was prepared to co-operate with Lloyd George in all that pertained to the reconstruction of Europe, eliminate the Anglo-French differences world-wide, refrain from broadening French naval and land arms, and even agree that on subjects such as Eastern Europe both sides would consult among themselves as to what steps should be taken. Under the pressure of the President, his Government colleagues and the Chamber of Deputies, Briand insisted that the pact should be a bilateral defensive alliance, that a German violation of the Versailles disarmament and demilitarization clauses should be treated on the same footing as a direct and unprovoked act of aggression against the soil of France, and that a military convention be signed between the two General Staffs, though he did not underscore this latter point. The sole prospect of finding a compromise was Briand's intense desire to reach an agreement.[181]

But before negotiations could continue, Briand had to resign. First he had to bear the uproar caused by the famous golf incident of 8 January. Photographs and reports showing the French leader taking instruction from the British Prime Minister symbolized to many Frenchmen the relationship of the two men of state, and the political relation between the two countries. Next, on 11 January, Briand was forced to return to Paris in order to reverse the decision of the President and Cabinet prohibiting him from reducing the amount of reparations, giving a moratorium to Germany, inviting Russia to the Genoa Conference, and granting world-wide concessions to Britain, *before* the conclusion of the pact. Thereafter he resigned, though he managed to carry the Chamber and Cabinet with him. His place was taken, on 15 January, by Poincaré.[182]

The negotiations faltered. Poincaré, in contrast to Briand, set stringent terms, which included, among others: a disinclination to scale down the sum of reparations;[183] a clause committing the two parties to act together on all questions likely to endanger the general order of things as set up under the Peace Treaties, and by implication to safeguard the integrity of France's East European Allies; and a demand to place at the head of the negotiations' scale of priorities a detailed military convention between the General Staffs.[184] Neither Poincaré nor the French Ambassador was prepared to retract the idea of policing the Continent, or the principle of an *entente constante* between the two General Staffs, when told that these two proposals were unlikely to be acceptable to Cabinet and Parliament.[185] The first proposal was, among other things, inconsistent with the principles of multilateral settlements of the League Covenant, which could only result in the reintroduction of the old policy of rival groupings of Great Powers; while the second proposal meant that the decision concerning the application of the pact rested not with the civilian authorities but with the General Staffs, which was hardly an appealing formula to the politicians.[186] In London, Curzon procrastinated with the Cabinet's approval, a deliberate procrastination on the question of the pact,[187] and put off dealing with France's Ambassador, time and again, during February to April.[188] His motives for this step were Hardinge's reports that Poincaré was in desperate need of the pact, and that if they acted with forbearance he would be prepared not only to alter the form of the pact, but also to settle beforehand the Anglo-French differences world-wide.[189] The outcome was that, until April, Poincaré would not grant Lloyd George the economic return he desired in exchange for the pact; namely, preparedness to reduce the sum of reparations, attend the Genoa Conference (10 April – 19 May 1922), and give support to the British moves for the reconstruction of Europe. Curzon, for his part, rejected Poincaré's pleas respecting the need to conclude the pact as soon as possible, observing that it was necessary first to settle all outstanding questions between them in other parts of the world.[190]

The Treaty of Rapallo

While the Allies were haggling over whether there was any point in the pact, the Germans and the Russians concluded one, on 16 April 1922, at Rapallo. The Treaty of Rapallo was, in part, the corollary of the Russian desires to win international recognition and to split the capitalist world. In part it was due to Germany's desire to emerge from her diplomatic isolation and enhance her bargaining power. Common hostility against the Allies – on the part of Germany because of the Treaty of Versailles, and on the part of Russia because of the Allied intervention in the Civil War, overcame Russo-German antagonism at various levels. The Social Democrats and President Friedrich Ebert had opposed the stand of Baron A. G. O. von Maltzan, Head of the Foreign Ministry's Eastern Department, and of Joseph Wirth, the Chancellor, that Germany's recovery, the wrecking of the Versailles Treaty, and the reinstatement of the common border between Russia and Germany in the east were contingent on Russo-German co-operation. In January 1922 Walther Rathenau, the Foreign Minister, thwarted the attempts of Georgi Chicherin, Russian Foreign Minister, Maxim Litvinov, Deputy Foreign Minister, and Maltzan, to reach a political agreement; it seemed that the President did not know that the Reichswehr, Junkers and Krupp firms were establishing armament factories in Russia and training troops there.[191]

At Genoa the Russians took advantage of the rumours that Lloyd George was striving towards an Anglo-Russian agreement at the expense of Germany, and threatened that they would, indeed, make a pact with Britain if Germany did not come to an agreement with them immediately. After the German Delegation had discussed the problem (wearing pyjamas), they drove, on 16 April, despite Rathenau's opposition, to Rapallo, where, together with the Russians, they worked out the final details. And in the evening they concluded the treaty. In this treaty Germany and Russia agreed to re-establish their diplomatic relations, waiver their right to claim from each other war damages, and accord each other, on the commercial level, a most-favoured nation status.[192]

The Treaty of Rapallo was received at the Foreign Office with mixed feelings. Some reasoned that the outcome was inevitable in view of the fact that on the eve of the Genoa Conference the Treasury and the Board of Trade had taken into their hands the formation of Britain's policy stands on such subjects as disarmament, peace, reparation, and the economic reconstruction of Europe.[193] Others pointed out that the Foreign Office representatives to the Genoa Conference did not play any role worthy of the name nor were they proficient in its details; and that those who remained in London had to rely on the fragmentary reports of the 'amateurs' who accompanied the Prime Minister. For example, following the report from Genoa of the Lord Chancellor, F. E. Smith, First Earl of Birkenhead, that progress made there under the Prime Minister's influence was 'wonderful', and that Lloyd George's personal position had never been stronger,[194] Tyrrell minuted ironically that, 'the Lord Chancellor's views are very comforting but the conclusion of the Russo-German agreement is rather a shock'.[195] While Curzon observed with sarcasm that, 'The Lord Chancellor, who only landed from his yacht on the morning on which he telegraphed, was very rapid in forming his conclusions.'[196] But in part there was also an expression of concern. Curzon observed to Chamberlain that:

> Germany undoubtedly guilty of a very shabby trick in making the agreement with Russia behind our backs. She has been caught in some bad deceptions about disarmament. She is probably more deeply involved with Russia than we know. She has not at all a clean bill of health.

Curzon further maintained that Germany should not be given any encouragement to persist with her political attitude towards Britain, by announcing that Britain favoured her inclusion in the League of Nations.[197]

Relations with France do not Improve

In as much as Britain and France were troubled by the Treaty, it did not serve to bring them together, or abate the deterioration that took place in the British Administration's relationship with Poincaré. There were, it appears, three principal reasons. First, Britain continued to maintain firmly that a pact should not be concluded with France until all outstanding questions between them, which included among others the economic reconstruction of Europe, peace with Turkey, and internationalization of the City of Tangier, had been liquidated first. In particular the British were annoyed that at first Poincaré had agreed to follow in Briand's footsteps and acquiesce to this stipulation, but that thereafter he retracted.[198] Curzon, on 18 March,[199] and Balfour, who replaced him during his illness in May and June,[200] made it clear to the French Ambassador – who enquired on behalf of Poincaré why no progress was being made in the negotiations over the pact, given the Russo-German co-operation – that if Poincaré modified his views 'now and then', Britain did not change her position. Gradually these differences of opinion deepened into a feeling that there was nothing more to talk about. For instance, at a clearing talk held in Paris on 16 June, Poincaré observed to Hardinge that, in a talk that he had held with Lloyd George, the question of the economic reconstruction of Europe had not been mentioned at all. He could only presume from this that Britain was striving to renege on her intention to conclude the pact. France was totally indifferent to the question of whether the pact, in the limited form preferred by Britain, was concluded or not, knowing that if France or Belgium were attacked by Germany, Britain would come to their aid, as occurred in 1914.[201] 'If this is not mere bluff', minuted the Head of the Western Department, 'we can abandon hope of coming to a settlement on any points while M. Poincaré remains in power'.[202]

Secondly, it was unfortunate that the personages who conducted most of the negotiations over the pact after the Cannes Conference were Poincaré and Curzon. The first was a formalist who liked quibbling.[203] For example, on 19 May, Poincaré expounded with the aid of ten different exegeses why Millerand's assurance of 1920, in which it was said that, in all inter-Allied questions raised by the execution of the Treaty of Versailles, France had no intention of acting save in accord with her Allies, did not bind him.[204] Poincaré saw everything through the spectacles of the German threat to Europe in general and France in particular, as had Briand, and treated Curzon's post-war aims as selfish and materialistic.[205] In contrast, Curzon was a haughty character who loved arguing, especially when faced with a worthy opponent. He saw, in the manner in which the French in general and Poincaré in particular deduced the future from the present, in their precision and lucidity, a parody of these qualities.[206] In fact, everything was subordinated by them to general considerations of expediency instead of to the ordinary rules of straightforward and loyal dealing.[207]

These personality traits caused bitter clashes between the two, which reached their climax at a meeting that took place on 22 September 1922, in Paris. This meeting was part of a British endeavour to crystallize a joint Anglo-French stand which would enable negotiations to be opened with the Turkish Nationalists, who had reached the Straits. However, Curzon, who wished to set the record straight, reproached the French repeatedly for having abandoned their British ally and taken their troops from the Straits. In response, Poincaré retorted that the British representatives in

Turkey had deliberately twisted the facts. To that, Curzon replied that there was no truth in the accusation; that the facts, as he portrayed them, were correct, and that France was, indeed, guilty of treachery:[208]

> This provoked the wrath of Poincaré and he suddenly lost his temper and shouted and screamed at Curzon, really in the most insulting manner, pouring out torrents of abuse and making the wildest statements with a flow of language like Niagara, which completely bowled over Curzon, who collapsed entirely. Curzon kept on saying to me, 'What am I to do, hadn't I better go home to London, I cannot go on, something must be done!' I thereupon took upon myself to interrupt Poincaré, saying that Curzon was not feeling well and that the Conference must be adjourned, Poincaré accepted and Curzon left the room with the rest of us.

Both sides refused to retract their words, but after strenuous efforts Hardinge managed to obtain Curzon's consent to say, on his behalf, to Poincaré, that he withdrew the charge. He then persuaded Poincaré to accept the withdrawal and to apologize to Curzon.[209] Hardinge saved the situation, but it did not gloss over the fact that Curzon and Poincaré were not the proper persons for establishing an Anglo-French understanding.

Thirdly, even if both sides could have reached an understanding on Imperial issues, on the reparation question they remained divided before and after the establishment of the Bonar Law Government, on 24 October 1922, in place of that of Lloyd George. There was no dispute among the Allies that Germany had not fulfilled her financial liabilities. Some Foreign Office officials pointed out that Germany had deliberately created the economic crisis in order to escape reparation payments. They argued that German industrialists had done their utmost to devalue the mark in order to evade tax-paying; and had transferred their capital abroad.[210] However, at meetings held on 19 June and 7–14 August 1922 at London, and also after the Reparation Commission formally declared Germany, on 26 December, in default on timber deliveries and telegraph poles,[211] the British maintained that since Germany had succeeded in destroying her currency she should be granted a full four-year moratorium on all reparation payments in order to facilitate financial reconstruction.

The French, on the other hand, objected to rewarding Germany for her evasions, and pointed out that France was in urgent need of reparations, for without them she could not balance her growing financial deficit caused by the costs of war, war debts and reconstruction of the devastated regions. French officials expressed a willingness to waive France's claim only if a comprehensive financial settlement for the payment of reparations, accompanied by 'productive guarantees', was forced through by the Allies. The British opposed the seizure of pledges in general, and the Ruhr in particular, arguing that this step would only impair German recovery, topple the German government, lead to internal anarchy and Bolshevism, without achieving the financial goals of the French. However, the French maintained that a moratorium without guarantees would mean the end of reparation payments, and that Germany would give in and comply with the obligations imposed upon her at the Paris Peace Conference only if coercive measures were exercised against her.[212]

Timber deliveries were not the only matter in which the Germans were in default. Since October 1920 they had defaulted regularly on coal deliveries, and it was expected that in January 1923 the thirty-fourth default would take place, not to mention financial defaults. In a last-minute effort, intended at overcoming the loss of French patience, the Allies and Germany met in Paris (2–4 January 1923), with each party (except Belgium) equipped with a plan of its own. The Germans, in an attempt at distracting attention from reparations default, offered to sign a non-aggression pact; the French and Italian plans called for limited economic sanctions on Germany, with France declaring that, in the absence of full unity, she would take more drastic steps. The British submitted an economic plan, based on a scheme drawn up previously by Bradbury, which was so excruciatingly complex that Carl Bergmann, a leading German reparations expert, grumbled that he would rather pay reparations than master the plan. And after discussing the plans the sides agreed to disagree.[213]

Crisis in the Ruhr

Against this background there occurred the final crisis during Curzon's tenure of office as Foreign Secretary – the Ruhr crisis. Although the possibility that France would occupy the Ruhr for the purpose of forcing Germany to meet her liabilities had been discussed, more than once, in the past, Poincaré had sought to avoid such a drastic step. However, the drawn-out failure of international attempts to reach a settlement that would enable the resumption of reparation payments; and the pressure of right-wing circles and the declaration of the Reparation Commission, on 9 January 1923, that Germany had again defaulted on coal deliveries, left him no other option.[214] On 11 January, French and Belgian forces entered the Ruhr. In consequence, Germany gave notice of the suspension of reparations, and within a few days developed in the Ruhr, with finance from the Berlin Government, passive resistance to the French and Belgians, which lasted eight whole months and severely impaired the German and French economies.

Reactions in Britain were mixed and ranged from outrage and dismay to covert sympathy. Left-wing opinion and the economic departments demanded vigorous action from the government before any irreparable damage was caused to European society and economy in general and to British interests in particular. For instance, the Labour Party strongly criticized what was termed by it as a French attack on the 'self-determination of the German people'.[215] The Treasury,[216] the Board of Trade,[217] and President of the Federation of British Industries and Association of Trade Protection Societies of the United Kingdom, Sir Eric Geddes,[218] underlined the economic need of finding a solution, even at the cost of a rupture with France. While Bradbury observed, with reference to the political reasons that underlay the French pressure to declare Germany in timber default, that: 'Since, in the tenth year of the war, Troy fell to the stratagem of the wooden horse history recorded no similar use of timber.'[219] He added that Britain had not fought for the removal of the German menace so that it be replaced with a French threat.[220]

The Prime Minister, Foreign Secretary and Foreign Office, on the other hand, refused to turn the dissensions into an open breach with France, but rather favoured a policy of 'benevolent neutrality' – i.e. the dissociation of the British Government from the separate French and Belgian action in the Ruhr, while interposing no obstacle to the successful prosecution of their understanding.[221] Bonar Law dreaded that, if they reached a breach with France, chaos would be created in Europe.[222] Curzon, who was negotiating with the French over the Turkish Peace Treaty at Lausanne, and the Central Department highlighted the immediate and long-term British interest of maintaining the *entente* with France, noting that she had a legitimate grievance against Germany in the matter of reparations.[223]

Crowe, for his part, pointed out that even if Britain had wanted to she would not have been able to impose her will on France.[224] While Tyrrell asserted that British aloofness was the only practicable policy at present,[225] a stand the Foreign Office repeated, with the approval of the Prime Minister, who acted for the absent Foreign Secretary, to British representatives, noting that this was not Britain's quarrel and that she preferred, at this juncture, to pursue a policy of wait and see.[226] In fact, Crowe believed that French policy in the Ruhr and Rhineland was illegal and he would have pursued a stronger line in Paris had Curzon agreed.

At a Cabinet meeting of 14 March, called to decide between these different opinions, the decision fell in favour of the Foreign Office's stand that was to shape Britain's policy throughout the Ruhr crisis. From this point on, Foreign Office policy stands carried increasing weight at Cabinet meetings as a result of the readiness of Bonar Law and his successor, Baldwin, in allowing the Foreign Office to take over the handling of foreign affairs, including the tasks of organizing international and

League of Nations conferences, which had previously been undertaken by the Cabinet Secretariat. In order to placate the critics, however, it was also agreed that, should the Germans show inclination to negotiate with France, Britain would do her best to encourage them.[227]

The first expressions of a negotiations trend were given by Curzon, on 14 and 21 March and 20 April, at his meetings with the German Ambassador,[228] the French Ambassador,[229] and in his speech in the House of Lords,[230] in which he called upon the Germans to make a 'firm offer' such as might ease the present stalemate. In response, the Germans proposed, on 2 May,[231] that the amount of reparations be reduced to 30 milliard marks, to be raised by loans on the international money markets, and that the whole reparation problem be submitted to an international tribunal should the foregoing payments be considered inadequate[232] – a suggestion first raised, on 7 December 1922, by Tyrrell[233] and the American Secretary of State[234] and then repeated in international discussions and Parliamentary debates.[235] Finally, to allay French security fears, Germany reiterated her offer to sign a non-aggression pact.[236] However, since the Ruhr was to be evacuated in advance, and in return for a plan whose real value was a mere 15 milliard marks, and given the fact that interest on the initial loan was to be paid out of the proceeds of the loan itself,[237] the French and Belgians turned it down on 6 May.[238] The Treasury and the Foreign Office did not enthuse over the plan either. Otto Niemeyer said that it was no use,[239] while Curzon termed it stupid and shadowy.[240] And in a note to the German Ambassador he demanded that the present proposal be converted to something clearer and more precise.[241]

On 7 June the Germans submitted a second plan, which no longer mentioned a lump sum, passive resistance or the illegality of the Ruhr occupation. Germany's acceptance of the idea that her capacity to pay, the amount and method, should be referred to an international meeting of experts was repeated, and concrete guarantees for the accomplishment of the final Reparation Settlement were offered.[242]

This latest offer met with a much more favourable reaction. Curzon and his Cabinet colleagues tended to see, in the idea of treating reparation as a business rather than a political problem, an honourable way out for all sides.[243] But unfortunately this idea was rejected on the spot by the French,[244] who reasserted during June–July, in response to British enquiries,[245] their refusal to treat with Germany until passive resistance ended, and insisted on the continued validity of the London Schedule of Payments.[246] The result was not only that the scheme was not adopted generally, but also that no reply was sent to the Germans.

Against the background of growing displeasure, Britain finally took the initiative. On 20 July Curzon submitted a definite British plan of action for resolving the Ruhr crisis. The purpose behind the plan was that it be preserved as an inter-Power pressure-measure, and as a means of forcing an international solution on Belgium and France.[247] In return for the abandonment of passive resistance and the resumption of civil administration of the Ruhr, the creditor Powers might agree to accept the decision of an impartial body of experts regarding Germany's capacity to pay reparations.[248] But the plan was rejected by France,[249] and so too, but less forthrightly, by Belgium.[250] A month later Curzon made a further effort. Crowe and Curzon induced the Cabinet to accept a full and strong statement of the British case, in the form of a Note to the French and Belgian Ambassadors, on 11 August.[251] Besides dealing with Franco-Belgian objections to the British suggestion of submitting the whole question of reparation to a committee of experts, it reviewed the whole past history and present elements of the controversy, reasoning that if any further debts were to be collected the whole French policy of making impossible demands for reparation and then enforcing these demands by purely destructive sanctions would have to be abandoned. The Note concluded by observing, with reference to members of the British Government, that:

> They are reluctant to contemplate the possibility that separate action may be required in order to hasten a settlement which cannot be much longer delayed without the gravest consequences to the recovery of trade and peace of the world.[252]

The Germans were delighted by the note, Poincaré was, however, unimpressed.[253] The Belgian lukewarm response,[254] and the abortive Baldwin–Poincaré meeting of 19 September,[255] depressed Curzon, who decided to refrain from pursuing this line any further.[256]

The turn came, on 26 September, when the new German Government, under Gustav Stresemann, overwhelmed by hyper-inflation and severe internal unrest,[257] declared its intention[258] to abandon passive resistance unconditionally and seek a settlement with France. This both reopened the question of reparation[259] and brought to a head the anxieties concerning the future of the European balance of power[260] and British economic interests on the Continent,[261] should Germany collapse or be forced into co-operation with the Franco-Belgian regime in the Ruhr and the Rhineland. Whereupon, pressure to draw up a new reparation plan, along with a scheme to reconstruct Germany financially and extract France and Belgium from the Ruhr, started to build up. This pressure came from a number of sources. It was expressed first in President Calvin Coolidge's declaration of 9 October that the United States Government rested on their proposal of December last;[262] and secondly, in the taking up of the offer by Curzon, on 12 October.[263] The Americans agreed to take part in an Advisory Economic Conference to consider Germany's capacity to pay reparations and to suggest a plan for securing payment or, alternatively, to send competent American citizens to serve on an advisory body appointed by the Reparation Commission.[264] In a forceful *démarche* made by Curzon, on 19 October, to Paris, Rome and Brussels he pressed for acceptance of one of the American offers,[265] and won the assent of Belgium[266] and Italy[267] to the second proposal. Poincaré attempted, it is true, to delay this process but France found herself increasingly isolated. In London, Baldwin rankled that he was misled into believing that Poincaré would negotiate with the Germans should they desist from their passive resistance.[268]

Crowe warned the French Ambassador that, in view of their repeated failures to carry out their subsequent promises to give way on the American offer and not to support the Separatist movement on the Rhine, Britain could no longer put any trust in the French Government's position.[269] Curzon expressed the fear that this signified a French bid to become the dominant Power on the Continent.[270] In Washington, the American Government categorically declined, on 9 November, to subscribe to the French specifications for the terms of reference of the proposed committee,[271] according to which the experts were not to question whether Germany could pay, only how they would pay for the next six years.[272]

Moreover, the increasing dependence of France on the assistance of Anglo-American markets over a wide spectrum of financial problems, caused by the precipitate fall of the franc, and, more broadly, the post-war mismanagement of the French economy,[273] reduced the bargaining power of France. The direct exploitation of the Ruhr became increasingly impractical as it would interfere with access to international capital markets and the preservation of the *entente*. Specific pressures against France included warnings by Baldwin, Crowe and Curzon against any French territorial disruption of the German State,[274] and the creation, through advantageous economic arrangements with the Ruhr industrialists, of a French-dominated industrial giant on the Continent.[275] Belgium and Italy edged closer to the position of Britain and the United States at the Reparation Commission.[276] Thus, under the prevailing circumstances, when Bradbury hit upon the plan of redetermining the practicability of the London Schedule of Payments under the guise of a comprehensive enquiry into all questions material to the restoration of German financial stability – including the flight of capital from Germany, as had been desired by the French – Louis Barthou, the French Delegate and Chairman of the Reparation Commission,[277] and, on 30 November, the Reparation Commission itself,[278] agreed to the plan despite Poincaré's subsequent attempt to put back the clock.[279]

On 26 December, the Reparation Commission appointed two committees of experts. One, under the chairmanship of General Charles G. Dawes, Chairman of the Board of the Central Trust Co., Director of the

Bureau of the Budget, was entrusted with considering the steps to be taken to balance the German budget and restore the currency. The other, under the chairmanship of Reginald McKenna, Chairman of the Midland Bank, formerly Chancellor of the Exchequer (1915–16), was to consider what measures could be adopted to ascertain the amount and secure the repatriation of exported German capital.[280]

The appointment of the committees was a major victory for the British stand, for by that an official stamp was given not only to her position of old, that reparation ought to be treated as a purely financial problem, but also to her view from mid-1922 onwards that the London Conference agreements of 1921 were no longer practicable.

But to return to October: when the French began perceiving that they had gained a Pyrrhic victory they endeavoured to salvage something from their very presence in the Ruhr by further encouraging the nationalist aspirations of the Separatist movements in the Rhineland, in order to establish a buffer between themselves and Germany. On 22 December 1923, the French announced the constitution of an 'Autonomous Government of the Palatinate' at Speyer, retrospectively from 11 November,[281] following the collapse of its predecessor, the 'Rhineland Republic', proclaimed at Aachen on 21 October.[282] This action created a major storm in Germany. It brought about increased activity by the militarist and extreme nationalist elements, who in any case enjoyed a marked rise in their prestige because of the Ruhr crisis, and provided the background to Hitler's 'Beer Hall Putsch' of 8–9 November.[283] In Britain, Curzon, who had just prevailed on the French not to occupy further German territory if Berlin failed to give satisfaction on the questions of resumption of military control and surrendering, or at least expulsion, of the ex-Crown Prince,[284] at once declared that the British Government would not recognize the new regime.[285] He also instructed Robert H. Clive, Consul-General at Munich, to proceed to the Palatinate to enquire into the actual stand of the population on this point.[286] On 19 January 1924, the Foreign Office received Clive's primary report,[287] and two days later it was presented to Parliament.[288] The report stated that the majority of the inhabitants of the Palatinate were strongly opposed to autonomy, that 75 per cent of the original promoters of the agitation were outsiders brought in for the purpose, and that the new regime could have neither come into existence nor survived without the support of the French army of occupation.[289] The publicity given to Clive's conclusions, the widespread public approval with which they were met in Britain and the rest of the States of Europe, and the violent outbreak of German nationalism in the Palatinate, inspired and led from unoccupied Germany, made the French position impossible and brought to naught their attempt to establish a buffer state between them and Germany.[290]

With the publication of the Clive Report, however, Curzon's career as Foreign Secretary was over. On 22 January 1924, a Labour Government was formed for the first time in Britain, under the Premiership of J. Ramsay MacDonald, and Curzon and the Conservatives resigned their posts. His tenure of office was one of the longest and stormiest that the Foreign Office knew after the war. One thing in which his period was distinguished, in all that pertained to Europe at least, was lack of consistency. He opined that European peace was contingent on the continuity of Anglo-French co-operation but was reluctant to accept the French logic that the element that should be neutralized, in order that the peace should not be undermined, was Germany. On the one hand he asserted that Britain was no longer able to refrain from active intervention on the Continent of Europe; and on the other hand he did his best to thwart agreements that would tie Britain to France. Curzon's trouble was that the desirable state of affairs, in his opinion, from the viewpoint of Britain, was her return to the policy of 'Splendid Isolation'. But under the new world situation that policy was no longer available. He temporized by taking half-measures, which led neither to 'Splendid Isolation' nor to serious involvement but created, together with Lloyd George's proceedings, internal inconsistencies and incessant friction with France from which the principal beneficiary was Germany.

The First Labour Government: MacDonald and the Foreign Office

The fall of the Conservatives and the coming of the Labour Party, for the first time in its history, into power, were accepted at the Foreign Office good-naturedly. In its modes of working, the Government resembled its predecessor.[1] Foreign policy, too, was conventional and did not defer to the anti-French sentiments of the left-wing of the Party.[2] Nor did Foreign Office officials find it especially difficult to put up with the departure of Curzon and the appointment of MacDonald as Foreign Secretary. 'It's odd', remarked Alexander George Montagu Cadogan, First Secretary of the Western Department,[3] referring to the difficulties encountered during Curzon's time,[4] 'that we should have had to wait for the Labour Party to give us a gentleman.'[5] Another senior official reacted likewise. In a conversation with Asquith he described the 'delight of that department at being relieved of the incubus of the "Archduke Curzon" and at starting daily relations with a mild mannered and so far quite rational Ramsay'.[6]

There was general satisfaction not only at the departure of the former Foreign Secretary but also at the fact that the entry of his successor enhanced the trend according to which the bulk of foreign-affairs issues was now back under Foreign Office domain, coupled with a significant uprise in the influence of officialdom over the political decisions adopted by the Government. This new trend stemmed from a number of factors, though the most important was undoubtedly the personality of the new Foreign Secretary himself.

MacDonald's Background, Personality and Methods

J. Ramsay MacDonald (1866–1937) was a somewhat mystic figure. He had a reputation for being rather woolly-minded, but he knew what he wanted and managed to attain a good deal of it. Although he had not been trained to be an administrator, and had neither the advantages of a privileged higher education nor the practical experience of business that could take the place of such training, he was nevertheless not as unsuited to be Prime Minister and Foreign Secretary as one might have thought. As Foreign Secretary he displayed good judgement. His extensive travels overseas while in opposition, his frequent meetings with foreigners and the fact that he had no prejudices against them developed the natural faculty that was harboured in him to show understanding towards the stands of European statesmen. He inspired them with the sense that he did, indeed, understand them, and usually reached an agreement with them.[7] On the governmental level MacDonald was not dependent upon his ministers. As Foreign Secretary, who was also Prime Minister, he carried out political moves without any debate in Cabinet, and on more than one occasion Ministers wondered what was said in his talks with foreign statesmen.[8] But if, in the era of Lloyd George, this had been a drawback, then in the days of MacDonald it was, from the respect of the Foreign Office, an advantage. MacDonald's method drastically reduced the intervention of neighbouring departments in foreign affairs and enabled Foreign Office officialdom to play a dominant role, and exert influence over the Foreign Secretary.

In their life histories, there could not have been a greater contrast between Curzon and his successor. MacDonald, illegitimate son of a Highland ploughman and a local mother, was born on 12 October 1866 in a two-roomed cottage in the fishing port of Lossiemouth, Scotland. His education was limited to the schooling he received at the local board school and that which he acquired by himself. The years of his youth were passed in relief work, self-education and activities in various socialist movements. In 1900 he became Secretary of the Labour Representation Committee, and then of the Labour Party itself, 1906–12, Treasurer, 1912–24, and its representative in Parliament. He actively opposed the South African War (1899–1902). After visiting India he wrote an excellent book on the country (1910), which together with his writings on socialism helped to consolidate his position as the most distinguished spokesman and thinker of the new Party. In 1914 he resigned the chairmanship of the Party because it declined to approve his proposal not to give the government public mandate to plunge Britain into war nor to collect £100 million for the war effort. Once the decision had been taken, however, he affirmed that, unnecessary as the war was, it had to be won; though he demanded that the ensuing peace be moderate and not vindictive, he was not actually a pacifist in the real sense of the term. In 1917 he welcomed the Russian Revolution, but when he saw where the Bolsheviks were headed he succeeded, in 1920, in persuading his Party to reject communism, and from there onwards established himself as the unquestionable leader of the Party until it took office in 1924.[9]

The division of functions between the new Foreign Secretary and his assistants was on the basis of a partnership, with the latter playing a more major role than hitherto. The new Foreign Secretary preferred not to take part in the various stages of departmental policy formulation that preceded the crystallization of final recommendations, on the ground that:

> details are overwhelming and ministers have no time to work out policy with officials as servants; they are immersed in pressing business with officials as masters. I must take care.[10]

Instead of putting his ideas in writing, in the form of departmental memoranda, a method preferred by Curzon, he developed the habit of giving his assistants a list of questions on which he wished them to prepare memoranda for his scrutiny, as well as particular points they should take into consideration. As a result, responsibility for policy planning now devolved on the Permanent Under-Secretary, Crowe. This method meant that a particular policy stand would evolve rather than be dictated from above. Thus, after discussion with officials at various levels, a common departmental stand would be co-ordinated and settled; differences of opinion with neighbouring departments would be solved, or at least delineated. This procedure exempted the Foreign Secretary from seeing a good number of the files which in the period of Curzon were dealt with directly by the Foreign Secretary. What turned Crowe into the right-hand man of the Foreign Secretary was the recognition of the practicality of his ideas, which now could be more fully implemented.[11] MacDonald's way of working brought about such a growth in Crowe's workload that he was forced to delegate more to the Assistant Under-Secretary of State and Heads of Departments, who became more autonomous than in the past. Crowe worked till the small hours every day. And on Saturdays and Sundays, too, he had to devote his time to Foreign Office work, including the formulation of answers, for the Foreign Secretary, to Parliamentary Questions.[12]

All the same, MacDonald was not idle, and neither was he considered as such in the eyes of his assistants.[13] Referring to his agenda, on 24 September 1924 he noted in his diary that:

> This long day's concerns, epitomises what distresses British statesmen have to face. First came the American Ambassador who is always welcomed . . . At 10 Sir Lee Stack and half a dozen officials discussed Sudan and Egyptian problems in view of my conversations with Zaghlul, at 11 Woolencott, editor of Indian Pioneer . . . ; 12 Turkish Minister on the fighting in Irak and the boundary discussion at League of Nations,

I lunched privately at the Savoy . . . ; 3 Governor of Bank of England and Mr. E. P. Morgan on the floating of the German loan interpretation of Versailles, occupation of the Ruhr, Cologne etc. etc.; 5 Inverpith on Mosul Irak boundaries etc. etc.; 8 half an hour at letters; 8.30 dispatch boxes from F.O. came up to the time of 11. Fine enjoyable and interesting work, but being ill, how I got through this day, I know not.[14]

It would be wrong to classify MacDonald as a policy legitimator, as just a rubber stamp to the policy suggested by his assistants.[15] Although he was not fond of putting his ideas in writing and did not, like Curzon, devoutly dedicate himself to mastering the details of all the files under his command,[16] MacDonald simply abstained from taking part in the early stages of the foreign-policy formulation process and instead put a premium on rapid decision-taking once a matter was brought to his attention.[17] MacDonald ran Cabinet meetings[18] and international conferences[19] with extreme competence.

Dealing with the French

MacDonald's major executive trait was that he knew what he wanted. He did not automatically adopt in full the final recommendation submitted to him but introduced, whenever necessary, modifications. For instance, on entering his post MacDonald turned to the Foreign Office with a question regarding what issues could be settled by Anglo-French negotiation and which ought to be handed over to the League of Nations. Bennett replied that putting an end to the attempts of France to attain a permanent presence in the Rhineland was contingent on resolving the security problem there, and this could be achieved by Anglo-French negotiation.[20] Nicolson minuted that this was also true regarding the reparation question, subject to the interlacing of other Powers, such as the United States, in the negotiations; adding that the question of the security of France could be solved under the auspices of the League of Nations, for the security question was an all-European problem.

As a first principle Nicolson recommended that Britain cease wrangling with the French over details and begin solving the fundamentals of outstanding problems[21] – a stand warmly approved by Crowe.[22] However, MacDonald was not content, for according to him it was a mistake to assume that the present situation was solely a consequence of the French feeling insecure. In his opinion the French were also motivated by 'historical craving'. Furthermore, MacDonald was convinced that if the mutual distrust and tendency to create inner alignments among the states of Europe should continue, it would lead to an arms race and inevitably to a renewed war.[23] Referring to the ideas of Bennett, he instructed the Foreign Office to turn to the War Office and Board of Trade requesting that they prepare memoranda on the military and economic aspects of French control over the Rhineland and the Ruhr, and to make sure that the full ramifications were jointly probed by the three departments before submitting to him their recommendations concerning final policy:

> I desire that this matter should be worked at so that we may be prepared with a comprehensive policy, because we cannot allow ourselves to drift. A half thought out policy or a dimly comprehended one will bring us to grief and what we finally adopt must be European in the full sense of the word.[24]

Thus, MacDonald took the first step towards what takes place in every modern Foreign Office, but had not happened in the days of Curzon – namely long-term policy planning.

The Foreign Office and MacDonald began to take primary measures to sort out the political confusion that ensued in Europe following the Ruhr crisis, and to moderate the polarization in relations between Britain and France. On 26 January, MacDonald turned to Poincaré with an appeal to clear the atmosphere that prevailed between the two states by a change of style. He suggested that, from this point forward, both sides should be frank, while taking pains to avoid hostility, and defend their interests without being at enmity with each other.[25] Also, he sent Poincaré a message of regret for the half-true allegations made by Lloyd George, in a Press interview, that Clemenceau and Wilson had signed a secret pact concerning the occupation of the Rhineland by France. The apology touched the dour Poincaré[26] and for a few months relations between MacDonald and Poincaré were better than under Poincaré and Curzon.

Future of the Allied Commission of Control

On 5 January, the Foreign Office embarked upon a series of consultations with the War Office concerning resumption of the activity of the Inter-Allied Commission of Control in Germany, which had been suspended during 1923. Troutbeck, Hurst (the Legal Adviser), Crowe and MacDonald adopted the recommendation of the War Office, which was to renew the activities of the Commission, even though its constitution was past the time assigned by the Treaty of Versailles. The War Office wanted Allied forces to return home as soon as possible, and an offer to Germany that thereafter control would be restricted to the Five Points enumerated to them in September 1922.[27] It was hoped to remove the Commission of Control, whose place would be taken by a Committee of Guarantee.[28] It was also decided, on the recommendation of D'Abernon,[29] that the British stand be that control over disarmament should not be resumed without an agreement with Germany, for fear it should lead to a crisis.[30] On 5 March, after negotiation, Britain's position was accepted by France,[31] but not by the Germans, who refused to acknowledge the right of the Commission of Control to remain active.[32] The question remained suspended until 30 June, when the Germans expressed their readiness to let the Commission make a general inspection provided that it subsequently remove itself from Germany.[33] In response, the Allies made clear that this would be done only if it turned out that the state of disarmament had reached an advanced stage.[34] The Germans did not accept the latter point but nevertheless expressed their agreement, on 14 and 16 August, that a general inspection should take place,[35] which began on 8 September.

Long-Term Policy Towards France and Germany

While these proceedings were taken as a matter of course by all sides in Whitehall, when they embarked – at the request of the Foreign Secretary – upon crystallizing a long-term foreign policy, strong differences of opinion emerged between the different departments, and between them and MacDonald. For instance, the sides were divided in their estimation regarding the expected developments in Europe during the next decade. Crowe,[36] Troutbeck and, presumably also, Tyrrell (though there is no evidence of it), continued to see in the traditional aspirations of Germany the true threat to the peace of Europe; while Lampson, Nicolson and Bennett called into question the German capability of challenging the *status quo*, given her great infirmity at present. Following a journey he had taken in Germany, Troutbeck became convinced that 'no people can be less qualified for democratic government than the Germans'; that they needed by nature to be led and drilled; that not long would pass before they found the leader they required; and following that, a new Germany would arise, which would indeed be reconstructed:

> but the new Germany that will arise will be an unpleasant neighbour. In her days of victory she trusts to force and nothing else, and her belief has been confirmed in defeat. Her only scruple will be to avoid the mistakes she made in 1914. The danger to this country will be just as great.[37]

Bennett, Lampson and Nicolson, on the other hand, rejected the assertions of the General Staff of 28 March and 23 September that war with France was impossible, while an Anglo-German war was inevitable, and that the danger would become tangible once the Rhineland was evacuated in 1935:[38]

> Under conditions as they are now and are likely to be for many years to come, renewed German aggression is the greatest danger that faces us and French security is our security.[39]

The rejection of this view was based on the assumption that France and not Germany was the dominant element on the Continent. And to reinforce this view officials pointed to the estimation of the Admiralty and Air Ministry that if France remained for any longer in the Rhineland it would affect the security of Britain,[40] – an opinion repeated, for economic reasons, by the Board of Trade,[41] and by the Department of Overseas Trade,[42] regarding the continuation of the French presence in the Ruhr.

Moreover, Nicolson and Lampson argued that the crux of the problem was not practical but psychological; namely, not how to neutralize Germany politically or militarily and secure France from an attack, but how to convince France that her fears were unfounded.[43] This was also the opinion of the Parliamentary Under-Secretary to the Foreign Office and President of the Council, Charles Alfred Cripps, First Baron Parmoor, and MacDonald. Cripps argued that it was a question of mutual trust and friendship, the more so as he had doubts whether the policy of France was propelled by a sense of fear;[44] while MacDonald requested his officials – following another General Staff memorandum, of June 1924, warning of a danger of war in Europe within ten years,[45] – to thank the War Office for their admirable note even though he could not at every point take their words as Holy Scripture. He noted that if war were to break out it would be because British policy manoeuvrings had not managed to put an end to the alliances policy that was dividing Europe into two camps,[46] and which would eventually bring to power the extreme Right,[47] who thought exclusively in military terms.[48] According to MacDonald:

> The good soldier of broad mind must help the politician, but do not draw these false distinctions between idealism & practicalism. The one thing that matters is psychology. All the sage materialists and self-styled realists will never be able to produce anything but wars. That is their natural offspring. Unless we change the qualities of our minds we had better arm to the teeth.[49]

The piqued Crowe minuted in response that it was not sufficient that the British Government should change its mind, supposing it was at all required. What was necessary was that other people should change their mind, notably the possible peace-breakers. As long as they were not prepared to do so, said Crowe, officials could not afford to neglect realistic views.[50]

The question of the manner by which peace and security could be established also raised differences of opinion in the Foreign Office. From the Paris Embassy, Eric Phipps recommended, on 18 February, the achievement of a peaceful solution by means of a buffer zone in the Rhineland and a guarantee, if necessary, that it would stay demilitarized by prolonging the presence there of Allied forces.[51] Bennett, on 23 February,[52] Joseph Addison, Counsellor to the Berlin Embassy, on 1 March,[53] and Lampson, on 16 April,[54] who thought that the root of the problem was the placating of France, favoured a non-aggression pact between Germany and her neighbours. Addison talked about this move as being just part of a much more general solution, while Lampson was to change his mind in September. Crowe, Troutbeck[55] and, from 2 September, also Lampson,[56] were, for their part, in favour of guaranteeing the boundaries of France against any external aggression, as a means of deterring Germany and placating France. While Nicolson,[57] Lord Parmoor[58] and MacDonald[59] stood for guaranteeing the well-being and security of the various states of Europe, including France, through attaining an all-European readiness to submit to the authority and regulations of the League of Nations, and providing the League, when necessary, with all the political and military instruments needed to enforce its authority on recusants. Nicolson explained:

> Whatever the French may say, they wish, under the guise of the League, to forge iron chains which will encircle Germany and keep her captive; whatever we may say, we wish while aiming at general pacification and disarmament, to avoid in any way committing ourselves to military intervention in Europe.[60]

MacDonald explained his categorical objection to any military agreement with France by saying that this was the worst possible way to provide security,[61] for it would return Europe to a vicious circle of pacts and counter-pacts, bring about an arms race and eventually result in war.[62] Even the assertions of Crowe that if Germany decided to go to war she would not pay regard to the regulations of the League of Nations[63] did not convince the pro-Leaguers.

As expected, on the question raised by the Dawes Plan, MacDonald had his way, but it did not mean, however, that Crowe lost his influence. Particularly in their opposition to French demands for sanctions, MacDonald and Crowe were in agreement on the eve of the London Conference, which was convened in mid-July for the purpose of discussing the implementation of the Dawes Plan. MacDonald let Crowe and his assistants work out, with the representatives of the Board of Trade and Treasury, the text of the Reparation Agreement to be offered to the sides in the coming Conference. MacDonald only came into the picture when the final agreement was submitted for his consideration, which he approved after putting in it a number of modifications.[64] On the eve of his meetings, on 2–3 May with the Belgian Premier and the Belgian Foreign Minister, and on 21 June with the French Premier, MacDonald took Crowe with him to Chequers:

> . . . so that if any technical points should arise with which he was not fully acquainted, he might be able to avail himself of Sir Eyre Crowe's assistance, if necessary.[65]

MacDonald also attached him to the British Delegation to the forthcoming London Conference, and prior to the conference, in July, took Crowe with him to Paris.[66]

Another example of the great esteem in which MacDonald held Crowe's professional capabilities was his request that he go over the draft letter written by him, on 14 May, to Poincaré. In this letter MacDonald expressed his regret that owing to the adverse results of the general election in France the latter was prevented from coming to visit MacDonald at his home in London, and his wish that until he left his office no change would take place in the intention to implement the Dawes Plan.[67] Crowe's opinion was asked as to whether it was advisable to send such a letter and whether any changes should be made in it.[68] Crowe replied that what the Prime Minister proposed was a good and judicious move, which would contribute to the stepping up of the implementation of the Dawes Plan. Nor did he think that the second paragraph, about which MacDonald had doubts, ought to be altered, for it could do no harm to be duly cautious.[69] And indeed, the letter was duly sent as it was.

Furthermore, on the question of the security of France, in correspondence he had with Poincaré between February and April, MacDonald began to clarify the solution of an all-European security settlement under the auspices of the League of Nations, including renewed demilitarization of zones and disarmament.[70] But, on the advice of Crowe, all further discussion of the security question was put off until the questions of reparations and French presence in the Ruhr had been disposed of.[71]

Immediate Priorities

The prioritization of those questions had two major aims. The first was the wish to avoid getting entangled in fresh arguments with the French over subsidiary problems such as the Ruhr, the Rhineland and the Palatinate but instead to concentrate on fundamentals, namely details of the reparation clause in the execution of the Treaty of Versailles,[72] with a view to finding a settlement that would enable a renewal of the payments of reparations by Germany and incidentally eliminate the need for continued French presence in the Ruhr.

The second aim was to bring about a normalization in Anglo-Franco-German relations. MacDonald believed that if the plan to compensate France for the damage caused to her territories during the War were to constitute part of an all-European reconstruction plan, including a loan from American sources,[73] and Germany could be persuaded to adopt it of her own free will, there was a reasonable chance that Germany would meet her liabilities.[74] It was hoped to bring about a German willingness to reconcile herself to the reparation clause of the Treaty of Versailles, and thereby to the Treaty as a whole. A positive response by Germany would, it was hoped, lead to a French willingness to exchange her self-centred policy towards Germany for a policy that would benefit all European countries.[75] MacDonald asserted that improved Anglo-French relations, and a normalization in relations with Germany, together with German co-operation, was contingent upon the goodwill of the Germans and Europeans in general. Upon this rested the future security of Europe.

> The more I read your history, the more I see that France has never found security by alliances. In reality, we have still to find out what security is. Neither a strong navy nor a powerful army nor numerous aerial forces have yet sufficed to give us the security which we seek in vain.[76]

The issuing, on 9 April, of the Dawes Plan gave priority to the reparation question. The Dawes Committee was requested to answer the question of how the German currency would be stabilized and a budgetary balance achieved as a pre-condition to the payment of reparations. The Committee tackled the question from political and financial angles. On the political level it determined that the attainment of these objectives was contingent upon the abrogation of French economic control over the Ruhr, the fiscal reamalgamation of the Ruhr with the rest of the German economy, and the reduction of the French military presence to a skeleton force despite the fact that the question of evacuation or non-evacuation of the Ruhr was not within its terms of reference.

On the financial level, the Committee called, as a preliminary measure before the payment of reparations and reorganization of German finances, for the gift of a large international loan to Germany and the appointment of an Agent-General in Berlin to oversee the complex supervisory structure. More specifically, the Committee determined that, in order that Germany should be able to raise the revenues necessary for financing the payment of reparations, the Government would have to mortgage German industry and state railways, reassume domestic indebtedness and increase the rate of taxes (which were much lower in Germany than in the victor Powers). However, the Committee did not attempt to form an estimate of the total sum of reparations owed to the Allies, and contented itself with determining that the primary sum called for from Germany would be one milliard marks a year, rising gradually. At the end of four years Germany would be called upon to increase her payments to the annual amount set by the London Conference of 1921, namely 2.5 milliard marks. This sum would be handed over to a special committee of bankers, whose duty would be to provide financial means to convert German wealth into foreign currencies without depreciating the mark, as had occurred in the past. Similarly, the Second Committee, under the chairmanship of McKenna, concluded that only restoration of confidence in the German currency could effectively discourage the flight of additional capital and induce the repatriation of the capital already abroad – tentatively estimated to be 6.75 milliard marks. No mention was made by the experts of foreign assets owned by German nationals as a means of footing Germany's bills.[77]

Despite recognizing that the Dawes Plan was serious and exhaustive, the Foreign Office had difficulty in deciding how sound the Plan was. Furthermore, the Foreign Office had already reconciled itself to the thought that politico-economic questions fell outside its jurisdiction and expertise, and that in this case:

> It was indisputable that it was the Treasury who must decide on the mechanism of the scheme and not the Foreign Office. It was for the Treasury first to pronounce whether the scheme was workable or not.[78]

The Foreign Office therefore turned to the Treasury for the purpose of evaluating the workability of the Plan, and whether it should be adopted in full.[79] After referring in detail to the positive and negative points of the Plan,[80] the Treasury replied affirmatively,[81] and as a result the Foreign Office evolved a stand according to which the Plan should be adopted in its entirety and given effect with the utmost despatch.[82]

Germany, Belgium, Italy and France accepted the Plan with evident lack of enthusiasm. On 11 April, Germany expressed willingness to adopt the Plan as a basis for negotiations.[83] Belgium and Italy, for their part, stipulated their willingness provided that the question of Inter-Allied war debts was resolved;[84] while France stated that she would release some of the productive pledges once the Plan was executed and imposed upon Germany.[85] But eventually, after the Foreign Office had warned the Germans that their reaction was bound to create an impression that they could not be trusted,[86] Crowe told the Belgians and Italians that Britain was tired of performing the functions of the milch cow.[87] MacDonald warned the French that should the Plan fail he would challenge the legality of the Ruhr occupation, the railway regime and the M.I.C.U.M. agreements,[88] and publish all documents in his possession to prove England's innocence.[89] on 16 April, Germany,[90] on 17 April, Italy,[91] on 24 April, Belgium,[92] and, on 14 May, France,[93] announced their willingness to accept the proposed scheme without reservations.[94]

The London Conference

From here on, matters moved swiftly. On 21–22 June, at his own invitation,[95] MacDonald met, at Chequers, the successor of the austere Poincaré and leader of the Cartel des Gauches, Édouard Herriot. MacDonald not only succeeded in establishing good relations with him,[96] but also won his agreement to an international conference in London at the end of July. The conference was to restrict itself to determining the application procedures of the Dawes Plan without going into other economic questions, such as Inter-Allied debts. Its conclusions would be codified in a protocol to be signed by Germany. Most important of all was that the dates of the application of the Plan be fixed and that only in case of flagrant German defaults should sanctions be imposed.[97] These ideas were suggested in part by Crowe and were the subject of discussions between Crowe and the legal advisers, and between Bradbury and the Permanent Under-Secretary at the Treasury, Fisher.[98]

In the meeting held at Chequers, and subsequently, on 8–9 July, in Paris, Herriot attempted to stipulate that the co-operation of France would depend on a number of factors: a gradual application of the Dawes Plan and a cancellation of war debts, a Mutual Guarantee Pact between Britain and France, the exclusion of Germany from the conference; and the principle that the body that would determine whether Germany was fulfilling faithfully the Dawes Plan would be the Reparation Commission. Crowe, who was present at the meetings, argued against each of these reservations. He emphasized that if the supreme executive authority was to be handed over to the Reparation Commission, France could, through the majority she had in the Commission, prevent the application of the scheme. Macdonald stressed that the Dominions and British public opinion would oppose a formal pact with France and that this would bring about the fall of his Government. He argued that military pacts had never solved the security problems of France in the past, but just gave birth to new wars. In the event of Germany declaring war upon France because of the reparation problem, he suggested Britain would regard this as a declaration of war upon her herself. Herriot retracted his demands and on 9 July both sides agreed to concentrate upon the application of the Dawes Plan and to invite Germany to the conference. It was accepted that the scheme would not diminish the authority of the Reparation Commission, but at the same time the application of subsidiary sanctions upon Germany would be carried out in the spirit of the recommendations included in the scheme. It was also determined that the discussion of the remaining points, about which there were differences of opinion, would be postponed until the coming international conference.[99]

These resolutions were reported by MacDonald, on 11 July, to the Cabinet, where he now, very much in the tradition of his predecessors, insisted that most of the negotiations on German questions were conducted mainly with France, while Germany was a side-line spectator. Accordingly it was the policy of France that prevented any reasonable settlement of the reparation question.[100]

Parallel to the inter-Power debates, the Treasury, Bradbury, Crowe and Hankey embarked upon the organization of the conference. With the consent of the Foreign Office, the Treasury took into its own hands the negotiations with the French over staffing the sub-committees of the coming conference.[101] The Treasury supplied the Foreign Office with a series of background memoranda, and advised the Foreign Office, together with Bradbury, to exclude France's East European allies from the conference lest the meetings become unnecessarily protracted[102] – a decision regretfully concurred with by MacDonald.[103] Crowe, who was co-opted by MacDonald onto the British delegation to the conference, centralized the reception of all necessary memoranda, held discussions with the Dominion representatives on MacDonald's behalf, and, with MacDonald, considered the composition of the British delegation,[104] and the delegation of the Foreign Office staff to take part in proceedings.[105] Lampson, under the supervision of Crowe, handled the subject of the lodgings of the various delegations.[106] The experienced Hankey was asked by the Foreign Office, despite past quarrels, to once again take responsibility for managing and co-ordinating the organization of the daily proceedings of the various sub-committees.[107]

The London Conference, under the chairmanship of MacDonald, opened on 16 July and was divided into two stages. At the first stage were present only the ex-Allies. After a good many talks, mostly concerned with the date for the evacuation of the Ruhr and the procedure for determining defaults in future, the Allies reached agreement among themselves. And on 2 August Germany was invited to join the Conference.[108]

On 5 August, the German Delegation came to London and its arrival opened the second stage of the Conference. This stage was imbued with frictions and disappointments. The French could not bear to be in the same room with the Germans.[109] Relations between the Belgians and the French could not be described as warm. On the British side, relations between Philip Snowden, Chancellor of the Exchequer, and MacDonald were somewhat shaky[110] because of the antipathy of the Chancellor to the general assumptions governing the Dawes Report, and because Snowden, seeking to make the political agreements subject to the conditions set down by the bankers, had taken a public stand contrary to that of the Prime Minister.[111] The intransigence of Snowden was so great that, in reaction to the remark of MacDonald that only Satan could separate Britain and France, the French Press published an article on Snowden headed 'Satan is here'.[112] The Germans, in exchange for their agreement, wanted an immediate evacuation of the Ruhr. The French held back, endeavouring to make their withdrawal dependent on payment of the first two Dawes annuities, the results of the prospective general inspection of the Control Commission, and the signing of a Franco-German commercial treaty.[113] The controversy between the sides was so fierce, and the rate of progress so slow, that MacDonald observed that writing his diary constituted his only source of pleasure.[114]

Despite these difficulties, on 16 August the sides initialled a common Protocol, signing the formal instruments, a fortnight later, in which arrangements were made to implement the Dawes Plan in full, to evacuate the Ruhr within twelve months, and to submit issues in dispute to arbitration.[115] This agreement, with all its implications, owed much to MacDonald. He showed considerable understanding towards the anxieties of the French and insisted on the application of the principle of arbitration whenever points came up that were liable to be in dispute in future. He insisted that Germany have an equal voice in choosing the arbitrators, and thus imbued the Germans with a feeling that he was on their side.[116] During the discussions he did not lose his calm, and always with humour, a good word and indulgence to the whims of his colleagues managed to lighten the atmosphere, obtain compromises, and wring a smile even from the severe countenance of the French Minister of War, General Charles Nollet.[117] Even Hankey, who had worked with other Prime Ministers, noted, after seeing Macdonald in action, that, 'Whether he remains in office or not, I shall always look back with pleasure to my association with him.'[118]

Financial Arrangements Settled

MacDonald was fortunate that parallel to Germany, France, too, suffered a financial crisis. In return for receiving loans from American bankers, who were accessories to the London Conference, the two Powers were prepared to make far-reaching concessions – the Germans because of their economic crisis, and the French because of the drastic decline of the franc. The French need for a loan was so great that they had to give way before the will of J. P. Morgan, who floated loans to both sides, and adopt the Dawes Plan as it stood; they also signed provisions making future French sanctions against German default virtually impossible.[119] Without the intervention of J. P. Morgan there would have been no Dawes agreement. With patience, understanding and a bit of luck, MacDonald brought the Conference to a successful end.

But, as with all political problems, the issues at stake were not resolved in their entirety. And the end of the 1920s and beginning of the 1930s revealed the difficulties that had not been fully addressed. The compliance of the Germans did not reconcile them to the reparations clause of the Treaty of Versailles in general and the Dawes annuities in particular.[120] So far as world opinion was concerned, the Dawes Plan constituted the beginning of four stable years, in the course of which the Germans, except for a slight default that from their respect was a point of honour, met all their liabilities.[121] The impression was created that the consequences of the First World War were finally overcome, and the reparation question could, temporarily at least, be taken off the agenda of Europe.

The Question of Security

With the Reparation obstacle behind him, MacDonald – under pressure from the French – now directed his attention to the question of security.[122] The possibility of this problem being resolved by a military pact with France[123] had been vigorously dismissed by him on the grounds that the Dominions and public opinion would not put up with it, and that pacts such as these constituted the most pregnant danger to peace.[124] For these very reasons he had formally rejected, on 5 July 1924, the Draft Treaty of Mutual Assistance of the League of Nations, according to which the League Council would be authorized to decide within four days of an outbreak of hostilities the identity of the aggressor and to make use of the armed forces and economic resources of the rest of the signatories for assisting the victim of aggression. The Draft was part of an endeavour gradually to bring about security and disarmament, and to reduce the military intervention of various states in the Continent of Europe.[125] According to MacDonald, the treaty – with its provision for partial agreements among signatories for self-defence by way of implementing regionally mutual assistance – would foreshadow in practice a return to the pre-war system of alliances. It would meet with fierce opposition from Labour circles, especially given its other ramification, i.e. the basing of security on existing armaments rather than on the general disarmament of the signatories; and preclude immediate and effective action against an aggressor:

> The Council will have great difficulty in reaching a unanimous decision, for no nation places its troops under a foreign command without very careful consideration. A system which involved prolonged delays before the first step in bringing military pressure to bear on an aggressor nation can be taken does not reach that standard of effectiveness which is essential.[126]

Furthermore, Foreign Office officials, such as the Head of the Western Department and Crowe, not to mention the military branches and the Dominions, threw their weight against the proposal, which they viewed as Utopian.[127] Crowe noted that the foundation of peace and the neutralization of hostile elements was contingent upon magnification of the individual military might of the Allies and their allies in Eastern Europe, and not upon the uncertainty of collective force.[128] It was also felt that too much weight was put on British naval and Imperial power.[129] Therefore, the sole hope of finding a solution to the problem, so the Prime Minister thought, was to find another formula, which would turn the League into an instrument of peace without those military provisions that aroused such wide opposition in Britain while meeting French demands.[130]

In his speech, on 4 September, before the Fifth Assembly of the League of Nations at Geneva (1 September – 2 October 1924), MacDonald turned down the pleas of Herriot, who was present with him at the convention, for military guarantees. The British statesman warned that military alliances would lead to a situation similar to that which prevailed in 1914, and to war. He urged the states who had not done so yet, to join the League of Nations; and proposed that the enforcement of peace be replaced by arbitration to remove the frictions menacing it.[131] His views did not meet any serious opposition – except, of course, from the French, who insisted that arbitration be backed by something more positive – and gave birth to a resolution by the League of Nations to examine the obligations contained in the Covenant in relation to guarantees of security and the settlement of disputes, which a resort to arbitration and a reduction of armaments might require. It was agreed to clarify the powers of the Permanent Court of International Justice. The Geneva Protocol, endorsed by the Assembly on 2 October, was, like the Draft Treaty of 1923 – a pledge of mutual assistance. But, in contrast to its predecessor, the present Treaty left the various states considerable discretion in deciding for themselves the scope of assistance they were able to make, something that would be determined according to their capability at the time and their geographical situation. Another novel feature was the striving towards prevention of wars and settlement of conflicts by peaceful means: first, by an acceptance by all signatories of the Optional Clause of the Statute of the Permanent Court; secondly, if neither party asked for arbitration, by authorization for the League Council to take charge of disputes and settle them through an agreed report; and finally, by the definition of an aggressor as any state that failed to carry out in good faith judicial sentences or arbitrated awards, and resorted to war, which would automatically lead to the application of sanctions by the League against the aggressor. However, a necessary condition for its realization was the readiness of all Powers to observe its principles and disarm. It was therefore stipulated that a world disarmament conference be convened, on 16 June 1925, in Geneva and that the Protocol was not to come into force until a disarmament plan had been adopted.[132]

The Protocol was accepted with evident lack of enthusiasm not only within the ranks of the Opposition,[133] the Fighting Services,[134] and the Dominions,[135] but among Ministers within the Labour Government as well[136] – despite the assurances of the British Government that the Protocol would not give the Council of the League or the Permanent Court any control whatsoever over the forces of Britain. The French, too, were not satisfied with the vagueness of the security clauses in the proposed Protocol and enquired as to what was the real meaning of the article according to which the signatory states would undertake 'to co-operate loyally and effectively' with each other in support of the Covenant, and in resistance to any act of aggression.[137] But before MacDonald could cope with these criticisms, which threatened to impede the ratification of the Protocol in Parliament, he was forced to devote his full attention to other matters, which reached their climax with the Zinoviev Letter.

The Perceived Russian Threat

First, the MacDonald Government had to face the fire which its policies towards Russia generated. The British public was far from being enthusiastic over the *de jure* recognition accorded, on 1 February 1924, by the Labour Government to Soviet Russia.[138] The Anglo-Soviet negotiations opened on 14 April and continued to 12 August.[139] Arthur Ponsonby, Parliamentary Under-Secretary of State for Foreign Affairs, whom MacDonald charged with negotiating with the Russians, managed, in August, to reach an agreement with the Russian Delegation according to which a commercial treaty would be signed between the two countries and negotiations would be opened over settlement of the pre-war debts of Russia in exchange for the giving of a British foreign loan to the Russians.[140] Eyre Crowe took no part in the Anglo-Soviet talks. MacDonald did not ask him to assist, knowing his views with regard to the Bolsheviks. Crowe told his wife that, 'he had put it formally and repeatedly on record that he entirely disapproved of and protested against the whole proceeding'.[141] The Conservatives fell upon this issue with glee. They perceived that at last there was an issue through which it would be possible to rouse the public against Labour and to remove it from office. Consequently, the Conservatives began a widespread public propaganda campaign in the course of which they described the agreements reached as a British pledge to finance the subversive activities of a state that was committed to bringing about the collapse of the British Empire.[142]

Nevertheless, it is highly improbable that these events by themselves would have enabled the Conservatives to convince public opinion that Labour was subject to orders from Moscow – especially since they themselves had signed, on 16 March 1921, a commercial treaty with the Russians – had it not been for the almost simultaneous media attention given to the Communist journalist John Ross Campbell. On 5 August, Campbell was arrested for inciting the armed forces to mutiny. In spite of the severity of the charge, Sir Patrick Hastings, the Attorney General, having been advised that a conviction was doubtful on the legal merits of the case, suddenly withdrew his indictment, on 13 August, with the approval of the Cabinet. Campbell was discharged. The Conservatives and the Press – though the Attorney General denied it – claimed that this step was taken under pressure of the Labour Left.[143] As Baldwin pointed out later, the feeling that the moves of the Government were being dictated by the extreme Left started to take hold among various groups of the public.[144]

The Russian treaties and the Campbell case together cost MacDonald the Liberal support on which his minority Government depended. On 8 October, the majority of Members of Parliament voted in favour of the establishment of a Select Committee to enquire into the Campbell case. Whereupon, the following day, Parliament was dissolved and MacDonald announced the holding of elections on 29 October.[145]

The pivot around which the election campaign centred was the 'Red Menace'. The main task of the Labour Party was to convince constituents that there was no truth to the claims of the Conservatives that it was being soft on Bolshevism.[146] Against this background the Zinoviev Letter fell on Labour as a bolt from the blue.

The Letter, dated 15 September, called on the British Communist Party to do its utmost for the ratification of the commercial treaty in Parliament on the grounds that tightening the ties between the two countries would give Russia the requisite scope first for beginning a propaganda campaign and then for bringing an armed insurrection against the capitalist order in Britain, Ireland and the Colonies.[147] At the bottom of the letter were the signatures of Grigori E. Zinoviev,

President of the Third Communist International; O. V. Kuusinen, Secretary of the International; and Arthur MacManus, Chairman of the British Communist Party. The letter instructed communist sympathizers to set up clandestine cells in military units and munition factories so that in the event of a strike or war against Russia the Army would side with them.[148] The Secret Intelligence Service, which reported the existence of the Letter on 9 October to the Foreign Office, and also supplied the Admiralty, Air Ministry, Scotland Yard and MI5 with copies, asserted that:

> 2. The document contains strong incitement to armed revolution, and clear evidence of intention to contaminate the Armed Forces; and it constitutes a flagrant violation of Article 16 of the Anglo-Russian Treaty signed on the 8th August.
> 3. The authenticity of the document is undoubted.[149]

However, these assertions, although convincing, did not satisfy Crowe. The subject was dealt with in the usual channels only after his having been informed, on 13 October, of the obtainment of corroborative proofs, two days before, that the Letter was, indeed, received by the addressee.[150]

The reactions amongst Northern Department officials were mixed. William Strang, the Second Secretary, was not over-enthusiastic over either of the options open to them, i.e. publishing the contents of the Letter or making a formal protestation to Moscow. Strang pointed out that the International would welcome the publishing of the Letter for it desired to discredit the British Government as much as possible; while Russian reaction to formal protestation would undoubtedly be that they were not responsible for the International and that if Britain objected to the doctrines of this body why did it not take action against the British Communist Party. Also, Strang did not see how Britain could, whether by withdrawal of her mission or by deportation of the Russian representatives from London, put a stop to secret correspondence between the International and the British Communist Party, or stop them making propaganda all over Britain. But if the Government had to do anything, concluded Strang, then of the two options he would rather have the Letter made public.[151]

In contrast, J. D. Gregory, Head of the Northern Department, minuted that:

> I very much doubt the wisdom of publication. The authenticity of the document would at once be denied and it would probably be the last of its kind which we should receive.[152]

Crowe responded that it was quite true that the Government had always felt there was no point in filing protests to the Soviet Government, simply because these quite shameless liars merely denied everything. However, it was not, in his opinion, sufficient to invalidate the view that the best and only defence against these proceedings was publicity. According to him, it did not seem fair to their own people that their knowledge of these Russian machinations should remain for ever concealed. The Establishment had not only every right but almost a duty to bring such cases to the notice of the Soviet Government, under the memorandum handed to her on 29 May 1923. Finally, Crowe dismissed the argument of Strang that officially the Soviet Government could disclaim responsibility for the actions of the International, noting both the support she extended to it and the formula respecting propaganda – i.e. not to support, with funds or in any other form, persons or bodies whose aim was to spread discontent or to foment rebellion in any part of the British Empire. The Soviet Government had been bound by this condition since 9 June 1923. Whereupon, he instructed that a formal note of protest be prepared with the intention of giving it to Khristian Rakovsky, Soviet Chargé d'Affaires, and at the same time he sent the file for MacDonald to study, who was campaigning in Manchester.[153]

But what Crowe failed to take into account, for the simple reason that he was unaware of it, was that forty-eight hours before he was appraised of the existence of the Letter the intelligence community had embarked on a course intended to ensure that the Letter not only be published,[154] but become public on the eve of the election, as a 'political bomb' beneath the feet of MacDonald and Labour.[155]

The tool selected for this purpose was Conrad Donald im Thurn, a former officer of MI5. From his diary, a copy of which was discovered by three correspondents amongst the papers of his close friend and confidant, Major Guy Kindersley, a former Conservative MP (1923–31), it turned out that on 8 October he had met with a mysterious informant [X], who told him that he had received news from Moscow to the effect that his [X's] old enemy Apfelbaum – Zinoviev was born Hirsch Apfelbaum – boasted of a great propaganda coup he was about to launch in Britain, and even sent instructions on this point, to be used as soon as the Trade Treaty was signed. On 13 October, the two met again, and Crowe was informed that the instructions had reached Britain.[156]

Over the years, evidence has accumulated to suggest, but not to prove, that an active part in ensuring that the Letter was publicized was played by that most celebrated master spy Sidney Reilly, born Sigmund Georgievic Rosenblum, illegitimate son of a Catholic Russian mother and a Jewish father, whose sometimes fantastic schemes against the Bolshevik regime have become a byword.[157] So perhaps X, who, according to Mrs im Thurn, was introduced by Boris Kadmozev, a White Russian aristocrat naval officer, and Reilly, who was well known in White Russian circles and whose enmity towards Zinoviev was of long standing, were one and the same. In which case, the 1928 financial claims of im Thurn on behalf of X were a blind for filching money from the Conservatives,[158] since in 1925 Reilly crossed into Russia from Finland never to be heard of again.[159] Unless, of course, X was, as stated in the 1928 letters of im Thurn, yet another involved party and a foreign one at that,[160] who was also an associate of the master spy. But that comes later.

A probable scenario was that in order to ensure that im Thurn would not content himself with bringing the facts to the notice of the authorities, but would take steps leading to the publication of the Letter, X specified to him, on 13 and 14 October, the contents of the Letter, and told him he thought MacDonald, Arthur Henderson, the Home Secretary, and James Maxton, leader of the Left Wing of the Labour Party, knew of the existence of the Letter. He further confirmed that these individuals had met, and acted in collusion with, Rakovsky and the British Communist leader – who was in fact in Moscow at the time – in order to hush the affair up. Im Thurn treated these last reports with scepticism, but having met one of his MI5 acquaintances, who hinted to him that the Letter did exist but was a strictly guarded secret, im Thurn became convinced that there was governmental intention to keep it under wraps. As an anti-Communist and former agent his duty seemed plain: to force the Government to expose the sensational information at its disposal.

Therefore, on 15 October, he met Kindersley, Lord George Younger, Treasurer of the Conservative Party, and Sir Stanley Jackson, Chairman of the Party, and unfolded before them the facts in his possession. At a second meeting that day it was agreed that in return for information corroborating the existence and the nature of the contents of the Letter, so that it would be possible to publish it in *The Times* and brand the Labour members as collaborators of the Bolsheviks, im Thurn – allegedly to recompense X – would receive £7,500 from the coffers of the Conservative Party.

From Thurn's point of view, he had an incentive to establish the existence of the Letter, discover its whereabouts, and ensure that it was circulated among as many departments as possible. This latter objective was particularly vital so far as the Conservatives were concerned, for its circulation would stamp it with a mark of authenticity, increase their chances of obtaining a copy, and give *The Times* an opportunity, through its own contacts, to confirm the story – an essential move if the information was to become public.[161]

Up to this point the story is (relatively) straightforward. Hereafter, a number of questions arise. Im Thurn claimed in his diary that, from a series of meetings he held during 16–20 October with Major-General Sir Wyndham Childs, Head of the Special Branch of Scotland Yard, his MI5 acquaintance, a Foreign Office Clerical Officer (woman), and Rear-Admiral Alan Geoffrey Hotham, Director of Naval Intelligence, he had learned that the Letter indeed existed. He claimed that Admiral Hugh Sinclair, Head of the Secret Intelligence Service, was avoiding circulating it among the various departments. He also gave a dramatic description of how, through camouflaged threats to publish the Letter (of which he had no copy), he had, on 21 October, extorted from Sinclair a promise to circulate the Letter, something which, indeed, was done on 23 October.[162]

From Foreign Office files and the evidence of various officials it turns out that copies of the Zinoviev Letter were circulated among the various departments on 9 October. Once officialdom became convinced that the Letter had been circulated and could no longer be kept under wraps, the Conservatives made final preparations to ensure that it won as wide a public resonance as possible and that Labour would be charged with trying to hush the affair up. In this way the 'Red Menace' would become an election issue. Im Thurn, for his part, attempted to get a copy of the Letter but without success. Consequently, on 22 October, he dictated the contents of the Letter orally to a correspondent of *The Times*, who got to work among his contacts. At the same time, the Conservatives, through the *Manchester Evening Chronicle*, dropped the first public hints of the ensuing political furore.[163]

While im Thurn and the Conservatives were trying to publicize the letter, on 16 October MacDonald instructed that:

> We must be sure that the document is authentic. I favour publication of such things, and the way to do it is to address a dispatch to M. Rakovsky. Prepare such and see how it looks. It must be so well founded and important it carries conviction of guilt. If not it will do harm.[164]

The Foreign Office carried out MacDonald's instructions.

The intelligence community, in all that related to giving the Letter publicity, refused to rely on the good intentions of Crowe and MacDonald, or on the conniving of im Thurn and the Conservatives. Therefore other branches of the intelligence community went into action for the purpose of making sure that the Letter would indeed be publicized, and in such a manner that the harm caused to the Labour Party would be maximal. First, on 23 October, Thomas Marlowe, editor of the *Daily Mail*, found on his writing table in his office a telephone message that had arrived the night before from 'an old and trusted friend', telling him of the existence of the Letter and of the fact that it had been circulated among the various departments. Then, on 24 October, he received two copies of the Zinoviev Letter – one sent by post by an unspecified gentleman, whom he had met the previous day; the second delivered to him by yet another friend, who came to his office to be advised on the best method of publication. Whereupon the *Daily Mail*, in collusion with *The Times* and the Conservative Central Office, who resigned themselves to being stolen a march on, made final preparations for printing.[165]

Evidence survives to show that the 'old and trusted friend' of whom Marlowe spoke was almost certainly Vice-Admiral Sir William Reginald 'Blinker' Hall, the legendary war-time Director of Naval Intelligence. The third source was very probably Lieutenant-Colonel Frederick Henry Browning, former Deputy Head of MI5.[166]

The motives of the intelligence community for wanting to bring Labour down from power are not hard to find. For one thing, they regarded Russia and Bolshevism as principal enemies of the British Empire, and tended to see in Leftist elements like Labour, who strove openly for improving the relations of Britain with Russia, a security risk. They also had a far more rational and concrete motive – namely, the future of the intelligence services should Labour remain in power. Ministers in the Labour Government demanded, more than once, the suspension of the Secret Service on the grounds that it was responsible, by its subversion, for European wars, but their wish was rejected. However, lately some people have thought that they detected a certain difference of emphasis in the attitude of MacDonald. Against this background it was only natural that the intelligence community – after becoming convinced of the seriousness of Labour's intentions on this point – should strive to bring the Labour Government down from power, for it judged, quite correctly, that by this the latter would be deprived of the ability to carry out its aim of reducing the power of the respective services.[167]

These facts inevitably raise the intriguing question of whether the Zinoviev Letter was authentic. The case is still open. Separate articles by Sybil Crowe and Christopher Andrew have demolished many of the arguments that it was a forgery but neither has proved that it was genuine. It is clear that Eyre Crowe believed the letter to be genuine and continued to do so even after the affair blew up.[168]

Evidence brought before governmental institutions in the course of October–November 1924 was, on paper at least, conclusive in favour of the Letter by Reilly, or so it would now seem.[169] Lastly, simultaneously with the Zinoviev Letter, a document containing instructions to the Balkan communists, and dated 7 August, circulated among intelligence officials in Europe. The similarity between the two in content, form and textual errors is so striking that there is little doubt that both documents emanated from the same source.[170]

Intelligence officials had some reservations about the Letter. The claim of Chamberlain, in December 1924, that there was no doubt among Government circles as to the authenticity of the document,[171] was denied by the Heads of the intelligence departments.[172] Childs, Head of Special Branch, noted in 1930 that, 'There was absolutely no reason to think that this particular effusion was genuine.'[173] In 1925 he assured the Principal

Private Secretary to the Foreign Secretary, Walford H. M. Selby, in confidence, that, 'he had come into possession of absolute proof that the Zinoviev letter was a forgery'.[174] Finally, the handwritten assertion of the SIS to Gregory attesting to the authenticity of the document reminds one of the 1921 'irrefutable' evidence[175] concerning Russian anti-British subversion,[176] which had been the basis of the Curzon protest and proved to be forgeries.[177]

After verifying to his satisfaction the authenticity of the letter, following the instructions of MacDonald on 16 October, Crowe directed the Northern Department, on the basis of MacDonald's minute that he favoured publication, to prepare a draft of protest[178] and sent it, after introducing revisions, on 21 October, for the Prime Minister's approval,[179] with the remark that:

> I now submit a draft to M. Rakovsky, which I think puts the case squarely. It can be published so soon as it has reached M. Rakovsky's hands.[180]

The draft reached MacDonald on 23 October. He was staying at Aberavon in South Wales, and returned it after studying it to the Foreign Office, without an initial but accompanied by substantial amendments to its content,[181] which were warmly welcomed by Crowe.[182] MacDonald wrote in his diary that despite the amendments introduced by him in the draft he still was not satisfied with it. Therefore he refrained from initialing it, with the purpose of studying it again after it had been copied by the Foreign Office.[183] Perhaps this was, indeed, his intention, but his refraining from initialing documents was not an infrequent omission.[184] So that when the handwritten amendments of MacDonald,[185] along with the previous remarks of the Permanent Under-Secretary that the draft could be published so soon as it reached the hands of Rakovsky,[186] were returned, albeit uninitialed,[187] to the Foreign Office, Crowe assumed quite justly that MacDonald sanctioned sending the amended protest draft to Rakovsky, and the publication of the 'Red' Letter and protest draft in the Press.[188]

This matter became particularly urgent so far as Crowe was concerned, given that he was informed that the *Daily Mail* had managed to get a copy of the Letter and was about to circulate it among its Press peers. Crowe judged that, should the issue be published in the *Daily Mail* before the official announcement, a misleading impression would be created in the public forum that the Foreign Office had been in possession of the incriminating document for some time but had refrained from all action on the instructions of the Labour Government and the Prime Minister and Foreign Secretary. He reasoned that the damage caused would be infinitely more serious than an early announcement about the existence of the Letter, the more so that the decision to publish it had already been taken.[189] On Crowe's directions the protest note signed by Gregory was handed over to Rakovsky, on 24 October,[190] in the presence of a reporter from the *Daily Express*, whose news editor had already received his copy.[191]

While Crowe managed to anticipate the publication by the Conservatives, the *Daily Mail* and the intelligence community, he was nevertheless unable to prevent the political storm that followed. Baldwin, Curzon and other Tory leaders charged MacDonald with attempting to cover up the existence of the Letter. At the same time the Conservative Party claimed that the Zinoviev Letter constituted dramatic evidence of the Bolshevist tendencies of the present Government. The bewilderment and inconsistent statements of Ministers and of MacDonald gave ostensible further substance to this charge.[192] From today's perspective it is clear that it was not the Zinoviev Letter which determined the results of the elections, though it certainly constituted the last nail in the coffin of the Labour Government.[193]

In the elections held on 29 October the Conservatives gained a considerable majority. And on 7 November a Conservative Government, under the Premiership of Baldwin, took the place of the Labour Government. J. Austen Chamberlain took the place of MacDonald as Secretary of State for Foreign Affairs.

6

The Second Baldwin Government: J. Austen Chamberlain and the Foreign Office

The departure of MacDonald, with whom the Foreign Office had excellent relations in spite of his descent and views, was viewed with regret by officialdom. However, officials soon found a common language with his successor. Chamberlain's willingness to develop informal and personal relationships with his subordinates,[1] his avoidance of petty details,[2] his attempts to boost the morale of staff by complementing them on their work[3] and, also, the kindness and consideration that he showed them,[4] were warmly accepted by officials[5] from whose mind the memory of the Curzon era had not yet been wiped out.

Foreign Office officials accepted with satisfaction Chamberlain's readiness to fight – though not always with success – against the intervention of the Chancellor of the Exchequer, Winston Churchill, in European[6] and Egyptian[7] foreign affairs, the Chancellor of the Duchy of Lancaster, Cecil, in matters related directly or indirectly to the League of Nations,[8] and of the Colonial Secretary, Amery, and Churchill in China's affairs.[9] Foreign Office members viewed with appreciation Chamberlain's intellectual faculties and the manner in which he represented their views in Cabinet. Chamberlain wrote to his sister:

> The Office is very happy. *We* have a policy for Egypt & the Sudan, approved by Cabinet after a stiff tussle, for Italy & for Albania . . . & for France, Germany & Security. They say that the Foreign Office view gets presented to the Cabinet & defended in Cabinet as it has not been for years, & that they are astonished at my 'grasp'. Indeed I have found they very constantly accept as improvements the major alterations that I have suggested in policy on despatches & telegrams.
>
> And the Diplomatic Corps sing the praises of their new 'Chief' . . . 'They like your frankness' says Crowe.[10]

The Foreign Office's satisfaction was understandable. Chamberlain's

Foreign Secretaryship not only heralded maintaining the good atmosphere prevailing in the days of MacDonald, but also further strengthened the Foreign Office monopoly over foreign affairs. The causes of this turn of events were diverse, but a prominent one was undoubtedly the personality of the new Foreign Secretary.

Chamberlain's Personality

Joseph Austen Chamberlain was a Victorian figure.[11] He was amazingly conscientious,[12] and took pride in the moralities and charities on which English society was founded. His unreserved loyalty to the prime minister, an efficient discharge of his public duties, consideration for his fellow-men, defence of his subordinates against public assaults, mediation by straight dealing and adaptation to the working arrangements of his department were typical of his nineteenth-century gentleman's code.[13]

Unlike his predecessor, Chamberlain benefited from the advantages of higher education. He also had an abundance of administrative and Parliamentary experience. He was educated first at Rugby, then at Cambridge University and finally, in 1885–6, at the École des Sciences Politiques in Paris. In 1887 he went to Germany for a year and attended regularly the lectures of Heinrich von Treitschke on Prussian history. However, Germany was not to his liking and the lectures he heard disquieted him. At the end of twelve months he left Germany, never to return.

In 1888, thanks to his father Joseph Chamberlain, Austen acquired his first pre-Parliamentary experience as prospective candidate for the Border Burghs. In 1892–3 he became Conservative MP for East Worcestershire. In 1895 he was made Civil Lord of the Admiralty by the Conservative Government, a post he held until 1900 when he became Financial Secretary to the Treasury. From 1903 until the resignation of the Conservatives in 1905 he served as Chancellor of the Exchequer, and from 1911 he became a regular member of the Conservative leadership. With the formation of the First Coalition Government, in May 1915, he was appointed Secretary of State for India. He resigned after criticisms were made of the Mesopotamian campaign, only to return to the Second Coalition, first as Minister without Portfolio (1918), and then again as Chancellor of the Exchequer (1919). He succeeded, from spring 1921 until the autumn of the following year, Bonar Law as Leader of the Conservative Party, and as a result ceased to act as Chancellor of the Exchequer. However, in October 1922 he surrendered his leadership post, refusing to participate in the Tory revolt against the leadership of Lloyd George which followed the latter's mishandling of the Greco-Turkish conflict. Only with the return of the Conservatives to power in 1924 did he once again head a department of state.[14]

Chamberlain's high principles, his inability to detach himself from views formed in young adulthood (views which more than once turned into prejudices),[15] and his concentration on the meticulous execution of matters while avoiding unconventional methods (which did not accord with his outlook on the world), created the impression that he was a dull and uninspired man.[16] But an assessment of his ideas suggests that he stuck only to the politically possible. Further, his command of French and Latin, his erudition in written literature in general and history in particular,[17] and his superior intelligence, counter-balanced some of the more negative aspects of his personality such as those prejudicial dispositions previously described. Lord George Riddell once said of him:

> Austen Chamberlain is a shrewd, capable Minister and although he looks superior is nothing of the sort. Indeed, he is modest, affable and kindly.[18]

With the aid of meticulous preparatory work and plain common sense he managed more than once to achieve results that less thorough persons failed to attain.

Chamberlain was aware that in foreign affairs he lacked the prestige and knowledge of his idol, Castlereagh.[19] But he made up for it by concentrating, first of all, on a sphere in which he did possess knowledge – Europe. Aided by the counsels of his assistants, while employing his own common sense as to whether to adopt recommendations or not, and totally relying on Foreign Office officials in spheres in which he admitted to being ignorant (such as the Far East), he presided competently over foreign-affairs policy.[20]

His private life was marked by simplicity. A devoted family-man, he liked to spend long hours in natural surroundings and the time he was able to allocate to nurturing his garden was an inexhaustible source of pleasure for him.[21] His wife, three children and two sisters were the refuge that enabled him to overcome his political disappointments.[22] His prolonged correspondences with them, in which he described at length political manoeuvrings, make marvellous background material, though they are not always accurate as to what actually took place behind the scenes.

Working Methods

The division of functions between Chamberlain and Foreign Office officials was, as in the MacDonald period, on the basis of partnership. The workload cast on the shoulders of the new Foreign Secretary was no less heavy. According to him he did not have the privilege of going to sleep before one, two and sometimes three o'clock in the morning.[23]

> I get two boxes of papers about 8.30 every night – one 'urgent' to be done before I go to bed & one which I may keep till the next afternoon if I like. But the Foreign Office & I have gone to bed at night with my papers finished on every night but one. Still it is the daily drip, drip, drip, the never being carefree that wears one out in the long run & makes one glad when the opposition becomes the government.[24]

While MacDonald had placed a premium on quick decision-making, Chamberlain belonged to that type of policy selector favoured by the Civil Service more than any other type, as Minister: namely, that of an intelligent laymen. Ministers who are intelligent laymen are supposed to bring to the sphere of which they are given charge new ideas, and to probe the weaknesses of assumptions that have hitherto guided officialdom.[25] And that was precisely what Chamberlain did. The role of Foreign Secretary was alien to him, but he rapidly reached a state in which he was able to react to memoranda that were submitted to him,[26] highlight the difficulties and bring the best out of his officials with probing questions.[27]

The minutes submitted by the various echelons for study by the new Foreign Secretary were extremely comprehensive, and often included opinions on the implications of the divergent views of all who had contributed to the issue to date. Chamberlain, for his part, took an active role. He would give an idea as to what line he intended to pursue, respond at length to the minutes, maintain close contact with Heads of Departments (such as Lampson),[28] and not hesitate to convene the senior assistants for consultations.[29]

For instance, when Foreign Office officials embarked, on the directions of Chamberlain, upon preparing a series of memoranda that were intended to make clear to themselves and others what precisely the Government was striving for in Europe,[30] it was done with the knowledge that the new Foreign Secretary saw, in providing security for France against a German war of revenge, a precondition for bringing about stability in Europe. This policy stand stemmed primarily from his pro-French leanings.[31] As he admitted on one occasion, 'I loved France as a woman, for her defects as well as for her qualities,'[32] and he wished to demonstrate a practical manifestation of his pro-French attitude.

The independence displayed by the new Foreign Secretary, and the immediate contact maintained between him and his assistants, might have weakened the position of the Permanent Under-Secretary. But in practice that was not the case. Chamberlain saw in Crowe the expert in foreign affairs and strategy,[33] consulted him often[34] and attached great weight to his opinion.[35] Functionally, Crowe continued to see all the papers read by the Foreign Secretary, and also those that were not read by the minister at all. He represented the Foreign Office on the Committee of Imperial Defence, and deputized for the Foreign Secretary during his absence. Chamberlain corresponded with Crowe whenever he stayed abroad and described at length to him the details of any talks that took place.[36] And when the health of Crowe began to fail, Chamberlain made special efforts to alleviate his workload.[37]

Policy Towards Germany

On assuming his new post, Chamberlain had to contend with the short-term question of whether to favour the evacuation of Allied forces on 10 January 1925 from the Cologne zone, as specified in the Treaty of Versailles, even though Germany had not fulfilled all her obligations, or to postpone it until the remaining obligations had been completed. The Powers were divided among themselves on this point, and the opinions submitted for the examination of the Foreign Secretary on the matter were not uniform either. The War Office, for example, was far from being enthusiastic over the evacuation, pointing out that Germany had not yet carried out all her disarmament obligations.[38] The War Office also asserted that the French, who treated the German infractions with great seriousness, would complain that the British were circumventing the Peace Treaty.[39] On 29 November, 20 December 1924 and 17 January 1925 there were repeated worryings that:

> . . . the danger of German aggression will increase with the evacuation of each successive zone, and plans for reinforcing our troops are therefore essential.[40]

The Foreign Office, on the other hand, was inclined to treat the German infraction with leniency and to evacuate the Cologne zone on time. It pointed out that the only material question at present outstanding against Germany, after the adoption of the Dawes Plan, was military control.[41] On the basis of past reports,[42] up to the end of 1924, that Germany's specific failures were not of great importance, the Foreign Office warned that unless Germany could be shown to be seriously in default in her Treaty obligations, failure to evacuate the Cologne zone would undo much of the good achieved at the recent London Conference.

In short, the Foreign Office felt that non-evacuation would cast doubt on the good faith of the Allies.[43]

Moreover, the Foreign Office dismissed the warnings of the War Office concerning the danger of Germany's future aggression.[44] Bennett claimed, on 23 December 1924, that the professed apprehensions of the War Office on this point were exaggerated, holding that one ought to take care not to interlock the Cologne evacuation with the French security question.[45] Lampson added that for the sake of rapid evacuation it was desirable not to turn a technical detail, such as the preparation of defence plans, into an international problem and that it could, if necessary, be dealt with by the forces remaining in the field.[46] Crowe and Chamberlain agreed;[47] the former minuted that it was important to avoid anything that might enable the French to say that they had been allowed by the British Government to work out plans for war against Germany.[48]

These policy stands were underlined with greater vigour by the Foreign Office in February 1925, in response to Marshal Ferdinand Foch's statements that once the Cologne zone was quit Germany could, within six months of the evacuation, invade Belgium successfully.[49] Bennett wrote condescendingly that this idea might pass muster among armchair strategists in a civilian club, but it was difficult to believe such a military absurdity. For an attack such as this to have the slightest chance of success, Bennett judged, the Germans would have to put up a force equal to at least three army divisions, and recruit a further force of similar size for the requisites of invasion. And for that purpose they would have to carry out general mobilization without the knowledge of the Allies, and to manufacture and accumulate modern arms, again without the knowledge of the other side:

> It is clear that neither of these two conditions is capable of fulfillment for a long time to come, and that while the Germans might possibly muster sufficient force to delay an allied attack on Germany, a German offensive is at present a practical impossibility.[50]

Lampson held that, as long as the evacuation of Cologne had not yet been decided upon, they should not be drawn into a discussion of these problems.[51] Crowe, too, thought that the picture drawn by Marshal Foch was childish and that the representatives of the War Office should be instructed to refrain from pursuing this discussion with the French.[52] Chamberlain also took this stand.[53]

All the same, at the beginning of December 1924, owing to German obstruction in the way of the Inter-Allied Military Commission of Control, the Commission was not able to submit its full report on time. Instead, the Commission was asked to submit a preliminary report.[54] On 17 December the Cabinet adopted the Foreign Office position that Germany should be notified that the date for evacuation of the Cologne zone was to be postponed until after the Commission of Control had completed its work,[55] and a proposal to that effect was circulated by Britain among the rest of the Allies.[56] However, following the negative character of the Commission's provisional report,[57] Chamberlain decided, on the grounds of the recommendations of Lampson[58] and Crowe,[59] to accept the French demand of 24 December that they include in the document to be sent to the Germans a list of their disarmament infractions, as detailed in the interim report of the Commission,

> I agree. The Germans are foolish. They have asked for it & they must get it.[60]

The issue was co-ordinated with the rest of the Allies and on 5 January 1925 a joint Note on this point was communicated to the Germans.[61] Germany, however, rejected it on 6 January.[62] Subsequently a second Note was communicated, on 26 January, making it clear that the Cologne zone would be evacuated only after the Germans had made good all that required rectification.[63] But again the Note was dismissed by the Germans on the grounds that they had fulfilled the disarmament clauses in full, and that in any case the date of the Cologne zone's evacuation did not rest on an Allied decision but on the terms of the Treaty of Versailles, which the Allies were bound to meet.[64] The Foreign Office was greatly irritated by the German statement.[65]

Bennett, however, was not happy with the firmness demonstrated towards the Germans:

> If we accept, as we can scarcely fail to do, that the right policy is to liquidate the war as speedily as possible and restore normality, it follows that so far as Germany is concerned we ought to do everything possible to encourage that part of German opinion which is working for the fulfillment of Germany's obligations and if possible give them something to show as the fruit of what must be a thankless and unpopular task.

Bennett's point was that the Allies should be careful not to provide material for those who fostered the idea of non-fulfilment and a war of revenge. Rather he sought to encourage economic intercourse between Germany and France as an incentive for the Germans to stick to peace. The decision to prolong the occupation of Cologne, asserted Bennett, tended in the opposite direction. In his opinion, Germany was disarmed and if the Allies were to insist on occupying the Rhineland because of a number of technical defaults, and as a guarantee of the payment of reparations, then it would undo all that was attained after the Ruhr crisis. Instead of achieving disarmament, it would encourage the Germans to accumulate arms secretly and hold secret training. Already, according to

Bennett, there were first signs of the negative consequences of the decision not to leave Cologne, which, in his opinion, was partly responsible for the nationalistic character of Hans Luther's new government. He also pointed out that the Allied stance had already led to a German stiffening in the commercial negotiations proceeding between them and Britain and France.[66]

The demand of Bennett that they should disregard the German infractions and evacuate Cologne did not evoke a positive echo from his superiors. Lampson minuted that they were not in Cologne for fun but to force the Germans to fulfil the Treaty of Versailles in general and disarmament in particular, and thus prevent a renewed war.[67] Crowe, too, took a serious view of the German disarmament infraction, and after reading the interim report of the Commission of Control he noted in a minute endorsed with a signature by Chamberlain:

> But to my mind the evidence strongly supports the theory that the German authorities show bad faith throughout and have no intention of loyally carrying out the treaty.[68]

Despite their doubts, the Heads of the Foreign Office did not conclude from the German evasions that the Cologne zone and the rest of the Rhineland territories were the sole guarantee of French security,[69] as was held, on 28 January, by Herriot.[70] They also rejected the later French stand of 3 March, according to which France was no longer opposed to the evacuation of Cologne, but demanded that the Germans would not only have to fulfil their obligations but must also demonstrate to the Allies that they had given up their aspirations of revenge and were no longer attempting to deceive the Allies on issues such as disarmament before they would move.[71] In the opinion of the Foreign Office, adherence to the Treaty of Versailles meant that the Cologne zone should be evacuated the minute the Commission of Control declared that Germany had rectified all that required rectification. If they allowed the French to link the evacuation of Cologne with the question of France's security there was no telling whether they would ever succeed in inducing France to agree to an evacuation.[72]

Long-Term Policy towards Europe

These dissensions gave birth to prolonged negotiations between England and the rest of the Allies in general, and France in particular: first, over the legal connection between the questions of Cologne, security and the future of the Commission of Control; and later, regarding the definition of what the Germans would be required to make good before the evacuation of the Cologne zone could begin. There were difficulties over the date of evacuation. The first part of the negotiations was tentatively resolved by France's acknowledgement of the separation between the evacuation of the Cologne zone and the fate of the Commission of Control, though not on the causality between them.[73] The second argument ended only on 2 June, when a detailed Note was communicated to the Germans on this point.[74] On the third point, on 16 November, it was concluded that the evacuation of the zone would begin on 1 December[75] and be completed by January 1926.[76]

Chamberlain had to consider his long-term European policy, and here it became clear to the Foreign Office managers, to their great joy, that there was a wide concord on this point between officialdom and the new Foreign Secretary. For instance, on 19 November 1924,[77] Chamberlain agreed with Campbell's stand of 10 November,[78] and with Crowe's stand of 17 November,[79] that the Geneva Protocol was too cumbersome an instrument to establish peace and stability in Europe, an opinion also held by the Committee of Imperial Defence and its Chairman, Curzon.[80] The Committee was unanimous in its estimations concerning expected developments by Germany in Europe during the next decade, if nothing was done to check its dictatorial aspirations.

On the basis of opinions he had built up over the years, and as a result of a chain of meetings that he held in Paris and Rome at the beginning of December, Chamberlain, even at the start of his tenure, expressed views amazingly similar to those voiced in the past by Crowe, Tyrrell, Troutbeck, Phipps and, to a certain degree, also Lampson – namely, that the security boundary of Britain was not the Channel but the eastern frontier of France, and that the dominant feature in European policy was the fear of invasion and wars.[81] As long as Germany had a glimmer of hope of success, she would be tempted to prepare for a *revanche*.

> Looking at Germany I see no chance of her settling down to make the best of new conditions unless she is convinced that she cannot hope to divide the Allies or to challenge them with any success for as long a time as any man can look ahead.

The general consensus was that a further European war was to be expected if they did not succeed in setting to naught German ambitions,[82] and did not prevent the intensification of French fears that might drive them to provocations against Germany.[83] It was recognized that if war did indeed break out, Britain, whether she liked it or not, would inevitably be involved in the fight against Germany. Chamberlain explained that Britain could not afford to see France crushed and to allow Germany to dominate the Low Countries.[84]

Moreover, Chamberlain contested D'Abernon's statement of 7 January 1925, that they should not fear the establishment of a Russo-German bloc, which would pose a military threat to Europe.[85] The British Ambassador's argument was part of an effort to persuade the Foreign Office that the old assumption that peace and stability were contingent on safeguarding Europe against Germany and her allies was obsolete.[86]

> I do not understand Lord D'Abernon's reason for thinking union between the Imperialist military leaders & the Communists of Russia 'unthinkable'. After all who started the Russian revolutionaries on their task, & why not again.[87]

Finally, besides his appreciation of the likelihood of such an alliance, Chamberlain foresaw, on 4 January, the possibility of Europe being mastered not by Germany alone but by a Russo-German combination,[88] although – and this must be stressed – he did not imagine that events would proceed so quickly. At best, he probably supposed a formidable Russo-German alliance in some 35 to 45 years' time. Meanwhile, in his opinion, the facts did not vindicate the French expectation that there was already reason to fear Germany in the immediate future.[89]

There was a concordance between Chamberlain and his senior assistants regarding the best way of establishing peace and stability in Europe. In Lampson's minute of 5 November 1924, he noted that if it was decided to cancel the Geneva Protocol it ought to be replaced by a guarantee pact like that of 1919.[90] Crowe minuted, at a later date, that the solution ought to be in the direction of an Anglo-French pact.[91] Chamberlain told the Committee of Imperial Defence, on 16 December, that if they wished to eliminate the component of fear then dictating the moves of the states of Europe, and to prevent another European war, they ought not to be content with the cancellation of the Geneva Protocol but to find a proper replacement. On this point Chamberlain refused to commit himself openly one way or another, though he did drop some broad hints when he mentioned that before the First World War he had pressed Grey to turn the *entente cordiale* into a formal alliance, that in 1919 he was for an Anglo-American guarantee to the security of France, and in 1922 was among those who, at the Cannes Conference, urged the Government to offer the French the Guarantee Pact. Moreover, according to Chamberlain, the French and the Belgians had made it clear that in any case they would not be satisfied by the Geneva Protocol alone, but would want it complemented by a security pact with Britain.

Chamberlain's ideas were general rather than specific. Thus the Committee of Imperial Defence accepted the proposal of Lord Curzon, to set up a sub-committee whose function would be to see how far the Geneva Protocol could be modified, and according to what principles the alternative proposals should be evolved. Originally, Curzon had suggested that Crowe be made chairman of the sub-committee, given his great authority. However, Chamberlain, who wished to alleviate Crowe's workload, explained that the Permanent Under-Secretary had too many duties. As a compromise it was agreed that Crowe be a member of the sub-committee, headed by Hankey, and that it be composed of representatives of the Colonial Office, India Office, Board of Trade, Treasury and the three Service departments.[92]

While the sub-committee was deliberating, Chamberlain embarked upon consolidating a clear-cut alternative scheme to the Geneva Protocol. On 4 January he instructed the Central and Western Departments to prepare a memorandum describing the present state of affairs in Europe, the need for securing the integrity of the various countries, and the bitter effects that would follow if that goal was not accomplished. At the same time, Headlam-Morley, the Historical Adviser to the Foreign Office, was requested by Chamberlain to review historically the position of Britain, and note the lessons that could be drawn from history concerning the present. Chamberlain minuted that all this did not answer the question of how the problem of the security of Europe could be solved. And to this end, according to him, he was at a loss:

> Can we propose an Anglo-Franco-Belgian pact of Guarantee to be followed by a Quadruple Pact embracing Germany? Or ought we to propose a unilateral declaration of British interests and of what we should regard as casus belli? Or again is there some third course?[93]

Chamberlain recognized the impossibility of defending vital British interests in the West without being dragged into a quarrel over Lithuania, Latvia, Poland or Bessarabia. And what ought Britain's policy to be in relation to those countries? He had no clear answers to that and wished to hear the thoughts of the Foreign Office on this point. For that purpose, he said, he was attaching Crowe's critical memorandum regarding the Geneva Protocol and his own statement, of 16 December, to the Committee of Imperial Defence. Finally, Chamberlain announced the holding of a departmental conference on 22 January, in the course of which his subordinates would be asked to express their opinions.[94] Thus, the new Foreign Secretary gave expression to his thoroughness and to the step-by-step system by which he set out to resolve problems.

After the departmental conference had taken place, Harold Nicolson drew up, under the supervision of Crowe, a brilliant rejoining memorandum, on 20 February, which was circulated by Chamberlain to the Cabinet saying that it reflected not only his own view but also the stand of the Foreign Office in general.[95] The memorandum described Europe as divided between the ex-Central Powers, dangerously angry at what they had lost, and Britain's late Allies, dangerously fearful of losing what they had won. According to Nicolson, despite the Austrian desire to fuse with

Germany, and the recalcitrants in Hungary and Bulgaria, the element that ought to worry England was nevertheless Germany; Germany in all certainty would recover and would then strive to restore to herself the Polish corridor and Upper Silesia. Moreover:

> If France were isolated, and British neutrality to be assured, she might also endeavour to attack France.

France, in contrast, the memorandum continued, found herself in a position where her request to turn the Rhine into a barrier between her and Germany was refused by the rest of the Allies in Paris; the Anglo-American guarantee offered as a substitute had not been put into effect; and the disarmament of Germany and the Allied military presence in the Rhineland were but provisional in duration. These facts, together with the bitter memories France had of Germany, understandably gave rise to deep apprehensions, and brought about France's determination to set up a 'Little Entente' with Eastern European States, which would inevitably increase tensions and was liable to drive France to despair, which in turn might provoke Germany to begin a war of revenge of which France stood in such terror. Therefore, if it was wished to deter Germany and moderate the attitude of France – two conditions without which it would not be possible to stabilize Europe – the Government ought to reach an understanding with Belgium and France that might entail, if necessary, a British guarantee of the integrity of these countries. Nicolson also turned to the question of Russia.

History and economics were pertinent enough reasons for it to be recognized that it was not possible to return to the policy of isolation. The giving of a guarantee to Belgium and France would prevent a repeat of past errors, namely that if, in 1914, Germany had known that the British Empire would come to France's assistance she would not have risked the Great War. The giving of a guarantee would also moderate France's political and economic attitude towards Germany, thus gradually enabling Europe to be brought from hostility to friendship and political and economic co-operation. Finally, although in the first stage the talk was of an Anglo-Franco-Belgian agreement, at a later stage it might be possible to include Germany also within the guarantees of security. To this memorandum was attached an historical paper, drawn up on 12 February by Headlam-Morley, which provided Nicolson's memorandum with the conclusion that no single Power should ever be allowed to control the opposite shores of the Channel and the North Sea. However, Headlam-Morley thought that a unilateral declaration on Britain's part concerning the position of France and the Low Countries would suffice,[96] although it did not altogether satisfy Crowe.[97] Headlam-Morley drew a parallel between Britain's attitudes to Western and Eastern Europe, respectively, something that was contested by Chamberlain.[98] Headlam-Morley therefore wrote a new memorandum from which it appeared that the solution lay in the guarantee pacts.[99]

Conflicting Schemes

What Chamberlain and the Foreign Office had no knowledge of, which was disclosed to them only later on, was that while they were busy consolidating guarantee pacts between Britain, France and Belgium, D'Abernon was working on the implementation of a different scheme altogether.

D'Abernon was much troubled by the non-evacuation of the Cologne zone and by the pro-French wind that blew from the Foreign Office. The basing of Britain's strategies on pre-1914 conceptions and the revival of war-time alignments constituted, from the Ambassador's viewpoint, a perpetuation of the Continental division into 'victors' and 'vanquished'; enhanced both France's non-conciliation and Germany's non-co-operation; and rendered, thereby, European pacification more unlikely. He was anxious, too, about the probable effect on the European balance of power, which, since the defeat of Germany, the disappearance of the Austrian and Russian Empires and the establishment of the 'Little Entente', had swung, in his view, altogether in favour of France.[100]

Whereupon, with the object of preventing a British commitment to France, bringing Germany into the Concert of Europe as an equal to the other Powers, and breaking old war alliances, D'Abernon – in continuation of his 1923–4 efforts to ward off exacting from Germany the execution of the Treaty of Versailles, either by an Allied–German mutual guarantee pact or through the neutralization of the Rhineland under the sovereignty of Germany[101] – suggested, on 29 December 1924, to Carl von Schubert, Secretary of State of the German Foreign Ministry, a revival of the Cuno initiative of December 1922,[102] i.e. a German–Allied non-aggression pact for a period of a generation,[103] a proposal repeated by the German Government in May and September 1923, and February and June 1924.

At first, Stresemann, Minister for Foreign Affairs, and other figures in the German Establishment treated the practicality of this suggestion with great scepticism. But with the prospect of increasingly stringent measures against infractions of the Peace Treaty's financial and military clauses,[104] should the solidification of the Anglo-French *entente* take place, it was decided to put D'Abernon's suggestion into effect.[105] Stresemann sent a note to Leopold von Hoesch, German Ambassador in Paris, on 15 January 1925, stating that it was scarcely possible that D'Abernon was not acting on the instructions of his Government.[106] On 20 January the

Germans put forward, through D'Abernon, an official proposal according to which Germany would be ready to enter, with other Powers having interests in the Rhine – but above all with England, France and Italy – into a solemn engagement for a lengthy period, with the Government of the United States acting as trustee, not to wage war.

In addition, Germany was prepared to conclude mutual arbitration treaties providing for the peaceful settlement of different issues, or alternatively to sign a special agreement obliging the signatories to guarantee the territorial *status quo* along the Rhine as well as demilitarization, thus ensuring that the questions of disarmament and evacuation were dealt with in friendly understanding.[107]

The reaction of the Foreign Office was cool. Bennett,[108] Lampson,[109] and even Crowe,[110] agreed that the proposal ought not to be rejected out of hand, but nevertheless regarded it not as a substitute for but as complementary to the initiatives evolved by the Foreign Office. Lampson explained in this connection that no treaty arrangement including France and Germany on equal terms was possible until the problem of security had been disposed of, as details like the demilitarization of the Rhineland already constituted part of the Treaty of Versailles.[111] Chamberlain, too, was unprepared to give up his initiative in favour of the German proposal. Under the chairmanship of Hankey, but drafted by Crowe, a memorandum was endorsed by the sub-committee recommending that the Geneva Protocol be replaced with a declaration, to be signed by Belgium, Britain and France, that Britain was prepared to fight if France or Belgium were in danger of seeing their North Sea ports pass into the hands of another Power.[112] Moreover, Chamberlain was annoyed because the Germans, who insisted on his keeping the details of the proposal from the French for the time being, on 9 February had made the same offer to France with the stipulation that it be kept from being brought to the knowledge of the British.[113]

In any case, Chamberlain was determined to get a green light from the Government for the execution of his policy, not to mention the fact that as far as he knew it was essentially supported by the Prime Minister[114] and the Services.[115] In a meeting that he held with the German Ambassador he explained to him that under present conditions the German initiative was premature.[116] But an unexpected hitch occurred, which was perhaps inevitable against the background of a growing involvement of various Ministers in foreign affairs. Chamberlain's policy was rejected. His arguments that only thus could Europe be stabilized and France be induced to evacuate the Rhineland at some time in the future, did not help. Curzon dismissed French feelings of insecurity as unjustified and even raised the suspicion that what stood behind France's manoeuvrings was a crude attempt at wringing political advantages from Britain. Balfour considered the French to be paranoid and stressed that the pact would be extremely unpopular. He maintained that Britain ought to wait until France stood in real danger. Churchill admitted that in the past he had been in favour of a guarantee pact, but noted that Britain's bitter experience with French political stances (primarily with Poincaré and to a lesser degree with Herriot) would result in Franco-German antagonism remaining in place and prevent the solidification of peace in Europe. Cecil, for his part, lamented the demise of the Geneva Protocol.

The Ministers did not content themselves with censure for its own sake and, as a substitute, suggested an Anglo-Franco-German pact, a proposal that was supported by Curzon. Birkenhead, Secretary of State for India, with the backing of Amery, Secretary of State for the Colonies, proposed a guarantee to Belgium. When Chamberlain saw that the majority was against him he observed – referring to the words of Churchill – that quite possibly his proposal could be exchanged for a quadrilateral pact which would guarantee the frontiers of France.[117] The following day he wrote to Britain's Ambassador in Paris that:

> . . . in the circumstances of today a guarantee of the eastern frontiers of France and Belgium by Great Britain would be rendered a much more practical policy if Germany was associated with it.[118]

Against that background the Committee of Imperial Defence decided that Chamberlain should submit proposals showing how an agreement between Britain, France, Belgium and Germany could be given effect.[119] And on 17 February a draft form of declaration to be signed by Belgium, France and Germany was, indeed, submitted by Crowe.[120] In response, D'Abernon, who followed the mood in London, noted with relief that the German initiative was no longer considered inopportune but was seen as an event of cardinal importance.[121]

With that, the lid was put on the initiative of Chamberlain and the Foreign Office, but still he did not give up. At a meeting of the Committee of Imperial Defence on 19 February, which discussed and also decided to reject the Geneva Protocol, the Foreign Secretary urged his colleagues to adopt the Foreign Office scheme. He rejected the observations of Sir Samuel Hoare, the Secretary of State for Air, that it would be better to wait several years until the position of France had softened somewhat – an opinion shared by Churchill, Birkenhead and Curzon – stating that pacts such as these increased the peril of war. He stressed the necessity of the plan, given the anticipated weakening of France and the current impoverishment of her population. He also dismissed the observations of Sir Hugh Trenchard, Air Chief Marshal, that the very implementation of the Foreign Office plan would turn conflict with Germany into a certainty. Chamberlain repeated that any further delay in settling the security question would cause serious international complications, such as French non-compliance with the evacuation of Cologne. He asserted that Germany would avoid aggression only if her position were untenable and if it were clear to her in advance that any attempt to alter the arrangements of Paris by force would end in failure. France would avoid any provocations that would bring about a German war of revenge, only if her security was guaranteed. At the same time he informed the Committee of the preparation of Nicolson's memorandum,[122] together with Lampson's[123] and Crowe's[124] endorsements, that the Government of Britain should be guided by the assumptions comprised in the document[125] and not the statements of Curzon and Churchill. On 20 February he circulated the memorandum among the members of the Cabinet. At a meeting of the Committee of Imperial Defence to discuss military and strategic matters, on 19 February, the Secretary of State for War and the Chief of the Imperial General Staff supported Chamberlain's stand. Even Balfour – albeit reluctantly, according to Chamberlain – changed his mind and now stood on the side of the Foreign Secretary.[126]

The Cabinet's final decision, on 2 March, was unfavourable with regard to the Foreign Office's initiative. Churchill, Balfour, Birkenhead and Curzon (who had changed his mind) totally rejected the proposal. The only positive support came from Lord Eustace Percy, President of the Board of Education, and, to a lesser extent, from Cecil. Furthermore, these individuals favoured renouncing the Geneva Protocol, rather than exchanging it for any other initiative.[127] But Baldwin, who beforehand had been impressed by Crowe's and Tyrrell's view that such a line could breed a definite breach with France,[128] threw his weight in favour of the Foreign Secretary's demand that he be allowed, at least, to tell the French at Paris and Geneva that Britain would not object to joining a West European security pact. And, indeed, that decision was taken. It was concluded that if the proposal was accepted by France, 'His Majesty's Government will, for their part, begin the necessary discussions with the Dominions, and in all their policy will endeavour to further the common cause.'[129]

This conclusion constituted a great blow to the policy of Chamberlain and the Foreign Office, but the affair was not ended yet. Curzon, 'dragging a reluctant Balfour with him', afterwards went round to the Foreign Office for the purpose of expressing his hostility to any agreement tying Britain to the Continent of Europe, including a quadrilateral pact.[130] On 4 March, apparently again on the instigation of Curzon and in the absence of Baldwin, the Cabinet empowered the Foreign Secretary, in his coming meetings with Herriot, only to inform him of the non-adoption of the Geneva Protocol and of the importance attached by Britain to the quadrilateral pact idea. Chamberlain was to encourage him to accept the German overture and to tell him, if the participation of the British Empire was desired by Germany, that his colleagues 'would use their best endeavours to secure that such a project should not fail for want of British concurrence'. But the Government was not in a position, given the state of public opinion, to agree to any particular formula nor to pledge the Dominions in advance.[131] Finally, on 11 March, a conference of Ministers rejected Crowe's recommendation that Chamberlain be empowered to propose to Herriot an Anglo-Franco-Belgian agreement provided that it would constitute one of the components of a quadrilateral pact with Germany.[132]

This last development came against the background of Chamberlain's reports from Paris regarding the severe effect that his announcement (concerning the impossibility of a separate Anglo-Franco-Belgian Pact, in view of public opinion both at home and in the Dominions) made on Herriot,[133] whose 'face turned very white and he looked suddenly a sick man'. Chamberlain feared that unless the French were reassured they would enforce their occupation of Cologne well beyond the stipulated period, which might well bring about a breach with Britain.[134] He thought that Herriot would agree to a two-part pact between Britain, France and Belgium, and between them and Germany.[135] Chamberlain resented the Cabinet's refusal to allow him to proceed with a separate Anglo-French-Belgian Pact, and asked Baldwin, through Crowe, on 8 March following the French Premier's threat to maintain the occupation of Cologne, for guidance.

Crowe, who shared Chamberlain's indignation, embarked, immediately on receiving Chamberlain's letter of 8 March, upon a vigorous campaign for the purpose of widening the manoeuvrability of his Foreign Secretary.[136] On 11 March he met Baldwin and impressed upon him the consequences for Anglo-French relations that would result from the Cabinet's uncompromising instructions. Crowe reminded the Prime Minister of the conversation which he [Crowe] and Tyrrell had had with him a week previously, in which they had forecast the present crisis with France if an appropriate substitute for the Geneva Protocol was not found. Crowe warned the Prime Minister that if Chamberlain was prevented from offering the French a pact with Britain, or at least a quadrilateral pact, the French presence in the Rhineland would be perpetuated and maximum friction would be created not only with Germany but also over the whole of Central Europe. He asked the Prime Minister whether he could not use the influence of his position to prevent such a catastrophe.

Baldwin gave Crowe the impression that he was in agreement. He said that in his absence the Cabinet had gone much further than he would have approved and that their instructions to Chamberlain had been far too rigid. Baldwin thought that Chamberlain 'ought to have been allowed a good deal more elasticity and should have not been compelled to make so drastic and categorical a statement, which was bound at once to put up the back of the French'. Asking advice on how the atmosphere could be improved, Baldwin was told that, frankly, Crowe did not like the Foreign Secretary's threat of withdrawal from the whole occupied provinces, leaving France and Belgium to face Germany alone in the event of their refusing the evacuation of Cologne. A far more restraining and effective method, to his mind, was to give France the pact she desired, provided that it was part of a comprehensive plan, of which Germany formed a part. Crowe suggested that as long as Britain was prepared to enter into a form of pact, it really did not matter whether it took the form of a multiple agreement or a series of linked pacts. Furthermore, from Chamberlain's Paris reports Crowe gathered that approaching the quadrilateral agreement through separate pacts would increase the chances of the whole plan's success, since it was compatible with Herriot's position. Baldwin, it seems, did not demur, but pointed out that he could not approve it without a fresh Cabinet decision, since it departed from what the Cabinet had previously laid down. Wherefore he summoned Crowe that afternoon to a conference of Ministers, which included Amery, Baldwin, Birkenhead, Cecil, Churchill, Hoare, William C. Bridgeman, First Lord of the Admiralty, James E. H. Gascoyne-Cecil, Fourth Marquess of Salisbury, the Lord Privy Seal and Sir Worthington-Evans, Secretary of State for War. Curzon was absent; he died on 20 March.

The Conference of Ministers

Crowe sent Chamberlain a private letter, in which he described in detail what had taken place in the Conference of Ministers the day before:

> I cannot describe to you the deplorable impression made upon me by this discussion, nor the feeling, I may frankly say, of indignation in which I felt it. I must give you, at the risk of being tedious, a short account of the sitting.[137]

Crowe having, at the request of the Prime Minister, reported to the Ministers the immediate question on which Chamberlain had asked for guidance, and described to them the wider issues that stood behind it as discussed between him and Baldwin for thirty minutes, a vague and inconclusive debate of one and a half hours developed, extending over every conceivable point – but mostly entirely irrelevant. According to Crowe, Churchill spoke first:

> . . . there was no reason why we should do anything at all; or why we should come to any arrangement with France, who could be left to stew in her own juice without its having any bad effect on anybody or

anything; that there was no immediate hurry either to take action or to make any decision; all we had to do was to go our own way and in a few years time we should see France on her knees begging for assistance and allowing us to impose anything whatever on her.

Churchill could not see why an inaction on their part should lead to a quarrel with France. It was quite untrue, in his opinion, that there was any danger of a breach with her. All that would happen, in his view, was that matters would remain as they were, namely that should France turn to Britain with requests, one should refuse them and everyone would be perfectly happy. Amery, as usual, continued Crowe, dilated on the impossibility of doing nothing, given the opposition of the Dominions. All that was required, according to him, was to avoid the danger of any talk of entanglements, and to restrict themselves to developing moral atmospheres by pacific methods, to the exclusion of anything to do with war, or disarmaments, or force or violence.

I confess I have never heard M. Ramsay MacDonald, in his most woolly-headed pronouncements, talk such utter rubbish as Mr. Amery poured forth.

Amery admitted that he had agreed to the Cabinet resolution that Britain might enter into a pact if Germany took part, but held that he did not really mean it. He saw resolution as a fatal policy, quite contrary to everything that was right, and announced his withdrawal of support.

Lord Birkenhead noted that though it was in British interests to prevent a foreign element from invading Belgium and France, it was quite impossible to put it into a formal document, for the public in Britain and the Dominions would repudiate any such undertaking. Birkenhead declared that he agreed entirely with both Churchill and Amery, and objected to any pact of any kind whatever. Moreover, in his opinion, the Cabinet had never sanctioned such an undertaking. Lord Salisbury expressed himself in vague terms in favour of a unilateral security declaration on the part of the Government as a limit beyond which Britain ought not to go, and his words won the support of Samuel Hoare.

The one person, according to Crowe, 'to say anything which had sense' was Lord Robert Cecil, who dissented from the light-hearted way in which Churchill, Amery and Birkenhead treated the latest development. Cecil confirmed Crowe's assessment of the seriousness of the situation. He noted that the British Empire was bound and was quite ready to fight for the maintenance of the *status quo* in the Channel Ports, and for the integrity of the eastern frontier of France, and urged his colleagues to sign a quadrilateral pact. However, his words did not elicit a positive response from the others present.

Birkenhead wound up the debate saying that Chamberlain's request for further instructions should not raise any real difficulty. All that needed to be said was that the Ministers entirely approved and highly appreciated the firm and clear statement of their policy which the Foreign Secretary had made to Herriot. With this his assignment would end, the Cabinet's instructions would have been carried out and the result would be exactly what was expected and was eminently satisfactory. Birkenhead stressed that it had been made clear to France that the British Government desired no pact and would commit themselves to nothing. If France could reach an agreement with Germany, Britain might consider an undertaking to enter into consultation in the event of any aggression.

Crowe was stunned by this conclusion. It became clear to him that the general trend was to cancel even the Cabinet resolution of 4 March to agree to a quadrilateral pact – intended to be offered to the French in place of the tripartite pact rejected by the Cabinet on 3 March. Therefore, when the Prime Minister, who sat mute throughout, at last enabled him to respond, he warned the Ministers against a breach with France and havoc in the states of Europe should their policy be put into effect. He also noted that if the interpretation put by Lord Birkenhead on the Cabinet's decision of 4 March was correct, then Chamberlain's statement to Herriot had been 'grossly misleading', because to an unprejudiced mind it could only be taken to convey that whilst the British Government could not enter into a pact of security with Belgium and France alone, they would endeavour to make their country and the Dominions enter into a quadrilateral pact. If, on the one hand, this was not a pledge of contemplated entry into a pact, concluded Crowe, it seemed to him 'that the words had no meaning'; and, on the other hand, he thought they were laying themselves open to an accusation of bad faith. The Prime Minister, who still kept his peace, thanked Crowe for his briefing and asked him to leave the room.

The next day, 12 March, Crowe sent the private letter to Chamberlain, who was greatly shocked. Baldwin's letter to him of the very same day did not improve his mood either. Baldwin conveyed his appreciation that Chamberlain was representing the views of the Government, telling him, however, that he could not refer to the pact question in the letter but intended to have a long talk over it with him when he returned to England.[138]

Chamberlain, thereby, felt that he could no longer function unless the Ministers left him to run foreign policy as he saw fit and stopped intervening in the affairs of his Office.[139] He therefore directed Crowe to meet Baldwin and inform him on his behalf that if it were not for the grave results that must immediately follow, he would have cancelled his engagement with Herriot and proceeded at once to London to place his resignation in the hands of the Prime Minister. Baldwin's letter gave no guidance and showed no support for him and his policy.

A Foreign Minister who does not possess the confidence of his colleagues is worse than useless to his country & the world, & I at any rate will not consent to hold the post if the policy of the Cabinet is to be changed every few days, if the whole effect of the conversations I have held since I left home is to be destroyed, & if my word is to be repudiated & I am to be dishonoured.[140]

What Crowe and Chamberlain did not know, because until the 15 March Baldwin did not take the trouble to bring them up to date,[141] was that Crowe's words in the conference of Ministers of 11 March had had a decisive effect. In the discussion that continued without his presence it was eventually decided 'to continue the policy of refusing any pact with France . . . unless a quadrilateral arrangement could also be made to include Germany'[142] – which put an end to the attempt of various Ministers to limit Chamberlain's offer. Thus, when Crowe conveyed, on 15 March, Chamberlain's latest message to Baldwin, it was made clear that there was no question of the Foreign Secretary not having his full confidence. Baldwin favoured entry into a quadrilateral pact and intended to convene the Cabinet in order to strongly support Chamberlain. He did not anticipate any serious difficulty in obtaining the Cabinet's assent to a policy of seeking a pact.[143] And that was precisely what took place. On 20 March, the Cabinet plenary, with the attendance of Chamberlain, took a decision to the effect that the German Government's proposals offered the best basis for European security as against reversion to a policy of isolation, the only result of which could be aggravation of the existing unrest on the Continent, leading ultimately to a fresh war.[144]

Negotiations for a Quadrilateral Agreement

With this resolution in their pocket, Chamberlain and the Foreign Office opened contacts with the French for the purpose of putting it into effect. But the negotiations proceeded laboriously, and only on 16 June did the French officially respond to the German initiative of 9 February.[145] The reasons were diverse. For instance, it took a month for the French to become gradually reconciled to the idea of a quadrilateral pact,[146] and even then Herriot saw no place for moving quickly. Also, France underwent changes of Governments. Herriot fell on 10 April, his positions as both Prime Minister and Foreign Secretary were taken by Paul Painlevé and Aristide Briand on 17 April. Only when Briand was established in office did things start moving once again.[147] A further reason was a pressure group that continued to oppose the agreement, seeing it as a trap[148] intended at driving a wedge between the Allies and limiting the military manoeuvring of France.[149]

The Foreign Office was not free from problems either. For instance, having begun to examine the possible ramifications of the quadrilateral pact, Chamberlain found himself face to face with the unpleasant possibility that Britain might have to declare war on France should she attempt, following a German assault on Poland, to transfer troops to Eastern Europe through the demilitarized Rhineland – a circumstance he did not like one little bit.[150] Another problem was whether they were justified in considering that future generations in Germany would hold themselves bound by any such agreement, something to which Sir William Max Muller, Envoy Extraordinary and Minister Plenipotentiary at Warsaw,[151] replied an emphatic 'No'.

Moreover, precisely at this decisive stage of the discussions, the Foreign Office sustained a heavy administrative shock. Sir Eyre Crowe, who had long been ailing and was finally persuaded to take an extended leave, became seriously ill and on 28 April he passed away. The new Permanent Under-Secretary, Sir William Tyrrell, was a worthy successor, but the heartbreak over the man who for so many years had served the Foreign Office as a model of professional conduct was universal. Chamberlain wrote to his sister:[152]

> I am in great distress, Sir Eyre Crowe, the head of my office is dying, if he be not at this moment already dead. He was a great public Servant, devoted to duty, delightful to work with, of immense knowledge & experience & proved judgement. He has been ill for a long time yet would not leave his work. At last I had got him to take 3 months leave of absence – too late alas! It ought to have been done 2 years ago. I did not think that I could have felt so much for a man whom I have only known intimately for a few months.

Life had to go on and, in response to the draft agreement of 12 May, and supplementary explanations transmitted by France for examination by Britain,[153] a counter-draft was transmitted to Paris on 28 May – following consultations held between Chamberlain and his assistants, and between them and the Cabinet.[154] Generally speaking, the Cabinet, Chamberlain[155] and his assistants found the spirit of the document eminently satisfactory. They had expected the French attempt to link up the Western with the Eastern settlements in order to 'tie Britain up' to both ends.[156] Consequently, Chamberlain acceded reluctantly to the stand of Sir Cecil B. Hurst, Legal Adviser to the Foreign Office,[157] Lampson[158] and Tyrrell[159] that the treaties contemplated would have to be regional in spirit and limited to the terms of the Covenant of the League.[160] Nevertheless, although the British Government was not prepared to assume, in respect of Central and Eastern Europe, obligations additional to those which were entailed on them as members of the League, yet they conceded the entitlement of France to do so, and to act upon it militarily without breaching the Western Pact, once all the methods to settle the dispute pacifically, as stipulated in Article 13 of the Covenant, were exhausted.[161]

On 8–12 June, Chamberlain and Briand met in Geneva, where they worked out France's formal reply to the German pact proposals.[162] During these discussions particularly warm relations were formed. They had long talks over the novels of Walter Scott, the poetry of Byron, the works of Voltaire and Rousseau. Chamberlain was deeply struck with Briand's willingness to do anything for furthering the Pact and his conviction that, if the negotiations could be brought to a successful conclusion, many matters would become of little or no importance.[163]

The French Note was, as aforementioned, presented to the Germans on 16 June.[164] It met with a chilly response.[165] The Germans rightly claimed that their original offer had been twisted out of all recognition. They argued that it was a French plot to encircle Germany by a system of alliances and that adopting the French draft would only make Germany's position worse for it would not only tie Germany to a stiff arrangement but also reserve France's right to impose sanctions under the Treaty of Versailles without submitting matters under dispute to arbitration as provided by the Pact. The Germans maintained that the draft would create a ridiculous situation according to which, by raising objection to internal matters, such as customs tariffs, France would be able to force Germany to submit them to arbitration.[166] Stresemann noted that giving France the right to guarantee a German–Polish arbitration treaty, and to intervene by force, was quite unacceptable.[167]

The German charges towards France, and indirectly towards Britain – given her taking an active part in drafting the French Note of June 16 – shook the Foreign Office in London. Chamberlain had drawn up in his own hand a letter to D'Abernon, in which he noted resentfully that never had France a Government so liberally inclined and never was there so much British readiness to better their relations with Germany. But instead of co-operating, Germany piled up obstacles that would justify every suspicion that their initiative was not the offspring of pure motives. According to Chamberlain:

I am now forced to ask myself whether I am being used as the dupe in a negotiation in which the German proposals were only put forward with the hope of creating dissension in the councils of the allies or to enhance the price which Germany might obtain from Russia in return for breaking them off.

Chamberlain warned the Germans that if the plan should fail because of their unbendingness there was no knowing when another opportunity would occur.[168] Lampson, for his part, asked Joseph Addison, Counsellor at the British Embassy in Berlin, to make clear to the Germans that rejection of the scheme would be perceived as a German move towards a new war.[169]

Yet in fact, so long as there was hope that the Germans would soften their position, and from the reports received there was place for optimism,[170] Chamberlain had no intention of discontinuing negotiations. He believed in his capacity to attain eventually the desired goal, and he and the Foreign Office would work enthusiastically for the possibilities latent in the Pact.[171] For instance, Chamberlain saw in the Pact an excellent expedient to avoid throwing Germany into the arms of Russia.[172] Other objectives, Bennett minuted, were to put an end to Franco-German rivalry, thereby stabilizing the peace and promoting European trade, and to turn the demilitarized territory of the Rhineland into the security frontier of Britain, thus removing still further from England's shores the line on which a future European war might take place. Britain would acquire a position of influence over the moves of France and Germany.[173] Tyrrell noted:

> If we can secure a clearly defined Rhine frontier the crossing of which would automatically constitute an act of aggression we should obtain ample protection for the Channel.[174]

Tyrrell's observations were later confirmed by Chamberlain.[175]

So, pending the official reply of Germany to the French Note of June 16, the Foreign Office embarked upon working the conclusions so far reached into a more formal draft treaty. On 1 July, Chamberlain submitted the final draft, the work of the Legal Adviser to the Central Department, for examination by the Committee of Imperial Defence. After it was approved by the Committee,[176] and by the Cabinet,[177] it was sent, with slight amendments, to France. The draft treaty was a brief thirteen-clause paper. Its main points were that all the contracting parties were to guarantee the maintenance and inviolability of the frontiers of Belgium, France and Germany; renounce war against each other; and conclude arbitration treaties; and to render assistance against aggression only on being appraised by the Council of the League that such a violation had indeed been committed.[178]

This last point caused dissension between Foreign Office officials and France, and consequently amongst the officials themselves. For example, officialdom, including the Foreign Secretary, were divided in their opinions as to how the principal French reservation to the 4 July draft treaty should be treated. Namely, that if Britain, in the event of a German army marching across the Rhineland or France proper, were to await the unanimous decision of the Council of the League as to whether there was an aggression, and until all methods to settle the rupture had been exhausted, the Germans might reach the gates of Paris before any response on the part of Britain would take place.[179] Bennett, Lampson and Cecil Hurst recommended nullifying the French stand that the British guarantee on this point would have to become operative prior to the decision of the Council of the League, while Tyrrell and Chamberlain tended to accept it. The first argued that Britain should not be committed to a condition forcing her to implement the guarantee whenever the French argued that there was a violation.[180] The British and German public would never be resigned to it. The Germans, on 20 July,[181] stated quite explicitly that the guarantee treaties would be put into operation only after it had been ruled by the Council that there had, indeed, been a breach.[182]

A further reason, in the opinion of Bennett and Lampson, for not accepting France's stand on this point, was that the treaties contemplated ought not to be regarded as directed against Germany.[183] Thus, they gave expression to the growing reluctance of the younger generation at the Foreign Office to reconcile themselves to the assumption that within a short while Germany would be capable of menacing once again the peace of Europe. Bennett asked:

> whether we were not proceeding too much on the assumption that it was always Germany who would act in bad faith, and that the French on their side would never act in bad faith. It seemed just as necessary to guard against the one as against the other.[184]

Tyrrell and Chamberlain, on the other hand, announced that they were in favour of the stand of France, and on this point adopted the legal formulations proposed by her. Tyrrell justified their stance on the grounds of having to allay the fears of France, while Chamberlain noted that he had no intention of creating a situation which rendered it impossible for Britain to come to the aid of France because of the inability of the Council of the League to rule on this point. As expected, the stand of the Foreign Secretary and Permanent Under-Secretary prevailed and it was agreed that the Legal Advisers of the British and French Foreign Offices would formulate an article laying down that in the case of a flagrant violation of the demilitarization of the Rhineland, or the outbreak of hostilities, the guarantee treaties would become operative even if the Council had not yet reached a decision.[185]

With this decision behind them the Foreign Office returned to the negotiating table with the French. After holding discussions in London, on 11–12 August, the sides evolved the final text of the proposed Pact. Among other things it was agreed that the guarantee was to operate prior to a decision by the Council only if all the contracting parties were agreed that the violation constituted an unprovoked act of aggression necessitating an immediate action.[186] Likewise, it was made clear to Briand by Chamberlain that the British Government could not take upon themselves any new obligation in the east of Europe, or put the guarantee into operation if, in reaction to force having been used against her on the 'Little Entente's' account, Germany was to invade France.[187]

However, while the sides were negotiating, Germany re-entered the picture. She accepted the idea of settling the points of contention face to face, qualified her agreement by restating past terms[188] and warned the French against sending any further counter-Notes.[189] Despite Chamberlain's great fury at what was termed by him 'the blindness of the German government',[190] he acceded to the stand of Bennett,[191] Lampson[192] and Tyrrell[193] that any Anglo-French attempt to confront Germany with a cut-and-dried scheme would be counter-productive. The draft Pact should, therefore, be used purely as a basis for discussions with the Germans. The readiness of Briand to negotiate with the Germans, and Germany's 26 August agreement to take part in a meeting of jurists – which was to evolve the text of the Pact, on the basis of the Anglo-French draft and the participants' reservations, into a more juridical document[194] – constituted a stimulus for Chamberlain to take this line.

The Conference of Jurists, which convened between 1 and 4 September in London and included representatives of Britain, France, Belgium, Italy and Germany, did, indeed, not go at all into issues such as disarmament and evacuation of the Cologne zone – notwithstanding Berlin's insistence, to please the German Nationalists, on these issues first being cleared before the Pact was concluded and Germany entered the League of Nations. The Conference also failed to evolve a conclusion concerning the proposed arbitration treaties between Germany and her eastern neighbours. All the same, they did reach a broad agreement on the question of the text of the Treaty of Mutual Guarantee in the West.[195]

The meeting somewhat eased the Germans' anxiety lest they should be confronted with an ultimatum, as occurred at Versailles. Consequently Luther and Stresemann managed to override the opposition of the extreme Right,[196] and joined the international conference that convened at Locarno[197] for the purpose of overcoming, or at least attempting to, the last items of contention between the parties, which to Chamberlain's mind were within easy reach of settlement.[198]

The Locarno Conference

The Locarno Conference extended from 5 to 16 October, and bar a few exceptions it proceeded smoothly since neither Briand nor Luther could afford to allow it to fail. Or, as Chamberlain defined it, quoting Lord Grey,[199] the success of the Conference was contingent on meeting each other on an equal footing and, if a concession was made, it was never exploited like a victory to press a further advance but used as a reason for urging moderation and concession.[200] For instance, France forewent her demand for an explicit guarantee of the arbitration treaties between Germany and her eastern neighbours, and that they be mentioned in the Western Pact,[201] in consideration of the Germans' professed anxieties against encirclement and British objections to any eastern commitments. Germany, for her part, agreed to incorporate in the Western Pact a list of cases where the use of military measures would not be illegitimate: i.e., action (1) in self-defence against aggression or a flagrant violation of the Rhineland demilitarized Peace Treaty clauses; (2) in pursuance of Article 15, paragraph 7, and Article 16 of the League Covenant;[202] and (3) as a result of a decision taken by the League. Germany affirmed the inviolability of the French–German–Belgian frontiers and the demilitarization of the left bank of the Rhine, as fixed in the Treaty of Versailles,[203] thus ostensibly giving up any desire for revenge. In contrast, Germany was not made, directly or indirectly, to recognize the inviolability of the Peace Treaty's territorial arrangements in Eastern Europe,[204] which was a sine qua non from the German point of view.

Likewise, responding to Russian pressure, Germany sought to be exempted from the Covenant, Article 16, obligation to engage in sanctions, either economic or military, against aggressors. In this connection Stresemann explained that even if Germany had wanted to, she would not have been able to participate in military action for she was disarmed. Specifically, he noted that Germany could not, in view of military deficiency, risk the hostility of elements such as Russia. Chamberlain maintained that if he had been a German he would have taken the same stand as the German Government.[205]

Germany was asked only to participate in any agreed sanctions to an extent compatible with her military situation and geographical position.[206]

The Treaty of Mutual Guarantee

The Pact was initialled on 16 October at Locarno[207] and ratified on 1 December 1925 by the Heads of State in London.[208] It comprised a Final Protocol to which were annexed a Treaty of Mutual Guarantee between Germany, Belgium, France, Great Britain and Italy; three Arbitration Conventions, between Germany and Belgium, between Germany and Poland, and between Germany and Czechoslovakia;[209] a Draft Collective Note to Germany regarding Article 16 of the Covenant of the League of Nations; and two separate Treaties, between France and Poland, and between France and Czechoslovakia,[210] which, though mentioned, were not part of the complex under discussion.[211]

Article 1 of the Treaty of Mutual Guarantee provided that all the contracting parties would guarantee the inviolability of the frontiers between Germany, France and Belgium, and observe the demilitarization of the Rhineland.

Article 2 embodied an undertaking by Belgium, France and Germany not to make war on each other except in special cases, and it was specified what they were.

Article 3 contained agreements by the three said parties for the settlement of all disputes by arbitration or conciliation.

Article 4 laid down that the decision as to whether any of the parties was entitled to military assistance would lie with the Council of the League, except in cases of flagrant violation of the above first two undertakings, in which case the signatories had to take steps in the matter independently.[212]

Article 5 dealt with the terms of the Arbitration Conventions.

Articles 6 and 7 stated that the Treaty of Locarno did not invalidate the Treaty of Versailles and its supplementary agreements, nor did it undermine the freedom of action of the League in this respect.

Article 8 detailed the circumstances and procedures under which the Pact could be terminated.

Article 9 ensured that the adherence of the British Dominions would be optional.

Article 10 provided that the Treaty should come into force after all ratifications had been deposited at Geneva, and after Germany had become a member of the League.[213]

There was widespread euphoria in Britain. It seemed to the Government and the public alike that German feelings of humiliation and the French sense of insecurity were now ended, and it was possible to embark upon the reconstruction of Europe.[214] Lampson, who claimed to be phlegmatic, displayed great enthusiasm over the Treaty and the business-like spirit that distinguished this Conference from other ones. Ex-enemies had turned into bosom friends.[215] Chamberlain reaped congratulations,[216] was given a K.C.G.B. and was made a Knight of the Garter and – the 1925 Nobel Peace Prize.[217] He convinced himself that the millennium had, indeed, arrived, and in November 1925 likened his Locarno achievements to the success with which Castlereagh had faced the rehabilitation problems of Europe after the Napoleonic Wars.[218]

Problems Continue

But the feeling of being in seventh heaven could not last forever, and slowly the Foreign Office started to come back to the grey reality of unceasing haggling with France and Germany over diverse issues. From a letter that Chamberlain sent to his sister in April 1926 it seems that he himself sensed it.

> I shall have a difficult & anxious year in my own sphere, for all the world is nervy & easily gets an attack of the jumps, & the Germans are prodigiously difficult to guide into or keep in the path of their own real interests.[219]

The first issue that the Foreign Secretary and his Office had to contend with on the eve of the new year was the disarmament problem. Namely, how could the assurances given at Locarno that great efforts would be made by both France and Great Britain to meet German desires in regard to the evacuation of Cologne and the withdrawal of the Inter-Allied Military Commission of Control out of Germany, possibly be squared with the latter's considered opinion that the Germans had not yet fully fulfilled their military obligations.[220] The Cologne evacuation issue had of course already been negotiated,[221] and by 31 January 1926 it was put into full effect.[222] This left open the Commission of Control question.

Here, Chamberlain tended to show firmness. Orme Sargent, Counsellor at the Central Department, Lampson and Tyrrell were inclined not to stand on what seemed to them to be trifles. Troutbeck thought that in any event the French complaint should not be dismissed.[223] The former three officials, on the other hand, dismissed the statements of the French that if the Commission of Control were to be taken out of Germany before it had completed its work it would no longer be possible to verify the state of disarmament there,[224] that the defaults were significant,[225] and that any concession would only be a prelude to other demands.[226] Their opinion was based on the grounds that Germany's state of disarmament could at any time be verified by the Allies through the League of Nations[227] and that insistence would merely give rise to unnecessary frictions and dissensions,[228] and while they admitted quite a few defaults had been committed, these did not justify, in their opinion, an extension of the Commission of Control's stay in Germany.[229] They found support in the reports of Major-General A. G. Wauchope, Chief of the British Section of the Commission of Control,[230] and of D'Abernon,[231] that Germany was effectively disarmed.[232] In May 1926, General Wauchope did warn that:

> On the other hand, after many conversations with Germans of moderate views, I am forced to believe that they will not remain content to accept permanently their present boundaries. Neither Germany nor, I suppose, Italy are among what I think Metternich called 'satisfied States', as are England and France.[233]

Yet this doubt did not stop either him or the War Office[234] from continuing to send papers confirming that Germany was disarmed, and that, therefore, the Commission of Control could be withdrawn, even though Germany had not completed her obligations in full.[235]

Even the Allied Military Committee of Versailles, which at first insisted upon the fulfilment of German obligations in full, on 30 January 1926 instructed the Commission of Control to be conciliatory towards the Germans.[236] When the War Office reported, in February, that members of the Commission of Control were continuing to be difficult over minor points, Chamberlain asked that the Commission be instructed by the Conference of Ambassadors, on the recommendation of the British Ambassador in Paris, to uphold their Governments' stands.[237]

This capitulation of the Foreign Secretary to the views of the Central Department officials was given more tangible expression on 7 April with his agreeing that they would have to press France to be more flexible over the question of disarmament in order that the Commission of Control be withdrawn as soon as possible from Germany.[238] However, the Central Department was prevented for many months from acting definitely on this point because of the reports reaching them that, even though Germany was generally disarmed, since February no effort had been made to correct previous infractions.[239] Lampson was forced to tell the German Ambassador, on 15[240] and 26 July,[241] that as long as no progress was made in these points a quick withdrawal of the Commission of Control would not be possible.

In September, the patience of Lampson's successor, Orme Sargent, finally ran out. The Conference of Ambassadors demanded on 26 August that the Germans fulfil their obligations.[242] The German reply, on 6 September, stated that their Government could not be made responsible for something for which the Commission of Control was, in their view, to blame, but that they would do their best to come to an understanding on the issue of disarmament.[243] In a draft letter to the War Office, submitted with the approval of his superiors, Sargent argued that, given enough time, they were bound always to find cases in which the Germans failed to meet the disarmament clauses in full and that these instances of non-compliance did not constitute a danger to British security. It was possible, in his opinion, to settle for the German proposal of resolving the questions of illegal recruitment, military associations and unauthorized military establishments by legislation. This left the non-dismantlement of the High Command and the General Staff questions. Sargent noted that there was no reason to delay the withdrawal of the Commission of Control because, in the War Office's considered opinion – as long ago as 1923 – Germany was effectively disarmed, and because of the need to avoid unnecessary frictions, against the background of the Locarno Treaty and of Germany's joining the League of Nations.[244]

His observations were accepted by his superiors, and the War Office, too, agreed to adopt this line.[245] The Foreign Office put together memoranda from Wauchope[246] and from the War Office[247] showing that militarily it was possible to withdraw the Commission of Control from Germany. These were sent to Paris so that the British Ambassador would be able to exert pressure upon the French and the Conference of Ambassadors on this point.[248] Direct negotiations were opened with the Belgians and the French, who, once the Thoiry scheme was abandoned, showed little enthusiasm for the British proposals.[249]

Following prolonged negotiation (5–12 December) between the British, French and Germans at Paris and Geneva, in the course of which Tyrrell threatened that the French would bear responsibility for a needless European crisis if the negotiation should fail,[250] an agreement was eventually reached on 11 December, and signed the following day at Geneva. It was agreed that the Commission of Control would be withdrawn from Germany by 31 January 1927 and the remaining points of contention – fortifications on the eastern frontier and the manufacture and export of war material – would be negotiated between Germany and the Conference of Ambassadors. If there was no agreement by the end of January, these matters were to be referred to the Council of the League. Article 213 of the Treaty of Versailles, authorizing the Council to inaugurate military investigations at Weimar, would be applied as from that date. In accordance with the Germans' demands, the investigations had to be limited to *ad hoc* instances, and no standing control agency of the League, as a replacement to the Commission of Control, was envisaged. Finally, each Allied Government was empowered to attach to its Embassy at Berlin a technical expert to deal with German authorities on disarmament matters after 31 January. Berlin first made certain, however, that these experts would be prohibited from conducting on-site investigations.[251] Thus special Allied rights to inspection came to an end.

Another problem that Chamberlain and the Foreign Office had to contend with was the question of Germany's admission to the League of Nations, and her receiving a permanent seat on the Council, under conditions that would do minimal damage to relations between Berlin and Moscow. The attainment of this objective was essential not only because until then it would formally not be possible to implement the Locarno Treaties, but also because the Treaties on the one hand, and the security guarantees provided by the Covenant of the League of Nations to the States of Europe on the other hand, were considered by Chamberlain to complement each other.[252]

The first step in this connection, as agreed at Locarno, was made by Germany. On 3 February 1926 the Foreign Affairs Committee of the Reichstag gave the Government full authority to take the necessary measures to enter the League of Nations,[253] and on 8 February Germany formally applied for a permanent seat on the Council[254] on account of her being accorded Great Power status.[255] France did not object but insisted on the admission to the Council of Poland as well.[256] This was violently opposed by Germany, who feared the move would derogate her international position.[257] Somewhat surprisingly, Belgium also objected.[258] France did not back down. On the contrary, France also began talking of the right of Brazil and Spain – two non-permanent members of the Council since 1920 – to permanent seats,[259] a proposal that again was rejected forcefully by Germany.[260]

Against this background the Foreign Office was required to determine the British standpoint. Troutbeck advocated a realistic approach, according to which the actual resolutions would be taken by the Great Powers, and by this criterion Poland had no place on the Council.[261] Tyrrell, on the other hand, favoured the admission of Poland to the Council,[262] as did Chamberlain, who pointed out that Germany was attempting to extort concessions from the Allies to which they had never agreed; namely, that after her own admission no other member was to be admitted to the Council.[263] But, according to Chamberlain, there were other grounds for more general admissions to the Council. First, already on 11 November 1925 the Cabinet had decided to support the admission of Spain. Secondly, the presence of Poland at the Council was, to his mind, essential.[264] According to Chamberlain, support for Poland in this respect was owed following Poland's co-operation during the Locarno Conference. Non-agreement would have created a disruption in relations between the two countries,[265] but instead a condition would be created that would allow the settlement of grievances – such as the Polish Corridor and Upper Silesia – between Germany and Poland, if Poland's status was made equal to that of Germany.[266] In this connection, Chamberlain quoted Briand's idealistic argument.[267] Tyrrell argued that the best means of achieving security was to admit Poland to the Council. Chamberlain recommended, on 17 February 1926, that Britain support the claim of France and look favourably on the Polish candidature, stressing that Poland's non-admission, or the giving of a temporary seat on the Council, would put Poland in an inferior position *vis-à-vis* Germany.[268]

But, as with his Tripartite Alliance initiative, Chamberlain did not win support within or outside the Government. The League of Nations Union, headed by Gilbert Murray, mounted a sustained pro-German campaign, which, in the ensuing weeks, won considerable press support.[269] Within the Government, Cabinet members – in particular Cecil, who had a special interest in the League of Nations[270] – criticized sharply the Foreign Secretary's approach, according to which Poland ought to be given a permanent seat on the Council just because she was embroiled with Germany. It was argued that the admission of Poland would result in the admission of scores of other small states. Such a policy would encumber the Council and cause resentment in Germany and Czechoslovakia. The view was also expressed that Poland's international behaviour to date did not entitle her to any reward.[271]

These criticisms were translated by the Cabinet into practice on 3 March, when it was determined not only that Poland was not entitled to more than a non-permanent seat on the Council – and that, too, only after Germany had received a permanent seat – but also that Chamberlain should not be empowered to handle these problems exclusively at the forthcoming Special League Assembly, called for 9 March at Geneva.[272] Instead, it was decided that the Foreign Secretary would be assisted on this point by Cecil, who would accompany him to Geneva.[273]

Though it was not to the liking of the Foreign Secretary[274] nor his Permanent Under-Secretary,[275] they accepted the ruling. However, Chamberlain, by way of expressing his dissatisfaction, did not invite Cecil to a series of meetings intended to reconcile the stands of Poland and Germany,[276] which he held at Geneva with the Allies and Germany. At first all sides showed obstinacy. But once France and Poland recognized that neither Germany nor Sweden would consent to Poland being given a permanent seat on the Council they backed down. Instead, they suggested that Poland be given a non-permanent seat for a period of three years. Eventually, this was agreed and Sweden – a non-permanent member – obligingly volunteered to resign. Even the efforts to persuade Czechoslovakia to give up her seat on the Council in favour of Holland, in order to compensate Germany for the replacement of neutral Sweden with hostile Poland, proved successful.[277]

But here, from Chamberlain's point of view, an unexpected hitch occurred. Spain and Brazil, who were deliberately excluded from the deliberations so long as the Polish problem had not been solved, threatened to put a spike in the wheels of the whole scheme unless they received a seat at the same time. Spain threatened to withdraw altogether from the League, while Brazil declared that she would veto Germany's candidature – a right vested to her by virtue of her being a non-permanent member of the Council. Chamberlain belatedly recognized his error in not having co-ordinated his moves in advance with these two states. However, all his and Briand's endeavours to persuade the two not to realize their threats came to naught; for the inflamed public opinion in Madrid and Rio de Janeiro, and the statements that flowed from their delegates in Geneva, meant that Brazil and Spain could not retreat totally.[278]

In these circumstances the British and French delegates, fearing the effect on Germany if a veto were actually employed, sought and obtained, on 16–17 March, an adjournment of the Special Assembly until September, when the annual League Assembly would be convened.[279]

A New Beginning?

The forthcoming General Assembly did not resemble its predecessor. The interval was utilized by the Powers and the League of Nations to appoint a committee – sitting at Geneva from 10 to 17 May, on which Germany was included – for the purpose of submitting suggestions concerning the future composition of the League Council. Its main recommendation – the product of the initiative of Lord Robert Cecil, the principal British delegate – was that the Council's membership be increased from ten to fourteen. The permanent seats were to be increased by one, in favour of Germany; three new semi-permanent seats would be created, whose occupants, unlike the non-permanent members, would be eligible to be re-elected after three years, provided any application was secured by a two-thirds vote in the General Assembly.[280]

This compromise eventually satisfied Poland, but not Brazil, which tendered her resignation on 14 June; nor Spain, whose resignation was sent in on 11 September, only to be rescinded before the mandatory two-year notice period expired in 1928. Thus, with the Brazilian threat removed, Germany could, on 8 September, join the League and become a regular permanent member of the Council.[281]

On the face of it, the conclusions reached concerning the evacuation of Cologne, the withdrawal of the Commission of Control from Germany and her admission to the League Council, constituted not only the start of putting the Locarno Pact into effect, but also the beginning of paving the way towards the pacification of Germany and the consolidation of peace in Europe. Yet, the result was the opposite. The prolonged contentions on these issues constituted a symptom of the beginning of the evaporation of the 'Locarno Spirit', which was to vanish altogether with the demise of the Weimar Republic and the advent of Adolph Hitler to the Chancellorship of Germany.

Appendix 1: The Establishment of the Department of Overseas Trade

The First World War led to an organizational and functional expansion of the departments of state in general and the Foreign Office in particular. Their work and staff increased dramatically, and there was a growing duplication of functions between the Foreign Office and the neighbouring departments. Inter-departmental overlapping was not just a momentary war-time phenomenon but an indication that, with the pace of modernization, the internal and external frameworks in which foreign policy had been conducted faced far-reaching changes. This process, which began in the nineteenth century, led to a redefinition of the relation between foreign and domestic policy and to a gradual dismantling of the traditional barriers that separated the two. Consequently, decisions taken in one sphere were bound to have important repercussions on others, thus making neighbouring departments into more active partners in foreign affairs than hitherto, and necessitating a new division of work between them. Among the developments, in the context of foreign policy in general and the status of the Foreign Office (as the chief adviser to the Cabinet on foreign affairs) in particular, was the politicization of economic and commercial affairs during the First World War.

Although economic relations were inseparable from foreign relations in the age of mercantilism, they were viewed for much of the nineteenth century, at the Foreign Office and in business circles, as the internal affair of the commercial community. The prevalent attitude was that so long as equal opportunities for British trade in the markets of the world were preserved, economic relations should, as much as possible, be divorced from the politics of foreign relations.[1] It is true that in 1865 a Commercial Department was established within the Foreign Office; and that, from the end of the nineteenth century onwards, the Foreign Office began to use political means to protect British commercial interests abroad.[2] Nevertheless, even in 1914 the extent of the Foreign Office's involvement in economic affairs was a restricted one. The Commercial Department, as reported by Victor Wellesley, the Controller of Commercial and Consular Affairs from 1916 until 1919, was no more than 'a post office for the Board of Trade in routine matters and a perfunctory editor of commercial reports'.[3] Further, the department that gave merchants their main support was not the Foreign Office but the Board of Trade, which was in charge of commercial policy. Individual diplomats had a key role in concession hunting in certain areas, i.e. Turkey, China, Morocco, Persia etc., but old attitudes died slowly in London until the shock of the First World War. But policy leads were limited as a result of the prevailing belief in the ultimate harm that would be caused to British trade should free trade be replaced by state intervention.[4]

The economic warfare waged against the Central Powers, and the interdependence that sprang up between Britain and her allies, made any continued divorce between economic and political relations impossible to maintain. Economic, financial and commercial affairs had become a central part of international relations.

An early move was the establishment of a Contraband Department within the Foreign Office, in November 1915, which, in connection with the blockade, took over certain duties originally performed by the Treaty and Commercial Department.[5] Commercial matters became inextricably interwoven with politics. During 1915 the department began to negotiate a network of contraband agreements with neutral countries, trading firms, shipping lines and even private individuals, aimed at cutting off the Central Powers from their sources of raw materials.[6] Each case posed different problems, and men who had never dealt with such matters rapidly became technical experts. For the Assistant Under-Secretary, Sir Eyre Crowe, the work of the new department was an outlet for his many talents.[7] Largely through his drive and ability, a new Ministry of Blockade was created under Lord Cecil.[8] The new ministry co-ordinated the activities of the various governmental agencies concerned with the blockade and was under the nominal control of the Foreign Office; the ministry used Foreign Office personnel to do the bulk of its work. Cecil, the minister in charge, had a Cabinet seat but remained Parliamentary Under-Secretary at the Foreign Office. Crowe, who was in fact, though not yet in name, the Permanent Under-Secretary of the new ministry, remained a Departmental Under-Secretary at the Foreign Office. Well over half the Foreign Office correspondence by 1916 was concerned with blockade matters.[9]

But there was pressure within and from outside the Foreign Office for more radical reforms. In March 1916, the Association of Chambers of Commerce demanded that the standard of service rendered by the Diplomatic and Consular Services to businessmen be raised, to enable the latter to compete successfully with the Central Powers' traders.[10] At the same time, members of parliament – prompted by the belief that

Germany intended to launch an economic offensive against the Allies once the war was over – urged the Foreign Office to reorganize itself, so that in the post-war period it would be able to offer the commercial community the same services furnished to it during the war.[11] In addition there were the recommendations of the 1914 Royal Commission on the Civil Service, for improving the staffing of the Consular Service and for expanding the commercial functions of the Diplomatic Service.[12] Within the Foreign Office, those who worked on blockade problems became convinced that the close relationship between commercial and political questions would intensify in the post-war period, and that the Foreign Office would lose its influence in Whitehall unless its officials were able to make a genuine contribution to the economic discussions. These men found in their work 'a more statisfying field of activity and a more tangible touch with realities and personalities than had been afforded by the political moves and countermoves of an older diplomacy, and they wished ardently to keep their place in the field'.[13]

Under steady pressure, the Heads of the Foreign Office reluctantly decided that the best way to forestall further interference in the affairs of the department would be to form their own committee. Thus a committee of officials was set up under Crowe, charged with the task of examining the best form of Government organization for the promotion and assistance of British trade. It submitted its report on 10 August 1916. Ironically, the committee came to the conclusion that the Government ought not, in future, to settle for the manipulation of the 'rules of the game' of international trade in favour of Britain, and recommended – with certain qualifications – that the German method of giving systematic diplomatic assistance to traders overseas be adopted. According to the committee, (1) this required the creation of a Foreign Trade Department, which would collect, co-ordinate and present all information relating to trade, industries and shipping overseas in such a way as to enable the government to neutralize dangerous commercial trends on the part of other powers and to furnish British traders with reliable information as to potential markets, local regulations, suitable agents for British firms in foreign countries etc.; and (2) it required better co-operation between foreign policy, economic policy and commerce. The attainment of these goals, the committee asserted, was dependent on the expansion of the Consular and Commercial Attaché Services, on the establishment of panels of advisers, made up of businessmen, on frequent visits of Foreign Trade Department officials to industrial centres, and on the setting-up of an inter-departmental committee to co-ordinate the work of the new department with that of the commercial departments. The committee submitted detailed organizational recommendations along these lines.

In order that the diplomatic support would correspond in scope to commercial trends in different countries, the committee recommended that the Foreign Trade Department and the Diplomatic Service be subordinate to the same authority, namely the Foreign Office. The committee came to the conclusion that it was incumbent on the Foreign Trade Department to keep in close touch with the business community – in other words, to provide the kind of assistance which interested the commercial community and the Consular and Commercial Attaché Services – rather than simply waste their time gathering useless information, as had been the case before the war. This would be secured, in the opinion of the committee, if the Board of Trade transferred to the Foreign Office the task of distributing commercial intelligence among British traders. The Crowe Committee report explained that:

> There cannot be a British foreign policy as regards commercial matters abroad separate from general foreign policy, of which it forms an integral and important part. There may be special agents charged more particularly with carrying on in a foreign country the commercial side of foreign policy. But if there is to be unity of foreign policy, such agents must derive their authority and receive their instructions from the same minister who directs the general foreign policy; and, as a general rule, unity and efficiency in policy will best be secured by the greatest possible consolidation of all such agents with the permanent foreign service.[14]

Thus the Crowe Committee anticipated one of the prime recommendations of the famous 1918 Report of the Machinery of Government Committee, which determined that a department of state should be organized on the basis of the service it provided (e.g., education, foreign policy) but that all questions relating to this service should be located in a single department.[15]

Conflict between the Foreign Office and the Board of Trade

Nothing was done immediately. The Foreign Secretary, Lord Grey, wavered between different courses. The Permanent Under-Secretary, Lord Charles Hardinge, was too much of a conservative to adopt so radical a reform,[16] and his reservations undoubtedly contributed to the delay. But the main factor that prevented the implementation of the reform was the fierce opposition of the Board of Trade to the establishment of the Foreign Trade Department within the framework of the Foreign Office. The department vigorously rejected the contention that assistance to traders, and collection and distribution of commercial intelligence, was an integral part of foreign affairs. It held that if those tasks were assigned to the Foreign Office it would blur the distinction between assistance to traders and commercial policy, and thus create a second Board of Trade within the Foreign Office.

The Board of Trade did not content itself with making protestations and, directly after the publication of the Crowe Committee report, it submitted a counter scheme to that of the Foreign Office, in which it demanded that the commercial functions of the embassies and legations be transferred to the Board of Trade. The Foreign Office viewed this suggestion as an attempt to undermine its status as the one institution responsible for whatever took place outside the boundaries of Britain. Grey hastened to tell the President of the Board of Trade, on 31 August 1916, that he was most anxious that 'the F.O. should not be abolished without a hearing'.[17]

The inability of the Foreign Office and the Board of Trade to agree whether the promotion of commercial interests overseas was part of foreign or commercial policy did not stem solely from considerations of prestige, but stemmed also from the fact that the politicization of economy and commerce, which till then had been part and parcel of domestic policy, blurred the distinction between foreign and domestic policies in general, and between foreign policy, commercial and economic policies in particular. In the past it had been presumed that the ends of foreign policy embodied the national interests of the whole of society, while those of domestic policy pertained to particular interests within the society. But from 1914 onwards – owing to the fact that the government took over the management of Britain's economy – economic and commercial goals became no less national than those of foreign policy. More important was the undermining of the premise that in order to achieve external goals any means might be utilized, while the use of diplomatic pressure for the purpose of achieving economic and commercial ends amounted to an intervention in the internal affairs of other countries.[18] The endeavour to break down the economic foundations on which the military might of the Central Powers was based 'proved' that, henceforth, any means could be invoked to achieve economic goals. This was reflected not only in the intrusion of the Foreign Office into the preserves of the domestic departments, but also in the pronounced effect that the Board of Trade's policies now had on foreign affairs.

No progress was made during the rest of 1916 in resolving this deadlock. In an attempt to find an agreement a committee was set up, on 5 December 1916, at the request of the Foreign Office, under Lord Alexander Faringdon MP. The committee consisted of Frank Dudley Docker (the representative of the Federation of British Industries), Sir John de F. Pennefather MP, Sir William Clark (Comptroller-General of the Commercial Intelligence Department of the Board of Trade) and Victor Wellesley. Unfortunately, its members could not agree among themselves about who in Whitehall should pool the commercial intelligence collected abroad and distribute it among traders. Docker, Pennefather and Wellesley were in favour of assigning the new department to the Foreign Office, which controlled British commercial representations overseas, while Lord Faringdon and Clark held that control over the proposed machinery for collecting and distributing commercial intelligence from abroad must be retained by the department charged by the executive branch with the promotion of British trade, namely the Board of Trade. Docker, Pennefather and Wellesley contended that the Foreign Office could receive information from abroad and distribute it much more quickly than the other departments, adding that:

> We think it is impossible to sever questions of commercial policy in foreign countries from general foreign policy, of which they are an integral part. International relations in general depend, and must tend to depend more and more in the future, upon commercial relations, and it is inevitable that foreign policy will consequently be affected to an ever increasing extent by commercial and industrial circumstances.[19]

In contrast, Lord Faringdon and Clark maintained that it was not possible to sever questions of general commercial policy, for which the Board of Trade was responsible, from the promotion of commercial interests overseas, and asserted that only the Board of Trade – the expert in commercial matters – could derive immediately, from the information gathered, the full commercial benefit.[20]

Out of this tussle emerged a compromise. Although not happy with the minority's report, Lord Robert Cecil considered, in contrast to Crowe and Wellesley, that the right attitude ought to be

> not how much can be got into the hands of the F.O., or even what is the ideally best organisation of Foreign Trade, but rather what can practically be accomplished having in view the opinions and existing organisation of the Board of Trade.[21]

Whereupon it was decided, following prolonged discussions between Cecil and the new President of the Board of Trade, Sir Albert Stanley, and Vice-Chairman of the Reconstruction Committee, Edwin Montagu, to adopt the approach that holds that common departmental functions necessitate the creation of a common department.

Cecil and Stanley therefore recommended to the Cabinet, on 14 August 1917, that a new department be set up under a Parliamentary Under-Secretary who would be responsible both to the Foreign Secretary and President of the Board of Trade. The department would take over from the Foreign Office the function of collecting commercial intelligence in foreign countries, and from the Board of Trade the function of distributing it among the business community.[22] On 15 August their suggestions were adopted by the Cabinet,[23] and on 14 September Sir Arthur Steel-Maitland was made Head of the new department.

The establishment of the new Department of Overseas Trade, as it was to be called, was preceded by a sharp argument as to whether the significance of the agreement was that the new department should promote overseas commercial interests and control the commercial services overseas, as Steel-Maitland held, or whether it should confine itself, until housed in the Foreign Office building, to collecting and distributing commercial information, as the Foreign Office officials maintained. Eventually, for the sake of peace and quiet, as Steel-Maitland put it,[24] a compromise was reached, on 11 January 1918, according to which Steel-Maitland would indeed supervise – as Parliamentary Under-Secretary for Foreign Affairs, subordinate to the Foreign Secretary – the work of the Commercial and Consular departments. They were not to be incorporated in the new department so long as the body was not housed with the political departments, but would remain an integral part of the Foreign Office.[25] The functions of the new department would be limited for the time being to export work and to the collection and distribution of commercial intelligence. All subjects relating to the promotion and protection of overseas commercial interests would remain – until the Department of Overseas Trade was housed in the Foreign Office building – in the care of the political departments.[26] In this way, the Foreign Office strove to integrate the new department as much as possible into the Foreign Office with the purpose of preventing it from becoming an independent body that would gradually take over traditional Foreign Office functions or become a Trojan Horse of the Board of Trade.[27] The Department of Overseas Trade was to be a common department for both the Board of Trade and the Foreign Office.

When Cecil and the President of the Board of Trade brought the question to the Cabinet, they observed that the creation of the Department of Overseas Trade represented a big step towards solving the duplication of functions between the two departments.[28] But, as later proved to be the case, this estimation was over optimistic. The politicization of domestic affairs, economy and commerce on the one hand, and the inability consequently to divide them among the different departments,[29] led to an overlapping of functions.[30] The Department of Overseas Trade, for its part, found it difficult not to handle, simultaneously with the Foreign Office and the Board of Trade and not always with their knowledge, matters such as promoting overseas commercial interests, instructing British representatives to approach foreign governments,[31] financial negotiations with foreign countries and giving advice to British industrialists.[32] On the personal level, Steel-Maitland found it difficult to co-operate with the heads of the Foreign Office. He told the Prime Minister, David Lloyd George, that as long as Cecil, who understood the circumstances of the Department of Overseas Trade, was at the Foreign Office, he (Steel-Maitland) could get his business properly done. 'As it is, however, as regards the commercial side of Foreign Affairs, Mr. Balfour frankly takes no interest and Lord Hardinge is hopelessly incompetent – as any capable person who knows the Foreign Office from within can testify.'[33]

These shortcomings drew sharp criticism from the Board of Trade and Foreign Office officials and from representatives of the commercial community, and led to the creation, in April 1919, of a committee under the former Home Secretary, Lord George Cave, for the purpose of examining the question of government machinery for dealing with trade and commerce. From 29 April until July, the committee heard the evidence of officials and examined witnesses from the business community, who pointed out the faults of the Department of Overseas Trade. The critics emphasized the number (as many as seven) of governmental agencies that dealt with commercial matters. In their view this led to confusion in inter-departmental co-ordination, with representatives overseas receiving instructions from different departments simultaneously.[34]

The solutions offered to the committee had a modern ring to them. In order that governmental domestic trade policy would tally with foreign trade policy, Board of Trade officials, headed by Auckland Geddes, the new President of the Board of Trade, recommended that domestic trade – for which they were responsible – be united with all aspects of foreign trade. The Board of Trade should be turned into a super-department, which could combine within itself all the governmental commercial agencies, including the Department of Overseas Trade, Consular Service and Commercial Attaché Service, though, with regard to the last two, control would be limited to their overseas activities.[35]

Foreign Office representatives, not surprisingly, recommended that foreign trade be merged with foreign policy – as far as it applied to the collection and distribution of commercial intelligence – by the union of the Department of Overseas Trade with the Foreign Office. The latter continued to insist, as Wellesley observed in his evidence, that whatever took place outside the boundaries of Britain was part of foreign affairs. If commercial intelligence gathered abroad reached Britain by means other than the Foreign Office, it would be cut off from a considerable amount of information relating to the system of inter-state relations. An absurd situation would be created whereby the Foreign Office would be responsible – from an administrative viewpoint – for services such as the Consular Service, while the authority that issued their practical instructions would be the Board of Trade. Another problem, according to Wellesley, was that for full information on the political and commercial aspects of any topic to be collected abroad, the resident representatives had to be aware of possible political implications. The services under the control of the Foreign Office were trained for this purpose, whilst the Board of Trade and its staff were not. Moreover, as observed by Crowe, who, on the advice of Wellesley, was summoned from the Paris Peace Conference in order to give added stress to the Foreign Office stand, though the Foreign Office argued that it had no intention of taking over from the Board of Trade the formulation of commercial policy, sooner or later the Foreign Office would be charged with the implementation of the commercial policy of the Board of Trade, and so it had to be well acquainted with the economic background of the problem. Crowe told the members of the committee:

> If the Foreign Office were to be a mere Post Office in the matter, they could know nothing about it, and would not understand its being though they would be responsible for the consul giving the information and would be blamed if he went astray.[36]

Representatives of the Federation of British Industries and Association of Chambers of Commerce also urged the committee not to separate foreign trade from foreign policy. In describing the commercial community's needs, they emphasized that the attainment of commercial goals was more than ever dependent on receiving diplomatic assistance, and observed that, of all the departments of state, the contribution of the Foreign Office was the most decisive. The general opinion was that the Foreign Office would lose the interest it had shown in commercial subjects during the war if responsibility for foreign trade were to be assigned to another department.[37]

The committee recommended that it favoured leaving the Department of Overseas Trade intact, so as to constitute a link between foreign policy, foreign trade and commercial policy. Moreover, in the committee's opinion, the administrative inconvenience that resulted from the existence of the Department of Overseas Trade was secondary to the public concern that there should be one governmental body which would dedicate all its time to promoting exports, a function that would not receive sufficient attention unless it became the sole duty of a specific department. However, the committee accepted the Foreign Office's plea that, should this solution be adopted, it would be incumbent upon the Department of Overseas Trade to occupy the same quarters as the Foreign Office. The Foreign Office, the Department of Overseas Trade and the Board of Trade ought to be housed in the same building, to promote the maximum co-ordination among them. The committee also recommended that the Consular and Commercial Departments of the Foreign Office be trasferred to the Department of Overseas Trade; that the Commercial Diplomatic Service – created on 1 April 1918 to replace the Commercial Attaché Service, which was disbanded – also be placed under the direction of the Department of Overseas Trade. The committee suggested that a

Foreign Office official be sent to the Department of Overseas Trade to specialize in commercial subjects and to serve as a liaison officer between the Department of Overseas Trade and the Foreign Office; and that an inter-departmental committee be set up to co-ordinate the work of the three departments.[38] The committee's recommendations were adopted on 23 July by the Cabinet, as a temporary arrangement, pending a few months' trial.[39] It eventually turned into a permanent settlement.

Unfortunately, this solution did more harm than good. True, it settled the conflict between the Foreign Office and the Board of Trade, under whose joint control the Department of Overseas Trade would fall, but it did not solve the duplication of functions among them. Instead of constituting a common department and a link between the two departments, the Department of Overseas Trade soon began to block 'both Foreign Office channels and the Board of Trade channels by refusing to accept the authority of either'.[40] The Consular Department, charged with the administration of the Consular Service, remained at the Foreign Office, while the supervision of the Consular Service's commercial functions overseas passed to the Department of Overseas Trade, which meant, according to H. Montagu Villiers, the Commercial Secretary at Madrid, that 'Ambassadors and the Foreign Office, under whom consuls always have been and always must be, got to know less than ever of what the consuls were doing commercially.'[41] Data on what was taking place in other countries, more often than not on the same subjects, was collected by the three departments simultaneously. Finally, negotiations with foreign elements were conducted sometimes by one department and sometimes by another, with an inevitable loss of consistency and bargaining position.[42]

The process of foreign policy formulation was no longer the sole preserve of the Foreign Office and Diplomatic Service. The Board of Trade also took part in it, as did the Department of Overseas Trade. In the Washington Embassy and South American missions, for example, the Commercial Diplomatic Service tended to take over the handling of politico-commercial matters from the Diplomatic Service, to encroach upon the authority of the Diplomatic Service members. This troubled the heads of the Foreign Office.[43] In London, from January until March 1920 – with Britain's de facto recognition of Bolshevik Russia – all contacts with the Russians were conducted through the Board of Trade, which also represented the Government, including the Foreign Office, in Moscow.[44]

In Janary 1922, on the eve of the Economic Conference for the Reconstruction of Europe, which was to take place in Genoa, a Board of Trade Committee attempted – despite vigorous opposition from Foreign Office officials – to discuss problems of peace and disarmament. In response, Crowe, who meanwhile had succeeded Hardinge as Permanent Under-Secretary, wrote indignantly:

This whole problem of disarmament is full of complexity and full of pitfalls and I cannot conceive why its study should be entrusted to the Board of Trade committee appointed to prepare economic agenda of the Genoa Conference as suggested in the minute on paper 6. The same objection applies to my mind to letting this committee advise on the wide and general question of 'the establishment of European peace' – whatever may have been intended by those who put this item into the Genoa agenda.[45]

It is true that during the years 1924–6 there was a decrease in the extent of the Board of Trade's intervention in purely political issues, largely due to the fact that J. Ramsay MacDonald served, in 1924, as Prime Minister and Foreign Secretary, and his successor, Austen Chamberlain, reasserted the authority of their department. Foreign Office interference, however, was not to disappear entirely.

Conflict between the Foreign Office and the Treasury

The arguments between the Foreign Office and the Commercial Departments over where economic and commercial policy ended and where foreign policy began were not limited to the Board of Trade and Department of Overseas Trade, but spilled over into relations between the Foreign Office and the Treasury. The main argument revolved around the question of who would control the British representative on the Reparation Commission and handle reparation issues. The Foreign Office maintained that as discussions would be conducted on an inter-state level outside the boundaries of England, and as they would have an effect on Anglo-Franco-German relationships, the issue was their responsibility. The Treasury, quite understandably, considered that the reparation question was basically a financial problem and that it was much better qualified to tackle the issue than the Foreign Office. An agreement was reached, on 2 February 1920, according to which the British Delegate to the Reparation Commission, Sir John Bradbury, would be directly responsible not to the Foreign Office, but to the Treasury. This agreement, and the Treasury circular of 4 March that confirmed it, also laid down that, although questions of financial policy were the principal concern of the Treasury, both the Minister and Bradbury would keep the Foreign Office fully informed on all reparation questions likely to interest it. On all questions of general foreign policy arising out of the reparation question, the Foreign Office would have the deciding vote.[46]

The compromise was good in theory but not in practice, for it caused constant friction between the two departments. For example, the Foreign

Office found itself – contrary to what had been agreed – often in ignorance of what was happening. The handling of its political and financial
aspects was concentrated during the years 1920–2 in the hands of the
Prime Minister, the Treasury and Bradbury. Copies of the proceedings of
the Reparation Commission were sent by Bradbury to the Foreign Office
only after considerable delay,[47] while, in London, the Treasury failed to
supply the Foreign Office with all the documents relating to various
developments regarding the reparation question. It was thus difficult
for the Foreign Office to gain a clear picture of what the Treasury's
stand was at any moment in the negotiations, which were central to
British–French–German relations. It was forced to agree in advance,
without any foreknowledge, to important steps taken by the Reparation
Commission and also to announcements made by the Treasury and
Bradbury which had political connotations.

The Foreign Office knew nothing of the notification given to the
German government regarding default in coal deliveries or about the
April 1921 talks concerning the fixing of the total German reparation
debt.[48] Following a protest made by the Foreign Office to the Treasury
on 7 May 1921,[49] there was an improvement in communications between
them, but in September 1921 the Foreign Office was shocked to discover
that the Belgians knew the details of conversations between British and
French ministers regarding the Allied Financial Agreement of August
1921, while the Foreign Office itself knew nothing of them.[50]

In November the Treasury published a White Paper which included
Bradbury's views on the Wiesbaden Agreement, without the Foreign
Office being consulted, even though it was apparent that it would
affect Anglo-French relations.[51] In this connection Wigram observed
that:

> We have long realised the danger of an acute Anglo-French press
> controversy arising on this question. . . . We have constantly during the
> last few months drawn attention to the danger of allowing the Treasury
> complete independence in these matters.[52]

The First Secretary of the Central Department, Sidney P. Waterlow, said
that those financial questions were in fact political and that unless they
restrained the Treasury, it would take over the Foreign Office's functions.[53] The Foreign Secretary, Curzon, also remarked that 'the Treasury
is a law unto itself and its wings must be clipped'.[54] Subsequently,
another vigorous protest was sent to the Treasury.[55] In October 1922,
at a time when Anglo-French relations were under severe strain over
the Chanak crisis, Bradbury published a note advocating a moratorium
on all German reparation payments, a policy to which France was firmly
opposed. The Foreign Office was not notified until it was too late, and,
despite last-minute attempts to head it off, the note was published.[56]

The Head of the News Department, Sir William Tyrrell, commented bitterly:

> I confess to a feeling of bitter resentment to seeing part of our foreign policy in the hands of Sir J. Bradbury. Apart from the evils of such a system, the individual, whom I have known for years, is wholly unfit to deal with foreign affairs or people.[57]

With the fall of Lloyd George's government and the rise of Andrew Bonar Law, it was decided that, all government correspondence with the Reparation Commission would be conducted via the Foreign Office and the Foreign Secretary. But the practical results of the acquisition of those powers were nullified in part, through Curzon's refusal to head international conferences on reparations,[58] and through the Ruhr crisis, which temporarily put an end to the work of the Reparation Commission. It is true that during the MacDonald–Chamberlain period the Foreign Office took a more active part in the reparations proceedings. Nevertheless, the stand of the Treasury and Bradbury had, and continued to have, an overwhelming importance in the eyes of the Foreign Office, which did not move a step without consulting them first.

The friction between the Foreign Office and the economic departments was accompanied by a considerable reduction, compared with the war period, in the scope of the economic and commercial functions of the Foreign Office and Diplomatic Service. It was not so much that the Foreign Office lacked adequate sources of commercial information, as the fact that the division of functions between the Foreign Office and the Board of Trade, and the Department of Overseas Trade and the Treasury, prevented its officials from solving commercial and economic questions on their own; this hardly contributed to the economic education of British diplomats abroad. Thus, economic matters, which constituted 75 per cent of the work of an average mission, were usually left to the Commercial Diplomatic Service, which reported directly to the Department of Overseas Trade.[59] Likewise, although a considerable percentage of the work of the Foreign Office political departments (which in the Western Department was as much as 50 per cent)[60] was commercial claims, i.e. against foreign governments, shipping customs, protection of British concessions, etc. They were, nine times out of ten, referred to the Board of Trade in order that it could advise the claimants as to what steps they should take. This procedure, according to the First Secretary of the Western Department, Ronald Campbell, was necessary because 'they have every commercial treaty at their finger-ends and a vast experience derived from continuity; they are in short the experts in these matter'.[61] Moreover, it was not considered that matters such as the negotiations with Germany in February 1924 regarding the German reparation payments in kind,[62] the exception taken by the Lancashire cotton industry to the possible signing of a Franco-German commercial agreement,[63] and the German request of 1925 to change the system of reparation collection,[64] which had been bread and butter to the Foreign Office officials during the war, were too technical to be handed over to the commercial departments.

Against this background, and given that all the data pertaining to reparations furnished to the Foreign Office by its sources of information – Treasury, Board of Trade, Bradbury, Department of Overseas Trade, and the British Ambassador in Berlin, Lord D'Abernon – throughout the 1920s indicated that Germany was practically bankrupt and that any idea of collecting large payments was fantastic,[65] it is hardly surprising that not a few Foreign Office officials came to the conclusion that the reparations were causing despair, bankruptcy, inflation and the destruction of the middle class in Germany. These officials accepted the Treasury–Bradbury–D'Abernon view that the problem was not the Germans' unwillingness to pay, but their inability to do so. As the Head of the Central Department, Miles Lampson admitted that because the Foreign Office lacked the requisite qualifications to assess Germany's capacity to pay reparations,[66] it too accepted without demur the trio's assertions that any attempt to extract large payments would lead to German bankruptcy, which would hinder the economic recovery of Europe and consequently of England.[67] Only a few – the most prominent of whom were Crowe and Tyrrell – suspected that, as indeed was the case,[68] the gloomy projections of the Treasury were groundless and much exaggerated, and that Germany was promoting the fall of the mark in order to avoid having to pay reparations, and thus bring about the revision of the Treaty of Versailles.[69]

The long-term effects of the changes of 1917–20 on the Foreign Office were no less devastating. The new political responsibilities taken on by the economic and financial departments, as a result of the increasing interdependence between economy, finance and foreign policy, reduced the influence of the Foreign Office further. While the whole question of Treasury influence in the 1930s is still the subject of political debate, there can be little doubt that an opportunity for radical reform after the Great War was missed. And in a period in which international relations considerably expanded in scope and foreign affairs increasingly intervened in the financial and economic spheres, the Foreign Office fielded few experts in these areas and was unable to increase its competence in this enlarged diplomatic canvas. It is true that an effort was made in 1931 to strengthen the Foreign Office on the economic side through the creation of an Economic Relations Section. However, this attempt also foundered in the face of inter-departmental rivalries,[70] a fact noted by Walford Selby and Frank Ashton-Gwatkin, and later used as a basis for charges that the Treasury and its Permanent Under-Secretary, Sir

Warren Fisher, had intervened improperly in the affairs of the Foreign Office.[71]

Consequently, the ascendancy of the economic departments, and particularly that of the Treasury, over the Foreign Office had become an accomplished fact. Under the leadership of the various Chancellors of the Exchequer (Winston Churchill 1924–9, Philip Snowden 1929–31, and above all, Neville Chamberlain 1931–7), foreign economic questions – such as negotiating with Germany over the abolition of the reparations in July 1932,[72] representing Britain in international conferences as the first and second Hague Conferences convened in 1929–30 to discuss the Young Plan,[73] and handling Dr Shacht's exchange manipulations in 1934[74] – were dealt with by the Treasury as it saw fit. The Treasury also examined the cost of policies suggested by the Foreign Office, and submitted to the Cabinet, whenever it deemed it necessary, alternative and less costly ones.

Differences between the two departments in subsequent years led some at the Foreign Office to lay the blame for the failure of British diplomacy on this weakness at the top of the administration. Wellesley was not the only Foreign Office official to believe that:

> if there had been co-ordination between British finance and British foreign policy in the first instance, to be followed by international agreement, the resurgence of Germany as the greatest military power would at least have been impeded if not entirely prevented.[75]

As the Treasury was responsible for overseeing all government spending, and as the Chancellor of the Exchequer in the crucial years, Neville Chamberlain, was one of the most powerful men in the Cabinet – and, from 1937 onwards, the energetic and dominant Prime Minister – the Treasury was in a strong position to ensure that its views were heard and heeded.

Appendix 2: The Formation of the Middle East Department

The Foreign Office and Colonial Administration

By the early twentieth century, although, colonial questions still remained at the heart of British Diplomacy, a division of responsibility had been established between the Foreign Office and the Colonial Office. Issues and responsibilities such as Britain's role in the Persian Gulf,[1] and the administration of Egypt, and British Central Africa, Somaliland and Uganda (until 1904), and Zanzibar (until 1913) were under Foreign Office control but were transferred to the Colonial Office. The extent of Foreign Office involvement in the external and internal affairs of the colonies was exceedingly limited.

Administration in War-time

The exact lines of demarcation were never clear but the events of the Great War were to blur the existing divisions of responsibility and to create new problems of jurisdiction that complicated personal and departmental relationships. It added to the confusion there was of overlapping and ill co-ordinated intelligence efforts involving many departments.

The disintegration of the Ottoman Empire and the addition of new territories to the British Empire threw into sharp relief the conflicting interests of the different ministries and departments. The question of whether territorial administration constituted a part of foreign policy came up on 11 February 1917, against the background of the pressure of the High Commissioner in Egypt, Sir Reginald Wingate, to replace the Egyptian self-administration with direct British management.[2] Robert Cecil commented that it seemed to him that the time had arrived to transfer the administration of Egypt from the Foreign Office to the

Colonial Office.[3] He explained that:

> The point of my suggestion is that the F.O. is not organised for administration but for diplomacy. Its officials are diplomats and not administrators and the point of view of the two professions is and ought to be entirely different. We have seen in Norway the disadvantage of a diplomacy which tries to administer. Administration by diplomatic methods is even more disastrous . . . And in spite of the genius of Lord Cromer . . . and the high ability of his two successors, everyone will agree that the system under which Egypt was governed before the war was deplorable just because it was and had to be diplomatic.[4]

Cecil's proposal met with strenuous opposition on the part of the Assistant Under-Secretary, Sir Ronald Graham,[5] and the Permanent Under-Secretary, then Lord Hardinge, an ex-Viceroy of India.[6] They rejected the traditional assumption that there existed a distinction between administrative and political issues,[7] and maintained that it was not possible to draw a dividing line between the two spheres, in view of the international character of Middle East problems. Moreover, according to them, the removal of Egypt from the Foreign Office preserve would be understood by the Moslem world as annexation. They would rise against Britain in sensitive places, such as India, severely damage prestige and prevent Egypt from becoming – in place of the Ottoman Empire – the centre of gravity of the Moslem world and the point from which Britain could influence important Mohammedan countries.[8] The parties to the opposing views would not bridge their differences, so the problem was laid before the Foreign Secretary, Balfour. As a compromise, it was concluded that for the time being the Foreign Office would continue to administer Egypt, but that a special Middle Eastern Department would be established in future that would devote itself to the administration of Egypt and Mesopotamia during their transition period from protectorates to sovereign states.[9]

Despite this decision the Foreign Office was forced to consider more radical reforms. The splitting of control of the Middle East among the Foreign Office, India Office, War Office and Colonial Office – so that the Foreign Office was responsible for western Arabia, northern Persia, Egypt and Palestine, the India Office for eastern Arabia, southern and eastern Persia and Mesopotamia,[10] the War Office constituted a co-partner for Mesopotamia and Palestine; and the Colonial Office a co-partner for Cyprus[11] – created administrative confusion and inconsistencies in the policies of the various departments towards the Middle East. Sir Mark Sykes related that he had had to turn to eighteen agencies and sub-agencies, in London and the Middle East, before he could put into effect a policy change.[12] It is true that efforts were made to group the whole complex of Middle Eastern interests, at first under the Mesopotamia Administration Committee, formed on 16 March 1917; and later under its successors, the Middle East Committee, formed in August

1917, and the Eastern Committee, formed on 21 March 1918.[13]

These committees, however, did not put an end to inter-departmental overlapping. Indeed, the vague definition of their terms of reference and the overbearing attitude of their chairman, Lord Curzon, resulted in their competing with the executive powers of the four departments, causing general resentment and bringing the demand that control over the Middle East be taken out of the hands of the committees and be transferred to some central authority.[14] Besides, the Financial Adviser to the Egyptian Government, Lord Edward Cecil, sent to London a comprehensive memorandum in which he described the existing flaws in the management of Egypt and called for the establishment of a new department, which would devote itself to the administration of the country.[15]

The Foreign Office was thus faced with a painful dilemma: whether to recruit the specialists needed for the purpose of taking charge of this new field, and to resign itself to the fact that the consequent increase in staff would mean the administration of the Foreign Office would become more bureaucratic and less personal, or whether to hand over the whole issue to some other body, which would inevitably diminish the influence of the Foreign Office over Middle Eastern foreign affairs. Robert Cecil subscribed to the second alternative. He rejected the notion that organizational expansion would enable the Foreign Office to cope successfully with administrative questions, and came to the conclusion that none of the present departments of state was competent to deal with the complex problems of the Middle East.[16] Cecil maintained that technical expertise would be secured by the establishment of an independent department, which would be composed of staff trained in administration relating to the various regions of the Middle East.[17]

In contrast to Cecil's view, Hardinge and Graham saw the problems of the Middle East in terms of political purpose[18] – a term which, in this case, implied that everything that pertained, even if indirectly, to foreign policy had to be handed over to the department whose upper echelons were committed to the pursuit of that goal, namely the Foreign Office. The international character of the problems of the Middle East was well emphasized by Hardinge on 6 September 1917:

> Egyptian affairs are affected by our relations with France and Italy in nearly all these countries. Whatever may be the future of Mesopotamia, Palestine, and Syria, Cairo tends to become more than ever the centre of British interests in the Near East, and the point from which we influence the rulers of Hedjaz and other important Mohammedan countries. Many of these Chiefs will be subjected to competing European influences in one way or another, and cannot, for a very long time ahead, be effectively dealt with by any other Imperial organisation than the Foreign Office.[19]

The argument that the Foreign Office could not cope with administrative problems was also dismissed. Hardinge maintained that the progress and development of Egypt during the past thirty-five years constituted one of the brightest pages in the record of the Foreign Office;[20] while Graham pointed out that the Permanent Under-Secretary of State for Foreign Affairs was none other than the ex-Viceroy of India.[21] In their view, the establishment of a new Middle Eastern Department within the Foreign Office did not contradict the need to avoid over-bureaucratization of the Foreign Office, for what they had in mind initially was a skeleton three-man department.[22] Finally, the two held, in contrast to Cecil, that – in view of the uncertainty as to the future status of the Middle East regions – it was necessary to put off the organization of the department until after the war.[23]

The persistent opposition of Graham and Hardinge, who won the backing of Balfour, annoyed Cecil. That his plans to suppress the committee system of control over Middle Eastern affairs should run afoul of the Middle East Committee was bad enough, but that his own department should obstruct him was intolerable. He resolved to enforce his stand on both agencies through the Cabinet. On 10 September 1917, he circulated among members of the Cabinet the memorandum of his brother, Lord Edward Cecil, on Egypt. Simultaneously, at the meeting of the War Cabinet, summoned on 14 September to discuss the recommendations of this memorandum, he emphasized the need to treat the Middle East as an organic whole and observed that it would not be possible to administer it efficiently if it should be under the Foreign Office or the Middle East Committee, for the first lacked administrative skills while the second was too cumbersome a body.[24]

The effort of Cecil to hand over the administration of the Middle East to an independent department and to abolish the Middle East Committee achieved only a partial success. At the War Cabinet meeting of 14 September the Secretary of State for War, Viscount Alfred Milner, agreed with him that it was not possible to continue administering Egypt on the present pattern.[25] Curzon concurred, commenting that developments in the Middle East indicated that perhaps there would be a need to set up a new department that would concentrate on the affairs of the Middle East.[26] In support, the new Secretary of State for India, Edwin Montagu, reported from Egypt on 4 November that he favoured treating the territories from India to Egypt as an organic whole; like Cecil, he, too, advocated the establishment of a Middle East Department with a Permanent Under-Secretary of its own.[27]

Unlike Cecil, however, Curzon considered that at any event it would not be possible to separate the new department, altogether, from the Foreign Office.[28] Montagu also objected to Cecil's suggestion that the Middle East Department should be independent, and rejected the Graham–Hardinge stand that it ought to be incorporated within the Foreign

Office. Instead he proposed that the newly-created department be a condominium of the Foreign and India Offices. In his opinion, the proper method of handling the affairs of the Middle East was by creating a new department, roughly analogous to the Department of Overseas Trade. He advocated splitting the control over Asian affairs (i.e. Chinese, Japanese, Tibetan and Mongolian questions), Near Eastern affairs (i.e. Persian, Afghan and Arabian questions) and Middle Eastern administrative matters among the Foreign Office, India Office and Middle East Department respectively, and bringing the three departments under the control of the Middle East Committee, in order that their activities would dovetail.[29]

A Cabinet committee was set up to examine Lord Edward Cecil's proposals. The committee, chaired by Balfour, and attended by Curzon and Milner, recommended, on 20 February 1918, that Egypt be left, for the time being, under the control of the Foreign Office, in view of the close link that existed in this country between diplomacy and administration. It further recommended that advantage be taken of the considerable skill of Graham in this field, and that personnel be co-opted into the Foreign Office from the Anglo-Egyptian Service, thus enabling the Foreign Office to cope more successfully with administrative problems. For the future, the committee favoured the establishment of a department that would take charge of Egypt, Hejaz, Sudan, Aden, Mesopotamia, Arabia and Palestine. As, in its opinion, it was too early to rule one way or another on this point, the committee determined that the decision be put off until after the war.[30] The report of the committee, and the shift of attention of the Cabinet from the Middle East to southern Russia, forced Cecil to give way and to content himself, for the time being, with the transfer of Sykes, on 4 January 1918, to the Foreign Office, and his appointment to the Palestine and the Hejaz desk under the superintendence of Hardinge.[31]

However, the capitulation of Cecil was merely tactical. On the return of Montagu from India, Cecil co-ordinated a policy stand with him in order to bring up anew the question of the efficiency of the administration of the Middle East.[32] On 5 July 1918, Montagu circulated among the members of the Eastern Committee – the successor of the Middle East Committee – a paper in which he again described in detail the existent duplication of functions in Persia and Mesopotamia among the Foreign Office, War Office and India Office, and the delays caused in putting various steps into effect, due to the necessity for awaiting the decisions of the Eastern Committee. On these grounds, he asserted that the Eastern Committee should refrain from dealing with executive matters and that the whole region be unified under a single authority.[33] On 20 July, Cecil circulated a note of his own in which he expressed his support for Montagu's assertion that the Eastern Committee was a valuable body so far as policy formulation towards the Middle East was concerned, but an inconvenient one in executive matters, adding that:

> 3. Further I agree with Mr. Montagu that it would be desirable for the Foreign Office, and, as far as I can judge, for the War Office also, to create a special department for dealing with Middle Eastern Affairs. Whether ultimately they should be put under the control of an entirely independent department or not may be left for future consideration, since we are all agreed it cannot be conveniently done during the war, but the lesser reform may be carried out at once.[34]

A few days earlier, on 15 July 1918, the Chief of the Imperial General Staff, Sir Henry Wilson, had also confirmed the deficiencies of the Eastern Committee as an executive body.[35] And, on 27 September, General Jan Christian Smuts remarked to his fellow committee members on the urgent need for a unity of direction in the eastern theatre.[36]

Despite the general consent regarding the need to unify the Middle East under a single executive authority, the attempt to change the present situation failed yet again, both because of the stand of the concerned parties that any decision ought to be postponed until after the war, and because of their opposition to Montagu's detailed proposals. These proposals included the surrender of the whole area from Palestine eastwards, or alternatively parts of it, to the administration of the Government of India, the establishment in India of a war-time Middle East Department – the function of which would be to act as a sub-department of the War Office – to administer the whole area and conduct the war in the East, and the giving of ultimate control over Middle Eastern affairs to an inter-departmental sub-committee, which would be created within the Eastern Committee in order to reduce the involvement of the latter in executive matters.[37] On 15 July, on behalf of the War Office, the Chief of the Imperial General Staff rejected the suggestion to split up the eastern theatre between the War Office and the Government of India, stating that there was a close link between the eastern and Russian theatres and the European theatre, and that they ought to be run by one body.[38]

Hardinge, on behalf of the Foreign Office, maintained that it was not possible at this phase of the war to restructure the machinery of government, especially as, in his opinion, the Eastern Committee proved to be an efficient instrument. Likewise, he opposed the transfer of control over the regions of the Middle East to the Government of India, stating that it would provoke a wave of anti-British feelings in Persia, given the traditional enmity between Persia and India. Such an action would thwart Arab nationalism – to which Britain pledged her support – in view of the fact that the Government of India was fundamentally anti-Arab and pro-Turk. As to the delays in the execution of various steps, this could be solved, in the opinion of Hardinge, by adopting the suggestion of Montagu to create an inter-departmental committee, which would deal with the executive side, while the Eastern Committee would concentrate on questions of high policy.[39]

Even Cecil, who adhered to the immediate implementation of the reform, agreed with the arguments of the War Office and of Hardinge as to why the Middle East should not be handed over to the Government of India. Instead, he proposed the immediate formation of a Middle Eastern Department in the Foreign Office, and deferment of the decision about to whom it should be subordinate until after the war.[40]

In a note circulated, together with the memoranda of Hardinge and Cecil, among the Eastern Committee, Balfour concluded tersely, on 27 July, that a distinction should be made between administrative issues and questions of high policy. The political arguments as to why not to hand over the region to the Government of India were overwhelming and in his opinion one ought not to change, until the end of the war, the *status quo*, but should allow the Eastern Committee to continue to run the Middle East areas.[41] At a meeting of the Eastern Committee, on 13 August, Curzon threatened to resign from the chairmanship of the committee rather than agree to the transfer of the political control for the Middle East to an inter-departmental sub-committee.[42]

It seemed, by Curzon's action, that the lid was put on the proposed reform, but Cecil did not give up. At the end of August, in a minute addressed to Balfour, Cecil made important concessions to the Graham–Hardinge–Curzon view. He no longer insisted on the immediate realization of the entire scheme; did not repeat earlier demands to abolish the Eastern Committee; and, as in his note of 20 July, did not advocate separating the Middle East Department from the Foreign Office. Instead, he concentrated on finding practical solutions to the current administrative problems of the Middle Eastern regions and, using the earlier Graham–Hardinge proposals as a lever, asked the leave of Balfour to form – as an interim stage – a special Eastern Department in the Foreign Office, which would be composed of persons with administrative experience.[43]

Permission to take these steps was granted,[44] but immediately gave birth to a fresh controversy, this time between Hardinge and Cecil, regarding the question of who would head the new Eastern Department. Hardinge promoted the candidacy of his protégé, Ronald Graham, while Cecil pressed Balfour to appoint Crowe, who worked under him in the Ministry of Blockade. Hardinge argued that, as one who worked in the Egyptian Administration and was intimately familiar with the Middle East, Graham possessed all the necessary qualifications,[45] Crowe, in his opinion, was fit to head the new Foreign Trade Department, contemplated to be set up after the war.[46] Cecil, on the other hand, held that Graham was a diplomat, while what was required in order to organize the department and to cope with its problems was an administrative talent, such as Crowe possessed.[47]

Hardinge not only believed that Graham would be the right man in the right place,[48] but also desired to keep Crowe out of the War Department, which dealt with all the main issues of foreign policy excluding blockade-related subjects during the war. He wanted to ensure that when he himself took over the Paris Embassy from Bertie, as Grey had promised, the only candidate to succeed him as Permanent Under-Secretary would be Graham, who was brought by him from Egypt for this very purpose.[49]

According to Hardinge, there was nobody in the Foreign Office or Diplomatic Service who could properly fill the post except for Graham.[50] In the opinion of Hardinge's circle, 'Crowe is far too much mixed up with middle-class Germans, and far too uncouth a creature to be the Permanent [Under-Secretary].'[51] It is true that, on 9 August 1917, he admitted to Colonel C. A. Repington that Crowe was qualified to take his place. However, he claimed that, owing to Crowe's German relatives, he could not, in the present state of public opinion, be given the post, while Graham, who was an excellent man, could do so with a little coaching.[52] 'For a long time after the war, the position of Head of the Foreign Office will be one of the greatest interest, and it will make me shudder to think, that a civil servant or an inferior diplomatist was in control of the office', he wrote to Graham.[53]

Cecil, on the other hand, was driven by sincere feelings of admiration for the personality of Crowe:

> He is rather angular, very clear and definite in his opinions and not always very careful of the language in which he expresses them. His knowledge and industry are prodigious, I know that it is the fashion to doubt his judgment: can only say that speaking generally I have found it very good. He has prejudices and they must be allowed for, but I think I now know what they are. He is senior to Graham and has worked devotedly in this office.[54]

Moreover, according to Cecil, though Crowe had declared, in modesty, that he knew very little about the Middle East, it was obvious, from conversation, that he knew more of the general aspects of it than Graham did. Also, in contrast to Hardinge and Graham, who thought the establishment of the new department to be a simple matter sufficing two officials to run it, Crowe realized instantly – in his conversation with Cecil – the organizational difficulties, the necessity for including sections of the department that would deal with the various countries, and the fact that its function would be, fundamentally, not political but administrative.[55] Eventually, the stand of Cecil was accepted, and on 28 August 1918 the appointment of Crowe was approved.[56] Crowe thus returned to the political arena after a forced exile, starting in the Contraband Department and subsequently in the Ministry of Blockade.

These settlements paved the way for the establishment of the new department, though, as detailed so far, they were preceded by contentiousness between Montagu and Cecil. Montagu, who was worried lest its creation would lead to a loss in the influence of the Government of India over Arabia, Mesopotamia and Persia,[57] wished, at first, to put off its establishment until after the war, and thereafter, wanted to place at its head the Permanent Under-Secretary of his office, Sir Arthur Hirtzel.[58]

Eventually, however, a compromise was reached, according to which Middle Eastern policy would be a shared concern of the new department and the Indian Office, and the post of Assistant Under-Secretary in charge of the new Middle East Department of the Foreign Office would go to Crowe.[59] The function of the new department, as agreed between Crowe and Hirtzel, was to control the Near East; though, in the field, the Government of India would continue to administer Mesopotamia and the Persian Gulf, and in London the Cabinet and the Eastern Committee would continue to decide high-policy subjects. The new department was to be organized into four sections: A. Egypt, Sudan and Ethiopia; B. Palestine and the Jews; C. Hejaz, the Arabs, Aden, Yemen and Mesopotamia; D. Persia and the Caucasus. Manpower would be recruited from members of the Foreign Office and the Consular Service.[60] The scheme was submitted to the heads of the two departments on 21 September 1918, and approved by the India Office,[61] by Hardinge, who was laid up at home with a broken leg,[62] and by Cecil, who observed:

> This is undoubtedly the best that can be done at present and on that ground I approve. But I should wish to record my opinion that at the earliest possible moment all the civilian affairs of Mesopotamia and the Persian Gulf should be transferred to the Middle East Dept., which ultimately should become an independent office.[63]

Despite these settlements, the War Office continued to exert control over Palestine and Mesopotamia; the Indian Office, over the Persian Gulf and eastern Arabia; the Foreign Office, over Egypt, Yemen, Persia, Hejaz and Syria; while Aden was administered by the three departments simultaneously.[64] In practice, what was created was not the all-controlling Middle East Department which Cecil desired, nor a condominium of the Foreign Office and India Office, as advocated by Montagu, but simply the Eastern Department of the Foreign Office, which was dismantled at the outbreak of war. The result was a displacement of goals, for the question of whether the much-discussed Middle East Department would be established, and what would be its organizational shape, was left in suspension until after the war.

Post-war Administration

The question of the Middle East control came up again on the agenda in 1920 following a series of diplomatic defeats suffered by British Middle Eastern policy, and in response to the growing public demand that the government rationalize its machinery with a view to reducing expenditure.[65] These defeats – which included the unrest that ensued in Egypt and Arabia in 1919, the Turkish Nationalists' defiance of the Treaty of Sèvres, the recognition of the French Middle Eastern interests at the San Remo Conference, the military withdrawal from Persia and the uprising in Mesopotamia – together with the contention that it was the end result of the inter-departmental dualism, brought the government to the conclusion that the Middle East ought to be placed under a single authority, in the hope that this would aid the re-establishment of the position of Britain in the Middle East.[66]

The government's conclusion brought about a renewal of the debate regarding the identity of this authority, a debate that brought to the surface the conflicts of views concerning the desirable division of functions between the Foreign Office and other departments in peace-time, and the goals of each side. The Foreign Office wanted to restore its pre-war status as the principal adviser to the Cabinet on foreign affairs, and to act as the main link between internal and external affairs.[67] The other departments dissented, and demanded to be equal partners, if not more, whenever subjects dealt with by them had implications on foreign policy, which occurred quite often in view of the growing interdependence between foreign policy and other spheres. Both the Foreign Office and the neighbouring departments claimed a right to the Middle East, even though the latter were embroiled in their own disagreement as to how the new department should be organized.

On 1 May 1920 the Secretary of State for War, Winston Churchill, who was later to become the Secretary of State for the Colonies, submitted a 2000-word memorandum to the Cabinet, which stated that, even though the subject in question was overseas territories, control over Mesopotamia should be turned over to the Colonial Office and its land control should be exchanged for aerial control, in order to reduce expenses. Churchill reasoned that economy and efficiency in territorial administration corresponded more with the functions of the Colonial Office than with those of the Foreign Office; as an example, he pointed out the great success of the Colonial Office in administering the protectorates of East Africa, as against the waste in resources and manpower caused by the Foreign Office in Uganda, Somaliland, Persia and Mesopotamia.[68]

In determining that Mesopotamia should be handed over to the Colonial Office, Churchill was guided by two principles: first, by the principle of division of functions according to process – a term which describes the special skills required in order to discharge a certain duty; second, by the principle of division of labour according to purpose – a term that portrays the major objective of one department or another.[69] With regard to the first, Churchill asserted that administrative skills were required in order to control Mesopotamia, and as to the question of which of the departments was already discharging functions in this sphere – a criterion that was suggested in 1918 by the Haldane Committee – he determined that it was the Colonial Office and not the Foreign Office. Churchill noted that:

> It is no reproach to the Foreign Office to say that they know nothing about administration. It is not their business . . . The Foreign Office is a great Department of State, the whole of whose experience and special aptitudes is devoted to the conduct of the relations of this country with foreign states and to mix up with this the administration of provinces is to impair the discharge of both functions.[70]

When various members of the Cabinet began expressing their opinion, most agreed with Churchill that the Foreign Office lacked the skills required for the purpose of administering the Middle East.[71] Montagu, in a memorandum circulated among Cabinet members on 2 July, observed that:

> Of all the practicable selections, that of the Foreign Office appears to be the least appropriate. The Foreign Office does not possess, and would not pretend to possess, administrative or quasi-administrative experience. It is not organised on an administrative basis; nor are its training and traditions such as to qualify it for administrative work.[72]

In contrast to Churchill, Montagu felt that the administration of the Middle East was a purpose that in itself justified the establishment of an independent department, which should be organized on a territorial basis – that is, the grouping of functions according to regional requirements.[73] As a second priority he was prepared to support its assignation to the Colonial Office.[74] Only the Secretary of State for the Colonies, Milner, was not enthusiastic over the new tasks that Churchill wished to thrust upon him, and in two counter memoranda, of 24 May and 17 June, he dismissed – in view of the Russian threat – not only the promise of Churchill that a reduction of the military forces in the Middle East was desirable,[75] but also his conclusion that the region ought to be turned over to the Colonial Office. Like Montagu, Milner preferred the establishment of an independent Middle East Department, and added that if, nevertheless, it must be assigned to an existent department, then the choice lay between the Foreign Office and the India Office.[76]

Simultaneously, at the Foreign Office, on 17 May, Major Hubert

Young, who was on temporary loan to the Eastern Department from the War Office, stated, in a memorandum which won the praises of his superiors,[77] that Middle East problems were in the main international and subordinate first and foremost, therefore, to the purpose named 'foreign policy'. Entrusting the region to the Colonial Office or to an independent department, he maintained, would be wrong from the administrative point of view. He warned that it would create inter-departmental duplications, delays in taking decisions, conflicts and confusion, and asserted that the most desirable solution was to let the Foreign Office have the exclusive responsibility for the administration of the area.[78]

His arguments were incorporated in a memorandum circulated amongst Cabinet members, on 8 June, by the Foreign Secretary, then Curzon, in which he stated that in theory he favoured the establishment of an independent Middle Eastern Department. He added that should it be decided, as he expected, that for budgeting reasons the Middle East ought to be assigned to an existing department, then it was imperative that it should be the Foreign Office. Curzon pointed out that Middle East elements that the government had to cope with were states with an ancient history and high conceit of their own importance, such as Persia; advanced, ambitious and cosmopolitan communities, such as Jews and Egyptians; and countries that were, and would continue to be, the focus of international rivalry and competition, such as Arabia and Syria. According to Curzon:

> It is safe to say that a lethal blow would be dealt at the pride of Egypt if it were to be placed under the Colonial Office; and that the mandated territories would utter a cry of rage if their conditions were, by even the implication of a misnomer, to be assimilated with that of British Colonies.

Subsequently Curzon criticized the preference of Churchill for the Colonial Office on the basis of division of functions according to process, stating that its Middle Eastern experience was limited to the administration of Cyprus, while the Foreign Office had in its service a very large number of administrators and experts possessing oriental experience and speaking oriental languages. He concluded by saying that in any event the Foreign Office would have to deal with complex interests of foreign powers in Palestine; to negotiate with foreign states on the future of Egypt; to be in sharp and immediate contact with the French in Syria and Feisal at Damascus; to be answerable to the League of Nations for mandatory matters; to oversee the execution of the Peace Treaty with Turkey – with its infinite opportunities for international complication; to be responsible for relations with Persia, owing to her becoming a sovereign state; to be responsible, in the future, for relations

with Afghanistan once it obtained independence and withdrew from the
occupying Government of India; and finally, to deal with Mesopotamia,
in view of the fact that problems of security, trade and frontier defence
would be closely bound up with those of sovereign Persia:

> Is it seriously proposed to leave Afghanistan under the divided control
> of the India Office and the Foreign Office, to leave Persia under the
> Foreign Office, but to place Mesopotamia under the Colonial Office?
> No present overlapping of functions or conflict of jurisdiction would
> be comparable with the friction produced by such a system or lack of
> system.

Curzon concluded that it was not possible to disconnect the Foreign
Office from Middle East affairs, and that therefore, to prevent inter-
departmental overlapping, it was necessary that the affairs of the region
be under the administration of the Foreign Office.[79]

When it became clear that what was to be established was a large
department with sections of its own, Foreign Office officials began to
seek means by which to separate it as much as possible from the Office.
The Head of the Eastern Department, John Tilley, expressed his anxiety
lest the heads of the Foreign Office would buckle under the burden of
an increased workload, and proposed to organize the new department
after the pattern of the Ministry of Blockade.[80] Curzon advocated using
the same method that was applied to the Department of Overseas Trade,
according to which the department would be run autonomically by an
official with the grade of Parliamentary Under-Secretary, who would
be subordinate to the Foreign Secretary. He also recommended – in
order to co-ordinate the interests and work of the various departments
concerned with the administration of the Middle East – to set up a
permanent co-ordinating committee, whose functions would be political
and deliberative, not administrative.[81] Even Hardinge, who fought so
determinedly for the turning of the department into an integral part of
the Foreign Office, now stressed that it must be borne in mind that its
ultimate purpose and eventual evolution was as an independent ministry
under its own Secretary of State.[82] Like Crowe,[83] he asserted that, in
order that the affair of the intrusion of the Department of Overseas
Trade into foreign affairs would not be repeated, it was necessary that
the department should limit itself to administrative matters, so that when
they separated there might be a minimum of disturbance for the Foreign
Office.[84]

In other words, Foreign Office officials demanded that the neighbouring departments transfer to the Foreign Office their Middle Eastern
interests, not so much in order to combine administrative subjects with
foreign affairs, as to prevent the consolidation of the right of the neighbouring departments to handle foreign affairs simultaneously with the

Foreign Office. By that they attempted to create a process of 'co-optation', which Philip Selznick has defined as 'the process of absorbing new elements into the leadership or policy-determining structure of an organization as a means of averting threats to its stability or existence'.[85]

The stands of the various departments were pretty clear, nevertheless, and the government was in no hurry to determine which of them would be adopted. Some ministers were undecided,[86] and preferred to defer their decision to a later date.[87] On the eve of the Cabinet adjournment for the summer holidays, both the contents and the conclusions of yet another circulated Foreign Office memorandum[88] – which proposed that the Middle East Department be placed in the Foreign Office; that it be created as from 1 October next; that a Parliamentary Under-Secretary be appointed as its head, and that he be given a non-Foreign Office staff[89] – were rejected.[90] Young[91] and Crowe[92] noted resentfully that it was mainly due to the opposition of Montagu that the question of the new department was not settled at this meeting. But in effect the main thing that prevented the taking of a prompt decision on this issue was the controversy raging in the Cabinet regarding Middle East policy in general, since on its consequences hung, to no small extent, the answer to the question of which department would be assigned the new agency.

What were the differences of outlook among members of the Cabinet? Curzon remained faithful to the ideas that dominated the British Establishment before the First World War, i.e. that British policy in the Middle East ought to be directed to the formation of a political *cordon sanitaire* round India.[93] Likewise he had no doubts that the best way to achieve this goal would be to recreate the pre-war buffer states system, but with two fundamental differences: first, the shattering of the assumption that the preservation of the integrity of the Ottoman Empire was a distinct British interest, and the abrogation of the Anglo-Russian Convention of 31 August 1907 necessitated the creation of substitute buffer states. Secondly, Britain, in Curzon's opinion, should abandon the strategy of safeguarding India and the imperial lines of communication through the navy, and switch to a direct military involvement in the Asian continent, until such time as these buffer states could stand on their own feet.[94]

As early as December 1918, in the framework of the Eastern Committee, Curzon defended his belief that, in place of the obsolete Ottoman Empire and southern Persia, a chain of vassal states should be created that would stretch from the Mediterranean to the Pamirs,[95] and that Britain must show a military presence there until such a time as those states could guard their own interests.[96] Curzon suggested – in order to prevent in advance any opposition on the part of oriental nationalism – to adopt the Wilsonian doctrine of self-determination, and to proclaim that Britain had no intention of annexing those territories, but meant just to act as trustee to them until such a time as they could be firmly established.[97]

Even while the British Delegation to the Peace Conference was parleying in Paris on the peace terms, without consulting the Delegation, Curzon signed, on 9 August 1919, the Anglo-Persian Agreement, whereby Persia would come under British tutelage and bar the way of Russia to India and the Gulf.[98]

Together with Milner, who displayed a close sympathy for his views,[99] Curzon continued to press the Cabinet, from December 1918 until August 1920, to leave a British presence in the Caucasus, Batum and northern Persia.[100] They even threatened to resign, on 6 August, should British troops be withdrawn from northern Persia.[101] Curzon and Milner claimed that if the Bolsheviks were allowed to occupy even a single British point, such as Batum, the rest of the Caucasus would succumb; the loss of the Caucasus would open the Bolshevik road to Persia; and their entrance into Persia would bring about the fall of India[102] – a view strikingly similar to the famous 'domino theory'.[103] 'You ask why should England do this? Why should Great Britain push herself in these directions? Of course the answer is obvious – India', Curzon told his colleagues on the Eastern Committee.[104]

The main opposition to this policy came from the War and India Offices. Churchill and the Chief of the Imperial General Staff had no time for the idea that Britain was able by military means to frame circumstances that would secure her dominance over all the Near East. The financial resources required for such a policy were too great and, as to manpower, even after the War Office had deployed all the troops at its disposal, British forces were still insufficient to meet the requirements of policies being pursued in various theatres. At best, Britain was in a position to control the different Mandated Territories, but only providing that it be done with a minimum of financial and military expense. In order that the region be administered economically, Churchill added, it should be assigned to the Colonial Office and the goodwill of the local population secured – which was of course conditional on the elimination of Moslem antagonism towards Britain. But to insist on the continued stationing of minute forces, which cost millions of pounds a year, too fragile to be useful against serious attack, in Ireland, Constantinople, Egypt, Palestine, Mesopotamia, the Caucasus, Persia and India, in the hope that their mere presence would frame circumstances required for the security of British interests, was, in his opinion, naïve and dangerous.[105] And, on 20 May 1920 – following the Russian takeover on 18 May of Enzeli on the Caspian Sea, while surrounding and taking prisoner the small British garrison there – he wrote indignantly to Curzon that, 'I must absolutely decline to continue to share responsibility for a policy of mere bluff.'[106]

For different reasons altogether, Montagu and the Government of India mourned the refusal of Curzon and Milner to be content with the setting up of a defence line on the Indian border. The way to bar the Bolshevik spectre, contended the Indian authorities, was not through the creation of a military *cordon sanitaire* but through the cultivation – by political means – of a national nucleus hostile to the basic tenets of Bolshevism. Furthermore, if Britain would not abandon its anti-Moslem policy in the Near East and continue to show there a military and economic presence, then she would take – in the eyes of the inhabitants of the region – the place of Russia as their arch-enemy.[107]

Moreover, at the Inter-Departmental Conference on Middle Eastern Affairs, Montagu was told that the Government of India was very short of funds and needed all it could get for domestic purposes. It would be better, therefore, to save the money Delhi had been asked to contribute to the upkeep of the British garrison in north-eastern Persia, and risk the possibility of a Bolshevik advance through Persia. For such an attack could be satisfactorily met on the Indian frontier itself.[108] This was a markedly different view from that taken in the pre-war period.

Against this background, Churchill, Wilson, Montagu and the Viceroy of India demanded an end to the British presence in a number of secondary theatres. Under their pressure the Cabinet sanctioned, on 5 May 1920, the withdrawal of British forces from north-eastern Persia,[109] and, on 11 June, the withdrawal from Batum on the Black Sea.[110] The representatives of the War Office were not satisfied with this and, on 9 and 18 June, the Chief of the Imperial General Staff warned that England would find itself inevitably committed to a process of gradual reinforcement which might entail unlimited liabilities and even war with Russia, unless the force from northern Persia was withdrawn.[111]

On 13 and 18 June and on 16 July, Churchill stressed, without much success, that if the government wished the War Office to meet its financial and military obligations it would be necessary to decide which of the British responsibilities in Turkey, Palestine, Mesopotamia and Persia was to be abandoned.[112] Churchill and Montagu asserted that the military forces at the disposal of Great Britain were insufficient to meet the requirements of the policies being pursued in the various theatres; that Britain would have to resign itself to the fact that it could not resist the Bolsheviks in the Caucasus, Trans-Caspia and Persia; and that it would be far better, before further disaster occurred, to retire from those areas and to set up a defence line on the Mesopotamian railheads and the Indian frontier.[113]

Ultimately, on 3 November, the Cabinet took a decision to withdraw from Persia in spring 1921.[114] The decision was approved on 18 November,[115] and although Curzon managed to prevent its early execution by delaying tactics, there was no disguising the fact that the Cabinet came close to accepting the Churchill–Montagu stand that Britain ought to limit its Middle Eastern activities to the administration of the Mandated

Territories, and the War Office appreciation that, 'Mesopotamia and North Persia were of no importance from the point of view of the defence of India, and that from that point of view the expenditure of a fraction of the cost would yield better results if spent on strategical railways and improving conditions in India.'[116]

Attitudes towards Turkey

Another aspect of the discussion, which hindered decision-making with regard to organizational matters, was the debate as to the desirable policy towards Turkey, the northern tip in the chain of buffer states planned by Curzon. Both sides agreed that, in order that the attempt to bring peace and stability to the area should not end in the embroilment of Britain in a new war, the national feelings of the Turks should be respected. They were also united in their opposition to the decision of the Council of Four in Paris to allow the Greeks to occupy Smyrna, on 15 May 1919, and to the consistent support of Lloyd George for the idea of parcelling Asia Minor between the Allies. During the Paris Peace Conference Curzon suggested, in order to put an end to Turkish imperialism, a solution stunning in its simplicity. Curzon proposed to dismantle the Ottoman Empire in Asia; to transfer control over the Straits to an international body; and, in order that it should win the blessing of Oriental nationalism, advised leaving in the hands of the Turks the sovereignty over the region which, since the Seljuks, had become their homeland, namely Asia Minor.[117] Churchill and Montagu not only supported the proposal, but also protested, in the course of the Paris Peace Conference, to Lloyd George against the intention of splitting Asia Minor among the Allies.[118]

However, the three Secretaries of State and their supporters in the Cabinet were divided among themselves on two fundamental points. Curzon, from the point of view of the need to secure the routes to India from any potential adversary, desired to suppress those elements which turned Turkey into the standard-bearer of pan-Islam.[119] Montagu, from the point of view of the need to secure the domestic peace in India, where a considerable part of the population was Moslem, wished to appease pan-Islamism and handle it with kid gloves.[120] While Churchill, from the point of view of the need to curtail the involvement of Britain in the region to a minimum, wanted to avoid any move likely to turn the Moslem world against Britain.[121]

Curzon wished to dispossess Turkey of her ability to pose as The Great Islamic Power, the origin of which, according to him, lay in the enormous strategical and political importance given to the power that held Constantinople. He held that it was highly desirable to get the Sultan out of Constantinople, the historic capital of the Eastern world, with its almost hieratic prestige. This would inflict a severe blow on the prestige of the Sultan, which might in the long run lose him the Caliphate – a severe blow to pan-Islamism.[122] Montagu wished to avoid damaging the prestige of the Caliphate for fear that it would lead to an explosion of Moslem feeling in the Near East.[123] Curzon was convinced, on the basis of his claimed intimate familiarity with Oriental psychology, that they would be reconciled to these changes, including the expulsion of the Turks from Europe, and content themselves with mere protestations.[124] Montagu, in contrast, who mistook the Khilafat movement in India for a demonstration of solidarity with the Turks, responded derisively that Curzon 'is led astray by his lifelong habit of knowing what people should think rather than considering what they do think'.[125]

Eventually, and despite the backing of Lloyd George and the French, Curzon was outvoted. In a Conference of Ministers held on 5 January 1920, and in a meeting of the Cabinet the day after, Montagu drew lurid pictures of the political consequences if a policy of eviction was pursued, while Wilson stated officially that an enormous army would be required in Turkey if the Sultan and his government were evicted and allowed to run in Anatolia unchecked. These views had such a devastating effect upon the Cabinet that it was decided, on 6 January, that the Sultan and his government be allowed to remain in Constantinople.[126]

From here the routes of Curzon, Churchill and Montagu separated. They continued to disagree not only about the future of Constantinople, but also about the essence of peace with Turkey. With regard to the handling of the Turkish problem, Curzon, it is true, shared the feeling of Hardinge[127] that:

> Nothing could have been more mismanaged and we are only at the begin-
> ning of our trouble with Turkey over the conditions of peace . . . The
> merest tyro [tyrant] who has lived in Turkey would know that the Turks
> would never agree to give up Smyrna and Adrianople to the Greeks
> whom they hate and despise.[128]

Hardinge concluded[129] that when the Turks realized that, in addition to being expelled from Europe and Asia, their Asian Minoric provinces were to be parcelled out among foreigners whom they despised, and that no corner of territory would be left to them which they could truly call their own, they would start a new war against the Greeks and Allies, which would turn Anatolia into a vast shambles, and end with the expulsion of the Greeks from the Smyrna vilayat.[130] Nevertheless, Curzon thought that, in view of the need to put an end formally to the state of war and to the dissensions among the Allies regarding the Middle East, there was no alternative but to accept the trends that took shape at the Paris Peace

Conference, First Conference of London (12 February–10 April 1920) and at the San Remo Conference (19–26 April), under the guidance of Lloyd George. Basically this involved the cession of Smyrna and Thrace to Greece, and the creation of Allied zones of influence in Asia Minor; and the pursuance of rectification policies in the future.[131]

Churchill, Montagu and the Chief of the Imperial General Staff refused to accept this compromise and bombarded the Cabinet with warnings regarding the disastrous political and military consequences that would ensue not only in Turkey but also in the Near East, the more so as the Allies were unable to implement these policies.[132] Eventually, the stand of Curzon, which was supported by Lloyd George, won the day. And, on 10 August, the representatives of the Government of Constantinople signed the draft treaty which had been drawn up by the Allies in the various conferences. The refusal of the Turkish Parliament, however, to ratify it, together with the armed opposition of the Turkish Nationalists – headed by Mustapha Kemal – to the Allies in general and the Greeks in particular, prevented the realization of the Treaty of Sèvres. The result was, from the end of 1920, to renew the controversy between Curzon, Churchill and Montagu.

Towards a Middle East Department

These fundamental policy dissensions, as aforesaid, set aside discussion on the organization of the Middle East Department. Only on 31 December 1920 was a decision taken on the matter. The decision resulted not from the sides reaching an agreement regarding the policy towards Turkey, but from Lloyd George throwing all his weight in favour of Montagu's demands of 18 November[133] and 22 December[134] to realize, as soon as possible, the organization of the new department. Although he had always supported the 'forward policy' strategy of Curzon – i.e. British military presence in the lands bordering India – Lloyd George became convinced, in view of the worsening financial state of Britain and the criticisms levelled against him from the rostrum of Parliament, that priority would have to be given to a curtailment of government expenditure in the region.[135] More important, Lloyd George determined, meanwhile, the identity of the Department of State to which the new department should be assigned – and it was not to be the Foreign Office, which of late had begun to be viewed by him as 'good for nothing'.

There was nothing unusual about this decision since Lloyd George, as a rule, showed little respect towards civil servants. More to the point is the fact that he began to display the first glimmerings of what was to become his private hobby-horse, namely to dwarf the aristocratic figure of his Foreign Secretary. On 16 November, as related by the Cabinet Secretary, Maurice Hankey, he 'bullied Curzon a good deal' about the need to evacuate British forces from Persia, and about 'trade with Russia to which Curzon is an obstacle'.[136] On 31 December, when the Cabinet met to decide to which department of state the Middle East territories be assigned, 'he had a sly thrust at Curzon over Persia, where – thanks almost entirely to Curzon – we have responsibilities, both civil and military, we ought never to have assumed, and which have cost us tremendous sums, and militated against the Russian trade agreement'.[137]

According to Hankey, Curzon made a most powerful plea for the Foreign Office – 'pointing out that the affairs of the two mandated territories are inextricably bound up with those of Egypt, Persia, Turkey, Syria, and Arabia; that Egypt in particular ought to be administered by the same Department as Palestine and Mesopotamia, that if Egypt were put under the Colonial Office, it would lead to a mutiny in Egypt'. 'The P.M., however, urged that the F.O. had never succeeded as an administering Department – in Egypt, Somaliland, or anywhere. Moreover, it distorted the perspective of the office, and induced them to attach too much importance to those regions which they were administering.'[138]

Even Montagu's last-minute reversal against supporting the Colonial Office did not sway the Prime Minister. Montagu's change of heart was based on the fact that taking such a step might be construed as annexation; that giving the task to the Foreign Office would ensure unity of control; and that withdrawing the British forces from Mosul to the Basra line, as demanded by Churchill, would demonstrate to the League of Nations and the inhabitants of Mesopotamia the inability of Britain to fulfil the terms of the Mandate.[139] In order to end the stalemate in the Cabinet, Lloyd George took the unconventional step of a general vote. The result was eight in favour of the Colonial Office, five in favour of the Foreign Office, and two abstentions.[140]

Curzon, in spite of his arrogance, was a sensitive man, and was very hurt when Lloyd George blamed him for the involvement of Britain in Persia, and by the decision to turn over responsibility for the Mandated Territories to the Colonial Office.[141] In a conversation with the Chancellor of the Exchequer, Austen Chamberlain, he said that he was considering resigning in view of the fact that the Prime Minister had treated him, more than once, with scant courtesy – almost with contumely – in the presence of his colleagues, and attached little weight to his opinion on matters directly within the sphere of his departmental responsibility: How, he asked, must his colleagues regard him and what use was he under such circumstances? Surely no previous Prime Minister had ever addressed one of his Secretaries of State in such a way or appealed to the votes of minor members of the Cabinet to overrule the advice of two Secretaries of State in the common affairs of their two offices.

Chamberlain, who was also shocked by the ferocity of the attacks of Lloyd George, did all he could to combat the depression of the poor man and to assure him that their esteem towards him had not changed. Simultaneously he asked Bonar Law to tell Lloyd George to desist from it.[142] But when Bonar Law turned, on this point, to Lloyd George, he turned down the request offhandedly, saying, 'Well he [Curzon] has no following', which brought Hankey to the conclusion that Lloyd George was suffering 'from a touch of a swelled head'.[143]

From here on, matters moved swiftly. On 1 January 1921, Lloyd George asked Churchill, who was celebrating the New Year with him at Lympne, whether he would be willing to succeed Lord Milner. The friend and ally of Curzon in the Middle East Department affair, Lord Milner had asked, on 27 November 1920, to be relieved of his duties[144] as Colonial Secretary. After some hesitation Churchill accepted, on 4 January, providing that his responsibilities as Colonial Secretary would include the civil and military administration of the Mandated Territories. This was accepted, on 7 January, by Lloyd George.[145]

On 11 January, in compliance with the Cabinet decision of 31 December 1920, an inter-departmental committee was constituted, chaired by the Assistant Secretary at the War Office – comprising Churchill and the future Permanent Under-Secretary of the Colonial Office, Sir James Masterton Smith – in order to determine the powers and functions of the new department. On 31 January, it submitted its report. It recommended that the new Middle East Department be responsible for Mesopotamia, Palestine, Trans Jordan, Arabia and Aden; and that the powers of the Middle East Department include the handling of questions such as the delimitation of boundaries, internal administration, military control, policing matters, civil and military expenditures, etc. Finally, the committee recommended that organizational solutions be found that would enable the Colonial, War and Foreign Offices to co-ordinate their policies towards the region.[146]

Lastly, on 14 February, the Cabinet met to consider the report of the Masterton Smith Committee. After what the President of the Board of Education, H. A. L. Fisher, described in his diary as a 'Long wrangle between Winston and Curzon as to spheres of interest',[147] the report was approved, with the supplementary observation that the Colonial Secretary would have to co-ordinate his policy towards Arabia with the Foreign Secretary.[148] Thus came to an end the Foreign Office attempt to secure its monopoly over Middle Eastern policies: the Colonial Office became senior partner to the process of formulating Middle Eastern policy.

The establishment of the Middle East Department in the framework of the Colonial Office solved the duplication of functions only partially. Responsibility for the Mandated Territories and Arabia was turned over to the Colonial Office, but control over the affairs of Egypt, Aden, Turkey and Syria remained in the hands of the Foreign Office. Likewise, Foreign Office staff continued to deal with regions which were transferred formally to the Colonial Office. The continuing duplication of functions was manifested not only by the profusion of files regarding the Middle East which flowed daily under the hands of Foreign Office officials, but also by (1) the composition of the Eastern Department of the Foreign Office, established in 1919 and comprising Turkey, Persia, the Middle East and Egypt, and (2) the composition of the Egyptian Department, formed in 1924 and comprising Egypt, Sudan, Ethiopia and the Italian Colonies in Africa.[149]

In response to a query by the Parliamentary Under-Secretary, Cecil Harmsworth, in reply to a Question in Parliament, the Foreign Office sent the following minute:

> It is the case that these responsibilities have been transferred from the Foreign Office to the Colonial Office, but the transfer does not completely terminate the work of the Foreign Office in connection with those portions of the Middle East, since the Foreign Office remains responsible for the diplomatic correspondence and negotiations with foreign Governments on questions of international concern arising in these countries.[150]

Although responsibility for the relations of Britain with sovereign states in the area was under the Foreign Office, the Colonial Office acquired the right to take part in formulating the policy towards these states. It based its claim on two grounds: first, its being part of the network of relations with states such as Turkey had an effect upon the Moslem population and was thus within the sphere of its responsibility; secondly, as the Colonial Office was responsible for Mesopotamia, for instance, it was its function to ensure that foreign developments did not cause unrest, bring about the return of Mosul to the bosom of Turkey (as was wished by the Turkish Nationalists) or result in the expulsion of Britain from the area.[151] Consequently, Churchill played a central role during the Anglo-Turkish crisis of 1921-2. The Colonial Office took part in the preparations for the Lausanne Conference – designed to put an end to this crisis – and its representatives participated in the discussions held among the representatives of the War Office, Treasury, Board of Trade and Foreign Office, regarding the content of the prospective peace treaty to be offered to Turkey,[152] including issues such as the border between Turkey and Mesopotamia and the fate of Kurdistan, etc.[153]

Relations with the Dominions

Finally, the growing identification of functions between the two departments found expression on a different level altogether: their relations with the Dominions. Before the war the Dominions had little or no voice in foreign policy; their step up from colonies to autonomous units of the British Commonwealth of Nations, however, and their decisive contribution – in resources and manpower – to the war effort, changed that.[154] 'Further, it must be remembered', Curzon told the French Ambassador in December 1921, ' . . . that British foreign policy was now not the policy of the Cabinet in Downing Street alone, but was the policy of the Empire, and the points of view of the Prime Ministers of our distant Dominions had also to be seriously considered.'[155]

Dominions representatives formed part of the British Delegations to the Paris Peace Conference, and Washington Conference on Naval Disarmament and Far Eastern Affairs. Signed separately, on 28 June 1919 and 13 December 1921, the Treaty of Versailles and Washington treaties were also accepted by the League of Nations. If there was a shadow of a doubt regarding the necessity to take the position of the Dominions into account, it was removed by the effects of the Chanak Crisis upon the Dominion Governments. The famous press communiqué of Churchill – issued on 16 September 1922 after consultation with Lloyd George, but published in the press of the Dominions before the secret telegram of Lloyd George, of 15 September, to the Dominion prime ministers had been deciphered by them – to the effect that the Dominions were asked to send contingents to Turkey, was their first intimation that they might have to go to war with the Turkish Nationalists. The communiqué was received with consternation in Australia and marked disapproval in Canada. Thereafter, the Dominions increased their watchfulness over the political moves of Britain; demanded that they be kept up to date on foreign affairs, and that they be consulted in advance; and refused to be bound by treaties signed by the British Government which had not been approved beforehand by them or by their Parliaments, as occurred with regard to the Locarno Treaty.[156]

These developments led the Colonial Office to begin sending the Dominions current information on foreign affairs. For its part the Foreign Office came into ever increasing contact with the Dominions, something that, until the outbreak of the First World War, was done via the Colonial Office. On the organizational level, it led to the establishment, on 11 November 1925, of a new extension of the Colonial Office, namely the Dominions Office,[157] and, on 28 June 1926, to the establishment of the Dominions Information Department of the Foreign Office,[158] which emphasized the politicization of the work of the Colonial Office, and the intrusion of the Foreign Office into inter-colonial matters, all the more.

Amongst its duties the Dominions Office handled foreign, defence and consular matters,[159] while the Dominions Information Department of the Foreign Office supplied the Dominions with information on foreign affairs, dealt with inter-imperial relations, so far as they affected the Foreign Office, and handled matters of protocol affecting the foreign relations of the Dominions.[160]

Abbreviations

AA	Auswärtiges Amt.
ADAPA	Akten zur deutsche Auswärtigen Politik, series A, series B.
ARWR	Akten der Reichskanzlei Weimarer.
BD	*British Documents on the Origins of the War, 1898–1914*, ed. G. P. Gooch and H. Temperley (London, 1926–38).
BOT	Board of Trade.
CBI Archives	Confederation of British Industry Archives.
CID	Committee of Imperial Defence.
CIGS	Chief of the Imperial General Staff.
Cmd	Command Paper.
CP	Conservative Party Archives.
CUL	Cambridge University Library.
DBFP	*Documents on British Foreign Policy, 1919–1939*, First Series, ed. E. L. Woodward and R. Butler (London, 1947–78).
EC	Eastern Committee.
FO	Foreign Office archives in the Public Record Office.
FRUS	Foreign Relations of the United States.
GC & CS	Government Code and Cypher School.
Le livre Jaune Français, 1924	Documents relatifs aux négotiations concernant les garanties de sécurité contre une aggression de l'Allemagne Ministère des Affaires Étrangères.
LP Archives	Labour Party Archives.
L/NOJ	League of Nations Official Journal.
MAE	Ministère des Affaires Étrangères, Europe 1918–1929, folder.
MEC	Middle East Committee.
MIIC	Predecessor of SIS; later military section of SIS.
MI5	Security Service.

OGPU	Predecessor archives of KGB, Soviet Intelligence Service, 1923–34.
PID	Political Intelligence Department of the Foreign Office.
PP	Parliamentary Papers.
PRO	Public Record Office, London.
RIIA	Royal Institute of International Affairs.
SIS	Secret Intelligence Service.
TC Deb 5	Parliamentary Debates (Hansard), Official Report 5th Series, House of Commons/House of Lords (London).
TL Deb 5	Parliamentary Debates (Hansard), Official Report 5th Series, House of Lords (London).
TUC	Trades Union Congress.
UF	Ursachen und Folgen, Vom deutschen Zusammenbruch 1918 und 1945 biz zur staatlichen Neuordnung Deutschlands in der Gegenwart Eine Urkundenune Dokumentensammlung zur Zeilgeschichte: vol. v (Berlin, 1958–); vol. vi (Berlin).
US National	US National Archives, Record Group 59, General Records of the Department of State, 1910–29.

Notes

The author's original organization of his material placed Appendices 1 and 2 prior to the current chapter 1, but at that stage he had not yet written an Introduction. Dr Steiner has written the introductory pages to chapter 1, up to 'Changes in Structure'; all the rest of the text was written by Ephraim Maisel and updated as determined in Dr Steiner's Preface.

For reasons of handwriting, and a card index system used by the author which was not fully understood by the editor, it has not been possible within the limitations of working on the text to determine the correct number assignation of a few of the last reference elements of some FO references. A great deal of care has been exercised in transcribing the handwritten notes to typeset text and the references given should be correct. However, researchers are forewarned.

Chapter 1
Organizational Changes at the Onset of Peace

1. See Appendices 1 and 2, 'The Establishment of the Department of Overseas Trade' and 'The Formation of the Middle East Department'.
2. FO 371/8291, W1475/14750/50, Minute by Villiers, 16 February 1922.
3. Ibid., Minute by Waterlow, 22 February 1922.
4. Hansard, Parliamentary Debates, 5th Series, H of L, vol. 326, cols 970–1, 30 March 1943.
5. For a discussion of the role of the Ministry of Information and the creation of the Political Intelligence Department, see Erik Goldstein, *Winning the Peace; British Diplomatic Strategy, Peace Planning, and the Paris Peace Conference, 1916–1920* (Oxford, 1991), pp. 57–90.
6. Ibid., p. 65.
7. See M. L. Sanders and P. M. Taylor, *British Propaganda during the First World War, 1914–1918* (London, 1982), p. 248; P. M. Taylor, *The Projection of Britain: British Overseas Publicity and Propaganda, 1919–1939* (Cambridge, 1981).
8. For a fuller discussion see Sir A. Willert, *Washington and Other Memories* (Boston, 1972).
9. For a further explanation of this see Christopher Andrew, *Secret Service: the Making of the British Intelligence Community* (London, 1985).
10. Curzon to Crewe, 13 October, 12 November, 12 December 1923, CUL Crewe MSS 12. Curzon to Baldwin, 9 November 1923, IOLR Curzon MSS Eur. F 112/320.

11. R. Warman, *The Foreign Office, 1916–1918: a Study of its Role and Functions* (New York, 1986), pp. 150–5; Z. S. Steiner and M. L. Dockrill, 'The Foreign Office Reforms, 1919–1926', *Historical Journal*, vol. 17, no. 1 (March 1974), p. 133.

12. Balfour MSS., Add. 49738, Cecil to Balfour, 8 January 1918; ibid., Add. 49744, Lord Derby to Balfour, 10 January 1919.

13. Hardinge MSS, vol. 39, Hardinge to Reginald Wingate, 28 November 1918.

14. Warman, *The Foreign Office 1916–1918*, pp. 152–3, 157, 159. See also Appendix 2.

15. Steiner and Dockrill, 'The Foreign Office Reforms', p. 134.

16. William Wallace, *The Foreign Policy Process in Britain* (London, 1977), pp. 27–8.

17. Public Record Office Handbooks, *The Records of the Foreign Office*, pp. 14, 24, 30.

18. Paul G. Lauren, *Diplomats and Bureaucrats: The First Institutional Responses to Twentieth-Century Diplomacy in France and Germany* (Stanford, 1976), pp. 88–90, 126–9.

19. FO 366/781, 19AI(e), Hardinge to Curzon, 26 September 1919.

20. See Appendix 1 and Chapter 4.

21. Public Record Office Handbooks, *The Records of the Foreign Office*, pp. 22–4; Steiner and Dockrill, 'The Foreign Office Reforms', p. 135. See also Appendix 2, 'The Formation of the Middle East Department'.

22. FO 366/781, 16AI(e), Tilley to Hardinge, 2 September 1919.

23. Ibid., 6AI(e), Hardinge to Montgomery, July 1919 (undated); ibid., 9AI(e), Montgomery to Hardinge, 30 July 1919.

24. Ibid., 12AI(e), Hardinge to Campbell, 25 August 1919.

25. Ibid., 21AI(e), Hardinge to Crowe, 30 October 1919.

26. Ibid., 19AI(e), Hardinge to Curzon, 26 September 1919.

27. Ibid., 20AI(e), Hardinge to Crowe, 22 October 1919. For further information on this point see Sybil Crowe and Edward Corp, *Our Ablest Public Servant: Sir Eyre Crowe, 1864–1925* (Devon, 1993), pp. 390–2.

28. FO 366/781, 19AI(e), Hardinge to Curzon, 26 September 1919; Steiner and Dockrill, 'The Foreign Office Reforms', p. 135.

29. Fo 366/781, 21AI(e), Hardinge to Crowe, 30 October 1919.

30. Ibid., 16AI(e), Tilley to Hardinge, 2 September 1919.

31. Ibid., 19AI(e), Hardinge to Curzon, 26 September 1919.

32. FO 366/781, 20AI(e), Hardinge to Crowe, 22 October 1919; Steiner and Dockrill, 'The Foreign Office Reforms', pp. 135–6.

33. FO 366/807, X9486/1272/504, Memorandum by Montgomery, 7 December 1923.

34. FO 366/826, X733/557/504, Sir Frank Bains to Montgomery, 22 October 1925.

35. FO 366/815, X4470/805/504, Crowe to MacDonald, 14 June 1924.

36. Ibid., Minute by MacDonald, 15 June 1924.

37. FO 366/789 X1495/990, Note by the Sub-committee on Amalgamation, 4 August 1920.

38. Ibid., Hardinge to Curzon, 6 August 1920, quoted in Steiner and Dockrill, 'The Foreign Office Reforms', p. 136.

39. FO 366/789 X1495/990, Note by the Sub-committee on Amalgamation, 4 August 1920; ibid., Hardinge to Curzon, 6 August 1920.

40. Ibid., Hardinge to Curzon, 6 August 1920.

41. FO 366/781, 5AI(b), Curzon to Hardinge, 13 August 1920.

42. FO 366/801, X7697/7275/504, Minute by Crowe, 19 August 1922; ibid., Minute by Crowe, 26 August 1922.

43. FO 366/790, X4790/1053/55, Minute by the Treasury, 1 December 1920.
44. FO 366/780, 1AI(b), Baldwin to Cecil, 20 February 1918. See also, Eunan O'Halpin, 'Sir Warren Fisher and the Coalition, 1919–1922', *Historical Journal*, vol. 24 (1981), pp. 909–10.
45. FO 366/790, X4790/1053/55, Sir N. F. Warren Fisher (Treasury) to Crowe, November 1920 (undated); Steiner and Dockrill, 'The Foreign Office Reforms', p. 137.
46. FO 366/790, X4790/1053/55, Hardinge to Sir Malcolm Ramsay (Treasury), 25 August 1920.
47. FO 366/801, X3578/3578/504, Minute by Crowe, 31 March 1922.
48. FO 366/789, AIV(c)10, Crowe to Sir Warren Fisher (Treasury), 3 December 1920.
49. FO 366/801, X7275/7275/504, Crowe to the Secretary to the Treasury, 20 July 1922.
50. Ibid., X4175/3578/504, R. R. Scott (Treasury) to Crowe, 25 April 1922; The Foreign Office List, 1939, p. 188.
51. FO 366/801, X7697/7275/504, Scott to Crowe, 11 August 1922.
52. Ibid., Circular by Curzon, 7 September 1922.
53. Cmd 8, Pts I–II; Cmd 7749, p. 15, Minutes of Evidence, QQ 38,751, 39,379, Appendixes 84, 87; Steiner, *The Foreign Office and Foreign Policy, 1898–1914* (Cambridge, 1969), pp. 19–20, 167–8, 183, 217–21.
54. It must be emphasized that the diplomatic establishment was not distinguished on this point from the Higher Civil Service and other Great Powers. See Raymond A. Jones, *The British Diplomatic Service, 1815–1914* (Waterloo, 1983), pp. 139–46.
55. Cmd 7749, Minutes of Evidence, QQ 39,381, 39,394–7, 40,658, 43,476, Appendix 87, pp. 167–8, 183–4, 198. See also A. J. A. Morris, *Radicalism Against War, 1906–14: The Advocacy of Peace and Retrenchment* (London, 1972), pp. 264–9; and A. J. P. Taylor, *The Troublemakers: Dissent Over Foreign Policy, 1792–1939* (London, 1964), pp. 97–9, 103, 118–19.
56. These features, too, were a common characteristic of most foreign ministries. See Zara S. Steiner (ed.), *The Times Survey of Foreign Ministries of the World* (London, 1982), pp. 16–18.
57. Ray Jones, *The Nineteenth-Century Foreign Office: An Administrative History* (London, 1971), pp. 111–35; Valerie Cromwell and Zara S. Steiner, 'The Foreign Office before 1914: A Study in Resistance', in Gillian Sutherland (ed.), *Studies in the Growth of the Nineteenth-Century Government* (London, 1972), pp. 82–6.
58. Jones, *The British Diplomatic Service*, pp. 164–7, 171–215, 219; Zara Steiner, 'The Foreign Office under Sir Edward Grey, 1905–1914', in F. H. Hinsley (ed.), *British Foreign Policy under Sir Edward Grey* (Cambridge, 1977), pp. 40–1.
59. Doreen Collins, *Aspects of British Politics, 1904–1919* (London, 1965), pp. 66–7; Jones, *The British Diplomatic Service*, p. 196.
60. PP 1890 (Cmd 6172), vol. 27, Fourth Report of the Royal Commission on Civil Establishments, recommendation 24.
61. Cmd 7748, Pt I, ch. 3, para. 14.
62. Lauren, *Diplomats and Bureaucrats*, pp. 44–77.
63. Collins, *Aspects of British Politics*, pp. 179, 181–2; M. Swartz, *The Union of Democratic Control in British Politics during the First World War* (Oxford, 1971), pp. 25, 42–3, 86, 95, 223–4, 230.
64. Gordon A. Craig, 'The British Foreign Office', in G. A. Craig and Felix Gilbert (eds), *The Diplomats, 1919–1939* (Princeton, 1953), pp. 21–2; Steiner and Dockrill, 'The Foreign Office Reforms', p. 131.

65. Lauren, *Diplomats and Bureaucrats*, pp. 99, 143; Steiner (ed.), *The Times Survey*.
66. Collins, *Aspects of British Politics*, p. 455.
67. FO 366/780, 14AI(b), Drummond to Hardinge, 4 April 1918.
68. The Commissioners considered that the lengthy unpaid probationary period followed by low wages, in addition to the property qualification, had both effectively closed the Diplomatic Service to all British subjects, however well-qualified, unless they possessed private means. It also discouraged transfers from the Foreign Office to the Diplomatic Service. Cmd 7748, Pt I, chs I–II.
69. Cmd 7748, Pt I, ch. III, paras 10, 13, 15, 17–19.
70. T1/12050, 26057/194027/16, Hardinge to the Secretary to the Treasury, 29 September 1916.
71. FO 366/786, 19763/101, Chief Clerk, Sir Robert Chalmers (Treasury) to Lord Hardinge, 24 January 1917.
72. Ibid., Walter Langley to the Secretary to the Treasury, 5 February 1917.
73. Christina Larner, 'The Amalgamation of the Diplomatic Service with the Foreign Office', *Journal of Contemporary History*, vol. 7 (1972), p. 117.
74. FO 366/780, 16AI(b), Cecil to Hardinge, 5 April 1918.
75. Ibid., 11AI(b), Fulton to Hardinge, 2 April 1918.
76. Ibid. 1AI(b), Stanley Baldwin (Treasury) to Cecil, 20 February 1918; ibid. 5AI(b), Draft Report on Mr Baldwin's letter, 15 March 1918.
77. PP 1914 (Cmd 7338), vol. 16, Fourth Report of the Royal on the Civil Service, ch. 9, para. 101; Cmd 9230, paras 15–20.
78. O'Halpin, 'Sir Warren Fisher', pp. 914–15.
79. Max Beloff, *New Dimensions in Foreign Policy: A Study in British Administrative Experience, 1947–1959* (London, 1961), p. 80.
80. FO 366/780, 5AI(b), Draft Report on Mr Baldwin's letter, 16 March 1918.
81. Ibid. 4AI(b), A Foreign Office official to Sir Maurice de Bunsen, 10 March 1918.
82. Ibid., 5AI(B), Draft Report on Mr Baldwin's letter, 16 March 1918.
83. Lord Gore-Booth, *With Great Truth and Respect* (London, 1974), p. 62.
84. Michael Hill, *The Sociology of Public Administration* (London, 1972), p. 189.
85. Sir Edward Bridges, 'Portrait of a Profession', in Richard A. Chapman and A. Dunsire (eds), *Style in Administration: Readings in British Public Administration* (London, 1971), p. 54.
86. Geoffrey K. Fry, *Statesmen in Disguise, The Changing Role of the Administrative Class of the British Home Civil Service, 1853–1966* (London, 1969), pp. 36–8, 49–60; Richard A. Chapman and J. R. Greenaway, *The Dynamics of Administrative Reform* (London, 1980), pp. 100–14.
87. FO 366/780, 21AI(b), Baldwin to Cecil, 18 April 1918.
88. PP 1930–1 (Cmd 3909), vol. 10, Report of the Royal Commission on the Civil Service, 1929–31, Minutes of Evidence, QQ 18,03–9, 18,852, 18,856.
89. Sir Edward Bridges, 'Administration: What it is? And how it can be learnt', in A. Dunsire (ed.), *The Making of an Administrator* (London, 1956), pp. 14–15.
90. FO 366/780, 3AI(b), Minute by Cecil, February 1918 (undated).
91. Ibid., Hardinge to Cecil, 25 February 1918.
92. Ibid., 9AI(b), Hardinge to Cecil, 1 April, 1918.
93. Sir John Tilley, *London to Tokyo* (London, 1942), p. 84.
94. FO 366/780, 5AI(b), Draft report on Mr Baldwin's letter, 16 March 1918.
95. Ibid., 11AI(b), Tufton to Hardinge, 2 April 1918.
96. Ibid., 8AI(b), Campbell to Hardinge, 22 March 1918.
97. Ibid., 15AI(b), Warner to Hardinge, 4 April 1918, quoted in Larner,

'The Amalgamation of the Diplomatic Service with the Foreign Office', pp. 119–20.

98. Ibid., 13AI(b), Minute by Wellesley, 3 April 1918.
99. Ibid., 9AI(b), Hardinge to Cecil, 1 April 1918.
100. Ibid., 16AI(b), Cecil to Hardinge, 5 April 1918.
101. Ibid., 20AI(b), Memorandum by Cecil, 16 April 1918.
102. Ibid.; ibid., 29AI(b), Treasury to Foreign Office, May 1918 (undated).
103. Ibid., 20AI(b), Memorandum by Cecil, 16 April 1918; ibid., 26AI(b), Cecil to Baldwin, 29 April 1918.
104. Ibid., 27 AI(b), Balfour to Bonar Law, Treasury, 6 May 1918.
105. Ibid., 30AI(b), Bonar Law (Treasury) to Balfour, 10 June 1918.
106. Ibid., 5AI(b), Draft Report on Mr Baldwin's letter, 16 March 1918.
107. Ibid., 1AI(b), Baldwin to Cecil, 20 February 1918.
108. These recommendations included the removal of the existing requirement that applicants should first obtain the permission of the Foreign Secretary to appear before the Board of Selection, which was thought to have already eliminated candidates who did not conform to a predetermined type; abolition of the property qualification; and assimilation of the entrance examination more closely to the Class I Civil Service examination for the purpose of bypassing the Etonian exclusivity and broadening the composition of the Board of Selection, which was felt to have been too departmental. Cmd 7748, Pt I, ch. II, paras 2, 3, 7, ch. III, para. 15.
109. Ibid., Pt I, chs I–II; Cmd 7749, Appendix 84, pp. 306–7; Steiner, *The Foreign Office*, pp. 217–21.
110. Comparative studies suggest, however, that in fact the Diplomatic Establishment was neither as socially exclusive as was presupposed, nor more aristocratic than the general run of the political élite, the Higher Civil Service and Great Powers services. Jones, *The British Diplomatic Service*, pp. 139–46; Richard A. Chapman, *The Higher Civil Service in Britain* (London, 1970), pp. 40–2.
111. D. C. Watt, *Personalities and Politics: Studies in the Formulation of British Foreign Policy in the Twentieth Century* (South Bend, 1965), pp. 187–8; Jones, *The British Diplomatic Service*, pp. 140–1, 144.
112. FO 366/780, 1AI(b), Baldwin to Cecil, 20 February 1918.
113. Ibid., 5AI(b), Draft Report on Mr Baldwin's letter, 16 March 1918; ibid., 39AI(b), Minute by Tilley, 13 August 1918.
114. Chapman and Greenaway, *Dynamics*, p. 39.
115. FO 366/780, 1AI(b), Baldwin (Treasury) to Cecil, 20 February 1918.
116. Ibid., 20AI(b), Memorandum by Cecil, 16 April 1918; ibid., 26AI(b), Cecil to Baldwin (Treasury), 29 April 1918.
117. Ibid., 21AI(b), Baldwin (Treasury) to Cecil, 18 April 1918.
118. Ibid., 33AI(b), Russell to Tilley, 14 June 1918.
119. Ibid., 39AI(b), Minute by Tilley, 13 August 1918.
120. Ibid., 10AI(B), Memorandum believed to be by Sir Eyre Crowe in connection with the Royal Commission (undated).
121. Ibid., 39AI(b), Minute by Drummond, 13 August 1918.
122. PP 1917–18 (Cmd 8657), vol. 8, Report of the Committee on the Scheme of Examination for Class I of the Civil Service; PP 1919 (Cmd 36), vol. 11, Third Interim Report of the Gladstone Committee on Recruitment to the Civil Service after the War; Fry, *Statesmen in Disguise*, pp. 75–83, 91–2, 95–6.
123. FO 306/780, 36AI(b), Report of the Committee on Conditions of Admission into the Foreign Office and Diplomatic Service, 1 July 1918; Steiner and Dockrill, 'The Foreign Office Reforms', p. 140.

124. FO 306/780, 41AI(b), Minute by Hardinge, 4 September 1918.
125. Ibid., 43AI(b), Minute by Crowe, 11 September 1918.
126. Ibid., 45AI(b), Minute by Tilley, September 1918 (undated).
127. Ibid., 48AI(b), Hardinge to Cecil, 11 October 1918.
128. Ibid., Cecil to Hardinge, 16 October 1918.
129. Ibid., 57AI(b), Hardinge to Heath, 28 December 1918.
130. Ibid., 1AI(d), Foreign Office Memorandum: 'The Foreign Service', January 1919 (undated).
131. Ibid., 58AI(b), Heath to Hardinge, 31 December 1918.
132. Ibid., 2AI(d), Cecil to Bonar Law (Treasury), 4 January 1919, quoted in Larner, 'The Amalgamation of the Diplomatic Service with the Foreign Office', pp. 123–4.
133. FO 306/780, 12AI(d), Draft letter, Curzon to Bonar Law (Treasury), 3 April 1919.
134. Ibid., 27AI(d), Graham to the Secretary to the Treasury, 23 June 1919; T.1, 12344/27573, Sir Malcolm Ramsay (Treasury) to Curzon, 30 June 1919.
135. FO 366/781, 6AI(f), Minute by Akers-Douglas, 6 November 1919.
136. FO 366/789, X1495/990, Hardinge to Curzon, 6 August 1920; Steiner and Dockrill, 'The Foreign Office Reforms', pp. 142–3.
137. FO 366/789 X1495/990, Note by the amalgamation and promotion sub-committee, 4 August 1920.
138. Ibid., Tilley to Hardinge, 29 July 1920, quoted in part in Larner, 'The Amalgamation of the Diplomatic Service with the Foreign Office', pp. 125–6.
139. FO 366/789, Hardinge to Curzon, 6 August 1920.
140. FO 366/781, 5AI(f), Curzon to Hardinge, 13 August 1920.
141. Hill, *Sociology of Public Administration*, pp. 90, 95–104. For further information on this point see Crowe and Corp, *Our Ablest Public Servant: Sir Eyre Crowe, 1864–1925*, pp. 392–7.
142. FO 366/832, X7026/7026/505, Minute by Montgomery, 26 October 1925.
143. FO 366/816, X4995/1179/504, Regulations for His Majesty's Diplomatic Service, 1 July 1924.
144. Ibid.
145. Sir Douglas Busk, *The Craft of Diplomacy: How to Run a Diplomatic Service* (London, 1967), pp. 187–9.
146. Sir Harold Nicolson, *Diplomacy* (London, 1969), p. 115; see also PP 1942–3 (Cmd 6420), vol. 11, Proposals for the Reform of the Foreign Service, para. 14. Printed in Lord Strang, *The Foreign Office* (London, 1955), p. 218.
147. Larner, 'The Amalgamation of the Diplomatic Service with the Foreign Office', pp. 115–16.
148. Cmd 7748, QQ 40,788, 40,791, 40,792; Steiner, *The Foreign Office*, p. 169.
149. Curzon MSS, Eur, F 112/213(a), Ian Malcolm to Curzon, 2, 8, 27 July 1919; Steiner and Dockrill, 'The Foreign Office Reforms', p. 142.
150. Ibid., 'The Foreign Office Reforms', p. 145.
151. Cmd 7449, Appendix 84, pp. 306–7, printed in Steiner, *The Foreign Office*, pp. 217–21.
152. In point of fact normal open competitions were not resumed after the First World War until 1925 when the new form of Civil Service examination was introduced. However, even when calculating for the years 1925–1929 alone, one still gets the same general picture, for the 1919–1924 figures did not differ materially from those of 1925–1929.
153. FO 366/787, 145545/1444a, Crowe to the Secretary to the Treasury, 26 August 1918.
154. Ibid., 189672/1923, Report of the Inter-Departmental Committee on the Re-organization of Foreign Office Registries, 14 November 1918.

155. Ibid., 145545/1444a, Crowe to the Secretary to the Treasury, 26 August 1918.
156. Ibid., 135088/1310, Report by the Sub-committee of Staffs on the FO and Ministry of Blockade, 29 July 1918.
157. Ibid., Minute by Crowe, 23 August 1918, quoted in Steiner and Dockrill, 'The Foreign Office Reforms', p. 147.
158. Ibid., Tilley to Crowe, 21 August 1918.
159. Ibid., 145545/1444a, Minute by Crowe, 23 August 1918.
160. Ibid. Hardinge to Bradbury, 24 August 1918.
161. Ibid., 135088/1310, Tilley to Crowe, 21 August 1918.
162. Ibid., 189672/1923, Report of the Inter-Departmental Committee on the Re-organization of Foreign Office Registries, 14 November 1918.
163. Public Record Office Handbooks, *The Records of the Foreign Office*, p. 67; Steiner and Dockrill, 'The Foreign Office Reforms', p. 148.
164. FO 366/789, 169809/CC, R. S. Meiklejohn (Treasury) to the Under-Secretary of State, 12 January 1920.
165. Ibid.
166. FO 366/796, X1616/266/505, M. F. Headlam to Montgomery, 21 February 1921.
167. Public Record Office Handbooks, *The Records of the Foreign Office*, pp. 68–70.
168. FO 366/847, X7851/4914/50, Minute by Leslie R. Sherwood, 11 January 1928.
169. FO 366/783, A111(f) 4, Memorandum by W. L. Dunlop, 7 June 1923; FO 366/806, X109/109/504, Memorandum on Foreign Office Shorthand Writers submitted by Florence Westwood, 30 October 1923; ibid., Memorandum by Miss M. Bell, 11 October 1923 – Crowe and Montgomery kept pressing the Treasury for promotions and pay rises for the clerical staff. But, with the exception of the raised salary of one Senior Assistant in 1920, and the creation of a number of new posts abroad, i.e. archivists, in 1923, they made very little headway. Indeed in 1922, following the report of the Cabinet Committee on National Expenditure, Crowe was forced to accept a reduction of 27 posts in the Registry. See FO 366/801, X9005/8565/504, Montgomery to the Secretary to the Treasury, 9 November 1922; ibid., X11679/8565/504, J. H. Craig to Montgomery, 15 December 1922; FO 366/807, X9297/1302/504, R. R. Scott (Treasury) to the Permanent Under-Secretary of State, 28 November 1923.
170. H. Nicolson, *Curzon: The Last Phase, 1919–1925* (New York, 1925), p. 74.
171. FO 371/7586, C14671/13/19, Minute by Curzon, 25 October 1922.
172. CAB 24/67, G.T. 6007, 'The Function of the Ministry of Information on the cessation of hostilities', by Lord Beaverbrook, 16 October 1918; Sir C. Stuart, *Secrets of Crewe House: The Story of a Famous Campaign* (London, 1920), p. 202; P. M. Taylor, *The Projection of Britain*, pp. 45–8.
173. FO 395/297, Tyrrell to Harry Batterbee (Colonial Office), 30 December 1918.
174. M. L. Sanders and P. M. Taylor, *British Propaganda during the First World War*, pp. 27–9, 47–50, 58–9, 61, 88, 252; Philip M. Taylor, 'The Foreign Office and British Propaganda during the First World War', *Historical Journal*, vol. 23 (1980), pp. 879, 882–7.
175. FO 395/301, Minute by Tilley, 31 January 1919, quoted in P. M. Taylor, *The Projection of Britain*, p. 49.
176. Ibid., Memorandum by S. A. Guest, 24 January 1919, quoted at length both in P. M. Taylor, *The Projection of Britain*, pp. 49–50 and in P. M.

Taylor, 'British Official Attitude towards Propaganda Abroad, 1918–1939', in Nicholas Pronay and D. W. Spring (eds), *Propaganda, Politics and Film. 1918–1945* (London, 1982), pp. 29–30.

177. Ibid., 'Policy and Propaganda', by George B. Beak, 2 December 1918.
178. FO 395/297, 'The Reconstruction of the Foreign Office', by Victor Wellesley, 30 November 1918.
179. FO 395/304, Minute by Cecil Harmsworth, 6 February 1919.
180. FO 395/297, Note of proceeding at a meeting held on 20 March 1919 to consider the future of British propaganda abroad; P. M. Taylor, *The Projection of Britain*, p. 51.
181. The Press Attachés were to be appointed to the more important missions abroad and maintain general liaison with the press of the countries in which they served.
182. FO 366/787, Chief Clerk 53464/682, 'Proposed News Department of the FO', by Curzon, 26 March 1919.
183. FO 366/790, X2800/2800/504, 'Activities of the News Department', by P. A. Koppel, June 1920 (undated); FO 366/805, X2397/226/505, Third Report of Committee on National Expenditure, Part 4, Annex A, Note by Sir W. Tyrrell, 27 February 1922.
184. FO 366/787, Chief Clerk 53464/682, Proposed News Department of the FO', by Curzon, 26 March 1919.
185. Ibid., Chief Clerk 32759/682.
186. Ibid.
187. Ibid., Sir Ronald Graham to the Treasury, 2 April 1919.
188. FO 395/297, Note of proceedings of a conference held at the Treasury, 14 May 1919, quoted in P. M. Taylor, *The Projection of Britain*, pp. 52–3.
189. FO 366/787, Chief Clerk 82638/1068, Sir. T. L. Heath (Treasury) to the Under-Secretary of State, Foreign Office, 31 May 1919.
190. FO 366/790, X2800/2800/504, Montgomery to the Secretary to the Treasury, 16 November 1920.
191. Sir Arthur Willert, *Washington and Other Memories*, p. 162.
192. FO 366/821, X8039, Minute by Wellesley, 2 November 1925; P. M. Taylor, *The Projection of Britain*, pp. 52–7.
193. Public Record Office Handbooks, *The Records of the Foreign Office*, pp. 22, 23.
194. FO 366/788, 49660/634, Montgomery to the Secretary to the Treasury, 11 August 1919.
195. Ibid., Minute by Warner, 14 May 1919.
196. Ibid., 40371/520, Minute by Tilley, 26 February 1919; ibid., 49660/634, Minute by Tilley, 14 May 1919.
197. Ibid., 40371/520, Minute by Tilley, 26 February 1919.
198. Ibid., 49660/634, Report of the Committee on the Amalgamation of King's Messengers and Cypher Depts, March 1919.
199. Ibid., Heath to the Under-Secretary of State at the Foreign Office, 19 August 1919.
200. Ibid., 118168/1475, The King's Messengers and Communications Department (undated).
201. Foreign Office List, 1920.
202. T.1 12460/1459, Montgomery to the Secretary to the Treasury, 9 January 1920.
203. FO 366/806, X426/270/504, FO Memorandum (undated), sent by Montgomery to G. G. Barnes, 26 May 1923. For further information, see Christopher Andrew, *Secret Service*, pp. 243–5.

204. FO 366/790, X2467/224/504, Memorandum by L. W. Stafford, 30 September 1920; ibid., 3909, Minutes of evidence, statement submitted by the Permanent Under-Secretary of State for Foreign Affairs, pp. 851–2.
205. Ibid.; FO 366/791, X4221/1116/505, Code of Instructions to Passport Control Officers, 1 December 1920.
206. See Appendix 2.
207. Public Record Office Handbooks, *The Records of the Foreign Office*, pp. 15, 17, 21.
208. FO 366/850, X2272/1840/505, Memorandum by Sir F. G. A. Butler, 1 April 1927.
209. In 1926 this number consisted of: Ambassadors – 10; Consular Officers etc. performing diplomatic duties – 11; Ministers – 32; Counsellors – 13; Clerical staff – 181; First Secretaries – 29; Local Subordinate Staff – 209; Second Third Secretaries – 37. FO 366/809, X3530/3530/503, Minute by S. K. Millar, 19 April 1923; FO 366/825, X2347/84/504, 'Diplomatic Service Position in March 1925', by Montgomery, 23 March 1925; FO 366/830, X6518/304/505, Estimates, Civil Services, 1926: Class II, 5, Diplomatic Service; FO 366/837, X3899/541/504, W. H. Robinson to E. E. Bridges, 28 June 1926; FO 366/850, X2272/1840/550, Memorandum by Butler, 1 April 1927.
210. CAB 24/5, G423, Scheme for the Reform and Development of the Consular and Commercial Diplomatic Services, 8 May 1919; FO 366/837, X3899/541/504, Robinson to Bridges, 28 June 1926.

Chapter 2
The Division of Functions between the Foreign Secretary and his Assistants

1. Charles Lindblom, *The Policy-Making Process* (Englewood Cliffs, N.J., 1968), pp. 13–17.
2. James Rosenau, 'The Study of Foreign Policy', in James N. Rosenau, Kenneth W. Thompson and Gavin Boyd (eds), *World Politics* (New York, 1976), p. 30.
3. Winston Churchill, *Great Contemporaries* (Chicago, 1973), p. 274.
4. John D. Gregory, *On the Edge of Diplomacy: Rambles and Reflections, 1902–28* (London, 1928), p. 250.
5. H. Nicolson, *Curzon: The Last Phase, 1919–1925* (New York, 1925), p. 193.
6. Kenneth Rose, *Superior Person: A Portrait of Curzon and his Circle in Late Victorian England* (London, 1969), p. 1.
7. Earl of Ronaldshay, *The Life of Lord Curzon* (London, 1928), vol. 1, pp. 26, 37; Kenneth Rose, *Curzon A Most Superior Person: a Biography* (London, 1985), pp. 25–6.
8. *Dictionary of National Biography, 1922–1930* (Oxford, 1961), pp. 228–230.
9. Lord Riddell, *Lord Riddell's War Diary, 1914–1918* (London, 1933), p. 229.
10. Bruce Heady, *British Cabinet Ministers: The Roles of Politicians in Executive Office* (London, 1974), pp. 69–72, 205.
11. H. Nicolson, *Curzon*, p. 193.
12. FO 366/781, 5AI(b), Curzon to Hardinge, 13 August; FO 371/7453, C16000/6/18, Minute by Curzon, 24 November.
13. The Earl of Swinton, *I Remember* (London, 1948), p. 76.

14. Trevor Wilson (ed.), *The Political Diaries of P. C. Scott, 1911–1928* (London, 1970), pp. 263–4.
15. The Marchioness Curzon of Kedleston, *Reminiscences* (London, 1955), p. 243.
16. Ronaldshay, *Curzon*, vol. 3, p. 384.
17. Sir Owen O'Malley, *The Phantom Caravan* (London, 1954), pp. 58–9.
18. H. Nicolson, *Curzon*, pp. 18–19, 89.
19. Ibid., pp. 40, 125, 128–9.
20. Sir Andrew Ryan, *The Last of the Dragomans* (London, 1951), p. 176.
21. FO 800/259, The Marquess of Crewe to Austen Chamberlain, 14 May 1926.
22. Ronaldshay, *Curzon*, vol. 3, p. 149.
23. Hardinge MSS, vol. 44, Curzon to Hardinge, 5 June 1921.
24. Ronaldshay, *Curzon*, vol. 2, p. 178; ibid., vol. 3, pp. 85, 149–150.
25. Curzon MSS, Eur F/112/319, 'Confidential Memo, written by me at Lausanne Dec. 1922–Jan. 1923 concerning the fall of the Lloyd George Govt. and other cognate matters. For the use of my biographer'.
26. H. Nicolson, *Curzon*, p. 20.
27. William Wallace, *The Foreign Policy Process in Britain* (London, 1977), pp. 51, 69.
28. FO 371/8302, W 8858/8858/50, Minute by Crowe, 21 October 1922.
29. FO 371/8184, N9662/573/38, Minute by Curzon, October 1922 (undated).
30. FO 371/7862, E4371/5/44, Minute by Curzon, 2 May 1922; FO 371/7863, E4466/5/44, Minute by Curzon, 4 May 1922.
31. FO 371/7877, E5158/19/44, Minute by Curzon, 22 May 1922.
32. Gregory, *On the Edge of Diplomacy*, pp. 247–8, 251.
33. Sir Arthur Willert, *Washington and Other Memories* (Boston, 1972), p. 165.
34. Sir David Kelly, *The Ruling Few: or the Human Background to Diplomacy* (London, 1952), p. 140.
35. Gregory, *On the Edge of Diplomacy*, p. 251.
36. Lord Vansittart, *The Mist Procession* (London, 1958), p. 262.
37. FO 371/5255, E5992/4870/44, 'Future Administration of the Middle East', Memorandum by the Secretary of State for Foreign Affairs, 8 June 1920; Lord Curzon of Kedleston, *Frontier* (Westport, 1976), pp. 9, 39–47, 56–7; H. Nicolson, *Curzon*, p. 13.
38. CAB 27/24, EC 42nd Minutes, Minutes of Meeting of the Eastern Committee, 9 December 1918.
39. John Darwin, *Britain, Egypt and the Middle East: Imperial Policy in the Aftermath of War, 1918–1922* (London, 1981), pp. 150–1, 159–60, 184–5, 204.
40. CAB 23/21, 30(20)(3), Conclusions of a Meeting of the Cabinet, 21 May 1920; CAB 23/22, 49(20), Appendix 1(1), Conclusions of a Meeting of the Finance Committee, 12 August 1920.
41. CAB 27/244, EC 41st Minutes, Minutes of a Meeting of the Eastern Committee, 5 December 1918.
42. H. Nicolson, *Curzon*, pp. 136–7.
43. Wm. Roger Louis, *British Strategy in the Far East, 1919–1939* (Oxford, 1971), p. 81.
44. CAB 2/3, Committee of Imperial Defence, Minutes of the 134th Meeting, 14 December 1920.
45. CAB 23/25, 43(21), Conclusions of a Meeting of the Cabinet, 30 May 1921.
46. CAB 27/24, EC 46th Minutes, Minutes of a Meeting of the Eastern Committee, 23 December 1918.
47. FO 371/4179, 46887/2863, Memorandum by Earl Curzon, 'A Note of Warning about the Middle East', 25 March 1919.

48. CAB 32/2, Pt 1, E. 4th Meeting, Stenographic Notes of a Meeting of Representatives of the United Kingdom, The Dominions and India (hereafter cited as Imperial Conference Minutes), 22 June 1921.
49. *DBFP*, 1st Series, vol. xvi, no. 768, p. 862, Memorandum by the Marquess Curzon of Kedleston on the question of an Anglo-French Alliance, 28 December 1921.
50. CAB 27/24, EC 40th Minutes, Minutes of a Meeting of the Eastern Committee, 2 December 1918.
51. H. Nicolson, *Curzon*, pp. 81, 88–9.
52. CAB 27/24, EC 40th Minutes, Minutes of a Meeting of the Eastern Committee, 2 December 1918.
53. CAB 32/3, Pt 1, E. 4th Meeting, Imperial Conference Minutes, 22 June 1921.
54. CAB 23/21, 38(20)(1), Conclusions of a Meeting of the Cabinet, 30 June 1920; Stephen Roskill, *Hankey, Man of Secrets* (London, 1972), vol. 2, pp. 209–10.
55. H. Nicolson, *Curzon*, p. 193.
56. CAB 32/2, Pt 1, E. 4th Meeting, Imperial Conference Minutes, 22 June 1921.
57. Brown, p. 140.
58. Lord Strang, *The Foreign Office* (London, 1957), p. 148.
59. Brown, p. 163.
60. Charles á Court Repington, *The First World War, 1914–1918* (London, 1920), vol. 2, p. 463.
61. CAB 29/1, W.C. 64, Suggested Basis for a Territorial Settlement in Europe, 7 August 1916, quoted for the major part in David Lloyd George, *Memoirs of the Peace Conference* (New York, 1972), vol. 1, pp. 11–23.
62. Hardinge MSS, vol. 44, Hardinge to Curzon, 28 April 1921, quoted in J. Douglas Goold, 'Lord Hardinge as Ambassador to France and the Anglo-French Dilemma over Germany and the Near East, 1920–1922', *Historical Journal*, vol. 21 (1978), p. 935.
63. FO 371/7000, W13420/12716/17, Notes respecting the possible conclusion of an Anglo-French Alliance by Crowe, 26 December 1921.
64. FO 371/6880, N13814/105/38, Minutes by Tyrrell, 19 November 1921.
65. FO 371/9802, C7897/737/18, Memorandum by J. M. Troutbeck, 15 May 1924.
66. FO 371/7000, W13420/12716/17, Notes respecting the possible conclusion of an Anglo-French Alliance by Crowe, 26 December 1921.
67. FO 371/8703, C848/313/18, Extract from an article by Hardinge in the *Sunday Pictorial*, 14 January 1923.
68. FO 371/7488, C16643/99/18, Central Department Memorandum on the Reparation Position, 5 December 1922.
69. Lord Strang, *Home and Abroad* (London, 1956), p. 269.
70. Christina Larner, 'The Organization and Structure of the Foreign and Commonwealth Office', in Robert Boardman and A. J. R. Groom (eds), *The Management of Britain's External Relations* (London, 1973), p. 58.
71. Strang, *Home and Abroad*, pp. 269–81.
72. D. Lloyd George, *Memoirs of the Peace Conference*, vol. 1, p. 132.
73. Zara S. Steiner, *Britain and the Origins of the First World War* (London, 1977), p. 181.
74. *The Foreign Office List*, 1923, pp. 234–5.
75. Lord Hardinge of Penshurst, *Old Diplomacy* (London, 1947), p. 250.
76. Sir Frank Oppenheimer, *Stranger Within* (London, 1960), p. 305.
77. Hardinge MSS, vol. 42, Hardinge to Harcourt Butler, 9 April 1920.

78. Ibid., vol. 43, Hardinge to Horace Rumbold, 13 July 1920.
79. R. Warman, *The Foreign Office, 1916–1918: a Study of its Role and Functions* (New York, 1986), p. 157.
80. Hardinge MSS, vol. 42, Hardinge to the Queen, 30 March 1920.
81. Ibid., Hardinge to Harcourt Butler, 9 April 1920.
82. Hardinge, *Old Diplomacy*, pp. 216, 243–4.
83. Vansittart, *Mist Procession*, p. 233.
84. FO 371/3914, 195387/40430/55, Minute by Hardinge, May 1920 (undated).
85. Hardinge MSS, vol. 42, Hardinge to Graham, 10 April 1920.
86. Zara S. Steiner, 'The Foreign Office and the War', in F. H. Hinsley (ed.), *British Foreign Policy under Sir Edward Grey* (Cambridge, 1977), p. 425.
87. Hardinge, *Old Diplomacy*, p. 243.
88. Hardinge MSS, vol. 22, Hardinge to Graham, June 1916 (undated).
89. Hardinge, *Old Diplomacy*, p. 226.
90. CAB 29/2, 'Europe', Foreign Office Memorandum (probably December 1918).
91. Hardinge, *Old Diplomacy*, p. 240.
92. Kenneth Calder, *Britain and the Origins of the New Europe, 1914–1918* (Cambridge, 1966), p. 124.
93. Hardinge, *Old Diplomacy*, pp. 240, 260.
94. Hardinge MSS, vol. 42, Hardinge to Harcourt Butler, 22 June 1920.
95. Hardinge, *Old Diplomacy*, pp. 238, 256–7.
96. Harold Nelson, *Land and Power* (London, 1963), pp. 6, 9, 96.
97. Lloyd George MSS, F/53/1/63, Hardinge to Lloyd George, 22 June 1921; FO 37/4725, C1079/2/18, Minute by Hardinge, July 1920 (undated).
98. Hardinge MSS, vol. 42, Hardinge to Graham, 10 April 1920.
99. CAB 32/2, Pt 2, E. 18th Meeting, Imperial Conference Minutes, 7 July 1921.
100. Hardinge, *Old Diplomacy*, p. 264.
101. FO 371/7470, C812/38/18, Minute by Curzon, 8 January 1922.
102. Goold, 'Lord Hardinge as Ambassador', pp. 915–16.
103. *Documents on British Foreign Policy, 1919–1939* (hereafter *DBFP*), 1st Series, vol. x, no. 174, p. 269, Note on the German Situation by Mr Waterlow, 5 July 1920; ibid., Minute by Hardinge, July 1920 (undated).
104. Hardinge MSS, vol. 42, Hardinge to Harcourt Butler, 22 June 1920.
105. Lloyd George MSS, F/53/1/63, Hardinge to Lloyd George, 22 June 1921.
106. FO 371/3781, 187976/4232/18, Minute by Hardinge, March 1920 (undated).
107. Hardinge MSS, vol. 44, Hardinge to Curzon, 25 and 28 November 1921; ibid., vol. 45, Hardinge to Curzon, 23 February 1922, quoted in Goold, 'Lord Hardinge as Ambassador', p. 921 and in C. J. Lowe and M. L. Dockrill, *The Mirage of Power: British Foreign Policy, 1914–22* (London, 1972), vol. 2, p. 352.
108. CAB 32/2, Pt 2, E. 18th Meeting, Imperial Conference Minutes, 7 July 1921.
109. *DBFP*, 1st Series, vol. xiii, no. 433, p. 487–8, Minute by Hardinge, 20 May 1920.
110. FO 371/4215, 132926/50535/44, Minute by Hardinge, September 1919 (undated).
111. Hardinge MSS, vol. 42, Hardinge to Sir Thomas Hohler, 27 January 1920.
112. Strang, *Home and Abroad*, pp. 284–5.
113. Sir Harold Nicolson, *Sir Arthur Nicolson, Bart, First Lord Carnock: A Study in the Old Diplomacy* (London, 1930), p. 328.
114. Strang, *Home and Abroad*, p. 273. For a very different view on this point see

Sybil Crowe and Edward Corp, *Our Ablest Public Servant: Sir Eyre Crowe, 1864–1925.*

115. Willert, *Washington and Other Memories*, p. 163.
116. Sir Lewis Namier, *Avenues of History* (London, 1952), p. 87.
117. Willert, *Washington and Other Memories*, p. 164.
118. Private information, provided by the late Miss Sibyl Crowe of St Hilda's College, Oxford, Crowe's daughter.
119. Willert, *Washington and Other Memories*, p. 163.
120. For further information on this point see Crowe and Corp, *Our Ablest Public Servant: Sir Eyre Crowe, 1864–1925.* This book provides essential information which was not available when the author wrote this chapter.
121. Zara S. Steiner, *The Foreign Office and Foreign Policy 1898–1914*, pp. 109–11; *Dictionary of National Biography, 1922–1930*, p. 219.
122. Steiner, *Britain and the Origins of the First World War*, p. 184. See also K. M. Wilson, 'Sir Eyre Crowe on the Origin of the Crowe Memorandum of 1 January 1907', *Institute of Historical Research Bulletin*, 1987, vol. 56, no. 134, pp. 240–42. See Crowe and Corp, *Our Ablest Public Servant: Sir Eyre Crowe, 1864–1925*, pp. 110–35.
123. Richard A. Cosgrove, 'The Career of Sir Eyre Crowe: A Reassessment', *Albion*, vol. 4 (1972), p. 195.
124. Hardinge MSS, vol. 93, Pt 2, Hardinge to Sir Arthur Nicolson, 8 June 1913.
125. Steiner, 'The Foreign Office and the War', in Hinsley (ed.), *British Foreign Policy*, p. 518; Edward Corp, 'Sir Eyre Crowe and the Administration of the Foreign Office, 1906–1914', *Historical Journal*, vol. 22 (1979), pp. 452–3. See Crowe and Corp, *Our Ablest Public Servant: Sir Eyre Crowe, 1864–1925*, 276–8.
126. Cosgrove, 'Career of Sir Eyre Crowe', pp. 200–1; H. Nicolson, *Curzon*, p. 428.
127. Viscount Cecil of Chelwood, *All the Way* (London, 1949), p. 138.
128. Vansittart, *Mist Procession*, p. 210.
129. FO 7944/2, Derby to Curzon, 8 December 1919.
130. Ibid., Norman to R. H. Campbell, 18 December 1919.
131. H. Nicolson, *Peacemaking, 1919* (London, 1938), pp. 173, 174, quoted in Steiner, *The Foreign Office*, p. 110.
132. Curzon MSS, Eur F/112/319, 'Confidential Memo, written by me at Lausanne Dec. 1922–Jan. 1923'.
133. Roskill, *Hankey*, vol. 2, pp. 22, 138–9, 182–3.
134. Curzon MSS, Eur F/112/319, 'Confidential Memo, written by me at Lausanne Dec. 1922–Jan. 1923'.
135. FO 800/253, Ty/21/1, Lancelot Oliphant to Horace Rumbold, 10 January 1921.
136. FO 371/10952, F 938/464/10, Minute by Crowe, 16 March 1925; Sir Ivone Kirkpatrick, *The Inner Circle: Memoirs of Ivone Kirkpatrick* (London, 1959), p. 32.
137. Willert, *Washington and Other Memories*, p. 164.
138. Kirkpatrick, *The Inner Circle*, p. 32.
139. Nigel Nicolson (ed.), Harold Nicolson, *Diaries and Letters* (London, 1966), vol. 1, p. 49.
140. Gregory, *On the Edge*, p. 260.
141. O'Malley, *Phantom Caravan*, p. 47, quoted with variants, in Cosgrove, 'Career of Sir Eyre Crowe', p. 202 and in Steiner, *The Foreign Office*, p. 117.
142. Sir Maurice Peterson, *Both Sides of the Curtain, Autobiography* (London, 1950), p. 50.

143. Nourah Waterhouse, *Private and Official* (London, 1942), p. 198.
144. Strang, *Home and Abroad*, p. 308.
145. FO 371/8617, C15704/15065/62, Minute by Curzon, 11 September 1923; John Vincent (ed.), *The Crawford Papers; The Journals of David Lindsay Twenty-seventh Earl of Crawford and Tenth Earl of Balcarres, 1871–1940; during the years 1892 to 1940* (Manchester, 1984), p. 422.
146. Kelly, *The Ruling Few*, p. 141.
147. O'Malley, *Phantom Caravan*, pp. 59–60.
148. Ibid., p. 60; Kirkpatrick, *The Inner Circle*, p. 33.
149. FO 371/700, W13420/12716/17, Notes respecting the possible conclusion of an Anglo-French Alliance by Crowe, 26 December 1921.
150. FO 371/4757, C11056/113/18, Minute by Crowe, 13 November 1920.
151. FO 371/7000, W13420/12716/17, Notes respecting the possible conclusion of an Anglo-French Alliance by Crowe, 26 December 1921.
152. *DBFP*, 1st Series, vol. xvi, no. 747, p. 828, Minute by Crowe, 30 November 1921.
153. FO 371/7000, W13420/12716/17, Notes respecting the possible conclusion of an Anglo-French Alliance by Crowe, 26 December 1921.
154. B.D., vol. 3, no. 445, App. A, p. 403, Memorandum on the Present State of British Relations with France and Germany by Mr. Eyre Crowe, 1 January 1907; Nelson, *Land and Power*, p. 95.
155. *DBFP*, 1st Series, vol. xvi, no. 747, p. 828, Minute by Crowe, 30 November 1921.
156. Ibid., vol. ix, no. 79, p. 121, Minutes by Crowe, 9 March 1920.
157. Ibid., vol. xvi, no. 747, p. 828, Minute by Crowe, 30 November 1921.
158. Nelson, *Land and Power*, pp. 24–5; Michael L. Dockrill and J. Douglas Goold, *Peace without Promise: Britain and the Peace Conferences, 1919–23* (London, 1981), pp. 108, 100.
159. Agnes Headlam-Morley, Russell Bryant and Anna Cienciala (eds), *Sir James Headlam-Morley, A Memoir of the Paris Peace Conference, 1919* (London, 1972), p. 187.
160. FO 371/7000, W13420/12716/17, Notes respecting the possible conclusion of an Anglo-French Alliance by Crowe, 26 December 1921.
161. FO 371/4757, C11056/113/18, Minute by Crowe, 13 November 1920.
162. Steiner, *The Foreign Office*, p. 114.
163. FO 371/6022, C6958/240/18, Minute by Crowe, 7 April 1921.
164. FO 371/9813, C3814/1346/18, Minute by Crowe, 13 March 1924.
165. FO 395/337, P1632/1632/150, Minute by Crowe, 28 December 1920.
166. *DBFP*, 1st Series, vol. ix, no. 79, p. 121, Minute by Crowe, 9 March 1920.
167. FO 371/3785, 196289/4232/18, Minute by Crowe, 7 May 1920; FO 371/4741, C7032/45/18, Minute by Crowe, 26 September 1920.
168. FO 371/3798, 185560/182853/18, Minute by Crowe, 10 March 1920.
169. FO 371/7518, C15069/333/18, Minute by Crowe, 3 November 1922.
170. FO 371/4741, C17032/45/18, The German Situation from a Military Aspect: Memorandum by the General Staff, 16 August 1920.
171. FO 371/3786, 202220/4232/18, Minute by Crowe, 7 June 1920.
172. FO 371/7518, C15069/333/18, Minute by Crowe, 3 November 1922.
173. H. Nicolson, *Peacemaking*, pp. 206, 278.
174. Steiner, *Britain and the Origins of the First World War*, p. 185. For a different view see Edward I. Corp, 'Sir William Tyrrell: The éminence grise of the British Foreign Office, 1912–1915', *Historical Journal*, vol. 25 (1982), pp. 707–8.
175. Willert, *Washington and Other Memories*, p. 164.

176. Viscount Grey of Falladon, *Twenty-Five Years* (London, 1925), vol. 1, p. xviii.
177. Kelly, *The Ruling Few*, p. 141; Thomas Jones, *A Diary with Letters, 1931–1950* (London, 1954), p. 123; Willert, *Washington and Other Memories*, pp. 142, 148, 150; Strachey MSS, S/19/14/11, Tyrrell to Strachey, 17 March 1922.
178. Steiner, *The Foreign Office*, p. 118.
179. Cynthia Gladwyn, *The Paris Embassy* (London, 1976), p. 206.
180. Tyrrell had a history of nervous collapse and was liable to turn to drink as a means of escape from the drudgery of departmental routine. Corp, 'Sir William Tyrrell', pp. 698, 706.
181. *Dictionary of National Biography, 1941–1950* (Oxford, 1959), p. 894.
182. FO 371/6915, N7647/5/38, Curzon to Tyrrell, 29 June 1921.
183. *DBFP*, 1st Series, vol. xiv, no. 212, pp. 221–7, Report of the Anglo-Japanese Committee, 21 January 1921.
184. FO 371/5618, A6719/18/45, Minute by Tyrrell, 19 September 1921.
185. FO 371/6464, E1207/1/44, Lord Hardinge (Paris) to Sir W. Tyrrell, 25 January 1921.
186. Sir Lewis Namier, *Avenues of History* (London, 1952), p. 87.
187. Keith Middlemas (ed.), *Thomas Jones, Whitehall Diary* (London, 1969), vol. 1, p. 243.
188. *Dictionary of National Biography, 1941–1950*, pp. 893–5; Steiner, *The Foreign Office*, p. 119.
189. FO 371/6880, N13814/105/38, Minute by Tyrrell, 19 November 1921.
190. CAB 29/1, W.C. 64, Suggested Basis for a Territorial Settlement in Europe, 7 August 1916. Quoted for the most part in D. Lloyd George, *Memoirs of the Peace Conference*, vol. 1, pp. 11–19.
191. FO 371/7486, C14819/99/18, Minute by Tyrrell, 26 October 1922.
192. Phipps MSS, 2/12, Tyrrell to Sir Eric Phipps (Paris), 18 December 1923.
193. FO 371/5843, C3340/3340/62, Memorandum by Crowe, 12 February 1921.
194. FO 371/7521, C11429/336/18, Minute by Tyrrell, 10 August 1922.
195. FO 371/10734, C8209/459/18, Minute by Tyrrell, 19 June 1925.
196. FO 371/7520, C5656/335/18, Minute by Tyrrell, 21 April 1922.
197. FO 371/10737, C9886/459/18, Minute by Tyrrell, 30 July 1925.
198. Hardinge, *Old Diplomacy*, p. 252.
199. Hardinge MSS, vol. 44, Curzon to Hardinge, 5, 9 December 1920.
200. Ibid., Hardinge (Paris) to Curzon, 6 December 1920; Hardinge, *Old Diplomacy*, p. 252.
201. FO 371/6515, E6411/143/44, Minute by Lindsay, 6 June 1921.
202. Brown, p. 140.
203. FO 371/47470, C812/38/18, Minute by Curzon, 18 January 1922.
204. FO 371/5035, E4698/2/44, Minute by Curzon, 16 May 1920.
205. See also, Strang, *The Foreign Office*, pp. 155–6.
206. Wallace, *The Foreign Policy Process*, p. 69.
207. Kelly, *The Ruling Few*, pp. 139–40.
208. Strang, *Home and Abroad*, pp. 284–5.
209. FO 800/253, Ty/21/1, Lancelot Oliphant to Horace Rumbold (Constantinople), 10 January 1921.
210. FO 371/4757, C11506/113/18, Minute by Crowe, 13 November 1920.
211. *DBFP*, 1st Series, vol. x, no. 174, pp. 268–9, Note on the German Situation by Mr Waterlow, 5 July 1920.
212. FO 371/9813, C5185/1346/18, General Staff Memorandum on the Military Aspects of the Future Status of the Rhineland, 28 March 1924.
213. Ibid., Minute by Bennett, 14 April 1924.
214. Ibid., Minute by Lampson, 16 April 1924.

215. FO 371/9819, C15288/2048/18, War Office to the Secretary of the C.I.D., 29 September 1924: enclosure – General Staff Memorandum on M. Herriot's Letter to Mr MacDonald, 11 August 1924.
216. Ibid., Minute by Nicolson, 15 October 1924.
217. Ibid., C13663/2048/18, Minute by Nicolson, 27 August 1924; FO 371/10736, C9784/459/18, Minute of a Departmental Conference by Bennett, 22 July 1925.
218. FO 371/3961, 145297/91/38, Minute by J. D. Gregory, 17 October 1920; FO 371/3914, 202067/40430/55, Minute by H. F. C. Crookshank, 7 July 1920.
219. FO 371/5361, F2200/199/23, Memorandum by Wellesley respecting the Anglo-Japanese Alliance, 1 September 1920.
220. FO 371/9397, W2558/2558/17, Minute by N. H. H. Charles, 18 June 1923.
221. FO 371/8251, W8928/50/17, Minute by R. H. Campbell, 28 October 1922.
222. *DBFP*, 1st Series, vol. xiv, nos. 443, 452, pp. 497–500, 513–16. The Marquess Curzon of Kedleston to Mr. Balfour (Washington), 23, 27 November 1921.

Chapter 3
Co-ordination of Foreign Policy in the Cabinet

1. E. Morse, *Foreign Policy and Interdependence in Gaullist France* (Princeton, 1973), pp. 5, 32.
2. E. Morse, 'Interdependence in World Affairs', in J. N. Rosenau, K. W. Thompson and Gavin Boyd (eds), *World Politics* (New York), p. 660.
3. FO 371/7000, W13420/12716/17, Notes respecting the possible conclusion of an Anglo-French Alliance by Crowe, 26 December 1921.
4. E. Morse, *Modernization and the Transformation of International Affairs* (London, 1976), pp. 83–4; Morse, *Foreign Policy and Interdependence*, pp. 14–15, 22–4.
5. Morse, *Modernization*, pp. 85–6, 88–90, 92–7; Morse, *Foreign Policy and Interdependence*, pp. 25–31, 41.
6. Sir Edward (later Lord) Bridges, *Portrait of a Profession* (Cambridge, 1950), p. 23, quoted in Brown, pp. 22–3.
7. Morse, *Modernization*, pp. 101–3.
8. Gordon A. Craig, 'The British Foreign Office' in G. A. Craig and Felix Gilbert (eds), *The Diplomats, 1919–1939*, p. 23.
9. A. L. Kennedy, *Old Diplomacy and New, 1876–1922* (London, 1923), pp. 364–5.
10. David Lloyd George, *War Memoirs* (London, 1938), vol. 1, pp. 27–30.
11. Lothian MSS, GD 40/17/1348, Kerr to Lloyd George, 15 November 1919.
12. H. Nicolson, *Diplomacy* (London, 1969), pp. 84–6.
13. H. Nicolson, *Curzon: The Last Phase, 1919–1925* (New York, 1925), p. 88.
14. Stephen Roskill, *Hankey, Man of Secrets* (London, 1972), vol. 2, pp. 69, 73, 142.
15. Ibid., p. 127.
16. Lloyd George MSS, F/12/2/11, Lloyd George to Curzon, 10 December 1919.
17. Lord Riddell, *Lord Riddell's Intimate Diary of the Peace Conference and After, 1918–1923* (London, 1933), p. 223.
18. Winston S. Churchill, *The World Crisis* (London, 1929), vol. 5, p. 374.

19. Curson MSS, Eur F/112/319, 'Confidential Memo, written by me at Lausanne Dec. 1922–Jan. 1923'.

20. Ibid., see also FRUS, The Paris Peace Conference, 1919, vol. 5, pp. 501–4, Notes of a Meeting of the Council of Four, 7 May 1919.

21. *DBFP*, 1st Series, vol. vii, nos 14, 26, pp. 122, 245, Notes of an Allied Conference held at 10 Downing Street, 18, 25 February 1920; ibid., no. 62, p. 512, Notes of Conference of Ambassadors and Foreign Ministers held in Lord Curzon's Room at the British Foreign Office, 16 March 1920.

22. *DBFP*, 1st Series, vol. viii, nos 4, 5, 7, 9, 10, 11, 12, 13, pp. 20–35, 35–45, 54–67, 83–93, 93–106, 107–19, 119–31, 132–43, Notes of a Meeting of the Supreme Council held at San Remo, 19, 20, 21, 22, 23 April 1920.

23. A. E. Montgomery, 'The Making of the Treaty of Sèvres of August 10th, 1920', *Historical Journal*, vol. 15 (1972), p. 777.

24. H. Nicolson, *Curzon*, pp. 81–93.

25. Roskill, *Hankey*, vol. 2, p. 149.

26. FO 371/5043, E 1297/3/44, Minute by Hardinge, March 1920 (undated).

27. FO 371/4215, 132926/50535/55, Minute by Vansittart, 6 September 1919.

28. FO 371/4179, 46887/2117/44, Memorandum by Earl Curzon, 'A note of Warning about the Middle East', 25 March 1919.

29. CAB 24/27, EC 46th Minutes, Minutes of a Meeting of the Eastern Committee, 23 December 1918; FO 371/5132, E 171/106/44, Minute by Hardinge, January 1920 (undated).

30. FO 371/4215, 132926/50535/44, Minute by Vansittart, 6 September 1919.

31. Hardinge MSS, vol. 41, Hardinge to Harcourt Butler, 5 July 1919.

32. FO 371/4215, 132926/50535/44, Minute by Curzon, 28 September 1919.

33. CAB 23/20, 1(20), Conclusions of a Meeting of the Cabinet, 6 January 1920.

34. Curzon MSS, Eur F/112/319, 'Notes on events attending break-up of Ll. G. Government written by me at end of Oct. 1922'.

35. Hardinge MSS, vol. 42, Hardinge to Harcourt Butler, 9 April 1920.

36. FO 371/4032, 172293/142549/38, The Earl of Derby (Paris) to Lord Hardinge, 22 January 1920. Quoted in *DBFP*, 1st Series, vol. ii, no. 76, note 5, p. 911.

37. *DBFP*, 1st Series, vol. ii, no. 71, pp. 867–75, Notes of a Meeting of the Heads of Delegations of the British, French and Italian Governments, 14 January 1920.

38. Ibid., no. 74, pp. 873–9, Notes of a Meeting of the Heads of Delegations of the American, British, French, and Italian Governments, 16 January 1920; ibid., no. 76, pp. 910–11, Notes of a Meeting, 16 January 1920.

39. Ibid., no. 71, p. 875, Notes of a Meeting of the Heads of Delegations of the British, French, and Italian Governments, 14 January 1920.

40. Richard H. Ullman, *Anglo-Soviet Relations, 1917–1921* (Princeton, 1968), vol. 2, pp. 99, 154, 308, 317–18.

41. *DBFP*, 1st Series, vol. iii, no. 664, pp. 804–5, Earl Curzon to Sir H. Rumbold (Warsaw), 27 January 1920.

42. Ullman, *Anglo-Soviet Relations*, vol. 2, p. 296.

43. FO 371/5433, N 2448/207/38, Russian Trade Negotiations: Memorandum by the Secretary of State for Foreign Affairs, 14 November 1920. Quoted in Ullman, *Anglo-Soviet Relations*, vol. 3, pp. 416–17.

44. Martin Gilbert, *Winston S. Churchill*, companion vol. 4, pt 2 (London, 1977), pp. 894–54.

45. Ullman, *Anglo-Soviet Relations*, vol. 2, pp. 8, 322, 448.

46. *DBFP*, 1st Series, vol. ii, no. 71, footnote 2, pp. 867–70, Economic Aspects of British Policy concerning Russia, 6 January 1920.

47. Milner MSS, 165/9, Curzon to Milner, 25 January 1920.
48. FO 371/4032, 172293/14259/38, The Earl of Derby (Paris) to Lord Hardinge, 22 January 1920; *DBFP*, 1st Series, vol. ii, no. 76, footnote 5, p. 911.
49. FO 371/4032, Lord Hardinge to the Earl of Derby (Paris), 20 January 1920.
50. Ibid., The Earl of Derby (Paris) to Lord Hardinge, 22 January 1920.
51. Ullman, *Anglo-Soviet Relations*, vol. 2, p. 326.
52. H. Nicolson, *Peacemaking, 1919* (London, 1933), p. 206.
53. FO 371/3961, 178568/91/38, Minute by Gregory, 19 February 1920; ibid., Minute by Hardinge, February 1920 (undated).
54. FRUS, The Paris Peace Conference, 1919, vol. 5, p. 501.
55. *DBFP*, 1st Series, vol. ii, nos 74, 76, pp. 873, 910–11.
56. R. Warman, *The Foreign Office, 1916–1918: a Study of its Role and Functions* (New York, 1986), pp. 137–8.
57. The 'Garden Suburb' was established in January 1917. For details of its war-time organization and activities, see John Turner, *Lloyd George's Secretariat* (Cambridge, 1980).
58. Frances Lloyd George, *The Years that are Past* (London, 1967), p. 101.
59. Ibid., pp. 101–2.
60. Lothian MSS, GD 40/17/1100/1, Hankey to Kerr, 31 March 1920; ibid., GD 40/17/1168/2, Kerr to Curzon, 20 May 1920.
61. J. R. M. Butler, *Lord Lothian (Philip Kerr), 1882–1940* (London, 1960), pp. 63–78; Turner, *Lloyd George's Secretariat*, pp. 67–82, 123–38, 157–66, 195–6.
62. Blanche E. C. Dugdale, *Arthur James Balfour, First Earl of Balfour* (London, 1939), vol. 2, p. 200.
63. FO 371/6933, N13752/12085/38, Minutes by Curzon, December 1921 (undated).
64. Curzon MSS, Eur F/112/319, 'Confidential Memo, written by me at Lausanne Dec. 1922–Jan. 1923'.
65. FO 371/3784, 194890/4232/18, Minute by Curzon, May 1920 (undated), quoted in Alan J. Sharp, 'The Foreign Office in Eclipse 1919–22', *History*, vol. 61 (1976), p. 207.
66. Lloyd George MSS, F/90/1/43, Minute by Kerr, 29 April 1921.
67. Ibid., F/90/1/42, Minute by Kerr, 29 April 1921; Sharp, 'The Foreign Office in Eclipse', p. 207.
68. Lothian MSS, GD 40/17/1311, Kerr to D'Abernon, 28 April 1921.
69. FO 800/154, Curzon to Sir Edward Grigg, 9 August 1922.
70. FO 371/7481, C11461/99/18, Minute by Wigram, 14 August 1922.
71. D. Lloyd George, *Memoirs of the Peace Conference* (New Haven, 1939), vol. 1, pp. 404–16.
72. Lloyd George MSS, F/90/1/18, Kerr to Lloyd George, 2 September 1920.
73. Lothian MSS, GD 40/17/1088, Minute by Kerr, 22 November 1919.
74. Riddell, *Intimate Diary*, p. 219.
75. Lloyd George MSS, F/24/3/9, Kerr to Hankey, 2 September 1920.
76. F. Lloyd George, *The Years that are Past*, p. 104; Kenneth O. Morgan, 'Lloyd George's Premiership: A Study in "Prime Ministerial Government"', *Historical Journal*, vol. 13 (1970), pp. 135–6, 146–7; Warman, *The Foreign Office, 1916–1918*, p. 138; Kenneth O. Morgan, *Consensus and Disunity: The Lloyd George Coalition Government, 1918–1922* (Oxford, 1979), p. 112.
77. Riddell, *Intimate Diary*, p. 223.
78. Public Record Office Handbooks, no. 11, *The Records of the Cabinet Office to 1922* (London, 1966), pp. 4, 20–3; John F. Naylor, *A Man and an Institution:*

Sir Maurice Hankey, the Cabinet Secretariat and the Custody of Cabinet Secrecy (Cambridge, 1984), pp. 27–9, 58–9, 90.

79. Public Record Office Handbooks, *The Records of the Foreign Office*, p. 29.
80. FO 800/243, Michael Palairet (for Sir Eyre Crowe) to Hardinge, 8 November 1919.
81. Lothian MSS, GD 40/17/1323/5, Kerr to Hankey, 21 July 1919.
82. Curzon MSS, Eur F/112/212, Hankey to Curzon, 7 November 1919.
83. Lothian MSS, GD 40/17/1323/5, Kerr to Hankey, 21 July 1919.
84. FO 371/5483, W523/523/98, Hankey to Lloyd George, 27 September 1920; ibid., Hankey to Hardinge, 1 October 1920; ibid., W774/523/98, Hankey to the Secretary to the Treasury, 4 October 1920; ibid., R. R. Scott to Hankey, 8 October 1920.
85. FO 371/8302, W8858/8858/50, Minute by Crowe, 21 October 1922; ibid., Minute by Curzon, 21 October 1922; ibid., Circular on the League of Nations by Crowe, 3 November 1922.
86. Roskill, *Hankey*, vol. 2, p. 301.
87. Ibid., pp. 171–2, 209, 230.
88. FO 371/5095, E4107/39/44, Forbes Adam to Tilley, 29 April 1920.
89. Riddell, *Intimate Diary*, p. 193.
90. Roskill, *Hankey*, vol. 2, p. 270.
91. Ibid., p. 222.
92. Curzon MSS, Eur F/112/319, 'Confidential Memo, written by me at Lausanne Dec. 1922–Jan. 1923'.
93. FO 371/6467, E3613/1/44, Minute by Crowe, 22 March 1921.
94. Briton C. Busch, *Mudros to Lausanne: Britain's Frontier in West Asia, 1918–1923* (Albany, 1976), pp. 319–27, 334–58; David Walder, *The Chanak Affair* (London, 1969), pp. 150–4.
95. Curzon MSS, Eur F/112/319, 'Confidential Memo, written by me at Lausanne Dec. 1922–Jan. 1923'.
96. Kenneth Morgan, *Consensus and Disunity*, pp. 113–14; Wm. Roger Louis, *British Strategy in the Far East, 1919–1939* (Oxford, 1971), pp. 45–8, 52–78, 107–8.
97. Ullman, *Anglo-Soviet Relations*, vol. 3, pp. 446–52.
98. FO 371/4683, C11940/343/19, Minute by Curzon, November 1920 (undated).
99. FO 800/243, Palairet (for Sir Eyre Crowe) to Hardinge, 8 November 1919.
100. Ibid.; Lord Gore-Booth, *With Great Truth and Respect* (London, 1974), p. 231.
101. Curzon MSS, Eur F/112/319, 'Confidential Memo, written by me at Lausanne Dec. 1922–Jan. 1923'; Joseph Davies, *The Prime Minister's Secretariat* (London, 1951), pp. 62–3; F. Lloyd George, *The Years that are Past*, p. 101.
102. Brown, pp. 205–6.
103. FO 800/243, Palairet (for Sir Eyre Crowe) to Hardinge, 8 November 1919.
104. Brown, pp. 22–4, 157–8, 201–6; Donald G. Bishop, *The Administration of British Foreign Relations* (Syracuse, 1961), pp. 121–30, 375–6.
105. Michael Crozier, *The Bureaucratic Phenomenon* (Chicago, 1964), pp. 111, 193–4, 253–4. See also Peter M. Blau and Richard W. Scott, *Formal Organizations: A Comparative Approach* (London, 1963), pp. 229–31.
106. Morse, *Modernization*, pp. 102–4.
107. FO 371/4223, 161686/70100/44, Minute by Hardinge, December 1919 (undated).
108. FO 371/4847, C10672/10613/18, Lord D'Abernon (Berlin) to Earl Curzon, 5 November 1920.
109. FO 371/48447, C10672/10613/18, Minute by Crowe, 14 November 1920.

110. Ibid.
111. FO 371/3782, 189420/4232/18, Report by General R. Haking: 'The Bolshevik Situation in Europe', 21 March 1920.
112. Ibid., Minute by Wigram, 3 April 1920.
113. Ibid., Minute by Phipps, 3 April 1920.
114. Ibid., Minute by Crowe, 4 April 1920.
115. FO 371/4741, C7032/45/18, 'The German Situation from a Military Aspect': Memorandum by The General Staff, 6 August 1920.
116. Minute by Waterlow, 24 September 1920.
117. Ibid., Minute by Crowe, 26 September 1920.
118. Ibid., Minute by Hardinge, September 1920 (undated).
119. Ibid., Minute by Curzon, 27 September.
120. FO 371/4847, C10672/10613/18, Minute by Waterlow, 13 November 1920.
121. Ibid., Minute by Crowe, 14 November 1920; ibid., Minute by Curzon, 14 November 1920.
122. Sharp, 'The Foreign Office in Eclipse', p. 214.
123. *DBFP*, 1st Series, vol. x, no. 312, pp. 421–2, Memorandum on the Execution by Germany of the Military Articles of the Peace Treaty of Versailles, 5 November 1920.
124. FO 371/5858, C15984/13/18, CP 3178, Memorandum by the Secretary of State for War: 'German Disarmament', 5 August 1921.
125. FO 371/4759, C15101/113/18, Minute by Waterlow, 1 January 1921; FO 371/7449, C4397/6/18, Lord D'Abernon (Berlin) to the Marquess Curzon of Kedleston, 20 March 1922; FO 371/7450, C7081/6/18, Lord D'Abernon (Berlin) to the Marquess Curzon of Kedleston, 9 May 1922.
126. FO 371/7450, C9596/6/18, Answer to a Question asked in the House of Commons, 4 July 1922.
127. FO 371/5864, C18725/37/18, Minute by Waterlow, 31 August 1921.
128. FO 371/7449, C4830/6/18, Minute by Waterlow, 4 April 1922.
129. Ibid., C5151/6/18, 'German Disarmament Position' by Wigram, 4 April 1922.
130. FO 371/7450, C5463/6/18, Minute by Lampson, 21 April 1922; ibid., Minute by Curzon, 21 April 1922.
131. Ibid., C8493/6/18, Minute by Lampson, 5 July 1922.
132. FO 371/7451, C10566/6/18, Minute by Wigram, 28 July 1922.
133. Brigadier-General J. H. Morgan, *Assize of Arms* (London, 1945), p. 34.
134. FO 371/4757, C11056/113/18, Minute by Crowe, 13 November 1920.
135. FO 371/4758, C12075/113/18, Minute by Crowe, 28 November 1920.
136. FO 371/7449, C5019/6/18, Minute by Crowe, 8 April 1922.
137. FO 371/8451, C9663/6/18, Minute by Wigram, 30 June 1922.
138. Ibid., Minute by Lampson, 30 June 1922.
139. FO 371/7451, C9663/6/18, Minute by Wigram, 30 June 1922.
140. Ibid., Minute by Crowe, 30 June 1922.
141. FO 371/5864, C18725/37/18, Minute by Crowe, 30 August 1921.
142. FO 371/9813, C5185/1346/18, 'General Staff Memorandum on the Military Aspect of the Future Status of the Rhineland', 28 March 1924; FO 371/9819, C15288/2048/18, 'General Staff Memorandum on M. Herriot's letter to Mr Ramsay MacDonald dated 11 April 1924 dealing with French security', 29 September 1924.
143. FO 371//9819, C15288/2048/18, Minute by Bennett, 13 October 1924.
144. FO 371/9813, C5185/1346/18, Minute by Bennett, 14 April 1924.
145. Ibid., Minute by Lampson, 16 April 1924.
146. FO 371/9819, C15288/2048/18, Minute by Nicolson, 15 October 1924.

147. FO 371/9833, C12724/4736/18, Minute by Lampson, 10 August 1924.
148. FO 371/10703, C8229/2/18, Lampson to Lt Colonel McGrath, 30 June 1920.
149. FO 371/11290, C310924/436/18, Minute by Troutbeck, 20 October 1926.
150. FO 371/11292, C11795/436/18, Minute by Chamberlain, 10 November 1926.
151. Jacques de Launay (ed.), *Louis Loucheur, Carnets Secrets, 1908–1932* (Brussels, 1962), pp. 185–8.
152. *DBFP*, 1st Series, vol. xv, nos 105, 106, 107, pp. 760–1, 764–5, 773, Notes of a Meeting between Mr Lloyd George and M. Briand, held at 10 Downing Street, 19, 20 December 1921.
153. Ibid., no. 111, Appendix 2, p. 799, 'Proposals for Re-establishing better Economic Conditions in Europe', 22 December 1921.
154. *DBFP*, 1st Series, vol. xix, no. 2, p. 9, Notes of a conversation between Mr Lloyd George and Signor Bonomi, held at Cannes, 4 January 1922.
155. FO 371/7000, W12716/12716/17, The Marquess Curzon of Kedleston to Hardinge (Paris), 5 December 1921.
156. *DBFP*, 1st Series, vol. xvi, no. 768, pp. 860–70, Memorandum by the Marquess Curzon of Kedleston on the question of an Anglo-French Alliance, 28 December 1921.
157. FO 371/9535, W5292/1585/17, Minute by Villiers, 2 July 1923.
158. Lord Vansittart, *The Mist Procession* (London, 1958), p. 281.
159. *DBFP*, 1st Series, vol. xix, nos 1, 3, 10, 17, pp. 1–7, 11–15, 56–8, 86–7, Notes of a conversation between Mr Lloyd George and M. Briand, held at Cannes, 4, 5, 8, 10 January 1922; ibid., nos 6, 19, 21, pp. 29–36, 90–99, 103–9, Notes of an Allied Conference, held at Cannes, 6, 10, 11 January 1922.
160. Georges Suarez, *Briand: sa vie, son œuvre avec son Journal et de nombreux documents inédits* (Paris, 1941), vol. 5, pp. 365–8, 371–3, 387–417. See also, *DBFP*, 1st Series vol. xix, nos 14, 19.
161. CAB 23/27, (93), Minutes of a Meeting of the Cabinet, 22 December 1921, printed in C. J. Lowe and M. L. Dockrill, *The Mirage of Power: British Foreign Policy, 1914–22* (London, 1972), vol. 3, pp. 733–8.
162. Hardinge MSS, vol. 44, Curzon to Hardinge, 23 December 1921.
163. Ibid., vol. 45, Curzon to Hardinge, 19 February 1922, quoted in J. D. Goold, 'Lord Hardinge as Ambassador to France and the Anglo-French Dilemma over Germany and the Near-East, 1920–1922', *Historical Journal*, vol. 21 (1978), p. 925.
164. Ibid., Hardinge to Curzon, 21 February 1922, quoted in Goold, 'Lord Hardinge as Ambassador', p. 925.
165. Ibid., Curzon to Hardinge, 22 February 1922; Lord Hardinge, *Old Diplomacy* (London, 1947), p. 270; Sharp, 'The Foreign Office in Eclipse', pp. 207–8.
166. Goold, 'Lord Hardinge as Ambassador', pp. 925–6; Leonard Mosley, *Curzon: The End of an Epoch* (London, 1961), pp. 218–20.
167. Hardinge MSS, vol. 45, Curzon to Hardinge, 1 March 1922.
168. Warman, *The Foreign Office, 1916–1918*, pp. 146–9.
169. CAB 23/20, 10(20), Conclusions of a Meeting of the Finance Committee, 30 January 1920.
170. Hardinge MSS, vol. 43, Hardinge to Rumbold, 13 July 1920.
171. Viscount D'Abernon, *An Ambassador of Peace: Pages from the Diary of Viscount D'Abernon* (London, 1929), vol. 1, pp. 53–4.
172. Arthur C. Murray, Viscount Elibank, *Reflections on Some Aspects of British Foreign Policy between the Two Wars* (London, 1946), p. 8.
173. Hardinge MSS, vol. 43, Hardinge to Rumbold, 13 July 1920.

174. Sir Maurice Peterson, *Both Sides of the Curtain, Autobiography* (London, 1950), p. 14.
175. Roskill, *Hankey*, vol. 2, p. 228.
176. Curzon, MSS, Eur F/112/319, 'Confidential Memo, written by me at Lausanne Dec. 1922–Jan. 1923'.
177. Winston S. Churchill, *Great Contemporaries* (Chicago, 1973), pp. 279–80.
178. Trevor Wilson (ed.), *The Political Diaries of P. C. Scott, 1911–1928* (London, 1970), p. 306.
179. H. Nicolson, *Curzon*, p. 34.
180. Riddell, *Intimate Diary*, p. 312.
181. Curzon MSS, Eur F/112/319, 'Notes on events attending break-up of Ll. G. Government written by me at end of Oct. 1922'; Mosley, *Curzon*, pp. 215–20.
182. H. Nicolson, *Curzon*, pp. 193, 269, 375.
183. F. Lloyd George, *The Years that are Past*, p. 102; H. A. L. Fisher MSS, Box 8A, Diary, 6 June 1920.
184. Roskill, *Hankey*, vol. 2, p. 210.
185. Vansittart, *Mist Procession*, p. 254.
186. See, for example, J. Vincent (ed.), *The Crawford Papers; The Journals of David Lindsay Twenty-seventh Earl of Crawford and Tenth Earl of Balcarres, 1871–1940; during the years 1892–1940* (Manchester, 1984), p. 416; Lord Beaverbrook, *The Decline and Fall of Lloyd George and Great was the Fall Thereof* (London, 1963), p. 40; H. Nicolson, *Curzon*, p. 214.
187. The Marchioness Curzon of Kedleston, *Reminiscences* (London, 1955), p. 223.
188. Ibid., p. 243; Earl of Ronaldshay, *The Life of Lord Curzon* (London, 1928), vol. 3, p. 256.
189. Vansittart, *Mist Procession*, p. 270.
190. Bonar Law MSS, 100/1/8. Austen Chamberlain to Bonar Law, 6 January 1921, reproduced in Beaverbrook, *Men and Power, 1917–1918* (New York, 1956), Appendix 4, no. 22, pp. 397–400.
191. Michael Kinnear, *The Fall of Lloyd George: The Political Crisis of 1922* (London, 1973), pp. 104–34; Busch, *Mudros to Lausanne*, pp. 340–60.
192. Robert Blake, *The Unknown Prime Minister: The Life and Times of Andrew Bonar Law, 1858–1923* (London, 1955), pp. 466, 501.
193. Naylor, *A Man and an Institution: Sir Maurice Hankey, the Cabinet Secretariat and the Custody of Cabinet Secrecy*, pp. 98–114.
194. Ibid., p. 98.
195. FO 371/7486, C15136/99/18, Minute by Crowe, 8 November 1922.
196. James Pope-Hennessy, *Lord Crewe, 1858–1945; The Likeness of a Liberal* (London, 1955), pp. 168–9.
197. A. J. P. Taylor (ed.), *My Darling Pussy: The Letters of Lloyd George and Frances Stevenson, 1913–1941* (London, 1975), p. 71.
198. John Ramsden (ed.), *Real Old Tory Politics: The Political Diaries of Sir Robert Sanders, Lord Bayford, 1910–35* (London, 1984), p. 199; Keith Middlemas and John Barnes, *Baldwin, A Biography* (London, 1969), p. 178.
199. Naylor, *A Man and an Institution: Sir Maurice Hankey, the Cabinet Secretariat and the Custody of Cabinet Secrecy*, p. 110.
200. The succession story has been told by others and is well documented. Blake, *The Unknown Prime Minister*, pp. 506–27; Middlemas and Barnes, *Baldwin*, pp. 158–68; Sir Harold Nicolson, *King George the Fifth: His Life and Reign* (London, 1953), pp. 375–9; H. Nicolson, *Curzon*, pp. 353–6; Robert Rhodes James, *Memoirs of a Conservative: J. C. C. Davidson's Memoirs and Papers 1910–37* (London, 1969), pp. 148–66; Kenneth Rose, *King George V*

(London, 1983), pp. 266–74; Sir Harold Nicolson relates that, on learning that the King's choice fell on Baldwin, Curzon broke down and wept like a child. 'A man of no experience. And of the utmost insignificance', he kept sobbing, 'the utmost insignificance.' H. Nicolson, *Curzon*, p. 355.

201. Roskill, *Hankey*, vol. 2, pp. 214–15; Blake, *The Unknown Prime Minister*, p. 482; Middlemas and Barnes, *Baldwin*, p. 178.

202. Busch, *Mudros to Lausanne*, pp. 365–84; Dockrill and Goold, *Peace without Promise*, pp. 236–243.

203. *DBFP*, 1st Series, vol. xxv, nos 41, 43, 46, 51–4, 63, 67–8, 71–2, 74, 78, 80, 86, 91, 93; Nicolson, *Curzon*, pp. 356–60.

204. Busch, *Mudros to Lausanne*, pp. 384–9; M. L. Dockrill and J. D. Goold, *Peace without Promise: Britain and the Peace Conferences, 1919–23*, pp. 243–7.

205. *DBFP*, 1st Series, vol. xxv, no. 94, pp. 162–6, M. Krassin to the Marquess Curzon of Kedleston, 9 June 1923; ibid., no. 100, pp. 172–3, Memorandum communicated by the Marquess Curzon of Kedleston to M. Krassin, 13 June 1923.

206. H. Nicolson, *Curzon*, p. 282; Ronaldshay, *Curzon*, vol. 3, pp. 325–6, 337–8, 355–6.

207. FO 371//7487, C16100/99/18, Minute by Curzon, 28 November 1922; CAB 23/45, 3(23), Conclusions of a Meeting of the Cabinet, 26 January 1923; *DBFP*, 1st Series, vol. xx, nos 120, 133, 136, 151; Blake, *The Unknown Prime Minister*, pp. 482–96.

208. Middlemas and Barnes, *Baldwin*, pp. 180–207.

209. Naylor, *A Man and an Institution: Sir Maurice Hankey, the Cabinet Secretariat and the Custody of Cabinet Secrecy*, pp. 113, 129.

210. *DBFP*, 1st Series, vol. xviii, no. 178, footnote 4, p. 255; Blake, *The Unknown Prime Minister*, p. 489.

211. Bonar Law MSS, 111/12/41, Curzon to Bonar Law, 6 December 1922, quoted in Blake, *The Unknown Prime Minister*, p. 489. See also Randolph S. Churchill, *Lord Derby: 'King of Lancashire'* (London, 1959), pp. 493, 498–9, 505, 517.

212. Blake, *The Unknown Prime Minister*, p. 489; Randolph Churchill, *Derby*, pp. 497–9, 511–21; Middlemas and Barnes, *Baldwin*, pp. 179, 187–8, 190–4.

213. Cecil MSS, vol. 7, Add. 51077, Curzon to Cecil, 2 June 1923; ibid., Cecil to Curzon, 7 June 1923.

214. Ibid., Curzon to Cecil, 14 June 1923; ibid., Cecil to Curzon, 16 June 1923; ibid., Curzon to Cecil, 18 June 1923; The Marchioness Curzon of Kedleston, *Reminiscences*, p. 193; Kenneth Rose, *The Later Cecils* (London, 1975), pp. 162–3.

215. Baldwin MSS, vol. , Curzon to Baldwin, 22 October 1923, quoted in G. M. Young, *Stanley Baldwin* (London, 1952), p. 50.

216. FO 371/8655, C16280/1/18, Communiqué to the Press, 19 September 1923, quoted in P. J. Grigg, *Prejudice and Judgement* (London, 1948), p. 165.

217. H. Nicolson, *Curzon*, p. 372. See also, The Marchioness Curzon of Kedleston, *Reminiscences*, pp. 185, 186. Curzon's resentment was, as the *procès-verbaux* of the conversation reveal, however, unjust. The Prime Minister did not exceed the bounds of the declared policy and the communiqué was designed to placate the French. See *DBFP*, 1st Series, vol. xxi, no. 567, pp. 529–35, Note on Conversation of 19 September 1923, between Mr Baldwin and M. Poincaré; Middlemas and Barnes, *Baldwin*, pp. 196–201.

218. H. Nicolson, *Curzon*, pp. 372–3.

219. The Marchioness Curzon of Kedleston, *Reminiscences*, p. 210.

220. Curzon MSS, Eur F/112/320, 'Intrigue between M. Poincaré, Comte de

St.-Aulaire (French Ambassador), H. A. Gwynne (Editor of *Morning Post*) to supplant Me at FO, October 1923'. Christopher Andrew, *Secret Service: the Making of the British Intelligence Community*, pp. 296–7.

221. Crewe MSS, C/12, Curzon to Crewe, 13 October 1923, quoted in Andrew, *Secret Service*, p. 297.
222. Ibid., Curzon to Crewe, 12 November 1923, quoted in Andrew, *Secret Service*, p. 297.
223. Randolph Churchill, *Derby*, pp. 515–16; Keith Middlemas (ed.), *Thomas Jones, Whitehall Diary* (London, 1969), vol. 1, pp. 243, 249, 261.
224. Curzon MSS, Eur F/112/320, Curzon to Baldwin, 9 November 1923, quoted in Andrew, *Secret Service*, p. 297.

Chapter 4
European Policy of the Prime Minister's Office

1. David Lloyd George, *Memoirs of the Peace Conference* (New Haven, 1939), vol. 1, pp. 266–7.
2. N. H. Gibbs, *Grand Strategy* (London, 1976), vol. 1, p. 38; Martin Gilbert, *The Roots of Appeasement* (London, 1966), pp. 18–22.
3. Brian Bond, *British Military Policy Between the Two World Wars* (Oxford, 1980), p. 12.
4. Ibid., pp. 1–2; N. H. Gibbs, 'British Strategic Doctrine, 1918–1939', in Michael Howard (ed.), *The Theory and Practice of War* (London, 1965), p. 194.
5. J. M. Keynes, *The Economic Consequences of the Peace* (London, 1920), p. 251, quoted in Gilbert, *The Roots of Appeasement*, p. 64.
6. J. M. Keynes, *A Revision of the Treaty* (London, 1920), pp. 96–100, 170, 187–8.
7. C. J. Lowe and M. L. Dockrill, *The Mirage of Power: British Foreign Policy, 1914–22* (London, 1972), vol. 2, pp. 350–2; M. Trachtenberg, *Reparation in World Politics: France and European Economic Diplomacy, 1916–1923* (New York, 1980), pp. 193–5.
8. *DBFP*, 1st Series, vol. xxi, no. 162, pp. 174–5, The Marquess Curzon of Kedleston to Sir S. Head (Berlin), 22 March 1923; G. Craig, 'The British Foreign Office from Grey to Austen Chamberlain', in G. A. Craig and Felix Gilbert (eds), *The Diplomats, 1919–1939*, pp. 22–3.
9. Gibbs, *Grand Strategy*, pp. 3–6.
10. Bond, *British Military Policy*, pp. 7, 14–22, 26–7, 30–4, 75.
11. CAB 32/9,1, E. 3rd Meeting, Imperial Conference Minutes, 5 October 1923.
12. Ibid.; *DBFP*, 1st Series, vol. xvi, no. 768, pp. 861–2, Memorandum by the Marquess Curzon of Kedleston on the question of an Anglo-French Alliance, 28 December 1921.
13. Hardinge MSS, vol. 44, Curzon to Hardinge, 15 June 1921; CAB 23/32,64 (22), Annex 4, Conclusions of a Meeting of the Cabinet, 1 November 1922.
14. Hardinge MSS, vol. 45, Curzon to Hardinge, 2 May 1922.
15. Ibid., vol. 44, Curzon to Hardinge, 15 June 1921.
16. CAB 32/2,1, E. 4th Meeting, Imperial Conference Minutes, 22 June 1921.
17. *DBFP*, 1st Series, vol. xvi, no. 768, p. 864, Memorandum by the Marquess

Curzon of Kedleston on the question of an Anglo-French Alliance, 28 December 1921.

18. Ibid., p. 862.
19. H. Nicolson, *Curzon: The Last Phase, 1919–1925* (New York, 1925), pp. 49–55.
20. Keith Middlemas (ed.), *Thomas Jones, Whitehall Diary*, (London, 1969), vol. 4, pp. 117, 178.
21. CAB 23/23, 80(20)(3), Conclusions of a Meeting of the Cabinet, 30 December 1920; Stephen Roskill, *Hankey, Man of Secrets* (London, 1972), vol. 2, p. 209.
22. Roskill, *Hankey*, vol. 2, p. 209.
23. Ibid., Middlemas (ed.), *Thomas Jones*, vol. 1, pp. 116–17; CAB 23/25, 40(21)(4), Conclusions of a Meeting of the Cabinet, 24 May 1921.
24. *DBFP*, 1st Series, vol. ix, no. 301, pp. 327–8, Memorandum by Sir E. Crowe on a conversation with M. de Fleuriau, 6 April 1920.
25. Ibid., no. 301, p. 328, Minute by Curzon, 6 April 1920.
26. CAB 23/29, 38(20(21), Conclusions of a Meeting of the Cabinet, 30 June 1920; Middlemas (ed.), *Thomas Jones*, vol. 1, p. 117.
27. Roskill, *Hankey*, vol. 2, p. 209.
28. *DBFP*, 1st Series, vol. xvii, no. 38, pp. 56–9; Memorandum by Sir E. Crowe, 12 February 1921.
29. CAB 32/2, 2, E. 18th Meeting, Imperial Conference Minutes, 7 July 1921.
30. CAB 32/2, 1, E, 4th Meeting, Imperial Conference Minutes, 22 June 1921.
31. Middlemas (ed.), *Thomas Jones*, vol. 1, pp. 108, 117; Lord Riddell, *Lord Riddell's Intimate Diary of the Peace Conference and After, 1918–1923* (London, 1933), pp. 188, 247.
32. Hartmut Pogge von Strandmann (ed.), *Walther Rathenau: Tagebuch, 1907–1922* (Dusseldorf, 1967), p. 266.
33. Riddell, *Intimate Diary*, pp. 196, 247.
34. Middlemas (ed.), *Thomas Jones*, vol. 1, pp. 117, 178.
35. CAB 32/2, 1, E. 7th Meeting, Imperial Conference Minutes, 27 June 1921.
36. FO 371/8703, C848/313/18, 'Putting the Screw on Germany: Whatever happens in the Ruhr the entente must be kept intact', by Lord Hardinge, *Sunday Pictorial*, 14 January 1923.
37. FO 371/4757, C11056/113/18, Minute by Crowe, 13 November 1920.
38. *DBFP*, 1st Series, vol. xvi, no. 747, footnote 7, p. 828, Minute by Crowe, 30 November 1921.
39. FO 371/7486, C14819/99/18, Minute by Tyrrell, 20 October 1922.
40. FO 371/7521, C11429/336/18, Minute by Tyrrell, 10 August 1922.
41. FO 371/8703, C848/313/18, 'Putting the Screw on Germany: Whatever happens in the Ruhr the entente must be kept intact', by Lord Hardinge, *Sunday Pictorial*, 14 January 1923; Lord Hardinge, *Old Diplomacy* (London, 1947), p. 260.
42. FO 371/4757, C11056/113/18, Minute by Crowe, 13 November 1920; FO 371/7000, W13420/2716/17, Notes respecting the possible conclusion of an Anglo-French Alliance by Crowe, 26 December 1921.
43. Hardinge MSS, vol. 44, Hardinge to Curzon, 1 December 1920.
44. FO 371/3780, 185720/4232/18, The Earl of Derby (Paris) to Earl Curzon, 16 March 1920. For reports from the field see F. L. Carsten, *Britain and the Weimar Republic: the British Documents* (London, 1984), pp. 34–41.
45. *DBFP*, 1st Series, vol. ix, no. 154, p. 185, Earl Curzon to the Earl of Derby (Paris), 18 March 1920.
46. Ibid., no. 150, p. 181, The Earl of Derby (Paris) to Earl Curzon, 18 March 1920.

47. FO 371/3780, 186016/4232/18, Minute by Waterlow, 18 March 1920.
48. Ibid., 185720/4232/18, Minute by Waterlow, 16 March 1920.
49. Ibid., 186016/4232/18, Minute by Crowe, 18 March 1920; FO 371/3781, 186567/4232/18, Minute by Crowe, 19 March 1920.
50. FO 371/3780, 185720/4232/18, Minute by Hardinge, March 1920 (undated).
51. *DBFP*, 1st Series, vol. vii, no. 64, pp. 542–7, British Secretary's Notes of an Allied Conference held at 10, Downing Street, London, 18 March 1920.
52. Ibid., no. 68, p. 586, British Secretary's Notes of Allied Conference held in Lord Curzon's room at the British Foreign Office, 22 March 1920; ibid., no. 70, p. 606, British Secretary's notes of a Conference of Ambassadors and Foreign Ministers held in Lord Curzon's room, 24 March 1920.
53. FO 371/3781, 187976/4232/18, Minute by Crowe, 24 March 1920.
54. Ibid., Minute by Hardinge, March 1920 (undated).
55. *DBFP*, 1st Series, vol. vii, no. 70, pp. 606–8, British Secretary's Notes of a Conference of Ambassadors and Foreign Ministers, held in Lord Curzon's room, 24 March 1920.
56. FO 371/3781, 186926/4232/18, Minute by Phipps, 22 March 1920.
57. ADAP, A⁵·, vol. 3, no. 86, pp. 147–8, Der Geschäftsträger in Paris an das Auswärtige Amt, 30 March 1920; *DBFP*, 1st Series, vol. ix, no. 254, pp. 287–93, Letter from the Earl of Derby (Paris) to Earl Curzon, 1 April 1920.
58. ADAP, A⁵·, vol. 3, no. 98, Aufzeichnung des Vorsitzenden der deutschen Friedensdelegation Goppert, 2 April 1920; *DBFP*, 1st Series, vol. ix, no. 276, pp. 307–9, Sir G. Grahame (Paris) to Earl Curzon, 9 April 1920.
59. ARWR, KAB. Muller, 1920, vol. 1, no. 10, pp. 21–4, Weisung des Reichskanzlers an die deutsche Friedensdelegation in Paris, 2 April 1920; ibid., no. 12, pp. 27–9, Interview des Reichskanzlers mit dem Korrespondenten des International News Service, Frank Mason, 5 April 1920.
60. *DBFP*, 1st Series, vol. ix, no. 322, pp. 346–7, Earl Curzon to the Earl of Derby (Paris), 8 April 1920.
61. Ibid., no. 298, pp. 324–5, Earl Curzon to the Earl of Derby (Paris), 6 April 1920.
62. Ibid., no. 318, pp. 340–4, Earl Curzon to the Earl of Derby (Paris), 8 April 1920.
63. Middlemas (ed.), *Thomas Jones*, vol. 1, pp. 108–11.
64. Riddell, *Intimate Diary*, p. 177.
65. Middlemas (ed.), *Thomas Jones*, vol. 1, pp. 108, 110.
66. Ibid., pp. 108–10.
67. CAB 23/21, 18(20), Conclusions of a Meeting of the Cabinet, 8 April 1920; *DBFP*, 1st Series, vol. ix, no. 322, p. 347, Earl Curzon to the Earl of Derby (Paris), 8 April 1920.
68. This subject has been studied in considerable depth by Anne Orde, *Great Britain and International Security, 1920–1926* (London, 1978); and *British Policy and European Reconstruction after the First World War* (Cambridge, 1990). For the subject of reparations, Bruce Kent, *The Spoils of War: the Politics, Economics and Diplomacy of Reparations, 1918–1932* (Oxford, 1989).
69. *DBFP*, 1st Series, vol. viii, no. 14, pp. 149–52, Notes of a Conversation held at the Hotel Royal, San Remo, 24 April 1920.
70. Ibid., nos 21, 22, pp. 258–61, 162–272, British Secretary's Notes of a Conversation between the Heads of the British and French Governments, held at Belcaire, Lympne, 15 May 1920.

71. Ibid., no. 31, Appendix, pp. 337–9, Note by the Financial Experts, 21 June 1920.

72. Ibid., no. 77, footnote 3, p. 6442, Earl Curzon to Lord Hardinge (London), 17 July 1920.

73. FO 371/4729, C9696/8/18, Minute by Sir Worthington-Evans, 22 October 1920.

74. *DBFP*, 1st Series, vol. viii, no. 88, pp. 776, Conclusions of a Conversation between Mr Lloyd George and Signor Giolitti at the Villa Haslihorn, Lucerne, 22 August 1920.

75. Ibid., no. 91, p. 793, Notes of a Conversation held at 10 Downing Street, SW1, between the British Prime Minister and the Belgian Prime Minister, 11 October 1920.

76. FRUS, 1920, vol. 2, p. 441, The Ambassador in France (Wallace) to the Secretary of State, 18 November 1920.

77. *DBFP*, 1st Series, vol. xxv, no. 8, p. 65, British Secretary's Notes on Allied Conference held in the Quai d'Orsay, Paris, 27 January 1921.

78. Ibid., no. 2, p. 9, British Secretary's Notes of an Allied Conference held in the Quai d'Orsay, Paris, 24 January 1921.

79. FO 371/5961/ C1876/386/18, Lord D'Abernon to Earl Curzon, 20 January 1921; Annex 1, Experts' Conference, Brussels, 18 January 1921; Roskill, *Hankey*, vol. 2, pp. 173–4.

80. *DBFP*, 1st Series, vol. viii, no. 21, p. 258, British Secretary's Notes of a Conversation between the Heads of the British and French Governments, held at Lympne, 15 May 1920; ibid., no. 32, pp. 341–2, Draft British Secretary's Notes of a Meeting of the Supreme Council, held at Boulogne, 21 June 1920.

81. Ibid., no. 36, pp. 371–2, British Secretary's Draft Notes of an Inter-Allied Conference, held at Boulogne, 22 June 1920.

82. Ibid., no. 50, pp. 480–1, British Secretary's Notes of a Meeting held at Spa, 8 July 1920; ibid., no. 77, footnote 3, p. 642, Earl Curzon to Lord Hardinge, 17 July 1920.

83. Ibid., vol. xv, no. 11, Appendix 1, pp. 101–4, Scheme for Payment of German Reparations, 29 January 1921.

84. Ibid., no. 29, Appendix, pp. 2445–6, German Reparation Proposals: Declaration drafted by the Financial Experts, 2 March 1921; ibid., no. 45, pp. 327–31, British Secretary's Notes of an Allied Conference, held at London, 7 March 1921.

85. Ibid., no. 446, Appendix, p. 334, Earl Curzon to the British High Commissioner, Rhineland, 7 March 1921; ibid., vol. xvi, no. 458, pp. 484–5, Earl Curzon to His Majesty's Representatives, 11 March 1921.

86. Ibid., vol. xv, no. 83, Appendix 2, pp. 566–9, Arrangement for the Discharge of Germany's Liability for Reparation under the Treaty of Versailles, 3 May 1921.

87. Lothian MSS, GD 40/17/1311, Kerr to D'Abernon, 28 April 1921.

88. CAB 23/25, 24(21), Conclusions of a Meeting of the Cabinet, 19 April 1921.

89. Lothian MSS, GD 40/17/1311, Kerr to D'Abernon, 28 April 1921.

90. *DBFP*, 1st Series, vol. xv, no. 85, Appendix 2, pp. 579–80, Allied Note to the German Government, 5 May 1921; Roskill, *Hankey*, vol. 2, p. 229. For further details respecting the aforementioned events see Trachtenberg, *Reparation*, chs 3–5, and Etienne Weill-Raynal, *Les Réparations allemandes et la France* (Paris, 1947), vol. 1, chs 14–17.

91. Sally Marks, 'The Myths of Reparations', *Central European History*, vol. 11 (1978), pp. 237–8.

92. CAB 23/30, 29 (22)(2), Conclusions of a Meeting of the Cabinet, 23 May 1922.
93. 'La Politique Française en 1922' (Paris, 1923), pp. 29–32, Text of Poincaré's speech at Bar-le-Duc, 24 April 1922. See also, *DBFP*, 1st Series, vol. xx, no. 19, pp. 40–1.
94. CAB 23/30, 44 (22), Conclusions of a Meeting of the Cabinet, 10 August 1922.
95. Trachtenberg, *Reparation*, pp. 240–50; Weill-Raynal, *Les Réparations allemandes*, vol. 2, pp. 148–65, 191–2.
96. *DBFP*, 1st Series, vol. xx, no. 51, pp. 120–1, British Secretary's Notes of an Allied Conference held at 10 Downing Street, 7 August 1922.
97. FO 371/6038, C22283/2740/18, Memorandum by Sir Basil Blackett, 22 November 1921.
98. FO 371/7516, C4852/333/18, Minute by Lampson, 6 April 1922; Lord Vansittart, *The Mist Procession* (London, 1958), p. 253.
99. FO 371/4730, C14119/8/18, Joseph Addison to S. P. Waterlow, 13 December 1920, quoted in Trachtenberg, *Reparation*, pp. 181–2. See also Carsten, *Britain and the Weimar Republic* pp. 78–80, 82–5.
100. *DBFP*, 1st Series, vol. xx, no. 102, p. 276, Sir J. Bradbury to Mr Bonar Law, 23 October 1922.
101. Ibid., vol. xxi, no. 212, pp. 252–3, Mr Niemeyer (Treasury) to Sir E. Crowe, 4 May 1929; CAB 24/161, CP 312, Memorandum by the President of the Board of Trade, Sir Philip Lloyd-Graeme, 6 July 1923.
102. FO 371/9813, C34420/1346/18, Waterlow to Lampson, 28 February 1924; ibid., C4218/1346/18, Board of Trade to Foreign Office, 12 March 1924.
103. FO 371/9740, C6080/70/18, Minute by Crowe, 9 April 1924. For the Secretary's reply, see *DBFP*, 1st Series, vol. xxvi, no. 430.
104. *DBFP*, 1st Series, vol. xxvi, no. 561, pp. 869–73, Board of Trade to Foreign Office, 4 September 1924.
105. *DBFP*, 1st Series, vol. ix, no. 50, footnote 4, p. 78, Minute by Waterlow, 24 February 1920; ibid., vol. 10, no. 174, p. 268, Note on the German situation by Mr Waterlow, 5 July 1920; H. Nicolson, *Curzon*, p. 220.
106. *DBFP*, 1st Series, vol. ix, no. 50, p. 75, Minute by Headlam-Morley, 20 February 1920; Agnes Headlam-Morley, Russell Bryant, Anna Cienciala (eds), *Sir James Headlam-Morley, A Memoir of the Paris Peace Conference, 1919* (London, 1972), pp. 44–5, 124–5, 162.
107. FO 371/7518, C15155/333/18, Minute by Lampson, 9 November 1922.
108. FO 371/6038, C22282/2740/18, Minute by Waterlow, 26 November 1921.
109. *DBFP*, 1st Series, vol. x, no. 174, p. 268, Note on the German Situation by Mr Waterlow, 5 July 1920.
110. FO 371/6022, C6958/2740/18, Minute by Waterlow, 7 April 1921.
111. FO 371/10713, C5750/35/18, Memorandum on the Election of Field-Marshal von Hindenburg as German President, by J. C. Sterndale Bennett, 27 April 1925.
112. FO 371/9833, C17812/4736/18, Minute by Austen Chamberlain, 26 November 1924.
113. FO 371/7516, C4852/333/18, Minute by Lampson, 6 April 1922.
114. *DBFP*, 1st Series, vol. x, no. 174, p. 267, Note on the German Situation by Mr Waterlow, 5 July 1920; FO 371/7487, C16100/99/18, Minute by Lampson, 27 November 1922.
115. *DBFP*, 1st Series, vol. xvi, nos 518, 521, 523, 555, 710, 727, 744, 747, 767; ibid., vol. xix, no. 111, Appendix; FO 371/7486, C15425/99/18, Memorandum on the History of the Negotiations with regard of Reparation and of Inter-Allied Debt from 5 May 1921, 2 November 1922;

FO 371/7488, C16643/99/18, Central Department Memorandum on the Reparation Position, 5 December 1922.

116. FO 371//7518, C15155/333/18, Minute by Lampson, 9 November 1922. For a modern development of this argument, see David Felix, *Walther Rathenau and the Weimar Republic: The Politics of Reparations* (Baltimore, 1971), pp. 25–40, 92–101, 105–11, 178–89; David Felix, 'Reparations Reconsidered with a Vengeance', *Central European History*, vol. 4 (1971), pp. 171–9.

117. FO 371/7486, C14819/99/18, Minute by Tyrrell, 26 October 1922.

118. FO 371/7518, C15069/333/18, Minute by Crowe, 3 November 1922.

119. Ibid.; *DBFP*, 1st Series, vol. xx, no. 16, pp. 27–9, Memorandum respecting the French budgetary situation, 4 April 1922.

120. FO 371/7487, C16100/99/18, Minute by Lampson, 27 November 1922; FO 371/7488, C16643/99/18, Central Department Memorandum on the Reparation Position, 5 December 1922; Trachtenberg, *Reparation*, pp. 275–89.

121. FO 371/5963, C3270/386/18, Minute by Headlam-Morley, 21 February 1921.

122. Sally Marks, 'Reparations Reconsidered: A Reminder', *Central European History*, vol. 2 (1969), pp. 356–65; Sally Marks, 'The Myths of Reparations', pp. 235–9, 245–6; Stephen A. Schuker, *The End of French Predominance in Europe: The Financial Crisis of 1924 and the Adoption of the Dawes Plan* (Chapel Hill, 1967), pp. 14–16, 19, 21–3, 78–9, 181–3, 384–5; Gerald D. Feldman, *Iron and Steel in the German Inflation, 1916–1923* (Princeton, 1977), pp. 112–13, 130–2, 187, 254–6, 282, 290–2; Walter A. McDougall, 'Political Economy versus National Sovereignty: French Structures for German Economic Integration after Versailles', *Journal of Modern History*, vol. 51 (1979), pp. 12–13. See the arguments in B. Kent, *The Spoils of War: the Politics, Economics and Diplomacy of Reparations, 1918–1932*, pp. 142–54, 170–2, 178–85, 194–7.

123. FO 371/7489, C16854/99/18, Mr Addison to the Marquess Curzon of Kedleston, 5 December 1922.

124. FO 371/7488, C16643/99/18, Central Department Memorandum on the Reparation Position, 5 December 1922.

125. *DBFP*, 1st Series, vol. xx, no. 151, pp. 349–50, 353, Memorandum, 24 December 1922; Herman J. Rupieper, *The Cuno Government and Reparations, 1922–1923: Politics and Economics* (The Hague, 1979), pp. 219–21; FO 371/9748, C7129/70/18, Minute by Crowe, 1 May 1924.

126. FO 371/4816, C5661/1621/18, Colonel Percival (Oppeln) to Earl Curzon, 31 August 1920; ibid., C6229/1621/18, Major Chas Macpherson to Colonel Percival, 1 September 1920.

127. FO 371/4816, C5661/1621/18, Colonel Percival (Oppeln) to Earl Curzon, 31 August 1920. For further reports from the field, see Carsten, *Britain and the Weimar Republic*, pp. 66–73.

128. FO 371/4816, C5930/1621/18, Minute by Phipps, 10 September 1920; ibid., Minute by Waterlow, 10 September 1920.

129. Ibid., C6221/1621/18, Minutes by Waterlow, 10, 15 September 1920.

130. Ibid., Minute by Crowe, 15 September 1920.

131. FO 371/4817, C6843/1621/18, Minute by Crowe, 22 September 1920.

132. FO 371/4816, C6221/1621/18, Minute by Hardinge, September 1920 (undated).

133. Ibid., C5661/1621/18, Minute by Curzon, 7 September 1920.

134. CAB 23/23, 70(20), Appendix 2(3), Conclusions of a Conference of Ministers, 29 November 1920.

135. FO 371/4820, C11655/1621/18, Minute by Waterlow, 19 November 1920.

136. FO 371/4823, C14725/1621/18, Minute by Crowe, 25 December 1920.
137. FO 371/4820, C11655/1621/18, Minute by Crowe, 19 November 1920.
138. FO 371/4823, C14725/1621/18, Minute by Crowe, 25 December 1920.
139. Ibid., Minute by Curzon, 26 December 1920.
140. FO 371/5889, C3293/92/18, B. B. Curtin to the Under-Secretary of State, 15 February 1921.
141. CAB 23/23, 80(20)(1)(2), Conclusions of a Meeting of the Cabinet, 30 December 1920.
142. FO 371/5889, C3464/92/18, Minute by Curzon, 19 February 1921.
143. Ibid., C9758/92/18, Minute by Crowe, 12 May 1921; FO 371/5900, C9910/92/18, Minute by Crowe, 15 May 1921.
144. FO 371/5900, C9910/92/18, Minute by Curzon, 16 May 1921.
145. CAB 23/25, 40(21), Conclusions of a Meeting of the Cabinet, 24 May 1921.
146. *DBFP*, 1st Series, vol. xvi, no. 1, p. 1, Earl Curzon to Colonel Percival (Oppeln), 22 March 1921.
147. FO 371/5900, C9874/92/18, Minute by Waterlow, 14 May 1921.
148. Lothian MSS, GD40/17/82, D'Abernon to Kerr, 20 October 1921.
149. *DBFP*, 1st Series, vol. xi, no. 123, pp. 146–7, Memorandum on the probable effects of the transfer of Upper Silesia to Poland, 29 December 1920; FO 371/5886, C420/92/18, Minute by Waterlow, 8 January 1921.
150. FO 371/5904, C10756/92/18, Minute by Crowe, 25 May 1921.
151. FO 371/5886, C420/92/18, Minute by Crowe, 8 January 1921.
152. Lloyd George MSS, F/53/1/63, Hardinge to Lloyd George, 22 June 1921.
153. *DBFP*, 1st Series, vol. xvi, no. 1, p. 1, Earl Curzon to Colonel Percival (Oppeln), 22 March 1921.
154. Ibid., no. 22, pp. 44–5, Colonel Percival (Oppeln) to Earl Curzon, 23 April 1921; Sarah Wambaught, *Plebiscites since the World War: With a Collection of Official Documents* (Washington, 1933), vol. 1, pp. 249–51; ibid., vol. 2, no. 88, p. 241.
155. Wambaught, *Plebiscites since the World War*, vol. 2, no. 89, pp. 242–3, The President of the Upper Silesian Plebiscite Commission to Mr Lloyd George, 30 April 1921.
156. FO 371/5892, C6194/92/18, Minute by Waterlow, 23 March 1921.
157. Wambaught, *Plebiscites since the World War*, vol. 2, no. 89, pp. 243–57, Reports and recommendations of the three members of the Commission regarding the frontier of Germany in Upper Silesia, 29, 30, April 1921.
158. F. Gregory Campbell, 'The Struggle for Upper Silesia, 1919–1922', *Journal of Modern History*, vol. 42 (1970), pp. 375–82.
159. *DBFP*, 1st Series, vol. xv, no. 102, p. 704, British Secretary's Notes of an Allied Conference held at Paris, 12 August 1921.
160. FO 371/5928, C19698/92/18, Recommendation of the Council of the League of Nations concerning the Delimitation of the Frontier in Upper Silesia, 12 October 1921; L/NOJ, vol. 2, December 1921, pp. 1223–32; James Barros, *Office without Power: General Sir Eric Drummond, 1919–1933* (Oxford, 1979), pp. 137–44.
161. Lloyd George MSS, F/13/2/47, Minute by Crowe, 23 September 1921.
162. Hardinge MSS, vol. 44, Curzon to Hardinge, 10 November 1921; *DBFP*, 1st Series, vol. vii, no. 434, p. 467, The Marquess Curzon of Kedleston to Sir G. Buchanan (Rome), 8 November 1921.
163. *DBFP*, 1st Series, vol. xiv, nos 442, 443, 452, pp. 495–7, 497–500, 513–16, The Marquess Curzon of Kedleston to Mr Balfour (Washington Delegation), 23, 27 November 1921; Martin Gilbert, *Winston S. Churchill*, companion vol. 4,

pt 3, pp. 1686–7, Winston S. Churchill: Note, 9 December 1921; Gibbs, *Grand Strategy*, pp. 46–8.

164. Austen Chamberlain MSS, AC 24/3/91, Saint-Aulaire to J. Austen Chamberlain, 9 December 1921.
165. FO 371/7000, W12716/12716/17, The Marquess Curzon of Kedleston to Lord Hardinge (Paris), 5 December 1921.
166. Ibid., W12728/12716/17, Lord Hardinge (Paris) to the Marquess Curzon of Kedleston, 7 December 1921; Hardinge, *Old Diplomacy*, pp. 263–4.
167. Le Livre Jaune Français, 1924, no. 17; M. Briand, Président du Conseil, Ministre des Affaires Étrangères, à M. de Saint-Aulaire, Ambassadeur de France à Londres, 4 December 1921; Hardinge MSS, vol. 44, Hardinge to Curzon, 16 September 1921; ibid., Hardinge to Curzon, 12 November 1921.
168. Le Livre Jaune Français, 1924, no. 19, p. 93, M. de Saint-Aulaire, Ambassadeur de France à Londres, à M. Briand, Président du Conseil, Ministre des Affaires Étrangères, 31 December 1921; CAB 23/29, 1(22), Conclusions of a Meeting of the Cabinet, 10 January 1922.
169. Austen Chamberlain MSS, AC 24/3/91, Saint-Aulaire to A. Chamberlain, 9 December 1921; Gilbert, *Churchill*, companion vol. 4, pt 3, pp. 1718–19, Winston S. Churchill: draft press statement, [9] January 1922.
170. FO 371/6999, C19270/19270/37, The Marquess Curzon of Kedleston to Sir H. Dering (Bucharest), 5 October 1921; *DBFP*, 1st Series, vol. xvi, no. 768, p. 869, Memorandum by the Marquess Curzon of Kedleston on the question of an Anglo-French Alliance, 28 December 1921.
171. *DBFP*, 1st Series, vol. xvi, no. 768, p. 864, Memorandum by the Marquess Curzon of Kedleston on the question of an Anglo-French Alliance, 28 December 1921.
172. Ibid., p. 860; ibid., vol. xv, no. 110, pp. 785–7, Notes of a Conversation between Mr Lloyd George and M. Briand at 10 Downing Street, 21 December 1921.
173. Hardinge MSS, vol. 44, Hardinge to Curzon, 26 December 1921; Hardinge, *Old Diplomacy*, p. 264; J. D. Goold, 'Lord Hardinge as Ambassador to France and Arab-French Dilemma over Germany and the Near East, 1920–1922', *Historical Journal*, vol. 21 (1978), p. 992.
174. Sir Ivone Kirkpatrick, *The Inner Circle: Memoirs of Ivone Kirkpatrick* (London, 1959), pp. 32–3.
175. FO 371/7000, W13240/2716, Notes respecting the possible conclusion of an Anglo-French Alliance by Crowe, 26 December 1921.
176. *DBFP*, 1st Series, vol. xvi, no. 768, pp. 868–9, Memorandum by the Marquess Curzon of Kedleston on the question of an Anglo-French Alliance, 28 December 1921.
177. Ibid., vol. xv, no. 110, pp. 786–7, Notes of a Conversation between Mr Lloyd George and M. Briand at 10 Downing Street, 21 December 1921; Roskill, *Hankey*, vol. 2, p. 255. This is further confirmed by the fact that Lloyd George brought with him to the Cannes Conference a draft of an Anglo-French security pact. See PP 1924 (Cmd 2169) vol. 26, papers respecting negotiation for an Anglo-French pact, no. 38, pp. 127–8.
178. Le Livre Jaune Français, 1924, no. 19, p. 93, M. de Saint-Aulaire, Ambassadeur de France à Londres, à M. Briand, Président du Conseil, Ministre des Affaires Étrangères, 31 December 1921. See also Carole Fink, *The Genoa Conference: European Diplomacy, 1921–1922* (Chapel Hill, 1984), pp. 33–8, and Donald Graeme Boadle, *Winston Churchill and the German Question in British Foreign Policy, 1918–1922* (The Hague, 1973), pp. 170–3. For additional information on Genoa, see Carole Fink, Alex

Frohn and Jürgen Heideking, *Genoa, Rapallo and European Reconstruction in 1922* (Cambridge, 1991), esp. chs 1, 2 and 3.

179. FO 371/9535, W5292/1585/17, Minute by Villiers, 2 July 1923.
180. CAB 23/29, 1 (22) Conclusions of a Meeting of the Cabinet, 10 January 1922; Curzon MSS, Eur F/112/319, 'Notes on events attending break-up of Ll. G. Government written by me at end of October 1922'.
181. Le Livre Jaune Français, 1924, no. 21, pp. 107–9, Exposé des vues du Gouvernement français sur les relations franco-britanniques, 8 January 1922, reproduced also in Cmd 2169, vol. 26, no. 35, pp. 121–6. See also *DBFP*, 1st Series, vol. xix, nos 10, 17, pp. 57–8, 86–7, British Secretary's Notes of a Conversation between Mr Lloyd George and M. Briand, vol. 5, pp. 352–86; Fink, *The Genoa Conference*, pp. 38–42, RIIA, Survey of International Affairs, 1924 (Oxford, 1926), pp. 9–10; Weill-Raynal, *Les Réparations allemandes*, vol. 2, pp. 109–18.
182. Georges Suarez, *Briand: sa vie, sa œuvre avec son Journal at de nombreux documents inédits* (Paris, 1941), vol. 5, pp. 387–417; Jules Laroche, *Au Quai d'Orsay avec Briand et Poincaré, 1913–1926* (Paris, 1957), pp. 151–7; A. J. Sylvester, *The Real Lloyd George* (London, 1947), pp. 71–4.
183. FO 371/8250, W614/50/17, Lord Hardinge (Paris) to the Marquess Curzon of Kedleston, 10 January 1922.
184. Le Livre Jaune Français, 1924, no. 23, pp. 113–14, M. Poincaré, Président du Conseil, Ministre des Affaires Étrangères, à M. de Saint-Aulaire, Ambassadeur de France à Londres, 23 January 1922. See also, Cmd 2169, vol. 26, no. 40, pp. 128–31; RIIA, Survey, 1924, pp. 11–12; Cmd 2169, vol. 26, no. 46, p. 164, The Earl of Balfour to Lord Hardinge (Paris), 30 May 1922.
185. FO 371/7000, W13226A/12716/17, British Secretary's Notes of a Conversation between Mr Lloyd George and M. Poincaré, held at Paris, 14 January 1922; Riddell, *Intimate Diary*, p. 349; Cmd 2169, vol. 26, no. 40, pp. 131–6, The Marquess Curzon of Kedleston to Lord Hardinge (Paris), 28 January 1922; Le Livre Jaune Français, 1924, no. 25, pp. 127–35, M. Poincaré, Président du Conseil, Ministre des Affaires Étrangères, à M. de Saint-Aulaire, Ambassadeur de France à Londres, 29 January 1922 – communicated to the Foreign Office on 1 February, Cmd 2169, vol. 26, no. 41, p. 131; FO 371/8250, W1379/50/17, The Marquess Curzon of Kedleston to Lord Hardinge (Paris), 9 February 1922, reported also in Le Livre Jaune Français, 1924, no. 26, p. 135.
186. Cmd 2169, vol. 26, no. 44, pp. 154–62, The Anglo-French Agreement: Memorandum by the Secretary of State for Foreign Affairs, 17 February 1922; F. S. Northedge, *The Troubled Giant: Britain Among the Great Powers, 1916–1939* (New York, 1967), pp. 227–9.
187. CAB 23/29, 2(22)(2), Conclusions of a Meeting of the Cabinet, 18 January 1922.
188. FO 379/9250, W1379/50/17, The Marquess Curzon of Kedleston to Lord Hardinge (Paris), 9 February 1922; ibid., [and *DBFP*, 1st Series, vol. xix, no. 50, pp. 240–4], W2448/50/17, The Marquess Curzon of Kedleston to Lord Hardinge (Paris), 19 March 1922; ibid., W2296/50/17, R. H. Campbell to Secretary of the Army Council, 18 April 1922.
189. Hardinge MSS, vol. 45, Hardinge to Curzon, 20 January, 1, 23 February 1922.
190. CAB 23/30, 22(3), Conclusions of a Meeting of the Cabinet, 23 May 1922; Goold, *Lord Hardinge as Ambassador*, pp. 923–8; Walter A. McDougall, *France's Rhineland Diplomacy, 1914–1924: The Last Bid for a Balance of Power in Europe* (Princeton, 1978), pp. 183–8.

191. Kurt Rosenblum, *Community of Fate: German–Soviet Diplomatic Relations, 1922–1928* (Syracuse, 1965), pp. 26–8, 40–1; Fink, *The Genoa Conference*, pp. 126–33, 162–4; Felix, *Rathenau*, pp. 127–38; William Manchester, *The Arms of Krupp, 1587–1968* (New York, 1970), pp. 365–6, 388–9.

192. FO 371/890, N4640/646/38, The Treaty of Rapallo, 16 April 1922; ARWR, KAB. Wirth, 1922, vol. 2, no. 246, pp. 705–6, Kabinettssitzung, 17 April 1922; Fink, *The Genoa Conference*, pp. 164–7, 172–4; Felix, *Rathenau*, pp. 138–44; see Peter Krüger, 'A Rainy Day, April 16, 1922: The Rapallo Treaty and the Cloudy Perspective for German Foreign Policy', in Carole Fink, Alex Frohn and Jürgen Heideking, *Genoa, Rapallo and European Reconstruction in 1922*, pp. 49–65.

193. For further information see also C. Fink, A. Frohn and J. Heideking (eds), *Genoa, Rapallo and European Reconstruction in 1922* (Cambridge, 1991).

194. FO 371/7426, C5473/458/62, J. D. Gregory to Crowe, 13 April 1922.

195. Ibid., Minute by Tyrrell, 18 April 1922.

196. Ibid., Minute by Curzon, 18 April 1922.

197. FO 371/7567, C6656/6347/18, Curzon to A. Chamberlain, 4 May 1922.

198. FO 371/8251, W4880/50/17, The Earl of Balfour to Lord Hardinge (Paris), 13 June 1922; ibid., Minute by Villiers, 31 May 1922.

199. *DBFP*, 1st Series, vol. xix, no. 250, pp. 242–5, The Marquess Curzon of Kedleston to Lord Hardinge (Paris), 19 March 1922.

200. FO 371/8251, W4880/50/17, The Earl of Balfour to Lord Hardinge (Paris), 30 May, 13 June 1922.

201. Ibid., W4995/50/17, Lord Hardinge (Paris) to the Earl of Balfour, 16 June 1922.

202. Ibid., Minute by Villiers, 17 June 1922.

203. Pierre Miquel, *Poincaré* (Paris, 1961), pp. 85–6, 308–12, 381–485; George Samme, *Raymond Poincaré: Politique et Personnel de la 3ᵉ· République* (Paris, 1933), pp. 7–9, 24, 77–8, 338, 376; Emmanuel de Perretti de la Rocca, 'Briand et Poincaré: Souvenirs', *Revue de Paris*, vol. 15, December 1936, pp. 775–88. For typical British opinions of Poincaré, see FO 371/9397, W2558/2558/17, France – Annual Report, 1922, by Lord Hardinge, 31 December 1922; Viscount D'Abernon, *An Ambassador of Peace: Pages from the Diary of Viscount D'Abernon* (London, 1929), vol. 1, pp. 317–18; ibid., vol. 2, pp. 21–3; Lord of Ronaldshay, *The Life of Lord Curzon* (London, 1928), vol. 3, p. 323.

204. FO 371/7477, C7526/99/18, Lord Hardinge (Paris) to the Marquess Curzon of Kedleston, 20 May 1922, enclosure: M. Poincaré to M. Cheetham, 19 May 1922. Reproduced for the bulk of it in *DBFP*, 1st Series, vol. xx, no. 23, footnote 2, pp. 49–51.

205. FO 371/9397, W2558/2558/17, France – Annual Report, 1922, by Lord Hardinge, 31 December 1922; *DBFP*, 1st Series, vol. xxi, no. 332, pp. 483–4, Mr Phipps (Paris) to the Marquess Curzon of Kedleston, 13 August 1923.

206. H. Nicolson. *Curzon*, pp. 19, 43–8, 194–5.

207. *DBFP*, 1st Series, vol. xvi, no. 768, p. 867, Memorandum by the Marquess Curzon of Kedleston on the question of an Anglo-French Alliance, 28 December 1921. In this connection, Curzon was particularly outraged by the evidence of secret dealings between Britain's French Allies and defeated Turkish Nationalists. Intercepted French telegrams not only re-convinced the Foreign Secretary of the justness of his belief of yore, that the French 'are not the sort of people one would go tiger-shooting with', but raised his suspicions to new heights. Christopher Andrew, *Secret Service: The Making of the British Intelligence Community*, p. 296; B. C. Busch, *Mudros to Lausanne: Britain's Frontier in West Asia, 1918–1923*

(Albany, 1976), pp. 225–6, 228–9, 234, 245, 310–11, 322–4, 335–8, 347–51; L. Mosley, *Curzon* (London, 1961), p. 210.

208. FO 371/7892, E9755/27/44, Lord Hardinge (Paris) to Sir Eyre Crowe, 22 September 1922; *DBFP*, 1st Series, vol. viii, nos 42, 48, pp. 51–3, 75–8, British Secretary's Notes of a Conference between the French President of the Council, The British Secretary of State for Foreign Affairs, and the Italian Ambassador in Paris, held at the Quai d'Orsay, 22 September 1922.

209. Hardinge, *Old Diplomacy*, pp. 272–3, partly quoted in Goold, *Lord Hardinge as Ambassador*, p. 932. Harold Nicolson's version of what took place in the adjoining room is even more dramatic. Curzon, he writes, 'collapsed upon a scarlet settee. He grasped Lord Hardinge by the arm. "Charley", he panted "I can't bear that horrid little man. I can't bear him. I can't bear him". He wept.' H. Nicolson, *Curzon*, pp. 273–4. For Curzon's version, see FO 371/7892, E9755/27/44, Lord Hardinge (Paris) to Sir E. Crowe following from Lord Curzon, 22 September 1922, partly reproduced in *DBFP*, 1st Series, vol. xviii, no. 48, footnote 26, p. 78, and John Vincent (ed.), *The Crawford Papers; The Journals of David Lindsay Twenty-seventh Earl of Crawford and Tenth Earl of Balcarres, 1871–1940; during the years 1892 to 1940* (Manchester, 1984), pp. 437–8.

210. FO 371/7486, C14819/99/18, Minute by Wigram, 26 October 1922; ibid., Minute by Tyrrell, 26 October 1922. See also B. Kent, *The Spoils of War: the Politics, Economics and Diplomacy of Reparations, 1918–1932*.

211. *DBFP*, 1st Series, vol. xx, no. 152, enclosure, p. 356, The Reparation Commission to the British Government, 26 December 1922.

212. Ibid., no. 38, pp. 68–79, British Secretary's Notes of a Meeting between Mr Lloyd George and M. Poincaré, held at London, 19 June 1922; ibid., no. 49, pp. 106–11, Sir J. Bradbury (Reparation Commission) to Treasury, 4 August 1922; ibid. nos 50–64, pp. 112–231, Proceedings of the Fifth Conference of London, 7–14 August, 1922; FO 371/7487, C16001/99/18, Minute by Wigram, 25 November 1922; ibid., Minute by Lampson, 25 November 1922; FO 371/7488, C16643/99/18, Central Department Memorandum on the Reparation position, 5 December 1922; FO 371/7491, C17445/99/1, Minute by Lampson, 19 December 1922; ibid., Minute by Crowe, 20 December 1922.

213. PP 1923 (Cmd 1812), vol. 24, pp. 68–70, 95–132, 140–2, 154–7, 188–93, Reports and Secretaries' Notes of Conversations at Inter-Allied Conferences on Reparations and Inter-Allied Debts, held in London and Paris, December 1922 and January 1923; Carl Bergmann, *The History of Reparations* (London, 1927), pp. 163–9; Marks, 'The Myths of Reparations', pp. 242–3; Rupieper, *The Cano Government*, pp. 43–77; Weill-Raynal, *Les Réparations allemandes*, vol. 2, p. 336. See B. Kent, *Spoils of War*, chs 5 and 6.

214. *DBFP*, 1st Series, vol. xxi, nos 11, 22, pp. 14, 26–7, The Marquess of Crewe (Paris) to the Marquess Curzon of Kedleston, 8, 11 January 1923; Jacques Bariety, *Les Relations Franco-Allemandes après la Première Guerre Mondiale: 10 Novembre 1918–10 Janvier 1925 de l'Exécution à la Négociation* (Paris, 1977), pp. 101–9; Marks, 'The Myths of Reparations', p. 244.

215. L.P. Archives, LP/IAC/2/270, Draft Manifesto, 11 January 1923. See also D. G. Williamson, 'Great Britain and the Ruhr Crisis, 1923–1924', *British Journal of International Studies*, vol. 3 (1977), pp. 73, 77.

216. *DBFP*, 1st Series, vol. xxi, no. 212, pp. 252–3, Mr Niemeyer (Treasury) to Sir E. Crowe, 4 May 1923; Baldwin MSS, vol. 125, Sir Warren Fisher to Baldwin, 11 June 1923; Keith Middlemas and John Barnes, *Baldwin, A Biography* (London, 1969), pp. 182–3.

217. CAB 24/161, CP 312, Memorandum by the President of the Board of Trade, Sir Philip Lloyd-Graeme, 6 July 1923.
218. CBI Archives, file nos 3521–899, British Industry and the Situation in Europe, May 1923; Williamson, 'Great Britain and the Ruhr Crisis', p. 77.
219. RIIA, *Survey of International Affairs, 1920–1923* (Oxford, 1925), pp. 191–2.
220. Baldwin MSS, vol. 126, Bradbury to Baldwin, 2 July 1923.
221. CAB 23/45, 1(23), Conclusions of a Meeting of the Cabinet, 11 January 1923; *DBFP*, 1st Series, vol. xxi, no. 161, enclosure, pp. 170–1, Memorandum communicated by the Secretary of State for Foreign Affairs to His Excellency the Comte de Saint-Aulaire, 20 March 1923.
222. T. Wilson (ed.), *The Political Diaries of P. C. Scott, 1911–1928* (London, 1970), p. 435.
223. CAB 24/158, The Marquess Curzon of Kedleston (Lausanne) to Mr Lampson, 31 January 1923; Hansard, Parliamentary Debates, 5th Series, H of L, vol. 53, cols 781–98, 20 April 1923; FO 371/8627 C5451/1/18, Foreign Office Memorandum, 10 January 1923; CAB 23/45, 3 (23), Conclusions of a Meeting of the Cabinet, 26 January 1923; *DBFP*, 1st Series, vol. xxi, nos 12, 15, 35, 47, 61, 74, 103, 120.
224. FO 371/8648, C13559/1/18, Memorandum by Crowe, 31 July 1923.
225. FO 371/8730, C8383/313/18, Minute by Tyrrell, 13 April 1923.
226. CAB 23/45, 1(23); ibid., 3(23), Conclusions of a Meeting of the Cabinet, 11, 26 January 1923; *DBFP*, 1st Series, vol. xxi, nos 20, 24, pp. 25, 28–9, The Marquess Curzon of Kedleston to Lord Kilmarnock (Coblenz), 11, 15 January 1923.
227. CAB 23/45, 15(23), Conclusions of a Meeting of the Cabinet, 14 March 1923. For further information on this point see Crowe and Corp, *Our Ablest Public Servant: Sir Eyre Crowe, 1864–1925*, pp. 433–40.
228. *DBFP*, 1st Series, vol. xxi, no. 151, pp. 158–9, The Marquess Curzon of Kedleston to Lord D'Abernon (Berlin), 14 March 1923.
229. Ibid., no. 161, pp. 169–70, The Marquess Curzon of Kedleston to Mr Phipps (Paris), 21 March 1923.
230. Hansard, Parliamentary Debates, 5th Series, H of L, vol. 53, cols 781–798, 20 April 1923.
231. Earlier on, at the beginning of April, the French, too, made unofficial feelers through Louis Loucheur, the Former Minister of the Liberated Areas, but his mission fell through. See *DBFP*, 1st Series, vol. xxi, nos 173, note 4, 176, 180, note 4; Jacques de Launay (ed.), *Louis Loucheur, Carnets Secrets, 1908–1932* (Brussels, 1962), pp. 117–21 and Rupieper, *The Cano Government*, pp. 136–7.
232. FO 371/8633, C7832/1/18, The German Embassy to the Foreign Office, 2 May 1923, reproduced in FRUS, 1923, vol. 2, pp. 57–60 and UF, vol. 5, no. 1044, pp. 121–4.
233. FO 800/243, vol. 2, Tyrrell to Crowe, 7 December 1922.
234. FRUS, 1922, vol. 2, pp. 199–202, The Acting Secretary of State to the Ambassador in France (Herrick), 29 December 1922. See also *DBFP*, 1st Series, vol. xx, no. 157, pp. 361–2.
235. See for example, Cmd 1812, pp. 112–19 and Hansard, Parliamentary Debates, 5th Series, vol. 162, cols 598, 609, 28 March 1923.
236. FO 371/8633, C7832/1/18, The German Embassy to the Foreign Office, 2 May 1923, reproduced in FRUS, 1923, vol. 2, pp. 59–60 and UF, vol. 5, no. 1044, p. 124.
237. *DBFP*, 1st Series, vol. xxi, no. 212, enclosure, pp. 253–4, Note by Sir J. Bradbury on the German Reparation Offer, May 1923 (undated).

238. FO 371/8634, C8044/1/18, Franco-Belgian Reply to the German Note, 6 May 1923, reproduced in UF, vol. 5, no. 1044, pp. 125–30.
239. *DBFP*, 1st Series, vol. xxi, no. 212, p. 252, Mr Niemeyer (Treasury) to Sir E. Crowe, 4 May 1923.
240. Ibid., no. 218, p. 263, The Marquess Curzon of Kedleston to the Marquess of Crewe (Paris), 5 May 1923.
241. FO 371/8635, C8311/1/18, The Marquess Curzon of Kedleston to the German Ambassador, 18 May 1923, reproduced in UF, vol. 5, no. 1044, pp. 130–2.
242. FO 371/8635, C8311/1/18, The German Ambassador to The Marquess Curzon of Kedleston, 7 June 1923, reproduced in FRUS, 1923, vol. 2, pp. 62–4 and UF, vol. 5, no. 1048, pp. 145–6.
243. CAB 23/46, 30(23), Conclusions of a Meeting of the Cabinet, 11 June 1923; *DBFP*, 1st Series, vol. xxi, no. 261, p. 337, The Marquess Curzon of Kedleston to the Marquess of Crewe (Paris), 11 June 1923; ibid., no. 286, note 5, pp. 386–7; Middlemas and Barnes, *Baldwin*, pp. 182–6.
244. *DBFP*, 1st Series, vol. xxi, no. 257, pp. 324–5, Record by Sir E. Crowe of conversations with French, Belgian and Italian representatives, 8 June 1923; ibid., no. 261, pp. 333–8, The Marquess Curzon of Kedleston to the Marquess of Crewe (Paris), 11 June 1923. The Belgians, it is true, did not veto the German plan. However, in deference to Poincaré, they demanded the cessation of passive resistance before negotiations could take place. See *DBFP*, 1st Series, vol. xxi, nos 257, 262, 286.
245. Ibid., no. 264, pp. 342–5, The Marquess Curzon of Kedleston to the Count de Saint-Aulaire, 13 June 1923.
246. FO 371/8639, C10185/1/18, Aide Mémoire by M. Poincaré, 11 June 1923; ibid., C13105/1/18, Note communicated by French Ambassador, 30 July 1923; *DBFP*, 1st Series, vol. xxi, nos 261, 266, 285, 287, 292, pp. 333–8, 346–7, 382–3, 387–90, 397–404, The Marquess Curzon of Kedleston to the Marquess of Crewe (Paris), 11, 15 June, 2, 3, 5 July 1923.
247. For details concerning the official reaction of the Government; far-reaching stands of persons such as Smuts, Hoare, Bradbury and Warren Fisher; plans laid down by the Foreign Office, Treasury and British members of the Reparation Commission for going independently ahead with an enquiry held by international experts and bringing – if necessary – into action financial and legal weapons to force its bindings on Belgium and France; soundings of Italy and the United States on this point; and deadlock reached in official negotiations, see CAB 23/46, 35–37(23), Conclusions of a Meeting of the Cabinet, 9, 11, 12 July 1923; Hansard, Parliamentary Debates, 5th Series, H of L, vol. 54, cols 992–6 and H of L, vol. 166, cols 1584–9, 12 July 1923; Middlemas and Barnes, *Baldwin*, pp. 182–9; Rupieper, *The Cano Government*, pp. 218–25.
248. *DBFP*, 1st Series, vol. xxi, no. 306, pp. 426–32, The Marquess Curzon of Kedleston to the French Ambassador, 20 July 1923.
249. FO 371/8645, C14306/1/18, Baron Moncheur to the Marquess Curzon of Kedleston, 30 July 1923, reproduced in PP 1923, Cmd 1943, vol. 25, no. 7, Correspondence with Allied Governments respecting Reparation Payments by Germany, June–August 1923.
250. Ibid., C13105/1/18, Count de Saint-Aulaire to the Marquess Curzon of Kedleston, 30 July 1923, reproduced in Cmd 1943, vol. 25, no. 6.
251. Nevertheless, this was preceded, it should be noted, by a heated inter-ministerial debate during which Crowe and Curzon's originally biting answer underwent – first by Baldwin and then by the more Francophile members of the Cabinet – certain revisions with the aim of softening

several passages which might have upset French public opinion. See L. S. Amery, *My Political Life* (London, 1929), vol. 2, pp. 267–8; R. S. Churchill, *Lord Derby 'King of Lancashire'* (London, 1959), pp. 613–16; Middlemas and Barnes, *Baldwin*, pp. 190–1. See, for a fuller account of Crowe's pressure on Curzon to depart from a policy of neutrality, Crowe and Corp, *Our Ablest Public Servant: Sir Eyre Crowe, 1864–1925*, pp. 428–38.

252. *DBFP*, 1st Series, vol. xxi, no. 330, pp. 467–82, The Marquess Curzon of Kedleston to Count de Saint-Aulaire and to Baron Moncheur, 11 August 1923.

253. FO 371/8649, C14380/1/18, M. Poincaré to the Marquess of Crewe, 20 August 1923, reproduced in Ministre des Affaires Étrangères, Documents Diplomatiques, Réponse du Gouvernement Français à la Lettre du Gouvernement Britannique du 11 Août 1923 sur les Réparations (20 Août 1923) (Paris, 1923). See also RIIA, *Survey*, 1924, pp. 336–8 and Weill-Raynal, *Les Réparations Allemandes*, vol. 2, pp. 444–6.

254. FO 371/8651, C14680/1/18, M. Jaspar to Sir G. Grahame, 27 August 1923; RIIA, *Survey*, 1924, p. 338; Weill-Raynal, *Les Réparations Allemandes*, vol. 2, pp. 456–7.

255. *DBFP*, 1st Series, vol. xxi, no. 367, pp. 529–34, Note on Conversation of 19 September 1923, between Mr Baldwin and M. Poincaré.

256. Ibid., no. 385, pp. 556–7, The Marquess Curzon of Kedleston to Sir G. Grahame (Brussels), 1 October 1923; ibid., no. 386, note 1, p. 557.

257. Feldman, *Iron and Steel*, pp. 393–405; Rupieper, *The Cano Government*, pp. 174–217, 231–3; Werner Weidenfeld, *Die Englandpolitik Gustav Stresemanns: Theoretische und praktische Aspecte der Außenpolitik* (Mainz, 1972), pp. 188–9.

258. ARWR, KAB. Stresemann, 1923, vol. 1, no. 81, pp. 361–72, Ministerrat, 25 September 1923. See also, ibid., nos 76–80 and K. P. Jones, 'Stresemann, the Ruhr Crisis, and Rhenish Separatism: A Case Study of Westpolitik', *European Studies Review*, vol. 7 (1977), pp. 312–15.

259. CAB 32/9, 1, E. 3rd Meeting, Imperial Conference Minutes, 5 October 1923.

260. Ibid., *DBFP*, 1st Series, vol. xxi, nos 271, 365, 367, 406, 422, 425.

261. *DBFP*, 1st Series, vol. xxi, nos 394, 396, 405.

262. President Coolidge's declaration of 9 October, *The Times*, 11 October 1923, p. 12.

263. *DBFP*, 1st Series, vol. xxi, no. 392, pp. 563–4, The Marquess Curzon of Kedleston to Mr Chilton (Washington), 12 October 1923.

264. FRUS, 1923, vol. 2, pp. 70–3, The Secretary of State to the British Charge (Chilton), 15 October 1923.

265. *DBFP*, 1st Series, vol. xxi, no. 403, pp. 574–6, The Marquess Curzon of Kedleston to the Marquess of Crewe (Paris), 19 October 1923.

266. FO 371/8658, C1838/1/18, M. Jaspar to Mr Wingfield (Brussels), 25 October 1923, and *DBFP*, 1st Series, vol. xxi, no. 409, pp. 585–6, Mr Wingfield (Brussels), 25 October 1923.

267. FO 371/9659, C18877/1/18, M. Mussolini to Sir R. Graham (Rome), 26 October 1923. See also *DBFP*, 1st Series, vol. xxi, no. 406 footnote 5, nos 405, 435.

268. Middlemas (ed.), *Thomas Jones*, vol. 1, p. 249.

269. MAE, Allemagne, 483, fols 26–32, M. de Saint-Aulaire, Ambassadeur de France à Londres, à M. Poincaré, Président du Conseil, Ministre des Affaires Étrangères, 8 November 1923. See also *DBFP*, 1st Series, vol. xxi, no. 472, p. 670, Record by Sir E. Crowe of a conversation with the French Ambassador, 16 November 1923, and Laroche, *Au Quai d'Orsay*, p. 182.

270. CAB 32/9, 1 E. 3rd Meeting, Imperial Conference Minutes, 5 October 1923.
271. FRUS, 1923, vol. 2, pp. 94–5, The Secretary of State to the Ambassador in France (Herrick), 9 November 1923. See also, ibid., pp. 87–9, 91–4, 97–8.
272. FO 371/8656, C18541/C18551/1/18, The Marquess of Crewe (Paris) to the Foreign Office, 27, 28 October 1923; ibid., C18866/1/18, Note pour l'ambassade de Grande-Bretagne, 31 October 1923, transmitted to the Foreign Office on 1 November; DBFP, 1st Series, vol. xxi, nos 432, 442, pp. 615–16, 624–6, The Marquess of Crewe (Paris) to the Marquess Curzon of Kedleston, 31 October, 3 November 1923; FRUS, 1923, vol. 2, pp. 86–95; RIIA, *Survey*, 1924, pp. 343–5.
273. Schuker, *The End of French Predominance*, chs 2–3 and pp. 89–100, 172–4, 196–7, 229–30, 384–6.
274. Mr Baldwin's speech of 25 October at Plymouth, *The Times*, 26 October 1924, pp. 12, 17; DBFP, 1st Series, vol. xxi, no. 472, pp. 670–1, Record by Sir E. Crowe of a conversation with the French Ambassador, 16 November 1923; ibid., nos 425, 465, pp. 609, 656–9, The Marquess Curzon of Kedleston to the Marquess of Crewe (Paris), 30 October, 10 November 1923. For further material respecting the French attempts to detach the Rhineland from Germany, and British and German reactions, see notes 287–90.
275. Mr Baldwin's speech of 19 November at Queen's Hall, London, *The Times*, 20 November 1923; DBFP, 1st Series, vol. xxi, no. 467, pp. 663–4, Record by Sir E. Crowe of a conversation with the German Ambassador, 12 November 1923; ibid., nos 394, 449, pp. 565–6, 638–9, The Marquess Curzon of Kedleston to the Marquess of Crewe (Paris), 15 October, 5 November 1923; ibid., no. 466, footnote 1, no. 510, footnote 7, pp. 660, 734, The Marquess Curzon of Kedleston to Lord D'Abernon (Berlin), 7 November 1923, 4 January 1924. For further details respecting the Franco-German industrial agreement see Feldman, *Iron and Steel*, pp. 405–27, and Trachtenberg, *Reparation*, pp. 325–9.
276. DBFP, 1st Series, vol. xxi, no. 416, p. 597, Sir J. Bradbury (Reparation Commission) to Sir E. Crowe, 26 October 1923; FO 371/8659, C18916/1/18, British Delegation, Reparation Commission, to Treasury, 30 October 1923; FO 371/8662, C20977/1/18, B. A. Kemball-Cook (Reparation Commission) to Sir J. Bradbury, 27 October 1923; Rupieper, *The Cano Government*, pp. 237–42.
277. DBFP, 1st Series, vol. xxi, no. 480, pp. 681–2, Minute by Sir E. Crowe, 27 November 1923; ibid., no. 488, p. 705, The Marquess Curzon of Kedleston to Mr Chilton (Washington), 6 December 1923.
278. Ibid., no. 481, footnote 2, p. 683, British Delegation, Reparation Commission, to Treasury, 30 November 1923; FRUS, 1923, vol. 3, p. 108, The Ambassador in France (Herrick) to the Secretary of State, 21 December 1923.
279. MAE, Allemagne, 483, fol. 100–11, 3 December 1923; FO 371/8663, C22245 / C22712 / C22294 / 1/18, Sir J. Bradbury (Reparation Commission) to Mr N. Chamberlain, 24, 27 December 1923, the latter reproduced in DBFP, 1st Series, vol. xxi, no. 515; Rupieper, *The Cuno Government*, pp. 246–7.
280. DBFP, 1st Series, vol. xxi, no. 481, footnote 2, p. 683, British Delegation, Reparation Commission, to Treasury, 30 November 1923; ibid., no. 511, pp. 734–6, Reparation Commission to British Delegation, Reparation Commission, 26 December 1923.
281. Ibid., no. 513, pp. 736–8, Lord Kilmarnock (Coblenz) to the Marquess Curzon of Kedleston, 27 December 1923; ibid., vol. 26, no. 313, p. 477, Lord Kilmarnock (Coblenz) to the Marquess Curzon of Kedleston, 2 January

1924; ibid., no. 396, pp. 508–9, The Marquess Curzon of Kedleston to Lord Crewe (Paris), 19 January 1924; ARWR, KAB. Marx, 1923–4, vol. 1, nos 12, 37, 55, 59, 63; Carsten, *Britain and the Weimar Rupublic*, pp. 331–2; H. Nicolson, *Curzon*, p. 376.

282.   *DBFP*, 1st Series, vol. xxi, no. 410, pp. 586–9, Lieut.-Colonel Ryan (Coblenz) to the Marquess Curzon of Kedleston, 25 October 1923, and nos 412, 418, 430, footnote 3, 433, 445–6, 494–5; ibid., vol. 26, no. 336, pp. 506–8, the Marquess Curzon of Kedleston to Lord Crewe (Paris), 9 January 1924; ARWR, KAB. Stresemann, 1923, vol. 2, nos 164–5, 179–180, 194, 198–9, 202–3, 210, 233–4, 240, 249–250, 263, 266–7; ibid., KAB. Marx, 1923–5, vol. 1, nos 6–7, 9, 17, 24, 28, 43; Carsten, *Britain and the Weimar Republic*, pp. 158–63; Karl Dietrich Erdmann, *Adenauer in der Rheinlandpolitic nach dem ersten Weltkrieg* (Stuttgart, 1966), chs 5–12; Jones, 'Stresemann', vol. 7, pp. 319–33; McDougall, *Rhineland*, pp. 190–206, 299–331, 346–53; H. Nicolson, *Curzon*, pp. 375–6; Weidenfeld, *Die England politik*, pp. 201–4.

283.   Der Hitler-Prozess vor dem Volksgericht in München, Teil 1–2 (Glashütten im Tanus, 1973); Jon Dornberg, *The Putsch that Failed, (Munich, 1923): Hitler's Rehearsal for Power* (London, 1982); Harold J. Gordon Jr, *Hitler and the Beer Hall Putsch* (Princeton, 1972).

284.   *DBFP*, 1st Series, vol. xxi, nos 633–5, 639–40, 643–5, 650, 652–3, 655–8, 660–1, 663.

285.   Ibid., vol. xxvi, no. 315, The Marquess Curzon of Kedleston to Lord Crewe (Paris), 4 January 1924.

286.   FO 371/9771, C509/91/18, The Marquess Curzon of Kedleston to Mr Clive (Munich), 9 January 1924.

287.   *DBFP*, 1st Series, vol. xxvi, no. 334, pp. 503–4, Lord Kilmarnock (Coblenz) to the Marquess Curzon of Kedleston, 19 January 1924.

288.   Hansard, Parliamentary Debates, 5th Series, H of L, vol. 169, cols 485–6, 21 January 1924.

289.   *DBFP*, 1st Series, vol. xxvi, no. 334, pp. 503–4, Lord Kilmarnock (Coblenz) to the Marquess Curzon of Kedleston, 19 January 1924. For Clive's full report, see, FO 371/9773, C1178/99/18, Mr Clive (Munich) to the Marquess Curzon of Kedleston, 22 January 1924.

290.   *DBFP*, 1st Series, vol. xxvi, no. 349, pp. 526–9, Mr MacDonald to Mr Phipps (Paris), 4 February 1924; ibid., no. 354, footnote 1, Mr Clive (Munich) to Mr MacDonald, 6 February 1924; ibid., nos 359, 360, 367, 375, 389, Lord Kilmarnock (Coblenz) to Mr MacDonald, 13, 19, 28 February, 15 March 1924; ARWR, KAB. Marx, 1923–5, vol. 1, nos 59, 63, 102, 107, 169; Carsten, *Britain and the Weimar Republic*, pp. 163–6; McDougall, *Rhineland*, pp. 333–5; Weidenfeld, *Die England Politik*, pp. 205–10.

Chapter 5
The First Labour Government:
MacDonald and the Foreign Office

1.   FO 800/218, Selby to W. D. Roberts, 26 May 1924.
2.   Martin Gilbert, *The Roots of Appeasement* (London, 1966), p. 105.
3.   Sir Arthur Willert, *Washington and Other Memories* (Boston, 1972), p. 166.
4.   PRO 30/69/8/1, MacDonald's Diary, vol. 1, 3 February 1924.
5.   Willert, *Washington and Other Memories*, p. 166.

6.   The Earl of Oxford and Asquith, *Memoirs and Reflections, 1857–1927* (London, 1928), vol. 2, p. 211.

7.   Richard Burdon Haldane, *An Autobiography* (London, 1929), p. 332; Sir Frederick Maurice, *Haldane, 1856–1928: The Life of Viscount Haldane of Cloan, K.T., O.M.* (London, 1939), vol. 2, p. 180.

8.   Margaret Cole (ed.), *Beatrice Webb's Diaries, 1924–1932* (London, 1956), p. 13.

9.   *The Dictionary of National Biography, 1931–1940* (Oxford, 1961), pp. 562–7; Lord Elton, *Life of James Ramsay MacDonald, 1866–1919* (London, 1939); David Marquand, *Ramsay MacDonald* (London, 1977), chs 1–13; H. Hessell and J. Filtman, *Ramsay MacDonald: Labor's Man of Destiny* (New York, 1929), chs. 1–8.

10. PRO 30/69/8/1, MacDonald's Diary, vol. 1, 3 February 1924.

11. FO 800/218, Selby to S. de Montille, 18 June 1924; FO 371/9744, C7953/70/18, Minute by Crowe, 14 May 1924.

12. FO 371/10562, W2845/2845/50, Note by Ponsonby, 24 March 1924; FO 371/9794, C11769/371/18, Minute by Crowe, 25 July 1924.

13. FO 371/9740, C6363/70/18, Minute by Crowe, 15 April 1924.

14. PRO 30/69/8/1, MacDonald's Diary, vol. 1, 24 September 1924. See also, FO 800/218, Selby to Archibald Clerk Kerr, 27 September 1924.

15. Bruce Heady, *British Cabinet Ministers: The Roles of Politicians in Executive Office* (London, 1974), p. 44.

16. Haldane, *An Autobiography*, pp. 332–3.

17. Heady, *British Cabinet Ministers*, pp. 70, 121–2, 143.

18. Maurice, *Haldane*, vol. 2, p. 160.

19. Philip Viscount Snowden, *An Autobiography* (London, 1934), vol. 2, p. 679.

20. FO 371/813, C2946/1346/18, Memorandum on British Foreign Policy in the Rhineland, 5 February 1924.

21. Ibid.; C2028/1346/18, Minute by Nicolson, 6 February 1924.

22. Ibid., Minute by Crowe, 9 February 1924.

23. Hansard, Parliamentary Debates, 5th Series, H of C, vol. 169, cols 772–4, 12 February 1924.

24. FO 371/9813, C2028/1346/18, Minute by MacDonald, 18 February 1924.

25. FO 800/218, Mr MacDonald to M. Poincaré, 26 January 1924, reproduced in *The Times*, 4 February 1924, p. 12, and translated in Etienne Weill-Raynal, *Les Réparations allemandes et la France* (Paris, 1947), vol. 3, p. 15.

26. FO 371/10359, W1188/631/17, Mr Phipps (Paris) to Sir E. Crowe, 7 February 1924.

27. For details, see *DBFP*, 1st Series, vol. xx, no. 270, footnote 5, p. 542.

28. FO 371/9723, C1461/9/18, Foreign Office Memorandum, 25 January 1924; ibid., C1725/9/18, War Office to Foreign Office, 31 January 1924; ibid., Minute by Troutbeck, 1 February 1924; ibid., Minute by Hurst, 7 February 1924; ibid., Minute by Crowe, 12 February 1924.

29. *DBFP*, 1st Series, vol. xxvi, no. 262, pp. 988–9, Lord D'Abernon (Berlin) to Mr MacDonald, 5 February 1924.

30. FO 371/9723, C2159/9/18, Mr MacDonald to Lord Crewe (Paris), 11 February 1924.

31. *DBFP*, 1st Series, vol. xxvi, no. 638, pp. 1004–6, Lord Crewe (Paris) to Mr MacDonald, 5 March 1924.

32. FO 371/9726, C12580/9/18, Central Department Memorandum, 12 August 1924.

33. Ibid., C10438/9/18, Lord Crewe (Paris) to Mr MacDonald, 30 June 1924.

34. Ibid., C11044/9/18, Lord Crewe (Paris) to Mr MacDonald, 9 July 1924.

35. Ibid., C13070/9/18, Minute by MacDonald, 14 August 1924; *DBFP*, 1st

Series, vol. xxvi, no. 692, footnote 7, p. 1100; ibid., no. 697, pp. 1105–6, Mr MacDonald to Lord D'Abernon (Berlin), 25 August 1924.

36. FO 371/9802, C7879/737/18, Minute by Crowe, 21 May 1924.
37. Ibid., Memorandum by J. M. Troutbeck, 15 May 1924.
38. FO 371/9813, C5185/1346/18, General Staff Memorandum on the Military Aspect of the Future Status of the Rhineland, 28 March 1924. See also, Stephen Schuker, *The End of French Predominance in Europe*, pp. 254–5.
39. FO 371/9819, C15288/2048/18, War Office to the Secretary of the CID, 29 September 1924, enclosure: General Staff Memorandum on M. Herriot's Letter to Mr MacDonald, dated 11 August 1924.
40. FO 371/9813, C5815/1346/18, Minute by Bennett, 15 April 1924; ibid., Minute by Lampson, 16 April 1924. See also, Schuker, *The End of French Predominance*, pp. 253–4.
41. Ibid., C4218/1346/18, Board of Trade to Foreign Office, 12 March 1924.
42. Ibid., C3420/1346/18, Waterlow to Lampson, 28 February 1924.
43. FO 371/9819, C14272/2048/18, Minute by Nicolson, 9 September 1924; ibid., Minute by Lampson, 10 September 1924.
44. FO 371/9813, C2028/1346/18, Minute by Lord Parmoor, 1 March 1924.
45. FO 371/9818, C10067/2048/18, General Staff Memorandum, 24 June 1924.
46. Ibid., Minute by MacDonald, 3 July 1924; Hansard, Parliamentary Debates, 5th Series, H of C, vol. 169, cols 772–4, 12 February 1924.
47. 'Mr MacDonald's Speech of April 19 at York', *Daily Telegraph*, 21 April 1924, p. 5.
48. FO 371/9802, C7897/737/18, Minute by MacDonald, 29 May 1924.
49. FO 371/9818, C10067/2048/18, Minute by MacDonald, 3 July 1924.
50. Ibid., Minute by Crowe, 4 July 1924.
51. FO 371/9813, C2946/1346/18, Memorandum by Mr Phipps, 18 February 1924.
52. Ibid., Minute by Bennett, 2–3 February 1924.
53. Ibid., C3814/1346/18, Mr Addison (Berlin) to Mr Nicolson, 1 March 1924. See also, Schuker, *The End of French Predominance*, pp. 252–3.
54. FO 371/9813, C5185/1346/18, Minute by Lampson, 16 April 1924.
55. FO 371/9819, C14272/2048/18, Note on French Security by J. N. Troutbeck, 9 September 1924.
56. FO 371/1813, C13819/1788/18, Minute by Lampson, 2 September 1924.
57. FO 371/9819, C14272/2048/18, Minute by Nicolson, 9 September 1924.
58. FO 371/9818, C10067/2048/18, Minute by Lord Parmoor, 22 July 1924.
59. Hansard, Parliamentary Debates, 5th Series, H of C, vol. 171, cols 1600–11, 27 March 1924.
60. FO 371/9819, C13663/2048/18, Minute by Nicolson, 27 August 1924.
61. FO 371/9818, C11164/2048/18, Minute by MacDonald, 17 July 1924.
62. Ibid., C2048/2048/18, Minute by MacDonald, 6 February 1924; Hansard, Parliamentary Debates, 5th Series, H of C, vol. 169, cols 772–4, 12 February 1924; FO 800/219, MacDonald to Albert Thomas, 16 April 1924; FO 371/9807, C3319/903/18, Extract from a speech by the Prime Minister in Wales, 28 April 1924; *DBFP*, 1st Series, vol. xxvi, no. 507, p. 758, Notes taken during the course of a meeting at the Quai d'Orsay, 8 July 1924; FO 371/9813, C13819/1288/18, Minute by MacDonald, 10 September 1924.
63. FO 371/9818, C8275/2048/18, Minute by Crowe, 24 May 1924.
64. FO 371/9743, 7169/70/18, Minutes of Meeting held at the Treasury, 1 May 1924; ibid., C7420/70/18, Minute by Crowe, May 1924 (undated); FO 371/9745, C8154/70/18, Minute by Lampson, 21 May 1924; ibid., Minute by Crowe, 21 May 1924; ibid., Minute by MacDonald, 25 May 1924.

65. FO 800/218, Selby to de Montille, 18 June 1924.
66. FO 371/9847, C10550/10193/18, Minute by Lampson, 25 June 1924, approved by MacDonald, 27 June 1924; *DBFP*, 1st Series, vol. xxvi, no. 507, pp. 749–50, Notes taken during the course of a meeting at the Quai d'Orsay, 8 July 1924.
67. FO 371/9744, C7953/70/18, Mr MacDonald to M. Poincaré, 14 May 1924, reproduced in *The Times*, 29 May 1924, p. 10. Saint-Aulaire, however, claims to have come upon evidence, corroborated by Tyrrell, that MacDonald was simply postponing decisions until after the French elections, and that the British Intelligence Service and the Labour Party were doing their best in the meantime to subsidize the Opposition in France. Compte de Saint-Aulaire, 'L'Angleterre et les élections de 1924', in *Écrits dès Paris*, (May 1959), pp. 9–19; Comte de Saint-Aulaire, *Confessions d'un Vieux Diplomate* (Paris, 1953), pp. 694–9.
68. FO 371/9744, C7953/70/18, Crowe to MacDonald, 14 May 1924.
69. Ibid.
70. FO 800/218, Mr MacDonald to M. Poincaré, 26 January 1924, reproduced in *The Times*, 4 February 1924, p. 13 and translated in Weill-Raynal, *Les Réparations allemandes*, vol. 3, p. 15; *DBFP*, 1st Series, vol. xxvi, no. 369, pp. 551–4, Mr MacDonald to M. Poincaré, 21 February 1924; Hansard, Parliamentary Debates, 5th Series, H of C, vol. 171, cols 1596–1611, 27 March 1924; FO 371/9804, C18950/737/18, Central Department Memorandum: Note on the MacDonald–Poincaré correspondence, 17 April 1924. For Poincaré's replies, see Weill-Raynal, *Les Réparations allemandes*, vol. 3, p. 15, and *DBFP*, 1st Series, vol. xxvi, no. 371.
71. FO 371/9828, C2842/2048/18, Minute by Crowe, 25 February 1924; Hansard, Parliamentary Debates, 5th Series, H of C, vol. 171, col. 1604, 27 March 1924.
72. FO 371/9813, C2028/1346/18, Minute by Nicolson, 6 February 1924; ibid., Minute by MacDonald, February 1924 (undated); *DBFP*, 1st Series, vol. xxvi, no. 369, p. 554, Mr MacDonald to M. Poincaré, 21 February 1924.
73. FO 371/9804, C18950/737/18, Central Department Memorandum: Note on the MacDonald–Poincaré correspondence, 17 April 1924.
74. *DBFP*, 1st Series, vol. xxvi, no. 507, pp. 754–5, Notes taken during the course of a meeting at the Quai d'Orsay, 8 July 1924.
75. PRO 30/69/8/1, MacDonald's Diary, vol. 1, 24 August 1924.
76. *DBFP*, 1st Series, vol. xxvi, no. 507, p. 758, Notes taken during the course of a meeting at the Quai d'Orsay, 8 July 1924.
77. PP 1924 (Cmd 2105), vol. 27, Reports of the Expert Committees appointed by the Reparation Commission, 9 April 1924. Reproduced in Charles G. Dawes, *A Journal of Reparations* (London, 1939), Appendix 3, pp. 278–511.
78. FO 371/9744, C7637/70/18, Foreign Office Memorandum, 10 April 1924.
79. FO 371/9742, C6913/170/18, Mr Niemeyer (Treasury) to Sir Eyre Crowe, 28 April 1924.
80. *DBFP*, 1st Series, vol. xxvi, no. 430, pp. 629–38, Memorandum by Mr Niemeyer on the Reparation Experts' Report, 14 April 1924.
81. Ibid., FO 371/9742, C6913/70/18, Mr Niemeyer (Treasury) to Sir Eyre Crowe, 28 April 1924.
82. *DBFP*, 1st Series, vol. xxvi, no. 490, pp. 724–30, Memorandum on the immediate steps to be taken to apply the Dawes Scheme, 19 June 1924.
83. Ibid., no. 421, p. 621, Lord D'Abernon (Berlin) to Mr MacDonald, 11 April 1924; ARWR, KAB. Marx, 1923–5, vol. 1, no. 174, pp. 552–4, Kabinettssitzung, 11 April 1924; ibid., no. 175, pp. 555–65, Besprechung

mit den Ministerpräsidenten der Länder, 14 April 1924; Hans Luther, *Politiker ohne Partei: Erinnerungen* (Stuttgart, 1960), pp. 270–1.

84. FO 371/9740, C6368/70/18, Minute by Crowe, 15 April 1924.
85. FO 371/9742, C6809/70/18, Lord Crewe (Paris) to Mr MacDonald, 25 April 1924.
86. *DBFP*, 1st Series, vol. xxvi, no. 428, p. 628, Mr MacDonald to Lord D'Abernon (Berlin), 14 April 1924; FO 371/9740, C6317/70/18, Mr MacDonald to Lord D'Abernon (Berlin), 15 April 1924.
87. FO 371/9740, C6368/70/18, Minute by Crowe, 15 April 1924.
88. FO 371/9743, C7427/70/18, Notes on Conversations held at Chequers, 2–3 May 1924.
89. FO 371/9742, C6758/70/18, Foreign Office Minute, 24 April 1924.
90. FO 371/9740, C6405/70/18, Lord D'Abernon (Berlin) to Mr MacDonald, 16 April 1924.
91. FO 371/9741, C6469/70/18, Marquis della Tarretta (London) to Mr MacDonald, 17 April 1924.
92. FO 371/9742, C6809/70/18, Lord Crewe (Paris) to Mr MacDonald, 25 April 1924.
93. FO 371/9745, C7960/70/18, M. Poincaré to Mr MacDonald, 14 May 1924.
94. In private, however, Poincaré held that this did not mean that France was to relinquish her military and economic control over the Ruhr and the Rhineland. For further details respecting the Germans' true intentions; the French early attempts to circumvent the Dawes Committee's political recommendations; the pressures exerted by the British; and the Belgian all-round effort at mediation, see Schuker, *The End of French Predominance*, pp. 186–219; B. Kent, *The Spoils of War*, p. 251.
95. FO 371/9843, C8510/8508/18, Mr MacDonald to Lord Crewe (Paris), 26 May 1924.
96. Georges Suarez, *Herriot, 1924–1932: Nouvelle Édition de 'Une Nuit chez Cromwell' suivie d'un récit historique de R. Poincaré* (Paris, 1932), pp. 18–19, 54.
97. FO 371/9845, C10114/9837/18, Summary of political telegrams for the secret information of the Dominion Governments, 24 June 1924.
98. FO 371/9742, C6913/70/18, Minute by Bennett, 29 April 1924; FO 371/9743, C7169/70/18, Minutes of interdepartmental meeting, 1 May 1924; FO 371/9747, C9529/70/18, Crowe to Bradbury, 17 June 1924; FO 371/9748, C9783/70/18, Bradbury to Crowe, 18 June 1924.
99. FO 371/9751, C11976/70/18, Notes of a conversation between Mr MacDonald and M. Herriot, 21–2 June 1924; *DBFP*, 1st Series, vol. xxvi, nos 507–8, pp. 749–87, Notes taken during the course of a meeting at the Quai d'Orsay, 8–9 July 1924. See also, Schuker, *The End of French Predominance*, pp. 237–45, 256–63, and Weill-Raynal, *Les Réparations allemandes*, vol. 3, pp. 29–65.
100. FO 371/9847, C11438/10193/18, Meeting of a Conference of Ministers, 11 July 1924.
101. FO 371/9747, C9073/70/18, Minute by Lampson, 6 June 1924; FO 371/9748, C9783/70/18, Minute by Crowe, 20 June 1924.
102. FO 371/9747, C10158/70/10, Minute by Crowe, 26 June 1924; FO 371/9847, C11033/10193/18, Minute by Lampson, 9 July 1924.
103. FO 371/9747, C10158/70/18, Minute by MacDonald, 27 June 1924.
104. FO 371/9847, C10550/10193/18, Minute by Crowe, 25 June 1924.
105. FO 371/9846, C10535/10193/18, Crowe to Bradbury, 30 June 1924.
106. FO 371/9847, C10596/10193/18, Minute by Lampson, 3 July 1924.
107. Stephen Roskill, *Hankey, Man of Secrets* (London, 1972), vol. 2, p. 368.

108. CAB 29/103, Proceedings of the London Reparation Conference, vol. 1, The Inter-Allied Conference; a more selective documentation, PP 1924 (Cmd 2270), vol. 27, Proceedings of the London Reparation Conference, July and August 1924, nos 1–26; *DBFP*, 1st Series, vol. xxvi, no. 526, pp. 807–13, Memorandum on the present position of the Inter-Allied Conference, 4 August 1924; Marquand, *Ramsay MacDonald*, pp. 342–5; Schuker, *The End of French Predominance*, pp. 300–44; Weill-Raynal, *Les Réparations allemandes*, vol. 3, pp. 66–96.

109. PRO 30/69/8/1, MacDonald's Diary, vol. 1, 8 August 1924.

110. Paul Hymans, *Mémoires* (Brussels, 1958), vol. 2, p. 588; Snowden, *An Autobiography*, vol. 2, pp. 528, 778–9; Marquand, *Ramsay MacDonald*, pp. 344–50.

111. CAB 29/103–4, Proceedings of the London Reparation Conference, vols 1–2; Schuker, *The End of French Predominance*, pp. 296–7, 300, 302–3, 307, 313, 315–16, 318, 344–5, 349, 379–80.

112. Roskill, *Hankey*, vol. 2, p. 372.

113. CAB 29/104, Proceedings of the London Reparation Conference, vol. 2, The International Conference; ARWR, KAB. Marx, 1923–5, vol. 2, Appendices 1–10, pp. 1283–399; *DBFP*, 1st Series, vol. xxvi, no. 535, pp. 825–6, Sir M. Hankey to Mr MacDonald, 11 August 1924; ibid., no. 537, pp. 828–30, Mr Finlayson to Mr Lampson, 12 August 1924; ibid., no. 540, pp. 833–5, Minute by Mr Finlayson to Mr Lampson, 14 August 1924; ibid., no. 690, pp. 1097–9, Memorandum by Sir C. Hurst on the question of the evacuation of occupied German territory, 12 August 1924; Shucker, *End of French Predominance*, pp. 327–39, 366–71.

114. PRO 30/69/8/1, MacDonald's Diary, vol. 1, 8 August 1924.

115. PP 1924 (Cmd 2259), vol. 27, Agreements Concluded between the Allied Governments and Germany, 16 August 1924. For further details on the negotiations and agreements, see Schuker, *The End of French Predominance*, pp. 344–93 and Weill-Raynal, *Les Réparations allemandes*, vol. 3, pp. 96–140.

116. Ibid.

117. PRO 30/69/8/1, MacDonald's Diary, vol. 1, 7 August 1924.

118. Roskill, *Hankey*, vol. 2, p. 373.

119. Schuker, *The End of French Predominance*, pp. 124–5, 140–68, 273–83, 289–94, 300–18, 349–52, 380–1.

120. When the Germans accepted the London Agreements they fully intended to ask for another reduction in reparations within three or four years, having viewed the Dawes Plan, from the start, as a temporary expedient to remove the French from the Ruhr, to overcome the economic problems attendant on a dearth of liquid capital, and to keep reparations minimal. Chancellor Marx's speech of 11 May at Cologne, *Frankfurter Zeitung*, 12 May 1924; FO 371/9745, C7967/70/18, Minute by Bennett, 20 May 1924; US National Archives 462.00 R, 296/376, Reception given in Western Germany to the Reports of the Committee of Experts by the Consul-General in Cologne (Emil Sauer), 7 June 1924; ARWR, KAB. Marx, 1923–5, vol. 2, pp. 766–855, Besprechung mit den Staats-und Ministerpräsidenten der Länder, 3 July 1924.

121. FO 371/15911, Waley (Treasury) to Foreign Office, 8 June 1924; *DBFP*, 2nd Series, vol. ii, Appendix 2, p. 487, Report of the Committee appointed on the Recommendation of the London Conference, 1931; Marks, 'The Myths of Reparations', pp. 249–50.

122. FO 371/9813, C13819/1288/18, Mr Phipps (Paris) to Mr MacDonald, 27 August 1924.

123. FO 371/9751, C1976/70/18, Notes of a conversation between Mr

MacDonald and M. Herriot, 22 June 1924; FO 371/9819, C12870/2048/18, M. Herriot to Mr MacDonald, 11 August 1924.

124. *DBFP*, 1st Series, vol. xxvi, no. 507, p. 758, Notes taken during the course of a meeting at the Quai d'Orsay, 8 July 1924; FO 371/9818, C11164/2048/18, Minute by MacDonald, 17 July 1924; William Rayburn Tucker, *The Attitude of the British Labour Party towards European and Collective Security Problems, 1920–1939* (Geneva, 1950), pp. 75–6, 94–9.

125. PP 1924 (Cmd 2200), vol. 27, Correspondence between His Majesty's Government and the League of Nations respecting the Proposed Treaty of Mutual Assistance, pp. 4–9, Draft Treaty of Mutual Assistance, 29 September 1923. For further details, see Lord Cecil of Chelwood, *A Great Experiment* (London, 1941), pp. 151–3, 157–9, and J. W. Wheeler-Bennett and F. E. Langermann, *Information on the Problem of Security, 1917–1926* (London, 1927), pp. 95–100, 240–5.

126. Ibid., pp. 101–4, Mr MacDonald to Sir E. Drummond (Geneva), 5 July 1924. See also, CAB 23/48, 35(24), Conclusions of a Meeting of the Cabinet, 30 May 1924; F. S. Northedge, *The Troubled Giant: Britain Among the Great Powers, 1916–1939*, pp. 236–7.

127. FO 371/9418, W1075/30/98, Minute by Villiers, 16 February 1924.

128. FO 371/9419, W5047/30/98, Memorandum by Crowe, 25 June 1924.

129. FO 371/10568, W637/134/98, Memorandum by the War Office, 14 March 1924; ibid., Memorandum by the Admiralty, 1924; RIIA, *Survey*, 1924, pp. 32–4.

130. Marquand, *Ramsay MacDonald*, pp. 354–5.

131. L/NOJ 1924, pp. 41–4, Records of the Fifth Assembly, Plenary Meeting, 4 September 1924.

132. Ibid., pp. 498–502, reproduced in Philip J. Noel-Baker, *The Geneva Protocol for the Pacific Settlement of International Disputes* (London, 1925), Annex 8, pp. 215–24.

133. CAB 2/4, Committee of Imperial Defence, Minutes of the 190th and 192nd Meetings, 4, 16 December 1924; CAB 24/172, CP 105, CID sub-committee report, 23 January 1925.

134. FO 371/10573, W9724/134/98, Joint Memorandum by the Chiefs of Staff, December 1924; Gibbs, *Grand Strategy*, pp. 37–41; Stephen Roskill, *Naval Policy Between the Wars* (London, 1968), vol. 1, pp. 430–1.

135. *The Times*, 16, 21 October 1924; PP 1925 (Cmd 2458), vol. 31, Protocol for the Pacific Settlement of International Disputes. Correspondence relating to the position of the Dominions; Gwendolen M. Carter, *The British Commonwealth and International Security: The Role of the Dominions, 1919–1939* (Toronto, 1947), pp. 118–23.

136. Maurice, *Haldane*, vol. 2, pp. 161–2, 167; Snowden, *An Autobiography*, Tucker, *Attitude of the British Labour Party*, pp. 101–9.

137. Mary Agnes Hamilton, *Arthur Henderson* (London, 1938), p. 247.

138. *DBFP*, 1st Series, vol. xxv, no. 208, pp. 333–4, Mr MacDonald to Mr Hodgson (Moscow), 1 February 1924; Gabriel Gorodetsky, *The Precarious Truce: Anglo-Soviet Relations, 1924–27* (Cambridge, 1977), pp. 7–13.

139. *DBFP*, 1st Series, vol. xxv, nos 275–97; Gorodetsky, *The Precarious Truce*, pp. 13–32.

140. *DBFP*, 1st Series, vol. xxv, no. 296, p. 617, Tenth Plenary Meeting of the Conference, held at the Foreign Office, 8 August 1924.

141. For further information on this point see Crowe and Corp, *Our Ablest Public Servant: Sir Eyre Crowe, 1864–1925*, pp. 456–7. For a further discussion of the Labour Government's relationship with the Soviet Union see Andrew J. Williams, *Labour and Russia: the Attitude of the Labour Party to the USSR,*

1924–1934 (Manchester, 1989).

142. Lewis Chester, Stephn Fay and Hugo Young, *The Zinoviev Letter* (London, 1967), pp. 32–3; Chris Cook, *The Age of Alignment: Electoral Politics in Britain, 1922–1929* (London, 1975), pp. 268–73, 299–305; Richard W. Lyman, *The First Labour Government, 1924* (London, 1957), pp. 196–207.
143. Hansard, Parliamentary Debates, 5th Series, H of C, vol. 177, cols 8–12, pp. 596–619, 8 October 1924; Cook, *Age of Alignment*, pp. 273–5; Sir Patrick Hastings, *The Autobiography of Sir Patrick Hastings* (London, 1948), pp. 238–42; Marquand, *Ramsay MacDonald*, pp. 364–70; Keith Middlemas (ed.), *Thomas Jones, Whitehall Diary* (London, 1969), pp. 287–90; F. H. Neward, 'The Campbell Case and the First Labour Government', *Northern Ireland Legal Quarterly*, vol. 20 (1969), pp. 19–42.
144. Hansard, Parliamentary Debates, 5th Series, H of C, vol. 215, col. 60, 19 March 1928.
145. Ibid., vol. 177, cols 619–23, 8 October 1924. See also Cook, *Age of Alignment*, pp. 275–8; Roy Douglas, *The History of the Liberal Party, 1895–1970* (London, 1971), pp. 178–80; Trevor Wilson, *The Downfall of the Liberal Party, 1914–1935* (Ithaca, 1966), pp. 275–9.
146. Cook, *Age of Alignment*, pp. 298–301, 303–5; Lyman, *First Labour Government*, pp. 248–57; Marquand, *Ramsay MacDonald*, pp. 378–81.
147. In the latter two cases a national uprising was involved as well.
148. PP 1927 (Cmd 2895), vol. xxvi, A Selection of Papers dealing with the relations between His Majesty's Government and the Soviet Government, 1921–1927, no. 3, pp. 30–2, The 'Zinoviev Letter', 15 September 1924, reproduced in Chester, Fay and Young, *The Zinoviev Letter*, pp. xi–xiii and RIIA, *Survey*, 1924, pp. 493–5.
149. FO 371/10478, N7838/108/38, SIS (Sir Hugh Sinclair) to Mr Gregory, 9 October 1924.
150. Ibid., N8105/108/38, Sir Eyre Crowe to Mr MacDonald, 26 October 1924.
151. Ibid., N7838/108/38, Minute by Strang, 14 October 1924.
152. Ibid., Minute by Gregory, 14 October 1924.
153. Ibid., Minute by Crowe, 15 October 1924, reproduced with modifications in *DBFP*, 1st Series, vol. xxv, no. 264, pp. 434–5.
154. Chester, Fay and Young, *The Zinoviev Letter*, pp. 72–3, 197.
155. Christopher Andrew, *Secret Service the Making of the British Intelligence Community*, p. 308, reprinted in Christopher Andrew, 'The British Secret Service and Anglo-Soviet Relations in the 1920s. Part I: From the Trade Negotiations to the Zinoviev Letters', *Historical Journal*, vol. 20 (1977), p. 705.
156. Chester, Fay and Young, *The Zinoviev Letter*, pp. 71–4, 196.
157. Michael Kettle, *Sidney Reilly: The True Story* (London, 1983), pp. 121–4; Robin Bruce Lockhart, *Ace of Spies* (London, 1967), pp. 172, 173; Hansard, Parliamentary Debates, 5th Series, H of C, vol. 215, col. 62, 19 March 1928.
158. Chester, Fay and Young, *The Zinoviev Letter*, pp. 183–91.
159. Robin Lockhart, *Ace of Spies*, chs 10–13; Kettle, *Sidney Reilly*, pp. 131–41.
160. Chester, Fay and Young, *The Zinoviev Letter*, pp. 188–9, 202–54.
161. Ibid., pp. 74–81, 84–5, 198–9, 201–5.
162. Ibid., pp. 83–91, 99, 198–200.
163. Ibid., *The Zinoviev Letter*, pp. 88, 90–3, 199–200.
164. FO 371/10478, N7838/108/39, Minute by MacDonald, 16 October 1924, reproduced with modifications in *DBFP*, 1st Series, vol. xxv, no. 264, p. 435.
165. *Observer*, Letter from Thomas Marlowe, 4 March 1928; Chester, Fay and

Young, *The Zinoviev Letter*, pp. 94–8.

166. Chester, Fay and Young, *The Zinoviev Letter*, pp. 95–6, 100–8.
167. Andrew, *Secret Service*, pp. 298–301, 308; Chester, Fay and Young, *The Zinoviev Letter*, p. 108; Sir Duncan Wilson, *Leonard Woolf: A Political Biography* (London, 1978), p. 160.
168. For a further discussion of the Zinoviev Letter see Sybil Crowe, 'The Zinoviev Letter: a reappraisal,' *Journal of Contemporary History*, vol. 10 (1975); and C. M. Andrew, 'The British Secret Service and Anglo-Soviet Relations in the 1920s. Part 1', *Historical Journal*, vol. 20 (1977). See also Crowe and Corp, *Our Ablest Public Servant: Sir Eyre Crowe, 1864–1925*, pp. 459–61.
169. *Sunday Times*, 15 February 1970; Crowe, 'The Zinoviev Letter', p. 414.
170. Natalie Grant, 'The Zinoviev Letter Case', *Soviet Studies*, vol. 19 (1967–8), pp. 272–3, reproduced in William Butler, *The Harvard Text*, Appendix 3, pp. 60–2.
171. Hansard, Parliamentary Debates, 5th Series, H of C, vol. 189, cols 673–4, 15 December 1924.
172. Ibid., vol. 215, cols 49–50, 19 March 1928. This hearsay evidence was presumably founded among other things on the examination of the Heads of the Secret Service Departments of the War Office, Admiralty, Air Ministry and Special Branch by the committee of the Labour Government appointed to go into the evidence about the Zinoviev Letter. Mrs Sidney Webb reported in her diary that 'all those departments regarded it as a fake and merely added it to the pile of suchlike documents'. While MacDonald minuted: 'We are inclined to regard it as forgery but no proofs one way or other. But plenty of evidence that forgers about. Scotland Yard thought so little about it that it put it on its file & took no action. Same with Service departments.' Cole (ed.), *Beatrice Webb's Diaries*, p. 49; PRO 30/69/8/1, MacDonald's Diary, vol. 1, 3 November 1924, quoted almost fully in Marquand, *Ramsay MacDonald*, p. 388.
173. Sir Wyndham Childs, *Episodes and Reflections* (London, 1936), p. 246, quoted in Chester, Fay and Young, *The Zinoviev Letter*, p. 93.
174. Letter from Philip Noel-Baker to *The Times*, 22 December 1966, p. 11.
175. *DBFP*, 1st Series, vol. xx, no. 414, p. 741, The Marquess Curzon of Kedleston to Mr Hodgson (Moscow), 7 September 1921.
176. Ibid., no. 379, pp. 691–5, Minutes of meeting of 1 July 1921, of Interdepartmental Committee on Bolshevik menace to the British Empire; ibid., no. 383, pp. 698–700, Minutes of meeting of 15 July 1921 of Interdepartmental Committee on Bolshevism. See also Andrew, *Secret Service*, pp. 279–80.
177. Ibid., no. 414, p. 741, enclosure: Note to Soviet Government, 7 September 1921; FO 371/6855, N11337/5/38, Minute by Curzon, October 1921 (undated).
178. FO 371/10478, N8105/108/38, Sir Eyre Crowe to Mr MacDonald, 25 October 1924.
179. Ibid., Sir Eyre Crowe to Mr MacDonald, 26 October 1924.
180. Ibid., N7838/108/38, Minute by Crowe, 21 October 1924, reproduced in *DBFP*, 1st Series, vol. xxv, no. 264, p. 435.
181. PRO 30/69/8/1, MacDonald's Diary, vol. 1, 31 October 1924; *DBFP*, 1st Series, vol. xxv, no. 264, p. 435.
182. FO 371/10478, N8105/108/38, Sir Eyre Crowe to Mr MacDonald, 25 October 1924.
183. PRO 30/69/8/1, MacDonald's Diary, vol. 1, 31 October 1924, quoted in Marquand, *Ramsay MacDonald*, pp. 383–4.

184. *DBFP*, 1st Series, vol. xxv, no. 264, p. 436, History of the Zinoviev Incident, 11 November 1924.
185. Ibid., p. 435; Crowe, 'The Zinoviev Letter', p. 424.
186. FO 371/10478, N7838/108/38, Minute by Crowe, 21 October, 1924.
187. FO 371/10479, N8567/108/38, Minute by Stephen Gaselee, 13 April 1938; *DBFP*, 1st Series, vol. xxv, no. 264, p. 436.
188. FO 371/10478, N8105/108/38, Sir Eyre Crowe to Mr MacDonald, 25 October 1924.
189. Ibid.
190. PP 1927 (Cmd 2895), pp. 28–9, Mr J. D. Gregory to M. Rakovsky, 24 October 1924, reproduced in RIIA, *Survey*, 1924, pp. 492–3.
191. John D. Gregory, *On the Edge of Diplomacy: Rambles and Reflections, 1902–28* (London, 1928), pp. 221–4.
192. Chester, Fay and Young, *The Zinoviev Letter*, pp. 12–13, 82–3, 87–8, 92–3, 130–4, 139–43; Lyman, *First Labour Government*, pp. 258–61; John Ramsden, *The Age of Balfour and Baldwin, 1902–1940* (London, 1978), pp. 200–4.
193. Charles Loch Mowat, *Britain Between the Wars, 1918–1940* (London, 1968), p. 190; Lyman, *First Labour Government*, pp. 264–70; Ramsden, *The Age of Balfour and Baldwin*, p. 206.

Chapter 6
The Second Baldwin Government:
J. Austen Chamberlain and the Foreign Office

1. FO 371/9834, C19457/4736/18, Minute by Lampson, January 1925 (undated).
2. FO 371/10728, C3246/459/18, Minute by Chamberlain, 18 March 1925.
3. FO 371/10703, C1785/2/18, Minute by Chamberlain, 4 February 1925; FO 371/10714, C1065/35/18, Minute by Chamberlain, 16 September 1925.
4. FO 371/10759, C13707/13120/18, Lampson to Colonel AS Pan, 4 November 1925.
5. Sir A. Willert, *Washington and Other Memories* (Boston, 1972), p. 168.
6. Martin Gilbert, *Winston S. Churchill* (London, 1976), vol. 5, 1922–1939, pp. 121–5; companion pt 1, pp. 244–6, 252–7, 279–80, 286–7, 338–404, 347–9, 358, 380, 390–7, 406, 413–17, 430–2; Douglas Johnson, 'Austen Chamberlain and the Locarno Agreements', *University of Birmingham Historical Journal*, vol. 8 (1961–2), pp. 68–9.
7. FO 371/10889, F143/3/16, CP 20 (25), Memorandum by the Secretary of State for Foreign Affairs on Egypt, 9 January 1925; Gilbert, *Churchill*, vol. 5, companion pt 1, pp. 247–9, 296–7, 317–21, 351–2.
8. CAB 23/52, 9(26)2, Conclusions of a Meeting of the Cabinet, 3 March 1926; Viscount Cecil of Chelwood, *All the Way* (London, 1949), pp. 189–91; Kenneth Rose, *The Later Cecils* (London, 1975), pp. 166–7.
9. CAB 23/50, 32 (025), Conclusions of a Meeting of the Cabinet, 1 July 1925; L. S. Amery, *My Political Life* (London, 1953–5), vol. 2, pp. 365, 475; Wm. Roger Louis, *British Strategy in the Far East, 1919–1939* (Oxford, 1971), pp. 120–37. For further details see Chapter 2.
10. Austen Chamberlain MSS, AC 5/1/347, J. Austen Chamberlain to Ida Chamberlain, 1 March 1925.
11. Ibid., AC 5/1/380, A. Chamberlain to Hilda Chamberlain, 25 April 1926.

12. Frances Lloyd George, *The Years that are Past* (London, 1967), p. 184.
13. Lord Eustace Percy, *Some Memories* (London, 1958), pp. 79–81.
14. *Dictionary of National Biography*, 1931–1940, pp. 163–6; David Dutton, *Austen Chamberlain, Gentleman in Politics* (Bolton, 1985), chs 1–6; Sir Charles Petrie, Bt, *The Life and Letters of the Right Hon. Sir Austen Chamberlain, K.G., P.C., M.P.* (London, 1939–40), vols 1 and 2, chs 1–5.
15. Margaret Cole (ed.), *Beatrice Webb's Diaries, 1924–1932* (London, 1956), p. 86.
16. Frances Lloyd George, *The Years that are Past*, p. 184.
17. *Dictionary of National Biography*, pp. 166, 168.
18. Lord Riddell, *Lord Riddell's Intimate Diary of the Peace Conference and After, 1918–1923* (London, 1933), p. 194.
19. Austen Chamberlain MSS, AC 5/1/370, A. Chamberlain to Ida Chamberlain, 28 November 1925.
20. FO 371/10939, F4633/190/10, Minute by Chamberlain, 26 September 1925. See also B. J. C. McKercher 'Austen Chamberlain's control of British Foreign Policy, 1924–1929,' *International History Review*, vol. 6 (1984); and Dutton *Austen Chamberlain: a Gentleman in Politics*.
21. Austen Chamberlain MSS, AC 5/1/379, A. Chamberlain to Ida Chamberlain, 18 April 1926.
22. *Dictionary of National Biography, 1931–1940*, p. 168.
23. Austen Chamberlain MSS, AC 5/1/346, A. Chamberlain to Ida Chamberlain, 15 February 1925.
24. Ibid., AC 5/1/343, A. Chamberlain to Ida Chamberlain, 28 December 1924.
25. Bruce Heady, *British Cabinet Ministers: The Roles of Politicians in Executive Office* (London, 1974), pp. 84–5, 142.
26. Austen Chamberlain MSS, AC 5/1/342, A. Chamberlain to Hilda Chamberlain, 23 December 1924.
27. *DBFP*, 1st Series, vol. xxvii, no. 180, pp. 255–8, Minute by the Secretary of State, 4 January 1925.
28. FO 371/9834, C19457/4736/18, Minute by Lampson, January 1925 (undated).
29. FO 371/10735, C9581/4549/19, Minute by Chamberlain, 21 July 1925.
30. FO 371/10939, F4633/190/10, Minute by Chamberlain, 26 September 1925.
31. *DBFP*, 1st Series, vol. xxvii, no. 180, pp. 255–8, Minute by the Secretary of State, 4 January 1925; FO 371/10703, C1849/2/18, Mr Chamberlain to Lord Crewe (Paris), 6 February 1925.
32. The Earl of Avon, *The Eden Memoirs: Facing the Dictators* (London, 1962), p. 7.
33. Austen Chamberlain MSS, AC 5/1/351, A. Chamberlain to Hilda Chamberlain, 25 April 1925.
34. FO 371/9834, C19483/4736/18, Minute by Chamberlain, 31 December 1924.
35. FO 371/10717, C1204/109/18, Minute by Chamberlain, 28 Januray 1925.
36. *DBFP*, 1st Series, vol. xxvii, no. 224, pp. 343–7, Mr Chamberlain to Sir E. Crowe, 7 March 1925.
37. CAB 2/4, Committee of Imperial Defence, Minutes of the 192nd Meeting, 16 December 1924.
38. *DBFP*, 1st Series, vol. xxvi, no. 708, pp. 1117–18, 'Memorandum on the Question of whether Germany has or has not carried out the Military Clauses of the Treaty of Versailles, with special reference to Articles 427–442 of that Treaty', 3 December 1924.

39. FO 371/9833, C19106/4736/18, War Office to Sir E. Crowe, December 1925 (undated).
40. FO 371/10702, C776/2/18, B. B. Cubitt to Sir E. Crowe, 17 January 1925.
41. *DBFP*, 1st Series, vol. xxvi, Appendix, p. 1157, Memorandum on the evacuation of the Rhineland, 29 November 1924.
42. Ibid., no. 701, p. 1110, Lord D'Abernon (Berlin) to Mr MacDonald, 17 September 1924; ibid., no. 705, p. 1114, Memorandum respecting German Disarmament, 7 November 1924.
43. Ibid., no. 588, p. 910, Memorandum on the evacuation of the Cologne Zone, 7 November 1924; ibid., Appendix 2, pp. 1157, 1158, Memorandum on the evacuation of the Rhineland, 29 November 1924.
44. FO 371/9833, C19106/4736/18, War Office to Sir E. Crowe, December 1924 (undated).
45. Ibid., Minute by Bennett, 23 December 1924.
46. Ibid., Minute by Lampson, 24 December 1924.
47. Ibid., Minute by Chamberlain, 24 December 1924.
48. Ibid., Minute by Crowe, 24 December 1924.
49. FO 371/10703, C1647/2/18, G. S. Clive to Lord Crewe (Paris), 26 January 1925.
50. Ibid., Minute by Bennett, 7 February 1925.
51. Ibid., Minute by Lampson, 9 February 1925.
52. Ibid., Minute by Crowe, 10 February 1925.
53. Ibid., Minute by Chamberlain, 10 February 1925.
54. *DBFP*, 1st Series, vol. xxvi, no. 710, p. 1121, Lord Crewe (Paris) to Mr Chamberlain, 12 December 1924; Michael Salewski, *Entwaffnung und Militarkontrolle in Deutschland, 1919–1927* (Munich, 1966), pp. 271–82.
55. CAB 23/49, 67(24), Conclusions of a Meeting of the Cabinet, 17 December 1924.
56. *DBFP*, 1st Series, vol. xxvi, no. 712, pp. 1122–5, Mr Chamberlain to Lord Crewe (Paris), 20 December 1924.
57. Ibid., no. 708, pp. 1117–18, 'Memorandum on the Question of whether Germany has or has not carried out the Military Clauses of the Treaty of Versailles, with special reference to Articles 427–429 of that Treaty', 3 December 1924; ibid., no. 717, pp. 1131–312, Lord D'Abernon (Berlin) to Mr Chamberlain, 24 December 1924.
58. FO 371/9834, C19288/4736/18, Minute by Lampson, 25 December 1924.
59. Ibid., Minute by Crowe, 25 December 1924.
60. Ibid., Minute by Chamberlain, 26 December 1924. For further details concerning the trends that prevailed in Weimar, see Salewski, *Entwaffnung*, pp. 284–8.
61. FO 371/10703, C1705/2/18, Allied Note to Germany, 5 January 1925. Printed in E.N., vol. 1, pp. 50–1 and in U.F., vol. 6, no. 1319/a.
62. *DBFP*, 1st Series, vol. xxvii, no. 556, pp. 899–901, Lord D'Abernon (Berlin) to Mr Chamberlain, 7 January 1925. Printed, also, in E.N., vol. 1, pp. 52–3 and in U.F., vol. 6, no. 1319/b.
63. E.N. vol. 1, pp. 54–5, Note de Alliierten Regierungen, 26 January 1925. See also, *DBFP*, 1st Series, vol. xxvii, nos 565, 570.
64. FO 371/10703, C1276/2/18, Lord D'Abernon (Berlin) to Mr Chamberlain, 27 January 1925. Printed in E.N., vol. 1, p. 56. For further details, see Salewski, *Entwaffnung*, pp. 288–91.
65. Minutes by Troutbeck, Lampson and Crowe, 28 January 1925.
66. Ibid., C1362/2/18, Memorandum by Bennett, 26 January 1925.
67. Ibid., Minute by Lampson, 27 January 1925.
68. FO 371/10707, C522/21/18, Minute by Crowe, 14 January 1925.

69. FO 371/10703, C1729/2/18, Minute by Lampson, 6 February 1925; ibid., Minute by Crowe, 7 February 1925; ibid., Minute by Chamberlain, 9 February 1925.

70. FO 371/10727, C1370/44549/18, Lord Crewe (Paris) to Mr Chamberlain, 29 January 1925.

71. FO 371/10709, C8109/21/18, Central Department Memorandum on the Evacuation of the Cologne Zone, 16 June 1925.

72. Ibid., see also *DBFP*, 1st Series, vol. xxvii, nos 225, 584, 600.

73. For details see FO 371/10702–10; John P. Fox, 'Britain and the Inter-Allied Military Commission of Control 1925–6', *Journal of Contemporary History*, vol. 4 (1969), pp. 149–54; Jon Jacobson, *Locarno. Diplomacy: Germany and the West, 1925–1929* (Princeton, 1972), pp. 47–67; Salewski, *Entwaffnung*, pp. 292–9; RIIA, *Survey of International Affairs, 1925* (Oxford, 1928), vol. 2, pp. 171–93.

74. E.N., vol. 1, pp. 6–47, Note Collectif des Gouvernements Alliés, 4 June 1925.

75. PP 1925 (Cmd 2527), vol. 30, 'Correspondence between the Ambassadors' Conference and the German Ambassador in Paris Respecting German Disarmament, Evacuation of Cologne Zone and Modifications in the Rhineland Regime', Paris, October–November 1925, no. 5, pp. 12–14, Note, 16 November 1925.

76. Foreign Office officials 1, no. 228, p. 378, Lord Kilmarnock (Coblenz) to Sir A. Chamberlain, 31 January 1926.

77. Cecil MSS, vol. 8, Add. MSS, 51078, A. Chamberlain to Cecil, 19 November 1924.

78. FO 371/10571, W10151/134/98, Minute by Campbell, 10 November 1924.

79. Ibid., Minute by Crowe, 17 November 1924.

80. CAB 2/4, Committee of Imperial Defence, Minutes of the 192nd Meeting, 16 December 1924.

81. Ibid.

82. *DBFP*, 1st Series, vol. xxvii, no. 180, pp. 256–7, Minute by the Secretary of State, 4 January 1925.

83. CAB 2/4, Committee of Imperial Defence, Minutes of the 192nd Meeting, 16 December 1924.

84. *DBFP*, 1st Series, vol. vii, no. 180 p. 256, Minute by the Secretary of State, 4 January 1925.

85. Ibid., no. 181, pp. 258–60, Lord D'Abernon (Berlin) to Mr Chamberlain, 7 January 1925.

86. In this connection D'Abernon added: 'As regards the future, no positive statement can be made, but it cannot be assumed with any probability that Europe will revert to the 1914 position, for a sudden development from non-army or minor secret arming by Germany to open and undisguised arming is impossible, since it would mean an immediate declaration of war by France, in which Germany would have no chance.' Ibid., p. 258.

87. Ibid., footnote 5, pp. 260–1, Minute by Chamberlain, 14 January 1925.

88. Ibid., no. 180, p. 256, Minute by the Secretary of State, 4 January 1925.

89. CAB 2/4, Committee of Imperial Defence, Minutes of the 195th Meeting, 13 February 1925.

90. FO 371/9820, C13819/1288/18, Minute by Lampson, 5 November 1924.

91. *DBFP*, 1st Series, vol. xxvii, no. 181, footnote 5, Minute by Crowe, 14 January 1925.

92. CAB 2/4, Committee of Imperial Defence, Minutes of 192nd Meeting, 16 December 1924.

93. FO 371/11064, W 312/9/98, Minute by Chamberlain, 4 January 1925.
94. *DBFP*, 1st Series, vol. xxvii, no. 205, footnote 1, p. 311, Minute by Chamberlain, 19 February 1925.
95. Ibid., no. 205, pp. 311–18, Memorandum by Mr Nicolson on British Policy considered in relation to the European Situation; prepared in pursuance of directions from the Secretary of State, 20 February 1925.
96. FO 371/11064, W1252/9/98, British Policy and the Geneva Protocol for the Pacific Settlement of International Disputes, 12 February 1925. Printed in Sir James W. Headlam-Morley, *Studies in Diplomatic History* (London, 1930), pp. 171–92.
97. Ibid., Minute by Crowe, 18 February 1925, quoted for the major part in Crowe, 'Sir Eyre Crowe and the Locarno Pact', p. 56.
98. Ibid., Minute by Chamberlain, 21 February 1925.
99. FO 371/11065, W2070/9/98, 'England and the Low Countries', 10 March 1925. Printed in Sir James W. Headlam-Morley, *Studies in Diplomatic History*, pp. 156–71.
100. FO 371/9704, C17265/737/18, Lord D'Abernon (Berlin) to Mr Chamberlain, 8 November 1924; *DBFP*, 1st Series, vol. xxvii, no. 181, pp. 258–60, Lord D'Abernon (Berlin) to Mr Chamberlain, 7 January 1925; Viscount D'Abernon, *An Ambassador of Peace: Pages from the Diary of Viscount D'Abernon* (London, 1929), vol. 2, p. 288; ibid., vol. 3, (London, 1930) pp. 122, 145, 151–2, 158–60; F. G. Stambrook, '"Das Kind" – Lord D'Abernon and the Origins of the Locarno Pact', *Central European History*, vol. 1 (1968), pp. 239, 244–6.
101. Stambrook, '"Das Kind"', pp. 238–46.
102. Stambrook, '"Das Kind"', p. 247.
103. FRUS, 1922, vol. 2, pp. 203–11; *DBFP*, 1st Series, vol. xx, no. 163 p. 369, Lord D'Abernon (Berlin) to the Marquess Curzon of Kedleston, 31 December 1922; ibid., vol. 21, no. 2, pp. 2–3, The Marquess of Crewe (Paris) to the Marquess Curzon of Kedleston, 2 January 1923.
104. That is, such as the annulment of the Dawes Plan, the non-evacuation of the Rhineland and the setting up of a permanent system of military control.
105. ARWR, KAB. Marx, 1923–5, vol. 2, no. 387, p. 1277, Ministerbesprechung, 6 January 1925; C. Gustav Stresemann, *Vermächtnis: der Nachlass in Drei Banden* (Berlin, 1932), vol. 2, pp. 13, 66, 112, 261; Stambrook, '"Das Kind"', pp. 249–53.
106. AA, 3123/642046–51, Der Reichsminister des Auswärtigen Stresemann an die Botschaft in Paris, 15 January 1925, quoted in Stambrook, '"Das Kind"', pp. 253–4.
107. *DBFP*, 1st Series, vol. xxvii, no. 189, enclosure, Memorandum, 20 January 1925.
108. FO 371/10726, C980/459/18, Minute by Bennett, 22 January 1925.
109. Ibid., Minute by Lampson, 22 January 1925.
110. Ibid., Minute by Crowe, 22 January 1925.
111. Ibid., Minute by Lampson, 22 January 1925.
112. *DBFP*, 1st Series, vol. xxvii, no. 191, pp. 286–9, Sir M. Hankey to Mr Lampson, 26 January 1925.
113. Ibid., no. 189, pp. 282–3, Lord D'Abernon (Berlin) to Mr Chamberlain, 20 January 1925; ibid., no. 200, p. 303, Mr Chamberlain to Lord Crewe (Paris), 16 February 1925.
114. FO 371/10727, C1143/459/18, Minute by Crowe, 27 January 1925.
115. FO 371/10726, C1063/459/18, Foreign Office Memorandum, 15 January 1925.

116. *DBFP*, 1st Series, vol. xxvii, no. 195, pp. 293–6, Mr Chamberlain to Lord D'Abernon (Berlin), 30 January 1925.
117. CAB 2/4, Committee of Imperial Defence, Minutes of the 195th Meeting, 13 February 1925, printed in part in Gilbert, *Churchill*, vol. 5, companion, pt 1, pp. 393–7.
118. *DBFP*, 1st Series, vol. xxvii, no. 200, p. 304, Mr Chamberlain to Lord Crewe (Paris), 16 February 1925, quoted in Crowe, 'Sir Eyre Crowe and the Locarno Pact', p. 61.
119. CAB 2/4, Committee of Imperial Defence, Minutes of the 195th Meeting, 13 February 1925.
120. FO 371/10731, C6582/459/18, Draft by Crowe, 17 February 1925.
121. D'Abernon, *An Ambassador of Peace*, vol. 3, pp. 141–2.
122. CAB 2/4, Committee of Imperial Defence, Minutes of the 196th Meeting, 19 February 1925.
123. FO 371/10727, C2201/459/18, Minute by Lampson, 17 February 1925.
124. Ibid., Minute by Crowe, 18 February 1925.
125. Ibid., Minute by Chamberlain, February 1925 (undated).
126. CAB 2/4, Committee of Imperial Defence, Minutes of the 196th Meeting, 19 February 1925; Crowe, 'Sir Eyre Crowe and the Locarno Pact', pp. 61–3.
127. Keith Middlemas and John Barnes, *Baldwin, A Biography* (London, 1969), pp. 30–1.
128. Austen Chamberlain MSS, AC 52/240, Crowe to A. Chamberlain, 12 March 1925.
129. CAB 23/49, 11(25), Conclusions of a Meeting of the Cabinet, 2 March 1925; Middlemas and Barnes, *Baldwin*, pp. 350–1.
130. Austen Chamberlain MSS, AC 39/2/35, A. Chamberlain to D'Abernon, 11 September 1930.
131. CAB 23/49, 4(25), Conclusions of a Meeting of the Cabinet, 4 March 1925, quoted in Middlemas and Barnes, *Baldwin*, p. 351 and in Crowe, 'Sir Eyre Crowe and the Locarno Pact', pp. 65–6.
132. Austen Chamberlain MSS, AC 52/240, Crowe to A. Chamberlain, 12 March 1925.
133. *DBFP*, 1st Series, vol. xxvii, no. 227, p. 353, Mr Chamberlain (Geneva) to Sir E. Crowe, 8 March 1925.
134. Ibid., pp. 354–5.
135. Ibid., no. 225, p. 348, Mr Chamberlain to Sir E. Crowe, 7 March 1925.
136. Ibid., no. 237, pp. 367–9, Sir E. Crowe to Mr Chamberlain (Geneva), 11 March 1925.
137. Austen Chamberlain MSS, AC 52/240, Crowe to A. Chamberlain, 12 March 1925, printed in part in Gilbert, *Churchill*, vol. 5, companion pt 1, pp. 430–2. See also Crowe, 'Sir Eyre Crowe and the Locarno Pact', pp. 67–70 and Middlemas and Barnes, *Baldwin*, pp. 352–4.
138. Austen Chamberlain MSS, AC 52/80, Baldwin to A. Chamberlain, 12 March 1925, quoted in Middlemas and Barnes, *Baldwin*, p. 354.
139. Ibid., AC 6/1/603, A. Chamberlain to Ivy Chamberlain, 15 March 1925, quoted in Crowe, 'Sir Eyre Crowe and the Locarno Pact', p. 71.
140. Ibid., AC 52/241, A. Chamberlain to Crowe, March 1925 (undated), printed in Petrie, *A. Chamberlain*, vol. 2, p. 264 and also for the major part in Crowe, 'Sir Eyre Crowe and the Locarno Pact', pp. 72–3.
141. Ibid., AC 52/244, Crowe to A. Chamberlain, 15 March 1925.
142. Bridgeman MSS, Political Notes, vol. 2, 11 March 1925, quoted in Middlemas and Barnes, *Baldwin*, p. 354.
143. Austen Chamberlain MSS, AC 52/244, Crowe to A. Chamberlain, 15

March 1925, quoted for the major part in Middlemas and Barnes, *Baldwin*, pp. 355–6.

144. CAB 23/49, 17 (25), Conclusions of a Meeting of the Cabinet, 20 March 1925.

145. PP 1925 (Cmd 2435), vol. 30, Papers respecting the proposals for a Pact of Security made by the German Government on 9 February 1925, no. 8, pp. 45–51, Réponse au Mémorandum Allemand, 16 June 1925.

146. *DBFP*, 1st Series, vol. xxvii, no. 291, pp. 452–3, Minute by Mr Sterndale Bennett, 7 April 1925.

147. Jules Laroche, *Au Quai d'Orsay avec Briand et Poincaré, 1913–1926* (Paris, 1957), p. 207.

148. *DBFP*, 1st Series, vol. xxvii, no. 277, pp. 428–9, Lord Crewe (Paris) to Mr Chamberlain, 29 March 1925; ibid., no. 279, pp. 430–1, Mr Phipps (Paris) to Sir E. Crowe, 30 March 1925.

149. Ibid., no. 279, p. 431, Mr Phipps (Paris) to Sir E. Crowe, 30 March 1925; ibid., no. 291, pp. 453–4, Minute by Mr Sterndale Bennett, 7 April 1925.

150. Ibid., no. 291, p. 453, Minute by Mr Sterndale Bennett, 7 April 1925.

151. FO 371/10732, C6984/459/18, Sir William Max Muller to Lampson, 23 May 1925.

152. Austen Chamberlain MSS, A 5/1/351, A. Chamberlain to Hilda Chamberlain, 25 April 1925.

153. *DBFP*, 1st Series, vol. xxvii, no. 318, pp. 492–5, Lord Crewe (Paris) to Mr Chamberlain, 13 May 1925; ibid., no. 328, pp. 512–44, Memorandum communicated informally by the French Embassy, 18 May 1925; ibid., nos 322, 340, pp. 501–4, 528–32, Mr Chamberlain to Lord Crewe (Paris), 14, 25 May 1925.

154. Ibid., no. 349, pp. 545–55, Mr Chamberlain to Lord Crewe (Paris), 28 May 1925.

155. FO 371/10731, C6493/459/18, Minute by Chamberlain, 14 May 1925.

156. *DBFP*, 1st Series, vol. xxvii, no. 321, pp. 499–501, Memorandum by Mr Chamberlain for the Cabinet, 14 May 1925.

157. Ibid., no. 317, p. 491, Notes by Sir C. Hurst on his suggested Draft of 12 May 1925, for a Security Pact, 13 May 1925. See also ibid., nos 315–16, 319, 347.

158. FO 371/10732, C7296/459/18, Minute by Lampson, 22 May 1925; ibid., C7297/45/28, Minute by Lampson, 18 May 1925.

159. Ibid., C7296/459/18, Minute by Tyrrell, 23 May 1925; ibid., C7297/459/18, Minute by Tyrrell, 18 May 1925.

160. Ibid., C7296/459/18, Minute by Chamberlain, 25 May 1925; *DBFP*, 1st Series, vol. xxvii, no. 349, pp. 545–50, Mr Chamberlain to Lord Crewe (Paris), 28 May 1925. See also ibid., nos 321, 330, 343, 347, 350.

161. Ibid., no. 363, p. 584, Mr Chamberlain to the French Foreign Minister, 8 June 1925.

162. FO 371/10733, C7884/459/18, French draft reply to Germany, agreed upon by Chamberlain and Briand, 11 June 1925.

163. *DBFP*, 1st Series, vol. xxvii, no. 364, pp. 586, 589, Mr Chamberlain (Geneva) to Sir W. Tyrrell, 8 June 1925.

164. Cmd 2435, PP 1925, no. 8, pp. 45–51, Réponse au Mémorandum Allemand, 16 June 1925, printed in U.F., vol. 6, no. 1337, pp. 364–7.

165. *DBFP*, 1st Series, vol. xxvii, nos 399, 401, pp. 650–1, 653–5, Lord D'Abernon (Berlin) to Mr Chamberlain, 29 June, 1 July 1925.

166. Ibid., no. 401, pp. 654–5, Lord D'Abernon (Berlin) to Mr Chamberlain, 1 July 1925; ARWR, KAB. Luther, 1925–6, vol. 1, no. 110, pp. 356–74,

Ministerbesprechung, 25 June 1925; ibid., no. 114, pp. 388–98, Besprechung mit den Ministerpräsidenten der Länder, 27 June 1925.

167. AA, Sicherheitspakt, vol. 4, Der Reichsminister des Auswärtigen Stresemann an die Deutschen Botschafts, 20 June 1925. Printed in Stresemann, *Vermächtnis*, vol. 2, pp. 206–8.
168. *DBFP*, 1st Series, vol. xxvii, no. 400, pp. 652–3, Mr Chamberlain to Lord D'Abernon (Berlin), 30 June 1925.
169. FO 8371/10735, C8804/459/18, Lampson to Addison, 6 July 1925.
170. Austen Chamberlain MSS, AC 5/1/357, A. Chamberlain to Ida Chamberlain, 27 June 1925.
171. Ibid., AC 5/1/370, A. Chamberlain to Ida Chamberlain, 28 November 1925.
172. *DBFP*, 1st Series, vol. xxvii, no. 430, p. 707, Mr Chamberlain to Sir W. Max Muller (Warsaw), 28 July 1925.
173. FO 371/10737, C9886/459/18, Minute by Bennett, 29 July 1925.
174. Ibid., Minute by Tyrrell, 30 July 1925.
175. Ibid., Minute by Chamberlain, 30 July 1925.
176. CAB 2/4, Committee of Imperial Defence, Minutes of the 200th Meeting, 1 July 1925.
177. CAB 23/50, 93(25), Conclusions of a Meeting of the Cabinet, 3 July 1925.
178. CAB 2/4, Committee of Imperial Defence, Minutes of the 201st Meeting, 1 July 1925; *DBFP*, 1st Series, vol. xxvii, no. 405, pp. 659–61, Mr Chamberlain to Lord Crewe (Paris), 4 July 1925.
179. *DBFP*, 1st Series, vol. xxvii, no. 408, pp. 663–5, The French Foreign Minister to the French Ambassador (London, 9 July 1925).
180. FO 371/10736, C9216/459/18, Minute by Lampson, 11 July 1925.
181. Ibid., C9784/459/18, Minute of a Departmental Committee by Bennett, 22 July 1925.
182. PP 1925 (Cmd 2468), vol. 30, Reply of the German Government to the Note handed to Herr Stresemann by the French Ambassador at Berlin on 16 June 1925, respecting the proposals for a Pact of Security. In continuation of Misc. no. 7. Printed in U.F., vol. 6, no. 1341, pp. 370–5.
183. FO 371/10736, C9784/459/18, Minute of a Departmental Committee by Bennett, 22 July 1925.
184. Ibid.
185. Ibid.
186. *DBFP*, 1st Series, vol. xxvii, no. 440, pp. 726–31, Proposed Treaty of Mutual Guarantee, 12 August 1925; ibid., nos 439, 443, pp. 723–5, 732–4, Mr Chamberlain to Lord D'Abernon (Berlin), 11, 13 August 1925.
187. CAB 23/50, 45(25) Conclusions of a Meeting of the Cabinet, 13 August 1925.
188. Stresemann, *Vermächtnis*, vol. 2, pp. 156–61, 163–6; *DBFP*, 1st Series, vol. xxvii, no. 427, pp. 701–4, Mr Chamberlain to Lord D'Abernon (Berlin), 28 July 1925.
189. *DBFP*, 1st Series, vol. xxvii, no. 427, p. 699, Lord D'Abernon (Berlin) to Mr Chamberlain, 28 July 1925.
190. Ibid., no. 431, p. 707, Mr Chamberlain to Lord D'Abernon (Berlin), 30 July 1925.
191. FO 371/ 10737, C9994/459/18, Minute by Bennett, 29 July 1925.
192. Ibid., Minute by Lampson, 29 July 1925.
193. Ibid., Minute by Tyrrell, 29 July 1925.
194. *DBFP*, 1st Series, vol. xxvii, no. 454, pp. 742–3, Lord D'Abernon (Berlin) to Mr Chamberlain, 26 August 1925.
195. Ibid., no. 466, pp. 754–60, Report by Sir C. Hurst on the Proceedings of

a Meeting of Jurists held at the Foreign Office, 1–4 September 1925, in connection with the proposed Treaty of Mutual Guarantee, 4 September 1925; ibid., no. 467, pp. 760–2, Note by Sir C. Hurst, 5 September 1925.

196. ARWR, KAB. Luther, 1925–6, vol. 1, nos 158–9, pp. 550–62, Minister-besprechung, 22–3 September 1925; ibid., no. 161, pp. 567–73, Kabinettsrat beim Reichspräsidenten, 24 September 1925; ibid., no. 162, pp. 574–9, Besprechung mit den Ministerpräsidenten der Länder, 25 September 1925; Stresemann, *Vermächtnis*, vol. 2, pp. 180–1.

197. Yet Germany's formal acceptance of the Allied invitation to attend a conference of Foreign Ministers at Locarno was marked by a last-minute attempt to stipulate the abolition of the notion of German war guilt, the prior settlement of the question of disarmament and the evacuation, first, of Cologne. *DBFP*, 1st Series, vol. xxvii, no. 493, pp. 789–92, Mr Chamberlain to Lord D'Abernon (Berlin), 26 September 1925; ibid., no. 507, p. 803, Mr Chamberlain to the German Ambassador, 29 September 1925; Jacobson, *Locarno Diplomacy*, pp. 57–9.

198. *DBFP*, 1st Series, vol. xxvii, no. 509, pp. 810–12, Memorandum by Mr Chamberlain on the Locarno Conversation, 2 October 1925.

199. Viscount Grey of Falladon, *Twenty-Five Years* (London, 1925), vol. 1, p. 273.

200. *DBFP*, 1st Series, vol. xxvii, no. 527, p. 853, Memorandum, 9 October 1925.

201. Harald von Rickhoff, *German–Polish Relations 1918–1933* (Baltimore, 1971), p. 109; Piotr S. Wandycz, *France and her Eastern Allies, 1919–1925: French–Czechoslovak–Polish Relations from the Paris Peace Conference to Locarno* (Minneapolis, 1962), p. 360; Stresemann, *Vermächtnis*, vol. 2, p. 235.

202. Regarding the right of the members of the League, in a case where the Council failed to reach a unanimous report, 'to take such action as they shall consider necessary for the maintenance of right and justice' [Article 15(7)]. And regarding their duty to subject any member that committed an act of war to economic, financial and military sanctions [Article 16]. Printed in James Barros, *Office without Power: General Sir Eric Drummond, 1919–1933* (Oxford, 1979), pp. 408–9.

203. *DBFP*, 1st Series, vol. xxvii, nos 531, 536–7, 550, pp. 861–4, 871–3, 888–93, Mr Chamberlain (Locarno) to Sir W. Tyrrell, 1, 14, 17 October 1925; ibid., Appendix, nos 6, 12, pp. 1097–1108, 1143–50, British Secretary's Notes of the Third and Seventh Meetings between the British, Belgian, French, Italian and German Delegations, held at Locarno, 7, 13 October 1925; Stresemann, *Vermächtnis*, vol. 2, p. 235.

204. *DBFP*, 1st Series, vol. xxvii, no. 520, pp. 828–31, Memorandum by Mr Lampson, 7 October 1925; ibid., no. 526, pp. 850–2, Record by Mr Chamberlain of a conversation with the Polish Foreign Minister, 9 October 1925; ibid., no. 550, p. 892, Mr Chamberlain (Locarno) to Sir W. Tyrrell, 17 October 1925; ibid., Appendix, no. 6, pp. 1097–1108, British Secretary's Notes of the Third Meeting between the British, Belgian, French, German and Italian Delegations, held at Locarno, 7 October 1925; ibid., no. 7, p. 1108, Notes of a Conversation between Members of the British Delegation in Mr Chamberlain's Room at the Grand Hotel, Locarno, 7 October 1925; ibid., nos 13, 15, pp. 1151–9, 1170–1, 1174–5, British Secretary's Notes of the Eighth and Ninth Plenary Meetings of the Locarno Conference, 15, 16 October 1925; PP 1925 (Cmd 2525), vol. 31, Final Protocol of the Locarno Conference, 1925 (and Annexes) together with Treaties between France and Poland and between France and Czechoslovakia, Locarno, 16 October 1925.

205. *DBFP*, 1st Series, vol. xxvii, nos 522–3, 532, 550, pp. 836–47, 865–6, 892, Mr Chamberlain (Locarno) to Sir W. Tyrrell, 8, 12, 17 October 1925; ibid., Appendix, no. 8, pp. 1110–21, British Secretary's Notes of the Fourth Meeting between the British, Belgian, French, German and Italian Delegations, held at Locarno, 8 October 1925.
206. Cmd 2525, PP 1925 , Annex F to item no. 1, Draft Collective Note to Germany regarding Article 16 of the Covenant of the League of Nations.
207. *DBFP*, 1st Series, vol. xxvii, no. 549, footnote no. 1, p. 887.
208. Ibid., 1st Series, vol. i, no. 122, footnote no. 1, p. 180, Record of a Meeting held in Mr Chamberlain's Room at the Foreign Office, 1 December 1925.
209. The Arbitration Treaties were analogous to the Arbitration Conventions, except that the Conventions mentioned in their preamble the Treaty of Mutual Guarantee between Germany and her Western neighbours, whereas the Treaties did not, thus both isolating Czechoslovakia and Poland from France and avoiding even the inference of a *status quo* policy in the East. Cmd 2525, PP 1925, Annexes B–E to item no. 1.
210. The terms of these treaties were identical. Cmd 2525, PP 1925, items nos 2 and 3 respectively.
211. *DBFP*, 1st Series, vol. xxvii, nos 549–50, pp. 887, 892, Mr Chamberlain (Locarno) to Sir W. Tyrrell, 17 October 1925. Again, not by inference either – i.e. neither from their having been annexed to the Final Protocol, nor specified as guarantee treaties nor initialled by other signatories owing to Germany's non-consent. For further details, see von Rickhoff, *German–Polish Relations*, pp. 110–11, and Wandycz, *France and her Eastern Allies*, pp. 362–3.
212. Yet, not having defined precisely if the discretion in judging the violations' severity was to be left to the aggrieved party or to the guarantors – and there can be no doubt that the British Government accepted the second meaning – this article, though theoretically committing Britain (and Italy) to act against any violation of a good deal of the existing territorial arrangements in Western Europe, did not in practice lead to the impairment of her traditional freedom of action. Contemporary sources in fact suggest that, for the British negotiators, the guarantee Britain gave could come into operation only in the event of an actual attack on such French and Belgian territory as would threaten the *status quo* of the Channel and the North Sea ports. Indeed, Chamberlain, quoting his assistants, told the 1926 Imperial Conference that he regarded British liabilities as reduced, not extended, by Locarno.

    DBFP, 1st Series, vol. xxvii, nos 530, 550, pp. 863–4, 890–1, Mr Chamberlain (Locarno) to Sir W. Tyrrell, 10, 17 October 1925; ibid.; 1st Series, vol. i, no. 1, p. 15, Foreign Office Memorandum respecting the Locarno Treaties, 10 January 1926; ibid., Appendix, pp. 848–9, Memorandum on the Foreign Policy of His Majesty's Government, with a List of British Commitments in their Relative Order of Importance, April 1926 (undated); N. H. Gibbs, *Grand Strategy* (London, 1976), pp. 42–3.
213. Cmd 2525, PP 1925, Annex A to item no. 1, Treaty of Mutual Guarantee between Germany, Belgium, France, Great Britain and Italy, 16 October 1925. Printed also in RIIA, *Survey*, 1925, vol. 2, Appendix no. 1, pp. 440–2, and in U.F., vol. 6, no. 1344, pp. 381–4.
214. Chamberlain MSS, AC 52/61, Amery to Chamberlain, 20 October 1925.
215. *DBFP*, 1st Series, vol. xxvii, pp. 858–60, Mr Lampson (Locarno) to Sir W. Tyrrell, 9 October 1925.
216. Ibid., no. 544, p. 882, Sir W. Tyrrell to Mr Chamberlain (Locarno), 15

October 1925; ibid., no. 550, footnote no. 6, p. 893, Conclusions of a Meeting of the Cabinet, 21 October 1925.

217. Petrie, *A. Chamberlain*, vol. 2, pp. 293, 314.
218. Austen Chamberlain MSS, AC 5/1/370, A. Chamberlain to Ida Chamberlain, 28 November 1925.
219. Ibid., AC 5/1/380, A. Chamberlain to Hilda Chamberlain, 25 April 1926.
220. FO 371/11286, C1683/436/18, Memorandum by J. M. Troutbeck: 'Present Position of German Disarmament', 6 February 1925.
221. On the threat of not proposing ratification of the Treaty to the Reichstag unless Cologne was evacuated, the Germans managed to have Allied insistence upon the meticulous execution by Germany of the whole list of unsatisfied disarmament points, contained in the Note of 4 June, changed to their being willing to settle for an indication of a true beginning on disarmament, in the form of a Note addressed to the Ambassadors' Conference by the German Government, showing in detail what points had already been liquidated, what points were on a fair way to settlement and those points on which agreement had not yet been reached. In return for which, if the position was found to be satisfactory, a date would be fixed for the evacuation. This was, indeed, addressed by Germany on 23 October, and accepted by both London and Paris, even though it did not give satisfaction on all counts. And on 16 November, Briand, acting as Chairman of the Ambassadors' Conference, notified the German Ambassador in Paris, Leopold von Hoesch, that the evacuation of Cologne would commence on 1 December 'without waiting for the execution to be entirely completed'. The evacuation was completed on 31 January 1926.
    Cmd 2527, *DBFP*, 1st Series, vol. xxvii, Annex, nos 11, 14; ibid., 1st Series, vol. i, nos 2, 28, 33, 50–1, 55, 59–64, 66, 68–73, 76–9, 85, 88, 92, 94–8, 100–3, 106, 228; Fox, 'Britain and the Inter-Allied Military Commission of Control 1925–6', pp. 153–4; Jacobson, *Locarno Diplomacy*, pp. 62–7; Salewski, *Entwaffnung*, pp. 300–28.
222. *DBFP*, 1st Series, vol. i, no. 228, p. 378, Lord Kilmarnock (Coblenz) to Sir A. Chamberlain, 31 January 1926.
223. FO 371/11286, C1683/436/18, Minute by Chamberlain, 8 February 1926.
224. *DBFP*, 1st Series, vol. 1, no. 174, pp. 306–10, Despatch from M. Briand to M. de Fleuriau, 11 January 1926.
225. Ibid., no. 395, pp. 571–2, The Marquess of Crewe (Paris) to Sir A. Chamberlain, 3 April 1926.
226. FO 371/11287, C4658/436/18, Minute by Troutbeck, 10 April 1926.
227. *DBFP*, 1st Series, vol. i, no. 199, p. 346, Letter from Sir W. Tyrrell to M. de Fleuriau, 23 January 1926. Though in private Lampson admitted that: 'We know in our heart of hearts that League investigation under Art. 213 is never going to be really effective.' Ibid., no. 201, footnote 2, p. 349, Minute by Lampson, 20 January 1926.
228. FO 371/11291, C11187/436/18, Minute by Sargent, 23 October 1926.
229. Ibid., Minute by Sargent, 16 September 1926.
230. FO 371/11286, C973/436/18, General Wauchope (Berlin) to Lord D'Abernon, 20 January 1926.
231. *DBFP*, 1st Series, vol. i, no. 265, pp. 423–4, Lord D'Abernon (Berlin) to Sir A. Chamberlain, 11 February 1926.
232. Ibid., no. 277, pp. 440–1, Sir A. Chamberlain to the Marquess of Crewe (Paris), 16 February 1926.
233. FO 371/11287, C6113/436/18, War Office to Foreign Office, 27 May 1925.
234. *DBFP*, 1st Series, vol. ii, no. 269, pp. 470–1, Sir A. Chamberlain to the Marquess of Crewe (Paris), 28 October 1926.

235. Ibid., no. 285, enclosure in no. 285, p. 500, Notes of Conversation between General Wauchope, General von Pawelsz and Dr Forster, 8 November 1926.
236. Ibid., no. 277, p. 440, Sir A. Chamberlain to the Marquess of Crewe (Paris), 16 February 1926.
237. FO 371/11286, C2016/436/18. Minutes by Troutbeck, 12 February 1926.
238. FO 371/11287, C4169/436/18, Minute by Chamberlain, 7 April 1926.
239. FO 371/11289, 8453/436/18, Annex 3, War Office to the Foreign Office: 'The State of Control in Germany', 22 July 1926.
240. Ibid., C8206/436/18, Memorandum by Lampson, 15 July 1926.
241. Ibid., C8453/436/18, Annex, Record of conversation between Mr Lampson and the German Ambassador, 26 July 1926.
242. FO 371/11290, 9473/436, Crowe (Paris) to Sir A. Chamberlain, 26 August 1926.
243. *DBFP*, 1st Series, vol. ii, no. 199, pp. 348–9, The Marquess of Crewe (Paris) to Sir A. Chamberlain, 10 September 1926. For further details see Salewski, *Entwaffnung*, pp. 326–42.
244. FO 371/11290, C9950/436/18, Minute by Sargent, 16 September 1926; ibid., Foreign Office to War Office, 21 September 1926.
245. Ibid., C10924/436/18, Minute by Troutbeck, 20 October 1926.
246. Ibid., C10593/436/18, Minute by Baxter, 5 October 1926.
247. FO 371/11291, C11460/436/18, War Office to Foreign Office, 29 October 1926.
248. FO 371/11292, C11795/436/18, Minutes by Tyrrell and Chamberlain, 10 November 1926.
249. Ever since October 1925 the Locarnites had looked for additional agreements to settle their economic, military and territorial differences. Eventually, Stresemann, taking advantage of France's financial difficulties and the Quai d'Orsay's pressure on its Chief to withdraw the Commission of Control in return for a large German sum, met Briand on 17 September 1926 at Thoiry, where they projected an extraordinary four-point scheme. First, the Saar district, which was governed by the League of Nations, would be restored to German sovereignty without the plebiscite of the local population provided for in the Treaty of Versailles, and Germany would repurchase the Saar coal mines from France at a cost of approximately 300 million marks. Secondly, the Commission of Control would be withdrawn and the remaining differences over disarmament between the Allies and the German Government would be settled at the Ambassadors' Conference. Thirdly, Paris would raise no further objections to German negotiations with the Belgians for the return to Germany of the Eupen-Malmedy district ceded to Belgium at the Peace Conference. Finally, the Rhineland would be completely evacuated within a year, and France would receive advance payment on her reparation account.

However, faced with the stabilization of the franc a short while after, and Berlin's fears lest any demonstration of an increased reparation-payment capacity should seriously weaken the Germans' case for a downward revision of the Dawes Plan – to which were added American and British resistance to loans to France and Germany, and French military opposition to an early end to military occupation – Paris and Berlin decided on 28 October not to follow up Thoiry with discussion among French and German experts. On 11 November, finally, Briand suspended all formal discussion of the Rhineland–reparation nexus.

ADAP, B Series, vols 1, 2, no. 88, pp. 188–91, Aufzeichnung des Reichsministers des Auswärtigen Stresemann, 17 September 1926; Stresemann, *Vermächtnis*, vol. 3, pp. 17–23; Georges Suarez, *Briand: sa vie, son œuvre avec son Journal at de nombreux documents inédits* (Paris, 1941), vol. 6, pp. 203–28; Jacobson, *Locarno Diplomacy*, pp. 84–90; Jon Jacobson and John T. Walker, 'The Impulse for a Franco-German entente: The origins of the Thoiry Conference, 1926', *Journal of Contemporary History*, vol. 10 (1975), pp. 157–75. *DBFP*, 1st Series, vol. ii, no. 283, p. 496, Sir G. Grahame (Brussels) to Sir A. Chamberlain, 12 November 1926. See discussion in B. Kent, *The Spoils of War*, pp. 271–2, 278.

250. Ibid., no. 304, p. 531, Letter from Sir W. Tyrrell to Mr Phipps (Paris), 26 November 1926.

251. Ibid., nos 333–54; ibid., no. 355, Annex to no. 355, p. 650, Agreement regarding military control in Germany, 12 December 1926, also printed in ADAP, B Series, vol. I, 2, no. 261, pp. 607–8; Fox, 'Britain and the Inter-Allied Military Commission of Control 1925–6', pp. 156–64; Jacobson, *Locarno Diplomacy*, pp. 91–8; Salewski, *Entwaffnung*, pp. 357–74.

252. *DBFP*, 1st Series, vol. xxvii, Appendix no. 8, pp. 1110–11, 1116–17, 1120, British Secretary's Notes of the Fourth Meeting between the British, Belgian, French, German and Italian Delegations, 8 October 1925; Johnson, 'Austen Chamberlain and the Locarno Agreements', p. 80.

253. FO 371/11262, C1393/71/18, Lord D'Abernon (Berlin) to Sir A. Chamberlain, 3 February 1926. See also *DBFP*, 1st Series, vol. i, no. 238, pp. 391–2.

254. ADAP, B Series, vol. I, 1, no. 88, p. 219, Der Reichsminister des Auswärtigen Stresemann an das Konsulat in Genf, 8 February 1926.

255. Ibid., no. 95, pp. 231, Der Reichsminister des Auswärtigen Stresemann an die Botschaft in London, 12 February 1926; *DBFP*, 1st Series, vol. i, no. 275, p. 436, Sir A. Chamberlain to Lord D'Abernon (Berlin), 15 February 1926.

256. *DBFP*, 1st Series, vol. i, no. 233, pp. 383–4, Memorandum by Sir A. Chamberlain respecting Poland and the Council of the League, 1 February 1926; FO 371/11263, C1994/71/18, Memorandum by Lampson, 15 February 1926.

257. ADAP, B Series, vol. I, 1, no. 90, pp. 222–4, Aufzeichnung des Vortragenden Legationsrats von Dirksen, 9 February 1926; ibid., no. 101, pp. 241–2, Der Botschafter in London Stohmer an das Auswärtige Amt, 15 February 1926; *DBFP*, 1st Series, vol. i, no. 265, p. 436, Sir A. Chamberlain to Lord D'Abernon (Berlin), 15 February 1926.

258. FO 371/11262, C1820/71/18, Sir G. Grahame (Brussels) to Sir A. Chamberlain, 12 February 1926.

259. ADAP, B Series, vol. I, 1, no. 99, pp. 236–8, Der Botschafter in Paris von Hoesch an das Auswärtige Amt, 13 February 1926; FO 371/11263, C1995/71/18, Minute by Tyrrell, 19 February 1926.

260. ADAP, B Series, vol. I, 1, no. 103, pp. 253–4, Aufzeichnung des Staatssekretärs des Auswärtigen Amts von Schubert, 16 February 1926; *DBFP*, 1st Series, vol. i, no. 276, pp. 438–9, Lord D'Abernon (Berlin) to Sir A. Chamberlain, 16 February 1926; ARWR, KAB. Luther, 1925–6, vol. 2, no. 299, pp. 1154–6, Ministerbesprechung, 24 February 1926; Christoph M. Kimmich, *Germany and the League of Nations* (Chicago, 1976), pp. 78–81.

261. FO 371/11263, C2144/71/18, Minute by Troutbeck, 19 February 1926.

262. Ibid., Minute by Tyrrell, 20 February 1926.

263. *DBFP*, 1st Series, vol. i, no. 275, p. 436, Sir A. Chamberlain to Lord D'Abernon (Berlin), 15 February 1926.

264. CAB 23/52, 6(26), Conclusions of a Meeting of the Cabinet, 17 February 1926.

265. *DBFP*, 1st Series, vol. i, no. 256, p. 412, Letter from Sir A. Chamberlain to Viscount Cecil, 9 February 1926.

266. CAB 23/52, 6(26), Conclusions of a Meeting of the Cabinet, 17 February 1926.

267. *DBFP*, 1st Series, vol. i, no. 233, pp. 383–4, Memorandum by Sir A. Chamberlain respecting Poland and the Council of the League, 1 February 1926.

268. Ibid., no. 257, p. 412, Letter from Sir A. Chamberlain to Viscount Cecil, 9 February 1926; CAB 23/52, 6(26), Conclusions of a Meeting of the Cabinet, 17 February 1926. See also, Chamberlain's Memorandum of his talks with Briand, *DBFP*, 1st Series, vol. i, no. 233, pp. 383–4.

269. Donald S. Birn, *The League of Nations Union, 1918–1945* (Oxford, 1981), pp. 62–4.

270. Lord Cecil of Chelwood, *A Great Experiment* (London, 1941), pp. 47–51; David Carlton, 'Disarmament with Guarantees: Lord Robert Cecil, 1922–1927', *Disarmament and Arms Control*, vol. 3 (1965), pp. 143–64.

271. CAB 23/52, 6(26), Conclusions of a Meeting of the Cabinet, 17 February 1926.

272. Though it avoided openly disowning the Foreign Secretary's policies in the House of Commons. Hansard, Parliamentary Debates, 5th Series, H of C, vol. 192, cols 1693–7, 4 March 1926.

273. CAB 23/52, 9(26), Conclusions of a Meeting of the Cabinet, 3 March 1926. For further details see David Carlton, 'Great Britain and the League Council Crisis of 1926', *Historical Journal*, vol. 11 (1968), pp. 355–8.

274. *DBFP*, 1st Series, vol. i, no. 330, pp. 498–9, Letter from Sir A. Chamberlain (Geneva) to Sir W. Tyrrell, 9 March 1926.

275. Ibid., no. 335, p. 504, Letter from Sir W. Tyrrell to Sir A. Chamberlain (Geneva), 11 March 1926.

276. Cecil, *All the Way*, p. 180.

277. ADAP, B Series, vol. I, 1, nos 145, 147, 153–4, 160–3, 169, Appendix I, pp. 712, 714, 716–17; *DBFP*, 1st Series, vol. i, nos 327–34, 336–8, 340–354; Barros, *Office without Power: General Sir Eric Drummond, 1919–1933*, pp. 184–5; Erik Lonnroth, 'Sweden: The Diplomacy of Osten Unden', in G. A. Craig and Felix Gilbert, *The Diplomats, 1919–1939*, pp. 92–9; Kimmich, *Germany and the League of Nations*, pp. 81–4.

278. ADAP, B Series, vol. I, 1, nos 147–8, 151, 153, 163; *DBFP*, 1st Series, vol. i, nos 333–4, 341, 346–7, 349–51, 353–60; Carlton, 'Britain and the League Crisis', pp. 360–2; F. P. Walters, *A History of the League of Nations* (Oxford, 1960), pp. 320–1.

279. ADAP, B Series, vol. I, 1, no. 166, pp. 402–4, Undatierte Aufzeichnungohne Unterschrift, 16 March 1926; *DBFP*, 1st Series, vol. i, no. 359, p. 531, Mr London (Geneva) to Sir W. Tyrrell, 16 March 1926; Petrie, *A. Chamberlain*, vol. 2, pp. 300–1; RIIA, *Survey, 1926, of International Affairs* (Oxford, 1927), pp. 45–7; Walters, *History of the League of Nations*, pp. 321–3.

280. ADAP, B Series, vol. I, 1, no. 222, pp. 531–4, Das Konsulat in Genf an das Auswärtige Amt, 16 May 1926; ibid., no. 224, pp. 537–541, Ministerialdirektor Gans an den Staatssekretär des Auswärtigen Amts von Schubert, z., Z. Marienbad, 20 May 1926; *DBFP*, 1st Series, vol. i, no. 517, footnote 1, p. 743, for documentary references of the minutes of this committee; ibid., no. 525, pp. 748–53, Viscount Cecil (Geneva) to Sir A. Chamberlain, 16 May 1926.

281. ADAP, B Series, vol. I, 1, nos 266, 282, 286, 295; ibid., vol. I, 2, nos 20–1, 28, 30, 45, 47, 50, 57, 61, 83; *DBFP*, 1st Series, vol. ii, nos 6, 37, 45, 46, 53, 57, 58, 114, 137, 145, 149, 194, 203, 204; Barros, *Office without Power: General Sir Eric Drummond, 1919–1933*, pp. 187–90; Elmer Bendiner, *A Time for Angels: The Tragicomic History of the League of Nations* (New York, 1975), pp. 216–223;

Kimmick, *Germany and the League of Nations*, pp. 87–91; RIIA, *Survey, 1926*, pp. 62–78; George Scott, *The Rise and Fall of the League of Nations* (London, 1973), pp. 150–8; Walters, *History of the League of Nations*, pp. 323–7.

Appendix 1
The Establishment of the Department of Overseas Trade

1. Edward Morse, *Modernization and the Transformation of International Affairs* (London, 1976), p. 19.
2. D. C. M. Platt, *Finance, Trade and Politics in British Foreign Policy, 1815–1914* (Oxford, 1968), pp. 207–17, 362–7, 372–4.
3. FO 371/8291, W1475/1475/50, Minute by Wellesley, 25 February 1922.
4. Platt, *Finance, Trade and Politics*, pp. xvii–xxix; D. C. M. Platt, *The Cinderella Service: British Consuls since 1825* (London, 1971), p. 91.
5. Public Record Office Handbooks, *The Records of the Foreign Office, 1782–1939* (London, 1969), p. 23.
6. Marian Siney, *The Allied Blockade of Germany, 1914–16* (Ann Arbor, MI, 1957), pp. 46, 52–6, 83–93.
7. Zara Steiner, 'The Foreign Office and the War', in F. H. Hinsley (ed.), *British Foreign Policy under Sir Edward Grey* (Cambridge, 1977), pp. 520–2. For further information on this point see Crowe and Corp, *Our Ablest Public Servant: Sir Eyre Crowe, 1864–1925*, pp. 278–86.
8. Zara S. Steiner, *The Foreign Office and Foreign Policy, 1898–1914* (Cambridge, 1969), pp. 165–6.
9. Arthur Marsden, 'The Blockade', in Hinsley (ed.), *British Foreign Policy*, p. 50; Siney, *The Allied Blockade of Germany*, ch. 8.
10. Hansard, Parliamentary Debates, 5th Series, H of C, vol. 80, col. 1809, 9 March 1916.
11. Ibid., vol. 85, cols 395–6, 2 August 1916; Rothwell, *British War Aims and Peace Diplomacy 1914–1918* (Oxford, 1971), p. 267.
12. FO 366/791, X3257/3257/505, Fifth Report of the Royal Commission on the Civil Service, 18 December 1914.
13. Lord Eustace Percy, *Some Memories* (London, 1958), p. 147, quoted in Doreen Collins, *Aspects of British Politics 1904–1919* (London, 1965), p. 257.
14. FO 368/854, 7785/2049, Report of the Foreign Office Committee, 10 August 1916.
15. Cmd 9230, Report of the Machinery of Government Committee, 1918.
16. Percy, *Some Memories*, pp. 147–8; Francis Oppenheimer, *Stranger Within* (London, 1960), p. 305.
17. Runciman MSS, Grey to Runciman, 31 August 1916, quoted in Keith Robbins, 'Foreign Policy, Government and Public Opinion', in Hinsley (ed.), *British Foreign Policy*, p. 541.
18. Edward Morse, *Foreign Policy and Interdependence in Gaullist France* (Princeton, 1973), p. 15.
19. FO 368/1854, 77853/2049, Report by Docker, Pennefather and Wellesley, 4 April 1917.
20. Ibid., Report by the Chairman and Sir W. H. Clark, 4 April 1917.
21. FO 368/1855, 141670/2049, Minute by Cecil (undated).
22. Ibid., 167887/2049, GT 1707, Joint Memorandum by the President of the Board of Trade and Lord Robert Cecil, 14 August 1917.

23. BT 61/4/4, Wellesley to Hardinge, 15 October 1919.
24. Cecil MSS, Add. 51094, Minute by Crowe, 6 September 1918.
25. FO 368/1855, 218677/2049, Minute by Hardinge (undated).
26. Ibid., 244751/2049, Wellesley to Clark, 11 January 1918.
27. Cecil MSS, Add. 51094, Steel-Maitland to Cecil, 28 August 1918.
28. FO 368/1855, 167887/2049, GT 1707, Joint Memorandum by the President of the Board of Trade and Lord Robert Cecil, 14 August 1917.
29. Morse, *Modernization and Transformation*, pp. 102–3.
30. FO 368/2253, 98230/79699/150, Minute of a Meeting of the Cave Committee, 28 May 1919.
31. Ibid., Minutes of a Meeting of the Cave Committee, 24 June 1919.
32. Ibid., Minutes of a Meeting of the Cave Committee, 28 May 1919.
33. Lloyd George MSS, F/36/3/3, Steel-Maitland to Lloyd George, 17 January 1919.
34. FO 368/2253, 98230/79699/150, Minutes of a Meeting of the Cave Committee, 22, 28 May, 4 June 1919.
35. Ibid., Minutes of a Meeting of the Cave Committee, 22, 27 May, 18 June 1919.
36. Ibid., Minutes of a Meeting of the Cave Committee, 27, 28 May, 2 July 1919.
37. Ibid., 103176/79699/150, Report of the Committee to Examine the Question of Government Machinery for Dealing with Trade and Commerce, 10 July 1919.
38. Ibid.
39. CAB 23/11, 598/6, Conclusions of a Meeting of the Cabinet, 23 July 1919.
40. Lothian MSS, GD 40/17/142/1, Kerr to Lloyd George, 2 February 1921.
41. H. Montagu Villiers, *Charms of the Consular Career* (London, 1925), pp. 165–6; quoted in Platt, *The Cinderella Service*, p. 92.
42. FO 371/8291, W4278/1475/50, Note on departmental co-ordination in the light of the Genoa Conference by Wigram, 16 May 1922.
43. FO 371/8432, A4220/3621/6, Report on Commercial Secretary's establishment by Wellesley (undated).
44. FO 371/6905, N3779/3779/38, Minute by Gregory, 21 March 1921; ibid., Minute by Gregory, 30 March 1921.
45. FO 371/7418, C1368/458/62, Minute by Crowe, 1 February 1922.
46. *DBFP*, 1st Series, vol. xvi, nos 544, 581, Memorandum by Mr Wigram on the position of Sir John Bradbury on the Reparation Commission, 29 April 1921; ibid., vol. x, ch. 2, 178, Introductory Note; Alan J. Sharp, 'The Foreign Office in Eclipse 1919–22', *History*, vol. 61 (1976), pp. 215–16.
47. FO 371/4728, C7375/8/18, Wigram to MacFadyean, 24 September 1920; FO 371/6027, C9289/2740/18, Minute by Tufton, 30 April 1921.
48. *DBFP*, 1st Series, vol. xvi, nos 544, 581–2, Memorandum by Mr Wigram on the position of Sir John Bradbury on the Reparation Commission, 29 April 1921; ibid., nos 546, 583–4, Memorandum by Mr Wigram on the Reparation Commission, 30 April 1921.
49. FO 371/6027, C9289/2740/18, Curzon to Robert Horne, 7 May 1921.
50. FO 371/6036, C18209/2740/18, Foreign Office to the Treasury, 22 September 1921; Sharp, 'The Foreign Office in Eclipse', p. 216.
51. *DBFP*, 1st Series, vol. xvi, nos 737, 810–11, Curzon to Horne, 19 November 1921; Sharp, 'The Foreign Office in Eclipse', p. 216.
52. FO 371/5978, C21403/416/18, Minute by Wigram, 10 November 1921.
53. Ibid., Minute by Waterlow, 11 November 1921.
54. Ibid., Minute by Curzon, 12 November 1921.

55. *DBFP*, 1st Series, vol. xvi, nos 737, 810–11, Curzon to Horne, 19 November 1921.
56. FO 371/7485, C1422/99/18, Hardinge to Curzon, 16 October 1922; Sharp, 'The Foreign Office in Eclipse', p. 216.
57. FO 371/7486, C14305/99/18, Minute by Tyrrell, 18 October 1922.
58. FO 371/8302, W8858/8858/50, Minute by Crowe, 21 October 1922; ibid., Minute by Curzon, 21 October 1922.
59. BT 61/27/8. E9471, Report on the work of the Commercial Secretaries in Europe by W. Glenny, 1 December 1926; Platt, *Finance, Trade and Politics*, pp. 394–5.
60. FO 371/9544, W613/613/50, Minute by Campbell, 28 January 1924.
61. Ibid.
62. FO 371/9768, C3104/81/18, Minute by Lampson, 29 February 1924.
63. FO 371/9867, C14280/14053/18, Minute by Lampson, 12 September 1924.
64. FO 371/10723, C4615/269/1666, Minute by J. C. Sterndale Bennett, 1 April 1925.
65. FO 371/6038, C22283/2740/18, Memorandum by Sir Basil Blackett, 22 November 1922; *DBFP*, 1st Series, vol. xx, nos 102, 275–7, Bradbury to Bonar Law, 23 October 1922; FO 371/9767, C1863/81/18, D'Abernon to MacDonald, 3 February 1924.
66. FO 371/7516, C4852/333/18, Minute by Lampson, 6 April 1922.
67. FO 371/6038, C22282/2740/18, Minute by Waterlow, 26 November 1921; FO 371/7518, C15153/333/18, Minute by Lampson, 9 November 1922; FO 371/9754, C15793/70/18, Minute by Bennett, 15 October 1924.
68. Sally Marks, 'The Myths of Reparations', *Central European History*, vol. 11 (1978), pp. 239, 246–7.
69. FO 371/6022, C6958/240/18, Minute by Crowe, 7 April 1921; FO 371/7486, C14819/99/18, Minute by Tyrrell, 26 October 1922.
70. Donald Boadle, 'The Formation of the Foreign Office Economic Relations Section, 1930–1937', *Historical Journal*, vol. 20 (1977), pp. 919–36.
71. Sir Walford Selby, *Diplomatic Twilight 1930–1940* (London, 1953), pp. 4–7, 10–11; Frank T. A. Ashton-Gwatkin, *The British Foreign Service* (Syracuse, 1950), pp. 24–7, 72.
72. W. N. Medlicott, 'Britain and Germany: The Search for Agreement, 1930–37', in David Dilks (ed.), *Retreat from Power, 1906–1936*, vol. 1 (London, 1981), p. 81.
73. David Carlton, *MacDonald versus Henderson* (London, 1970), pp. 48–9, 60.
74. Medlicott, 'Britain and Germany', p. 85.
75. Sir Victor Wellesley, *Diplomacy in Fetters* (London, 1944), p. 91.

Appendix 2
The Formation of the Middle East Department

1. Frankel, op. cit., p. 5.
2. FO 371/2926, 61053/24699/16, Sir Reginald Wingate (Cairo) to Mr Balfour, 11 February 1917.
3. Ibid., Minute by Cecil, March 1917 (undated).
4. Ibid., Minute by Cecil, 17 March 1917.
5. Ibid., Memorandum by Sir Ronald Graham on future British policy with regard to Egypt, 2 March 1917.
6. Ibid., Minute by Hardinge, 16 March 1917.

7. Michael Hill, *The Sociology of Public Administration* (London, 1972), pp. 197, 200.
8. FO 371/2926, 61053/24699/16, Minute by Hardinge, March 1917 (undated).
9. FO 371/2932, 162105/158543/16, Minute by Hardinge, 3 September 1917.
10. Cecil MSS, Add. 51094, Cecil to Montagu, 13 September 1918.
11. CAB 21/186, CP 1372, Lord Winterton to Hankey, 26 May 1920.
12. Briton C. Busch, *Britain, India and the Arabs, 1914–1921* (Berkeley, 1971), pp. 205–6.
13. CAB 23/2, 98, Conclusions of a Meeting of the War Cabinet, 16 March 1917; CAB 23/5, 369, Conclusions of a Meeting of the War Cabinet, 21 March 1918.
14. Balfour MSS, Add. 49738, Cecil to Balfour, 8 January 1918; Helmut Mejcher, 'British Middle East Policy 1917–1921', *Journal of Contemporary History*, vol. 8 (1973), p. 89.
15. FO 371/2932, 162105/258543/16, Memorandum by Lord Edward Cecil respecting the future government of Egypt, 10 September 1917.
16. CAB 27/22, MEC 11th Minutes, Minutes of a Meeting of the Middle East Committee, 12 January 1918.
17. Cecil MSS, Add. 49738, Cecil to Balfour, 23 August 1918.
18. Brown, op. cit., pp. 205–6.
19. FO 371/2932, 162105/15843/16, Minute by Hardinge, 6 September 1917.
20. Ibid.
21. Ibid., Minute by Graham, 2 Septemebr 1917.
22. Cecil MSS, Add. 49738, Cecil to Balfour, 23 August 1918.
23. FO 371/2932, 162105/158543/16, Minute by Graham, 2 September 1917; Minute by Hardinge, 6 September 1917.
24. CAB 23/4, 233(17), Conclusions of a Meeting of the War Cabinet, 14 September 1917.
25. Ibid.
26. Ibid.
27. CAB 24/34, G.T. 2837, Memorandum by E. S. Montagu, 4 November 1917; Mejcher, 'British Middle East Policy', pp. 91–2.
28. CAB 23/14, 233(17), Conclusions of a Meeting of the War Cabinet, 14 September 1817.
29. CAB 24/34, G.T. 2837, Memorandum by E. S. Montagu, 4 November 1917; Mejcher, 'British Middle East Policy', pp. 91–2.
30. CAB 24/46, Report of the Egyptian Administration Committee by Lord Curzon, 20 February 1918.
31. FO 371/3388, Minute by Hardinge, 4 January 1918.
32. Mejcher, 'British Middle East Policy', p. 93.
33. CAB 27/28, EC 718, 'The War in the East' by Montagu, 5 July 1918.
34. CAB 27/29, EC 978, Note by Cecil, 20 July 1918.
35. Ibid., EC 809, 'The War in the East', Note by the Chief of the Imperial General Staff, 15 July 1918.
36. CAB 27/24, EC 31st Minutes, Minutes of a Meeting of the Eastern Committee, 17 September 1918; Aaron Klieman, *Foundations of British Policy in the Arab World: The Cairo Conference of 1921* (Baltimore, 1970), pp. 85–6.
37. CAB 27/28, EC 718, 'The War in the East', by Montagu, 5 July 1918.
38. CAB 27/29, EC 809, 'The War in the East', Note by the Chief of the Imperial General Staff, 15 July 1918.
39. Ibid., EC 978, 'The War in the East', Departmental Note by the Foreign Office on Mr Montagu's paper, 17 July 1918.
40. Ibid., Note by Cecil, 20 July 1918.

41. Ibid., Note by Balfour, 27 July 1918.
42. CAB 27/24, EC 24a Minutes, Minutes of a Meeting of the Eastern Committee, 13 August 1918.
43. FO 371/3481, W50/161561/161560, Minute by Cecil, August 1918 (undated).
44. FO 800/207, Cecil to Montagu, 29 August 1918.
45. FO 794/6A, Minute by Hardinge, 12 September 1918.
46. Repington, *The First World War, 1914–1918* (London, 1920), vol. 2, pp. 25–6.
47. Cecil MSS, Add. 49738, Cecil to Balfour, 23 August 1918. For further information on this point see Crowe and Corp, *Our Ablest Public Servant: Sir Eyre Crowe, 1864–1925*, pp. 291–4.
48. Hardinge MSS, vol. 29, Hardinge to George Allen, 31 January 1917.
49. Ibid., vol. 22, Hardinge to Graham, June 1916 (undated).
50. Ibid.
51. FO 800/163, Reginald Lister to Bertie, 12 December 1905.
52. Repington, *The First World War*, vol. 2, pp. 25–6.
53. Hardinge MSS, vol. 22, Hardinge to Graham, June 1916 (undated).
54. Cecil MSS, Add. 49738, Cecil to Balfour, 21 August 1918.
55. Ibid., Cecil to Balfour, 23 August 1918.
56. FO 794/6A, Minute by Hardinge, 12 September 1918.
57. FO 800/207, Montagu to Cecil, 3 September 1918; Mejcher, 'British Middle East Policy', pp. 95–6.
58. Ibid., Cecil to Montagu, 29 August 1918; ibid., Montagu to Cecil, 3 September 1918; ibid., Cecil to Montagu, 5 September 1918; FO 371/3481, W50/161561/161560, Montagu to Cecil, 11 September 1918; Cecil MSS, Add. 51094, Cecil to Montagu, 13 September 1918.
59. Cecil MSS, Add. 51094, Montagu to Cecil, 17 September 1918.
60. FO 371/3481, W50/161561/151560, Memorandum by Crowe, 21 September 1918.
61. Ibid., Hirtzel to Crowe, 23 September 1918.
62. Ibid., Crowe to Cecil, 23 September 1918; ibid., Minute by Hardinge, September 1918 (undated).
63. Ibid., Minute by Cecil, September 1918 (undated).
64. DBFP, 1st Series, vol. xiii, no. 250, p. 267, Memorandum on the future control of the Middle East by Major H. W. Young, 17 May 1920.
65. Klieman, *Foundations of British Policy*, pp. 83–5.
66. Helmut Mejcher, *Imperial Quest for Oil: Iraq, 1910–1928* (Oxford, 1976), pp. 71–5; Klieman, *Foundations of British Policy*, pp. 11–12, 83–85.
67. Zara S. Steiner, *The Foreign Office and Foreign Policy, 1898–1914*, pp. 86, 209–13.
68. CAB 24/106, CP 1320, Memorandum by the Secretary of State for War, Mesopotamian Expenditure, 1 May 1920; also, in Martin Gilbert, *Winston S. Churchill* (London, 1977), companion vol. 4, pt 2, pp. 1078–82.
69. Brian C. Smith and Jeffrey Stanyer, *Administering Britain* (Glasgow, 1976), p. 139; D. N. Chester and F. M. G. Wilson, *The Organisation of British Central Government, 1914–1956* (London, 1957), pp. 402–3, 405–13.
70. CAB 24/106, CP 1320, Memorandum by the Secretary of State for War, Mesopotamian Expenditure, 1 May 1920; also, in Gilbert, *Churchill*, companion vol. 4, pt 2, pp. 1079–80.
71. CAB 23/22, 49(20), Conclusions of a Meeting of the Cabinet, 17 August 1920.
72. CAB 24/107, CP 1402, 'Mesopotamia and the Middle East: Question of future Control', Memorandum by the Secretary of State for India, 1 June 1920.

73. Smith and Stanyer, *Administering Britain*, p. 139.
74. CAB 24/107, CP 1402, 'Mesopotamia and the Middle East: Question of future Control', Memorandum by the Secretary of State for India, 1 June 1920.
75. CAB 24/106, CP 1337, 'Mesopotamian Expenditure', Memorandum by the Secretary of State for the Colonies, 24 May 1920.
76. CAB 24/108, CP 1512, Memorandum by the Secretary of State for the Colonies, 17 June 1920.
77. FO 371/5255, E4870/4870/44, Minute by Hardinge, 29 May 1920.
78. Ibid., and in DBFP, 1st Series, vol. xiii, no. 250, pp. 260–9, Memorandum on the future control of the Middle East by Major H. W. Young, 17 May 1920.
79. Ibid., E5992/4870/44, 'Future Administration of the Middle East', Memorandum by the Secretary of State for Foreign Affairs, 8 June 1920.
80. Ibid., E4870/4870/44, Minute by Tilley, 18 June 1920.
81. Ibid., E5992/4870/44, Future Administration of the Middle East, Memorandum by the Secretary of State for Foreign Affairs, 8 June 1920.
82. Ibid., E7636/4870/44, Minute by Hardinge, July 1920 (undated).
83. Ibid., Minute by Crowe, 22 July 1920.
84. Ibid., Minute by Hardinge, July 1920 (undated).
85. Philip Selznick, *TVA and The Grass Roots, A Study in the Sociology of Formal Organisation* (Los Angeles, 1949), p. 13.
86. CAB 24/117, CP 2343, Memorandum by the Secretary of State for India, 22 December 1920.
87. CAB 27/71, Minutes of a Meeting of the Cabinet Finance Committee, 3 August 1920.
88. CAB 24/110, CP 1777, 'A Middle-Eastern Department', by Earl Curzon, 16 August 1920.
89. Ibid.
90. CAB 23/22, 49(20), Conclusions of a Meeting of the Cabinet, 17 August 1920.
91. FO 371/5255, E14522/4780/44, Minute by Young, 23 November 1920.
92. Ibid., Minute by Crowe, 24 November 1920.
93. CAB 27/24, EC 39th Minutes, Minutes of a Meeting of the Eastern Committee, 27 November 1918; ibid., EC 42nd Minutes, Minutes of a Meeting of the Eastern Committee, 9 December 1918; John Darwin, *Britain, Egypt and the Middle East: Imperial Policy in the Aftermath of War, 1918–1922* (London, 1981), p. 246.
94. Richard H. Ullman, *Anglo-Soviet Relations, 1917–1921* (Princeton, 1972), vol. 3, pp. 320–4; Darwin, *Britain, Egypt and the Middle East*, pp. 19, 157–8, 246.
95. Sir Harold Nicolson, *Curzon: The Last Phase, 1919–1925* (London, 1937), pp. 121–2.
96. CAB 27/24, EC 40th Minutes, Minutes of a Meeting of the Eastern Committee, 2 December 1918; ibid., EC 42nd Minutes, Minutes of a Meeting of the Eastern Committee, 9 December 1918; ibid., EC 45th Minutes, Minutes of a Meeting of the Eastern Committee, 19 December 1918.
97. Ibid., EC 5th Minutes, Minutes of a Meeting of the Eastern Committee, 24 April 1918; EC 21st Minutes, Minutes of a Meeting of the Eastern Committee, 18 July 1918.
98. DBFP 1st Series, vol. iv, no. 710, pp. 1119–22, Memorandum by Earl Curzon on the Persian Agreement, 9 August 1919; H. Nicolson, *Curzon*, p. 133.

99. Sir Henry Wilson: Diary, 21 May 1920, in Gilbert, *Churchill*, companion vol. 4, pt 2, p. 1104.
100. CAB 23/42, 45th Minutes, Minutes of a Meeting of the Imperial War Cabinet, 23 December 1918; CAB 23/22, 49(20), Appendix 1, Minutes of a Meeting of the Cabinet Finance Committee, 12 August 1920.
101. Sir Henry Wilson: Diary, 6 August 1920, in Gilbert, *Churchill*, companion vol. 4, pt 2, p. 1164.
102. Ibid., p. 1104, Sir Henry Wilson: Diary, 21 May 1920; CAB 23/21, 30(20), Conclusions of a Meeting of the Cabinet, 21 May 1920.
103. Ullman, *Anglo-Soviet Relations*, vol. 3, p. 324.
104. CAB 27/24, EC 42nd Minutes, Minutes of a Meeting of the Eastern Committee, 9 December 1918.
105. Gilbert, *Churchill*, companion vol. 4, pt 4, pp. 1030, 1078–82, 1105, 1116, 1117, 1119–20, 1123, 1164–5, 1261–2.
106. Ibid., p. 1103.
107. CAB 24/115, CP 2164, The Viceroy of India to the India Office, 23 November 1920; DBFP, 1st Series, vol. xiii, no. 624, pp. 675–6, footnote 3, Lord Chelmsford to Mr Montagu, 6 December 1920; ibid. no., 662, pp. 704–6, Lord Chelmsford to Mr Montagu, 22 January 1921; Ullman, *Anglo-Soviet Relations*, vol. 3, pp. 328–9.
108. FO 371/E3706/11/44, I.D.C.E. 37th Minutes, Minutes of a Meeting of the Inter-Departmental Conference on Middle Eastern Affairs, 13 April 1920.
109. CAB 23/21, 24(20), Conclusions of a Meeting of the Cabinet, 5 May 1920.
110. CAB 23/21, 24(35), Conclusions of a Meeting of the Cabinet, 11 June 1920.
111. CAB 24/107, CP 1467, British Military Liabilities, Note by Sir Henry Wilson, Chief of Imperial General Staff, 9 June 1920; CAB 23/21, 38(20), Appendix 1, Statement by the Chief of Imperial General Staff.
112. Gilbert, *Churchill*, companion vol. 4, pt 2, pp. 1119–20, 1123, 1143, 1144.
113. Ibid., pp. 1123, 1161, 1164–5, 1169, 1170, 1197.
114. CAB 23/23, 59(20), Conclusions of a Meeting of the Cabinet, 3 November 1920.
115. CAB 23/23, 62(20), Conclusions of a Meeting of the Cabinet, 18 November 1920.
116. CAB 23/23, 69(20), Conclusions of a Meeting of the Cabinet, 13 November 1920.
117. FO 371/4179, 46887/2117/44, Memorandum by Earl Curzon, 'A Note of Warning about the Middle East', 25 March 1919; FO 371/4180, 60671/2117/44, Note respecting the Middle Eastern Question by Earl Curzon, 22 April 1919.
118. Sir Harold Nicolson, *Peacemaking, 1919* (London, 1945), pp. 278, 280, 281; S. D. Waley, *Edwin Montagu: A Memoir and an Account of his Visits to India* (London, 1964), pp. 240–1.
119. FO 371/4179, 3636/2117/44, Earl Curzon to Earl of Derby (Paris), 5 March 1919.
120. Lloyd George MSS, F/40/2/58, Montagu to Curzon, 14 August 1919; CAB. 23/20, 1(2), Appendix 1, Minutes of a Conference of Ministers, 5 January 1920.
121. Gilbert, *Churchill*, companion vol. 4, pt 2, pp. 1054–5, 1114–16, 1198–1200.
122. CAB 27/24, EC 46th Minutes, Minutes of a Meeting of the Eastern Committee, 23 December 1918; FO 371/4179, 3636/2117/44, Earl Curzon to the Earl of Derby (Paris), 5 March 1919.
123. Lloyd George MSS, F/40/2/59, Montagu to Lloyd George, 20 August 1920; CAB 23/20, 1(20), Minutes of a Conference of Ministers, 5 January 1920.

124. Ibid.; CAB 27/24, EC 46th Minutes, Minutes of a Meeting of the Eastern Committee, 23 December 1918.
125. Lloyd George MSS, F/40/2/59, Montagu to Lloyd George, 20 August 1919.
126. CAB 23/20, 1(20), Appendix 1, Minutes of a Conference of Ministers, 5 January 1920; ibid., Conclusions of a Meeting of the Cabinet, 6 January 1920.
127. FO 371/5043, E1297/3/44, Minute by Curzon, 11 March 1920; Lothian MSS, GD40/17/208/268, Curzon to Kerr, 12 March 1920.
128. Hardinge MSS, vol. 42, Hardinge to Harcourt Butler, 9 April 1920.
129. FO 371/4215, 132926/50535/44, Minute by Hardinge, September 1919 (undated); FO 371/5043, E1297/3/44, Minute by Hardinge, March 1920 (undated).
130. FO 371/4180, 60671/2117/44, Note respecting the Middle Eastern Question by Earl Curzon, 22 April 1919; FO 371/4215, 132926/50535/44, Minute by Curzon, 28 September 1919; Lothian MSS, GD40/17/208/268, Curzon to Philip Kerr, 12 March 1920.
131. FO 371/5043, E1297/3/44, Minute by Hardinge, March 1920 (undated); ibid., Minute by Curzon, 11 March 1920; FO 371/5044, E1917/3/44, Minute by Hardinge, March 1920 (undated); ibid., Minute by Curzon, 25 March 1920.
132. FO 371/5042, E946/3/44, Earl Curzon to Vice-Admiral Sir J. de Robeck, 6 March 1920; CAB 24/115, CP 2193, 'The Greek Position', Note by Lord Curzon, 27 November 1920; FO 371/5103, E56/56/44, The Secretary of the Army Council B. Cubitt to the Under-Secretary for Foreign Affairs, 12 February 1920; DBFP, 1st Series, vol. xiii, no. 23, pp. 26–8, Situation in Turkey, 15 March 1920; ibid., no. 40, pp. 54–7, General Staff Memorandum on the Turkish Peace Treaty, 1 April 1920; FO 371/5109, E5650/56/55, F. W. Duke, India Office, to the Under-Secretary of State, Foreign Office, 1 June 1920; FO 371/5142, E8753/139/44, The Secretary of State for India to the Viceroy, 21 July 1920.
133. Curzon MSS, Eur F/112/281, 'Mesopotamian Affairs', by Montagu, 18 November 1920.
134. CAB 24/117, CP 2343, Memorandum by the Secretary of State for India, 22 December 1920.
135. Klieman, *Foundations of British Policy*, p. 85.
136. Stephen Roskill, *Hankey, Man of Secrets* (London, 1972), vol. 2, p. 210.
137. Ibid., p. 202.
138. Ibid.
139. CAB 24/117, CP 2343, Memorandum by the Secretary of State for India, 22 December 1920; ibid., CO 2356, 'Mesopotamia', Note by the Secretary of State for India, 24 December 1920.
140. Roskill, *Hankey*, vol. 2, p. 202; CAB 23/23, 82(20) Conclusions of a Meeting of the Cabinet, 31 December 1920.
141. Roskill, *Hankey*, vol. 2, p. 214.
142. Bonar Law MSS, 100/1/8, Chamberlain to Bonar Law, 6 January 1921, in Lord Beaverbrook, *Men and Power, 1917–1918* (New York, 1956), Appendix 4, pp. 397–400.
143. Roskill, *Hankey*, vol. 2, pp. 214–15.
144. Klieman, *Foundations of British Policy*, p. 90.
145. Martin Gilbert, *Winston S. Churchill* (London, 1975), vol. 4, pp. 507–9.
146. CAB 24/119, CP 2571, Report of the Inter-departmental Committee: Middle East, 31 January 1921.
147. Fisher MSS, Box 8A, Diary, 14 February 1921.

148. CAB 23/24, 7(21), Conclusions of a Meeting of the Cabinet, 14 February 1921.
149. Public Record Office Handbooks, *Records of the Foreign Office*, p. 15.
150. FO 366/795, X5635/520/504, Minute approved by Harmsworth, June 1921 (undated).
151. CAB 23/29, 19(22), Conclusions of a Meeting of the Cabinet, 20 March 1922.
152. FO 371/7952, E11372/10102/44, Minute by Crowe, 19 October 1922.
153. FO 371/7907, E11785/27/44, Forbes Adam to Crowe, 24 October 1922.
154. FO 372/2197, T5885/5885/384, Memorandum on Consultation with and Communication of Information to the British Dominions on Foreign Policy by Percy A. Koppel, 16 January 1926.
155. FO 371/7000, W12716/12716/17, The Marquess Curzon of Kedleston to Lord Hardinge (Paris), 5 December 1921, quoted in Gordon A. Craig, 'The British Foreign Office from Grey to Austen Chamberlain', in G. A. Craig and Felix Gilbert (eds), *The Diplomats, 1919–1936*.
156. FO 372/2197, T5885/5885/384, Memorandum on Consultation with and Communication of Information to the British Dominions on Foreign Policy by Percy A. Koppel, 16 January 1926; G. A. Craig, 'The British Foreign Office from Grey to Austen Chamberlain', in Craig and Gilbert (eds), *The Diplomats, 1919–1936*.
157. R. F. Holland, *Britain and the Commonwealth Alliance, 1918–1939* (London, 1981), pp. 40–6; Sir George V. Fiddes, *The Dominions and the Colonial Offices* (London, 1926), pp. 277–81.
158. FO 372/2198, T16434/5885/384, Memorandum by Tyrrell, 28 June 1926.
159. Fiddes, *The Dominions and the Colonial Offices*, p. 281.
160. FO 372/2198, T16435/5885/384, Draft Memorandum on the Duties of the Dominions Information Department 1926 (undated); Public Record Office Handbooks, *Records of the Foreign Office*, p. 23.

Bibliography

Primary Sources: Government Archives

Public Record Office, London (Kew)

Board of Trade Papers
BT 61/   Estimate and Accounts Papers

Cabinet Papers
CAB 2/   Committee of Imperial Defence
CAB 21/   Registered Cabinet Office Files: Miscellaneous Papers
CAB 23/   Cabinet Minutes
CAB 24/   Cabinet Memoranda
CAB 27/   Cabinet Committees: General Series
CAB 29/   Proceedings of the London Reparations Conference
CAB 32/   Imperial Conference Minutes
CAB 103/   Cabinet Office Historical Section, Registered Files

Foreign Office Papers
FO 366/   Chief Clerk's Department: Archives
FO 368/   General Correspondence: Commercial
FO 371/   General Correspondence: Political
FO 372/   General Correspondence: Treaty
FO 395/   General Correspondence: News
FO 794/   Private Office: Individual Files
FO 800/   Private Collections:
            FO 800/147–58   Curzon of Kedleston Papers
            FO 800/243   Sir Eyre Crowe Papers

Treasury Papers
T1/   Treasury Board Papers, 1557–1920

PRO 30
The Diary of J. Ramsay MacDonald

Command Papers, 9230 [1918], 8, 7748, 7749 [1914], 8657 [1917–18], 2169, 1812 [1922], 2270 [1924], 2527 [1925], 2468 [1925], 3909 [1929–31].

Primary Sources: *Private Collections and Personal Papers*

Balfour Papers, PRO, London
Baldwin Papers, Cambridge University Library
Bonar Law Papers, House of Lords Record Office
Cecil of Chelwood Papers, British Library
Lord Curzon Papers, Indian Office Library, British Library
Austen Chamberlain Papers, Birmingham University Library
Hardinge Papers, University Library, Cambridge
Lloyd George Papers, House of Lords Record Office
Lothian Papers (Philip Kerr), Scottish Record Office, Edinburgh
H. A. L. Fisher Papers, Bodleian Library, Oxford

The Labour Party Archives, LP/1AC/2/27
The CBI Archives, File numbers 3521–899

Primary Sources: *Published Documents*

Documents on British Foreign Policy, 1919–1939, First Series, vols i, ii, iii, iv, vii, viii, ix, x, xiii, xiv, xv, xvi, xviii, xix, xx, xxv, xxvi (London, 1947–69)

British Documents on the Origins of the First World War, 1898–1914, 11 vols, (London, 1927)

Hansard, Parliamentary Debates, House of Lords and House of Commons, 5th Series.

The League of Nations, Official Journal, 1920–1925
The Foreign Office List, 1920 (London, 1920)
The Foreign Office List, 1939 (London, 1939)
The Public Records Office Handbooks Series, *The Records of the Foreign Office, 1782–1939* (London, 1969)

The Foreign Relations of the United States (FRUS) Series:

1919, vol. 15 (Washington, 1946)
1920, vol. 2 (Washington, 1936)
1922, vol. 2 (Washington, 1938)
1923, vol. 2 (Washington, 1938)

A Journal of Reparations (London, 1939)

The Times, 1928
Pravda, 1924

Secondary Sources: Biographical Accounts, Monographs and Articles

D'Abernon, Viscount, *An Ambassador of Peace: Pages from the Diary of Viscount D'Abernon*, vols 1 and 2 (London, 1929).

Amery, L. S., *My Political Life*, 3 vols (London, 1953–5).

Andrew, Christopher, 'The British Secret Service and Anglo-Soviet Relations in the 1920s. Part I: From the Trade Negotiations to the Zinoviev Letters', *Historical Journal*, vol. 20 (1977).

Andrew, Christopher, *Secret Service: The Making of the British Intelligence Community* (London, 1985).

Armhen, R., Birke, M. and Howard, M. (eds), *The Quest for Stability: Problems of West European Security, 1918–1957* (Oxford, 1993).

Ashton-Gwatkin, Frank T. A., *The British Foreign Service* (Syracuse, 1950).

Avon, The Earl of, *The Eden Memoirs: Facing the Dictators* (London, 1962).

Bariety, Jacques, *Les Relations Franco-Allemandes après la Première Guerre Mondiale: 10 Novembre 1918 – 10 Janvier 1925 de l'Exécution à la Négociation* (Paris, 1977).

Barnes, Trevor, 'Special Branch and First Labour Government', *Historical Journal*, vol. 22 (1979).

Barros, James, *Office without Power: General Sir Eric Drummond, 1919–1933* (Oxford, 1979).

Beak, George B., 'Policy and Propaganda', 2 December 1918.

Beaverbrook, Lord, *The Decline and Fall of Lloyd George and Great was the Fall Thereof* (London, 1963).

Beloff, Max, *New Dimensions in Foreign Policy: A Study in British Administrative Experience, 1947–1959* (London, 1961).

Bendiner, Elmer, *A Time for Angels: The Tragicomic History of the League of Nations* (New York, 1975).

Bergmann, Carl, *The History of Reparations* (London, 1927).

Birn, Donald S., *The League of Nations Union, 1918–1945* (Oxford, 1981).

Bishop, Donald G., *The Administration of British Foreign Relations* (Syracuse, 1961).

Blake, Robert, *The Unknown Prime Minister: The Life and Times of Andrew Bonar Law, 1858–1923* (London, 1955).

Blau, Peter M. and Scott, Richard W., *Formal Organizations: A Comparative Approach* (London, 1963).

Boadle, Donald Graeme, *Winston Churchill and the German Question in British Foreign Policy, 1918–1922* (The Hague, 1973).

Boadle, Donald, 'The Formation of the Foreign Office Economic Relations Section, 1930–1937', *Historical Journal*, vol. 20 (1977).

Bond, Brian, *British Military Policy between the Two World Wars* (Oxford, 1980).

Bridges, Sir Edward, 'Portrait of a Profession', in Richard A. Chapman and A. Dunsire (eds), *Style in Administration: Readings in British Public Administration* (London, 1971).

Bridges, Sir Edward (later Lord), *Portrait of a Profession* (Cambridge, 1950).

Bullen, R. (ed.), *The Foreign Office, 1782–1982* (Maryland, 1984).

Busch, Briton C., *Britain, India and the Arabs, 1914–1921* (Berkeley, 1971).

Busch, Briton C., *Mudros to Lausanne: Britain's Frontier in West Asia, 1918–1923* (Albany, 1976).

Busch, B. C., *Hardinge of Penshurst: A Study in the Old Diplomacy* (New York, 1980).

Busk, Sir Douglas, *The Craft of Diplomacy: How to Run a Diplomatic Service* (London, 1967).

Butler, R., 'Sir Eyre Crowe', *World Review*, no. 50 (1953), pp. 8–13.

Butler, J. R. M., *Lord Lothian (Philip Kerr), 1882–1940* (London, 1960).

Butler, William E., 'The Harvard Text of the Zinov'ev Letter', *Harvard Library Bulletin*, vol. 18 (1970).

Calder, K. J., *Britain and the Origins of the New Europe* (Cambridge, 1966).

Calhoun, Daniel F., *The United Front: The TUC and the Russians, 1923–1928* (Cambridge, 1976).

Campbell, F. Gregory, 'The Struggle for Upper Silesia, 1919–1922', *Journal of Modern History*, vol. 42 (1970).

Carlton, David, 'Disarmament with Guarantees: Lord Robert Cecil, 1922–1927', *Disarmament and Arms Control*, vol. 3 (1965).

Carlton, David, 'Great Britain and the League Council Crisis of 1926', *Historical Journal*, vol. 11 (1968).

Carlton, D., *MacDonald vs. Henderson: The Foreign Policy of the Second Labour Government* (London, 1970).

Carr, Edward Hallett, *Socialism in One Country, 1924–1926* (London, 1964), vol. 3, pt 1, no. 5.

Carsten, F. L., *Britain and the Weimar Republic: the British Documents* (London, 1984).

Carter, Gwendolen M., *The British Commonwealth and International Security: The Role of the Dominions, 1919–1939* (Toronto, 1947).

Chelwood, Lord Cecil of, *A Great Experiment* (London, 1941).

Chelwood, Viscount Cecil of, *All the Way* (London, 1949).

Chapman, Richard A. and Greenaway, J. R., *The Dynamics of Administrative Reform* (London, 1980).

Chapman, Richard A., *The Higher Civil Service in Britain* (London, 1970).

Chester, D. N. and Wilson, F. M. G., *The Organisation of British Central Government, 1914–1956* (London, 1957).

Chester, Lewis, Stephen Fay and Hugo Young, *The Zinoviev Letter* (London, 1967).

Childs, Sir Wyndham, *Episodes and Reflections* (London, 1930).

Churchill, Randolph S., *Lord Derby: 'King of Lancashire'* (London, 1959).

Churchill, Winston S., *The World Crisis* (London 1929).

Churchill, Winston S., *Great Contemporaries* (Chicago, 1973).

Cole, Margaret (ed.), *Beatrice Webb's Diaries, 1924–1932* (London, 1956).

Collins, Doreen, *Aspects of British Politics, 1904–1919* (London, 1965).

Cook, Chris, *The Age of Alignment: Electoral Politics in Britain, 1922–1929* (London, 1975).

Corp, Edward T., 'Sir Eyre Crowe and the Administration of the Foreign Office, 1906–1914', *Historical Journal*, vol. 22 (1979).

Corp, Edward I., 'Sir William Tyrrell: The Eminence Grise of the British Foreign Office, 1912–1915', *Historical Journal*, vol. 25 (1982).

Cosgrove, Richard A., 'The Career of Sir Eyre Crowe: A Reassessment', *Albion*, vol. 4 (1972).

Craig, Gordon A. and Gilbert, Felix (eds), *The Diplomats, 1919–1939* (Princeton, 1953).

Craig, Gordon A., 'The British Foriegn Office from Grey to Austen Chamberlain', in Craig, Gordon A. and Gilbert, Felix (eds), *The Diplomats, 1919–1939* (Princeton, 1953).

Crozier, Michael, *The Bureaucratic Phenomenon* (Chicago, 1964).

Cromwell, Valerie and Steiner, Zara S., 'The Foreign Office before 1914: A Study in Resistance', in Gillian Sutherland (ed.), *Studies in the Growth of the Nineteenth-Century Government* (London, 1972).

Crowe, Sybil, 'The Zinoviev Letter: a reappraisal', *Journal of Contemporary History*, vol. 10 (1975).

Crowe, Sybil, 'Sir Eyre Crowe and the Locarno Pact', *English Historical Review*, vol. 87 (1972), pp. 49–74.

Crowe, Sybil and Corp, Edward, *Sir Eyre Crowe: Our Ablest Public Servant, 1864–1925* (Devon, 1993).

[Curzon] Kedleston, Lord Curzon of, *Frontier* (Westport, 1976).

[Curzon] Kedleston, The Marchioness Curzon of, *Reminiscences* (London, 1955).

Darwin, John, *Britain, Egypt and the Middle East: Imperial Policy in the Aftermath of War, 1918–1922* (London, 1981).

Davies, Ernest, 'The Foreign and Commonwealth Services', in William A. Robson (ed.), *The Civil Service in Britain and France* (London, 1956).

Davies, Joseph, *The Prime Minister's Secretariat* (London, 1951).

Degras, Jane (ed.), *Soviet Documents on Foreign Policy 1925–1932*, vol. 2 (London, 1952).

Dockrill, M. L. and Goold, Douglas J., *Peace without Promise: Britain and the Peace Conferences, 1919–1923* (London, 1981).

Dornberg, Jon, *The Putsch that Failed (Munich, 1923): Hitler's Rehearsal for Power* (London, 1982).

Douglas, Roy, *The History of the Liberal Party, 1895–1970* (London, 1971).

Dugdale, Blanche E. C., *Arthur James Balfour, First Earl of Balfour*, vol. 2 (London, 1939).

Dutton, David, *Austen Chamberlain: a Gentleman in Politics* (Bolton, 1985).

Egerton, G. W., *Great Britain and the Creation of the League of Nations: Strategy, Politics and International Organisation, 1914–1919* (London, 1979).

Elton, Lord, *Life of James Ramsay MacDonald, 1866–1919* (London, 1939).

Erdmann, Karl Dietrich, *Adenauer in der Rheinlandpolitik nach dem ersten Weltkrieg* (Stuttgart, 1966).

Feldman, Gerald D., *Iron and Steel in the German Inflation, 1916–1923* (Princeton, 1977).

Felix, David, *Walther Rathenau and the Weimar Republic: The Politics of Reparations* (Baltimore, 1971).

Felix, David, 'Reparations Reconsidered with a Vengeance', *Central European History*, vol. 4 (1971).

Ferris, John, 'The Greatest Power on Earth: Great Britain in the 1920's', *International History Review*, vol. 13 (1991).

Fiddes, Sir George V., *The Dominions and the Colonial Offices* (London, 1926).

Fink, Carole, *The Genoa Conference: European Diplomacy, 1921–1922* (Chapel Hill, 1984).

Fink, Carole, Frohn, Alex and Jürgen Heideking, *Genoa, Rapallo and European Reconstruction in 1922* (Cambridge, 1991).

Fisher, H. A. L., MSS, Box 8A, Diary, 6 June 1920.

Fischer, Louis, *The Soviets in World Affairs: A History of the Relations between the Soviet Union and the Rest of the World, 1917–1929*, vol. 2 (Princeton, 1951).

Fox, John P., 'Britain and the Inter-Allied Military Commission of Control 1925–6', *Journal of Contemporary History*, vol. 4 (1969).

Frederick, Sir Maurice, *Haldane, 1856–1928: The Life of Viscount Haldane of Cloan, K.T., O.M.*, vol. 2 (London, 1939).

Fry, Geoffrey K., *Statesmen in Disguise, The Changing Role of the Administrative Class of the British Home Civil Service, 1853–1966* (London, 1969).

Gibbs, N. H., 'British Strategic Doctrine, 1918–1939' in Michael Howard (ed.), *The Theory and Practice of War* (London, 1965).

Gibbs, N. H., *Grand Strategy*, vol. 1 (London, 1976).

Gilbert, Martin, *The Roots of Appeasement* (London, 1966).

Gilbert, Martin, *World in Torment: Winston S. Churchill*, vol. 4: *1917–1922* (London, 1975).

Gilbert, Martin, *Prophet of Truth: Winston S. Churchill*, vol. 5: *1922–1939* (London, 1976).

Gilbert, Martin, *Winston S. Churchill*, companion vol. 4, pt 1, pt 2, vol. 5 (London, 1977).

Gilbert, Martin, *Churchill, a Life* (London, 1992).

Gladwyn, Cynthia, *The Paris Embassy* (London, 1976).

Goldstein, Erik, *Winning the Peace: British Diplomatic Strategy, Peace Planning and the Paris Peace Conference, 1916–1920* (Oxford, 1991).

Golinkov, David, *The Secret War against Russia* (Moscow, 1982).

Goold, J. D., 'Old Diplomacy: The Diplomatic Career of Lord Hardinge, 1910–1922' (Ph.D. Thesis, University of Cambridge, 1976).

Goold, J. D., 'Lord Hardinge as Ambassador to France and the Anglo-French Dilemma over Germany and the Near East, 1920–1922', *Historical Journal*, vol. 21, no. 4 (1978), pp. 913–937.

Gordon, Harold J. Jr, *Hitler and the Beer Hall Putsch* (Princeton, 1972).

Gore-Booth, Lord, *With Great Truth and Respect* (London, 1974).

Gorodetsky, Gabriel, *The Precarious Truce: Anglo-Soviet Relations, 1924–27* (Cambridge, 1977).

Grant, Natalie, 'The Zinoviev Letter Case', *Soviet Studies*, vol. 19 (1967–68).

Gregory, John D., *On the Edge of Diplomacy: Rambles and Reflections, 1902–28* (London, 1928).

Grey, Viscount of Falladon, *Twenty-Five Years*, vol. 1, (London, 1925).

Grigg, P. J., *Prejudice and Judgement* (London, 1948).

Haldane, Richard Burdon, *An Autobiography* (London, 1929).

Hamilton, Mary Agnes, *Arthur Henderson* (London, 1938).

Lord Hardinge of Penshurst, *Old Diplomacy* (London, 1947).

Hastings, Sir Patrick, *The Autobiography of Sir Patrick Hastings* (London, 1948).

Heady, Bruce, *British Cabinet Ministers: The Roles of Politicians in Executive Office* (London, 1974).

Headlam-Morley, Sir James W., *Studies in Diplomatic History* (London, 1930).

Headlam-Morley, Agnes, Russell Bryant and Anna Cienciala (eds), *Sir James Headlam-Morley, A Memoir of the Paris Peace Conference, 1919* (London, 1972).

Hemery, J. A., 'The Emergence of Treasury Influence in British Foreign Policy, 1914–1921' (Ph.D. Thesis, Univeristy of Cambridge, 1988).

Hessell, H. and Filtman, J., *Ramsay MacDonald: Labor's Man of Destiny* (New York, 1929).

Hill, Michael, *The Sociology of Public Adminstration* (London, 1972).

Hinsley, F. H. (ed.), *British Foreign Policy under Sir Edward Grey* (Cambridge, 1977).

Holland, R. F., *Britain and the Commonwealth Alliance, 1918–1939* (London, 1981).

Hymans, Paul, *Memoires*, vol. 2 (Brussels, 1958).

Jacobson, Jon, *Locarno. Diplomacy: Germany and the West, 1925–1929* (Princeton, 1972).

Jacobson, Jon and Walker, John T., 'The Impulse for a Franco-German Entente: The origins of the Thoiry Conference, 1926', *Journal of Contemporary History*, vol. 10 (1975).

James, Harold, *The German Slump: Politics and Economics, 1924–1936* (Oxford, 1986).

James, Robert Rhodes, *Memoirs of a Conservative: J. C. C. Davidson's Memoirs and Papers 1910–37* (London, 1969).

Johnson, Douglas, 'Austen Chamberlain and the Locarno Agreements', *University of Birmingham Historical Journal*, vol. 8 (1961–2).

Jones, K. P., 'Stresemann, the Ruhr Crisis, and Rhenish Separatism: A Case Study of Westpolitik', *European Studies Review*, vol. 7 (1977).

Jones, Ray, *The Nineteenth-Century Foreign Office: An Administrative History* (London, 1971).

Jones, Raymond A., *The British Diplomatic Service, 1815–1914* (Waterloo, 1983).

Jones, Thomas, *A Diary with Letters, 1931–1950* (London, 1954).

Kelly, Sir David, *The Ruling Few: or the Human Background to Diplomacy* (London, 1952).

Kennedy, A. L., *Old Diplomacy and New, 1876–1922* (London, 1923).

Kent, Bruce, *The Spoils of War: the Politics, Economics and Diplomacy of Reparations, 1918–1932* (Oxford, 1989).

Kettle, Michael, *Sidney Reilly: The True Story* (London, 1983).

Keynes, J. M., *A Revision of the Treaty* (London, 1920).

Keynes, J. M., *The Economic Consequences of the Peace* (London, 1920).

Kimmich, Christoph M., *Germany and the League of Nations* (Chicago, 1976).

Kinnear, Michael, *The Fall of Lloyd George: The Political Crisis of 1922* (London, 1973).

Kirkpatrick, Sir Ivone, *The Inner Circle: Memoirs of Ivone Kirkpatrick* (London, 1959).

Klieman, Aaron, *Foundations of British Policy in the Arab World: The Cairo Conference of 1921* (Baltimore, 1970).

Kuusinen, Aino, *Before and After Stalin* (London, 1974).

Larner, Christina, 'The Amalgamation of the Diplomatic Service with the Foreign Office', *Journal of Contemporary History*, vol. 7 (1972).

Larner, Christina, 'The Organization and Structure of the Foreign and Commonwealth Office', in Robert Boardman and A. J. R. Groom (eds), *The Management of Britain's External Relations* (London, 1973).

Laroche, Jules, *Au Quai d'Orsay avec Briand et Poincaré, 1913–1926* (Paris, 1957).

de Launay, Jacques (ed.), *Louis Loucheur, Carnets Secrets, 1908–1932* (Brussels, 1962).

Lauren, Paul G., *Diplomats and Bureaucrats: The First Institutional Responses to Twentieth-Century Diplomacy in France and Germany* (Stanford, 1976).

Lees-Milne, J., *Harold Nicolson: a Biography*, vol. 1 (London, 1980).

Lees-Milne, J., *Harold Nicolson: a Biography*, vol. 2 (London, 1981).

Leventhal, F. M., *Arthur Henderson* (Manchester, 1989).

Lindblom, Charles, *The Policy-Making Process* (Englewood Cliffs, 1968).

Lloyd George, D., *Memoirs of the Peace Conference*, vol. 1 (New Haven, 1939).

Lloyd George, David, *War Memoirs*, vol. 1 (London, 1938).

Lloyd George, Frances, *The Years that are Past* (London, 1967).

Lockhart, Robin Bruce, *Ace of Spies* (London, 1967).

Lonnroth, Erik, 'Sweden: The Diplomacy of Osten Unden', in Gordon Craig and Felix Gilbert, *The Diplomats* (London, 1966).

Louis, Wm. Roger, *British Strategy in the Far East, 1919–1939* (Oxford, 1971).

Ludowa Spółdzielnia Wydawnicza (ed.), *Maciej Rataj, Pamiętniki, 1918–1927* (Warsaw, 1965).

Luther, Hans, *Politiker ohne Partei: Erinnerungen* (Stuttgart, 1960).

Lyman, Richard W., *The First Labour Government, 1924* (London, 1957).

McDougall, Walter A., *France's Rhineland Diplomacy, 1914–1924: The Last Bid for a Balance of Power in Europe* (Princeton, 1978).

MacDougall, Walter A., 'Political Economy versus National Sovereignty: French Structures for German Economic Integration after Versailles', *Journal of Modern History*, vol. 51 (1979).

Manchester, William, *The Arms of Krupp, 1587–1968* (New York, 1970).

Mantoux, Étienne, *The Carthaginian Peace or The Economic Consequences of Mr. Keynes* (London, 1946).

Marks, Sally, 'Reparations Reconsidered: A Reminder', *Central European History*, vol. 2 (1969).

Marks, Sally, 'The Myths of Reparations', *Central European History*, vol. 11 (1978).

Marsden, Arthur, 'The Blockade', in Hinsley (ed.), *British Foreign Policy under Sir Edward Grey* (Cambridge, 1977).

Marquand, David, *Ramsay MacDonald* (London, 1977).

McKercher, B. J. C., 'From Enmity to Co-operation: The Second Baldwin Government and the Improvement of Anglo-American Relations, November 1928 to June 1929', *Albion*, vol. 24 (1982).

McKercher, B. J. C., *The Second Baldwin Government and the US, 1924–29, Attitudes and Diplomacy* (Cambridge, 1984).

McKercher, B. J. C. and Moss, D. J. (eds), *Shadow and Substance in British Foreign Policy, 1895–1939. Essays Honouring C. J. Lowe* (Edmonton, 1984).

McKercher, B. J. C., 'Austen Chamberlain's Control of British Foreign Policy, 1924–1929', *International History Review*, vol. 6 (1984).

McKercher, B. J. C., *Esme Howard: a Diplomatic Biography* (Cambridge, 1989).

Medlicott, W. N., 'Britain and Germany: The Search for Agreement, 1930–37', in David Dilks (ed.), *Retreat from Power, 1906–1936*, vol. 1 (London, 1981).

Mejcher, Helmut, 'British Middle East Policy 1917–1921', *Journal of Contemporary History*, vol. 8 (1973).

Mejcher, Helmut, *Imperial Quest for Oil: Iraq, 1910–1928* (Oxford, 1976).

Middlemas, Keith (ed.), *Thomas Jones, Whitehall Diary*, vol. 1 (London, 1969).

Middlemas, Keith and Barnes, John, *Baldwin, A Biography* (London, 1969).

Miquel, Pierre, *Poincaré* (Paris, 1961).

Montgomery, A. E., 'The Making of the Treaty of Sèvres of August 10th, 1920', *Historical Journal*, vol. 15 (1972).

Morgan, Austen, *James Ramsay MacDonald* (Manchester, 1987).

Morgan, Brig.-Gen. J. H., *Assize of Arms* (London, 1945).

Morgan, Kenneth O., 'Lloyd George's Premiership: A Study in "Prime Ministerial Government"', *Historical Journal*, vol. 13 (1970).

Morgan, Kenneth O., *Consensus and Disunity: The Lloyd George Coalition Government, 1918–1922* (Oxford, 1979).

Morris, A. J. A., *Radicalism against War, 1906–14: The Advocacy of Peace and Retrenchment* (London, 1972).

Morse, E., *Foreign Policy and Interdependence in Gaullist France* (Princeton, 1973).

Morse, Edward, *Modernization and the Transformation of International Affairs* (London, 1976).

Morse, Edward, 'Interdependence in World Affairs', in James Rosenau, K. W. Thompson and Gavin Boyd (eds), *World Politics* (New York, 1976).

Mosley Leonard, *Curzon: The End of an Epoch* (London, 1961).

Mowat, Charles Loch, *Britain between the Wars, 1918–1940* (London, 1968).

Murray, Arthur C., Viscount Elibank, *Reflections on Some Aspects of British Foreign Policy between the Two Wars* (London, 1946).

Namier, Sir Lewis, *Avenues of History* (London, 1952).

Namier, Julia, *Lewis Namier, A Biography* (London, 1971).

Naylor, John F., *A Man and an Institution: Sir Maurice Hankey, the Cabinet Secretariat and the Custody of Cabinet Secrecy* (Cambridge, 1984).

Nelson, Harold, *Land and Power* (London, 1963).

Neward, F. H., 'The Campbell Case and the First Labour Government', *Northern Ireland Legal Quarterly*, vol. 20 (1969).

Nicolson, H. G., *Sir Arthur Nicolson, Bart., First Lord Carnock: A Study in the Old Diplomacy* (London, 1930).

Nicolson, H., 'Allen Leeper', *The Nineteenth Century and After* (October 1935), pp. 473–83.

Nicolson, Sir Harold, *Curzon: The Last Phase, 1919–1925* (New York, 1925).

Nicolson, Sir Harold, *Peacemaking, 1919* (London, 1933).

Nicolson, Sir Harold, *King George the Fifth: His Life and Reign* (London, 1953).

Nicolson, Sir Harold, *Diplomacy* (London, 1969).

Nicolson, Nigel (ed.), Harold Nicolson, *Diaries and Letters*, vol. 1 (London, 1966).

Northedge, F. S., *The Troubled Giant: Britain Among the Great Powers, 1916–1939* (New York, 1967).

O'Halpin, Eunan, 'Sir Warren Fisher and the Coalition, 1919–1922', *Historical Journal*, vol. 24 (1981).

O'Malley, Sir Owen, *The Phanton Caravan* (London, 1954).

Oppenheimer, Sir Frank, *Stranger Within* (London, 1960).

Orde, Anne, *British Policy and European Construction after the First World War* (Cambridge, 1990).

Oxford and Asquith, The Earl of, *Memoirs and Reflections, 1857–1927*, vol. 2 (London, 1928).

Parker, James G., *Lord Curzon, 1859–1925: a Bibliography* (Connecticut, 1991).

Percy, Lord Eustace, *Some Memories* (London, 1958).

de Perretti de la Rocca, Emmanuel, 'Briand et Poincaré: Souvenirs', *Revue de Paris*, vol. 15, December 1936.

Petrie, Sir Charles, Bt, *The Life and Letters of the Right Hon. Sir Austen Chamberlain, K.G., P.C., M.P.*, vols 1 and 2 (London, 1940).

Peterson, Sir Maurice, *Both Sides of the Curtain, Autobiography* (London, 1950).

Platt, D. C. M., *Finance, Trade and Politics in British Foreign Policy 1815–1914* (Oxford, 1968).

Platt, D. C. M., *The Cinderella Service: British Consuls since 1825* (London, 1971).

Pope-Hennessy, James, *Lord Crewe, 1858–1945; The Likeness of a Liberal* (London, 1955).

Post, Gains J. R., *The Civil-Military Fabric of Weimar Foreign Policy* (Princeton, 1973).

Pronay, N. and Spring, D. W. (eds), *Propaganda, Politics and Film, 1918–1945* (London, 1982).

Ramsden, John, *The Age of Balfour and Baldwin, 1902–1940* (London, 1978).

Ramsden, John (ed.), *Real Old Tory Politics: The Political Diaries of Sir Robert Sanders, Lord Bayford, 1910–35* (London, 1984).

Repington, Charles á Court, *The First World War, 1914–1918*, vol. 2 (London, 1920).

Richardson, D., *The Evolution of British Disarmament Policy in the 1920s* (London, 1989).

Riddell, Lord, *Lord Riddell's War Diary, 1914–1918* (London, 1933).

Riddell, Lord, *Lord Riddell's Intimate Diary of the Peace Conference and After, 1918–1923* (London, 1933).

Ronaldshay, Earl of, *The Life of Lord Curzon*, vols 1, 2 and 3 (London, 1928).

Rose, Kenneth, *Superior Person: A Portrait of Curzon and his Circle in Late Victorian England* (London, 1969).

Rose, Kenneth, *The Later Cecils* (London, 1975).

Rose, Kenneth, *King George V* (London, 1983).

Rose, Kenneth, *Curzon, A Most Superior Person: a Biography* (London, 1985).

Rosenau, James, 'The Study of Foreign Policy', in James N. Rosenau, Kenneth W. Thompson and Gavin Boyd (eds), *World Politics* (New York, 1976).

Rosenblum, Kurt, *Community of Fate: German–Soviet Diplomatic Relations, 1922–1928* (Syracuse, 1965).

Roskill, Stephen, *Hankey, Man of Secrets*, vol. 2 (London, 1972).

Roskill, Stephen, *Naval Policy Between the Wars*, vol. 1 (London, 1968).

Rothwell, *British War Aims and Peace Diplomacy 1914–1918* (Oxford, 1971).

Rupieper, Herman J., *The Cuno Government and Reparations, 1922–1923: Politics and Economics* (The Hague, 1979).

Ryan, Sir Andrew, *The Last of the Dragomans* (London, 1951).

Saint-Aulaire, C. de B., *Confessions d'un Vieux Diplomate* (Paris, 1953).

Salewski, Michael, *Entwaffnung und Militarkontrolle in Deutschland, 1919–1927* (Munich, 1966).

Samme, George, *Raymond Poincaré: Politique et Personnel de la 3ee République* (Paris, 1933).

Sanders, M. L. and Taylor, Philip M., *British Propaganda during the First World War, 1914–18* (London, 1982).

Schuker, Stephen A., *The End of French Predominance in Europe: The Financial Crisis of 1924 and the Adoption of the Dawes Plan* (Chapel Hill, 1967).

Scott, George, *The Rise and Fall of the League of Nations* (London, 1973).

Selby, Sir Walford, *Diplomatic Twilight 1930–1940* (London, 1953).

Selznick, Philip, *TVA and The Grass Roots, A Study in the Sociology of Formal Organisation* (Los Angeles, 1949).

Seton-Watson, H. and Seton-Watson, C., *The Making of a New Europe: R. W. Seton-Watson and the Last Years of the Austria-Hungary* (London, 1981).

Sharp, Alan J., 'The Foreign Office in Eclipse 1919–22', *History*, vol. 61 (1976).

Sharp, Alan, *The Versailles Settlement: Peacemaking in Paris, 1919* (Basingstoke, 1991).

Shay, Robert, *British Rearmament in the Thirties: Politics and Profits* (Princeton, 1977).

Siney, Marian, *The Allied Blockade of Germany, 1914–16* (Ann Arbor, Mich, 1957).

Smith, Brian C. and Stanyer, Jeffrey, *Administering Britain* (Glasgow, 1976).

Snowden, Viscount Philip, *An Autobiography*, vol. 2 (London, 1934).

Stambrook, F. G., '"Das Kind" – Lord D'Abernon and the Origins of the Locarno Pact', *Central European History*, vol. 1 (1968).

Steiner, Zara S., *The Foreign Office and Foreign Policy, 1898–1914* (Cambridge, 1969).

Steiner Z. S. and Dockrill, M. L., 'The Foreign Office Reforms, 1919–1921', *Historical Journal*, vol. 17, 1 (March 1974).

Steiner, Zara S., *Britain and the Origins of the First World War* (London, 1977).

Steiner, Zara S., 'The Foreign Office and the War', in F. H. Hinsley (ed.), *British Foreign Policy under Sir Edward Grey* (Cambridge, 1977).

Steiner, Z. S. and Dockrill, M. L., 'The Foreign Office at the Paris Peace Conference of 1919', *International History Review*, vol. 2 (1980).

Steiner, Zara S., 'The Foreign Office under Sir Edward Grey, 1905–1914', in F. H. Hinsley (ed.), *British Foreign Policy under Sir Edward Grey* (Cambridge, 1977).

Steiner, Zara S. (ed.), *The Times Survey of Foreign Ministries of the World* (London, 1982).

Lord Strang, *The Foreign Office* (London, 1955).

Lord Strang, *Home and Abroad* (London, 1956).

Stresemann, C. Gustav, *Vermächtnis: der Nachlass in Drei Banden* (Berlin, 1932).

Stuart, Sir C., *Secrets of Crewe House: The Story of a Famous Campaign* (London, 1920).

Suarez, Georges, *Herriot, 1924–1932: Nouvelle Édition de 'Une Nuit chez Cromwell' suivie d'un récit historique de R. Poincaré* (Paris, 1932).

Suarez, Georges, *Briand: sa vie, son œuvre avec son Journal et des nombreux documents inédits*, vols 5 and 6 (Paris, 1941).

Swartz, M., *The Union of Democratic Control in British Politics during the First World War* (Oxford, 1971).

Sylvester, A. J., *The Real Lloyd George* (London, 1947).

Swinton, The Earl of, *I Remember* (London, 1948).

Taylor, A. J. P., *The Troublemakers: Dissent Over Foreign Policy, 1792–1939* (London, 1964).

Taylor, A. J. P. (ed.), *My Darling Pussy: The Letters of Lloyd George and Francis Stevenson, 1913–1941* (London, 1975).

Taylor, Philip M., *The Projection of Britain: British Overseas Publicity and Propaganda, 1919–1939* (Cambridge, 1981).

Taylor, Philip M., 'The Foreign Office and British Propaganda during the First World War', *Historical Journal*, vol. 23 (1980).

Temperley, A. C., *The Whispering Gallery of Europe* (London, 1939).

Tilley, Sir John, *London to Tokyo* (London, 1942).

Toynbee, A., *Acquaintances* (London, 1967).

Trachtenberg, M., *Reparation in World Politics: France and European Economic Diplomacy, 1916–1923* (New York, 1980).

Trevor-Roper, Hugh, *The Philby Affair: Espionage, Treason and Secret Services* (London, 1968).

Tucker, William Rayburn, *The Attitude of the British Labour Party towards European and Collective Security Problems, 1920–1939* (Geneva, 1950).

Turner, John, *Lloyd George's Secretariat* (Cambridge, 1980).

Ullman, Richard H., *Anglo-Soviet Relations, 1917–1921*, vols 2 and 3 (Princeton, 1972).

Vansittart, Lord, *The Mist Procession* (London, 1958).

Vincent, John (ed.), *The Crawford Papers; The Journals of David Lindsay Twenty-seventh Earl of Crawford and Tenth Earl of Balcarres, 1871–1940; during the years 1892 to 1940* (Manchester, 1984).

von Rickhoff, Harald, *German–Polish Relations 1918–1933* (Baltimore, 1971).

von Strandmann, Hartmut Pogge (ed.), Walther Rathenau: Tagebuch, 1907–1922 (Dusseldorf, 1967).

Walder, David, *The Chanak Affair* (London, 1969).

Waley, S. D., *Edwin Montagu: A Memoir and an Account of his Visits to India* (London, 1964).

Wallace, William, *The Foreign Policy Process in Britain* (London, 1977).

Walters, F. P., *A History of the League of Nations* (Oxford, 1960).

Wambaught, Sarah, *Plebiscites since the World War: With a Collection of Official Documents*, vols 1 and 2 (Washington, 1933).

Wandycz, Piotr S., *France and her Eastern Allies, 1919–1925: French-Czechoslovak-Polish Relations from the Paris Peace Conference to Locarno* (Minneapolis, 1962).

Ward, A. and Gooch, G. P., *The Cambridge History of British Foreign Policy*, vol. 3, (Cambridge, 1923).

Ward, Stephen R., *James Ramsay MacDonald: a Low Born Among High Brows* (New York, 1990).

Warman, R. M., 'The Erosion of Foreign Office Influence in the Making of Foreign Policy, 1916–1918', *Historical Journal*, vol. 15, no. 1 (1972).

Warman, R., *The Foreign Office, 1916–1918: a Study of its Role and Functions* (New York, 1986).

Waterfield, G., *Professional Diplomat: Sir Percy Lorraine of Kirkharle, Bt* (London, 1973).

Waterhouse, Nourah, *Private and Official* (London, 1942).

Watt, D. C., *Personalities and Politics: Studies in the Formulation of British Foreign Policy in the Twentieth Century* (South Bend, 1965).

Weidenfeld, Werner, *Die Englandpolitik Gustav Stresemanns: Theoretische und praktische Aspecte der Außenpolitik* (Mainz, 1972).

Weill-Raynal, Étienne, *Les Réparations allemandes et al France*, vols 1 and 2 (Paris, 1947).

Wellesley, Sir Victor, *Diplomacy in Fetters* (London, 1944).

Wheeler-Bennett, J. W. and Langermann, F. E., *Information on the Problem of Security, 1917–1926* (London, 1927).

Willert, Sir Arthur, *Washington and Other Memories* (Boston, 1972).

Williams, Andrew J., *Labour and Russia: The Attitude of the Labour Party to the USSR, 1924–1934* (Manchester, 1989).

Williamson, D. G., 'Great Britain and the Ruhr Crisis, 1923–1924', *British Journal of International Studies*, vol. 3 (1977).

Wilson, Sir Duncan, *Leonard Woolf: A Political Biography* (London, 1978).

Wilson, K. M., 'Sir Eyre Crowe on the Origin of the Crowe Memorandum of 1 January 1907', *Institute of Historical Research Bulletin*, no. 134 (1987), pp. 240–2.

Wilson, K. M., 'The Protocols of the Morning Post, 1919–1920', *Patterns of Prejudice*, vol. 19, no. 3 (1985).

Wilson, Trevor, *The Downfall of the Liberal Party, 1914–1935* (Ithaca, 1966).

Wilson, Trevor (ed.), *The Political Diaries of P. C. Scott, 1911–1928* (London, 1970).

Wrigley, C., *Arthur Henderson* (Cardiff, 1990).

Yearwood, Peter, 'The Foreign Office and the Guarantee of Peace through the League of Nations, 1916–1925' (Ph.D. Thesis, University of Sussex, 1980).

Yearwood, Peter, 'On the Safe and Right Lines: The Lloyd George Government and the Origins of the League of Nations, 1916–1918', *Historical Journal*, vol. 32, no. 1, (1989), pp. 131–55.

Index

Abyssinia, 30
Accounting Officer, 11–12, 39
Adam, Eric Forbes, 67
Addison, Joseph, 110, 137, 177
Aden, 209, 225, 226
Administrative Division of News
    Department, 27, 28
Admiralty: FO, conflict with, 2;
    Government Code and Cypher School,
    6, 29; Rhineland, 136
Adrianople, 65, 66, 222
Advisory Economic Conference, 126
Afghanistan, 209, 217
African Department of FO, 8, 46
African Empire, 35, 115
Air Ministry, 136
Akers-Douglas, Aretas, 21
Allied Military Committee of
    Versailles, 184
Alsace-Lorraine, 42
Ambassadors, appointment of, 82–3
American Department of FO, 7, 8, 10,
    30, 54; and France, 59; Head of
    Department, role of, 57–8
Amery, Leopold Charles Maurice
    Stennett, 155, 169, 172, 173
Anatolia, 72, 222
Andrew, Christopher, 152
Anglo-American guarantee, 164, 166
Anglo-Belgian relations, 164
Anglo-Franco-Belgian Alliance:
    proposal (1916), 55; proposal (1920), 92;
    proposal (1925), 165, 166, 167, 168–9,
    171, 174 see also Quadrilateral Pact
Anglo-Franco-German pact (1925), 169
Anglo-French alliance proposals: with
    Baldwin government, 164, 171–2, 175;
    with Lloyd George government, 52, 53,
    56, 80–1, 89, 91–2, 96, 114–19, 120–1;
    with MacDonald government 137, 141,
    144, 145
Anglo-French entente, 59, 89, 115, 167;
    Crowe on, 51, 52, 53, 92–3, 94; Curzon
    on, 53, 91, 92, 93, 129; Hardinge on,
    42, 43, 94; Lloyd George on, 91, 93;

pressure on, 69, 114
Anglo-French relations, 80, 91, 92; Africa,
    115; Baldwin–Poincaré meeting (1923),
    87; Cologne, evacuation of, 159, 162–3,
    171; Commission of Control, 135, 163,
    183, 184, 185; Curzon–Poincaré clashes,
    121–2, 134; Egypt, 207; French press and
    public attitude, 114; German reparations,
    96–106, 108, 109, 118, 119, 122–3, 133,
    200; MacDonald government, 133–4,
    135–8, 139; Middle East, 43, 53, 92, 115,
    214, 216; Rhineland, 133, 134, 136, 138;
    Ruhr and Franco-Belgian occupation
    (1923), 109, 110, 123–9, 134, 138; Ruhr
    crisis (1920), 95–6; Turkish conflict,
    91, 92, 114, 120, 121–2, 124; Upper
    Silesia conflict, 111–14; Washington
    Conference, 92, 114
Anglo-French war, risk of, 58, 59, 79, 135,
    176
Anglo-German entente, 51, 94
Anglo-German relations, 39, 58–9;
    Chamberlain's period of office, 159–
    62, 163; MacDonald's government,
    135–8, 139
Anglo-German war, risk of, 38, 58, 79, 135
Anglo-Italian relations, 207
Anglo-Japanese Alliance, 35–6, 47, 55, 73
Anglo-Persian Agreement (1919), 35, 219
Anglo-Russian Convention (1907), 35, 218
Anglo-Russian relations: MacDonald
    government, 146–54; Middle East issues,
    220
Anglo-Russian trade, 6, 54, 66–7, 84, 224
Anglo-Turkish crisis (1921–2), 226, 227
Apfelbaum, Hirsch, 149
Arab nationalism, 210
Arabia, 206, 209, 213, 216, 224, 225;
    Colonial Office control, 226; unrest in
    (1919), 214
Ashton-Gwatkin, Frank, 202
Asia see Far East
Asia Minor: Crowe on, 53; Curzon on, 36,
    223; Greeks in, 72, 73; Hardinge on, 44,
    222; Lloyd George on, 64–6, 67, 221, 223

Association of Chambers of Commerce, 190, 197
Australia, 227
Austrian Empire: British policy during war, 47; dismantlement of, 42, 51, 56; post-war Austria, 75, 165

balance of power, 58, 62, 126; Crowe on, 50, 51–2; D'Abernon on, 167; French threat to, 53; Hardinge on, 42, 43; Lloyd George on, 91, 93; Tyrrell on, 38
Baldwin, Stanley: and Curzon, 85, 87, 88; on Europe, 170, 171–2, 174–5; FO, relationship to, 124; FO reform, 15–16; on France, 127, 168–9; on MacDonald government, 147; and Poincaré, 87, 126, 127; becomes Prime Minister (1923), 85; becomes Prime Minister (1924), 154; propaganda issue, 28; recruitment policy, 18, 19; Ruhr crisis, 86; and Tyrrell, 55; Zinoviev Letter, 154
Balfour, Arthur James: on Egypt, 206; on Europe, 170; and FO, 7; FO reform, 14, 18; on France, 92, 121; and Kerr, 69; and Lloyd George, 1; on Middle East, 208, 209, 211; trade/commercial issues, 196
Balkans, 91
Ball, Joseph, 204
Barthou, Louis, 127
Battenberg, Prince Louis, 47
Batum, 219, 220
Bavaria, 91
Beak, George, 26
Beaverbrook, William Maxwell Aitken, 1st Baron, 4, 26
'Beer Hall Putsch' (1923), 128
Belgium: Allied Financial Agreement, 200; Commission of Control withdrawal, 185; Curzon on, 37, 92; Dawes Plan, 140, 141; guarantees to, 169; German reparations, 97, 126, 127; German threat to, 121, 160; Hardinge on, 42; Mutual Guarantee, Treaty of, 181–2; Polish seat on Council of League, 186; Ruhr occupation (1923), 86, 109, 123–9; Ruhr proposed occupation (1920), 97; and War Office, 76–7; Washington Conference, 60 *see also* Anglo-Belgian relations; Anglo-Franco-Belgian Alliance; Franco-Belgian relations
Belloc, Hilaire, 111
Benes, Eduard, 81
Bengal, partition of, 41
Bennett, J.C. Sterndale, 58, 79; on Europe, 137, 178, 179, 180; German-Allied non-aggression pact, 168; German reparations, 108; on Germany, 135, 160, 161–2;

Rhineland, 133
Bergmann, Carl, 123
Berthelot, Phillipe, 8
Bertie, Francis, 42, 82, 212
Bessarabia, 165
Bingham, Francis, 79
Birkenhead, Frederick Edwin Smith, 1st Earl of, 169, 170, 172, 173–4
Board of Selection, 17, 18, 19, 21, 23
Board of Trade: commercial policy role, 190; Dawes Plan, 107, 137; FO, conflict with, 2–3, 192–9, 201; at Genoa Conference (1922), 120, 198–9; German disarmament, 198–9; German reparations, 202; Ruhr crisis (1923), 107, 124, 134, 136; Turkish crisis, 226
Bolshevik revolution, 6, 51, 89, 131
Bolsheviks, fear of, 6, 44, 52, 66–7, 92, 93; in Germany, 43, 52–3, 58, 76, 95, 108, 123; by intelligence community, 152; and MacDonald government, 147, 150, 154; in Middle East, 219, 220
Bonar Law, Andrew, 225; and Curzon, 85, 88; FO, relationship to, 124; FO reform, 14, 18; acts as Foreign Secretary (1922–23), 86; on France, 91, 124; and League of Nations, 72; Prime Minister (1922), 85; Ruhr crisis (1920), 95; salary proposals, 19–20
Boulogne Conference (1920), 97, 98
Bradbury, Sir John: Dawes Plan, 107, 141; German reparations, 3, 43, 98, 107, 108, 123, 127, 199, 200, 202; London Conference (1924), 142; Registry reorganization, 25; Ruhr crisis (1923), 124; Tyrrell on, 201
Brazil, 186, 187, 188
Briand, Aristide, 80, 81; Anglo-French alliance, proposed, 115, 116, 117–18; and Chamberlain, 177; Council of the League issue, 186, 187; German threat, 121; Locarno Conference (1925), 180; Quadrilateral Pact, 175, 177, 179, 180
Bridgeman, William C., 172
Bridges, Edward, 16, 61
British Central Africa, 205
British Communist Party, 147, 148
British Delegation to Peace Conference, 2, 25, 47, 54
British Empire, Curzon's policy on, 35, 62
Browning, Lieutenant-Colonel Frederick Henry, 152
Brussels Conference (1920), 97–8
buffer states: Europe, 51, 56, 59, 128, 137; Near East, 35, 36; Russia, 67
Bulgaria, 166
bureaucratization, 62, 73, 74, 207, 208

Burnett-Stuart, Major-General J.T., 86
Butler, Sir Frederick, 12

Cabinet, 74
Cabinet Secretariat, 68, 71–3, 85, 86; League
    of Nations conflict with FO, 34, 71–2,
    125
Cable and Wireless Section of FO, 10,
    27, 28
Cadogan, Alexander George Montagu, 130
Campbell, John Ross, 147
Campbell, Ronald: Board of Trade, 201; FO
    reform, 16; Geneva Protocol (1924),
    163; hierarchy changes, 8
Canada, 227
Cannes Conference (1922), 80–1, 117, 164
Caucasus, 35, 77, 213, 219, 20
Cave, Lord George, 196
Cecil, Lord Edgar Algernon Robert: on
    Crowe, 47, 212; and Curzon, 86–7;
    at Eastern Department, 212; on Egypt,
    205–6; on Europe, 172, 173; FO reform,
    14, 15, 16, 17; Geneva Protocol (1924),
    169; and Hardinge, 7; League of Nations,
    86, 155, 186, 187, 188; on Middle East,
    207, 208, 209–10, 211–13; at Ministry
    of Blockade, 47, 190; propaganda issue,
    26; recruitment policy, 18, 19; salary
    proposals, 19–20, 21; trade/commercial
    issues, 194, 195
Cecil, Lord Edward, 207, 208, 209
Central America, 58
Central Department of FO, 7, 10, 50;
    Commission of Control, 184; and Curzon,
    57, 77; and Europe, 165; and France,
    124; German reparations, 43, 111; Head
    of Department, role of, 58
Chalmers, Sir Robert, 20
Chamberlain, Sir J. Austen: Anglo-French
    alliance, proposed (1921), 115, 164;
    Board of Trade influence, 199; and
    Briand, 177; Cologne zone, evacuation
    of, 160; Commission of Control
    withdrawal, 80, 183–5; conflict with
    Ministers, 155; Council of the League
    issue, 185–8; and Crowe, 155, 159, 176;
    and Curzon, 84–5, 224–5; on Europe,
    158, 162–88; FO, relations with, 155–6;
    becomes Foreign Minister, 154; on
    France, 92, 158, 164; Geneva
    Conference (1927), 49; German-Allied
    non-aggression pact, 168; German
    disarmament, 161, 162, 183–5; German
    reparations, 201; on Germany, 156,
    159–62, 163; Locarno Conference (1925),
    180–2; personality, 156–7; Quadrilateral
    Pact, 175–80; Ruhr crisis (1920), 96;

Russo-German alliance, 164; working
    methods, 155, 157–9; Zinoviev Letter,
    152
Chamberlain, Neville, 203
Chanak débâcle, 85, 200, 227
Chancery Service, 25
Chicherin, Georgi, 119
Chief Clerk's Department of FO, 11, 12, 17
Childs, Major-General Sir Wyndham, 150,
    152–3
China, 36, 60, 155, 209
Churchill, Sir Winston: Anglo-French
    alliance, proposed, 115; and
    Chamberlain, 155; Chancellor of the
    Exchequer (1924–9), 203; Colonial
    Secretary appointment (1921), 225;
    and Dominions, 227; on Europe, 169,
    170, 172; on France, 91, 92, 172–3; on
    Germany, 76; on Kerr, 71; on Lloyd
    George, 63; and Middle East, 214–15,
    216, 219, 220, 224; on Turkey, 221, 222,
    223, 226; Upper Silesia conflict, 113
cipher work, 28–9
Civil Service: entrance examination, 21, 23;
    and FO, 14–17, 19; and Lloyd George,
    48, 223; salaries, 20
Clark, Sir William, 193–4
Clemenceau, Georges, 48, 64, 134
Clerk, George, 29, 75
Clive, Robert H., 128
coding work, 28–9
Cologne zone, evacuation of, 79, 159–63,
    170, 171, 183
colonial administration: post-war, 214–21;
    war-time, 205–13
Colonial Office, 205; Cyprus, control of,
    206, 216; and Dominions, 227; Egypt,
    control of, 205–6, 216, 224; Mesopotamia,
    control of, 214–15, 217, 226; Middle
    East, control of, 215–16, 219, 224, 225–6
commerce, 3–4, 8, 189–99, 201–2 *see also*
    Anglo-Russian trade; Board of Trade;
    Commercial Department of FO;
    Department of Overseas Trade
Commercial Attaché Service, 191, 192,
    196, 197
Commercial Department of FO, 3, 4, 8,
    189–90, 195, 197
Commercial Diplomatic Service, 197, 198,
    201
Commission of Control, 78, 79, 134–5,
    160–1, 162, 163, 183–5
Committee of Guarantee, 135
Committee of Imperial Defence, 163, 164,
    165, 169, 170, 178
Committee on Conditions of Admission,
    19
communism, *see* Bolsheviks, fear of; British

Communist Party; Third Communist International

Conference of Ambassadors (1926), 184, 185

Conference of Jurists (1925), 180

Conference of Ministers (1925), 172–5

Conservatives: and Bolshevism, 147; gain power (1924), 154; lose power (1924), 129, 130; Zinoviev Letter, 150, 151, 154

Consular Department of FO, 8, 11, 46, 195, 197, 198

Consular Service, 24, 196, 213; commercial role, 190, 191, 192, 197; staffing, 30, 191

Contraband Department of FO, 3, 47, 190, 213

Coolidge, Calvin, President, 126

Cripps, Charles Alfred, First Baron Parmoor, 136, 137

Cromer, Lord, 206

Crowe, Sir Eyre A.B.W., 45–54; African Department, 46; amalgamation of FO with Diplomatic Service, 21; Anglo-French alliance, proposed, 116–17; Anglo-French pact, 164; assistant under-secretary, 44–5; Board of Trade dispute, 198–9; Central Department, 50; and Chamberlain, 155, 159, 176; Cologne zone, evacuation of, 160; Commission of Control, 134; Conference of Ministers (1925), 172–5; Consular Department, 46; Contraband Department, 3, 47, 190, 213; and Curzon, 50, 53, 55, 69, 83; Dawes Plan, 107, 137, 140, 141; death of, 176; on Europe, 46, 49, 50–3, 136–7, 163, 165, 166, 168, 169, 170, 171–5; FO building extension, 10; FO reforms, 16, 46; on France, 46, 51, 52, 53, 79, 92–3, 94, 124, 127, 133; German-Allied non-aggression pact, 168; German disarmament, 50, 53, 79, 161, 162, 198–9; German reparations, 52, 53, 108–9, 110, 111, 116, 126, 202; on Germany, 38, 46, 50–3, 78–9, 94, 135; on Greece, 53, 67; and Grey, 47; Hardinge on, 47; hierarchy changes, 8, 9, 10, 11; and League of Nations, 8, 34, 71; and Lindsay, 56; and Lloyd George, 1–2, 48, 63; London Conference (1924), 142; and MacDonald, 132, 138; on Middle East, 53, 211–12, 213, 217, 218; Ministry of Blockade, 3, 47, 190, 213; Nicolson on, 45, 48; Paris Peace Conference, 7, 47–8, 52; permanent under-secretary, 38, 48–54; personality, 49; Press Department, 5; recruitment policy, 19; Registry reorganization, 24, 25; Ruhr crisis (1920), 95; Ruhr crisis, (1923), 94, 124; on Russia, 50, 51–2, 54, 146–7, 148–9; salary proposals,

20; San Remo Conference (1920), 96; security issues, 144; trade/commercial issues, 191, 192, 194, 197; and Tyrrell, 45, 56, 87; Upper Silesia conflict, 112, 113; Versailles Treaty, 51, 52; and War Office, 75, 76, 77, 78; war-time role, 3; Western Department, 46, 50; Zinoviev Letter, 148–9, 151, 152, 153–4

Crowe, Sybil, 152

cryptography, 5–6

cultural services, 26–8

Cuno initiative, 167

Curzon, George Nathaniel, 1st Marquess Curzon of Kedleston, 31–7; and Africa, 35, 115; amalgamation of FO with Diplomatic Service, 22; Ambassador appointments, 82–3; Anglo-French pact, proposed, 114, 115, 116, 117, 118, 119, 121; anti-Bolshevism, 54, 67; and Baldwin, 85, 87, 88; on Belgium, 37, 92; and Bonar Law, 85, 88; British Empire, 35, 62; and Cecil, 86–7; and Chamberlain, 84–5; character and abilities, 31–3; and Crowe, 50, 53, 55, 69, 83; death of, 172; and Dominions, 35, 227; on Europe, 36–7, 42, 43–4, 52, 53, 62, 163, 164–5, 169, 170; Foreign Secretary appointment, 7; on France, 36–7, 53, 59, 73, 80, 81, 87, 90–1, 92–3, 124, 127, 129, 168; 'Garden Suburb', 68, 69–70; German reparations, 36, 125–6, 127, 201; on Germany, 36, 58, 73, 78, 87, 90–1, 120; on Greece, 64, 69–70, 73; and Hankey, 73; and Hardinge, 33, 41, 81–2; and Heads of Departments, 57; hierarchy changes, 8, 9, 11; intelligence services, 6; and Kerr, 69, 73; League of Nations, 33, 86; and Lindsay, 56; and Lloyd George, 2, 32, 36, 48, 63, 69, 73, 81, 82, 84, 88, 224–5; on Middle East, 36, 53, 115, 207, 208, 209, 211, 216–17, 218–19, 220, 223, 224; on Near East, 35, 36, 37, 41, 44, 65–6, 80; Nicolson on, 26, 84, 87; and Poincaré, 6, 82, 87, 121–2, 134; policies, 35–7; propaganda issue, 27; Registry reorganization, 25–6; resignation, 129; Rhineland, 36, 128; Ruhr crisis (1920), 95, 96; Ruhr crisis (1923), 34, 86, 124, 125–6; on Russia, 66–7, 73, 81, 84, 85–6, 153; Russo-German alliance, 120; salary proposals, 21; San Remo Conference (1920), 65, 96; on the Treasury, 200; on Turkey, 36, 53, 65–6, 73, 221–3; and Tyrrell, 54, 55, 87; Upper Silesia conflict, 112–13; and War Office, 77; working methods, 33–4, 130, 132, 133; Zinoviev Letter, 154

Cyprus, 206, 216
Czechoslovakia, 51, 75, 181, 187

D'Abernon, Lord Edgar Vincent:
    Ambassador in Berlin, 77, 82–3;
    Commission of Control, 78; German-
    Allied non-aggression pact, 167–8,
    169; German disarmament, 135, 183;
    German reparations, 98, 107, 108, 202;
    Russo-German alliance, 164; Upper
    Silesia conflict, 113
Daily Express, 154
Daily Mail, 151, 153, 154
Davies, J.T., 68
Dawes, Charles G., 127–8; Dawes Plan,
    107, 111, 137, 138, 139–44, 159
Decyphering Branch, 5
Department of Information, 4
Department of Overseas Trade, 3, 8, 9,
    201, 217; establishment of, 4, 189–99;
    German reparations, 202; propaganda
    issue, 28; Ruhr crisis (1923), 107, 136
Derby, Lord Edward George Villiers, 2, 47,
    82, 86, 96, 112
diplomacy by conference, 33, 62
Diplomatic Service, 12–13; amalgamation
    with FO, 13, 16, 19, 21–4, 192; clerical
    duties relief, 25; commercial role, 190, 191,
    198, 201; and Europe, 75; permanent
    under-secretary role, 39; property
    qualification, 13, 16, 17; recruitment, 12,
    18; salaries, 20; staff numbers, 30
diplomats: Lloyd George on, 1, 62; public
    distrust of, 2
disarmament: Conference (1925), 145;
    Crowe on, 50, 53, 79, 161, 162, 198–9;
    Curzon on, 120; France and, 92; German
    rate of, 98, 135, 159, 161–2, 183–5;
    Hardinge on, 43; War Office on, 75,
    78–80; Waterlow on, 58
Docker, Frank Dudley, 193–4
domestic policies, relation to foreign
    policy, 60–1, 73, 74, 75, 189, 193, 196
Dominions: Anglo-French pact, 141, 144;
    Curzon on, 35, 227; and European pact,
    170, 171, 173, 174; Geneva Protocol
    (1924), 146; Mutual Assistance, Draft
    Treaty of, 144; Mutual Guarantee,
    Treaty of, 182; relations with, 227–8
Dominions Information Department of FO,
    30, 227, 228
Dominions Office, 227, 228
Dormer, Cecil, 29
Drummond, Sir Eric, 13, 19

East Africa, 214

Eastern Committee, 207, 209–10, 211, 213,
    218
Eastern Department of FO, 7, 211, 213, 226;
    and Curzon, 57; Egypt, loss of, 29–30;
    and France, 59; and Hardinge, 9, 10, 42;
    and Lindsay, 56
Eastern Empire, Curzon's policy on, 35, 37
Eastern Europe, French Allies in, 115, 116,
    117, 118, 166
Ebert, Friedrich, 119
economic affairs *see* commerce
Economic Relations Section of FO, 202
Egypt: Anglo-French conflict, 91; Colonial
    Office/FO conflict, 205–6, 207–8, 209,
    213, 216, 224, 226; military presence in,
    219; unrest in (1919), 214
Egyptian Department, 29, 226
entente cordiale see Anglo-French *entente*
Ethiopia, 58, 213, 226
Europe: Chamberlain on, 158, 162–88;
    Churchill on, 169, 170, 172; Crowe on,
    46, 49, 50–3, 136–7, 163, 165, 166, 168,
    169, 170, 171–5; Curzon on, 36–7, 42,
    43–4, 52, 53, 62, 163, 164–5, 169, 170;
    economic recovery, 60, 80, 90, 93, 107,
    117, 119, 120, 121, 139; Hankey on,
    72, 165, 168; Hardinge on, 42–4, 52;
    Kerr on, 70; Lloyd George on, 80, 81,
    91–4, 117; MacDonald on, 135–8, 139;
    Nicolson on, 58, 137, 165, 166, 170;
    Tyrrell on, 163, 170, 171, 176, 178, 179,
    180; and War Office, 75–80 *see also*
    Eastern Europe
Evans, Sir Arthur, 13
examinations, entrance, 14, 18, 19, 21, 23;
    cipher work, 29

Facilities Division of News Department,
    27, 28
Far East, 35–6, 115, 209
Far Eastern Department of FO, 7, 8, 10,
    54; German question, 59; Head of
    Department, role of, 58
Faringdon, Lord Alexander, 193–4
Federation of British Industries, 197
financial matters, 11–12, 13–14
Fisher, Herbert A.L., 83, 225
Fisher, Sir Warren, 11, 16, 141, 203
Foch, Marshal Ferdinand, 160
Fontainebleau Memorandum, 70
Foreign Office building, 9–10
Foreign Trade Department *see* Department
    of Overseas Trade
France: and Chamberlain, 92, 158, 164;
    Churchill on, 91, 92, 172–3; Cologne
    zone, evacuation of, 159, 162, 163, 170,
    171; Commission of Control, 135, 163;

Crowe on, 46, 51, 52, 53, 79, 92–3, 94, 124, 127, 133; Curzon on, 36–7, 53, 59, 73, 80, 81, 87, 90–1, 92–3, 124, 127, 129, 168; Dawes Plan, 140, 141, 144; financial crisis (1924), 143–4; German reparations, 3, 43, 52, 92, 116, 118, 119, 122–3, 125–6, 200; Germany, agreement with, 174; Germany, economic recovery of, 42, 52, 56, 116; Germany, fear of, 38, 43, 51, 91, 92, 114, 117, 121, 136, 163–4, 166; Germany, relations with, 174, 178; Hardinge on, 42–3, 92, 94; Kerr on, 69; and Lloyd George, 80–2, 91–4; and MacDonald, 133–4, 137, 139, 140–1, 143; Middle East, 43, 53, 92, 115, 214, 216; Mutual Guarantee, Treaty of, 181–2; Nicolson on, 133; post-war attitude to, 89–91; Ruhr crisis (1920), 95–6; Ruhr evacuation (1925), 79; Ruhr occupation (1923), 86, 87, 107, 109, 110, 111, 123–9, 134; Ruhr proposed occupation (1921), 69, 97, 98; San Remo Conference (1920), 64; security question, 133, 138, 139, 146, 158, 160, 162; Turkish policy, 222; Tyrrell on, 56, 92, 94; and War Office, 76–7; Washington Conference, 60 *see also* Anglo-Franco-Belgian Alliance; Anglo-Franco-German pact; Anglo-French alliance; Anglo-French *entente*; Anglo-French relations; Anglo-French war

Franco-Belgian relations, 142
Franco-German commercial treaty, 143

'Garden Suburb', 4, 40, 63, 68–71, 73, 85
Gascoyne-Cecil, William C., 172
Geddes, Auckland, 83
Geddes, Sir Eric, 124
Geneva Conference (1927), 49
Geneva Protocol (1924), 145–6, 163, 164–5, 168, 169, 170, 171
Genoa Conference (1922), 72, 81, 117, 118, 119, 120, 198–9
German-Allied non-aggression pact, 137, 167–8
German-Polish arbitration treaty, 177, 181
Germany: arms concealment, 38, 79; Chamberlain on, 156, 159–62, 163; Commission of Control, 135, 184; Communist revolution threat, 43, 52–3, 58, 76, 95, 108, 123; Crowe on, 38, 46, 50–3, 78–9, 94, 135; Curzon on, 36, 58, 73, 78, 87, 90–1, 120; Dawes Plan, 140, 141, 144, 159; economic recovery, 42, 52, 56, 60, 80, 116; Far Eastern fleet, 47; France, relations with, 174, 178; Geneva Conference (1927), 49; Hankey on, 72;

Hardinge on, 38, 42–3, 94; Kerr on, 69; and League of Nations, 115, 120, 180, 182, 184, 185–8; letter to Pope, 70; and Lloyd George, 80, 81, 91–4; London Conference (1924), 142; and MacDonald, 139, 143; Mutual Guarantee, Treaty of, 181–2; Nicolson on, 79, 135, 136; post-war attitude to, 89–91; potential threat of, 38, 42, 43, 50–1, 117, 121, 136, 163–4, 166, 169–70, 179, 203; Ruhr crisis (1920), 95–6; Tyrrell on, 38, 55–6, 78, 94, 135; Upper Silesia conflict, 111–14; and War Office, 75, 76–80, 160; Waterlow on, 58, 76–7, 78 *see also* Anglo-Franco-German pact, Anglo-German *entente*; Anglo-German relations; Anglo-German war; disarmament; Franco-German commercial treaty; Quadrilateral Pact; reparations; Russo-German relations; Russo-German-Japanese Alliance

Glasgow, George, 13
Government Code and Cypher School, 6, 29, 54
Government Hospitality Fund, 28
Graham, Sir Ronald, 20; on Egypt, 206, 209; Hague, appointment to, 38, 48; on Middle East, 207, 208, 211–12
Greco-Turkish conflict, 72, 73, 114, 157, 222–3
Greco-Turkish relations, 65, 66, 222
Greece: Crowe on, 53, 67; Curzon on, 64, 69–70, 73; Hankey on, 72; Kerr on, 70; Lindsay on, 56; Lloyd George on, 64–6, 72, 157; Nicolson on, 67 *see also* Smyrna
Gregory, J.D., 67, 148, 153, 154
Grey, Edward, 1st Viscount Grey of Fallodon: Ambassador to Washington, 83; and Crowe, 47; diplomacy in war-time, 1; and Hardinge, 42, 212; trade/commercial issues, 192, 193; and Tyrrell, 54
Grigg, Sir Edward, 69, 70, 71
Guest, S.A., 26
Gwynne, H.A., 87

Habsburg Empire *see* Austrian Empire
Haking, Richard, 76
Haldane Committee (1918), 215
Hall, Vice-Admiral Sir William Reginald, 152
Hankey, Sir Maurice, 34, 67, 71–3, 224; and Curzon, 73; on Europe, 72, 165, 168; on Germany, 72; and Greco-Turkish conflict, 72; and Lloyd George, 48, 68,

72, 225; London Conference (1924), 142; on MacDonald, 143

Hardinge, Lord Charles: amalgamation of FO with Diplomatic Service, 21, 22; Ambassador appointments, 82–3; Ambassador to France, 5, 38, 41, 42, 48, 56, 83; Anglo-French pact, proposed, 116, 117, 118, 121, 122; Anglo-Russian trade, 67; bottle-neck system, 40, 48; career, 40; and Cecil, 7; cipher work, 29; on Crowe, 47; and Curzon, 33, 41, 81–2; on Egypt, 206, 207–8; on Europe, 42–4, 52; FO reform, 14, 16, 17; on France, 42–3, 92, 94; on Germany, 38, 42–3, 94; and Grey, 42, 212; hierarchy changes, 8, 9, 10, 11; in India, 41; and League of Nations, 71; on Middle East, 53, 207, 208, 209, 210–12, 213, 217; on Near East, 41, 44, 64–5, 66, 207; at Paris Peace Conference, 41, 47; permanent under-secretary, 40–4, 63; Political Intelligence Department, 4, 5; Registry reorganization, 25; on reparations, 43; Ruhr crisis (1920), 95; on Russia, 42, 67; salary proposals, 20; structural changes, 8; trade/commercial issues, 192, 196; on Turkey, 41, 64–5, 66, 222; Upper Silesia conflict, 112, 113; and War Office, 75, 77

Harmsworth, Cecil, 27, 226

Hastings, Sir Patrick, 147

Headlam-Morley, J.W., 5, 108, 109, 165, 166–7

Heads of Departments, 9, 20, 57–8

Heath, Sir Thomas, 20

Hejaz, 207, 209, 213

Henderson, Arthur, 150

Herriot, Édouard, 141, 145, 162, 169; Quadrilateral Pact, 170, 171, 172, 174, 175

hierarchy, changes in, 8–12

Hirtzel, Sir Arthur, 213

Historical Section, 5, 30

Hoare, Sir Samuel, 169, 172, 173

Hoesch, Leopold von, 167

Holland, 42, 187

Holtzendorff, Admiral Henning von, 46

Home Propaganda Organization, 27

Hong Kong, 35

Horne, Sir Robert, 81

Hotham, Rear-Admiral Alan Geoffrey, 150

Hughes, Charles Evans, 83

Hungary, 51, 91, 166

Hurst, Sir Cecil B., 134, 176, 179

Hythe Conference (1920), 72, 97, 98

India: anti-Arab views, 210; Curzon's policy on, 35, 36, 218, 219; domestic peace of, 221; French guarantee of borders, 115; Khilafat movement, 222; Middle East control by Government, 210, 211, 213, 217; security of, 221, 223; Soviet threat, 6, 44, 219–20; threat of uprising, 206; War Office on, 221

India Office: and Afghanistan, 217; FO, conflict with, 2–3; Middle East issues, 206, 209, 213, 215, 219

information services, 26–8

Intelligence Bureau, 4–5

intelligence community, Zinoviev Letter, 149, 151–2, 154

intelligence functions, 4, 5–6

inter-departmental relationships, 61–2, 73–5, 189, 202; Middle East, 205, 207, 214, 216, 217 see also Board of Trade; Cabinet Secretariat; Colonial Office; 'Garden Suburb'; India Office; Treasury; War Office

Ireland, 219

Italy: African colonies, 226; Conference of Jurists (1925), 180; Dawes Plan, 140, 141; German reparations, 97, 123, 127; Mutual Guarantee, Treaty of, 181; San Remo Conference (1920), 64; Smyrna, occupation of, 65; Upper Silesia conflict, 114 see also Anglo-Italian relations

Jackson, Sir Stanley, 150

Japan, 209; Curzon's policy on, 35–6; Germany, influence of, 59; Washington Conference (1921–22), 60 see also Anglo-Japanese Alliance; Russo-German-Japanese Alliance

Jews, 213, 216

Jones, Thomas, 55

Kadmozev, Boris, 149

Kalogerpoulus, Nikolaos, 72

Kapp Putsch, 95

Kerr, Philip, 11th Marquess of Lothian, 68–71

Keynes, John Maynard, 90

Kindersley, Major Guy, 149, 150

King's Messengers and Communications Department of FO, 11, 29

Koppel, Percy A., 28

Kurdistan, 226

Kuusinen, O.V., 148

Labour Party: balance of power policies, 62; and Bolsheviks, 147, 150, 154; and intelligence community, 152; power, gain of (1924), 129, 130; power, loss of (1924), 154; Ruhr crisis (1923), 124; security issues, 144, 146; Zinoviev Letter, 149, 151, 153

Lampson, Miles: and Chamberlain, 158; Cologne zone, evacuation of, 160, 162; Commission of Control, 78, 183, 184; on Europe, 163, 164, 170, 176, 180; German-Allied non-aggression pact, 137, 168; German disarmament infractions, 79, 161, 162; German reparations, 108, 202; German war threat, 58, 79, 135, 136, 179; on Germany, 177; London Conference (1924), 142; Mutual Guarantee, Treaty of (1925), 182

Lansdowne, Lord, 115

Latvia, 165

Lausanne Conference (1922–23), 85–6, 124, 226

League of Nations, 155, 227; Cabinet Secretariat/FO conflict, 34, 71–2, 85, 125; creation of, 60; and Crowe, 8, 34, 71; Curzon on, 33, 86; Genoa Conference (1922), impact of, 81; German disarmament, 183, 185; and Germany, 115, 120, 180, 182, 184, 185–8; Locarno Conference (1925), 181; Middle East issues, 216, 224; Mutual Guarantee, Treaty of, 182; security issues, 133, 137, 138, 144, 145, 146, 176, 178–9; Upper Silesia conflict, 114

Leathes, Sir Stanley, 19

Liberal Foreign Affairs Committee, 12

Liberals: balance of power policies, 62; withdraw support for MacDonald, 147

Library Department of FO, 11

Liddell Hart, Captain Basil, 90

Lindley, Francis, 75

Lindsay, Sir Ronald, 56

Lithuania, 165

Litvinov, Maxim, 119

Lloyd George, David: Ambassador appointments, 42, 82–3; Anglo-French alliance, proposed, 115, 116, 117, 119, 121; on Asia Minor, 64–6, 67, 221, 223; and Balfour, 1; and Civil Service, 48, 223; and Crowe, 1–2, 48, 63; and Curzon, 2, 32, 36, 48, 63, 69, 73, 81, 82, 84, 88, 224–5; on diplomats, 1, 62; and Dominions, 227; on Europe, 80, 81, 91–4, 117; and FO, 1–2, 62–3, 73, 74, 80–2, 131; foreign policy, 64–8; and France, 80–2, 91–4; 'Garden Suburb', 4, 63, 68–71; at Genoa Conference (1922), 120; German disarmament, 98; German reparations, 80, 81, 94, 96, 97, 98, 106; on Germany, 80, 81, 91–4; on

Greece, 64–6, 72, 157; and Hankey, 48, 68, 72, 225; and Kerr, 68–9, 70; and League of Nations, 71, 81; leaves office (1922), 85; on Middle East, 53; Middle East Department, 223, 224; Paris Peace Conference, 3, 64, 66, 67; and Poincaré, 81, 82; Rhineland occupation, 134; Ruhr and French proposed occupation (1921), 98; Ruhr crisis (1920), 95, 96; on Russia, 54, 66–7, 80, 81, 115, 117, 119; trade/commercial issues, 195; on Turkey, 64–6, 72, 157, 222; Upper Silesia conflict, 113; Versailles, Treaty of, 1, 114

Locarno Conference (1925), 180–2, 227

London Conference (1920), 64, 65, 223

London Conferences (1921), 72, 98, 108, 128, 140

London Conference (1924), 137–8, 141–3

London Reparation Conference (1922), 86

London Schedule of Payments, 109–10, 125, 127

Loucheur, Louis, 80

Luther, Hans, 162, 180

MacDonald, J. Ramsey: background, 131–2; Board of Trade influence, 199; Commission of Control, 134; and Crowe, 132, 138; Dawes Plan, 137, 140–1; and Europe, 135–8, 139; FO, relations with, 155, 156; FO building extension, 10; and France, 133–4, 137, 139, 140–1, 143; German reparations, 201; and Germany, 139, 143; Hankey on, 143; and Herriot, 141; intelligence community, 152; London Conference (1924), 142, 143; personality, 130–1; and Poincaré, 134, 138; becomes Prime Minister, 129; on Russia, 54; security issues, 144, 145; and Snowden, 143; working methods, 132–3, 158; Zinoviev Letter, 149, 150, 151, 153, 154

Machinery of Government Committee Report, (1918), 192

MacManus, Arthur, 148

Malcolm, Ian, 19

Maltzan, Baron A.G.O. von, 119

Manchester Evening Chronicle, 151

Marlowe, Thomas, 151, 152

Masterton Smith, Sir James, 225

Maxton, James, 150

McKenna, Reginald, 128, 140

Mesopotamia: Anglo-French conflict, 91; Curzon on, 35, 217; departmental control of, 206, 209, 213, 214–15, 217, 224, 225, 226; FO waste of resources

in, 214; Hardinge on, 41, 207; military
    presence, 219, 220; uprising in, 214; War
    Office on, 221
Mesopotamia Administration Committee,
    206
Middle East, 205–21; Anglo-French
    relations, 43, 53, 92, 115, 214, 216;
    Balfour on, 208, 209, 211; Cecil on,
    207, 208, 209–10, 211–13; Churchill
    on, 214–15, 216, 219, 220, 224; Crowe
    on, 53, 211–12, 213, 217, 218; Curzon
    on, 36, 53, 115, 207, 208, 209, 211, 216–
    17, 218–19, 220, 223; Graham on, 207,
    208, 211–12; Hardinge on, 53, 207, 208,
    209, 210–12, 213, 217; Montagu on, 208–
    10, 211, 213, 215, 218, 223, 224;
    post-war administration, 214–21;
    war-time administration, 205–13
Middle East Committee, 206, 208, 209
Middle Eastern Department of FO, 3;
    formation of, 205–21, 223–6
Millerand, Alexander, 95, 121
Milner, Viscount Alfred, 208, 209, 215,
    219, 225
Ministry of Blockade, 3, 8, 47, 190, 213
Ministry of Information, 4, 26
Mongolia, 209
Montagu, Edwin: Department of Overseas
    Trade, 194; Middle East, military
    presence in, 219–20; Middle East
    Department, 208–10, 211, 213, 215,
    218, 223, 224; on Turkey, 221, 222, 223
Montgomery, C. Hubert: amalgamation
    of FO with Diplomatic Service, 21;
    Assistant Under-Secretary, 11, 12;
    cipher work, 29; hierarchy changes,
    8, 10
Morel, E.D., 13
Morgan, J.H., 79
Morgan, J.P., 144
Morocco, 80, 91
Muller, Sir William Max, 176
Murray, Arthur C., 83
Murray, Gilbert, 186
Mustapha Kemal, 72, 223
Mutual Assistance, Draft Treaty of, 144
Mutual Guarantee, Treaty of, 180, 181–2

Namier, Lewis B., 13
Near East, 223; anti-Moslem policy, 220,
    222; Curzon on, 35, 36, 37, 41, 44, 65–6,
    80; departmental control of, 209, 213;
    French policy, 92; Hardinge on, 41, 44,
    64–5, 66, 207; military presence in, 219
New Europe, 13
News Department of FO, 4, 10, 27, 28, 54
Nicolson, Sir Arthur, 47

Nicolson, Harold: on Crowe, 45, 48;
    on Curzon, 84, 87; on Curzon's filing
    system, 26; on Europe, 58, 137, 165,
    166, 170; on France, 133; German
    reparations, 133; on Germany, 79,
    135, 136; on Greece, 67; on selection
    procedures to FO, 23; on Tyrrell, 45
Niemeyer, Otto, 125
Nollet, Charles, 143
Norman, Herman, 47
North America, 58
Northern Department of FO, 7, 56; German
    question, 59; and Hardinge, 9, 10,
    42; Head of Department, role of, 58;
    Zinoviev Letter, 148, 153
Norway, 206

Office of the Works, 10
organizational changes, 7–8
Ottoman Empire, dismantlement of, 35, 36,
    44, 64, 205, 218, 221

Paget, Sir Ralph, 38, 54
Painlevé, Paul, 175
Palatinate *see* Rhineland
Palestine: Anglo-French conflict, 92;
    Curzon on, 35, 216; departmental
    control of, 206, 209, 213, 216, 224,
    225; Hardinge on, 207; military
    presence in, 219, 220
Palewski, Gaston, 54
Palmer, W.W., 2nd Earl of Selborne, 19
Paris Conference (1921), 98, 117
Paris Peace Conference: Asia Minor,
    44, 64, 65; Crowe's role, 47–8, 52;
    Curzon, 65, 67; Hardinge's role, 41,
    47; Lloyd George, 3, 64, 66, 67;
    Political Intelligence Department
    influence, 5
Paris Reparation Conference (1923), 86, 123
Parker, Alwyn: Registry reorganization, 24,
    25; war-time role, 3
Parliament, control of foreign policy, 62
Parliamentary Department of FO, 28, 29
Passport Control Department of FO, 29
Passport Control Officer system, 6
Passport Office of FO, 29
Pennefather, Sir John de F., 193–4
Percival, Colonel, 112
Percy, Lord Eustace, 13, 170
Permanent Court of International Justice,
    145, 146
Persia: anti-British feeling, 210; Curzon's
    policy on, 35, 37; departmental control
    of, 206, 209, 213, 216, 217, 224, 226; FO
    waste of resources in, 214; Hardinge

on, 77; military presence in, 214, 219, 220, 224; War Office on, 221 *see also* Anglo-Persian Agreement
Peterson, Maurice, 49
Phipps, Eric: on Europe, 163; Rhineland buffer zone, 137; Ruhr crisis (1920), 95; Upper Silesia conflict, 111; and War Office, 76
Plebiscite Commission, 111, 114
Poincaré, M. Raymond: Anglo-French pact, proposed, 118–19, 120–1, 169; and Baldwin, 87, 126, 127; and Curzon, 6, 82, 87, 121–2, 134; German reparations, 106, 109, 126, 127; and Lloyd George, 81, 82; and MacDonald, 134, 138; Ruhr occupation (1923), 123
Poland: Anglo-French conflict, 92; buffer zone, 51, 56, 59; Chamberlain on, 165; Council of League seat, 186–7, 188; France, treaty with, 181; Germany, threat of, 166, 176; Upper Silesia conflict, 111–14 *see also* German-Polish arbitration treaty
Political Intelligence Department, 4–5, 8, 27, 54
Ponsonby, Arthur, 13, 146
Portugal, 60
Press Department, 5
press relations, 4, 27, 28
Prime Minister's Office, 73–4
Private Secretariat *see* 'Garden Suburb'
Promotion Board, 21, 24, 39
propaganda, 4, 26–8
property qualification, 13, 14, 16, 17, 21
Public Accounts Committee, 11
public suspicion of FO, 62

Quadrilateral Pact: Baldwin on, 175; Cecil on, 173; Chamberlain on, 165, 169; Crowe on, 171, 172, 174; Curzon on, 170; negotiations for, 175–80

Radolin, Prince Hugo, 55
Rakovsky, Khristian, 149, 150, 151, 153, 154
Rapallo, Treaty of, 119–20
Rathenau, Walther, 119
record keeping, 24–6
recruitment, 12, 17–19
Registry of FO, 10, 24–6
Reilly, Sidney, 149, 150, 152
Reparation Articles, 43
Reparation Commission: Dawes Plan, 141, 142; FO/Treasury conflict, 199, 200, 201; German default on payment, 122, 123; German financial stability, 127–8; German liability, scale of, 97, 98,

108; moratorium on payment (1922), 106
reparations, 96–111; Anglo-French differences, 96–106, 108, 109, 118, 119, 122–3, 133, 200; Conferences, 86, 96–8; Crowe on, 52, 53, 108–9, 110, 111, 116, 126, 202; Curzon on, 36, 125–6, 127, 201; Dawes Committee, 139–40, 144; FO/Treasury conflict, 3, 107, 199–202, 203; Genoa Conference (1922), impact of, 81; German payment delays, 38–9, 56, 94, 106–11, 122; German suspension due to Ruhr crisis, 123; Hardinge on, 43; Kerr on, 69, 70; Lloyd George on, 80, 81, 94, 96, 97, 98, 106; Nicolson on, 133; Tyrrell on, 108–9, 110, 111, 125, 202; Waterlow on, 43, 108
Repington, Charles á Court, 212
Rhineland: Allied occupation, 89, 137; Curzon on, 36, 128; demilitarization, 179, 181, 182; evacuation (1935), 58, 79, 136; and France, 92, 116, 133, 134, 136, 138, 168, 171; Locarno Conference (1925), 181; Mutual Guarantee, Treaty of, 182; Separatist movement, 127, 128
Rhineland Commission, 8
Riddell, Lord George, 157
Ridley Commission (1890), 13
Rosenblum, Sigmund Georgievic, 149
Royal Commission on Civil Service (1914), 13–14, 191; recruitment policy, 12, 17, 18
Ruhr: Communist counter revolution (1920), 95–6; French proposed occupation (1921), 69, 97, 98
Ruhr crisis (1923): Crowe on, 94, 104; and Curzon, 34, 86, 124, 125–6; and Dawes Committee, 139; Franco-Belgian occupation, 107, 109, 110, 111, 123–9, 134, 138–9; London Conference (1924), 143
Rumbold, Sir Horace, 83
Russell, O. Theophilus V., 19, 29
Russia: Crowe on, 50, 51–2, 54, 146–7, 148–9; Curzon on, 66–7, 73, 81, 84, 85–6, 153; Genoa Conference (1922), 118; Hardinge on, 42, 67; Kerr on, 70; Lloyd George on, 54, 66–7, 80, 81, 115, 117, 119; Near East, presence in, 35; subversion, 6, 54, 70 *see also* Anglo-Russian Convention; Anglo-Russia relations; Anglo-Russian trade; Bolshevik Revolution; Bolsheviks, fear of
Russo-German-Japanese Alliance, 59
Russo-German relations, 59, 181; and Chamberlain, 164, 178; and Crowe, 53; and Curzon, 36; German admission

to League of Nations, 185; and Lloyd George, 93; Rapallo, Treaty of, 119–20; and Waterlow, 43, 76–7
Ryan, Sir Andrew, 33

Saar, 42, 107
Saint-Aulaire, Auguste F. de, Comte, 80, 87, 114
salaries, 13–14, 16, 17, 19–21
Salisbury, 173
San Remo Conference (1920), 64, 65, 66, 72, 96, 214, 223
Sargent, Orme, 183, 184
Sasoon, Philip, 73
Schubert, Carl von, 167
Schüler, F. Edmond, 8
Scott-Eden reforms, 24
Secret Intelligence Service, 6, 27, 72; and Labour Government, 152; Zinoviev Letter, 148, 153
security issues, 139, 144–6, 165 *see also* France; League of Nations
Selby, Walford H.M., 153, 202
Selznick, Philip, 218
Seton-Watson, R.W., 13
Sèvres, Treaty of, 64, 73, 214, 223
Shacht, Dr H., 203
Silesia, 42, 91, 166, 186; Upper Silesia conflict, 70, 98, 111–14
Sinclair, Admiral Hugh, 151
Smith, F.E., First Earl of Birkenhead, 120
Smuts, General Jan Christian, 210
Smyrna, 221; Crowe on, 53; Hardinge on, 64–5, 66, 222; Kerr on, 70; Lloyd George on, 64, 65, 223
Snowden, Philip, 143, 203
Somaliland, 205, 214, 224
Soviet Union *see* Russia
Spa Conference (1920), 72, 97, 98
Spain, 186, 187, 188
Spartacists' revolt, 92, 96
Sperling, Roland, 58
Stanley, Sir Albert, 194
Stanley, Edward G.V., 42
state centralization, 60
Steel-Maitland, Sir Arthur, 195–6
Strang, William, 24, 148, 149
Stresemann, Gustav, 126, 167, 177, 180, 181
structural changes, 7–8
Stuart, Sir Campbell, 26
Sudan, 209, 213, 226
Suez Canal, 35
Supreme Council, 2, 8, 47, 48, 66, 114
Sweden, 187
Sykes, Sir Mark, 206, 209
Sykes–Picot Agreement, 35
Syria, 92, 207, 213, 216, 224, 226

Tangier, 91, 92, 120
Ten Year Rule, 90
Thelwall, John, 113
Third Communist International, 148, 149
Thoiry scheme, 185
Thrace, 223
Thurn, Conrad Donald im, 149, 150, 151
Tibet, 209
tiger-shooting, British Foreign Secretary would decline to go, with French, 262
Tilley, John: cipher work, 29; FO/Diplomatic Service amalgamation, 1, 22; hierarchy changes, 8; Middle East Department, 217; recruitment policy, 19; Registry reorganization, 25; Rio de Janeiro posting, 56; salary proposals, 20
Times, The, 150, 151
trade *see* commerce
Trans-Caspia, 35, 220
Trans Jordan, 225
Treasury: Berlin Embassy, 77; British rearmament, 203; cipher work, 29; Dawes Plan, 107, 137, 140; FO, conflict with, 2, 3, 199–204; FO/Diplomatic Service amalgamation, 22; FO, financial control of, 11–12; FO reform, 7, 14–19; at Genoa Conference (1922), 120; German reparations, 3, 43, 106–7, 108, 109, 110, 125, 199–202, 203; hierarchy changes, 8, 9; London Conference (1924), 142; Political Intelligence Department closure, 5; Press Department, 5; propaganda issue, 27–8; Registry reorganization, 24–5; Ruhr crisis (1923), 107, 111, 124; salary agreement with FO, 19–20, 21; Turkish crisis, 226
Treaty Department of FO, 11, 190
Treitschke, Heinrich von, 156
Trenchard, Sir Hugh, 169
tripartite pact *see* Anglo-Franco-Belgian Alliance
Troutbeck, J.M.: Commission of Control, 134, 183; Council of the League issue, 186; France, guarantee of boundaries, 137; Germany, threat of, 38, 135, 163
Tufton, Charles, 57
Turkey, 64–6, 221–3; Anglo-French conflict, 91, 92, 114, 120, 121–2, 124; British policy during war, 47; Curzon on, 36, 53, 65–6, 73, 221–3; departmental control of, 224, 226; Hankey on, 72; Hardinge on, 41, 64–5, 66, 222; Lindsay on, 56; Lloyd George on, 64–6, 72, 157, 222; military withdrawal from, 220; Montagu on, 221, 222, 223; Peace

Treaty, 66, 85, 86, 216; Treaty of Sèvres, defiance of, 214 *see also* Greco-Turkish conflict; Greco-Turkish relations

Tyrrell, Sir William G.: assistant under-secretary, 38, 44–5, 54–6, 63; and Baldwin, 55; on Bradbury, 201; on commercial functions, 4; Commission of Control withdrawal, 183, 185; Council of the League issue, 186; and Crowe, 45, 56, 87; and Curzon, 54, 55, 87; on Europe, 163, 170, 171, 176, 178, 179, 180; FO/Diplomatic Service amalgamation, 21; on France, 56, 92, 94; German reparations, 108–9, 110, 111, 125, 202; on Germany, 38, 55–6, 78, 94, 135; and Grey, 54; hierarchy changes, 10; Nicolson on, 45; Permanent Under-Secretary, 176; Political Intelligence Department, 5; propaganda issue, 27, 28; Ruhr crisis (1923), 124; Russo-German alliance, 120; salary proposals, 20

Uganda, 205, 214

under-secretaries of FO: assistant, 8, 10, 11, 17, 20, 44–5; permanent, 39–40

Union of Democratic Control, 13, 62

United States, 60, 73, 92, 93; German-Allied pact guarantor, 168; German reparations, 126–7, 133 *see also* Anglo-American guarantee

Vansittart, Sir Robert, 65, 84, 92

Venizelos, Eleutherios, 64

Versailles, Treaty of: Crowe on, 51, 52; Genoa Conference (1922), impact of, 81; German attitude to, 52, 110, 114, 119; Hardinge on, 42; Keynes on, 90; and Lloyd George, 1, 114; Russo-German Alliance threat, 53; and War Office, 76

Villiers, Gerald, 3, 22, 57

Villiers, H. Montagu, 198

war: impact on FO, 1–6; public revulsion of, 89

War Cabinet: disbandment of, 63; Middle East issues, 208

War Department of FO, 7, 47

War Office: Cologne zone, evacuation of, 159; Commission of Control, 134, 184, 185; Europe, danger of war, 136; FO, conflict with, 2, 75–80, 83; future German aggression, 160; Middle East issues, 206, 209, 210, 211, 213, 219, 220, 221, 225; Rhineland and Ruhr, French control of, 134; Turkish crisis, 226; Upper Silesia conflict, 112, 113

Warren, George, 16

Washington Conference (1921–22), 55, 60, 92, 114, 227

Waterlow, Sidney P.O.: economic matters, 4; German reparations, 43, 108; and Germany, 58, 76–7, 78; Ruhr crisis (1920), 95; on the Treasury, 200; Upper Silesia conflict, 111, 112, 113, 114

Wauchope, Major-General A.G., 183, 186

Wellesley, Victor: FO/Treasury conflict, 204; on Germany, 59; propaganda, 26–7; salary proposals, 17; trade/commercial issues, 190, 193–4, 196–7

Western Department of FO, 7, 8; and Crowe, 10, 46, 50; and Europe, 165; France, *entente* with, 59; and Hardinge, 9; Head of Department, role of, 57

Wigram, Ralph, 70, 76, 78, 79, 200

Willert, Arthur, 5, 28

Wilson, Sir Henry, 68, 210, 220, 222

Wilson, Woodrow, President, 2, 64, 134

Wingate, Sir Reginald, 205

Wirth, Joseph, 119

Wise, E.F., 67, 68

Worthington-Evans, Sir, 81, 172

Yemen, 213

Young, Sir George, 13

Young, Major Hubert, 215–16, 218

Younger, Lord George, 150

Yugoslavia, 56

Zanzibar, 205

Zinoviev, Grigori E., 147, 149; Letter, 147–54